PHOTOSELECTIVE CHEMISTRY

PART 1

ADVANCES IN CHEMICAL PHYSICS

VOLUME XLVII

PHOTOSELECTIVE CHEMISTRY

PART 1

Edited by

JOSHUA JORTNER
Tel-Aviv University

RAPHAEL D. LEVINE
Hebrew University of Jerusalem

STUART A. RICE
University of Chicago

ADVANCES IN CHEMICAL PHYSICS
VOLUME XLVII

Series editors

Ilya Prigogine
University of Brussels
Brussels, Belgium
and
University of Texas
Austin, Texas

Stuart A. Rice
Department of Chemistry
and
The James Franck Institute
University of Chicago
Chicago, Illinois

AN INTERSCIENCE® PUBLICATION

JOHN WILEY AND SONS
New York · Chichester · Brisbane · Toronto

An Interscience® Publication

***Library of Congress Catalog Card Number:* 58–9935**

ISBN 0-471-06275-8

Printed in the United States of America

10 9 8 7 6 5 4 3 2 1

INTRODUCTION

Few of us can any longer keep up with the flood of scientific literature, even in specialized subfields. Any attempt to do more, and be broadly educated with respect to a large domain of science, has the appearance of tilting at windmills. Yet the synthesis of ideas drawn from different subjects into new, powerful, general concepts is as valuable as ever, and the desire to remain educated persists in all scientists. This series, *Advances in Chemical Physics*, is devoted to helping the reader obtain general information about a wide variety of topics in chemical physics, which field we interpret very broadly. Our intent is to have experts present comprehensive analyses of subjects of interest and to encourage the expression of individual points of view. We hope that this approach to the presentation of an overview of a subject will both stimulate new research and serve as a personalized learning text for beginners in a field.

ILYA PRIGOGINE
STUART A. RICE

PREFACE

Photoselective chemistry is concerned with the influence of selective optical excitation on the acquisition, storage, and disposal of energy and on the reactivity of molecules, in both gaseous and condensed phases. The very considerable recent progress in this field is largely due to the introduction of lasers, both as pumping and as probing devices, and to the intense theoretical effort stimulated by the many new and intriguing experimental findings. Photoselective chemistry is an interdisciplinary research area, blending concepts and techniques from a wide variety of fields. The articles assembled in these volumes describe many of the theoretical and experimental results now in hand, with the goal of contributing to the synthesis of a conceptual framework with which one can understand a broad spectrum of photophysical and chemical processes. The field is too young, and too little is known, to permit compilation of a definitive treatise. Instead, the contributions in these volumes reflect our opinion of the kinds of information that must be accumulated before it is possible to develop an integrated approach to the interpretation of observations already made and to the development and exploitation of these for new approaches to photochemistry. The same (opinionated) point of view guided the Organizing Committee [J. Jortner (Chairman), S. Kimel, A. Levin, R. D. Levine] of the Laser Chemistry Conference, which took place on December 15–22, 1979 at Ein Bokek, Israel, under the auspices of the National Council for Research and Development, Jerusalem, Israel. Indeed the decision to compile these volumes was triggered by the intensive and exhaustive problem-oriented discussions that took place during that Conference. In addition to contributions by participants at the Ein Bokek meeting, we have included many others, the overall goal being to provide an up-to-date set of authoritative reviews spanning the various aspects of photoselective chemistry.

The first article is meant to serve as an introduction to the entire field; it provides an overview of the relevant concepts, problems, ideas, and experiments described in the following papers. This introductory article was written after receipt of the other contributions so that cross-references could be made. The following articles have been organized in topical groups. Where it was desirable and practical, an introductory article to the given topic is placed first. The general organization of the material is as follows:

1. Aspects of Intramolecular Dynamics
2. Multiphoton Induced Chemistry
3. Studies of Collision Effects
4. Studies in Condensed Media
5. Other Aspects of Photoselective Chemistry

We have followed the general policy of the *Advances in Chemical Physics* in that the authors have been given complete freedom, our point of view being that the person who pioneered the topic is the best judge of the appropriate mode for its presentation. We believe that the results have more than vindicated our approach and hope that the reader concurs. These volumes offer the newcomer a review of the entire field, yet in each and every direction reach the forefront of the current research effort and even attempt to explore the perspectives and future of photoselective chemistry.

We are grateful to the participants of the Ein Bokek Conference and to numerous colleagues and friends, whose lively and probing discussions convinced us of the merits of this project, and to the staff of Wiley-Interscience for welcoming and supporting it. We thank the authors for their willingness to contribute to this endeavor and for their adherence to a timetable, which enabled us to send the manuscripts to the publisher in the Fall of 1979. The wide range of subjects touched on in these volumes bears witness to the scope of photoselective chemistry and to the contagious enthusiasm of its practitioners.

JOSHUA JORTNER
R. D. LEVINE
STUART A. RICE

Tel-Aviv, Israel
Jerusalem, Israel
Chicago, Illinois
January 1981

CONTRIBUTORS TO VOLUME XLVII, PART 1

ABRAHAM BEN-REUVEN, Institute of Chemistry, Tel-Aviv University, Tel-Aviv, Israel

J. A. BESWICK, Laboratoire de Photophysique Moléculaire, Université de Paris Sud, Orsay, France

R. G. BRAY, Exxon Research and Engineering Company, Corporate Research Laboratories, Linden, New Jersey

PAUL BRUMER, Department of Chemistry, University of Toronto, Toronto, Canada

C. D. CANTRELL, Center for Quantum Electronics and Applications, The University of Texas at Dallas, Richardson, Texas

D. M. COX, Exxon Research and Engineering Company, Corporate Research Laboratories, Linden, New Jersey

YEHUDA HAAS, Department of Physical Chemistry, The Hebrew University, Jerusalem, Israel

R. B. HALL, Exxon Research and Engineering Company, Corporate Research Laboratories, Linden, New Jersey

J. A. HORSLEY, Exxon Research and Engineering Company, Corporate Research Laboratories, Linden, New Jersey

PAUL L. HOUSTON, Department of Chemistry, Cornell University, Ithaca, New York

JOSHUA JORTNER, Department of Chemistry, Tel-Aviv University, Tel-Aviv, Israel

A. KALDOR, Exxon Research and Engineering Company, Corporate Research Laboratories, Linden, New Jersey

G. M. KRAMER, Exxon Research and Engineering Company, Corporate Research Laboratories, Linden, New Jersey

R. D. LEVINE, Department of Physical Chemistry and Institute for Advanced Studies, The Hebrew University, Jerusalem, Israel

DONALD H. LEVY, The James Franck Institute, Department of Chemistry, University of Chicago, Chicago, Illinois

W. H. LOUISELL, Department of Physics, University of Southern California, Los Angeles, California

E. T. MAAS, JR., Exxon Research and Engineering Company, Corporate Research Laboratories, Linden, New Jersey

A. A. MAKAROV, Institute of Spectroscopy, USSR Academy of Sciences, Moscow, USSR

SHAUL MUKAMEL, Department of Chemistry, William March Rice University, Houston, Texas

YITZHAK RABIN, Institute of Chemistry, Tel-Aviv University, Tel-Aviv, Israel

P. RABINOWITZ, Exxon Research and Engineering Company, Corporate Research Laboratories, Linden, New Jersey

HANNA REISLER, Departments of Electrical Engineering, Physics, and Chemistry, University of Southern California, Los Angeles, California

STUART A. RICE, The Department of Chemistry and The James Franck Institute, The University of Chicago, Chicago, Illinois

AVIGDOR M. RONN, Department of Chemistry, Brooklyn College of the City University of New York, Brooklyn, New York

MARTIN L. SAGE, Department of Chemistry, Syracuse University, Syracuse, New York

CURT WITTIG, Departments of Electrical Engineering, Physics, and Chemistry, University of Southern California, Los Angeles, California

CONTENTS

PHOTOSELECTIVE CHEMISTRY
By Joshua Jortner and R. D. Levine 1

Section 1. Aspects of Intramolecular Dynamics **115**

AN OVERVIEW OF THE DYNAMICS OF INTRAMOLECULAR TRANSFER OF VIBRATIONAL ENERGY
By Stuart A. Rice 117

INTRAMOLECULAR ENERGY TRANSFER: THEORIES FOR THE ONSET OF STATISTICAL BEHAVIOR
By Paul Brumer 201

THE INFORMATION THEORETIC APPROACH TO INTRAMOLECULAR DYNAMICS
By R. D. Levine 239

BOND MODES
By Martin L. Sage and Joshua Jortner 293

VAN DER WAALS MOLECULES
By Donald H. Levy 323

INTRAMOLECULAR DYNAMICS OF VAN DER WAALS MOLECULES
By J. A. Beswick and Joshua Jortner 363

Section 2. Multiphoton-Induced Chemistry **507**

REDUCED EQUATIONS OF MOTION FOR COLLISIONLESS MOLECULAR MULTIPHOTON PROCESSES
By Shaul Mukamel 509

N-LEVEL MULTIPLE RESONANCE
By Abraham Ben-Reuven and Yitzhak Rabin 555

LASER EXCITATION OF SF_6: SPECTROSCOPY AND COHERENT PULSE PROPAGATION EFFECTS
By C. D. Cantrell, A. A. Makarov, and W. H. Louisell 583

Initiation of Atom–Molecule Reactions by Infrared Multiphoton Dissociation
By Paul L. Houston 625

Infrared Laser Chemistry of Complex Molecules
By R. B. Hall, A. Kaldor, D. M. Cox, J. A. Horsley, P. Rabinowitz, G. M. Kramer, R. G. Bray, and E. T. Maas, Jr. 639

Luminescence of Parent Molecule Induced by Multiphoton Infrared Excitation
By Avigdor M. Ronn 661

Electronic Luminescence Resulting from Infrared Multiple Photon Excitation
By Hanna Reisler and Curt Wittig 679

Electronically Excited Fragments Formed by Unimolecular Multiple Photon Dissociation
By Yehuda Haas 713

Author Index **735**

Subject Index **763**

PHOTOSELECTIVE CHEMISTRY

PART 1

ADVANCES IN CHEMICAL PHYSICS

VOLUME XLVII

PHOTOSELECTIVE CHEMISTRY*

JOSHUA JORTNER

Department of Chemistry, Tel-Aviv University, Tel Aviv, Israel

and

R. D. LEVINE

Department of Physical Chemistry, The Hebrew University, Jerusalem, Israel

CONTENTS

I. Prologue . . . 2
  A. Energy Acquisition . . . 2
  B. Energy Storage . . . 9
  C. Energy Disposal . . . 12
II. Experimental Observables . . . 13
  A. Time-Resolved Observables . . . 13
  B. Energy-Resolved Observables . . . 14
  C. Observables Pertaining to Coherent Optical Effects . . . 15
III. Intramolecular and Intermolecular Relaxation . . . 16
IV. Predissociation . . . 19
V. Nonreactive Molecular Processes . . . 29
VI. Electronic Relaxation . . . 40
VII. Inverse Electronic Relaxation . . . 46
VIII. Intramolecular Energy Flow . . . 51
IX. Unimolecular Reactions . . . 56
X. Collisional Processes . . . 58
XI. Coherent Optical Effects for Pedestrians . . . 65
XII. High-Order Multiphoton Molecular Processes . . . 70
XIII. Relaxation and Dephasing in Condensed Phases . . . 82
XIV. Multiphonon Processes . . . 90
XV. Digression on Biophysics . . . 98
XVI. Epilogue . . . 102
  References . . . 104

*Work supported by the U.S.–Israel Binational Science Foundation, Grant 1404, and the U.S. Air Force of Scientific Research (AFOSR), Grant 77-3135.

I. PROLOGUE

Photoselective chemistry is the exploration of the consequences of selective optical excitation on the acquisition, storage, and disposal of energy in molecules and in condensed phases. The outstanding progress achieved in this field during the last decade stems from two directions. First, extensive and exhaustive theoretical studies in the area of intermolecular and intramolecular dynamics provided a conceptual framework for the general understanding of a variety of nonreactive and reactive molecular photophysical and chemical processes, thereby establishing a firm basis for the elucidation of the nature of energy storage and disposal on the molecular level.[1–17] Second, the advent of laser sources had a remarkable impact on experimental progress in the field.[18–22] Although significant information on excited-state dynamics was obtained in the past utilizing conventional optical excitation sources, the introduction of laser sources surpassed and eclipsed previous experimental work in providing new ways and means for energy acquisition as well as in advancing novel methods for the interrogation of the basic processes of energy storage and disposal. The degree of detail currently available is such that methods for the compaction and correlation of the basic data have to be introduced.

Photoselective chemistry can be considered as a truly interdisciplinary field, providing blending and integration of modern experimental methods and of theoretical concepts for a wide spectrum of fields, such as radiation theory,[15] quantum electronics,[23] scattering theory,[24] molecular spectroscopy,[25–27] intramolecular dynamics,[1–17] as well as solid-state and condensed phase physics and chemistry.[28, 29] A multitude of experimental and theoretical methods from all these areas has been adopted to unveil the nature of excited-state photophysical and photochemical processes. It is our opinion that the basic approach to the understanding of photoselective chemistry should be problem-oriented rather than technique-oriented, so that the variety of the new, sophisticated, and exciting experimental methods advanced in this field should be considered, as well as the elucidation of the basic microscopic processes. Accordingly, we shall proceed to discuss the many-faceted nature of energy acquisition, storage, and disposal in molecules and in condensed phases as experimentally investigated and as explored theoretically from a microscopic point of view.

A. Energy Acquisition

The aspect of energy acquisition pertains to the central feature of photoselective excitation, i.e., the "preparation" of well-defined "initial" states by optical excitation.[13-17] During the past decade lasers have been extensively utilized in this area, taking advantage of many of the unique features of these optical excitation sources, which will now briefly be considered.

1. *Spectral Range.* Currently, available lasers span a broad energy region from 10 eV to 0.1 eV. Typical sources include vacuum ultraviolet lasers in the range 8.0–9.2 eV operating by third harmonic generation in metallic vapors,[30] ultraviolet excimer lasers emitting at several energies in the range 6.3–4.0 eV,[31-33] ultraviolet and visible solid-state Neodymium glass and Ruby lasers together with their second and third harmonics,[34] ultraviolet and lasers based on second harmonic generation from dye lasers and visible lasers spanning the energy range of 6.0–1.5 eV,[34] as well as infrared laser sources such as chemical lasers[35] and the popular CO_2 laser at ~0.12 eV.[34]

2. *Tunability.* Some laser sources can be tuned over quite a broad energy range. Vacuum ultraviolet lasers are currently tunable over a range of 9.2–8.0 eV.[30] A combination of a variety of ultraviolet and visible dye lasers and their second harmonic results in tunability over the broad range of 6.0–1.5 eV.[34] Tunability in the infrared is currently restricted to low-power narrow-range CW diode lasers.[36]

3. *High Power.* The power output of some laser sources is remarkably high, e.g., 10^9 W cm^{-2} for mode-locked Nd glass and Ruby lasers,[19, 37] a power output of $\sim 10^{12}$ W cm^{-2} being accomplished for the iodine near infrared laser,[38] while for infrared CO_2 laser the accessible power is 10^6–10^9 W cm^{-2}.[39-43] High-power laser sources were utilized to achieve high-order electronic and vibrational multiphoton excitation in atoms, molecules, and solids. These unconventional excitation processes involve two- and three-photon excitation studies of molecular states,[44] high-order multiphoton atomic ionization,[45, 46] and high-order molecular multiphoton vibrational excitation,[40-43] fragmentation,[39-43] isomerization,[47, 48] and ionization.[49]

4. *High Energy.* Photoselective excitation of extremely weak absorption bands can be accomplished by intracavity one-photon absorption, providing a novel way for the interrogation of chemical consequences of high vibrational excitations.[40-43] High-order multiphoton excitation of very large molecules, such as S_2F_{10} [51] or the esoteric molecule $UO_2(hfacac)_2THF_3$,[52, 53] is determined by the laser energy fluence[40] and can be conducted using CW infrared lasers.[52, 53]

5. *Ultrashort Duration.* The temporal duration of some pulses can be extremely short, that is, in the picosecond region. Mode-locked solid-state lasers in the visible deliver pulses in the time domain of 6–30 psec,[19, 37] mode-locked dye lasers yield pulses in the range >1–10 psec[19] in the visible and near ultraviolet, while 30-psec pulses were obtained from an infrared CO_2 laser.[54] These ultrashort pulses provide a powerful tool for the interrogation of intramolecular and intermolecular dynamics on the picosecond time scale.

6. *Coherence Effects.* Single-mode laser sources can be utilized for excitation of molecular ensembles by $(\pi/2)$ or π optical pulses.[55, 56] One can study the interesting process of driving of the system by an electromagnetic field, exploring coherent transient effects in electronically excited states, and interrogating the retention of phase coherence between the molecular ground state and the excited states.

Lasers have played a leading role as optical excitation sources in the energy range 10–0.1 eV. At higher energies above 10 eV, photoselective, tunable, energy-resolved, and time-resolved excitation can be accomplished using synchrotron radiation sources, which show great promise for the use in high-energy intramolecular dynamics.[57] We shall confine the present discussion to energy acquisition below 10 eV where the characteristics of lasers were utilized for energy-resolved and time-resolved excitation, for high-order multiphoton excitation as well as for establishing phase coherence between the excited state and the ground state in a variety of molecular and condensed-phase systems. Studies of intramolecular dynamics focus attention on nonradiative relaxation processes occurring in "isolated" collision-free molecules and how these intramolecular processes are affected by coupling between a molecule and an external medium. The systems utilized for the exploration of intramolecular dynamics of isolated molecules essentially involve molecules in the bulb, as well as molecules in the thermal beams and in supersonic beams, which will now be considered.

a. Molecules at low pressures. Optical excitation of molecules in the low-pressure gas phase can be conducted at pressure down to 10^{-3}–10^{-4} torr, where the time between gas–kinetic collisions is 10^{-4} sec. Considerably higher collision cross-sections were reported, however, for rotational–vibrational relaxation in electronically–vibrationally or vibrationally excited states. For example, the cross-section for collisional rotational relaxation of electronically excited benzene in the $^1B_{2u}$ state by a ground-state benzene molecule is 500 $Å^2$,[58] while the cross-section for SF_6–SF_6 collision involving a vibrationally excited molecule is 300 $Å^2$.[59] Thus, in bulb experiments collision-free conditions at 10^{-3} torr can be accomplished on a time scale of $\sim 10^{-5}$ sec, and the excited-state intramolecular dynamic process can be considered to occur in the "isolated" molecule which is collision-free on the time scale of microseconds, or so. Such bulb experiments led to pertinent information on electronic relaxation in electronically–vibrationally excited states of large "isolated" molecules[9–17] and were crucial in establishing the features of collisionless, multiphoton molecular photofragmentation and isomerization on the ground-state potential surface.[40–43]

b. Molecules in thermal molecular beams. To accomplish electronic and/or vibrational excitation of an "isolated" collision-free molecule, a laser beam was crossed with an effusive, thermal, molecular beam.[60,61] The intramolecular dynamics in the first electronically excited singlet state of several large molecules, i.e., pentacene and benzophenone, were investigated in thermal beams.[60,61] Studies of multiphoton molecular photofragmentation in beams established the collisionless nature of this class of phenomena and contributed to the understanding of the decomposition mechanism by the determination of the translational energy of the fragments.[62,63] The intrinsic limitation of probing the consequences of laser excitation of large molecules in thermal molecular beams is due to thermal inhomogeneous broadening effects (TIB), that is, thermal rotational broadening and vibrational sequence congestion (see Table I), which preclude truly photoselective excitation of an ensemble of large molecules at room temperature.

c. Molecules in supersonic beams. A powerful way to overcome TIB effects rests on the use of isentropic–nozzle beam expansions. The low translational, rotational, and vibrational temperatures achieved in supersonic-free expansions are sufficient to eliminate all rotational broadening effects in small molecules, and to avoid all vibrational sequence congestion effects in large molecules. The use of seeded molecular beams provides a novel experimental approach for genuine photoselective excitation of large

TABLE I
Origins of Inhomogeneous Broadening

Systems	Conserved Quantity	Type of Inhomogeneous Broadening
Isolated molecules	Linear momentum	Doppler
	Angular momentum	Rotational
	Vibrational state	Sequence congestion
Medium-perturbed molecules	Electronic–vibrational state	Site splitting and statistical distribution of trapping sites

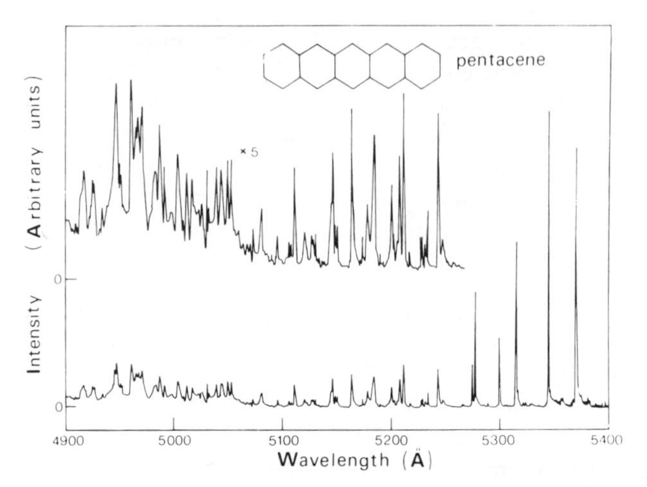

Fig. 1. Fluorescence excitation spectrum of gas-phase isolated pentacene ($C_{22}H_{14}$) molecule cooled in a supersonic expansion. The data are reproduced from the work of Amirav, Even, and Jortner (Ref. 76). Pentacene (at pressure of 5×10^{-2} torr) was seeded into Ar and expanded from a pressure of 210 torr through a 150-μm nozzle. The exciting dye laser (spectral bandwidth 0.3 cm^{-1}) crossed the supersonic beam at 5 mm down the nozzle. The wavelength scale corresponds to that of the laser. The intensity scale monitors the total fluorescence, normalized to the laser intensity. The fluorescence excitation spectrum in the range 5400–4900 Å was assigned[76] to the $S_0(^1A_{1g}) \rightarrow S_1(^1B_{2u})$ transition, the electronic origin being located at 5368 Å.

molecules.[64–69] As is evident from Figs. 1 and 2 laser spectroscopy of large molecules seeded in supersonic beams allows for an increase of spectral resolution of about three orders of magnitude over that possible with room-temperature bulb experiments. The dynamics of several isolated "ultracold" large molecules, such as formaldehyde,[70] benzene,[71] naphthalene,[72, 73] phtalocyanine,[74] tetracene,[69, 75] pentacene,[75, 76] and ovalene,[77] in their lowest excited singlet states in a supersonic beam, provided central information on interstate electronic relaxation and on intrastate vibrational energy redistribution in excited states of large molecules. Not only conventional molecules can be excited in supersonic beams but supersonic-free expansion has also been extensively and fruitfully utilized to prepare a wide variety of weakly bound van der Waals molecules[66, 67] (Chapter 5), whose intramolecular dynamics is of considerable interest.

We shall now proceed to consider the basic systems which were explored in the context of medium-perturbed, intramolecular dynamics and nonradiative relaxation processes in condensed phases:

d. Selective collisional effects. Collisional effects have constantly plagued the experimentalists engaged in studies of intramolecular dy-

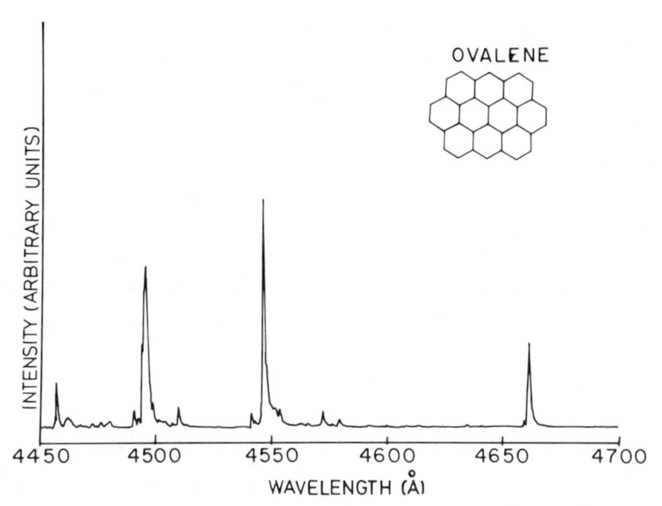

Fig. 2. Fluorescence excitation spectrum of the ovalene ($C_{32}H_{14}$) molecule cooled in a supersonic expansion. These data are reproduced from the work of Amirav, Even, and Jortner (Ref. 77). Ovalene (at pressure of 5×10^{-2} torr) was seeded at Ar and expanded through a 200-μm ceramic nozzle. Experimental conditions similar to that of Fig. 1. The excitation spectrum in the range 4700–4450 Å corresponds[77] to the $S_0(^1A_{1g})\rightarrow S_1(^1B_{3u}^-)$, the lowest energy spectral feature at 4662 Å being attributed to the electronic origin.

namics in "isolated" molecules. It is extremely important to understand the interplay between intramolecular dynamics and intermolecular perturbations. In this context one should attempt to extract microscopic information regarding the selective effects of collisions, which was obtained from energy-resolved emission studies from electronically excited states of collisionally perturbed large molecules ("Collision Induced Intramolecular Energy Transfer in Electronically Excited Polyatomic Molecules," Chapter 23). The thermal distribution of the energies of the colliding molecules limits the information content emerging from such experiments. The ultimate goal in this area will be the exploration of the consequences of monoenergetic collisions on intramolecular dynamics. Significant progress in that direction was accomplished recently (Chapter 23) by exploiting some features of supersonic beam expansions, which made it possible to study energy-resolved collisional effects in the proximity of the nozzle source.[78, 79] Other recent studies are discussed in Chapters 22–26.

e. Matrices and Mixed Crystals. A traditional way to eliminate TIB effects in the spectroscopy of large molecules involves the study of low concentration guest molecules in a low-temperature solid, glass, or in a mixed crystal. Even at low temperatures the effects of phonon broadening, manifested in the appearance of multiphonon side bands, can be severe in spreading the intensity of the vibronic molecular state over a wide energy range. These phonon-broadening effects can be reduced by a proper choice of the host, so that the molecule–lattice nuclear equilibrium configurations are not distorted on electronic excitation and the zero-phonon lines dominate the spectrum, as in the case for Shpolskii matrices[80] and for some mixed organic crystals.[81] Even in such favorable systems site-splitting effects are exhibited, and each of the zero-phonon lines is severely broadened due to the statistical distribution of trapping sites, i.e., intrinsic inhomogeneous broadening effects. A visual demonstration of such complications is presented in Fig. 3. Spectroscopic studies of zero-phonon lines

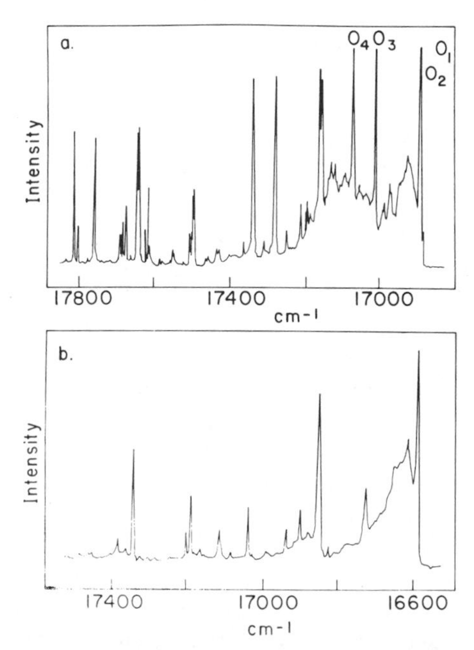

Fig. 3. Absorption spectrum of the pentacene molecule in host molecular crystals at 1.4 K. These data are reproduced from the work of Lambert and Zewail[334] (a) Pentacene in *p*-terphenyl host. (b) Pentacene in naphthalene host. These low-temperature guest spectra reflect sharp features corresponding to zero-phonon lines (ZPO) of the vibrational transitions of the $S_0 \rightarrow S_1$ electronic excitation. The lowest energy ZPO corresponds to the electronic origin. Four complications inherent in these spectra are evident: (1) Site splitting. The features 0_1, 0_2, 0_3 and 0_4 of pentacene/*p*-terphenyl correspond to different trapping sites. (2) Lattice phonon excitation. The broad absorption on the high energy side of each ZPO reflect excitation of lattice phonons, induced by molecule–lattice coupling. (3) Inhomogeneous broadening. The width of the lowest ZPO originates from a statistical distribution of trapping sites. (4) Vibrational relaxation. The broadening of some of the ZPOs corresponding to vibrationally excited states has a large contribution from medium-induced vibrational relaxation, cf. Section XIII.

provide useful data regarding the energetics of electronically–vibrationally molecular excitations. Unfortunately, such studies of matrix isolated molecules provide only very limited information regarding intramolecular dynamics, as selective excitation of vibrationally excited molecular states results in medium-induced vibrational relaxation, which masks the intramolecular effects. Nevertheless, some useful information regarding intramolecular dynamics in the vibrationless electronic origin of an electronically excited state of a molecule in a low-temperature glass or crystal can be obtained.

f. Solutions. Studies of excited state dynamics of molecules in solutions provide a traditional research area in physical chemistry. The interpretation of the rich and diverse information regarding excited-state dynamics of molecules in solution, usually conducted at room temperature, is complicated due to TIB effects, phonon broadening, as well as severe inhomogeneous broadening effects. Of particular interest are dynamic processes in dense polar and also in nonpolar fluids, which are specific to the liquid phase. These can be grouped into excited-state or ground-state processes. In the first category we shall consider excited-state chemical processes, such as intermolecular and intramolecular proton transfer in polar solvents ("Laser Studies of Proton Transfer," Chapter 33), which provide interesting information regarding the dynamics of electronically excited states.[82] In the second category we shall mention the interesting process of the dynamics of the localization of an excess electron in polar solvents[83] ("Picosecond Dynamics of Electron Relaxation Processes in Liquids," Chapter 30), which utilizes the excess electron, produced by photoionization of a guest molecule or ion in solution, as a probe for the local and the long-range structure in liquids.

g. Biological systems. Photoselective chemistry is expected to provide fundamental contributions toward the elucidation of some of the basic phenomena in photobiology,[84] unveiling the basic mechanism of photosynthesis[85] and of the vision processes[86] (Chapter 32). During the last few years several studies on model systems of the photosynthetic systems[87–90] and of rhodopsin[91] were performed, as well as investigation of the chemical and biochemical consequences of photoselective excitation of systems in vitro was conducted. The utilization of photoselective ultrafast excitation at room temperature, as well as low temperature systems,[86, 92] is expected to provide a fruitful blending between the techniques and the concepts of chemistry, photophysics, photobiology, and biophysics.

B. Energy Storage

The problem of energy storage pertains to the basic aspects of intramolecular and intermolecular dynamics of "isolated" molecules, of collisionally perturbed molecules and of molecules in condensed phases. A systematic description of dynamic processes should address the following three aspects.

1. *Characterization of the molecular level structure.* Excited-state level structure in vibrationally–electronically excited states of molecules may include:
   a. Sparse bound level structure in diatomics, in small polyatomics, as well as in some low-lying vibrationally excited states of large molecules.
   b. An intramolecular quasicontinuum consisting of a dense vibrational manifold of bound vibronic levels, which is characterized by a high density of states.
   c. Dissociative continua consisting of unbound states of the fragments.
   d. Ionization continua corresponding to unbound electron-positive ion pairs.

The molecular structure is specified in terms of the eigenstates of the entire molecular (nuclear–electronic) Hamiltonian for the system. There is a basic physical distinction between bound level structures, such as *a* and *b*, where the molecular electronic–nuclear states are localized, and a continuous spectrum, where the electronic–nuclear states are extended. Above the onset of the lowest-lying dissociative or ionization continuum, the molecular states correspond to extended levels. There can be no peaceful coexistence between bound and continuous states above the onset of the lowest-lying continuum. When resonances are exhibited in the continuous spectrum, the relevant molecular states can be described in terms of a superposition of discrete and continuum states. However, a finite contribution from continuum states will always prevail above the onset.

2. *Characterization of the excitation amplitudes.* The energetic spread of the transition moments connecting the ground state of lower-lying states with the excited-state level structure determines the accessibility of these excited states to optical excitation. The distribution of the transition moments between the molecular eigenstates of the total Hamiltonian is expected to exhibit sharp structure in range *a*, to show a quasicontinuous structure characterized by resonances in range *b* and to reveal a smooth distribution with some superimposed resonances in ranges *c* and *d*. We note in passing that the characterization of the excitation amplitudes is more general than the specification of the distribution of the oscillator strengths, as it makes possible the description of interference effects. The resonance structure and the

interference effects exhibited in congested and in continuous spectra can adequately be described in terms of zero-order states, a problem of considerable technical importance which will subsequently be considered in Sections V and VII. What is important for the purpose of the present general discussion is that the distribution of the transition amplitudes to the molecular eigenstates determines accessibility and photoselectivity by optical excitation.

3. *Specification of the initial conditions.* A variety of excitation methods can be applied to molecular systems, e.g., short-time excitation, energy-resolved excitation, high-power multiphoton excitation, a coherent excitation, just to mention a few examples. It has been realized since the early days of quantum mechanics, and is now well established in the area of chemical physics, that a stationary state of the molecular Hamiltonian exhibits only radiative decay, and that only nonstationary states of the molecular Hamiltonian are metastable with respect to intramolecular dynamics. The metastable molecular states "prepared" by an optical excitation are determined by the excited-state level structure, by the energetic spread of the transition amplitude, as well as by the energetic temporal intensity and coherence properties of the excitation light source. Thus, the excitation conditions govern the subsequent dynamic time evolution of a molecular system. We can conclude that the intramolecular dynamics of a given system provide the signature of the initial conditions.

Information concerning excited-state level structure is provided by the well-established discipline of molecular spectroscopy. Basic data concerning vibrational level structure on the ground-state potential surface, important for the understanding of multiphoton excitation[40–43] were obtained from high-resolution infrared spectroscopy,[41] while laser spectroscopy of large molecules in supersonic beams lead to interesting data regarding sparse and quasicontinuous bound molecular level structure in low-lying electronic states.[69, 76] Spectroscopic data also provide information on the distribution of the transition amplitudes. Conventional spectroscopy establishes the energy distribution of the oscillator strength from the ground vibrational–electronic state to excited states. Low resolution spectroscopy of vibrationally excited molecules[93] yield important information regarding the distribution of intensity between highly excited states, which is crucial for the understanding of high-order multiphoton vibrational excitation.

The basic information concerning energy storage should address the question of the mechanisms of intramolecular energy exchange and the time scales for such processes.

C. Energy Disposal

The problem of energy disposal is concerned with the basic microscopic mechanisms of the damping of the excitation energy. It should be borne in mind that the radiative decay channel provides just one possible energy decay route and that a variety of alternative energy disposal mechanisms should be considered. Exploration of the energy disposal phenomena requires the interrogation of the decay channels. This information essentially involves the determination of all possible cross-sections but has to be singled out in view of its primary chemical significance.

1. Identification of reactive and nonreactive channels in molecular systems. This is a simple matter in diatomics but becomes a rather difficult and important task with increasing size of the molecule.
2. Specification of reactive and nonreactive channels in condensed phases. Here the media affect the intramolecular processes and, in addition, new intramolecular and/or intermolecular electronic and vibrational energy transfers have to be considered.
3. Determination of branching ratios between various parallel channels.
4. Characterization of parallel and consecutive decay channels.
5. Investigation of possible interference effects between different channels.
6. Determination of internal energy (electronic, vibrational, rotational) content of the fragments.

Laser sources in conjunction with modern spectroscopic techniques have been extremely useful for the investigation of decay channels. Some of the modern techniques are:

1. Doppler spectroscopy of fragments ("Doppler Spectroscopy of Photofragments," Chapter 20).
2. Fluorescence of fragments resulting from one-photon excitation[94] Chapter 19 and from multiphoton excitation[95, 96] (Chapters 12–14).
3. Laser spectroscopy of highly excited vibrational states reached by multiphoton excitation, (Chapter 7–9).
4. Laser spectroscopy of fragments.[97] ("Laser Diagnostics of Reaction Product Energy Distributions," Chapter 19).
5. Double resonance techniques.[98]
6. Identification of products by conventional chemical techniques, such as gas chromatography and mass spectrometry.[62]
7. Identification of products by laser spectroscopy[97] (Chapter 19).
8. Identification of products by spectroscopic methods, such as resonance Raman or CARS.[99]
9. Studies of isotope separation resulting from multiphoton excitation,[42] (and "Infrared Laser Chemistry of Complex Molecules," Chapter 11).

Lasers have also been extensively employed (Chapters 10, 15–17) to generate reactive excited states and radicals.

This brief introduction has been concerned with some aspects of the basic physical and chemical phenomena of acquisition, storage, and disposal of energy, with a special emphasis on the role of laser photoselective excitation in energy acquisition and on the interrogation of the rise and fall of excited molecular and condensed-phase excited states. It is the purpose of this chapter to serve as an overview of the problems, concepts, experiments, and ideas described in the present volumes. The prominence of the microscopic point of view in the description of excited-state dynamics, which is emphasized in the present and in the following chapters, reflects the blending between the development of theoretical concepts and of sophisticated experimental methods and provides a firm conceptual framework for the understanding of intramolecular and intermolecular dynamics. It is our goal to expose the present state of the art in this field and to explore the perspectives and future of photoselective chemistry.

II. EXPERIMENTAL OBSERVABLES

We shall first proceed to examine the experimental information pertaining to various aspects of energy acquisition, storage, and disposal. This will provide background for the subsequent discussion of the conceptual framework for the rationalization and interpretation of many of these diverse phenomena.

The basic information concerning intramolecular and intermolecular nonradiative relaxation dynamics in molecules and in condensed phases emerges from the following general classes of experimental observables.

A. Time-Resolved Observables

Direct information regarding the temporal decay of excited states is obtained from:

1. Monitoring radiative decay by time-resolved photon counting.
2. Monitoring populations of reactive and nonreactive decay channels by time-resolved absorption methods.

This direct information on the rise and fall of rotational–vibrational–electronic molecular states and vibrational–electronic excitations in condensed phases can be classified according to their different decay modes:

1. *Exponential decay.* This is the common decay mode, which was well documented in the low-energy (say, below 5 eV) range, and which corresponds to the decay times of metastable states and to the rates of population of all decay channels.

2. *Nonexponential decay.* This consists of a superposition of exponentials. Such effects may originate from two sources:
   a. Inhomogeneous broadening of molecular states (Table I), which results in excitation of an incoherent superposition of states, and whose independent decay may be characterized by a distribution of lifetimes. These effects complicate the elucidation of the intrinsic dynamics.
   b. Genuine intramolecular effects originating from intrastate and interstate intramolecular coupling may give rise to such interesting intrinsic effects.[16, 100, 101]
3. *Quantum beats in the decay.* Such a pattern originates from interference effects between closely-lying excited levels.[102] These interesting manifestations of interrelationships between the phases of excited states were reported for externally perturbed (Zeeman and Stark shifted) levels of atoms and diatomics. The more interesting situation of coherent decay of closely spaced indistinguishable closely lying states of polyatomic molecules resulting from mixing between two electronic configurations was not yet experimentally explored. A recent study[103] has reported the observation of molecular quantum beats in the decay of the singlet manifold of biacetyl, which is scrambled with the triplet background manifold, so that coherent effects can be exhibited by closely spaced singlet–triplet molecular eigenstates.

B. Energy-Resolved Observables

A variety of cross-sections for optical absorption, photon scattering, etc., yield basic data regarding reactive and nonreactive molecular processes. These studies sacrifice the time resolution for the sake of energy resolution, providing a blending of spectroscopic data with basic information concerning intramolecular dynamics. These observables are:

1. *Absorption cross-sections.* In the case of a simple isolated resonance in reactive and nonreactive processes, the homogeneous width of the Lorentzian line shape is determined by the decay rate of the metastable state. In more complex situations, when the background manifold, i.e., the continuum or the quasicontinuum, carries oscillator strength from the ground-state, Fano-type interference effects in absorption will be exhibited.[104–106] Finally, interference effects in absorption to overlapping resonances[107] provide interesting information on interstate and intrastate coupling, which is complementary to that which will be obtained from time-resolved quantum beat experiments.
2. *Photon-scattering cross-sections.* These involve conventional off-resonance fluorescence and Raman scattering,[108] near-resonance

Raman scattering from dissociative continua and from ionization continua.[109–111] The information obtained is complementary to but not identical with that obtained from absorption data.

3. *Cross-sections for population of reactive and nonreactive decay channels.* The modern techniques of picosecond spectroscopy in absorption and in emission, both in the ultraviolet, visible, and infrared regions[19] have been extremely useful in this context.
4. *Angular distribution of products,* i.e., neutral fragments and/or ions, resulting from excitation by polarized radiation.[112–114] For direct molecular ionization and dissociation processes, such experiments provide spectroscopic evidence concerning spatial orientation of optical transition moments in polyatomic molecules. For indirect reactive processes, such as autoionization and predissociation, the lifetimes of metastable states can be extracted from the angular distribution.[114]

C. Observables Pertaining to Coherent Optical Effects

Optical excitation by coherent light pulses can establish definite phase relationships between the ground-state and excited doorway states.[115–129] The temporal persistence of the phase relationships can be interrogated by studies of optical nutation, free-induction decay, and photon-echo experiments. These phase relationships are destroyed by dephasing phenomena, which are of three distinct types. First, intramolecular dephasing in reactive processes is equivalent to T_1 level depletion processes. Second, intramolecular dephasing in interstate and intrastate nonreactive relaxation in large isolated molecules and can be regarded as intramolecular T_2 processes.[130, 131] Third, erosion of phase coherence by medium perturbations provide important chemical and physical information on intermolecular T_2 processes, which pertain to the consequences of the coupling of excited molecular states with the host medium.[56] Most of the coherent effects in the optical region have been limited to two-level systems.[55] The studies of retention of phase relationships in multilevel systems[132–134] will be of considerable interest for the elucidation of multiphoton excitation of large molecules.

From the point of view of general methodology, the basic physical and chemical information emerging from the experimental observables falls into two broad classes:

I. *Populations of individual excited states and/or populations of discrete (quasicontinuum) or continuous decay channels.* Temporal and energy-resolved information is obtained from some of the time-resolved and energy-resolved observables. This information regarding the population of excited and ground electronic–vibrational–rotational states of the parent molecule and/or the fragments is of central chemical importance.

II. *Phase relationships.* Information on phase relationships between excited states is obtained from the analysis of interference effect exhibited in the time-resolved observables and in the energy-resolved variables. Phase relationships between the ground and excited doorway states can be monitored by studies of coherence effects.

The major goal of the experimental and theoretical studies in the area of intramolecular and intermolecular dynamics is twofold. First, one has to elucidate the nature of the various intermolecular and intramolecular decay channels, the interaction between channels and the sequence of photo-physical reactive and nonreactive processes in excited states. Second, the consequences of phase relationships between excited states, as well as between the ground state and the excited doorway state, have to be explored in order to understand intramolecular interference effects, intramolecular dynamical processes and medium perturbations on excited states.

III. INTRAMOLECULAR AND INTERMOLECULAR RELAXATION

An outstanding goal of research in the area of molecular and condensed phase excited-state dynamics is the elucidation of a wide class of radiationless processes involving a large variety of intermolecular and intramolecular phenomena, such as electronic–vibrational exchange, (Chapter 26) vibrational–vibrational exchange, (Chapter 22) dissociation (and ionization) in direct processes and in the form of indirect processes, that is, predissociation and autoionization, as well as more complex chemical phenomena. These intramolecular and intermolecular processes can be classified as follows in the order of increasing complexity of the system involved:

1. Basic molecular processes.
2. Complex molecular processes.
3. Relaxation processes in external electric and magnetic fields.
4. Relaxation in intense electromagnetic fields.
5. Medium-induced processes.
6. Relaxation in condensed phases.

The basic intramolecular relaxation (Table II) processes fall into two distinct classes:

1. *Reactive processes,* e.g., autoionization[135] and predissociation,[25, 27] which result in ionization or fragmentation.

TABLE II
Classification of Basic Intramolecular Relaxation Processes

	Nature of Coupling	
Nature of decay channel	Intrastate (1 electronic configuration)	Interstate (2 electronic configurations)
Reactive	Rotational predissociation	Autoionization
	Vibrational predissociation	Electronic predissociation
Nonreactive	Intramolecular vibrational energy redistribution in large molecules	Electronic relaxation (internal conversion and intersystem crossing) in large molecules

2. *Nonreactive processes,* e.g., intramolecular electronic relaxation[7–17] and intramolecular vibrational redistribution[14] in large molecules.

Another useful classification of the basic molecular processes separates those (intrastate) processes occurring on a single electronic potential surface from the (interstate) processes involving at least two distinct electronic configurations:

1. Intrastate dynamics involving some forms of rotational predissociation, as well as intramolecular vibrational energy redistribution which occurs between bound vibrational levels of large molecules.
2. Interstate dynamics incorporating the processes of autoionization, electronic predissociation, as well as electronic relaxation, i.e., internal conversion and intersystem crossing between bound states of large molecules.

The basic molecular processes we have been concerned with provide the framework for the elucidation of the nature of a variety of interesting chemical phenomena. Complex chemical processes in large molecules can be described in terms of several parallel and/or sequential basic molecular relaxation processes. While molecular photofragmentation of diatomic molecules involves just direct photodissociation or predissociation, the photofragmentation process becomes more complex with increasing size of the molecule. For triatomic molecules the effects of coupling various dissociative channels, that is, final-state interactions,[136] and effects of intramolecular energy exchange have to be incorporated. For the case of molecular photofragmentation of large polyatomic molecules one should consider electronic relaxation, intramolecular vibrational energy redistribution, as well as vibrational predissociation and electronic predissociation. Photochemical rearrangements, such as isomerizations of large molecules,[137] involve contributions from a variety of basic processes, for example, electronic relaxation and intramolecular vibrational energy redistribution, and may also be strongly influenced by medium-induced steric hindrances and vibrational relaxation. The elucidation of the variety of decay channels contributing to these complex processes will provide firm grounds for the understanding of molecular photochemistry and radiation chemistry from a unified point of view.

Some of the intramolecular processes we have just considered can be affected by external fields. Control of intramolecular dynamics by external fields is of considerable interest, as such effects will provide a new variable to probe dynamic molecular processes. Little is presently known concerning the effects of external magnetic and electric fields on molecular relaxation processes. Only the effect of the magnetic field on some cases of electronic predissociation of diatomics, where the Zeeman term enhances the predissociation rate, is well documented.[138] The effects of external fields on nonreactive electronic relaxation have not yet been explored.[139] Another interesting feature of the effects of external fields involves field-induced decay processes, such as the ionization of highly excited atomic and molecular Rydberg states in intense electric fields.[140] This effect is of interest in relation to high-order multiphoton ionization.

Of considerable current interest are the consequences of radiative coupling with intense electromagnetic fields on molecular and condensed phase dynamics. Problems of intramolecular dynamics in intense radiation fields fall into two major categories:

1. Effects of intense fields on energy acquisition, as is the case of multiphoton atomic and molecular ionization,[45, 46] as well as multiphoton molecular photodissociation and isomerization.[39–43]
2. Effects of intense fields on energy storage and disposal, which is exhibited in the novel and interesting field of radiative collisions[141, 142] and electronic energy transfer induced by intense radiation fields.[143]

Up to this point we were concerned with intramolecular dynamics in isolated molecules. The effects of medium perturbations on intermolecular and intramolecular dynamics are of importance in the understanding of the effects of collision-induced processes and medium perturbations in excited state reactivity in condensed phases. Collisional processes can be explored on the microscopic level by considering the role of van der Waals bonding on vibrational predissociation,[144] which results in vibrational relaxation.[145] The induction of collision-induced electronic relaxation[146] can be investigated by a similar approach. Turning now to some relaxation phenomena in condensed phases, we shall briefly consider separately intermolecular and intramolecular processes. Intermolecular processes involve electron transfer[148–150] from ground-state or electronically excited states, electron-hole recombination in semiconductors[150] and in amorphous solids,[151] group transfer processes,[152] and intermolecular electronic energy transfer.[153, 154] Intramolecular processes correspond to nonradiative electronic relaxation in ionic centers[155] and in insulators.[156] These processes can be described in terms of the decay of resonances, that is, metastable states of the entire system involving the reactive centers (electron or energy donor and electron or energy acceptor) coupled to the phonon field of the medium. One should notice the conceptual similarity between these phenomena and electronic relaxation in large molecules. The basic operative tools in the theory of intramolecular electronic relaxation phenomena in large molecules draws heavily on the concept of the weighted density of states, which bears a close analogy to the customary approach to elementary excitations in solids.

We shall now proceed to the discussion of intramolecular and intermolecular dynamics with an emphasis on the development of this field, not by defining it in the historical order but in the order of increasing complexity on the molecular level.

IV. PREDISSOCIATION

Unimolecular dissociation is the basic reactive process in an isolated, energy-rich molecule. In this section we pay particular attention to smaller molecules where the process is best described as one of predissociation.[25, 27] For larger polyatomics, intramolecular energy migration plays an important role and needs to be discussed in the first instance (Sections V and VIII). We shall then return to unimolecular processes in Section IX.

It is convenient to classify the predissociation processes in accordance with which degrees of freedom undergo a change of state (are "not adiabatic") during the dissociation. The basic distinction is between interstate and intrastate processes, the latter taking place on a single electronic potential energy surface. The simplest example of interstate electronic predissociation is the decay of a metastable, discrete vibronic level to another, repulsive, electronic configuration. As early as 1927 Wentzel[157] used the quantum-mechanical time-dependent perturbation theory to derive the familiar expression for the decay rate, which is nowadays usually referred to as the Fermi golden rule.[158] With appropriate changes the same expression applies to all other predissociation processes in which a change of state of any type takes place.

To describe the decay of an isolated metastable state n, it is convenient to assign to its energy ϵ_n an imaginary part

$$\epsilon_n = E_n - \frac{i\Gamma_n}{2} \tag{4.1}$$

The time-dependent part of the wavefunction, $\exp(-i\epsilon_n t/\hbar)$, is thus of diminishing amplitude. When excited, such a metastable state will exhibit a simple exponential decay, $|\exp(-i\epsilon_n t/\hbar)|^2 = \exp(-\Gamma_n t/\hbar)$ with a decay rate $\Gamma_n/\hbar$. It is the purpose of this section to identify the processes that give rise to a finite value of the "width" Γ_n and to examine the dependence of Γ_n on the relevant molecular parameters. We shall also comment on "inverse" predissociation or association processes which are important in recombination and other relaxation phenomena.

Intrastate processes cover the entire range from those where all degrees of freedom change adiabatically throughout to those where they are all strongly coupled. The latter are typical of thermally induced unimolecular processes in polyatomics, as discussed in Section IX. The simplest example of the former is rotational predissociation in diatomic molecules (Fig. 4). Consider a potential energy curve that can support bound states and where its long-range attractive part decreases typically as $C_6 R^{-6}$. For a finite value of orbital angular momentum J, the centrifugal barrier is repulsive and of longer range $[\hbar^2 J(J+1)R^{-2}]$. The resulting potential has a hump that can trap the system in the region of the well. Escape is by tunneling through the hump. The required theory in connection with α decay in nuclei was developed earlier by Condon and Gurney[159] and by Gamow.[160]

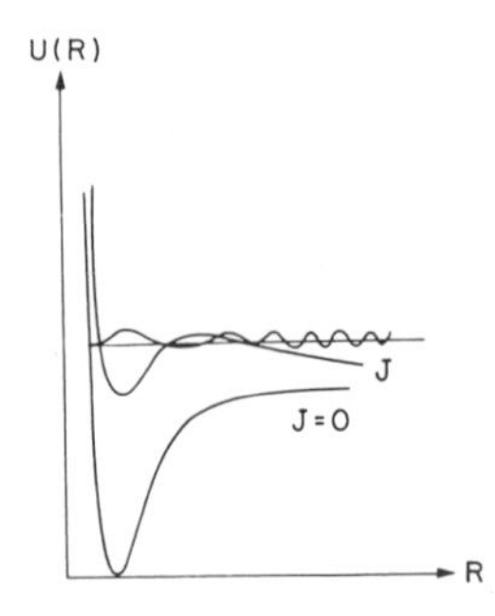

Fig. 4. Rotational predissociation in a diatomic molecule. The potential curve at finite values of the orbital angular momentum may support a metastable state, which escapes by tunneling through the hump in the potential.

The resulting expression for the decay rate is of the form

$$\frac{\Gamma_n}{\hbar} = \nu P \tag{4.2}$$

Here ν can be interpreted as the vibration frequency inside the well (i.e., the number of times per second that the system reaches the barrier), and P is the tunneling probability via the classically forbidden region.

An alternative approach is to solve the Schrödinger equation. If one does so, it is found that for most energies the wavefunction is very largely confined to the unbound region to the right of the barrier. At (and within a range of about Γ_n) certain energies the radial function has a large component within the well. These are the energies of the predissociating states. They can be approximated by the bound state energies in a problem where the barrier is made impenetrable (e.g., by making it much thicker). About such energies the density of states has a Lorentzian component

$$A_n(E) = \frac{\Gamma_n}{(E - E_n)^2 + (\Gamma_n/2)^2} \tag{4.3}$$

The delay time,[24] given by $\hbar A_n(E)$, is also maximal at E_n. These characteristics, that is, a large localized component of the radial wavefunction at an energy in the continuum, or equivalently an increase in the density of states/delay time, are usually referred to as a "resonance" or a quasibound state.

Such rotational predissociation is common in diatomics, providing the elementary dissociation channel on the ground-state potential curve. This process is expected to prevail for high-order multiphoton photofragmentation of diatomics, which can be accomplished only at extremely intense radiation fields,[39–43] and which has not yet been experimentally documented. Evidence for rotational predissociation in polyatomic molecules is scarce, some evidence being reported for the occurrence of this process in HNO.[27] Of considerable interest is the inverse rotational predissociation (IRP) process involving a collision of the two constituents with angular momentum $\hbar J$ and energy $\sim E_n$, which results in the emission of an infrared photon to a lower bound state of the diatomic molecule. This process can be envisioned as the decay from a translational continuum via an intermediate state to a radiative continuum. Such IRP processes (e.g., the recombination of $He + H^+$) are of considerable astrophysical interest.[161] IRP of diatomics involves essentially a radiative recombination process. For IRP involving atom–diatom collisions, with the diatom being in a vibrationally excited state, the decay channel can involve nonradiative vibrational predissociation. Rice[78] and Herschbach et al.[79] have recently suggested that the low-temperature vibrational relaxation of $I_2(B^3\Pi, v)$ by He occurs via IRP, with an orbiting resonance of $HeI_2(B^3\Pi, v)$ decaying into the dissociation channel $He + I_2(B^3\Pi, v-1)$.

All other predissociation processes on a given potential energy surface involve deexcitation of either rotational or vibrational degrees of freedom or both.[162] Vibrational predissociation involves the decay of a metastable vibrational level of a polyatomic molecule into the dissociative continuum corresponding to the same electronic configuration (Fig. 5). This process was treated by Rosen's 1933 theory[163] of decomposition of metastable molecules. The appropriate classical description of vibrational predissociation involves the Lissajous motion of an image point on a multidimensional potential surface to the dissociative region. Unambiguous experimental evidence for vibrational predissociation in conventional polyatomic molecules was not well documented, as the dividing line between this intrastate process and the interstate electronic predissociation is vaguely defined.[27] From the theoretical point of view, a complete picture of vibrational predissociation of ordinary molecules requires the description of the nuclear level structure at highly excited vibrational states, and the elucidation of intramolecular intrastate vibrational energy flow in a strongly scrambled dense manifold, which in turn is coupled to a dissociation continuum.

Of considerable current interest are the intrastate predissociation processes that involve rotational or vibrational deexcitation of weakly bound complexes. Here the presence of an attractive well plays a key role. As the simplest example, consider a collision between an atom and a diatomic molecule at an energy not sufficient to, say, excite the vibration of the molecule. As the two approach, the attraction increases the available kinetic energy with the result that the diatomic can be excited, while the atom is now bound by the attractive force. The resulting species (see Fig. 6) is only quasibound, since it can predissociate by the process of transfer

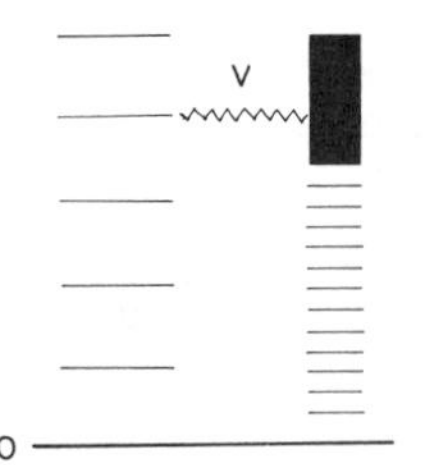

Fig. 5. Vibrational predissociation in a polyatomic molecule. Zero-order vibrational states located above the threshold of the dissociative continuum, corresponding to the same electronic configuration, are metastable, exhibiting an intrastate reactive relaxation process.

of the vibrational excitation of the diatomic back to the translation (V–T). Such quasibound states can also be formed by optical excitation from lower-lying states[66, 67] and can dissociate by either V–T or R–T processes, or both of them. This mechanism is quite general, since even in the absence of chemical attraction the longer range van der Waals forces suffices to support bound states, and hence to give rise to such "subexcitation" quasibound states. In a predissociation, via an intramolecular V–T or R–T process, the energy required to promote the translation to a dissociative state is provided by the deexcitation of either the vibration or the rotation. Since even van der Waals wells are tens of wavenumbers or more deep, R–T predissociation requires a deexcitation of a rotational

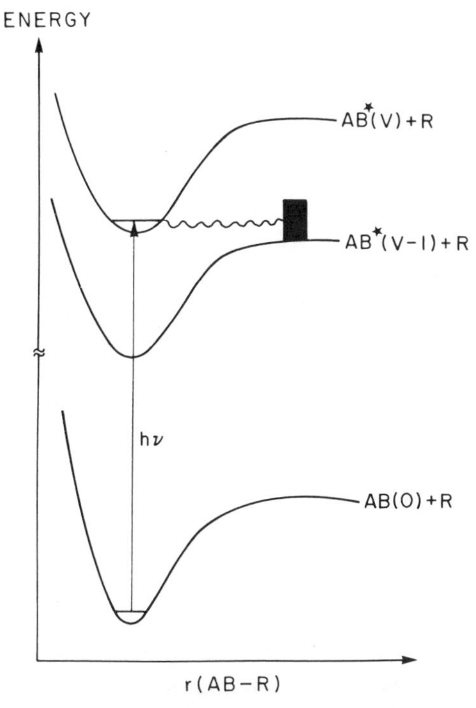

Fig. 6. Vibrational predissociation of the triatomic van der Waals molecule AB–R, where AB is a conventional diatomic molecule and R represents a rare-gas atom. Optical excitation from the ground electronic–vibrational state $AB(v=0)-R$ of the complex to the electronically–vibrationally excited state $AB^*(v)-R$ results in a quasibound state. This metastable state can decay via V–T into the translational continuum $AB^*(v')+R$, with $v'<v$, corresponding to lower vibrational states of the diatomic molecule. The VP process $AB^*(v)-R \rightarrow AB^*(v-1)+R$ is denoted by a horizontal wiggly line.

mode with a high B constant and would thus be typical of, for example, rare-gas-hydrogen-halides van der Waals bound pairs.[165]

Recently, photofragmentation of the van der Waals molecules I_2R, ($R = He, Ne, Ar$) upon excitation of the I_2 molecule, has been demonstrated.[66, 67, 164] Here the quasibound state is prepared by optical excitation to a vibrationally excited state of the $^3\Pi$ electronic state of I_2 and decays by a V–T transfer. For the simple case of dissociation of AB–R by direct intramolecular V–T transfer (e.g., as is the case in I_2He), the rate of dissociation can be semiquantitatively described in terms of an exponential gap law[144]

$$\Gamma = h\nu \exp\left[-\pi d\left(\frac{2\mu\epsilon}{\hbar}\right)^{1/2}\right] \tag{4.4}$$

Here [cf. (4.2)] ν is the effective frequency for the bound relative motion (of AB–R), and the second factor can be interpreted as the probability of the V–T transfer. d is the range of the AB–R interaction and μ is the reduced mass. ϵ denotes the released kinetic energy of the fragments, which is determined by the mismatch between the frequency of the AB vibration and the binding energy of the AB–R mode. Such probability factors have indeed also been extensively discussed by V–T transfer processes in AB–R collisions.[2]

The exponential gap relation (4.4), which was derived[144] from the collision theory analysis is one member of a family of such relations. Equation 4.4 is isomorphous to the conventional Gamow formula (4.2) for tunneling of a particle of mass μ over a barrier of height ϵ and spatial extent d, providing a semiclassical description of intrastate overlap between bound and continuum nuclear states. The variation of the lifetimes of various triatomic AB–R complexes, with respect to the V–T vibrational predissociation process, exhibits a strong energy gap dependence and can vary over 20 orders of magnitude from the picosecond domain up to lifetimes comparable to the age of the universe.[144] This admittedly oversimplified description, which disregards rotational effects as well as the rate of the R–AB vibrational bending mode, constitutes the essential conceptual framework for the understanding of vibrational predissociation in small model systems. Furthermore, these studies provide compelling evidence that the intramolecular vibrational energy flow in the AB–R complex is not necessarily fast, in contrast to the intuitive arguments.

The energy-gap law addresses the dynamics of a direct vibrational predissociation process, where a single discrete state is coupled to a dissociative continuum, as is the case for HeI_2. More complex V–T processes in triatomic complexes can be exhibited when the van der Waals bond is relatively strong. For example, the ArI_2 is characterized by a dissociation energy of 220 cm^{-1}, the van der Waals potential well supporting a large number of bound states.[168] In the rich level structure, accidental degeneracies will be encountered between bound states corresponding to different vibrational states of the molecular modes (Fig. 7). This state of affairs will result in accidential vibrational predissociation, a process which bears a close analogy to electronic accidental predissociation.[25]

More complex dynamic processes are exhibited in van der Waals molecules formed from molecular building blocks. Of considerable interest are homodimers, such as the Cl_2–Cl_2 complex, whose vibrational predissociation was experimentally interrogated.[169] Intramolecular dynamics in such homodimers involves both nonreactive intramolecular vibrational energy flow between the two molecular bonds as well as reactive vibrational energy exchange between the molecular bond and the van der Waals bond. The interplay between the strength of near-resonant coupling between discrete states and the widths of the reactive channels, which are both very sensitive to the potential parameters, determine the relative role of the nonreactive and the reactive energy flow processes. For the $(F_2)_2$ linear dimer the direct interbond coupling is expected to be efficient on the time scale of fragmentation; this vibrationally excited dimer will play a game of musical chairs, oscillating $\sim 10^8$ times between the bond modes before slow fragmentation occurs.[170] On the other hand, the $(Cl_2)_2$ dimer is characterized by exceedingly weak discrete–discrete coupling and no oscillatory energy exchange between the two bond modes is exhibited on the time scale of the reactive process.[170] In both homodimers the reactive V→T process is expected to be rather slow, occurring on the 10^{-3} sec time

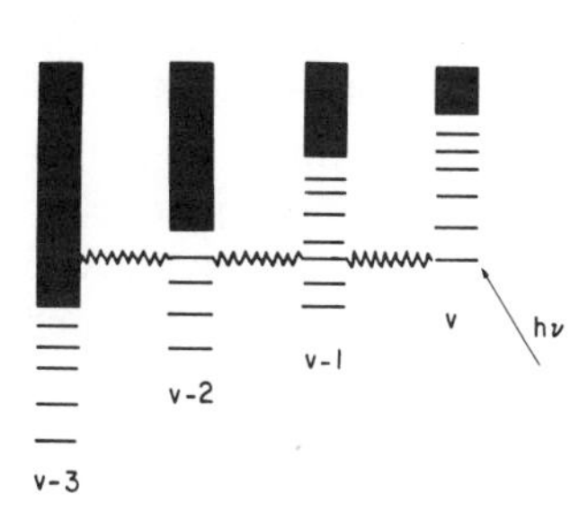

Fig. 7. A schematic description of accidental vibrational predissociation in AB–R van der Waals complex, which is characterized by high binding energy relative to the frequency of the diatomic molecule. The van der Waals potential supports a large number of discrete quasibound states. The channel corresponding to the one quantum jump $v \rightarrow (v-1)$ of the initially excited state is closed and the first open channel corresponds to a three quantum jump $v \rightarrow (v-3)$. Accidental degeneracies between discrete states result in VP intermediated by coupling between nearly discrete levels.

scale,[170] which is in accord with the energy gap law (4.4), and is in agreement with the scarce experimental information.[169] An interesting counterintuitive conclusion emerges at this point concerning inefficiency of intramolecular vibrational energy exchange in small complexes, such as van der Waals homodimers.

While the reactive channel in homodimers involves solely the V→T channel, new avenues open up for reactive processes in heterodimers AB–CD, consisting of a pair of distinct diatomic molecules. In these systems an intermolecular V→V+T process can be exhibited, with the vibrational energy of one molecular bond being interconverted into the vibrational energy of the second bond, while the energy balance is made up by the translational energy. Accordingly, the energy gap ϵ in (4.4) is considerably reduced, whereupon the vibrational predissociation rate can be significantly enhanced. These examples are sufficient to demonstrate how investigations of the fragmentation of small van der Waals complexes provide a rich and diverse range of tests of novel experimental and theoretical concepts in the area of vibrational (V–T) predissociation. The basic understanding of these reactive processes in such small molecular systems rests on quantum-dynamic models, while statistical models, such as the celebrated RRKM approach, are inadequate. It is thus apparent that for these basic unimolecular fragmentation processes chemical intuition has to be supplemented and complemented by the results of a microscopic theory of intramolecular dynamics.

Of considerable interest is the "transition" from the small molecular case, where a detailed dynamic picture is required, to the "statistical limit" for vibrational predissociation dynamics in a large system (Section V). Such a transition from small to "statistical" systems can be explored by the study of vibrational predissociation of I_2R_n complexes, with a gradual increase of the coordination number n. With increasing size of the complex, the number of the vibrational modes involving the rare-gas atoms increases linearly with n. There is now a substantial number of torsional and bending modes of the complex, which do not lead to dissociation. When the size of the complex increases, the density of states of these modes increases. For large complexes these will act as accepting modes for vibrational energy flow from the molecular mode. Such intramolecular, nonreactive, V–V transfers will delay the reactive V–T transfer. Consequently, the dissociation process in a large "statistical" I_2R_n complex may be retarded, as compared to the I_2R small complex. The results of Levy[171] on the dynamics of I_2Ne_n ($n=1,2\cdots 6$) complexes are in accord with such a picture. It will be extremely intriguing and interesting to establish the crossing point between the reactive dynamics of small systems and of the statistical ones. Preliminary observations are discussed in Section VI.

We now depart from the intrastate rotational and vibrational predissociation processes and proceed to consider interstate electronic predissociation. Electronic predissociation is traditionally envisioned in terms of decay of metastable bound states, which belong to some electronic configuration into a dissociative continuum corresponding to another electronic state. This state of affairs (Fig. 8) is well known for diatomics, where the rate Γ_{ER} for electronic predissociation is expressed in terms of the celebrated Golden Rule[157]

$$\Gamma_{ER} = \frac{2\pi}{\hbar}|\langle n|V|f\rangle|^2\rho_f \qquad (4.5)$$

where $|n\rangle$ and $|f\rangle$ are the electronic nuclear wavefunctions for the discrete and continuum states, respectively, while ρ_f is the density of continuum states, and V is the residual perturbation term which induces the nonradiative electronic transition. This decomposition mechanism prevails in the multiphoton photodissociation of the triatomic OCS molecule,[172] the smallest system which was decomposed via a collisionless high-order multiphoton process. The dissociation energy of the ground-state OCS ($^1\Sigma$) yielding CO($^1\Sigma$) and excited S(^1D) is 97.6 kcal mole^{-1}. There exists a nonbonding triplet state dissociating into CO($^1\Sigma$)+S(^3P) with a threshold at 68 kcal mole^{-1} (Fig. 8). The interstate nonadiabatic coupling is provided by spin–orbit coupling V_{SO} and the electronic relaxation rates of the

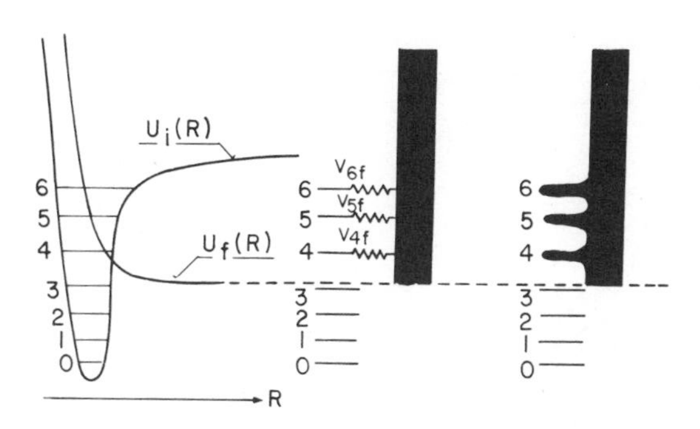

Fig. 8. Electronic predissociation resulting from interstate coupling between two potential surfaces. (a) One-dimensional potential curves. (b) Zero-order levels corresponding to the two electronic states. Near-resonance coupling between discrete and continuum zero-order states is designated by arrows. (c) Nonoverlapping resonances appear above the continuum threshold. This description is adequate for electronic predissociation of diatomics and was applied to electronic predissociation of the linear molecules N_2O, CO_2 and OCS (ref. 173) in a one-dimensional world.

metastable states are characterized by the vibrational state $\mathbf{v}$

$$\Gamma = \frac{2\pi}{\hbar} |V_{SO}|^2 |\langle \mathbf{v} | f \rangle|^2 \rho_f \tag{4.6}$$

leading to a conventional microscopic rate, which is determined by a product of an electronic term, and a vibrational overlap Franck–Condon factor.[173]

Of considerable interest is the process of inverse electronic predissociation (IEP), which involves radiative recombination intermediated by a molecular resonance. The initial state in inverse predissociation is a wavepacket of translational states coupled by intrastate interaction to a zero-order bound state, which in turn is coupled to a radiative continuum originating from spontaneous one-photon decay to the ground electronic state (Fig. 9). This process can be described within the framework of scattering theory,[24] the transition probability at resonance for IEP of a diatomic molecule being given by $(2\pi/\hbar\rho_d)\Gamma_V\Gamma_R(\Gamma_V+\Gamma_R)^{-2}$, while the branching ratio for radiative recombination is Γ_R/Γ_V, where ρ_d is the density of states in the dissociative continuum, while Γ_R and Γ_V are the radiative width and the predissociative width, respectively. The recent observation of inverse electronic relaxation (Section VII) induced by multiphoton excitation in a bound level structure of a large molecule bears some analogy to the IEP process. However, the latter proceeds via a dense superposition (wavepacket) of continuum translational states.

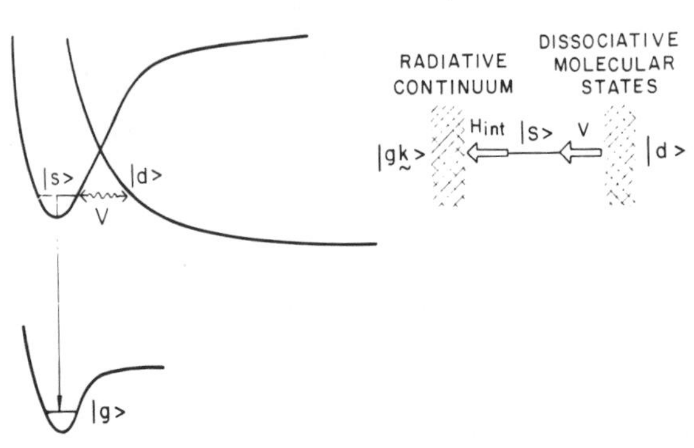

Fig. 9. Inverse electronic predissociation of a diatomic molecule. A dissociative state $|d\rangle$, or rather a wavepacket of such dissociative states, is coupled by the nonadiabatic interaction V to a discrete zero-order $|s\rangle$ state, which in turn is radiatively coupled to the ground state $|g\rangle$.

V. NONREACTIVE MOLECULAR PROCESSES

Two notable features of the unimolecular decomposition processes, with which we were just concerned, should be emphasized. First, all the predissociation processes are essentially irreversible, as they involve the decay of a metastable state into a dissociative channel. Second, the density of states in the dissociative channel, as well as the intrastate and interstate coupling terms, vary smoothly and slowly with energy, whereupon the reactive intramolecular processes involve irreversible relaxation into a smooth dissipative continuum. In addition to these reactive processes, electronically–vibrationally excited states of large molecules also exhibit intramolecular nonreactive phenomena, which involve intrastate vibrational energy redistribution (IVR), as well as interstate electronic relaxation (ER). These nonreactive processes differ quantitatively and qualitatively from the reactive predissociation phenomena with respect to the following three major features. First, the molecular level structure is located below any dissociation (or ionization) threshold, so that intramolecular dynamics is exhibited within a bound molecular level structure. This bound intramolecular level structure is congested, being characterized by a high density of states.[174] Nonreactive relaxation occurs within an intramolecular quasicontinuum, in contrast to predissociation which takes place into a continuous spectrum. Second, the level spacing between adjacent states within the quasicontinuum varies in an irregular way and the notion of density of states can be defined only in terms of a coarse-grained quantity, while in the case of a dissociative channel the density of states varies smoothly with energy. Third, the interstate or intrastate coupling terms exhibit a wild and wide variation with respect to the quantum numbers specifying the zero-order states, in contrast to the case of predissociation where the coupling terms exhibit a slow and smooth energy dependence. We can assert that intrastate and interstate nonreactive relaxation processes involve time evolution within a "bumpy" quasicontinuum.

There is a close formal analogy between the general features of interstate ER and intrastate IVR in large molecules, which can be described within a unified conceptual framework. Both processes manifest the consequences of time evolution within a bound quasicontinuum which, according to the discussion of Section (I.B), should now be described in terms of the characterization of the level structure and the excitation amplitudes, as well as the specification of the initial conditions for intramolecular dynamics. At this stage we have to address the important technical problem of the choice of zero-order basis for the proper description of intramolecular dynamics. The traditional separation of the molecular Hamiltonian[8]

$H_M = H_{MO} + V$, into a zero-order part H_{MO} and a perturbation term V rests on the following guidelines:

A. Similarity Criteria

1. The energies of the zero-order states should be close to those of the true levels.
2. Small off-resonance coupling. The zero-order basis set is chosen to minimize off-resonance interactions. The effect of off-resonance interactions can either be entirely disregarded or incorporated as perturbative correction terms.

The similarity criteria imply that the spectrum of H_{MO} is close to that of H_M. From the point of view of general methodology, these criteria make it possible to use a truncated basis set, e.g., a two-electronic level structure for interstate coupling. Next, we have to invoke an additional criterion for the choice of a zero-order basis which pertains to the "preparation" of the initial metastable state.

B. Accessibility Criterion

Only a single (or a small number of) doorway state(s) is (are) accessible to optical excitation. In the case of one-photon excitation only the doorway state carries oscillator strength from the ground state, while for multiphoton excitation a small number of zero-order states are radiatively coupled in each energy region.

The accessibility criterion is not necessary for the choice of the zero-order molecular basis. Nevertheless, the concept of a doorway state is essential for a meaningful characterization and specification of time-resolved observables.

These general criteria for a practical dissection of the molecular Hamiltonian pertain both to interstate ER and to intrastate IVR in large molecules, whose features are summarized in Table III. In the case of interstate coupling between vibronic levels corresponding to two distinct electronic manifolds, Born–Oppenheimer pure-spin states usually provide the adequate zero-order description of the level structure. For intrastate coupling the choice of the zero-order basis is less transparent and may involve X—H bond modes in molecules containing hydrogens ("Bond Modes," Chapter 4) or normal vibrational modes, such as the ν_3 mode in the celebrated SF_6 molecule.[40, 93, 175]

The level structure for a bound spectrum, where a single doorway state $|s\rangle$ is accessible to one-photon excitation from the ground state, is portrayed in Fig. 10. Such a ladder diagram was first advanced about ten years ago[176] for the case of interstate coupling but is also applicable for the

TABLE III
Interstate and Intrastate Level Scrambling

Feature	Electronic Relaxation	Intramolecular Vibrational Energy Redistribution
Coupling between zero-order states V	Nuclear kinetic energy or Spin–orbit coupling	Kinetic energy Anharmonicity
Single doorway state	Born–Oppenheimer state	C—H bond mode ν_3 normal mode of SF_6 (?)
Sequential decay widths of zero-order manifold γ_l	Infrared decay or Optical decay	Infrared decay
Molecular eigenstates	Mixed BO states of 2 electronic configurations	Mixed anharmonic states of 1 electronic configuration
Occurrence of intramolecular relaxation (or dephasing)	1. In statistical limit 2. Short t excitation of intermediate level structure	1. In statistical limit 2. Short t excitation of intermediate level structure

case of intrastate scrambling. The time evolution within such discrete spectrum should be envisioned in terms of the dynamics of wavepackets of bound states. For the simple level structure of Fig. 10, the molecular eigenstates[7, 8, 174] of H_M are given in terms of the superposition of the doorway state[8, 13] $|s\rangle$ and background manifold $\{|l\rangle\}$

$$|m\rangle = a_s^m|s\rangle + \sum_l b_l^m|l\rangle \tag{5.1}$$

with energies E_m. For characterization of the time-resolved experiments the dynamics is completely specified in terms of the state of the system at the time t

$$\psi(t) = \sum_m A_m \exp\left(\frac{-iE_m t}{\hbar}\right)|m\rangle \tag{5.2}$$

where A_m are the preparation amplitudes. The relevant observables, such as the probability $P_0(t)$ for finding the system in its initial state

$$P_0(t) = |\langle\psi(0)|\psi(t)\rangle|^2 = \left|\sum_m |A_m|^2 \exp(-iE_m t/\hbar)\right|^2 \tag{5.3}$$

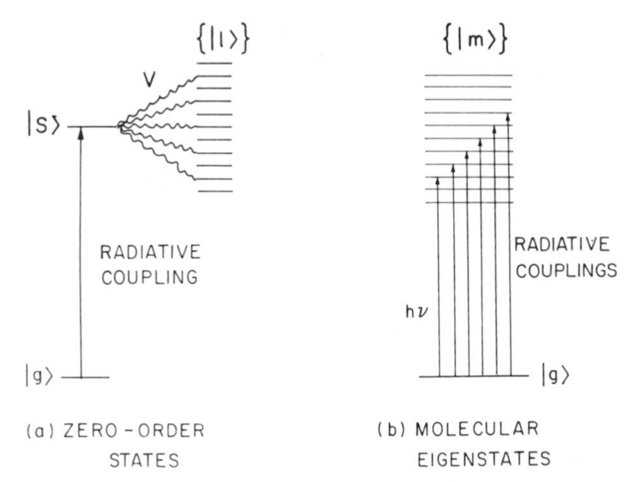

Fig. 10. A molecular energy-levels model used to discuss coupling and nonreactive relaxation in a bound-level structure in excited states of large molecules. This model was originally introduced by Bixon and Jortner[176] to discuss interstate coupling and electronic relaxation, but is also adequate to describe intrastate coupling and intramolecular vibrational energy redistribution. The zero-order molecular levels $|g\rangle$, $|s\rangle$, and $\{|l\rangle\}$ are Born–Oppenheimer states for interstate dynamics or appropriate nuclear zero-order states for intrastate dynamics. They correspond, respectively, to the ground state $|g\rangle$, the (one-photon) optically accessible doorway state $|s\rangle$ and the background manifold $\{|l\rangle\}$. The radiative coupling connects $|g\rangle$ and $|s\rangle$, while the transition moments from $|g\rangle$ to the $\{|l\rangle\}$ manifold are assumed to vanish. The wiggly arrows represent the intramolecular interstate or intrastate coupling. The molecular eigenstates $|m\rangle$ diagonalize the molecular Hamiltonian and are all radiatively coupled to the ground state.

or the decay possibility $P_D(t)$, e.g., spontaneous emission of the system,

$$P_D(t)=\left|\sum_m A_m B_m \exp\left(\frac{-iE_m t}{\hbar}\right)\right|^2 \tag{5.4}$$

where $\{B_m\}$ represent the decay amplitudes of the $\{|m\rangle\}$ states, being explicitly expressed in terms of Fourier sums. Analysis of such Fourier sums has to consider the energetic spread of the energies $\{E_m\}$ and the specification of the excitation and decay amplitudes.[13, 15, 16] Before doing so, it is imperative to extend the concept of molecular eigenstates incorporating genuine decay processes, such as spontaneous (infrared or optical) emission or intramolecular decomposition[13] by assigning the decay widths γ_s and $\{\gamma_l\}$ to the zero-order states (Fig. 11). The resulting (complex) molecular eigenstates $|m\rangle$, usually referred to as independently decaying

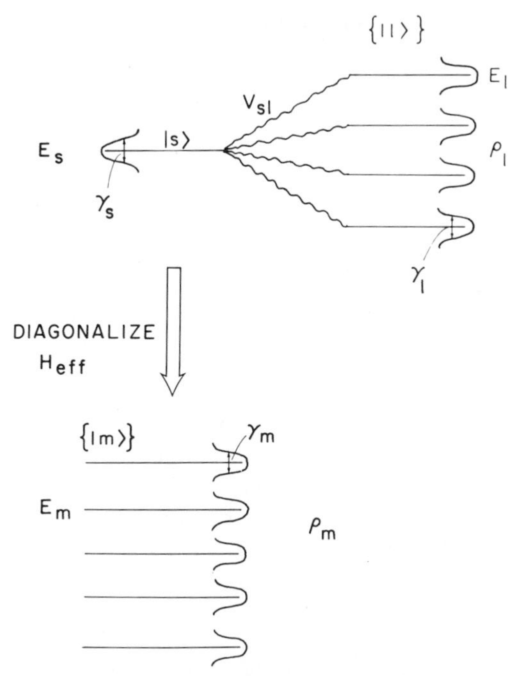

Fig. 11. The effective Hamiltonian formalism. The zero-order states $|s\rangle$ and $\{|l\rangle\}$ are characterized by the energies E_s and $\{E_l\}$, respectively, and by the decay widths γ_s and $\{\gamma_l\}$. V_{sl} represents the intramolecular (interstate or intrastate) coupling between the doorway state $|s\rangle$ and the $\{|l\rangle\}$ manifold, which is characterized by the density of states ρ_l. Diagonalization of the effective Hamiltonian results in a set of independently decaying levels $\{|m\rangle\}$, i.e., generalized molecular eigenstates, characterized by energies $\{E_m\}$, decay widths $\{\gamma_m\}$, and density of states ρ_m.

levels, are obtained from the diagonalization of the effective Hamiltonian[13–17] $H_{\text{eff}} = H_{\text{M}} - (i/2)\gamma$, where γ is the decay matrix (See Fig. 11). The $\{|m\rangle\}$ states are now characterized by the complex energies

$$\epsilon_m = E_m - \frac{i}{2}\gamma_m \tag{5.5}$$

where the decay widths, γ_m, are obtained in a self-consistent manner. These complex energies (5.5) provide an extension of relation (4.1) from the case of a single resonance to handle the decay of a manifold of metastable states, constituting the generalization of the Fermi golden rule. The relevant time-resolved observables (5.3) and (5.4) are readily obtained by substitution of the complex energy (5.5), being given in the form

$$P_0(t)=\left|\sum_m |A_m|^2 \exp\left(\frac{-iE_m t}{\hbar}-\frac{\gamma_m t}{2\hbar}\right)\right|^2 \tag{5.3a}$$

$$P_D(t)=\left|\sum_m A_m B_m \exp\left(\frac{-iE_m t}{\hbar}-\frac{\gamma_m t}{2\hbar}\right)\right|^2 \tag{5.3b}$$

which constitute Fourier sums damped by real decay components. These general results provide a unified framework for the understanding of all nonreactive processes for IVR in vibrationally excited states and for IVR and ER in vibrationally-electronically excited states of large molecules.

We can obtain significant molecular information from the classification of the level structure, which pertain to the effective intrastate or interstate scrambling, i.e., $V_{sl}\rho_l \gg 1$. The interstate scrambling reflects the breakdown of the Born–Oppenheimer separability, while intrastate mixing can be described in terms of many-level-Fermi resonances. We shall now focus attention on simple excitation conditions, which for the time being will be restricted to one-photon optical excitation in weak radiation fields. The three limiting situations for intramolecular dynamics are:[8, 13–17]

1. The Small Molecule Limit

Here the density of states in the $\{l\}$ manifold is sparse, i.e., $\rho_l\gamma_l \ll 1$ and the molecular eigenstates are well separated, their energetic spread being relative to the coherent energetic width of any available light sources. In general, each molecular eigenstate acts on its own in absorption and in emission so that only photoselective excitation of individual molecular eigenstates is feasible. The resulting decay is expected to be exponential and is determined by the decay rate[13, 16, 101]

$$\gamma_m \simeq \gamma_l + \frac{\gamma_s}{N} \tag{5.6}$$

where $N=\pi^2|V_{sl}|^2\rho_l^2$ is the dilution factor and $N \gg 1$. For interstate coupling $\gamma_m \ll \gamma_s$, a dilution of the radiative and nonradiative decay probabilities of the doorway state will be exhibited. This general pattern of a "diluted" exponential decay can be violated by thermal rotational congestion and vibrational congestion effects. For intrastate coupling $\gamma_s \simeq \gamma_l \equiv \gamma$ for all zero-order states and $\gamma_m \simeq \gamma(1+1/N)$ for all molecular eigenstates. In this limit a single molecular eigenstate is photoselected, exhibiting exponential decay. However, subtle effects may originate from accidental near-degeneracy of a small number of levels which can then be coherently excited. In the latter case, oscillatory quantum beats are expected to be exhibited in the decay, although such effect has not yet been experimentally documented. The small molecule limit exhibits the consequences of intramolecular scrambling but no relaxation or dephasing effects are expected to be exhibited in the isolated molecule.

2. *The Statistical Limit*

In this case, effective intramolecular coupling and relaxation prevail. The density of states is very large relative to the sequential (infrared or optical) decay widths of the background states

$$\gamma_l \rho_l \gg 1 \tag{5.7}$$

Furthermore, a wavepacket of all the molecular eigenstates contaminated by the doorway state can be coherently excited so that in the short time domain, $\gamma_m t \ll 1$, the decay mode of the wavepacket assumes the familiar form of a Fourier sum of the absorption strengths. On the time scale shorter than the characteristic recurrence time,[174] $t \ll h\rho_l$, the exponential decay law is obtained

$$P_0(t) \propto \exp\left[-\frac{(\Gamma_s + \gamma_s)t}{\hbar}\right] \tag{5.8}$$

where the nonradiative rate is given by

$$\Gamma_s = 2\pi \langle V_{sl}^2 \rho_l \rangle \tag{5.9}$$

with the average being taken over the $\{|l\rangle\}$ manifold. In the statistical limit Γ_s is the only intramolecular nonradiative experimental observable. One is lead to the familiar results concerning (nearly) Lorentzian broadening of the doorway state in the absorption spectrum, (nearly) exponential decay mode, shortening of the experimental lifetime relative to the pure radiative lifetime and decrease of the quantum yield below unity. The width Γ_s (5.9) can be considered to constitute a genuine intramolecular decay rate as a condition (5.7) for overlapping resonance, together with the long recurrence time $h\rho_l$, insure irreversible decay in the bound level structure which corresponds to the statistical limit. The nonreactive relaxation in the statistical limit in an isolated molecule retains the molecular excitation in the energy range corresponding to that of the doorway state. Two supplementary and complementary points of view can be advanced to understand such intramolecular relaxation, that is, intramolecular E–V for ER or intramolecular V–V for IVR, in an isolated large molecule. First, one can adopt the conventional attitude, considering the depletion of the doorway state $|s\rangle$ as a genuine intramolecular relaxation process, which can be interrogated by time-resolved photon counting. Second, the time evolution of the coherently excited wavepacket can be considered (see Section XI) in terms of an intramolecular dephasing process, which can be experimentally explored by the study of coherent optical effects. Identical physical information stems from conventional photon counting and from the fancy coherent optical probes conducted on a statistical level structure of an isolated molecule.

3. *Intermediate Level Structure*

This situation, intermediate between the small molecule and the statistical limit, is characterized by moderately large density of states. However, these states are still well separated relative to their decay widths, i.e., $\gamma_l \rho_l < 1$. Under these circumstances, the coarse-graining procedures inherent in the description of the statistical limit are inapplicable. When a coherent excitation of the wavepacket is possible two decay components are expected to be exhibited,

$$P_0(t) \propto \exp[-(\gamma_s + \Gamma_s)t] \qquad t < \rho_l^{-1}, \gamma_l^{-1} \tag{5.10}$$

$$P_0(t) \propto \sum_m |A_m|^4 \exp(-\gamma_m t) \qquad t > \rho_l^{-1}, \gamma_l^{-1} \tag{5.11}$$

The short-time component (5.10) exhibits the dephasing of the initially excited wavepacket and, as in the statistical limit, can be interrogated by photon counting or by probing of coherent optical effects. The long-time decay component (5.11) reveals a sum of exponentials when all phase relationships between the molecular eigenstates were eroded, and each molecular eigenstate decays on its own. The long-time decay rates γ_m exhibit the dilution effect relative to the radiative and nonradiative decay widths of the doorway state.

The classification of intramolecular level structure was advanced several years ago[8] for interstate coupling and ER in electronically excited states of large molecules. For the sake of general methodology, it is important to emphasize that this approach provides a comprehensive, complete, and general description of both interstate ER and intrastate IVR intramolecular dynamics in excited states of polyatomic molecules. Several illustrative examples from the areas of intrastate and interstate coupling may be of some interest. The small molecule case for interstate coupling is dramatically demonstrated in the optical absorption of NO_2 in a supersonic beam[64, 65] (Fig. 12), which reveals in the spectral range 7000–5000 Å about 150 molecular eigenstates originating from scrambling between a small number of electronically excited vibronic states and the electronic ground-state manifold. An analogous example for intrastate coupling in the small

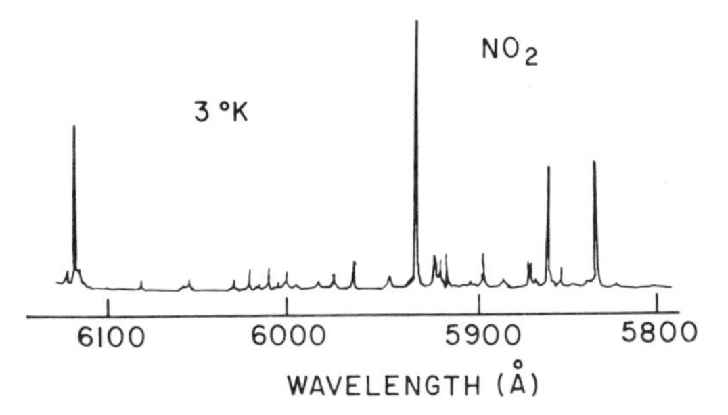

Fig. 12. Spectroscopic implications of interstate coupling in the small molecule case. The fluorescence excitation spectrum of NO_2 cooled in a supersonic free expansion to a rotational temperature of 3 K was taken from the work of Smalley, Wharton, and Levy (Refs. 65, 66). This electronic spectrum reveals a rich vibrational structure originating from interstate mixing between a small number of electronically excited vibronic levels and the ground-state vibrational manifold.

molecule limit is the $5\nu_1$ absorption band of NH_3 at 6475 Å, which is exhibited in the atmospheres of Jupiter and Saturn.[177] The rich structure of molecular eigenstates (Fig. 13) reflects the spectroscopic implications of intrastate scrambling. Proceeding to the intermediate level structure for interstate coupling, which can be exhibited in some electronically excited states of large molecules that satisfy two conditions. First, the upper electronic state is characterized by a small electronic energy gap. Second, the sequential decay widths of the background states are small, for example, the background states correspond to a triplet manifold, or to singlet states whose radiative decay to the ground state is symmetry-forbidden. Figure 14 reproduces the experimental results of van der Werf and Kommandeur[178] for the time-resolved decay of the S_1 state of the "iso-

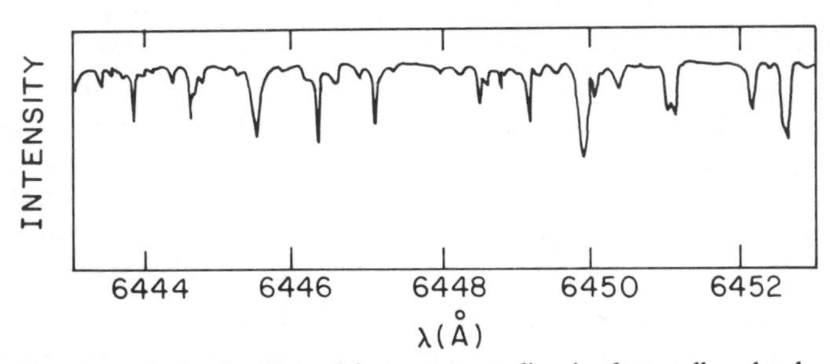

Fig. 13. Spectroscopic implications of intrastate coupling in the small molecule case. The room temperature absorption spectrum of the $5\nu_1$ overtone of the ammonia molecule was taken from the work of Giver, Miller, and Boese [177]. This spectrum reveals a rich structure originating from intrastate scrambling.

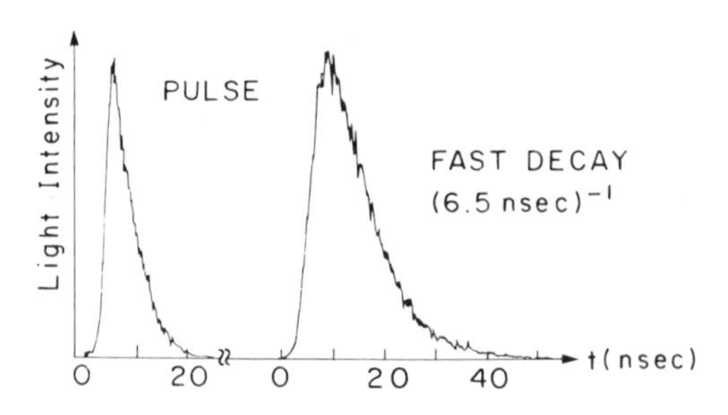

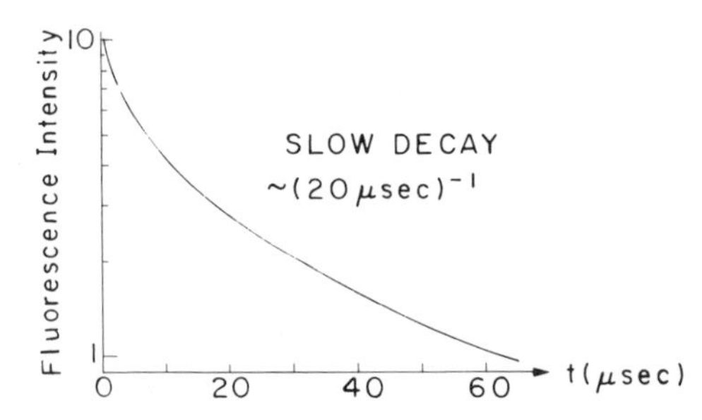

Fig. 14. Time-resolved emission spectrum of the first singlet state of the biacetyl molecule excited at 3725 Å in a low-pressure bulb (0.3 torr) at room temperature. Data taken from the work of van der Werf and Kommandeur[178]. The laser profile is denoted by "Pulse." The short-time decay component is exponential with a lifetime of 6.5 nsec, which presumably reflects "dephasing" of the initially excited wavepacket of S_1-T_1 mixed molecular eigenstates. The nonexponential decay on the time scale of ~20 μsec reflects the independent decay of the molecular eigenstates. These data demonstrate the characteristics of decay resulting from "coherent excitation" of an intermediate level structure for interstate coupling.

lated" biacetyl molecule, where the S_1-T_1 coupling presumably corresponds to the intermediate level structure and which exhibits the two-component decay, as predicted by (5.10) and (5.11). The only difficulty inherent in the interpretation of these low-pressure room temperature data involves the effects of vibrational sequence congestion, resulting in severe thermal inhomogeneous broadening, which may erode the effects of coherent excitation. The "transition" from the small molecule limit to the statistical limit may be exhibited in the same molecule for different electronic configurations in the case of interstate coupling and at different levels of vibrational excitation within the same electronic manifold for the case of intrastate scrambling. A good example for the novel area of intrastate coupling is provided by the absorption lineshapes for the C—H vibrational overtones of the "isolated" benzene molecules studied by Bray

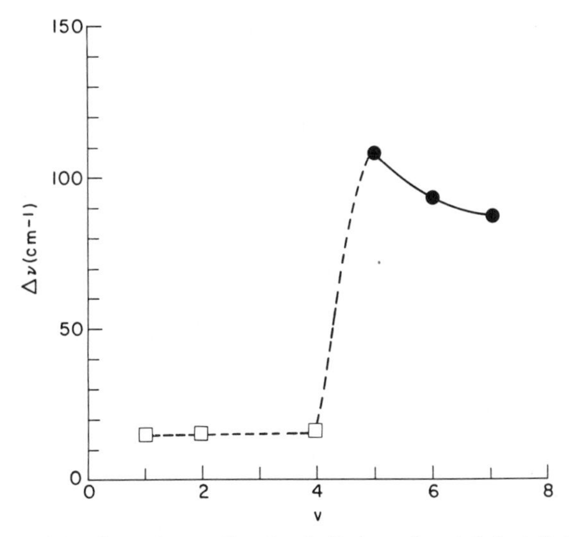

Fig. 15. The transition from the small molecule limit to the statistical limit for intrastate coupling. The dependence of the halfwidth $\Delta\nu$ of the C—H overtone spectra of the benzene molecule in the low-pressure gas phase on the vibrational quantum number v are taken from the work of Bray and Berry[179]. The low v data represent an upper limit for the homogeneous line broadening. The transition to large values of $\Delta\nu$ at $v>4$ reflects the transition to the statistical limit, where a C—H bond mode decays to a quasicontinuum of skeleton modes.

and Berry.[179] These C—H bond modes constitute adequate doorway states for one-photon excitations. The low C—H overtones exhibit the effects of Fermi resonances, manifesting intrastate scrambling, while the high v C—H excitations correspond to the statistical limit, exhibiting a Lorentzian absorption lineshape. The dependence of the absorption linewidths on the vibrational overtone v (Fig. 15) demonstrates the "transition" from the small molecule case to the statistical limit for intrastate coupling.

From the foregoing discussion it is apparent that the "two cultures" involving interstate and intrastate coupling and dynamics can be reconciled, being described within a unified conceptual framework. These central ideas based on time evolution of wavepackets of bound states, which is well established and accepted for interstate coupling and ER,[8–17] have been extremely useful and fruitful in this general area loosely referred to as "electronic radiationless transitions." The application of the same concepts to intrastate coupling and IVR is conceptually straightforward, and its practical implementations are extremely useful, supplementing and even replacing some traditional current approaches, which are based on classical mechanics, to the important problem of intrastate intramolecular vibrational energy flow.

As was discussed in Section IV, intrastate vibrational energy redistribution also takes place during the predissociation of an energy rich polyatomic molecule. Such reactive processes are further discussed in Section VIII. It should, however, be emphasized that the phenomena of IVR, as discussed in Sections V and VIII, is one and the same. In particular, during the up-pumping of a molecule by a multi(infrared)-photon process, the energy is below the dissociation threshold and the IVR process is precisely the one discussed in this section. Indeed, the "second" type of energy selectivity discussed in Section VIII is essentially the phenomena of IVR as considered here. The loss of memory of the initial mode of excitation is the decline of $P_0(t)$ with time, e.g., (5.8).

VI. ELECTRONIC RELAXATION

Interstate electronic relaxation in the statistical limit constitutes a common example for excited-state dynamics in electronically excited states of large molecules. We would like now to dwell on the physical description and then proceed to consider some applications and implications of ER. The nonradiative decay rate, Γ_s, of the doorway state into the bumpy quasicontinuum can be rigorously expressed in terms of the level shift operator[6, 13–17, 24] in the vicinity of the energy $E \simeq E_s$,

$$\Gamma_s = -2\,\mathrm{Im}\sum_l \frac{|V_{sl}|^2}{E_s - E_l + i\eta} \qquad \eta \to 0^+ \tag{6.1}$$

For the simple level structure we are concerned with (see Fig. 10) this result incorporates the wide and wild variations of the level spacings and the interstate coupling terms. From the formal point of view (6.1) is

$$\Gamma_s = 2\pi \sum_l |V_{sl}|^2 \delta(E_s - E_l) \tag{6.2}$$

yielding the Fermi golden rule rate (4.5). A reasonable description of the molecular parameters, which determine the dynamics, rests on the notion of maximum ignorance, invoking a random coupling model,[180–188] where V_{sl} terms and the spacings between adjacent l levels are taken to be random variable characterized by well defined first and second moments of their distribution. The results of numerical computer simulations,[188] some of which are portrayed in Fig. 16, concur with the exponential decay law, resulting in the simple kinetic result for the statistical limit. The notion of random interstate or intrastate coupling introduced herein can be extended to consider random coupling between intramolecular quasicontinua, a problem which does not render an exact explicit solution for the

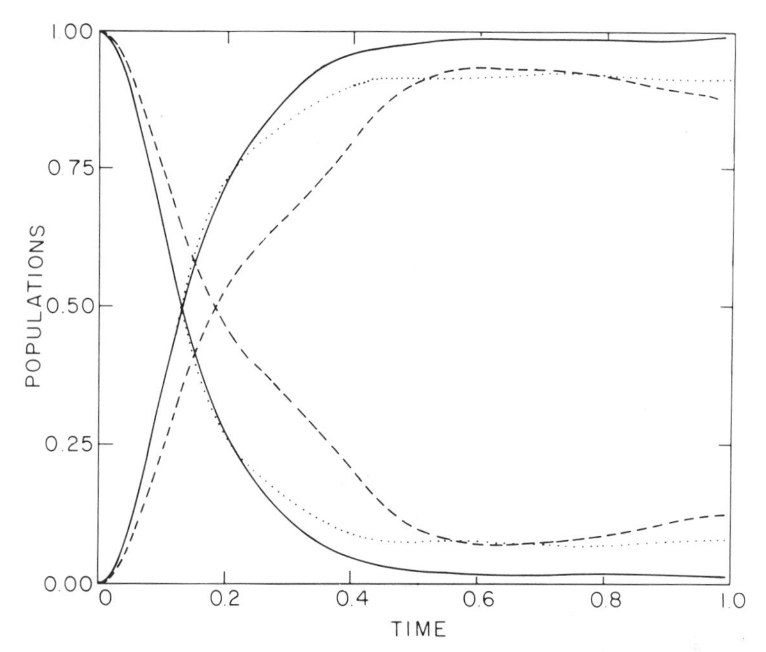

Fig. 16. Numerical computer simulations of the dynamics of the decay of a single doorway state $|s\rangle$ into a manifold of levels $\{|l\rangle\}$, characterized by a constant level spacing ρ^{-1}. The coupling terms V_{sl} are taken to be random. The decay (dephasing) width is $\Gamma_s = 2\pi\langle V_{sl}^2\rangle\rho = 6.28$ and $\langle V_{sl}\rangle = 0$. The initial state at $t=0$ is $|s\rangle$ so the three curves with unit populations at $t=0$ correspond to the $|s\rangle$ state, while the three curves with zero populations at $t=0$ represent the total occupancy of the background manifold. Full line: 150 levels, $\rho=3$, single trajectory. Dashed line: 50 levels, $\rho=1$, single trajectory. Dotted line: 50 levels, $\rho=1$, average over 15 trajectories.

dynamics and which is of considerable interest for the elucidation of excited-state unimolecular processes[184] and high-order multiphoton excitation of large molecules.[186–188]

From the point of view of the experimentalist, the formal theory has to be applied to handle real-life situations. A useful approximation for the decay rate from the electronic origin of an excited state to a lower electronic configuration is given in terms of the energy gap law[16, 17, 189]

$$\Gamma_s \propto \exp\left(\frac{-\gamma\Delta E}{\hbar\omega_M}\right) \tag{6.3}$$

where ΔE is the electronic energy gap, $\hbar\omega_M$ represents a characteristic molecular frequency close to the maximum frequency, while $\gamma \simeq 1-3$ is a numerical coefficient. This relation, which is expected to hold for large values of $\Delta E/\hbar\omega_M$, was systematically derived for a model system involving ER between two displaced, but otherwise identical, potential surfaces. However, model calculations[190] indicate that this energy gap law applies also to a more realistic system, which is characterized by configurational displacements and frequency changes between the two potential surfaces. The admittedly oversimplified energy gap law is extremely useful for the correlation of a vast amount of experimental data.[16, 17] This semiquantitative result has to be complemented and supplemented by detailed calculations for specific molecular systems, which boil down to the evaluation of the interstate nonadiabatic and/or spin-orbit coupling and to the estimates of the vibrational overlap Franck–Condon factors. All the detailed calculations of the ER rate in large molecules performed up-to-date adopted a harmonic model for the nuclear potential surfaces. The most comprehensive calculation[190] of this type was performed for the ${}^3B_{1u}\rightarrow{}^1A_{1g}$ intersystem crossing from the lowest triplet state to the ground state of the benzene molecule. This calculation utilized extensive spectroscopic input information on configurational and frequency changes for all the molecular normal modes. These detailed calculations conducted within the framework of the harmonic model[190] confirm the energy gap law and the well-known deuterium isotope effect on the intersystem crossing in benzene. However, the absolute value of the ER rate is lower by two to three orders of magnitude from the experimental results. This failure reflects the limitations inherent in the use of the harmonic approximation for quantitative calculations of the ER rates. This conclusion is not surprising as the ground-state potential surface of benzene at excess vibrational energy of $\sim$30,000 cm^{-1}, which acts as the dissipative quasicontinuum for the ER process, is by no means harmonic. From the point of view of general methodology, it is important to realize that, in principle, an ER process will be exhibited in a system where all nuclear states are harmonic and only the quantitative details of the rates will be affected by anharmonicity corrections. These general characteristics are in marked contrast to the features of intrastate IVR processes, which are induced by anharmonicity effects and which will not be exhibited in a harmonic model system.

A central question which directly bears on photoselective chemistry, is the dependence of the nonradiative decay rate $\Gamma_s(E_V)$ on the excess vibrational energy E_V in an isolated large molecule. Numerical model calculations based on the harmonic model for the interstate decay of individual vibronic levels of large molecules were performed,[16, 17] resting on the basic assumption that E_V is sufficiently low to prohibit any IVR in the electronically excited state, whereupon the molecule does not act as its own heat bath. Under these circumstances, several general trends for the dependence of $\Gamma_s(E_V)$ on E_V can be exhibited.[16] (1) For a large electronic

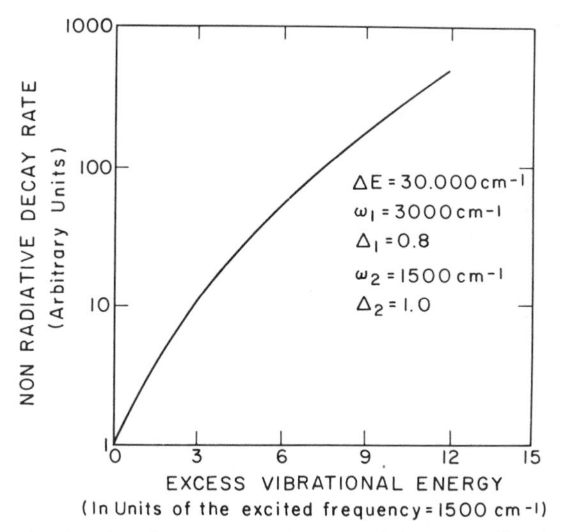

Fig. 17. Electronic relaxation decay rate as a function of the excess vibrational energy for a large electronic energy gap. The model molecule is characterized by two accepting modes with the frequencies $\omega_1 = 3000\ \text{cm}^{-1}$ and $\omega_2 = 1500\ \text{cm}^{-1}$, the reduced displacements $\Delta_1 = 0.8$ and $\Delta_2 = 1.0$, and the electronic energy gap $\Delta E = 30000\ \text{cm}^{-1}$. ω_2 is taken to the optically excited frequency.

energy gap relative to the characteristic vibrational frequencies, $\Gamma_s(E_V)$ exhibits a fast, nearly-exponential increase with increasing E_V, as is evident from Fig. 17. (2) For moderate energy gaps, $\Gamma_s(E_V)$ exhibits a moderate, close-to-linear increase with increasing E_V, as demonstrated in Fig. 18. (3) For small electronic energy gaps $\Gamma_s(E_V)$ initially decreases with increasing E_V. This retardation of the nonradiative decay rate, demonstrated in Fig. 19, is due to the decrease of the vibrational overlap at small ΔE with decreasing E_V. These theoretical predictions can be confronted with experiment for an isolated molecule which satisfied two conditions. First, the excitation spectrum is free from vibrational sequence congestion effects and, second, IVR effects can be disregarded. A molecular electronic excitation satisfying these two conditions is the $^1B_{2u}$ state of the benzene molecule and extensive experimental and theoretical studies of the energy dependence of the rate for the $^1B_{2u} \rightarrow {}^3B_{1u}$ intersystem crossing at $E_V = 0$–$3000\ \text{cm}^{-1}$, which corresponds to case (2), were conducted.[16, 17] At $E_V \geqslant 3000\ \text{cm}^{-1}$ an effective so-called channel three ER process of the $^1B_{2u}$ state sets in,[16, 17] presumably due to the opening of the $^1B_{2u} \rightarrow {}^1A_{1g}$ internal conversion route to the ground state. This process corresponds to case (1), for which $\Gamma_s(E_V)$ will exhibit a fast exponential increase with increasing E_V. For molecules larger than benzene, the interpretation of optical studies

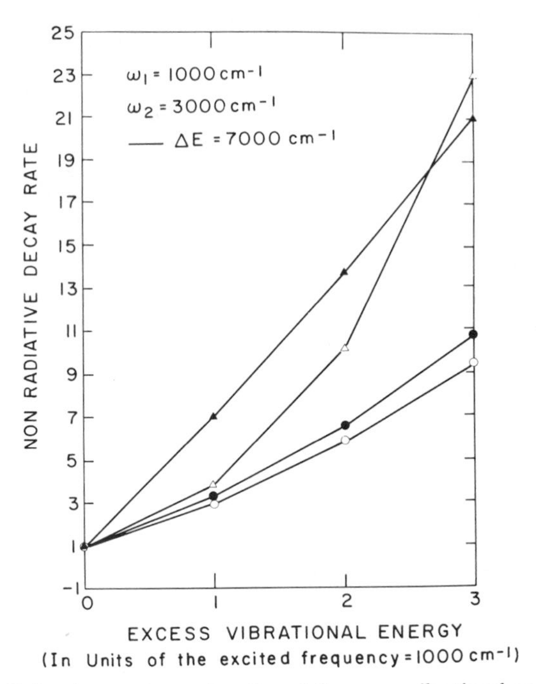

Fig. 18. Nonradiative decay rate as a function of the excess vibrational energy for moderate electronic energy gaps. The model molecule is characterized by two accepting modes; $\omega_1 = 1000\ \mathrm{cm}^{-1}$, $\omega_2 = 3000\ \mathrm{cm}^{-1}$. ω_1 is taken to be the optically excited frequency. $\Delta E = 7000\ \mathrm{cm}^{-1}$. ○: $\Delta_1 = 0.24$; $\Delta_2 = 0.071$; ●: $\Delta_1 = 0.24$; $\Delta_2 = 0.6$; △: $\Delta_1 = 0.1$; $\Delta_2 = 0.071$; ▲: $\Delta_1 = 0.9$; $\Delta_1 = 0.071$.

conducted in the low-pressure gas phase at room temperature, or at higher temperatures, is severly hampered by technical difficulties originating from extensive vibrational sequence congestion and by conceptual difficulties regarding the consequences of IVR within the "initially excited" electronic manifold on the ER process. For a large molecule cooled in a supersonic expansion, the effects of vibrational sequence congestion can be eliminated and the dynamics of genuinely photoselected states can be explored.[64–77] The spectrum of pentacene seeded in a supersonic beam displayed in Fig. 1[76] provides a good illustrative example for the gross features of the vibrational level structure in the $S_1(^1B_{2u})$ manifold, which is directly relevant to intramolecular interstate and intrastate dynamics. Two energy regions can be distinguished in the optical spectrum of the ultracold large molecule.[69, 76]

Range A. A sparse vibrational level structure in the $^1B_{2u}$ state of pentacene is exhibited in the range $E_V = 0-1000\ \mathrm{cm}^{-1}$. Individual spectral

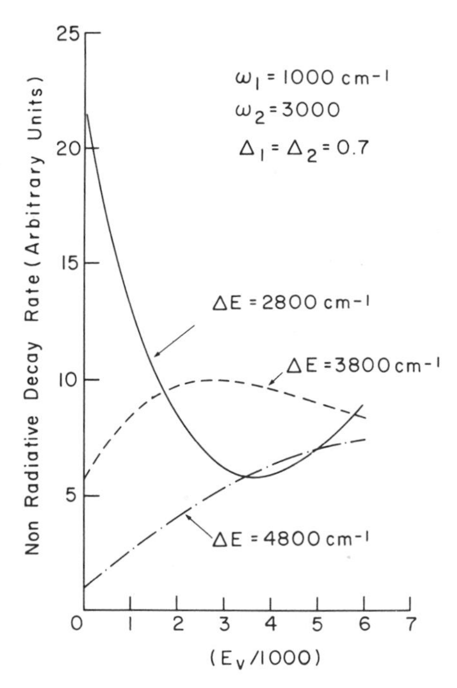

Fig. 19. Reversal effect of the nonradiative decay rate as a function of excess vibrational energy. Model calculations performed for a model molecule with $\Delta_1 = \Delta_2 = 0.7$, $\omega_1 = 1000$ cm^{-1} $\omega_2 = 3000$ cm^{-1}. —$\Delta E = 2800$ cm^{-1}, --- $\Delta E = 3800$ cm^{-1} and -·-· $\Delta E = 4800$ cm^{-1}.

features correspond to distinct vibronic levels. In this range each vibrational state can be photoselectively and individually excited and its subsequent intramolecular decay proceeds only via ER, which can be accounted for in terms of the simple theoretical framework previously considered.

Range B. A congested vibrational level structure is exhibited at higher excess vibrational energies, which for pentacene $^1B_{2u}$ state are exhibited at $E_V > 1000$ cm^{-1}. This congested level structure corresponds to overlapping vibrational resonances, where not only the fundamental frequencies and their combinations are optically active. Here effective intrastate anharmonic mixing between close-lying vibronic levels, which can be envisioned as originating from many-level-Fermi resonances, induces optical absorption to a multitude of vibronic levels.

In range *B* the simple theory of optical selection requires some considerable modifications.

A recent experimental demonstration of the dynamic implications of optical selection within the S_1 manifold of a very large molecule is given in

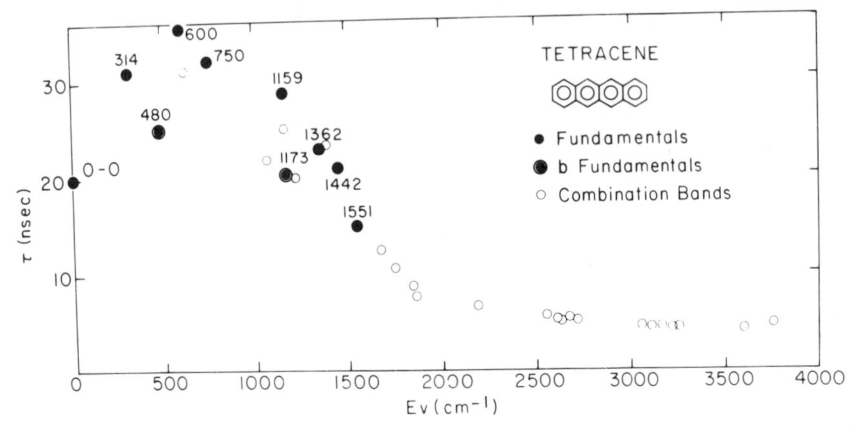

Fig. 20. Fluorescence decay lifetimes resulting from photoselective excitation of an isolated tetracene molecule cooled in a supersonic beam, taken from the work of Amirav, Even, and Jortner[69, 75]. Tetracene was expanded in a 210 torr Ar through a 150 μm nozzle and crossed 5 mm down the nozzle with the nitrogen pumped dye laser (spectral bandwidth 0.3 cm^{-1}, pulse width FWHM 3 nsec). The decay was found to be exponential with lifetimes τ depending on the excess vibrational E_V above the electronic origin. Note the retardation of the nonradiative decay at low E_V and the weak dependence of τ on E_V at high E_V. It is important to realize that photoselective excitation of such an isolated large molecule can be accomplished only in a supersonic beam.

Fig. 20, where we have portrayed the fluorescence decay lifetimes, τ, resulting from photoselective excitation within the vibronic manifold which corresponds to the $S_0 \rightarrow S_1$ transition of the tetracene molecule cooled in a supersonic beam.[69, 75] Here range A spans the energy region $E_V =$ 0–2000 cm^{-1}. The energy dependence of τ in the ultracold isolated molecule in range A exhibits a surprising retardation of the ER rate with increasing E_V. This retardation of the ER rate can be tentatively accounted for in terms of S_1-T_1 intersystem crossing, which corresponds to case (3). The weak dependence of τ and of the ER rate in range B at $E_V > 2000$ cm^{-1} was tentatively blamed on the effects of effective intrastate anharmonic scrambling which affect the interstate nonreactive relaxation process.[69] Such an interplay between intrastate and interstate dynamics is of considerable interest and undoubtedly many examples of this type will be unveiled in the future. We shall proceed to consider such a case which pertains to electronic excitation induced by multiphoton vibrational excitation.

VII. INVERSE ELECTRONIC RELAXATION

The consequences of interstate coupling and ER in electronically excited states of large molecules were extensively interrogated by one-photon excitation. A different optical excitation mode of a collision-free molecule involves high-order multiphoton excitation (MPE) on the ground state potential surface (Section XII and Chapter 12). Such MPE may populate molecular eigenstates, which result from interstate scrambling between ground state vibronic levels and vibronic levels corresponding to an electronically excited state. These molecular eigenstates will exhibit one-photon spontaneous radiative decay. The observation of fluorescence from the first singlet excited state of CrO_2Cl_2 [191] and possibly of F_2CO[192] induced by MPE, which will be discussed in Chapter 12, provides an experimental demonstration of such an up-conversion process, which was referred to as "inverse electronic relaxation" (IER).[193, 194] This process essentially involves one-photon radiative decay of molecular eigenstates, which are reached by MPE. The physical features of the IER are of considerable interest because of two reasons. First, it provides another demonstration of the generality and usefulness of the concept of molecular eigenstates in a bound level structure. Second, this novel phenomenon provides a blending between intrastate dynamics which affects the features of the MPE and interstate scrambling which determines the nature of the mixed molecular eigenstates. The concept of IER is compatible with the ideas of irreversibility inherent in the conventional theory of ER as the existence of the final radiative continuum for spontaneous one-photon decay insures the irreversibility of the IER for any molecular level structure, provided that effective interstate scrambling prevails. It is thus legitimate to invoke the concept of IER even for an isolated diatomic molecule, in contrast to the conventional phenomenon of ER in an isolated molecule, where practical irreversibility is insured only in the statistical limit.

The accessibility of an electronically excited configuration via MPE on the ground-state potential surface can be accounted for in terms of two radiative coupling mechanisms:

1. Interstate radiative coupling between high vibrational levels $\{|g'\rangle\}$ of the ground state and an electronically excited configuration $|s\rangle$. The dipole moment $\mu_{g's}$ for this coupling is determined by the electronic coordinates.[191, 194]
2. Intrastate radiative coupling within the high vibrationally excited levels of the ground-state $\{|g'\rangle\}$ excited levels and the ground-state manifold $\{|g\rangle\}$, which is quasidegenerate with $|s\rangle$. The dipole moment $\mu_{g'g}$ for this coupling is determined by the nuclear coordinates.[194]

There is no dichotomy between these two mechanisms which can peacefully coexist. We have to consider the radiative coupling between the ground-state manifold $\{|g'\rangle\}$ located at the energy $\sim\hbar\omega$, which corresponds to that of the infrared photon, below the mixed molecular eigenstates $|m\rangle = \alpha_s^m|s\rangle + \Sigma_g \beta_g^m|g\rangle$. The transition moment being $\mu_{g'm} = \alpha_s^m \mu_{g's} + \Sigma_g \beta_g^m \mu_{g'g}$, where the first term provides the electronic contribution of mechanism 1, while the second term yields the nuclear contribution of mechanism 2. At present the relative contributions of the two mechanisms cannot be assessed and both have to be incorporated in the theory of IER. Mechanism 1 provides photoselection of those molecular eigenstates which are grossly contaminated by $|s\rangle$, while mechanism 2 provides essentially a democratic excitation of the molecular eigenstates. On the basis of a careful comparison of the decay lifetimes excited by conventional one-photon absorption and by MPE the relative contributions of these two mechanisms can be elucidated.

We now address the interesting question of how do the features of the IER depend on the intramolecular level structure, with a special reference to the nature of the MPE and the characteristics of the radiative decay. The simplest molecular system involves level mixing in a diatomic molecule, (Fig. 21), where accidential degeneracy prevails between a pair of rotational–vibrational levels corresponding to two distinct electronic configurations. The MPE involves coherent excitation of a sparse level structure. Such MPE process of an isolated diatomic molecule involves large anharmonicity defects, and requires very large intensitites of the radiation field.[195] This serious difficulty may be overcome by MPE of the diatomic in a dense medium. It may be possible to considerably enhance the

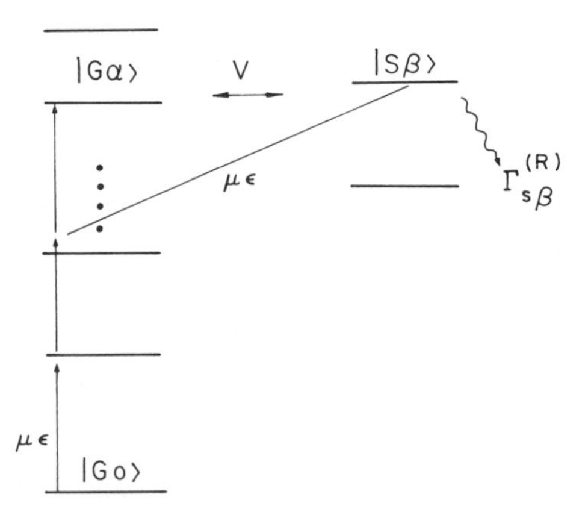

Fig. 21. IER in a diatomic molecule. In the isolated molecule case a close matching between two vibrational levels ($|G\alpha\rangle, |S\beta\rangle$) belonging to the two electronic manifolds is necessary.

efficiency of multiphoton absorption via T_2 line broadening in the absence of an appreciable T_1-type vibrational relaxation process by embedding the diatomic molecule in a low-temperature simple solid or liquid.[194] Vibrational relaxation of diatomics in such media are slow, while proper dephasing is fast, so that IER following MPE will be efficient. We shall next consider IER in a small polyatomic molecule, where two sparse electronic configurations are heavily scrambled (Fig. 22). The MPE in isolated small polyatomics can be described in terms of a coherent MPE of a sparse ground state manifold until a range of contaminated molecular eigenstates is reached. The latter states are active in one-photon emission. The radiative decay rate is characterized by "diluted" radiative decay times. The radiative lifetimes should be similar to those obtained by conventional one-photon excitation. Very recently, $S_1 \rightarrow S_0$ optical emission of the SO_2 molecule was induced by MPE with a CO_2 laser,[196] providing the first example for IER resulting from interstate coupling in the small molecule limit. Finally, we shall dwell on IER in a level structure of a large molecule where a ground state is effectively coupled to an electronically excited configuration. The MPE of a congested dense vibrational manifold on the ground state potential surface is incoherent (Section XII and

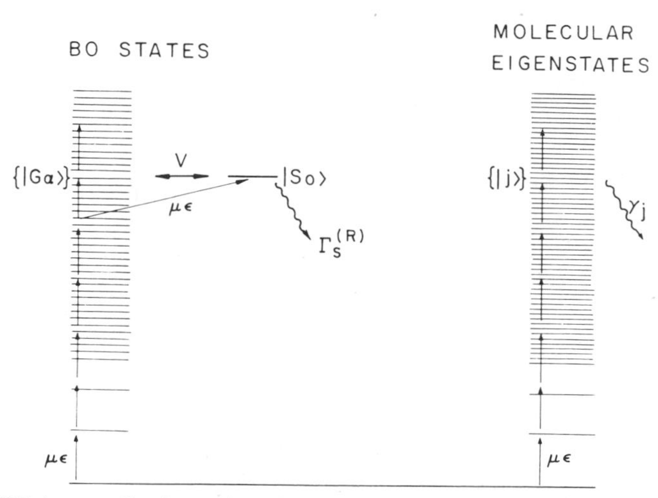

Fig. 22. IER in a small polyatomic molecule. The ground state manifold is dense but not necessarily statistical. The excited state manifold is discrete in the interesting energy region. Alternatively, the molecular eigenstates basis may be utilized. The same picture applies for *range A* of large molecules.

"Reduced Equations of Motion for Collisionless Molecular Multiphoton Processes," Chapter 7) and results in excitation of the contaminated molecular eigenstates. Regarding these emitting molecular eigenstates, we can (see Section VI) distinguish between two energy regions:[194]

1. *Range A* (Fig. 22), which is characterized by sparse nonoverlapping resonances, each having its parentage in a single $|s\rangle$ state, and which are separated by "black holes" of uncontaminated $|g\rangle$ states.
2. *Range B* (Fig. 23), which is characterized by high density of $|s\rangle$ states, where the "black holes" disappear and the level structure consists of overlapping contaminated molecular eigenstates.

The IER in the large molecule constitutes the radiative decay of incoherently excited molecular eigenstates in range A or in range B. The decay lifetimes of the molecular eigenstates can be expressed in terms of the radiative width of a zero-order state $|S\rangle$ diluted by the number of effectively coupled levels (5.6) for range A, or by a statistical dilution factor in range B. The IER is expected to be amenable to experimental observation in range A of a large molecule, characterized by an intermediate level structure and in accord with the experimental results for IER in the CrO_2Cl_2 molecule.[191]

The phenomenon of IER is of interest within the theoretical framework of intramolecular dynamics, providing a nice demonstration for the characteristics of interstate scrambled molecular eigenstates within a single

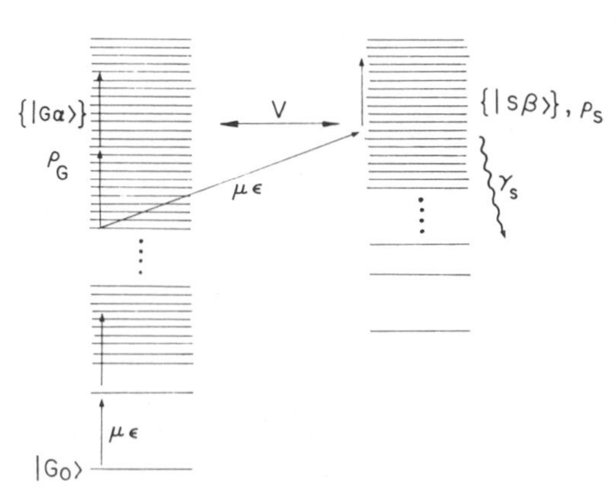

Fig. 23. IER in *range B* of a polyatomic molecule. Both ground and excited state manifolds are statistical.

molecule which can be excited by the novel high-order infrared MPE. This brings us to the end of the exposition of the conceptual framework required to describe the implications and applications of the basic reactive and nonreactive intramolecular processes. We shall now proceed to examine some more complex phenomena, which rest on the interplay between nonreactive and reactive processes, and on the role of reactive collisions.

VIII. INTRAMOLECULAR ENERGY FLOW

Increasing the energy of the initial state can have a considerable qualitative effect on its reactivity. Thus, promoting the reagents to an excited electronic potential energy surface will often open up new reaction pathways. Classical photochemistry has extensively explored such possibilities. The availability of lasers (high photon flux within a narrow frequency interval) and, in particular, of high intensity infrared lasers has posed a new challenge for photoselective chemistry: Can one achieve selectivity on a given potential energy surface and, at higher energies, can one selectively cross among different surfaces?

The central current issue for photoselective chemistry is therefore the nature and evolution of the laser-induced energy distribution in polyatomic molecules. It is important to note from the very beginning that there are really two related but distinct problems. The first is the distribution of the total energy. For a given pumping process and intramolecular dynamics, what is the fraction of molecules with total energy in any given range? It is worthwhile to emphasize that photoselectivity can be manifested already at this level. Consider, for example, a multiphoton absorption. If the up-pumping by the laser is fast compared to the rates of unimolecular dissociation one can reach exceedingly energy-rich states that would be inaccessible by a slower (e. g., collisional) pumping process. It is the competition between the up-pumping and the decay process (which depletes energy-rich molecules and thus reduces the extent of pumping to even higher energies), which governs this type of (energy) selectivity.

It may be the case that the extensive fragmentation of polyatomics noted in multiphoton ionization (MPI) at high laser powers,[197–199] as is illustrated in Fig. 24, will provide an example of energy selectivity. Of course, the use of visible (or UV) photons in MPI studies implies that, other things being equal, the rate of up-pumping is faster than that possible using infrared photons. The preliminary results are however that the major trends can be explained on statistical grounds.

The second problem in photoselective chemistry is the distribution of the given total energy among the possible modes. There are two traditional

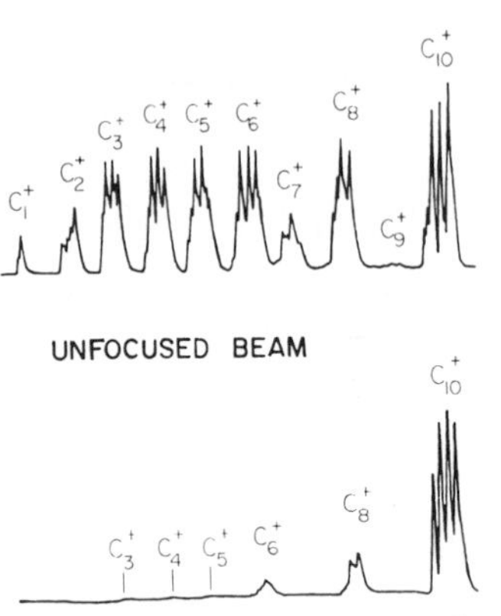

Fig. 24. Fragmentation pattern of naphthalene at 281 nm (adapted from Ref. 198). See Fig. 7 of Chapter 3 for the MPI mass spectra of benzene. The fragments are identified by the number of carbon atoms and different peaks correspond to different numbers of hydrogen atoms. Note the extensive fragmentation at the higher laser power.

points of view associated with the names of Slater[200] and of Rice, Ramsperger, Kassel, and Marcus, so called RRKM,[201, 202] respectively. In the Slater approach the molecule is represented as a set of normal (and hence noninteracting) modes. The energy distribution is thus completely specified by its initial partitioning among these modes. No subsequent energy exchange is possible. Precisely the opposite point of view is adopted by RRK[201] and RRKM.[202] The rate of intramolecular energy flow is assumed to be rapid compared to the rates of collisional up/down pumping and to the rate of unimolecular breakdown. For chemical purposes, the energy-rich molecules is then fully characterized by its energy content. Any selectivity due to the initial mode of excitation is thus lost on the time scale of chemical interest. This point of view has dominated chemical kinetics and a considerable body of experimental backing has recently been summarized.[203, 204]

The early results on bond cleavage following infrared multiphoton dissociation[205, 206] were interpreted to be in accord with the RRKM picture. This was pleasing, since it was consistent with the "quasicontinuum" model of multiphoton absorption (cf. Section XII). If the anharmonic coupling between the higher vibrational states is sufficiently strong so as to smear them into a continuum then, on the time scale of the laser pumping, the intramolecular energy transfer rate must be fast and no selectivity due to selective deposition of the excitation energy is possible.

More recent evidence suggests, however, that this initial pessimism was premature. Four center multiphoton induced eliminations,[207–209] e.g.,

$$CH_3CCl_3 \rightarrow CH_2 = CCl_2 + HCl \tag{8.1}$$

cannot be fully accounted for in terms of the RRKM theory. At the same time improved theoretical models of infrared multiphoton absorption[210] and the interpretation of overtone spectra[211] suggest that the quasicontinuum is not quite as smooth as the simplest picture would suggest. Rather, the intramolecular energy transfer processes can be quite selective —the relaxation can be rapid among a particular group of states, but much slower among different groups.

Another potentially photoselective process is the single photon excitation to an energy-rich or to a dissociating state. If the rate of intramolecular energy transfer is comparable or slower than the rate of removal of the energy-rich species by chemical means (dissociation or bimolecular reaction), then selectivity is possible. In this case, the reaction rate will again depend not only on the energy content but also on the initial mode of excitation. Figure 25 shows recent results[212] for the isomerization of allyl isocyanide. By exciting different bands in the C—H overtone spectra it is possible to prepare AIC in an initially highly nonstatistical state. The bandwidths show that the initial intramolecular vibrational energy redistributions occur on subpicoseconds time scales.[211, 212] Yet marginal selectivity remains on the chemical (nanoseconds, cf. Fig. 25) time scale. This is primarily evident in the nonmonotonic variation of the isomerization rate constant with the energy content.

The possibility of bottlenecks for intramolecular energy transfer is also suggested by the two laser-induced reactions of cyclopropane[213] There are two reaction channels, isomerization (which has a lower threshold) and fragmentation. Their relative yield depended on the mode which was excited. Multiphoton excitation of the C—H asymmetric stretch induces essentially pure isomerization; excitation of the (lower frequency) CH_2 wag produces roughly equal yields of fragmentation and isomerization products.[213] If these results reflect incomplete randomization between the higher and lower frequency modes then the effect of collisions should be to enhance the intramolecular relaxation. Indeed, for the CH stretch excitation the effect of collisions is to increase the yield of fragmentation.[213]

State-selective chemistry (i.e., selectivity at a given total energy, reflecting different pathways for different initially excited states) depends there-

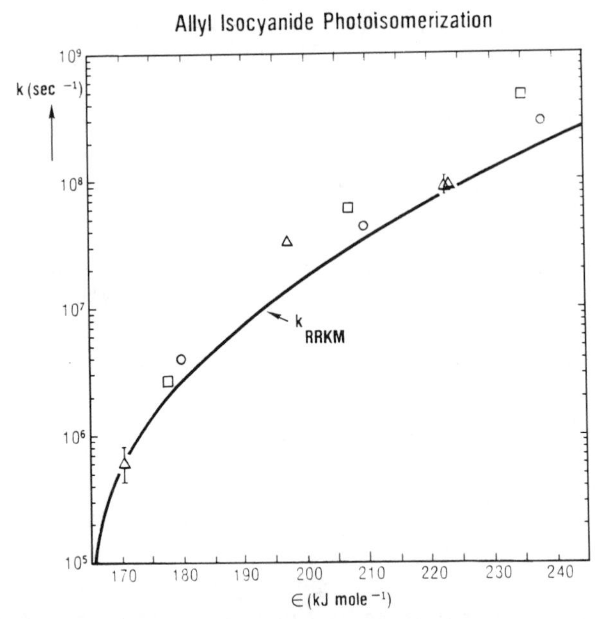

Fig. 25. Experimental (symbols) and calculated (RRKM, solid curve) rate constant for photoisomerization of allyl isocyanide ($CH_2{=}CH{-}CH_2NC$) vs. the energy content. (Adapted from Ref. 212.) The symbols refer to different CH overtone bands: (○) the terminal olefinic group which is most removed from the isomerization site; (□) the central (=CH—) group; and (Δ) the methylenic group, which is directly bonded to the NC group.

fore on the rate at which an isolated energy-rich molecule appears to relax toward equilibrium.

A formulation of the relaxation of a system to equilibrium in mechanical terms was first attempted over a century ago by Boltzmann.[214] The active debate following his work[215, 216] continues to this very day. The problem is that very strictly speaking an isolated mechanical system does not relax. However, and as has been pointed out earlier on by Einstein,[217] experience teaches us that the observable magnitudes do relax. The physics of the problem consists therefore in the identification of what are the observable properties for a system governed by reversible equations of motion. The accepted interpretation is that these are average values defined over an ensemble of systems. Over the last two decades equations of motion for such averaged or reduced quantities have been derived by a variety of techniques[218–222] and applications to multiphoton pumping are discussed in Chapter 7. As long as these equations of motion are exact they are necessarily equivalent to the original, reversible equation of motion. The relaxation toward equilibrium is only manifest when approximations are introduced into these equations. Such approximations are valid only when the reduced properties are chosen with due regard to the physics of the problem. One rule of thumb is to use those properties with the longest relaxation time, the so-called "slow" variables. In particular, both experimental and theoretical evidence suggests that for highly vibrationally excited polyatomics these are not necessarily the normal modes made familiar by spectroscopic studies of the low-lying vibrational states. Rather, the concept of the local modes[223, 224] has recently received considerable attention in this context (Chapters 4).

How can one understand the apparent relaxation of reduced quantities? One possibility[225] is to examine the trajectories (in phase space) for different initial conditions, ("Intramolecular Energy Transfer", Chapter 2. See also Chapters 1 and 3). It is then found that at higher energies and for model anharmonic potentials, trajectories which originate at neighboring points rapidly diverge. Another way of viewing this is to note that a given trajectory cannot be confined to a sub-phase-space of low dimensionality. The property one is after is best characterized by what mathematicians term "mixing."[226] Consider two finite and nonintersecting regions A and B of phase space. Consider now the transformation of the region A under the equations of motion while region B is held in place. As time progresses the region A(t) that evolved from A may begin to intersect B. The time evolution is mixing if, for long times, the intersection of A(t) and B is simply proportional to the size of B. In other words, for long times the spreading of A is uniform for any finite region B which is used to sample this spreading.

It should, however, be remembered that there are several time scales in the present problem. It is not enough to know that if left on its own the system will be mixing. The question is whether it will do so on the time scales (pumping or dissociation) of interest. This is not meant as an idle caveat. There is both experimental and computational evidence (mentioned in Chapter 3, Section I) that systems that are potentially mixing may well show considerable selectivity on time scales of molecular interest.

Many experiments on MPD have, however, been discussed on the basis of the assumption that the relaxation of the vibrational energy is fast compared to both pumping and dissociation. This assumption is being increasingly questioned[227, 228] and is being gradually replaced by models where energy relaxation is rapid among limited groups of states with a slower inter-group relaxation rate.

The information theoretic approach, Chapter 3, leads to a similar point of view. It shows that the exact state of the system can be described as statistical, subject to constraints. These constraints are determined both by the excitation process and by the system's Hamiltonian. All regions of phase space which correspond to the same values for the constraints are uniformly populated. In this sense, the approach is statistical and similar to the RRKM point of view. Due to the presence of the constraints, regions of phase space which correspond to different values of the constraints are not equally populated. The constraints reflect not only the systems's Hamiltonian (as in the Slater model) but also the preparation process. Indeed, they serve as a signature of the initial state. Why do the constraints not confine the trajectory to a subspace? Indeed they do.[229] However, the constraints are time dependent constants of the motion and hence the subspace continuously deforms in time and is not evident by inspection. The RRKM results obtained when there are no constraints imposed by the preparation nor by the dynamics (beside conservation of flux).[230]

IX. UNIMOLECULAR REACTIONS

Much of the recent work on laser-induced unimolecular reactions was aimed at clarifying the mechanism of the MPD process. Some of these studies were already mentioned in Sections VII and VIII. In particular, the systematics of the rate constant for dissociation can be used to demonstrate the competition between the up-pumping by the laser photons and the removal of energy-rich molecules by unimolecular breakdown. For the purpose of the present, qualitative, discussion it is sufficient to employ the RRK approximation.[201, 232] At a given total energy E the rate constant is

$$k(E)=\nu\left[\frac{E-E_0}{E}\right]^{s-1} \tag{9.1}$$

s is the number of vibrational modes, ν is the (mean) vibrational frequency, and E_0 is the threshold for dissociation. At a given E and E_0, smaller molecules (lower s) will therefore dissociate faster. If significant dissociation occurs when it can compete with the rate of up-pumping, then it will require a higher excess energy to observe dissociation in the larger polyatomic molecules. The dissociation products will then be internally excited and could therefore be more readily fragmented upon further absorption of laser photons. Due to the possibility of pumping of larger polyatomics past E_0 one can also reach dissociation or elimination processes with higher threshold energies. One example of a system with multiple pathways is ethyl vinyl ether[233]

$$
\begin{array}{ccccccccc}
 & O & & H & & O & & H \\
CH_2 & & C & & & CH_2 & & C & \\
| & & \| & & \rightarrow & \| & + & | & \\
CH_3 & & CH_2 & & & CH_2 & & CH_3 &
\end{array}
$$

$$
\begin{array}{ccccccccc}
 & O & & H & & O & & H \\
CH_2 & & C & & & \dot{C}H_2 & & C & \\
| & & \| & & \rightarrow & | & + & | & \\
CH_3 & & CH_2 & & & CH_3 & & \dot{C}H_2 &
\end{array}
\qquad (9.2)
$$

where the radicals produced in the second path will proceed to undergo secondary processes.

Traditionally, the dynamics of unimolecular dissociation processes were probed using collisions with a buffer gas.[234] Collisional deactivation is competing with and hence serves as a built-in clock for the dissociation process. Similar studies were reported for MPD processes (see, e.g., Ref. 235). As is the case for chemical activation,[204, 236] collisional deactivation is particularly probing when used for molecules with multiple reaction pathways such as (9.2) or[237]

$$
\begin{array}{ccccc}
CH_2 - CHCl & & CH_2 - CH & \\
| \qquad\quad | & \rightarrow & | \qquad\quad \| & +HCl \\
CH_2 - CH_2 & & CH_2 - CH &
\end{array}
$$

$$
\begin{array}{ccccc}
CH_2 - CHCl & & CH_2 \quad CHCl & \\
| \qquad\quad | & \rightarrow & \| \quad + \quad \| & \\
CH_2 - CH_2 & & CH_2 \quad CH_2 &
\end{array}
\qquad (9.3)
$$

$$
\begin{array}{ccccc}
CH_2 - CHCl & & CH_2 - \dot{C}H & \\
| \qquad\quad | & \rightarrow & | \qquad\quad | & +Cl \\
CH_2 - CH_2 & & CH_2 - CH_2 &
\end{array}
$$

where the radicals produced in the third pathway proceed to undergo secondary processes.

An important diagnostic of the intramolecular dynamics is the energy distribution among the reaction products. In the simplest, RRK, approach, the relative rate of dissociation for molecules with translational energy in the range E_T, $E_T + dE_T$ along the reaction coordinate is[232]

$$(\text{relative rate}) \propto \left[(E - E_0) - E_T \right]^{s-2} dE_T \tag{9.4}$$

When there is no significant repulsion among the products past the transition state, (9.4) is also the fraction of products with translational energy in the range E_T, $E_T + dE_T$. Equation (9.4) is a rapidly decreasing function of E_T, while the range of E_T determines the excess energy, $(E - E_0)$, in the energy rich molecule. The functional dependence (9.4) is often in accord with dissociation processes requiring a simple bond fission,[205, 206, 231] except when considerations of conservation of angular momentum need be introduced,[238] or when the excess energy is not small so that the lifetime is short. Three or four center elimination processes, where there is a significant barrier to the recombination process, will qualitatively deviate from the predictions based on (9.4). Additional discussion can be found in "The Information Theoretic Approach to Intramolecular Dynamics," Section I.

Single photon induced dissociation has been studied in considerable detail, particularly because of the high photon flux and wavelength selection made possible by the use of lasers. Products-state analysis is usually carried out either via the technique of photofragment spectroscopy[229, 240] where the fragment velocity is being monitored or via laser-induced fluorescence,[240, 241] which determines the internal states. Both methods have been used in a recent study of the photodissociation of CS_2 using an excimer laser.[242] The vibrational state distribution of CS (and its surprisal) for two different pathways is shown in Fig. 26.

Laser diagnostics of products velocity ("Doppler Spectroscopy of Photofragments," Chapter 20) and state ("Laser Diagnostics of Reaction Product Energy Distributions," Chapter 19) distributions are increasingly employed and have stimulated an extensive theoretical effort.[243–249]

X. COLLISIONAL PROCESSES

Much of our current understanding of the role of internally excited reagents in bimolecular collision processes results from studies of energy disposal in exoergic chemical reactions.[250–254] Reactions such as

$$\mathrm{F} + \mathrm{HR} \rightarrow \mathrm{HF}(v, J) + \mathrm{R} \tag{10.1}$$

tend to preferentially release the energy into internal excitation of the HF

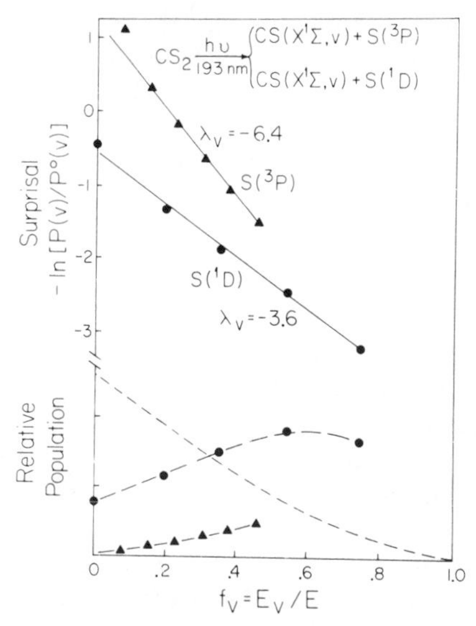

Fig. 26. Bottom panel: The distributions of CS vibrational states accompanying the sulphur atom in the ^3P and ^1D electronic states plotted vs. the fraction of the available energy which is in the CS vibration. The dashed line is the vibrational state distribution when all final quantum states are equally probable. Top panel: The surprisal (cf. Chapter 3) of the two vibrational distributions, which is the measure of the deviance of the observed from the purely statistical vibrational energy disposal. (Adapted from Ref. 242.)

molecule (cf. Fig. 4 of Chapter 3) and this remains the case even when R is a large organic radical.[253] It follows from considerations of detailed balance[255, 256] that in the reversed,

$$HF(v, J)+R \rightarrow RH+F \tag{10.2}$$

reaction, internal excitation of HF will be particularly effective in promoting the reaction. It should be noted, however, that the collision of R with HF(v, J) can also lead to a nonreactive, vibrational energy transfer. Such processes will diminish the yield of the reaction and may, in fact, offer a severe competition.

Excited reagents prepared in an exoergic prereaction have indeed been used in studies of both reactive (e.g., Ref. 257) and inelastic (e.g., Ref. 258) collisions. In particular, chemical laser emission can be used to monitor the secondary collisional transfer of the internal excitation released by the primary chemical reaction.[259] Laser pumping of reagents is discussed in Chapters 15–17.

For direct collisions, of short duration, the rough rule of thumb is that the transition rate is exponentially small in the energy released into or required from the translational motion. This "exponential gap rule" accounts for the considerable enhancement of the rates of endoergic reactions by vibrational excitation. One of several examples where laser-pumped reagents were used[260] is

$$Br + HCl(v) \rightarrow HBr + Cl \tag{10.3}$$

The rates for this and other reactions of halogen atoms with HCl molecules in a definite vibrational state are shown in Fig. 27. The exponential rise of the reaction rate with vibrational excitation is quite evident.

As long as the reagent internal excitation is below the energetic barrier of the reaction it diminishes the amount of translational energy required to overcome the barrier and hence serves to diminish the exponential gap. when an excess vibrational energy is employed or when there is no significant barrier to reaction, the effects of vibrational excitation on the reaction rate is much more moderate (Fig. 27).

The exponential gap rule also implies that any excess reagent internal excitation will be preferentially channelled into product vibrational excita-

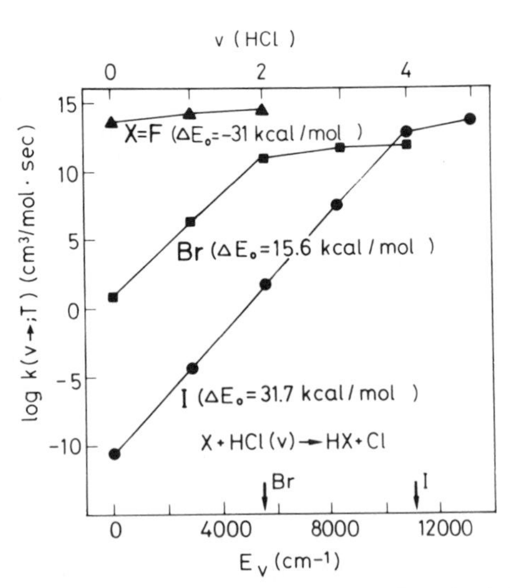

Fig. 27. The dependence of the rate constant for the reaction of a halogen atom with HCl molecules in a given vibrational state on the amount of HCl vibrational energy. The arrows indicate the point beyond which the reaction is exothermic. (Figure adapted from Ref. 256.)

tion. This has been demonstrated both by direct experiments (e.g., Ref. 261),

$$Ba + HF(v=1) \rightarrow H + BaF(v') \tag{10.4}$$

using laser pumped HF and by trajectory computations.[262, 263]

A given initial energy can be partitioned in different ways between the degrees of freedom of the reagents. In the "prior" limit[264] all of these different possible initial states (of the same total energy) are taken to react with the same rate. The dependence of the prior reaction rate (at a given temperature) on the (reduced) energy gap is shown in Fig. 28. For very endoergic processes the increase is exponential,

$$\left(\frac{\partial \ln k^\circ}{\partial E_v}\right)_T = \frac{1}{kT} \tag{10.5}$$

Observed (or computed) rate constants (e.g., Fig. 28) tend, however, to increase with a higher slope[264]

$$\left(\frac{\partial \ln k}{\partial E_v}\right)_T = \frac{1-\lambda_v}{kT} \tag{10.6}$$

where λ_v is negative [and is about equal to -0.2 for the reaction (10.3)]. The surprisal (the logarithmic deviance of the actual rate from the prior limit) for the state-to-state process

$$Cl + HCl(v) \rightarrow H + Cl_2(v') \tag{10.7}$$

is shown in Fig. 29. As v increases, the process changes from an endother-

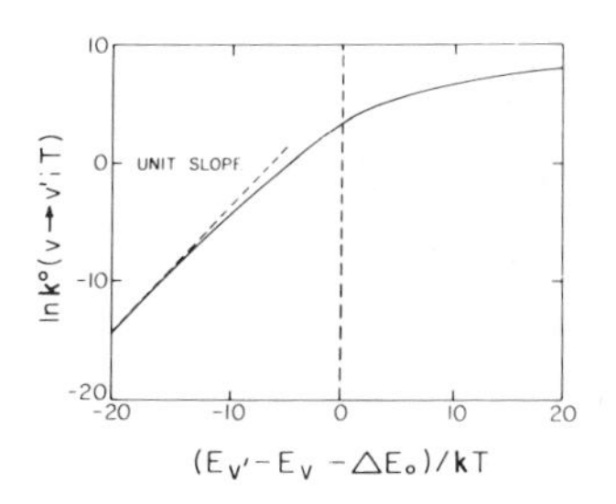

Fig. 28. The dependence of the "prior" reaction rate on the energy gap (the difference between the internal energy of the products and reactants), in units of kT where T is the translational temperature. (Adapted from Ref. 264.) The prior rate applies when all possible isoenergetic states of the reagents react with the same rate (and, by detailed balance, all isoenergetic states of the products are formed at the same rate).

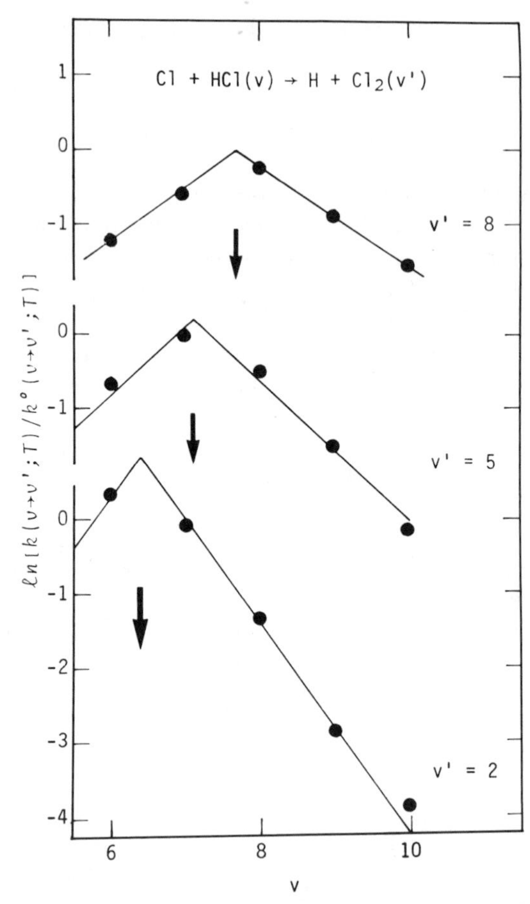

Fig. 29. Surprisal of the state-to-state rate constants of the reaction (10.7) for three values of v'. (Adapted from Ref. 263.) The surprisal is a quantitative expression of the gap, which is seen to be minimal at the v and v' pair for which the energy gap is minimal.

mic to an exothermic one, and the preference for final states for which the gap is minimal is quite evident.

Reagent vibrational excitation can also affect the branching ratio between two reaction pathways. Figure 30 shows the surprisal of the two branches in the reaction

$$HBr(v)+H \begin{array}{l} \nearrow H_2+Br \\ \searrow H+BrH \end{array} \tag{10.8}$$

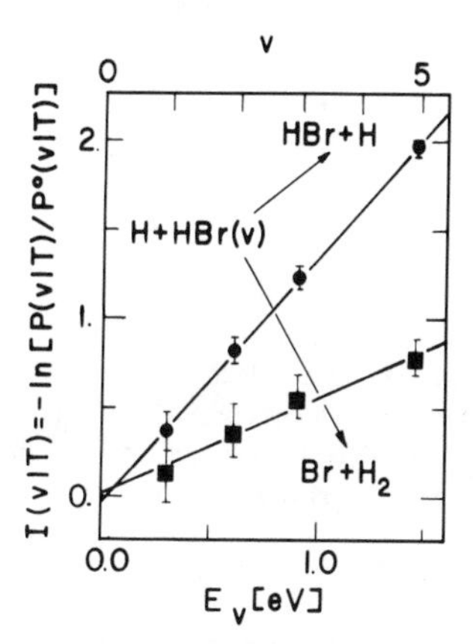

Fig. 30. Surprisal analysis of the effect of HBr vibrational excitation on the two branches of the reaction $H+HBr(v)$. Shown is the surprisal (the logarithmic deviance of the actual from the prior rate) vs. the vibrational excitation of HBr. (Adapted from Ref. 264, using trajectory computations of J. M. White and D. L. Thompson.)

"Enhancement of Chemical Reactions by Infrared Lasers" is discussed in Chapter 15. An interesting example is

$$O_3 + NO \rightarrow O_2 + NO_2 \qquad (10.9)$$

where either O_3 [265] or NO[266] were vibrationally excited. A process under active study is the multiphoton pumping of polyatomic molecules to be used as reagents in bimolecular processes. Preliminary results for the

$$Na + SF_6 \rightarrow NaF + SF_5 \qquad (10.10)$$

reaction are reported in Chapter 17.

Rotational energy is typically not sufficient to significantly change the energy gap except that for hydrides it can be used to fine-tune the balance. There is, however, a dynamical effect that has been demonstrated using laser pumped HCl in the

$$K + HCl(v=1, J) \rightarrow KCl + H \qquad (10.11)$$

reaction.[267] The center of mass of HCl is quite near to the Cl atom. As the molecule is rotating the H atom spans a sphere about the Cl atom. Increasing the rotational state of HCl causes a shielding of the Cl atom and the observed reaction probability indeed declines (Fig. 31). A similar

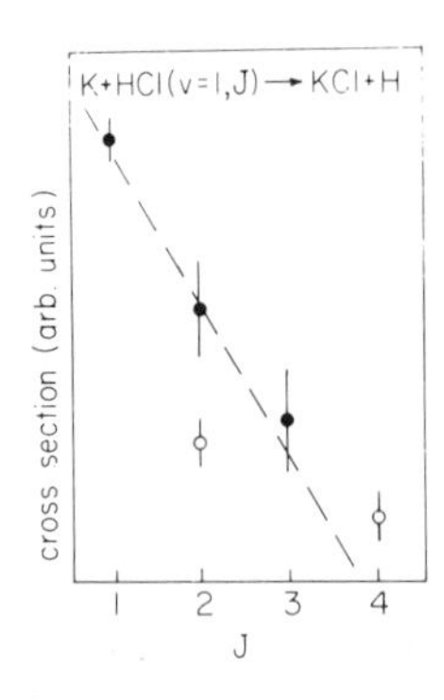

Fig. 31. The decline of the cross-section for the reaction K + HCl as a function of the rotational quantum number of HCl. (Adapted from Ref. 267, where the difference between the dots and circles is discussed.)

explanation has previously been proposed[268] to account for the dependence of the branching ratio in the

$$F + HD(J) \begin{matrix} \nearrow HF + D \\ \searrow DF + H \end{matrix} \tag{10.12}$$

reaction on the rotational state of HD.

A finer probe of the steric requirements of chemical reactions is provided by the use of oriented reagents.[269] Using polarized laser light it is possible to preferentially orient those molecules that have absorbed. As an example,[270] Sr essentially does not react with HF molecules in the ground vibrational state. Sending a beam of Sr atoms through a cell of very low pressure HF pumped by a polarized output of an HF laser one can preferentially orient the HF molecules in the $v = 1$ state with respect to the Sr beam. The SrF product can then be probed using laser-induced fluorescence. Its vibrational distribution is found to depend on rotation and orientation of the HF reactant. In particular, the results suggest that the sideways approach of Sr to HF may be energetically preferable. For both In and Tl reaction with I_2^* (cf. reaction 10.14), it is found that a collinear approach is preferred.

Lasers have been used to pump both atomic[271]

$$Sr(^3P_1) + HF \rightarrow H + SrF \tag{10.13}$$

and molecular[272]

$$In + I_2(B\,^3\Pi) \rightarrow InI + I \tag{10.14}$$

reagents to electronically excited states.

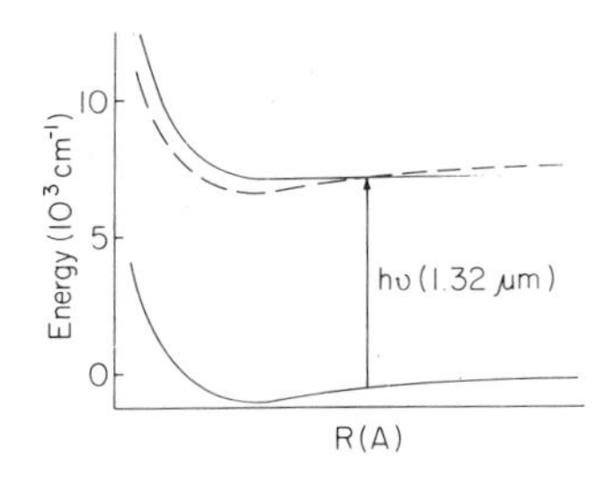

Fig. 32. Laser-assisted bridging of the energy gap in an atomic collision (schematic). The potential energy curve of the ground state plus a photon of a Nd-glass laser (the "dressed" state), shown as a dashed curve, can cross the potential energy curve for the excited atoms. Because of this crossing, transitions (in the presence of the field) can take place without requiring changes in the momenta. See Ref. 278 for an actual experimental example.

The use of excited reagents in bulk systems is necessarily accompanied by side collisional processes which tend to degrade the energy. Considerable attention was therefore given to the study of energy transfer pathways for laser pumped species.[273–277] Several aspects of this work are discussed in Chapters 22 and 26. Again, for direct collisions of simple species, the rule of thumb is the exponential gap rule—the energy release into translation tends to be minimal. Hence, and particularly for hydrides, vibrational deactivation of diatomic molecules by atoms converts a significant fraction of the energy into rotation. In a diatom–diatom collision, vibration–vibration exchange is a preferred mode. For polyatomics, collision-induced intramolecular energy transfer (Chapter 23) is an important pathway. Such near-resonant processes tend to increase the time required to degrade the initial excitation into heat, cf. Chapter 18.

In view of the primary role of the translational energy gap for direct collisions it is of obvious interest to try and bridge it using laser photons.[278, 279] The essential idea is shown in Fig. 32. The required translational energy is being supplied by the photons or, in other words, the potential energy of the initial state is increased (it is being dressed) by the photon to a point where it equals the potential energy of the final state. When that happens, efficient transitions are possible. The process has been demonstrated for transfer of electronic energy in atom–atom collisions.[277] It should however be noted that the field intensity required is quite high and that for molecular systems the dressing of the excited states may lead to unimolecular multiphoton dissociation and ionization.

XI. COHERENT OPTICAL EFFECTS FOR PEDESTRIANS

Up to this point we have been concerned with experimental observables pertaining to the populations of discrete molecular energy levels, intramolecular continua and quasicontinua. Another class of interesting, useful and informative experimental observables is concerned with the interrogation of the retention of phase relationships between the ground-state and excited doorway states. Studies of coherent transient effects[55, 56, 115–129] in electronically excited states of molecules in the gas phase and in low-temperature solids, reviewed in Chapter 27, involve the observation on a macroscopic scale of the dynamics of a pulsed interaction between the radiation field and a molecular ensemble. The relevant new physical information emerging from these studies pertains to the destruction of phase coherence via T_2 dephasing processes. Extensive experimental and theoretical studies focused attention on dephasing processes in a two-level system[55, 56] induced by medium perturbations, due to gas phase collisions, phonon coupling and electronic energy transfer in solids. A different class of problems in this novel area pertains to coherent optical effects in electronically excited states of large polyatomic molecules (Fig. 33).[130, 131] A cursory examination of this complex level structure indicates that the molecular eigenstates are unevenly spaced and exhibit irregular variations in the radiative coupling with the ground state $|g\rangle$, so that at first sight it seems hopeless to observe phase coherence effects in such a system. The possibility of observing photon echoes from an excited electronic state, which corresponds to the statistical limit, was considered[130, 131] and it was demonstrated that under certain conditions the problem can be reduced to the familiar two-level case. The destruction of phase coherence in an excited electronic state of a large molecule will then originate from the following major causes:

1. Intramolecular effects originating from the congested level structure.[130, 131]
2. Inhomogeneous broadening originating from Doppler broadening in the gas phase or distribution of trapping sites in a host solid (Fig. 33d).[55, 127–129, 280]
3. Intermolecular dephasing, which is collisionally induced in the gas phase or originating from (nonlinear) phonon coupling processes in a solid.[55, 56, 127–129]
4. Long-range intermolecular resonance coupling between impurity molecules.
5. Spectral diffusion originating from electronic transfer to other guest molecules.[282]

The interesting effects of medium perturbation are considered in Chapter 29, while we shall address the question of intramolecular dephasing, which is of considerable interest in the area of interstate and intrastate intramolecular dynamics.

The concept of $\pi/2$ and π optical pulses, widely utilized to induce and probe coherent optical effects in two-level systems,[55, 56] has to be more

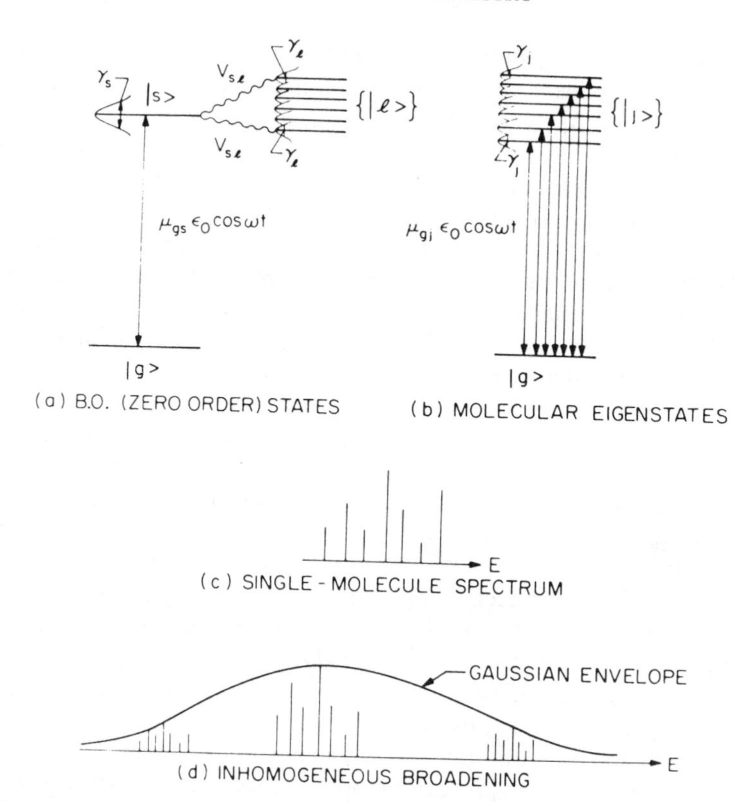

Fig. 33. Schematic representation of an inhomogeneously broadened level structure of a large molecule. (a) Zero-order manifold. Horizontal lines represent energy levels, while γ_s and γ_l denote the corresponding widths. (b) The eigenstates $\{|j\rangle\}$ of the effective Hamiltonian. These molecular eigenstates are characterized by energies ϵ_j and widths γ_j. (c) A schematic spectrum of a single molecule. The height of each vertical line represents the (relative) peak intensity in absorption. Note the irregular level structure. (d) The effects of inhomogeneous broadening resulting in a Gaussian spread of the individual spectra of (c).

carefully specified for a congested level structure, where many levels are driven by the field. Coherent optical effects, such as photon echoes[55] or free-induction decay,[55] can be conducted under short-time excitation, which spans the entire congested excited state spectrum.[130, 131] This "coherent" molecular excitation mode corresponds to optical selection of the doorway state $|s\rangle$, being specified by the condition that the pulse length Δt_p is considerably shorter than the decay time $\hbar\Gamma_s$, (5.9) and (6.2), of the wavepacket of coupled molecular eigenstates,

$$\Delta t_p \ll \hbar\Gamma_s^{-1} \tag{11.1}$$

Consider first the problem of optical free induction decay (OFID) of an ensemble of "collision-free" polyatomic molecules, which interact on the time scale $0 \leqslant t \leqslant \Delta t_p$ with a short laser pulse, and which satisfies condition (11.1). Following the pulse the sample radiates, the characteristics of the OFID being determined by intramolecular dephasing processes as well as by dephasing due to inhomogeneous broadening, which is specified in terms of a Gaussian distribution. The polarization decay after termination of the pulse can be separated into a product of two terms corresponding to dephasing due to inhomogeneous broadening and to intramolecular effects.[131] In the photon echo (PE) experiment two short light $\pi/2$ and π pulses, characterized by the durations Δt_1 and Δt_2, respectively, and which are separated by the time τ, impinge on the ensemble of the isolated molecules. We again consider short-time excitation conditions for these two light pulses. The temporal behavior of the PE is essentially determined by the inhomogeneous dephasing contribution at $t' = \tau$, while the amplitude of the echo at $t' = \tau$ is determined by the contribution of the intramolecular dephasing. The aspects of intramolecular dephasing are determined by the intramolecular decay amplitude of the doorway state, as is the case for free induction decay. At this stage we would like to comment on the relation between the decay of the OFID and of the PE, which are both exhibited in the direction of the laser beam, and the conventional, incoherent, isotropic fluorescence excited by a short-pulse excitation. The intensity (i.e., the photon counting rate) of the conventional fluorescence at time t, when the doorway state $|s\rangle$ can be optically selected at $t = 0^+$, is given by

$$I_F(t) \propto |C_{ss}(t)|^2 \tag{11.2}$$

where $C_{ss}(t)$ is the population probability of the doorway state at time t. In the assembly of isolated molecule the intensities of the OFID and of the PE are[131]

$$I_{\text{OFID}}(\tau) \propto I_F(\tau)$$

$$I_{\text{PE}}(t' = \tau) \propto [I_F(\tau)]^2 \tag{11.3}$$

Thus, in principle, physical information emerging from the study of coherent optical effects in isolated large molecules can be obtained from the interrogation of the fluorescence decay. For a level structure which corresponds to the statistical limit, the intramolecular interstate or intrastate relaxation within the isolated molecule retains the excitation energy within the energy range corresponding to the doorway state. The time evolution of the "coherent" wavepacket, "prepared" according to condition (11.1), signifies a T_2 process. The dephasing width is just

$$\frac{\hbar}{T_2} = \Gamma_s \tag{11.4}$$

The total dephasing rate, $(T_2^{\prime})^{-1}$), of the doorway state incorporates also a level depletion contribution due to the radiative decay rate γ_s^r, being given by

$$\hbar / T_2^{\prime} = \Gamma_s + \gamma_s^r \tag{11.5}$$

Experimental interrogation of dephasing in electronically excited states of large molecules was conducted on large molecules embedded in a low-temperature host crystal.[123–129] Medium perturbations can be completely switched off when the dephasing of the electronic origin of an electronically excited state of a large molecule is studied at extremely low temperatures, $T < 2$ K, so that vibrational relaxation effects are absent and all the effects of molecule medium phonon coupling are eliminated. For the electronic origin of the S_1 state of the pentacene molecule embedded in *p*-terphenyl at $T < 2$ K, the experimental data clearly exhibit the basic relation $T_2^{\prime} = 2T_1$, where $T_2^{\prime}$ is the experimental dephasing lifetime obtained from PE and OFID techniques, while T_1 corresponds to the decay lifetime monitored by conventional photon counting.[123–129] The electronic origin of this S_1 state is coupled to the ground state and the level structure corresponds to the statistical limit. The experimental observation[123–129] of the equivalence of intramolecular interstate decay and intramolecular dephasing provides compelling evidence for the validity of the physical picture (Section V) for the statistical limit in interstate coupling and ER.

The notion of intramolecular dephasing in a congested, bound, level structure pertains both to interstate and intrastate processes. While all the available experimental data pertain to interstate coupling and ER, a related problem, which can be explored by coherent optical techniques, involves IVR in vibrationally or electronically–vibrationally excited states of large isolated molecules.[131] The dephasing of the wavepacket of nuclear states on a single potential surface can be directly interrogated providing central information regarding intramolecular vibrational energy flow in large molecules. Such information will be of considerable interest for the assignment of the threshold of the vibrational quasicontinuum that plays a central role in determining the features of high-order multiphoton excitation of large molecules, which we shall now discuss.

XII. HIGH-ORDER MULTIPHOTON MOLECULAR PROCESSES

The nature and characteristics of the photophysical and chemical processes induced by high-order infrared multiphoton excitation (MPE) of collision-free, isolated molecules triggered hectic experimental and theoretical activity and generated a considerable amount of interest[39–43, 186–188, 283–314] and excitement because of several reasons. First, MPE processes provide a novel technique for high-order multiphoton photoselective excitation on the ground state potential surface of a large molecule. Second, these MPE processes open up a new research area in photophysics and photochemistry, which is concerned with the interaction of isolated polyatomic molecules with intense radiation fields. Third, this excitation process may provide an avenue for specific excitation of some nuclear modes, resulting in unconventional photochemical consequences. Fourthly, diverse and interesting chemical phenomena were induced already by MPE and deserve further study. These include reactive processes, such as photofragmentation resulting in ground state fragments[39–43] ("Laser Excitation of SF_6: Spectroscopy and Coherent Pulse Propagation Effects," Chapter 9 and "Infrared Laser Chemistry of Complex Molecules," Chapter 11), as well as in radicals in their electronically excited states[95, 96] ("Electronic Luminescence Resulting from IR Multiple Photon Excitation," Chapter 13 and "Electronically Excited Fragments Formed by Unimolecular Multiple Photon Dissociation," Chapter 14) and nonreactive processes, e.g., intramolecular isomerization[47, 48] and the production of electronically excited parent molecule[191] (Section VII, "Luminescence of Parent Molecule Induced by Multiphoton Infrared Excitation", Chapter 12). In Sections VIII and IX we discussed some of the problems pertaining to the chemical consequences of molecular MPE, which were extensively experimentally explored during the last few years. On the other hand, our current understanding of the mechanisms of high-order MPE of a molecular level structure is still incomplete and the relevant experimental and theoretical evidence accumulates quite slowly. The elucidation of the mechanism of energy acquisition by MPE is crucial for a better understanding of the specific and nonspecific photochemical process induced by this photoselective excitation mechanism. We shall now address the basic mechanism of molecular MPE.

The gross features of these high-order multiphoton molecular processes on the ground state potential surface are best described in terms of three distinct energy regions[283, 289] (Fig. 34), which will now be specified in terms of the level structure and the nature of the high-order radiative coupling effects. The low energy range (range I) is characterized by a sparse level

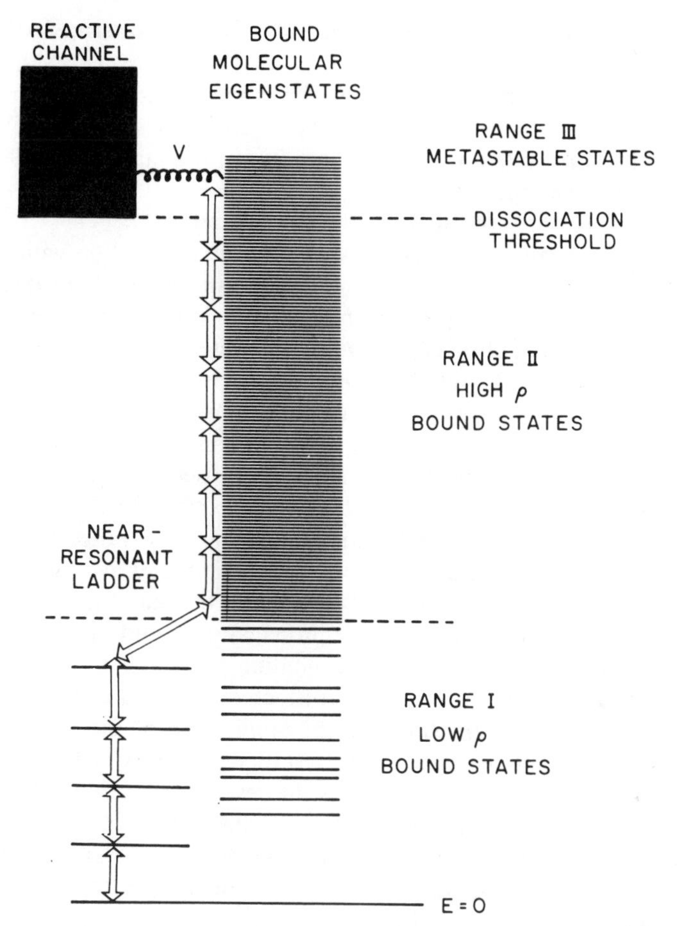

Fig. 34. Traditional energy level scheme for infra-red multiphoton excitation of an isolated large molecule. The three energy ranges, corresponding to the sparse bound level structure (range I), the bound quasicontinuum (range II) and the reactive region (range III), are marked. These molecular levels shown as horizontal lines correspond to the eigenstates of the molecular Hamiltonian. Radiative couplings to a photon field of frequency $\hbar\omega$ are marked by vertical arrows. In range I near-resonant radiative coupling within a discrete ladder is exhibited, while in range II radiative coupling between bunches of quasicontinuum states prevails. This schematic representation of radiative coupling should not be taken to imply that MPE occurs via consecutive one-photon absorption events.

structure of bound states. Range II corresponds to the intermediate energy region which consists of a dense bound level structure, i.e., a quasicontinuum of bound states. Finally, range III is the continuum region where, in the case of dissociation or predissociation, channels (Section IV) open up, while the nonreactive isomerization process in a large molecule results in a configurational change.

The general features of range I are well understood.[41, 195, 290, 303–306] Here the MPE process involves near-resonant radiative interaction with a sparse manifold of levels, where anharmonicity splitting effects[41] in degenerate states and rotational compensation effects[287] presumably insure near-resonant radiative coupling of the multilevel system. Range I processes are responsible for the celebrated isotopic selectivity[40–43] and should reveal some conventional features of near-resonant multiphoton excitation, such as a power dependence of the yield at low intensities followed by saturation effects at higher intensities of the radiation fields.[41] Finally, and most interesting, excitation in range I is expected to exhibit the characteristics of coherent, phase-preserving excitation, such as the appearance of multiphoton resonances (exhibited in the curves of the yield and average excitation energy vs. the laser frequency),[289, 290, 303–306] and the manifestation of some esoteric effects, such as coherent wave propagation[134] and photon echoes[132] in a multilevel system. The coherent excitation processes in range I can be handled in terms of a multilevel optical Bloch equation. The basic simplification involved in such treatment is the application of the rotating wave approximation, which is useful for a ladder of states, where radiative coupling prevails only between adjacent levels and provided that the photon energy considerably exceeds the characteristic Rabi frequencies. When these conditions break down the Magnus expansion[306] or a brute force solution of the equations of motion can be applied to improve upon the rotating wave approximation.

The radiative coupling in range II involves resonant interactions between quasicontinuum manifolds of bound states.[40, 283] The intramolecular quasicontinuum, which plays a central role in determining the features of nonreactive interstate and intrastate dynamics in large molecules (Section V), dominates the characteristics of intrastate MPE of such isolated molecules. Experimental evidence[206, 310] for MPE of a moderately large molecule, such as SF_6, led to the conclusion that the excitation process in range II essentially consists of a sequence of incoherent consecutive one-photon processes, which can be described in terms of a conventional kinetic master equation.[206, 314] Subsequent theoretical work[307–309, 311–314, 186–188] addressed the nature of intramolecular erosion of phase coherence effects in the MPE of the quasicontinuum. Concerning the "transition" from range I to range II, two comments should be made. First, regarding the molecular level structure, the "transition" from range I to range II with increasing excess vibrational energy bears a close analogy to the "transition" from the small molecular case to the statistical limit in the theory of intramolecular dynamics (Section V), the transition region may be fuzzy corresponding to the intermediate level structure. Second, regarding the radiative coupling between range I and range II, this may be described in terms of a "decay" of the upper level $|M\rangle$ in range I into the quasicontinuum.[306, 315] The coupling responsible for this escape process involves the radiative interaction $\mu_{Mq}\epsilon$, where μ_{Mq} is the transition dipole between $|M\rangle$ and $|q\rangle$ is any of the quasicontinuum states characterized by the density of states ρ_c, while ϵ represents the amplitude of the electromagnetic field. The escape rate being given by the golden rule

$$\Gamma_M = 2\pi(\mu_{Mq}\epsilon)^2\rho_c \tag{12.1}$$

Following the general classification of intramolecular energy ranges it will be useful to provide a cursory and superficial discussion of the implications for MPE of molecular systems, considered in the order of increasing complexity. The only molecular system whose bound level structure belongs entirely to range I is a diatomic molecule which, because of large anharmonicity defects, can be excited only at extremely high intensities and whose MPE and multiphoton photofragmentation were not yet experimentally documented.[195] For a triatomic molecule, it is an open question whether range II exists at all and, in any case, the low density of molecular eigenstates will make the properties of range I dominate the high-order MPE process. The experimental observation[172] of multiphoton photofragmentation of COS should be understood essentially in terms of high-order MPE in range I to the energy range where the electronic predissociation channel (Section IV) opens up. In a large molecule, such as SF_6, the energy span of range I is narrow, i.e., 4–6 levels in the ν_3 ladder[40] so that, although some characteristics of range I excitation will still be exhibited, the features of range II dominate the MPE process at reasonably high energies of the CO_2 laser. For huge molecules, such as S_2F_{10}[51] or the 44-atom molecule UO_2(hfacac)THF_3[52, 53] the onset of the quasicontinuum is very low, being located at an excess energy of about a single vibrational quantum.[52] As the truly novel and not completely understood central feature of molecular MPE involves high-order multiphoton pumping of the quasicontinuum in large and huge molecules, we shall discuss the problems pertaining to the identification of the onset of the quasicontinuum, the physical mechanisms underlying MPE of the quasicontinuum and the relevance of IVR to the multiphoton pumping of this congested, bound, level structure of an isolated molecule.

Theoretical estimates of the onset of the vibrational quasicontinuum in the ground electronic configuration of large molecules start from two complementary descriptions, that of zero-order states coupled by anharmonic interactions V_A or that of molecular eigenstates. These estimates rest on:[40, 186, 188, 195, 310]

1. Intramolecular criteria of strong intrastate coupling, i.e., $V_A\rho \gg 1$, where ρ is the density of these states, as well as on efficient sequential decay, i.e., $\gamma_{IR}\rho \gg 1$, and where γ_{IR} corresponds to the infrared decay width of the zero-order vibrational states.
2. Field dependent criteria implying that the Rabi frequency $\mu\epsilon$ for radiative coupling between molecular eigenstates exceeds their mean level spacing, i.e., $\mu\epsilon\rho \gg 1$.

Experimental criteria for the onset of the vibrational quasicontinuum stem from several sources:

1. Infrared picosecond spectroscopy.[54] These time-dependent data for SF_6 were interpreted[54] in terms of an ultrafast (<30 psec) intramolecular vibrational relaxation at excess vibrational energy of $1\nu_3$($\sim$940 cm^{-1}).
2. High resolution infrared spectroscopy.[41] The interpretation of these spectroscopic data for SF_6[41] indicates that at $1\nu_3$ the level structure is still sparse.
3. Photon echoes at a single vibrational excitation.[316] The photon echo experiments conducted at the level of $1\nu_3$ excitation of SF_6 yield conclusive evidence that at this excess vibrational excitation of 940 cm^{-1} intramolecular dephasing does not occur, and only intermolecular dephasing is exhibited, which originates from collisional effects.
4. Intermolecular V–V transfer between polyatomic molecules at excess vibrational energies of 2000–3000 cm^{-1} indicates that the vibrational level structure is sparse.[317]
5. Single-level collisional effects in the first excited electronic state of some large molecules, such as substituted benzenes,[318] provide strong evidence that at excess vibrational energies of 2000–4000 cm^{-1} the level structure is still sparse.

We note that the interpretation of the picosecond time-resolved data for SF_6 (method 1)[54] is inconsistent with observations 2–5. We believe that the latter four independent sources of information are more reliable in assigning the onset of the quasicontinuum. The experimental interrogation methods 2–5 provide compelling evidence that for large molecules, such as SF_6, the onset of the vibrational quasicontinuum is located at excess vibrational energies above $\sim 3000\ cm^{-1}$.

The most reliable methods for the identification of the onset of the vibrational quasicontinuum rests on the study of coherent optical effects to interrogate dephasing of a wavepacket of nuclear states, which mark intrastate IVR in a bound level structure. Studies of photon echoes in SF_6 [316] provide some useful negative evidence, setting a lower limit for the onset of the quasicontinuum. These photon echoes experiments pertain to the conventional case of coherent effects in a two-level system. To study the location of the onset of the vibrational congested level structure, which is located at $\sim 3000\ cm^{-1}$ in a large molecule by a CO_2 laser, one has to consider coherent optical effects in multilevel systems. The problem of coherent effects in some multilevel systems were studied recently, both theoretically[132] and experimentally,[133] but were not yet applied to the problem at hand. Recently, Cantrell has suggested[134] a new method based on coherent pulse propagation in a multilevel system to probe the onset of the vibrational quasicontinuum, which will be of considerable interest.

Up to this point we have been concerned with large molecules, such as SF_6. For huge molecules many of the techniques just considered will be inapplicable in view of extensive thermal rotational–vibrational inhomogeneous broadening effects. These difficulties may be overcome by the cooling of huge molecules in supersonic beams.[64–67] Two new methods recently advanced for the identification of the onset of the vibrational quasicontinuum in the first electronically excited state of huge molecules are:

6. Identification of the onset of a congested level structure in the optical spectrum interrogated by laser spectroscopy in seeded supersonic beams.[69–76]
7. The observation of an energetic threshold for splitting[319] and broadening[74, 77] of the emission lines for the $S_1 \rightarrow S_0$ fluorescence spectrum of a large molecule in a supersonic beam marks the threshold for extensive intrastate mixing.

Although these methods were applied to interrogate the vibrational level structure in the first electronically excited state of some huge molecules, such as tetracene,[69, 75] pentacene,[76] and ovalene,[77] some reliable conclusions can be inferred regarding the level structure in the ground electronic state, as the configurational changes between these two electronic states are not appreciable. Application of these spectroscopic methods 6 and 7 to some huge molecules characterized by 84–132 vibrational degrees of freedom,[76] located the onset of the vibrational quasicontinuum above the electronic origin of the S_1 state at the excess vibrational energy of 2000–1000 cm^{-1}. This spectroscopic information concurs with the physical picture, which implies that the onset of the quasicontinuum in huge molecules, such as $UO_2(hfacac)_2THF_3$, is located around the first vibrational excitation around $\sim 1000\ cm^{-1}$.

The pumping of the quasicontinuum via MPE is underlined by intramolecular erosion of phase coherence effects. On the basis of experimental evidence this high-order process is described in terms of a kinetic master equation for the populations. Several theoretical approaches, both on the "working hypothesis" level[40, 283, 310, 311] and on the formal level,[307–309, 186–188, 312–314] were advanced to account for the erosion of coherence effects and to account for the structure of the kinetic equations. The latter fall into the following categories:

1. *Weak coupling approach*.[307–309, 314] The zero-order molecular modes are subdivided into a "relevant" group consisting of a radiatively active mode and a "bath" containing all other modes. The system-bath coupling is due to intramolecular anharmonic coupling. The intramolecular bath provides the dephasing source. The rapid intramolecular dephasing limit is expressed by the inequality

$$\frac{\Gamma_{(W)}}{\hbar} \gg \bar{\mu}\epsilon \tag{12.2}$$

where $\Gamma_{(W)}/\hbar$ is the rate associated by IVR of the optically active mode, while $\bar{\mu}$ is the transition moment connecting the optically active states. Under these circumstances, the populations of the relevant modes obey a Markoffian master equation within the relevant subspace. The main advantage of this treatment is that "it fits," being useful for a reasonable description of the experimental facts of life. Nevertheless, such an approach suffers from several serious drawbacks. First, it invokes the applicability of the weak coupling master equation[324] (Table IV) to describe the MPE. The radiative coupling may be quite strong being just restricted by condition (12.2), so that the conventional weak coupling master equation may be irrelevant. Second, the role of the intramolecular bath in inducing dephasing of the interesting mode is somewhat obscure as, when the bath itself is not excited, it can induce dephasing only via population relaxation. Third, a hidden assumption underlying this treatment involves the hypothesis of selective radiative coupling between the special zero-order states, while no radiative coupling is assumed to occur between the reservoir states. This assumption is inadequate for a large molecule where no distinction can be made between relevant and reservoir states with respect to near-resonant radiative coupling.

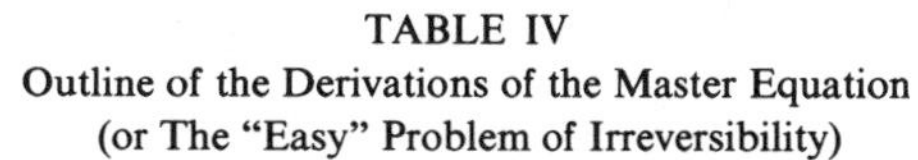

TABLE IV
Outline of the Derivations of the Master Equation
(or The "Easy" Problem of Irreversibility)

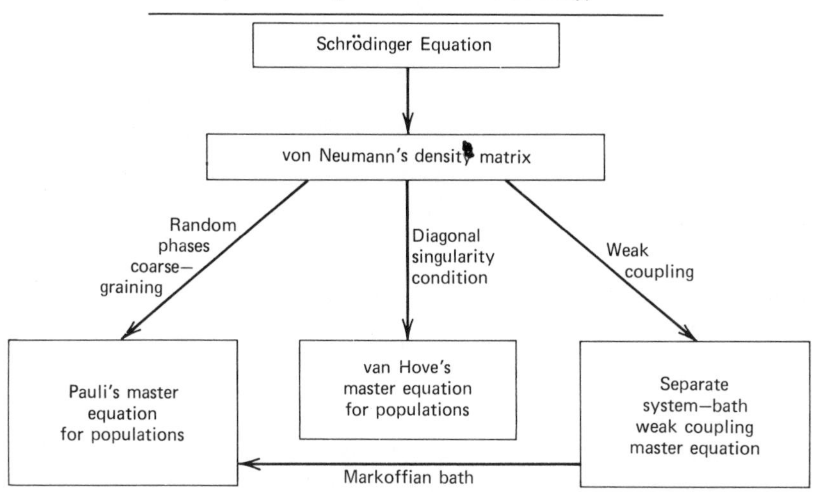

2. *Probabilities for MPE.*[186–188, 312, 313] The natural approach to the problem of MPE of the quasicontinuum should rest on considering the total energy content of the molecule expressed in terms of the number of photons absorbed. The following sensible assumptions are invoked: (1) Monochromatic excitation source. (2) Spontaneous infrared emission is disregarded. (3) The Rabi frequency $\mu\epsilon$ between quasicontinuum states separated by the laser frequency $\hbar\omega$ is small, relative to $\hbar\omega$. (4) the condition

$$\mu\epsilon \ll \hbar\omega \tag{12.3}$$

together with assumed coupling between adjacent distinct bunches of states, (Fig. 35), justifies the use of the rotating wave approximation to handle the MPE. Each manifold of "dressed" molecule-radiation field states is characterized by a manifold of molecular levels $\{|\alpha\rangle\}$ and by absorbed K photons. The population $P_K(t)$ of the manifold $\{|K\alpha\rangle\}$ is the probability that the molecule absorbs K photons at time t.

3. *Reduction schemes.*[312, 313] A reduction scheme was developed (Chapter 7) leading to the equations of motion for the populations $\{P_K\}$, which is valid for arbitrary radiative coupling strength. Invoking the notion of intramolecular dephasing, which originates from IVR, it was concluded (Chapter 7) that provided IVR is fast relative to radiative Rabi frequency (12.2) a Markoffian kinetic equation emerges for the $\{P_K\}$. It is still an

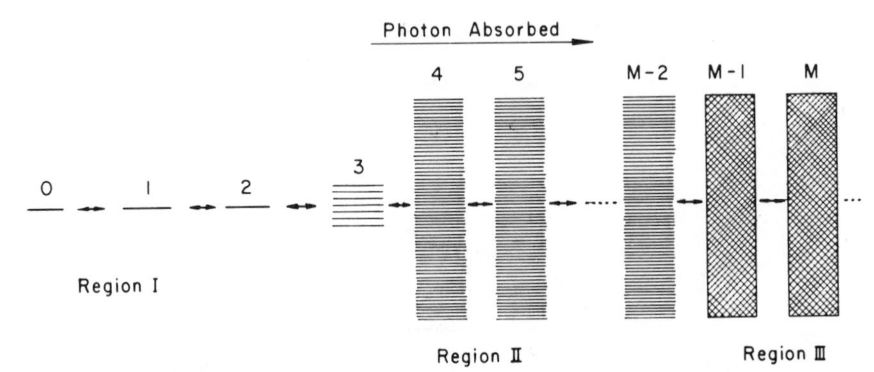

Fig. 35. Dressed states of a larger molecule in an intense radiation field. The energies of each of these states correspond to (energy of molecular level)$+(n-L)\hbar\omega$, where n is the initial number of photons, L denotes the number of photons absorbed and $\hbar\omega$ is the laser frequency. This dressed states picture is valid within the rotating wave approximation.

open question whether the Markoffian assumption (12.2) constitutes a sufficient condition for the validity of a simple master equation for the populations without invoking additional assumptions concerning the nature of the radiative coupling. This brings us to the last approach, which attempts to provide physical insight on the basis of minimal physical information.

4. *The random coupling model* (*RCM*).[186–188] The variation of the near-resonant radiative coupling terms μ_{lm} within the quasicontinuum are assumed to be random functions of the level indices, being characterized by the distribution

$$\langle \mu_{lm} \rangle = 0$$

$$\langle \mu_{lm}\mu_{l'm'} \rangle = \langle \mu_{lm}^2 \rangle \delta_{ll'}\delta_{mm'} \tag{12.4}$$

The validity of the RCM is restricted by condition (12.2) for rapid intramolecular IVR (or dephasing). The RCM, which rests simultaneously on the implicit validity of (12.2) and the explicit applicability of relation (12.4), results in the kinetic equations for the populations $\{P_K\}$, which involve absorption and stimulated emission. It should be emphasized that in contrast to weak coupling methods usually used to derive the Pauli master equation the RCM results in kinetic equations under the conditions of strong radiative coupling $\langle \mu_{lm}^2 \rangle \rho_l^2 \gg 1$.

The RCM for MPE of the intramolecular quasicontinuum rests on the notion of random radiative coupling between nuclear molecular eigenstates, evading the issue of IVR.[186–188] It is interesting to establish the

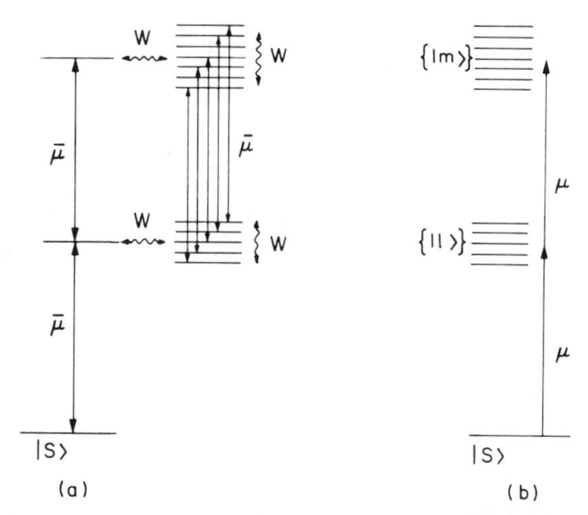

Fig. 36. A comparison between two molecular models for high-order multiphoton molecular excitation: (a) Zero-order molecular levels with a single optically active mode and with anharmonic interactions. (b) Exact molecular eigenstates with diluted random radiative couplings.

relation between this picture of radiative RCM and the alternative approach, which starts from a zero-order molecular basis and considers the radiative coupling between a small number of zero-order modes, i.e., bond modes or normal modes (Section V), which in turn are coupled to the rest of the vibrational zero-order manifold. In Fig. 36 we compare the two complementary models. The radiative interaction terms $\bar{\mu}\epsilon$ in the zero-order basis couple all states, which correspond to a change of one quantum in the optically active mode with no change in the other modes. Accordingly, for each consecutive manifolds the $\bar{\mu}\epsilon$ terms vary weakly with the level indices. On the other hand, the intramolecular anharmonic perturbation terms W vary widely and wildly with the level indices and can reasonably be assumed to be random. Thus the zero-order molecule model (Fig. 36a) for MPE involves random intramolecular coupling and approximately constant radiative coupling, in contrast to the RCM, which considers just random radiative coupling between the genuine molecular eigenstates. To establish the relation between the two approaches one can consider the dilution of the constant radiation coupling terms $\bar{\mu}\epsilon$, which combine zero-order states among the molecular eigenstates. This dilution effect is essentially determined by the IVR width, first appearing in (12.2), which is given explicitly by

$$\Gamma_{(W)} = 2\pi\langle W^2\rangle\rho \tag{12.5}$$

The dilution factor for such interstate coupling[188]

$$D_{(W)} = \pi^2 \langle W^2 \rangle \rho^2 = \frac{\pi \Gamma_{(W)} \rho}{2} \tag{12.6}$$

in analogy to relation (5.6) and the dilution factor N for interstate scrambling. The dilution of the radiative coupling is now obtained from the conservation of absorption intensities, so that the dilution effect is exhibited for $(\bar{\mu}\epsilon)^2$ being given explicitly by[188]

$$\langle \mu^2 \rangle = \frac{\bar{\mu}^2}{D_{(W)}} = \frac{2\bar{\mu}^2}{\pi \Gamma_{(W)} \rho} \tag{12.7}$$

This relation establishes the connection between radiative coupling and intramolecular intrastate dynamics. On the basis of numerical simulations[188] and analytical solutions,[320] it is established that the random coupling in multiphoton excitation of the quasicontinuum can be taken to involve either random intramolecular or random radiative interactions, depending on the choice of the basis set. As long as the Markoffian condition (12.2) holds, both approaches are equivalent. Finally, we would like to draw attention to the breakdown of the Pauli master equation when the Markoffian condition (12.2) is no longer valid. This state of affairs can, in principle, be accomplished at very high fields, when

$$\bar{\mu}\epsilon \gtrsim \Gamma_{(W)} \tag{12.8}$$

As $\Gamma_{(W)} \sim 10\ \text{cm}^{-1}$ or so,[40] one demands that $\bar{\mu}\epsilon > 10\ \text{cm}^{-1}$. In that case the radiative RCM approach breaks down, while the intramolecular RCM still applies. In such extremely high radiation fields coherent effects in the excitation of the special bond mode may be exhibited within the congested quasicontinuum. As condition (12.8) implies that the radiative driving is effective relative to the IVR "dephasing" rate, selective excitation of a single bond mode may be realized at extremely high fields. This novel effect, which was not yet subjected to an experimental test, provides an example of the perspectives of the novel area of photochemistry in intense radiation fields.

Finally, it will be useful to consider some further experimental implications and applications of the theory. The "working hypothesis" for MPE of large molecules rests on the simple kinetic model for consecutive one-photon absorption, which is applied to the quasicontinuum, evading the issue of coherent effects in MPE of range I. The celebrated experimental evidence[40] for the dependence of the photofragmentation yield of SF_6 on the laser fluence, F, is only approximate, as at low values of F some admittedly weak but definite intensity dependence is clearly exhibited.[310, 321] Several recent experimental studies of MPE in which the laser pulse length was varied at constant F have revealed definite intensity dependence of the yield.[322, 323] The results indicate that the role of coherent excitation in range I cannot be disregarded. This conclusion is, of course, not surprising for MPE of small molecules like COS,[172] where under well-controlled excitation conditions range I spans a wide energy region. However, even for large molecules, such as SF_6, the dogmatic "working hypothesis" for MPE has to be modified. The general scheme for treating coherent excitation in range I, followed by leaking to the quasicontinuum where incoherent excitation prevails, has been developed[186–188] and can be applied for model calculations which will be useful and interesting. Before undertaking such an endeavor, we would like to address the cardinal question of whether the experimental observation of the dominating effect of the fluence on the yield of high-order multiphoton processes in large molecules, does solely reflect the intramolecular erosion of phase coherence effects within the quasicontinuum. In this context it will be useful to consider the following three major sources of dephasing in an isolated, collision-free molecule:

1. Leaking from range I.[315] At sufficiently high fields when Γ_M (12.1) exceeds all the Rabi frequencies $\bar{\mu}\epsilon$ and the off-resonance energy mismatches within range I, the coherent oscillations within range I will be damped due to the effective excitation from range I to range II. A rough estimate indicates that erosion of coherence in range I will be realized when $\Gamma_{(W)} \ll \bar{\mu}\epsilon$, being exhibited only at extremely high fields, when the excitation of range II is no longer incoherent.
2. Intramolecular phase erosion due to MPE of the quasicontinuum, which was considered in some detail.
3. Fluctuations in the laser field. The lasers currently utilized for MPE are by no means ideal, single-mode sources. The radiative interaction and the MPE process may be affected by the fluctuations in the phase of the electromagnetic field which, under certain circumstances, may result in extrinsic erosion of phase coherence effects. The complete elucidation of the diverse and exciting aspects of intensity and coherence effects in MPE and multiphoton photochemistry require a more profound understanding of the effect of the mode structure of the radiation field on the molecular target.

Notable accomplishments in the area of intramolecular and intermolecular excited-state dynamics include the elucidation of some of the photophysical and photochemical consequences of diverse photoselective excitation processes, the merger between the small molecule and the large molecule point of view, the unification and extension of the description of reactive molecular processes, and the introduction of a unified conceptual framework for interstate and intrastate coupling and dynamics in large molecules. Moreover, dynamics of isolated large molecules bridge the gap between intramolecular dynamics and dynamics in condensed phases, which we shall now briefly consider.

XIII. RELAXATION AND DEPHASING IN CONDENSED PHASES

Collisional perturbations and coupling of a molecule to a dense medium introduce two new central features of excited-state dynamics, which involve erosion of phase coherence, i.e., T_2-type processes, and medium-induced relaxation phenomena, resulting in population transfer, i.e., T_1-type processes, (Chapters 28 and 29).

The most interesting and subtle medium effects on a photoselected state of a guest molecule in a host medium pertain to the destruction of phase coherence between the ground state and the electronically excited state due to molecule-medium coupling. These dephasing processes retain the molecule in the initially excited energy range, so that no genuine relaxation of the guest molecule has occurred, but phase destruction did take place. The oldest spectroscopic observations of such phenomena dates back to 1895 when Michelson demonstrated the effect of collision broadening of spectral lines.[325] Since then sophisticated and fancy techniques of coherent optical effects were advanced[55, 56] to attack this problem. The information emerging from homogeneous broadening of spectral lines and from studies of coherent optical effects is identical, except that the modern studies probe dephasing in an inhomogeneously broadened system. These T_2 processes belong to two major categories. (1) Intramolecular dephasing already considered in Section XI, and (2) intermolecular dephasing essentially due to the modulation of the energy levels by molecule-medium coupling, e.g., elastic scattering in collisional effects,[56, 326] molecule-medium phonon coupling,[123–129] impurity–impurity long-range interactions[281] and energy transfer.[281] The experimental results for intermolecular T_2 processes fall into two broad classes, which involve dephasing of vibrational excitations[327] (Chapter 29) and of electronic excitations, which will now be briefly discussed. Four techniques were recently applied to study intermolecular T_2-type processes in electronically excited states, which involve

(1) photochemical hole burning,[328–330] (2) spectral line broadening, with an attempt to eliminate inhomogeneous broadening,[129, 331] (3) time-resolved and energy-resolved observables for resonance fluorescence and for near-resonant Raman scattering,[332, 333] and (4) coherent optical effects (Section XI and Chapter 27). On the basis of general arguments, one can assert that in the limit of zero-temperature, medium-induced elastic scattering effects are negligible, so that

$$(T_2)^{-1} = (2T_1)^{-1} \qquad (T \rightarrow 0) \tag{13.1}$$

Thus, for the electronic origin of a large statistical guest molecule in a host medium only radiative decay and intramolecular dephasing prevail, so that the experimental T_2 rate is

$$(T_2)^{-1} = \gamma_s^r + \Gamma_s \qquad (T \rightarrow 0) \tag{13.2}$$

For higher vibrationally excited states of an electronically excited configuration, a new type of medium-induced vibrational relaxation rate is exhibited, being characterized by the rate γ_{VR}, the zero temperature dephasing rate being

$$(T_2)^{-1} = \gamma_s^r + \Gamma_s + \gamma_{\mathrm{VR}} \qquad (T \rightarrow 0) \tag{13.3}$$

It is important to realize that in the zero-temperature limit the information obtained from the direct study of T_2 processes and from the interrogation of T_1 processes is identical, as is evident from (13.1). Thus, in the zero-temperature limit both intramolecular relaxation in the statistical limit, as well as medium-induced relaxation, e.g., vibrational relaxation, can be studied either by the conventional methods of probing the level population or by the hard way of interrogating T_2-type processes. This is not the case at finite temperatures, when a medium-induced contribution T_2^* to the dephasing, due to "elastic scattering" events, sets in. Thus, for the electronic origin

$$(T_2)^{-1} = \gamma_s^r + \Gamma_s + (T_2^*)^{-1} \tag{13.4a}$$

while for a vibrationally excited state

$$(T_2)^{-1} = \gamma_s^r + \Gamma_s + \gamma_{\mathrm{VR}} + (T_2^*)^{-1} \tag{13.4b}$$

where γ_{VR} (and also Γ_s) may be temperature dependent.

The most extensive studies of the effects of intramolecular dephasing in electronically excited state of a large molecule were conducted for the S_1 state of pentacene in host crystals of p-terphenyl[128, 129] and of naphthalene[334, 335] ("Electronic Transitions of Large Molecules in the Condensed Phase," Chapter 27). For the electronic origin of S_1 at low temperatures, $T < 1.5$ K, relation (13.1) is obeyed (Section XI). The medium-induced elastic contribution $(T_2^*)^{-1}$ dominates the dephasing process of the electronic origin at higher ($T=2-10$ K) temperatures, exhibiting an Arrhenius-type temperature dependence $(T_2^*)^{-1} = A\exp(-\Delta E/kT)$, with $\Delta E \simeq 13$ cm^{-1}.[124] Such Orbach-type[29] dephasing process reflects coupling with a local phonon mode ("Electronic Transitions of Large Molecules in the Condensed Phase," Chapter 27). At even higher temperatures ($T>20$ K) optical line broadening data for the electronic origin indicate that $(T_2^*)^{-1} \propto T^7$ marking a Raman phonon process.[29] Next, we shall mention some studies of vibrationally excited states of this system.[331] In contrast to the features of the electronic origin the dephasing of the vibrationally excited levels in the temperature range 0–14 K is dominated by the medium-induced vibrational relaxation rate, while the role of the medium-induced elastic scattering $(T_2^*)^{-1}$ is negligible. Line broadening, together with picosecond time-resolved population monitoring, result in $T_2 = 7$ psec at 1.5 K for the lowest optically allowed vibrational state at 261 cm^{-1}, while both photon echo and population monitoring give $T_2 = 66$ psec at 1.5 K for the 750 cm^{-1} vibration.[331] At higher temperatures (3–14 K), the dephasing rate and the vibrational relaxation rate obey (13.1), so that $(T_2^*)^{-1}$ is negligible. These rates exhibit an Arrhenius-type activation energy $\gamma_{VR} = A\exp(-\Delta E/kT)$ with $\Delta E = 16.6 \pm 1.5$ cm^{-1}. This activation energy for vibrational relaxation of a vibronically excited state[331] is identical to that obtained[124] for elastic dephasing of the electronic origin, indicating the role of low frequency local phonon modes in both processes.[331] From the foregoing exposition it is apparent that studies of dephasing processes yield valuable information on medium-induced level depletion processes. We shall now turn to consider such T_1 level depletion processes in electronically–vibrationally excited states of molecules in a dense medium starting from medium-induced vibrational relaxation and proceeding to medium effects on electronic relaxation.

The two dominant factors which determine the dynamics of medium-induced vibrational relaxation (VR), reviewed in "Vibrational Population Relaxation in Liquids," and "Experimental Studies of Nonradiative Processes in Low Temperature Matrices," Chapters 28 and 29, are the vibrational level structures of guest and of the host environment. The simplest case is VR on the ground-state potential curve of a diatomic molecule embedded in a rare-gas solid.[335, 336] The guest vibrational frequency is $\hbar\omega$ while the vibrational spectrum of the host is simple, being characterized by an acoustic phonon spectrum with the Debye frequency in the range $\hbar\omega_D = 45\ \text{cm}^{-1}$ for Xe to $\hbar\omega_D = 65\ \text{cm}^{-1}$ for Ar. The simplest mechanism for VR involves dumping of molecular vibrational energy to lattice phonons via high-order multiphonon process, which can be described by an exponential energy-gap law,[337]

$$\gamma_{VR} = A\exp\left[-\alpha\left(\frac{\hbar\omega}{\hbar\langle\omega\rangle}\right)\right]\left(1+\frac{1}{\exp(\hbar\langle\omega\rangle/kT)-1}\right)^{(\hbar\omega/\hbar\langle\omega\rangle)} \tag{13.5}$$

where $A \sim 10^{12}\ \text{sec}^{-1}$ is a numerical constant, $\alpha \simeq 1-3$ is another number, while $\langle\omega\rangle$ is a mean-medium phonon frequency, $\langle\omega\rangle \leqslant \omega_D$. This relation predicts strong exponential dependence of γ_{VR} on the guest frequency, a moderate dependence of γ_{VR} on the host, a dependence of γ_{VR} on the isotopic composition of the guest, the rate increasing for the heavier isotope, a temperature independent rate at $kT \ll \hbar\langle\omega\rangle$ and a strong activated temperature dependence of the rate for $kT \gg \hbar\langle\omega\rangle$. This simple and general relation has not yet been validated experimentally.[335] VR in electronically excited states of Cl_2, NO, and of O_2 may correspond to multiphonon VR.[338]

Relation (13.5) disregards coupling to local modes of the guest molecule. VR of diatomics containing a light atom, such as hydrogen and deuterium, which are characterized by a high moment of inertia, reveal that local modes associated with hindered (or free) rotation of the guest provide important accepting modes.[339, 340] The VR process can then be best described in terms of a medium-induced vibration to rotation (V–R) energy exchange.[335, 341, 342]

The relation between VR process in dense media, such as solid rare gases and gas-phase intermolecular dynamics, is of considerable interest. The VR of HX and DX molecules in simple solids via a medium-induced V–R process bears a close analogy to collisionally induced V-R relaxation of such molecules in the gas phase.[2] This useful analogy is reinforced by the observation that in binary mixtures of Kr and Ar host, specific Kr—NH pair interaction have a profound effect on the VR process,[343] so that the chemist's intuitive picture of the role of guest–host interactions, presumably of van der Waals type, dominate over the physicist's symmetry arguments for the trapping site. Not only the medium-induced V–R processes but also the multiphonon VR process bears close analogy to gas-phase reactive relaxation in small molecular systems. Multiphonon VR of a diatomic guest in a rare-gas host may involve V–V transfer intermediated by a specific guest-host pair interaction.[338] Such a VR process bears a close analogy to vibrational predissociation of van der Waals molecules (Section IV, "Van der Waals Molecules," and "Intramolecular Dynamics of van der Waals Molecules," Chapters 5 and 6). The VP in O_2 ($c^1\Sigma_u^-$) and in NO ($a^4\Pi$) were proposed to occur via such mechanism.[338] These considerations are of considerable interest in providing a unified description of gas-phase intramolecular relaxation phenomena and relaxation processes in condensed phases.

At the next level of complexity of VR one should consider medium-sized polyatomic molecules in rare-gas hosts. Medium-induced VR of the lowest vibrational frequency of the polyatomic guest should resemble the VR features of "heavy" diatomics, which proceed via the solid-state multiphonon mechanism, as described by (13.5) or by vibrational predissociation of a van der Waals bond. Following photoselective excitation at a higher vibrational frequency of the guest, the medium-induced VR process may involve cascading down the vibrational ladder. The high vibrational frequency being interconverted into a lower molecular vibrational mode, the energy balance being taken up by the host medium. This medium-induced intramolecular V–V exchange bears a close analogy to the intramolecular V–V vibrational predissociation process of van der Waals complexes of polyatomic molecules ("Intramolecular Dynamics of van der Waals Molecules," Chapter 6). Finally, one has to consider VR in molecular hosts. An important difference between rare-gas solids and molecular solids is the existence of host vibrational states in molecular solids, which can act as accepting modes for the VR process. These include, in addition to acoustic modes, the optical lattice modes of the solid, as well as intramolecular vibrational modes (vibrons) of the molecules constituting the molecular host. Coupling with optical lattice phonons may result in a decrease of the energy gap for multiphonon VR, making this mechanism much more efficient. Of considerable interest are the consequences of coupling of the vibrational states of the guest molecule to the vibron states of the host. The VR process may result in interconversion of the vibrational quantum of the guest to intramolecular vibration of a host molecule, the (small) energy mismatch being taken up by the lattice phonons. This intermolecular V–V process in VR bears analogy to intermolecular V–V in the vibrational predissociation of van der Waals complexes, consisting of a pair of polyatomic molecules.[170] As the current stage of theoretical understanding of medium-induced VR is still in the embryonic stage, we shall find comfort not in alluding to further conjectures but rather by considering some modern techniques to probe vibrational population changes:

1. Time-resolved antistokes scattering from liquids[327] result in VR times on the ground-state potential surfaces. It was found that γ_{VR} varies from >10 sec for liquid N_2[335, 344] to a few psec for polyatomic

liquids.[327] No information on the fate of the excited state was obtained and the information pertains just to γ_{VR} of the photoselected state.

2. Two-photon infrared-visible consecutive excitation. This technique[345] was applied to probe medium-induced VR (in solution T=300 K), as well as IVR in an isolated molecule (T=580 K), and was applied to the huge Coumarin 6, 42-atom molecule, with the result that the lifetimes for VR and for IVR at excess vibrational energies of 2700 cm^{-1} and 6000 cm^{-1}, respectively, are a few picoseconds. The interpretation of these results is hampered by severe sequence congestion effects.
3. Hot luminescence studies[346] based on measurements of relative intensities of hot luminescence bands when the molecule cascades down the vibrational ladder. The quantum yields for the unrelaxed states are[346] $10^{-3}-10^{-4}$ and can readily be detected.
5. Picosecond absorption recovery experiments of photoselected vibrational states of large molecules in low-temperature mixed crystals.[331]
5. Picosecond photon echo experiments.[280]
6. Hole burning experiments.[330, 331]

The modern methods 3–5 were recently applied to VR of some large molecules in low-temperature host solids, where the effects of sequence congestion could be overcome. The following interesting conclusions emerge from these studies:

a. Dependence on excess vibrational energy. VR does not increase with increasing of the excess vibrational energy. For pentacene in naphthalene[331] at 1.5 K, γ_{VR} = 3.3 psec for the 261 cm^{-1} vibration, while γ_{VR} = 33 psec for 750 cm^{-1} vibrations. Similar irregular dependence of γ_{VR} on excess vibrational energy was reported for various vibrational excitations in the S_1 state of perylene in Ne.[346]
b. Dependence on host matrix. The lowest allowed 350 cm^{-1} mode of perylene exhibits a lifetime of 35 psec in heptane and 55 psec in solid Ne, reflecting the possible role of intermolecular V–V transfer in the polyatomic host.[346]
c. Intramolecular cascading. Anthracene in solid Ne shows an 18 psec lifetime for the 400 cm^{-1} mode. This level becomes significantly populated by the decay of higher vibrational states.[346] For the higher 1400 cm^{-1} level the lifetime was independent of host composition for monoatomic and molecular hosts. These results indicate in the S_1 state of anthracene medium-induced vibrational cascading prevails.
d. Intermolecular V–V. Vibrational relaxation times for various vibrational levels of the S_1 state of azulene in naphthalene at 2 K

were in the range 1–10 psec.[347] The 384 cm^{-1} lowest optically allowed mode was not populated by the decay of higher vibrational levels. These results were interpreted[347] by asserting that intermolecular V–V transfer to the host naphthalene vibron states provides the dominating mechanism for dissipation of vibrational energy in this system.

We have limited our discussion to VR dynamics of low-temperature well-defined systems, where genuine photoselection can be achieved and the detailed pathways of intramolecular and intermolecular vibrational energy degradation will be amenable for study. Although a lot of chemical information pertains to large molecules embedded in room-temperature solids and liquids, the interpretation of these results is hampered by some inhomogeneous broadening, ultrafast dephasing and sequence congestion, all of which will obscure the details of the dynamics. Accordingly, we have not discussed such systems in the context of VR and we will provide only a cursory examination of medium-induced ER in solution, focusing attention on the sparse available information which pertains to electronic processes in well-defined medium-perturbed systems.

The first question we address in relation to electronic processes of a guest molecule in a dense medium is: What are the configurational changes of the medium in the immediate vicinity of the electronically excited molecule? Some information can be obtained from the analysis of the optical lineshapes in low-temperature systems, where the effects of inhomogeneous broadening are not severe.[348] The appearance of zero-phonon lines marks small configurational changes, as is the case for intravalence excitation of aromatic molecules in some solid rare gases[346] and in Shpolskii matrices.[80] While molecular medium-configurational changes are expected to be small for intravalence excitation, large configurational changes are expected to be exhibited for extravalence excitations.[156] The emission spectrum NO in solid Ar was assigned to a medium relaxed Rydberg state.[349] Similar effects were previously observed for the radiative decay of extravalence excitations of atomic impurities in solid and in liquid rare gases,[350] where the emission which exhibits a large Stokes shift is very close to the free atom transition. These medium relaxations were described in terms of "cavity formation around excited states".[349, 350] Search for emission from Rydberg states of diatomic and polyatomic molecules in a dense medium will be of considerable interest. This simple and transparent example of medium relaxation around excited states provide examples for a broad class of much more complex phenomena of solvent reorganization, which involve orientation of a polar solvent around a giant dipole,[351] as well as the dynamics of solvation of excess electrons in fluids, (Chapter 30).[83]

We now proceed to a cursory examination of medium effects on ER. In doing so, it will be useful to follow the traditional classification of Section V according to the intramolecular level structure. For excited states of polyatomic molecules, which correspond to the small molecule limit and to the intermediate level structure, ER is not exhibited in the isolated molecule and can only be induced by medium perturbations.[8, 16, 17] Obvious examples are to be found with atoms and diatomic molecules in inert matrices which exhibit ER, degrading electronic energy ΔE_l into lattice phonon energy. This nonradiative medium-induced ER process can presumably be described in terms of an exponential energy-gap law, the rate being proportional to $\exp(-\gamma\Delta E_l/\hbar\langle\omega\rangle)$, so that small step cascading is preferred. An interesting example in this category is the medium-induced ER of $CN(A^2\Sigma)$ in solid Ne, where the excited states relax through intervention of ground state levels.[352] Proceeding to the small molecular case, we note that the isolated SO_2 molecule exhibits fluorescence[353] with unity quantum yield, while solid SO_2 emits only phosphorescence,[354] marking medium-induced ER. Next, we consider medium-induced ER in an electronically excited state of a large molecule, which corresponds to the intermediate level structure. For example, the S_1 state of the benzophenone molecule, which is separated by 2000 cm^{-1} from the origin of T_1, exhibits a fluorescence radiative decay time of $\sim$1 μsec in a molecular beam,[61] while the decay lifetime of S_1 in solution is $\sim 10^{-11}$ sec.[355] These five orders of magnitude change in the lifetime of S_1 of the isolated benzophenone molecule and of this state in solution reflects the effect of medium-induced ER on the intermediate level structure. Medium-induced ER in the small molecule case and for the intermediate level structure can be described in terms of intrastate VR within the background manifold intermediated by interstate nonadiabatic coupling ("Collision Induced Intersystem Crossing," and "Collisional Effects in Electronic Relaxation," Chapters 24 and 25).

Finally, we shall consider medium effects in the statistical limit. In a molecular level structure, corresponding to the statistical limit, the ER characteristics of the electronic origin are invariant with respect to perturbation excited by an "inert" solvent, which does not modify the energy levels or the intermolecular nonadiabatic coupling.[16, 17] A notable example in the category of active solvent perturbations on ER, which enhance S_1–T_1 or T_1–S_0 interstate spin–orbit coupling, involves the external heavy atom effect on intersystem crossing.[356] Solution and matrix isolation data on this phenomenon provide only the gross features of this effect. Supersonic beams provide an excellent means for the production of weakly bound van der Waals molecules and clusters involving aromatic molecules and rare-gas atoms. Selective gradual and controlled solvation of large aromatic molecules and rare-gas atoms can be accomplished, making it possible to study solvent effects on ER in a well-defined system.[356–359] The observation of the heavy atom effect in intersystem crossing in tetracene–Kr_n and tetracene–Xe_n complexes provided a clue for the nature of specific intermolecular interactions which modify the spin–orbit coupling.[360]

The evidence for minimal perturbations of an "inert" solvent on ER dynamics in the statistical limit is sparse. Most of the available discussions[16] compare low-pressure room temperature $T > 300$ K dynamics of a molecule suffering from severe sequence congestion to room-temperature solution spectra, which we have decided not to analyze in detail. An esoteric system, which suffers from all these shortcomings, is the ultrafast relaxation dynamics of the S_1 state of azulene.[361] The lifetimes reported are 4 ± 3 psec for 530 nm excitation in solution, 3 ± 2 psec at 625 nm in solution and 4 ± 3 psec in a collision-free sample.[362] Additional data are the directly measured lifetime of 1.9 ± 0.2 psec at 605 nm excitation[363] and the estimate of 3.1 ± 0.3 psec for the electronic origin deduced from spectroscopic data.[348] From these results it was tentatively asserted[362] that (1) the $S_1 \rightarrow S_0$ internal conversion depends weakly on excess vibrational energy, and (2) that the ER rate is rather insensitive to perturbations induced by an inert medium, this conclusion being in accord with the theoretical predictions for the statistical limit.

The ultrashort psec lifetimes of the S_1 state of azulene raise the distinct possibility that the intramolecular ER rates are faster or comparable with medium-induced VR processes. All conventional studies of multiphonon processes addressed the situation where the VR process overwhelms the rate of the electronic process, making it possible to define two distinct time scales for the relaxation process.[364] This problem of the competition between electronic processes and medium-induced VR is of considerable interest and in the forthcoming discussion we shall demonstrate explicitly the necessity to extend the theory of conventional multiphonon processes to handle this situation.

XIV. MULTIPHONON PROCESSES

There have been extensive experimental and theoretical studies of nonradiative electronic processes (NREP) in solids and in liquids. The origins of this field date back to the work of Seitz[365] on the configurational diagram description of radiative and nonradiative processes in impurity centers in solids and to the quantification of these ideas by Huang and Rhys[366] and by Kubo[367, 368] in relation to the theory of electron-hole recombination in semiconductors. These NREPs span a broad spectrum of intermolecular behavior in condensed phases. Several of the relevant phenomena are small polaron motion,[369] electron transfer between ions in solutions[147–149] and donor (D)–acceptor (A) pairs in solids, electron-hole recombination in semiconductors[150] and in amorphous solids,[151] electronic energy transfer between donor (D)–acceptor (A) pairs involving molecules or ions in solids and liquids.[153, 154] Several interesting biophysical phenomena (Section XV) fall also into this category, typical examples being intermolecular electron transfer in the photosynthetic center[85, 370, 376] and group transfer in the low-temperature recombination of dioxygen and carbon monoxide with hemoglobin.[152, 377]

The elucidation of the nature of these intermolecular NREPs in condensed phases operates on a different level of depth, as compared to the foregoing discussion of medium-induced molecular VR and ER. While in the latter cases we have carefully distinguished between T_1- and T_2- type processes and an elaborate approach was adopted for the initial photoselected states, the characterization of the dynamics of intermolecular NREP rests on the change in the population of an entire electronic manifold, adopting a traditional approach of conventional chemical kinetics with just considering a single observable, i.e., the thermally averaged rate of the process. A unified theoretical framework exists at present for the understanding of all these diverse phenomena. All these NREPs can be described in terms of relaxation processes between two potential surfaces corresponding to two different zero-order electronic configurations,[365] and where energy conservation is insured by the absorption and emission of phonons[368] (Fig. 37). These intermolecular multiphonon processes bear a close analogy to intramolecular ER, where population or depletion of vibrational levels, i.e., intramolecular phonons, prevail.[189] To be more specific we have listed in Table V some examples for such intermolecular processes. The rates of these nonradiative processes are calculated by considering two electronic–vibrational manifolds of the entire system consisting of the (energy or electron) donor–acceptor pair, together with the entire phonon bath of the medium. The residual coupling induces the nonradiative process in the solid (Table V) from a single initial vibronic state $|I\rangle$ to the manifold $\{|J\rangle\}$ of final states quasidegenerate to it. The residual coupling $V=V_e\langle I|J\rangle$ is usually expressed within the framework of the Condon approximation in terms of a product of an electronic term V_e and a nuclear Franck–Condon vibrational overlap factor $\langle I|J\rangle$ between the initial and final states. The microscopic rate constant, W_I, for the $|I\rangle\rightarrow\{|J\rangle\}$ transition can be described in terms of nonadiabatic formalism provided that the residual coupling is weak relative to the characteristic vibrational frequencies of the system. The nonadiabatic microscopic rate is given in terms of the golden-rule analogous to (6.2),

$$W_I=\frac{2\pi}{\hbar}|V_e|^2\sum_J|\langle I|J\rangle|^2\delta(E_I-E_J) \qquad (14.1)$$

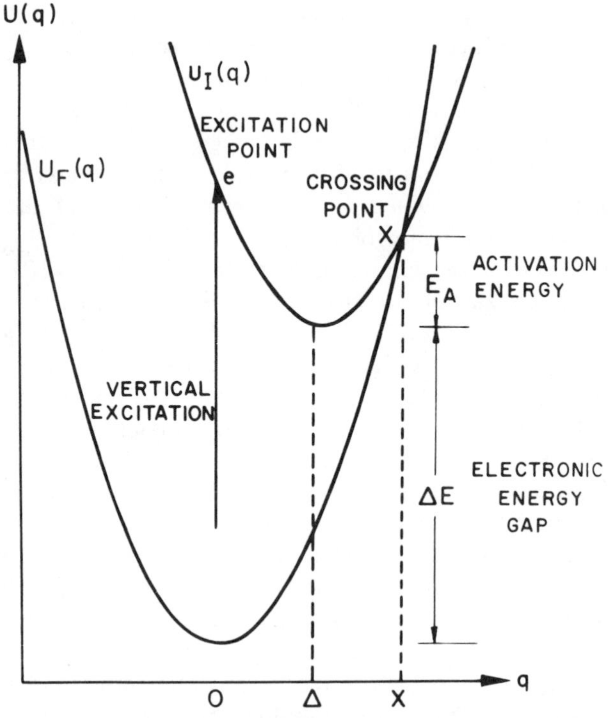

Fig. 37. One-dimensional configurational diagram for a multiphonon process in a dense medium. The initial zero-order states belong to the potential surface $U_I(q)$, while the final zero-order states belong to the potential surface $U_F(q)$. q denotes the nuclear coordinate(s).

The basic assumption underlying most of the studies is that medium-induced vibrational relaxation and vibrational excitation processes are fast on the time scale of the NREP itself.[364] One can then perform a separation of time scales for the "fast" vibrational relaxation (VR) and "slow" NREP, so that the transition probability W for the NREP is then expressed in terms of a thermal average of the microscopic rates

$$W = \sum_I \exp\left(\frac{-E_I}{kT}\right) W_I \Big/ \sum_I \exp\left(\frac{-E_I}{kT}\right) \tag{14.2}$$

Before alluding to any, necessarily simplified, specific models it will be useful to consider two limiting cases for the general nonadiabatic multiphonon rate:

1. The low-temperature limit is exhibited when the thermal energy is considerably lower than all the relevant characteristic vibrational fre-

TABLE V
Intermolecular Nonradiative Processes in Condensed Phases

Process	Electronic States	V_e Electronic Coupling
Small polaron motion in solids	$A^-A \rightarrow AA^-$ A = neutral molecule A^- = Negative ion	Two-center one-electron electrostatic interaction
Electron transfer in solids, liquids, and biological systems	$DA \rightarrow D^+A^-$ D = electron donor A = electron acceptor	Two-center one-electron electrostatic interaction
Electron-hole recombination in semiconductors	$D^+\|k\rangle \rightarrow D^+\|b\rangle$ $\|k\rangle$ = free electron $\|b\rangle$ = electron bound to D^+ D^+ = positive ion	Nuclear momentum
Electronic energy transfer in solids, glasses, and liquids	$D^*A \rightarrow DA^*$ D = energy donor A = energy acceptor	Intermolecular electrostatic interaction dipole–dipole monopole–monopole, also electron exchange
Group transfer in hemoglobin	$Fe(S=2)+CO \rightarrow Fe(S=0)+CO$	Spin–orbit coupling

quencies of the system. The low temperature rate is

$$W(T \rightarrow 0) = W_{I=0} \tag{14.3}$$

being determined by the microscopic rate from the vibrationless nuclear level of the initial-state potential surface. This low-temperature limit corresponds to nuclear tunnelling from the lowest zero-point state of the initial nuclear configurations to the final vibronic states, which are nearly degenerate with this level.

2. The high-temperature limit, where the thermal energy considerably exceeds the variation of the potential energy of the initial state within an average de Broglie wavelength.[189, 368] The high temperature rate expression

$$W \propto \exp\left(\frac{-E_A}{kT}\right) \tag{14.4}$$

assumes an Arrhenius-type form with the activation energy E_A, being given by the potential energy of the initial electronic state at the lowest intersection point of the two potential surfaces. This activated rate equation is obtained without invoking the concept of the activated complex.[189]

The general nonadiabatic multiphonon rate equation (14.2) passes from a finite temperature independent nuclear tunneling rate at low temperatures to an activated rate at high temperatures. Considerable effort was devoted toward the systematic simplification of the nonadiabatic multiphonon rate expression, which rests on the following assumptions. First, the nuclear motion is taken to be harmonic. Second, only linear electron phonon coupling terms corresponding to configurational distortion of the origin of the potential surfaces are considered, while the role of frequency changes are neglected. These two assumptions are sufficient to derive the exponential energy-gap law for low temperature exoergic multiphonon processes[189]

$$W \propto \exp\left(\frac{-\gamma \Delta E}{\hbar \langle \omega \rangle}\right) \qquad T \to 0 \tag{14.5}$$

where ΔE is again an electronic energy gap for the intermolecular process and one requires that $\Delta E / \hbar \langle \omega \rangle \gg 1$. This energy gap law for intermolecular NREPs was subjected to an experimental test for electronic energy transfer[378] and for electron transfer.[379, 380] Equation 14.5 is isomorphous with a broad class of exponential energy gap laws encountered in diverse areas, such as intramolecular ER in the statistical limit (6.3), ER in ionic centers in solids, medium-induced processes such as VR, (13.5), and medium-induced ER of small molecules in a dense medium.[353, 354] All these intramolecular and intermolecular processes can be handled in terms of a unified physical description, which just pertains to a reasonable though approximate description of the nuclear overlap factor for the vibrationless level of the initial-state potential surface. An interesting alternative description of the energy gap law for exoergic multiphonon processes can be obtained by utilization of the semiclassical WKB approximation to calculate the vibrational overlap Franck–Condon factor. For a model system of two one-dimensional harmonic potential surfaces whose minima are displaced by d, each of which being characterized by frequency ω and reduced mass μ, the low-temperature rate can be recast in the form[152, 381]

$$W \simeq \frac{2\pi |V_e|^2}{\hbar \omega} \exp\left[\frac{-\sqrt{2}\, d(\mu E_A)^{1/2}}{\hbar}\right] \tag{14.6}$$

and E_A being the activation energy, i.e., the barrier height. This result establishes the equivalence between the nuclear contribution to the low temperature multiphonon rate and the Gamow tunneling formula.[159, 160] On the other hand, the preexponential factor in the low-temperature exoergic rate expression cannot be expressed in terms of Gamow's frequency factor, but rather it is determined by the square of the electronic coupling $|V_e|^2$. The simplified multiphonon rate expression (14.6), which describes an NREP between two bound states in a dense medium, where irreversibility is insured by dispersion of the phonon frequencies and by medium-induced sequential vibrational relaxation, is akin to the tunneling expressions which characterize reactive molecular relaxation processes, such as rotational predissociation (4.2) and vibrational predissociation (4.4). This formal analogy pertains to the form of the nuclear contributions to the rates of these diverse processes.

The alert reader will note that the conventional multiphoton NREPs we were concerned with do not correspond to nonradiative processes occurring from a photoselected initial state, as fast medium-induced vibrational relaxation, implicit in the derivation of (14.2), erodes all memory regarding the initially excited state. Interesting effects of radiative coupling on NREP in condensed phases can be exhibited when such processes will be interrogated in intense radiation fields. For example, intermolecular electronic energy transfer between a pair of molecules in a dense medium, whose energy levels are off-resonance, can be induced by simultaneous intermolecular coupling and radiative interaction with an intense radiation field, the energy conservation being insured by the absorption of an additional optical photon.[143] A similar state of affairs holds for inelastic atomic collisions induced by an intense radiation field,[142] except that in the molecular case Franck–Condon nuclear vibrational overlap has to be incorporated. These intermolecular processes in intense radiation fields belong to a broad class of radiative collisions (Section X) which are of considerable interest.

Up to this point we have been concerned with intermolecular NREP occurring from a thermally equilibrated vibrational manifold. When ultrafast electronic relaxation processes are exhibited, one encounters the situation when the assumption of "fast" VR cannot be taken for granted, and the separation of time scales between VR and NREP is no longer applicable. Under these circumstances, the coupling between VR and NREP has to be explicitly considered.[367] We envisage the system starting from a highly vibrationally excited state sliding down by VR and simultaneously undergoing and NREP. Such situations can be realized under the

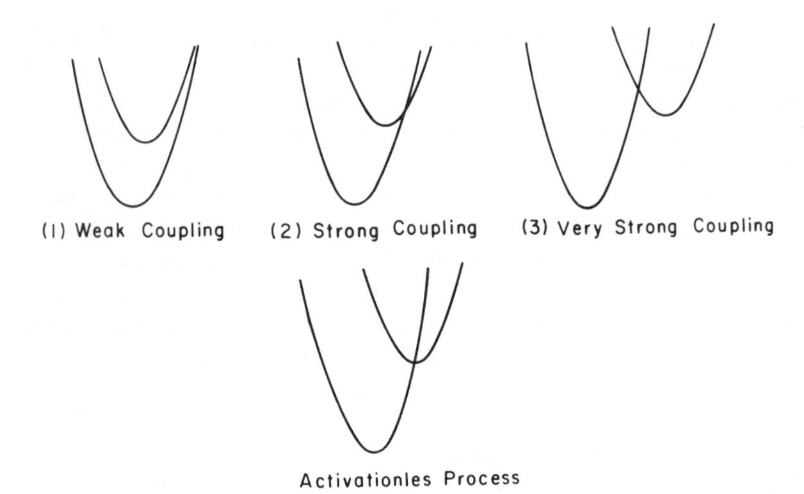

Fig. 38. Nuclear coupling schemes for multiphonon processes portraying (1) the weak coupling limit, (2) the strong coupling situation and, (3) the very strong coupling case. The activationless exoergic process corresponds to a borderline case between situations (2) and (3).

following circumstances, which are schematically portrayed in Fig. 38:

1. In the weak electron-phonon coupling limit [Fig. 38(1)] the rate of the NREP is enhanced at higher vibrational levels (Section VI), so that, provided the electronic coupling is sufficiently strong, the competition between NREP and vibrational relaxation will result in nonexponential decay.
2. In the strong electron–phonon coupling limit [Fig. 38(2)] the crossing point of the potential surfaces is close to the minimum of the initial surface. At the crossing point the nuclear vibrational overlap is large and the microscopic rate for NREP in this energy range may be appreciable for reasonable values of the electronic coupling. The system initially excited to the energy E_l (Fig. 39), will start to slide downward by VR. If the initially excited configuration E_l is located below the crossing point, the microscopic rates for NREP are usually sufficiently low to enable the system to relax vibrationally before NREP occurs (Fig. 39). However, when the initially excited configuration E_l is located above the crossing point of the potential surfaces (Fig. 39), the system will pass on its way downward through the crossing point where efficient NREP occurs and efficient nonradiative decay is exhibited before vibrational equilibration is established. This efficient mechanism for NREP, which is determined by the nature of

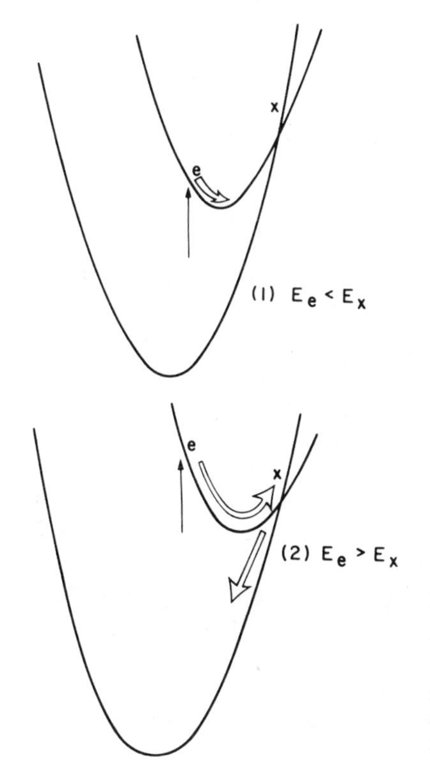

Fig. 39. Schematic description of initial excitation followed by vibrational relaxation, which is designated by arrows. When $E_e < E_x$ the system usually equilibrates thermally. When $E_e > E_x$ the system can undergo efficient NREP at the crossing point x before equilibration has been accomplished.

the initially photoselected state, was proposed for fluorescence quenching of impurity centers in solids,[382, 383] but has much wider applicability.

3. In the very strong coupling situation when the minimum of the initially-excited potential surface is located outside the crossing point of the two surfaces [Fig. 38(3)] then energy storage of some of the excited state population, which survived the crossing, will be exhibited.[365]

Three theoretical methods were applied recently to describe the dynamics of the competition between NREP and VR in condensed phases, which involve a semiclassical relaxation model based on the Landau–Zener treatment,[364, 384] a stochastic model resulting in a master equation for the populations[385] and a quantum mechanical treatment.[364] The elucidation of the features of these coupled processes is of interest because of three reasons. First, they provide an important example for the consequences of photoselective excitation in condensed phases. Second, theoretical and experimental studies of such phenomena will provide a conceptual framework for the understanding of ultrafast processes, occurring on a picosecond and shorter time scale, in condensed phases. Third, these ultrafast processes may be relevant for the establishment of the nature of some primary events in photobiology, which will now briefly be considered.

XV. DIGRESSION ON BIOPHYSICS

The conceptual framework developed for the understanding of intermolecular nonradiative relaxation processes in condensed phases is expected to contribute significantly toward the elucidation of some of the basic phenomena in biophysics and photobiology. Electron transfer processes[370–376] and atom or molecule transfer processes[152, 377] play a central role in a variety of biological systems. Many of the basic reactions can be described in terms of intermolecular nonradiative relaxation processes, which are amenable to a theoretical description in terms of nonadiabatic multiphonon formulism outlined in Section XIV. Thus, a broad spectrum of condensed phase electron transfer and atom transfer reactions in the areas of physical chemistry, solid-state physics and biophysics can be described within the framework of a unified approach.

The central biological phenomena pertaining to excited-state dynamics involve the basic mechanisms of photosynthesis[87–89, 370–376] (Chapter 31) and of vision ("Proton Transfer: A Primary Picosecond Event," Chapter 32). Consecutive electron transfer reactions constitute the key steps in the primary events of photosynthesis, while proton translocation is a primary process in visual pigments.[86] Both processes can be viewed as nonradiative relaxation processes in a biological system. During the last decade the dynamics of some of these elementary photobiological processes were investigated over a broad temperature range from cryogenic temperatures around 2 K up to room temperature. Notable examples are the light-induced oxidation of cytochrome c in the photosynthetic bacterium *Chromatium*[374–376] (Fig. 40) and the production of perlimirhodopsin from, electronically excited rhodopsin[86] ("Proton Transfer: A Primary Picosecond Event," Chapter 32). In these two cases the rate was found to be finite and temperature independent at low temperatures, then changing within a narrow temperature region into an Arrhenius temperature dependence, characterized by a finite activation energy. Without alluding to any specific model, one can assert that these processes can be described in terms of multiphonon nonradiative phenomena, where the finite temperature-independent rate exhibited at low temperatures manifests the effects of nuclear tunneling, while at higher temperatures an activated multiphonon process is exhibited.[152, 375, 376] This temperature behavior (Fig. 40) is

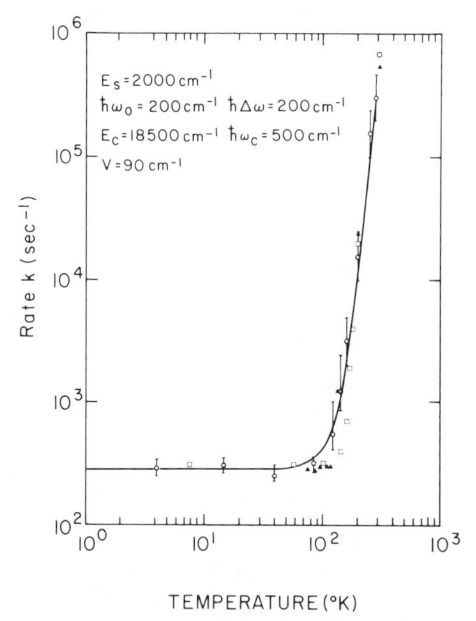

Fig. 40. The temperature dependence of the rate of the Chance–DeVault reaction (Ref. 374) for electron transfer from cytochrome c to the photooxized reaction center in the photosynthetic bacterium *Chromatium*. The experimental data (points) exhibit the effects of nuclear tunneling at low temperature, resulting in a temperature independent rate in the range 4–70 K. It is important to realize that nuclear tunneling rather than electron tunneling is involved in the low-temperature electron transfer process. The high-temperature data reflect activated process. The solid curve represents a theoretical fit using multiphonon theory [Buhks, Bixon, and Jortner, (unpublished)], which incorporates the effects of both long-range coupling with optical medium modes and short-range coupling with molecular vibrational modes of the donor and acceptor centers. The latter coupling effects dominate the nuclear contribution to the rate.

general for a broad class of multiphonon processes in condensed phases, as discussed in Section XIV. Before considering the consequences of a microscopic description of these processes it is imperative to distinguish between two classes: (1) Adiabatic processes that proceed on a single potential surface, e.g., proton translocation,[86] and (2) nonadiabatic processes involving a transition between two zero-order potential surfaces which correspond to electronic configurations. The theory of nonadiabatic multiphonon processes (Section XIV) is applicable to handle these processes. For electron transfer processes in biological systems,[87–89, 370–376] the residual electronic coupling between the donor and acceptor centers is sufficiently weak to warrant the utilization of nonadiabatic electron transfer theory.[375, 376]

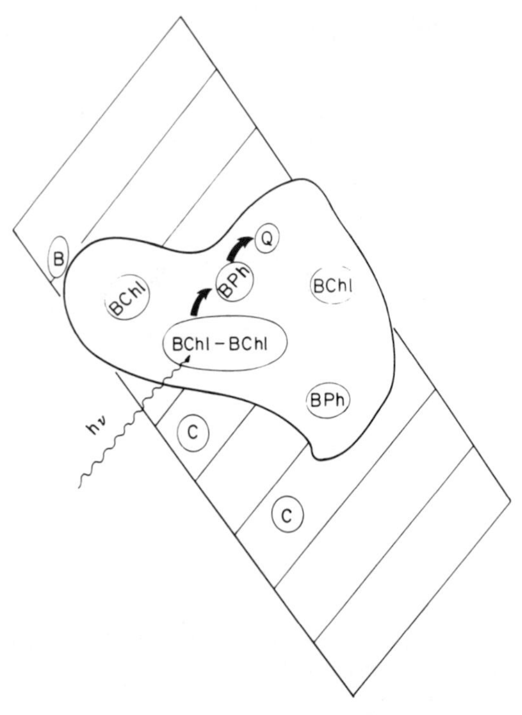

Fig. 41. Artist's view of the isolated reaction center in bacterial photosynthesis. BChl, bacteriochlorophyll-a; BPh, bacteriopheophytin; Q, ubiquinone-10; B, C secondary cytochrome; Ladder, membrane. Black arrows designate initial steps of charge separation, resulting from optical excitation of the bacteriochlorophyll dimer.

The understanding of the primary events in bacterial photosynthesis[87–89, 370–373] have been extremely helpful in the elucidation of the fundamental aspects of the photosynthetic process. The reaction centers of photosynthetic bacteria play the same role in biophysics as the hydrogen atom does in physics and chemistry, as they serve as basic, sufficiently simple, model systems for the applications of new theories and novel experimental techniques. The bacterial reaction center of photosynthetic bacteria[87–89] (Fig. 41) consists of a bacteriochlorophyll-a dimer $(BChl)_2$, with optical absorption bands peaking at 865 and 605 nm, two bacteriochlorophylls-a that absorb light at 800 and 595 nm, two bacteriopheophytins (BPh) absorbing at 760 and 535 nm and ubiquinone-10 (Q). Recent experimental studies utilizing the techniques of picosecond spectroscopy have established the initial processes of bacterial photosynthesis are[370–373]

$$[(BChl)_2BPh]Q \xrightarrow{h\nu} [(BChl)_2BPh]^*Q \tag{15.0}$$

$$[(BChl)_2BPh]^*Q \to [(BChl)_2^+BPh^-]Q \tag{15.1}$$

$$[(BChl)_2^+BPh^-]Q \to [(BChl)_2^+BPh]Q^- \tag{15.2}$$

Process (15.0) involves direct optical excitation of the $(BChl)_2$. The first chemical process, (15.1), corresponds to electron transfer (ET), which is characterized by: (1) Lifetime (reciprocal rate), τ_1, shorter than 10 psec, $\tau_1 \leqslant 10$ psec. (2) The upper limit for τ_1 is temperature independent over the temperature range 300–4 K. The second chemical process, (15.2), again constitutes ET and its features are: (3) The lifetime is $\tau_2 \cong 150$ psec at 300 K. (4) τ_2 is temperature independent over the temperature range 300–4 K. The general features of the lifetimes τ_1 and τ_2, which characterize the two primary chemical events in bacterial photosynthesis, differ both qualitatively and quantitatively from the celebrated DeVault–Chance reaction which involves ET from cytochrome c to the reaction center[374] (Fig. 40). The understanding of the primary events in bacterial photosynthesis has to provide a proper physical picture for the ultrafast rates and the temperature independence of these rates over a broad range.

From the point of view of general methodology, one should inquire at this stage whether conventional nonadiabatic multiphonon theory (Section XIV) is indeed directly applicable for the ultrafast processes (15.1) and (15.2). As we have pointed out, the basic hidden assumption underlying the applicability of (14.2) is that medium-induced VR is fast on the time scale of microscopic electron transfer.[364] The rate of reaction (15.2) is longer than characteristic medium-induced VR rates (Section XIII), so that this process has to occur from a thermally equilibrated nuclear configuration and conventional theory is adequate. The temperature independence of reaction (15.2) cannot, however, be accounted for in terms of low-temperature nuclear tunneling through a nuclear barrier, as such a mechanism will imply an unrealistically high characteristic molecular frequency ($>1000\ \mathrm{cm}^{-1}$) for electron–vibration coupling within the electron donor and electron acceptor centers.[386, 387] The temperature-independent rate of reaction (15.2) can adequately be explained in terms of specific intramolecular distortions of the equilibrium configurations of the donor and acceptor centers and of the external medium, which accompany the change in the charge distribution. For an exoergic reaction, where the nuclear potential surfaces for the initial and final states cross at the equilibrium configuration of the initial state (Fig. 38), the electron transfer process is activationless, revealing a temperature independent rate over a wide temperature range[386–388] and exhibiting a decrease of the rate at higher temperatures, i.e., a slight apparent negative activation energy. Reaction (15.2) was assigned to an activationless process,[386, 387] where the absolute lifetime $\tau_2 = 150$ psec could be adequately accounted for in terms of a reasonable estimate of the residual electronic coupling V_e, which corresponds to a reasonable donor-acceptor spacing of ~10 Å. On the other hand, the ultrafast primary process (15.1) can be definitely fast on the time scale of medium-induced VR, whereupon the separation of time scales for the electronic process and for VR may be inapplicable.[364] It was suggested that this primary process (15.1) occurs from a nonthermally equilibrated nuclear configuration.[364,386,389] The system is envisioned to be initially excited above the crossing point of the two potential surfaces and efficient electron transfer takes place at the crossing point. This ultrafast process brings up a significant and interesting example for the far-reaching consequences of photoselective excitation in a complex biological system. A crucial test for the proposed mechanism will emerge from the study of the dependence of the yield of process (15.1) on the initial excitation energy when the electron transfer is expected to be retarded at lower excitation energy below the crossing point.[389] From this brief survey it is apparent that a unified picture for dynamic processes in photobiology is currently emerging.

We have begun the saga of intramolecular dynamics in Section IV by a discussion of nuclear tunneling, which results in rotational predissociation in diatomic molecules. We have then progressed to consider intramolecular and intermolecular dynamics in more complicated systems, being finally led to consider the features of nuclear tunneling in a biological molecule characterized by a molecular weight of 10^5 daltons, i.e., the reaction center in bacterial photosynthesis. In recording this development, we have attempted to emphasize the unification and integration of concepts in the many-faceted area of photoselective chemistry.

XIV. EPILOGUE

The last two decades have seen an impressive progress in the understanding of dynamical processes on the molecular level. In the beginning was the alkali age of both elastic and reactive molecular collisions, and the earlier trajectory computations on realistic potential energy surfaces. Chemiluminescence from the products of elementary exoergic reactions led to the development of chemical lasers and universal detectors opened up new possibilities for molecular beam research. Larger molecules were being considered, and systematic approaches to intramolecular radiationless transitions were discussed. The separate developments centered about smaller and larger molecules were finally brought together by laser-induced chemistry. The experimental progress stimulated corresponding theoretical developments, and one is now finally on a threshold where understanding of elementary intramolecular and intermolecular dynamical events has been achieved. Much of the activity of these two decades can thus be characterized as the study of the "isolated molecule" limit. This molecule may have just a fleeting existence as in a direct collision or may be quite long-living with attention centered on the intramolecular energy transformation. The point of view, however, was decidedly microscopic. Simultaneously there was considerable progress in the study of what are "macroscopic" systems, e.g., dynamic processes in condensed phases, molecular crystals, disordered materials, surfaces, interphases, etc. Moreover, several of the techniques, both experimental (e.g., probing with lasers) and theoretical (e.g., computer simulation), are common to both the microscopic and the macroscopic domains.

What are the potential future developments? One possibility is to argue that the microscopic point of view will increasingly influence our approach, both experimental and theoretical to macroscopic systems. Indeed, increasingly finer probes, improved resolution and theoretical interpretation on the molecular level, can be and often already are employed. Even so, one may want to be more specific. What general areas will best benefit from an infusion of the molecular–microscopic point of view? For example, is it the case that the time is ripe to examine, both experimentally and theoretically, reactions on surfaces with the same degree of detail previously reserved for isolated collisions of simple species? However, there is also a flow in the opposite direction. Macroscopic points of view are increasingly influencing our approach to molecular dynamics. Is it therefore necessarily obvious that the detailed molecular picture should serve as the ultimate goal? In the area of collision dynamics we still do not have *ab initio* potential energy surfaces for such elementary processes as $H+ICl$ or even $H+DCl$. This despite the fact that there is a real wealth of experimental information on the branching ratios among the different arrangement channels on such surfaces, which could and should be interpreted in fundamental microscopic terms. In the field of intramolecular radiationless transitions we are not aware of a single quantitative calculation of the rate for electronic relaxation in the statistical limit which could stand the cold scrutiny of experiment. Are such bottlenecks important in principle? Are they important in practice? Should it influence our choice of a point of view?

The dichotomy between the microscopic and macroscopic points of view will gradually be bridged. It will become increasingly evident that these opposite approaches are complementary and supplementary. The merging between the microscopic and macroscopic points of view will undoubtedly be reflected in establishing the nature of the gradual "transition" from the intramolecular dynamics of an isolated molecule to the intramolecular and intermolecular processes underlying molecular relaxation in a condensed phase. The experimental techniques for preparation and identification of molecular clusters in supersonic beams[390–392] together with the development of computer simulation methods[393] will extend the interesting research areas concerned with the electronic structure, nuclear and electronic excitations, and excited state dynamics to large finite systems. In this context one will inquire: What are the optical properties of a large molecule solvated in finite cluster of rare-gas atoms or polar molecules? What is the phonon spectrum of a finite cluster and how does it affect vibrational energy flow? What is the nature of medium-induced VR and ER in finite clusters? What is the nature of surface states and surface dynamics and clusters? These and other questions will help to bridge the conceptual gap between the isolated molecule and condensed phase dynamics.

There are, moreover, other exciting new directions. Fundamental processes in biophysics are a good example. Applications of excited-state dynamics to astrophysics and astrochemistry are developing.[394] Even within traditional chemistry there are several areas, such as organic photochemistry, photoelectrochemistry, catalysis, etc., which are beginning to rely on modern developments in molecular dynamics. It is therefore likely that one key activity in the next decade will be integration: The synthesis of elementary steps into a description of a complex, multistage process.

References

1. R. D. Levine and J. Jortner, eds., *Molecular Energy Transfer*, Wiley, New York, 1975.
2. R. D. Levine and R. B. Bernstein, *Molecular Reaction Dynamics*, Clarendon Press, Oxford, 1974.
3. M. S. Child, *Molecular Collision Theory*, Academic Press, New York, 1974.
4. E. E. Nikitin, *Theory of Elementary Atomic and Molecular Processes in Gases*, Clarendon Press, Oxford, 1974.
5. E. C. Lim, ed., *Excited States*, Vols. I, II, and III, Academic Press, New York, 1974, 1975, 1977.
6. R. D. Levine, *Quantum Mechanics of Molecular Rate Processes*, Clarendon Press, Oxford, 1969.
7. B. R. Henry and M. Kasha, *Ann. Rev. Phys. Chem.*, **19**, 161 (1968).
8. J. Jortner, S. A. Rice, and R. M. Hochstrasser, *Adv. Photochem.*, **7**, 149 (1969).
9. E. W. Schlag, S. Schneider, and S. F. Fischer, *Ann. Rev. Phys. Chem.*, **22**, 465 (1971).
10. K. F. Freed, *Top. Curr. Chem.* **31**, 105 (1972).

11. B. R. Henry and W. Siebrand, *Org. Mol. Photophys.* **1**, 153 (1973).
12. G. W. Robinson, in *Excited States*, Vol. I, E. C. Lim, ed., Academic Press, New York, 1974, p. 1.
13. J. Jortner and S. Mukamel, in *The World of Quantum Chemistry*, R. Daudel and B. Pullman, eds., Reidel, Boston, Mass., 1974, p. 225.
14. S. A. Rice, in *Excited States*, Vol. II, E. C. Lim, ed., Academic Press, New York, 1975.
15. W. Heitler, *The Quantum Theory of Radiation*, Clarendon Press, Oxford, 1954.
16. J. Jortner and S. Mukamel, in *MTP Series in Science*, Vol. I, *Theoretical Chemistry*, C. A. Coulson and A. D. Buckingham, eds. (1976).
17. K. F. Freed, in *Radiationless Transitions in Molecules and Condensed Phases*, F. K. Fong, ed., Springer-Verlag, Berlin, 1976, p. 23.
18. C. B. Moore, ed., *Chemical and Biochemical Applications of Lasers*, Vols. 1, 2, and 3, Academic Press, New York, 1976, 1977, 1978.
19. C. V. Chank, E. I. Ippen, and S. L. Shapiro, eds., *Picosecond Phenomena*, Springer-Verlag, Berlin, 1979.
20. K. L. Kompa and S. D. Smith, eds., *Laser Induced Processes in Molecules*, Springer-Verlag, Berlin, 1979.
21. A. H. Zewail, ed., *Advances in Laser Chemistry*, Springer-Verlag, Berlin, 1978.
22. A. BenShaul, Y. Haas, K. L. Kompa, and R. D. Levine, *Lasers and Chemical Change*, Springer-Verlag, Berlin, 1980.
23. (a) R. H. Pantell and H. E. Puthoff, *Fundamentals of Quantum Electronics*, Wiley, New York, 1969.
    (b) M. Sargent, M. O. Scully, and W. E. Lamb, *Laser Physics*, Addison-Wesley, Reading, Mass. 1974.
24. M. L. Goldberger and K. M. Watson, *Collision Theory*, Wiley, New York, 1964.
25. G. Herzberg, *Spectra of Diatomic Molecules*, Van Nostrand, Toronto, 1950.
26. G. Herzberg, *Infrared and Raman Spectra*, Van Nostrand, Toronto, 1945.
27. G. Herzberg, *Electronic Spectra of Polyatomic Molecules*, Van Nostrand, Toronto, 1966.
28. K. K. Rebane, *Impurity Spectra of Solids*, Plenum Press, New York, 1970.
29. B. Di Bartolo, *Optical Interactions in Solids*, Wiley, New York, 1968.
30. S. F. Wallace and G. Zdasink, *App. Phys. Lett.*, **28**, 449 (1976).
31. J. M. Hoffman, A. K. Hayes, and G. G. Tisone, *App. Phys. Lett.*, **28**, 538 (1976).
32. T. M. McKee, B. P. Stoichoff, and S. C. Wallace, *App. Phys. Lett.*, **30**, 278 (1977).
33. S. D. Rockwood, in *Laser Induced Processes in Molecules*, K. L. Kompa and S. D. Smith, eds., Springer-Verlag, Berlin, 1979, p. 3.
34. F. T. Arecchi and E. O. Schultz-Dubois, eds., *Laser Handbook*, Vol. 1, North-Holland, Amsterdam, 1972.
35. R. W. F. Gross and J. F. Bott, eds., *Handbook of Chemical Lasers*, Wiley, New York, 1976.
36. J. N. Walpole, A. R. Calawa, T. C. Harman, and S. H. Groves, *App. Phys. Lett.*, **28**, 552 (1978).
37. D. Huppert and P. M. Rentzepis, in *Molecular Energy Transfer*, R. D. Levine and J. Jortner, eds., Wiley, New York, 1975, p. 270.
38. K. L. Kompa, private communication.
39. N. R. Isenor and M. R. Richardson, *App. Phys. Lett.*, **18**, 225 (1971).
40. N. Bloembergen and E. Yablonovitch, *Phys. Today*, **31**, 23 (1978).
41. C. D. Cantrell, S. M. Freund, and J. L. Lyman, in *Laser Handbook*, Vol. III(b), North-Holland, Amsterdam, 1979.
42. V. S. Lethokov and C. B. Moore, in *Chemical and Biochemical Application of Lasers*, C. B. Moore, ed., Vol. 3, Academic Press, New York, 1977, p. 1.

43. R. V. Ambartzumian and V. S. Lethokov, in *Chemical and Biochemical Applications of Lasers*, C. B. Moore, ed., Vol. 3, Academic Press, New York 1977, p. 167.
44. J. M. Womlock, in *Laser Handbook*, Vol. 2, p. 1323, North-Holland, Amsterdam, 1972.
45. N. R. Isenor, in *Multiphoton Processes*, J. H. Eberly and P. Lambropoulos, eds., Wiley, New York, 1978, p. 179.
46. J. J. Wyne, J. A. Armstrong, and P. Esherick, in *Multiphoton Processes*, J. H. Eberly and P. Lambropoulos, eds., Wiley, New York, 1978, p. 215.
47. A. Yogev and R. M. J. Benmair, *Chem. Phys. Lett.* **46**, 290 (1977).
48. I. Glatt and A. Yogev, *J. Am. Chem. Soc.*, **98**, 7087 (1976).
49. A. Harrman, S. Leutwyler, E. Schumacher, and L. Wöste, *Chem. Phys. Lett.* **52**, 418 (1977).
50. D. L. Feldman, R. K. Lengel, and R. N. Zare, *Chem. Phys. Lett.*, **52**, 413 (1977).
51. J. L. Lyman, in *Laser Induced Processes in Molecules*, K. L. Kompa and S. D. Smith, eds., Springer-Verlag, Berlin, 1979, p. 131.
52. (a) R. B. Hall, A. Kaldor, J. A. Horsley, D. M. Cox, and E. B. Priestley, Paper 151, Presented at the 178th ACS National Meeting, Washington, D. C. (1979).
    (b) A. Kaldor, R. B. Hall, D. M. Cox, J. A. Horsley, P. Rabinowitz, and G. M. Kramer, *J. Am. Chem. Soc.*, **101**, 4465 (1979).
53. R. L. Woodim, D. S. Bornse, and J. L. Beauchamp, *J. Am. Chem. Soc.*, **100**, 3248 (1978).
54. H. S. Kwok and E. Yablonovitch, *Phys. Rev. Lett.*, **41**, 745 (1978).
55. L. Allen and J. H. Eberly, *Optical Resonance and Two-Level Atoms*, Wiley, New York (1975).
56. R. G. Brewer, in *Frontiers in Laser Spectroscopy*, R. Baliam, S. Haroche, and S. Liberman, eds., North-Holland, Amsterdam, 1977, p. 341.
57. J. Jortner and S. Leach, *J. Chim. Phys.*, **77**, 7 (1980).
58. J. H. Callamon, V. M. Anderson, J. R. Christie, and A. R. Lacey, Paper 25 at General Discussion Meeting on Radiationless Processes, Schliersee, September (1974).
59. J. I. Steinfeld, private communication.
60. R. K. Sander, B. Soep, and R. N. Zare, *J. Chem. Phys.*, **64**, 1242 (1976).
61. R. Naaman, D. M. Luban, and R. N. Zare, *Chem. Phys.*, **32**, 17 (1978).
62. (a) M. J. Coggiola, P. Schultz, Y. T. Lee, and Y. Shen, *Phys. Rev. Lett.*, **38**, 17 (1977).
    (b) D. J. Krajnovich, A. Giardini-Guidon, Aa. S. Subdø, P. A. Schultz, Y. R. Shen, and Y. T. Lee, in *Laser Induced Processes in Molecules*, K. L. Kompa and S. D. Smith, eds., Springer-Verlag, Berlin, 1979, p. 176.
63. A. Kantrowitz and T. Grey, *Rev. Sci. Instrum.* **22**, 328 (1951).
64. (a) K. P. Huber and A. E. Douglas, Quarterly Progress Report, National Research Council, Ottawa, Canada, Division of Pure Physics, Reports 161 (1963), (1964), and (1965).
    (b) M. P. Sinha, A. Schultz, and R. N. Zare, *J. Chem. Phys.*, **58**, 549 (1973).
65. (a) R. E. Smalley, B. L. Ramakrishna, D. H. Levy, and L. Wharton, *J. Chem. Phys.*, **61**, 4363 (1974).
    (b) R. E. Smalley, L. Wharton, and D. H. Levy, *J. Chem. Phys.*, **63**, 4977 (1975).
66. D. H. Levy, L. Wharton, and R. E. Smalley, in *Chemical and Biochemical Applications of Lasers*, Academic Press, New York, 1977, Vol. 2, p. 1.
67. D. H. Levy, L. Wharton, and R. E. Smalley, *Acc. Chem. Res.*, **10**, 139 (1977).
68. P. S. H. Fitch, L. Wharton, and D. H. Levy, *J. Chem. Phys.*, **69**, 3424 (1978).
69. A. Amirav, U. Even, and J. Jortner, *J. Chem. Phys.*, **71**, 2319 (1979).
70. E. W. Schlag, private communication.
71. S. M. Beck, M. G. Liverman, D. L. Monts, and R. E. Smalley, *J. Chem. Phys.*, **70**, 232 (1979).

72. S. M. Beck, D. L. Monts, M. G. Liverman, and R. E. Smalley, *J. Chem. Phys.*, **70**, 1062 (1979).
73. F. M. Behlen, N. Mikami, and S. A. Rice, *Chem. Phys. Lett.*, **60**, 364 (1979).
74. P. S. H. Fitch, L. Wharton, and D. H. Levy, *J. Chem. Phys.*, **70**, 2019 (1979).
75. A. Amirav, U. Even, and J. Jortner, *Chem. Phys. Lett.* **72**, 21 (1980).
76. A. Amirav, U. Even, and J. Jortner, *Opt. Comm.*, **32**, 266 (1980).
77. A. Amirav, U. Even, and J. Jortner, *Chem. Phys. Lett.*, **69**, 15 (1979).
78. J. Tusa, M. Sulkes, and S. A. Rice, *J. Chem. Phys.*, **70**, 3136 (1979).
79. G. M. McClelland, K. L. Saenger, J. J. Valentini, and D. R. Herschbach, *J. Phys. Chem.*, **83**, 947 (1979).
80. E. V. Shpolskii, *Sov. Phys. Uspekhi*, **6**, 411 (1963).
81. R. M. Hochstrasser and P. Prasad, in *Excited States*, Vol. 1, E. C. Lim, ed., Academic Press, New York, 1974, p. 79.
82. J. Goodman and L. E. Brus, *J. Am. Chem. Soc.*, **100**, 7472 (1978).
83. P. M. Rentzepis, R. P. Jones, and J. Jortner, *J. Chem. Phys.*, **59**, 766 (1973).
84. G. Wald, *Science*, **162**, 230 (1968).
85. B. Chance, H. Fraunfelder, and R. Schriefer, eds., *Tunnelling in Biological Systems*, Academic Press, New York, 1979.
86. K. Peters, M. L. Applebury, and P. M. Rentzepis, *Proc. Natl. Acad. Sci. USA*, **74**, 3119 (1977).
87. P. L. Dutton, R. C. Prince, and D. M. Tlide, *Photochem. Photobiol.*, **28**, 939, (1978).
88. D. Holten and M. W. Windsor, *Ann. Rev. Biophys. Biochem.*, **7**, 189 (1978).
89. R. K. Clayton and R. T. Wang, *Meth. Enzymol.*, **23**, 696 (1971).
90. G. Feher, *Photochem. Photobiol.*, **14**, 373 (1971).
91. T. G. Manger, R. R. Alfano, and R. H. Callender, *Biophys. J.*, **27**, 105 (1979).
92. D. De Vault and B. Chance, *Biophys. J.*, **6**, 825 (1966).
93. W. Fus and J. Hartman, in *Laser Induced Processes in Molecules*, K. L. Kompa and S. D. Smith, eds., Springer-Verlag, Berlin, 1979, p. 128.
94. K. R. Wilson, in *Symposium on Excited States Chemistry*, J. N. Pitts, Jr., ed., Gordon and Breach, New York, 1970.
95. R. V. Ambartzumian, N. V. Chekalin, V. S. Lethokov, and E. A. Ryabov, *Chem. Phys. Lett.*, **36**, 301 (1975).
96. Y. Hass and G. Yahav, *Chem. Phys. Lett.*, **48**, 63 (1977).
97. R. N. Zare and P. J. Dagdigian, *Science*, **185**, 739 (1974).
98. T. F. Deutch and R. J. Brueck, *Chem. Phys. Lett.*, **54**, 258 (1978).
99. J. P. Boquillon, in *Laser Induced Processes in Molecules*, K. L. Kompa and S. D. Smith, eds., Springer-Verlag, Berlin, 1979, p. 43.
100. J. Jortner, *J. Chim. Phys.*, Special Issue "*Transitions Non-Radiative dans les Molecules*", p. 1 (1970).
101. A. Nitzan, J. Jortner, and P. M. Rentzepis, *Proc. Roy. Soc.*, **A327**, 367 (1972).
102. M. Bixon, Y. Dothan, and J. Jortner, *Mol. Phys.*, **17**, 109 (1969).
103. J. Chaiken and J. D. McDonald, *Chem. Phys. Lett.*, **61**, 195 (1979).
104. U. Fano, *Phys. Rev.*, **A124**, 1866 (1961).
105. U. Fano, *Phys. Rev.*, **A137**, 1364 (1964).
106. A. F. Starace, *Phys. Rev.*, **A16**, 231 (1977).
107. A. Nitzan and J. Jortner, *J. Chem. Phys.*, **57**, 2870 (1972).
108. P. F. Williams, D. L. Rousseau, and S. H. Dworetsky, *Phys. Rev. Lett.*, **32**, 196 (1974).
109. W. Keifer and H. J. Bernstein, *J. Chem. Phys.*, **57**, 3017 (1972).
110. M. Jacon, M. Brejot, and L. Bernard, *J. Phis.*, **32**, 517 (1971).
111. S. Mukamel and J. Jortner, *J. Chem. Phys.*, **61**, 227 (1974).
112. R. N. Zare, *Mol. Photochem.*, **4**, 1 (1972).

113. R. Bersohn and S. H. Lin, *Adv. Chem. Phys.*, **16**, 67 (1969).
114. C. Jonah, *J. Chem. Phys.*, **55**, 1915 (1971).
115. N. A. Kurmit, I. D. Abella, and R. Hartman, *Phys. Rev. Lett.*, **13**, 567 (1964); *Phys. Rev*, **141**, 391 (1966).
116. R. G. Brewer and R. L. Shoemaker, *Phys. Rev. Lett.*, **27**, 631 (1971).
117. R. L. Shoemaker and R. G. Brewer, *Phys. Rev. Lett.*, **28**, 1430 (1972).
118. R. G. Brewer and R. L. Shoemaker, *Phys. Rev.*, **A6**, 2001 (1972).
119. R. G. Brewer and A. Z. Gemack, *Phys. Rev. Lett.*, **36**, 959 (1976).
120. A. Z. Gemack, R. M. Macfarlane, and R. G. Brewer, *Phys. Rev. Lett.*, **37**, 1078 (1976).
121. F. A. Hopf and M. O. Scully, *Phys. Rev.*, **134**, A1492 (1964).
122. F. A. Hopf, R. F. Shea, and M. O. Scully, *Phys. Rev.*, **A7**, 2105 (1973).
123. T. J. Aartsma and D. A. Wiersma, *Phys. Rev. Lett.*, **36**, 1360 (1976).
124. T. J. Aartsma and D. A. Wiersma, *Chem. Phys. Lett.*, **42**, 520 (1976).
125. A. H. Zewail, T. E. Orlowski, and D. R. Dawson, *Chem. Phys. Lett.*, **44**, 379 (1976).
126. A. H. Zewail, *Chem. Phys. Lett.*, **45**, 399 (1977).
127. A. H. Zewail, Proc. *SPIE Unconvent. Spect.*, **82**, 43, (1976).
128. H. de Vries and D. A. Wiersma, *J. Chem. Phys.*, **70**, 5807 (1979).
129. T. E. Orlowski and A. H. Zewail, *J. Chem. Phys.*, **70**, 1390 (1979).
130. R. Lefebvre and J. Savolamer, *Chem. Phys. Lett.*, **3**, 449 (1969).
131. J. Jortner and J. Kommandeur, *Chem. Phys.*, **28**, 273 (1978).
132. R. G. Brewer and E. L. Hahn, *Phys. Rev.*, **A11**, 1641 (1975).
133. A. Amirav, U. Even, J. Jortner, and L. Kleinman, *J. Chem. Phys.*, **73**, (1980).
134. C. D. Cantrell, W. H. Louisell, and J. F. Lam, in *Laser Induced Processes in Molecules*, K. L. Kompa and S. D. Smith, eds., Springer-Verlag, Berlin, 1979, p. 138.
135. G. Herzberg, *Commentarii Pontificia Academia Scientiarum*, Vol. II-49, p. 1, 1972.
136. O. Atabek, J. A. Beswick, R. Lefebvre, S. Mukamel, and J. Jortner, *J. Chem. Phys.*, **65**, 4035 (1976).
137. R. N. Greene, R. M. Hochstrasser, and R. B. Weisman, *J. Chem. Phys.*, **70**, 1247 (1979).
138. J. Vigue, H. Bourger, and J. C. Lehman, *J. Chem. Phys.*, **62**, 4441 (1975).
139. H. G. Kuttner, H. L. Selzle, and E. W. Schlag, *Israel J. Chem.*, **16**, 264 (1977).
140. V. A. Kovarskii and N. F. Perel'man, *Sov. Phys. JETP*, **34**, 738 (1972).
141. K. M. Watson, in *Multiphoton Processes*, J. H. Eberly and P. Lambropoulos, eds., Wiley, 1978, p. 385.
142. S. E. Harris and D. B. Lidow, *Phys. Rev. Lett.*, **33**, 674 (1974); **34**, 172 (1975).
143. J. Jortner and A. Ben Reuven, *Chem. Phys. Letts.*, **41**, 401 (1976).
144. J. A. Beswick and J. Jortner, *J. Chem. Phys.*, **69**, 512 (1978).
145. G. Ewing, *Chem. Phys.*, **29**, 253 (1978).
146. A. Tramer, private communication.
147. R. A. Marcus, *J. Chem. Phys.*, **24**, 966 (1956).
148. V. G. Levich, in *Advances in Electrochemistry and Electrochemical Engineering*, Vol. 4, P. Delahay, ed., Wiley-Interscience, New York, p. 249.
149. N. R. Kestner, J. Logan, and J. Jortner, *J. Phys. Chem.*, **78**, 2148 (1974).
150. C. H. Henry and D. V. Lang, *Phys. Rev.*, **B15**, 989 (1977).
151. N. F. Mott, E. A. Davis, and R. A. Street, *Phil. Mag.*, **32**, 961 (1975).
152. J. Jortner and J. Ulstrup, *J. Am. Chem. Soc.*, **101**, 3744 (1979).
153. T. Forster, *Naturwissenschaften*, **33**, 166 (1964).
154. D. L. Dexter, *J. Chem. Phys.*, **21**, 836 (1953).
155. M. D. Sturge, *Phys. Rev.*, **B8**, 6 (1973).
156. J. Jortner, in *Vacuum Ultraviolet Radiation Physics*, E. E. Koch, R. Haensel, and C.

Kunz, eds., Pergamon, Vieweg, 1974, p. 263.
157. G. Wenzel, *Z. Phys.*, **43**, 524 (1927).
158. E. Fermi, *Nuclear Physics*, University of Chicago Press, Chicago, Ill., 1949.
159. R. W. Gurney and E. U. Condon, *Nature*, **122**, 439 (1928).
160. G. Gamow, *Zeit. Phys.*, **51**, 204 (1928).
161. G. Herzberg, private communication.
162. R. D. Levine, *Acct. Chem. Res.*, **3**, 273 (1970).
163. N. Rosen, *J. Chem. Phys.*, **1**, 319 (1933).
164. (a) R. E. Smalley, D. H. Levy, and L. Wharton, *J. Chem. Phys.*, **64**, 3266 (1976); (b) M. S. Kim, R. E. Smalley, and L. Wharton, *ibid.*, **65**, 1216 (1976); (c) R. E. Smalley, D. H. Levy, and L. Wharton, *ibid.*, **66**, 2750 (1977).
165. R. D. Levine, J. T. Muckerman, B. R. Johnson, and R. B. Bernstein, *J. Chem. Phys.*, **49**, 56 (1968).
166. L. Landau and E. Teller, *Phys. Z. Sowjun*, **10**, 34 (1936).
R. D. Levine and R. B. Bernstein, *J. Chem. Phys.*, **56**, 2281 (1972).
167. J. D. Lambert, *Vibrational and Rotational Relaxation in Gases*, Clarendon Press, Oxford, 1977.
168. D. H. Levy (to be published).
169. D. A. Dixon and D. R. Herschbach, *Ber. Bunsenges. Phys. Chem.*, **81**, 145 (1977).
170. J. A. Beswick and J. Jortner, *J. Chem. Phys.*, **71**, 4737 (1979).
171. D. H. Levy (to be published).
172. D. Proch and H. Schröder, *Chem. Phys. Lett.*, **61**, 426 (1979).
173. H. Gebelein and J. Jortner, *Theor. Chim. Acta*, **25**, 143 (1972).
174. M. Bixon and J. Jortner, *J. Chem. Phys.*, **48**, 715 (1968).
175. W. Fus and J. Hartmann, *J. Chem. Phys.*, **70**, 5468 (1979).
176. M. Bixon and J. Jortner, *Mol. Cryst.*, **213**, 237 (1969).
177. L. P. Giver, J. H. Miller, and R. W. Boese, *Icarus*, **25**, 34 (1975).
178. R. Van der Werf and J. Kommandeur, *Chem. Phys.*, **16**, 125, 161 (1976).
179. R. G. Bray and M. J. Berry, *J. Chem. Phys.*, **71**, 4904 (1979).
180. E. J. Heller and S. A. Rice, *J. Chem. Phys.*, **61**, 936 (1974).
181. B. Carmeli and A. Nitzan, *Chem. Phys. Lett.*, **58**, 310 (1978).
182. J. M. Delory and C. Tric, *Chem. Phys.*, **3**, 54 (1974).
183. C. Tric, *Chem. Phys.*, **14**, 189 (1976).
184. K. G. Kay, *J. Chem. Phys.*, **61**, 5205 (1974).
185. W. M. Gelbart, D. F. Heller, and M. L. Elbert, *Chem. Phys.*, **7**, 116 (1975).
186. I. Schek and J. Jortner, *J. Chem. Phys.*, **70**, 3016 (1979).
187. B. Carmeli and A. Nitzan, *J. Chem. Phys.*, **72**, 2054 (1980).
188. B. Carmeli, I. Scheck, A. Nitzan and J. Jortner, *J. Chem. Phys.*, **72**, 1928 (1980).
189. R. Engelman and J. Jortner, *Mol. Phys.*, **18**, 145 (1970).
190. A. Nitzan and J. Jortner, *Theor. Chim. Acta*, **30**, 217 (1973).
191. Z. Karny, A. Gupta, R. N. Zare, S. T. Lim, J. Nieman, and A. M. Ronn, *Chem. Phys.*, **37**, 15 (1979).
192. J. W. Hudgens, J. L. Durant, B. J. Bogan, and R. A. Coveleskie, *J. Chem. Phys.*, **70**, 5906 (1979).
193. A. Nitzan and J. Jortner, *Chem. Phys. Lett.*, **60**, 1 (1979).
194. A. Nitzan and J. Jortner, *J. Chem. Phys.*, **71**, 3524 (1979).
195. J. Jortner, in ref. 21.
196. G. W. Flynn, private communication.
197. P. M. Johnson, *J. Chem. Phys.*, **62**, 4562 (1975); *ibid.*, **64**, 4143 (1976).
198. L. Zandee and R. B. Bernstein, *J. Chem. Phys.*, **71**, 1359 (1979).

D. M. Lubman, R. Naaman, and R. N. Zare, *J. Chem. Phys.*, **72**, 3034 (1980).
199. D. H. Parker, J. O. Berg, and M. A. El-Sayed, *Advances in Laser Chemistry*, A. H. Zewail, ed., Springer-Verlag, Berlin, 1978.
200. N. B. Slater, *Theory of Unimolecular Reactions*, Cornell University Press, Ithaca, N. Y., 1959.
201. L. S. Kassel, *Kinetics of Homogeneous Gas Reactions*, Chemical Catalog, New York, 1932.
202. R. A. Marcus, *J. Phys. Chem.*, **20**, 359 (1952).
203. M. Quack and J. Troe, *Gas Kinetics and Energy Transfer*, Vol. II, Chemical Society, London, 1977, p. 175.
204. I. Oref and B. S. Rabinovitch, *Acct. Chem. Res.*, **12**, 166 (1979).
205. M. J. Coggiola, P. A. Schulz, Y. T. Lee, and Y. R. Shen, *Phys. Rev. Lett.*, **38**, 17 (1977).
206. J. G. Black, E. Yablonovitch, N. Bloembergen, and S. Mukamel, *Phys. Rev. Lett.*, **38**, 1131 (1977).
207. Aa. S. Sudbǿ, P. A. Schulz, Y. T. Lee, and Y. R. Shen, *J. Chem. Phys.*, **68**, 1306 (1978).
208. G. A. West, R. E. Weston, Jr., and G. W. Flynn, *Chem. Phys. Lett.*, **35**, 275 (1978).
209. E. Zamir and R. D. Levine, *Chem. Phys. Lett.*, **67**, 237 (1979).
210. J. R. Ackerhalt and H. W. Galbraith, *Laser Spectroscopy IV*, H. Walther and K. W. Rothe, eds., Springer-Verlag, New York (1979); J. A. Horsley, J. Stone, M. F. Goodman, and D. A. Dows, *Chem. Phys. Lett.*, **66**, 461 (1979).
211. R. G. Bray and M. J. Berry, *J. Chem. Phys.*, **71**, 4909 (1979).
212. K. V. Reddy and M. J. Berry, *Chem. Phys.Lett.*, **66**, 223 (1979).
213. R. B. Hall and A. Kaldor, *J. Chem. Phys.*, **70**, 4027 (1979).
214. L. Boltzmann, *Wein. Ber.*, **63**, 397 (1871), *ibid.*, **76**, 373 (1877).
215. T. S. Kuhn, *The Emergence of the Quantum Discontinuity*, Clarendon Press, Oxford, 1979.
216. M. J. Klein, *Acta Phys. Austr., Suppl.*, **X**, 12 (1973).
217. A. Einstein, *Ann. Phys.*, **11**, 170 (1903).
218. R. Zwanzig, *J. Chem. Phys.*, **33**, 1338 (1960); K. S. J. Nordholm and R. Zwanzig, *J. Stat. Phys.*, **13**, 347 (1975).
219. H. Mori, *Prog. Theor. Phys.*, **33**, 423 (1965).
220. L. S. Garcia-Colin and J. L. del Rio, *J. Stat. Phys.*, **16**, 235 (1977).
221. Y. Alhassid and R. D. Levine, *Phys. Rev.*, **C20**, 1775 (1979).
222. R. Ramaswarmy, S. Augustin, and H. Rabitz, *J. Chem. Phys.*, **69**, 5509 (1978).
223. B. R. Henry, *Acct. Chem. Res.*, **10**, 207 (1977).
224. A. C. Albrecht, *Advances in Laser Chemistry*, A. H. Zewail, ed., Springer-Verlag, Berlin, 1978, p. 235.
225. S. A. Rice, *Advances in Laser Chemistry*, A. H. Zewail, ed., Springer-Verlag, Berlin, 1978, p. 2.
226. V. I. Arnold and A. Avez, *Ergodic Problems of Classical Mechanics*, Benjamin, New York, 1968.
227. J. R. Ackerhalt and H. W. Galbraith, *Laser Spectroscopy IV*, H. Walther and K. W. Rothe, eds., Springer-Verlag, New York, 1979.
228. E. Thiele, M. F. Goodman, and J. Stone, *Opt. Eng.*, **19**, 10 (1980).
229. Section V.C. of Chapter 3
230. E. Pollak and R. D. Levine, *J. Chem. Phys.*, **72**, 2990 (1980).
231. Aa. S. Sudbǿ, D. J. Karjnovich, P. A. Schulz, Y. T. Shen, and Y. T. Lee, in *Multiphoton Excitation and Dissociation of Polyatomic Molecules*, C. D. Cantrell, ed., Springer-Verlag, New York, 1979.
232. P. J. Robinson and K. A. Holbrook, *Unimolecular Reactions*, Wiley, New York, 1972.

233. R. N. Rosenfeld, J. I. Brauman, J. R. Barker, and D. M. Golden, *J. Am. Chem. Soc.*, **99**, 8063 (1977).
234. W. Forst, *Theory of Unimolecular Reactions*, Academic Press, New York, 1973.
235. R. B. Knott and A. W. Pryor, *J. Chem. Phys.*, **71**, 2946 (1979).
236. D. C. Tardy and B. S. Rabinovitch, *Chem. Rev.*, **77**, 369 (1977).
237. J. Francisco and J. I. Steinfeld, A.C.S. March Meeting, 1980.
238. S. A. Safron, N. D. Weinstein, D. R. Herschbach, and J. C. Tully, *Chem. Phys. Lett.*, **12**, 564 (1972).
239. G. E. Busch and K. R. Wilson, *J. Chem. Phys.*, **56**, 3626 (1972).
240. R. Bersohn, *Israel J. Chem.*, **14**, 111 (1975).
241. J. L. Kinsey, *Ann. Rev. Phys. Chem.*, **28**, 349 (1977).
242. S. C. Yang, A. Freedman, M. Kawasaki, and R. Bersohn, *J. Chem. Phys.*, **72**, 4058 (1980).
243. M. Shapiro, *Chem. Phys. Lett.*, **46**, 442 (1977).
244. M. D. Morse, K. F. Freed, and Y. B. Band, *J. Chem. Phys.*, **70**, 3620 (1979).
245. M. J. Berry, *Chem. Phys. Lett.*, **29**, 329 (1974).
246. S. Mukamel and J. Jortner, *J. Chem. Phys.*, **65**, 3735 (1976).
247. O. Atabek and R. Lefebvre, *J. Chem. Phys.*, **67**, 4983 (1977).
248. K. E. Holdy, L. C. Klotz, and K. R. Wilson, *J. Chem. Phys.*, **52**, 4588 (1970). M. Shapiro and R. D. Levine, *Chem. Phys. Lett.*, **5**, 499 (1970). A. D. Wilson and R. D. Levine, *Mol. Phys.*, **27**, 1197 (1974).
249. M. Tamir, U. Halavee, and R. D. Levine, *Chem. Phys. Lett.*, **25**, 38 (1974).
250. J. C. Polanyi and J. L. Schreiber, *Physical Chemistry*, V. 6A, H. Eyring, D. Henderson, and W. Jost, eds., Academic Press, New York, 1974.
251. S. H. Bauer, *Chem. Rev.*, **78**, 147 (1978).
252. R. D. Levine and R. B. Bernstein, *Acct. Chem. Res.*, **7**, 393 (1974).
253. B. E. Holmes and D. W. Setser, in *Physical Chemistry of Fast Reactions*, I. W. M. Smith, ed., Plenum, N.Y., 1980.
254. I. W. M. Smith, *Gas Kinetics and Energy Transfer*, Vol. 2, P. G. Ashmore and R. J. Donovan, eds., Chemical Society, London, 1977.
255. H. Kaplan, R. D. Levine, and J. Manz, *Chem. Phys.*, **12**, 447 (1976).
256. R. D. Levine, *New World of Quantum Chemistry*, B. Pullman and R. Parr, eds., Reidel, Holland, 1976.
257. B. A. Blackwell, J. C. Polanyi, and J. J. Sloan, *Chem. Phys.*, **24**, 25 (1977).
258. A. M. G. Ding and J. C. Polanyi, *Chem. Phys.*, **10**, 39 (1975).
259. E. Cuellar and G. C. Pimentel, *J. Chem. Phys.*, **71**, 1385 (1979).
260. J. Wolfrum, *Ber. Bunsenges. Phys. Chem.*, **81**, 114 (1977).
261. Z. Karny and R. N. Zare, *J. Chem. Phys.*, **68**, 3360 (1978).
Also R. N. Zare, *Faraday Disc. Chem. Soc.*, **67**, 7 (1979).
262. A. M. Ding, L. J. Kirsch, D. S. Perry, J. C. Polanyi, and J. L. Schreiber, *Faraday Disc. Chem. Soc.*, **55**, 252 (1973).
263. E. Pollak and R. D. Levine, *Chem. Phys. Lett.*, **39**, 199 (1976).
264. R. D. Levine and J. Manz, *J. Chem. Phys.*, **63**, 4280 (1975).
265. E. Bar Ziv, J. Moy, and R. J. Gordon, *J. Chem. Phys.*, **68**, 1013 (1978);
K. K. Hui and T. A. Cool, *J. Chem. Phys.*, **68**, 1022 (1978).
266. J. C. Stephenson and S. M. Freund, *J. Chem. Phys*, **65**, 4303 (1976).
267. H. H. Dispert, M. W. Geis, and P. R. Brooks, *J. Chem. Phys.*, **70**, 5317 (1979).
268. J. T. Muckerman, *J. Chem. Phys.*, **54**, 1155 (1971);
See also D. J. Douglas and J. C. Polanyi, *Chem. Phys.*, **16**, 1 (1976).
269. P. R. Brooks, *Science*, **193**, 11 (1976).

270. Z. Karny, R. C. Estler, and R. N. Zare, *J. Chem. Phys.*, **69**, 5199 (1978).
271. R. W. Solarz, S. A. Johnson, and R. K. Preston, *Chem. Phys. Lett.*, **57**, 514 (1978).
272. R. C. Estler and R. N. Zare, *J. Am. Chem. Soc.*, **100**, 1323 (1978).
273. C. B. Moore and P. F. Zittel, *Science*, **182**, 541 (1973); K. Bergman, S. R. Leone, R. G. MacDonald, and C. B. Moore, *Israel J. Chem.*, **14**, 105 (1975).
274. H. Pummer, D. Proch, U. Schmailzl, and K. L. Kompa, *Opt. Comm.*, **19**, 273 (1976).
275. J. H. Smith and D. W. Robinson, *J. Chem. Phys.*, **68**, 5474 (1978).
276. H. M. Lin, M. Seaver, K. Y. Tang, A. E. W. Knight, and C. S. Parmenter, *J. Chem. Phys.*, **70**, 5442 (1979).
277. I. Shamah and G. W. Flynn, *J. Am. Chem. Soc.*, **99**, 319 (1977).
278. W. R. Green, J. Lukasik, J. R. Wilson, M. D. Wright, J. F. Young, and S. E. Harris, *Phys. Rev. Lett.*, **42**, 970 (1979).
279. T. F. George, I. H. Zimmermann, J. M. Yuan, J. R. Laing, and P. L. De Vries, *Acct. Chem. Res.*, **10**, 449 (1977).
280. W. H. Hesselink and D. A. Wiersma, *Chem. Phys. Lett.*, **56**, 227 (1978).
281. D. E. Cooper, R. W. Olson, R. D. Wieting, and M. D. Fayer, *Chem. Phys. Lett.*, **67**, 41 (1979).
282. K. Godzik and J. Jortner, *J. Chem. Phys.*, **72**, 4471 (1980).
283. N. Bloembergen, *Opt. Commun.*, **15**, 416 (1975).
284. M. F. Goodman and E. Thiele, *Phys. Rev.*, **A5**, 1358 (1972).
285. J. Stone, E. Thiele, and M. F. Goodman, *J. Chem. Phys.*, **59**, 2909 (1973); *ibid.*, **63**, 2936 (1975).
286. M. F. Goodman, J. Stone, and E. Thiele, *J. Chem. Phys.*, **59**, 2919 (1973); *ibid.*, **63**, 2929 (1975).
287. D. M. Larsen and N. Bloembergen, *Opt. Commun.*, **17**, 254 (1976).
288. F. H. M. Faisal, *Opt. Commun.* **17**, 247 (1976).
289. S. Mukamel and J. Jortner, *Chem. Phys.Lett.*, **40**, 150 (1976).
290. S. Mukamel and J. Jortner, *J. Chem. Lett.*, **65**, 5204 (1976).
291. T. P. Cotter W. Fuss, K. L. Kompa, and H. Stafast, *Opt. Commun.*, **18**, 220 (1976).
292. N. Bloembergen, C. D. Cantrell, and D. M. Larsen, in *Tunable Lasers and Applications*, Springer-Verlag, Berlin, 1976, p. 162.
293. D. M. Larsen, *Opt. Commun.* **19**, 404 (1976).
294. V. S. Letokhov and A. A. Makarow, *Opt. Commun.*, **17**, 250 (1976).
295 M. F. Goodman, J. Stone, and D. A. Dows, *J. Chem. Phys.*, **65**, 5052 (1976).
296. M. F. Goodman, J. Stone, and D. A. Dows, *J. Chem. Phys.*, **65**, 5062 (1976).
297. C. D. Cantrell and H. W. Galbraith, *Opt. Commun.*, **18**, 513 (1976).
298. C. D. Cantrell and H. W. Galbraith, *Opt. Commun.*, **21**, 374 (1977).
299. V. I. Gorchakov and V. N. Sazanov, *JETP*, **43**, 241 (1976).
300. D. P. Hodgkinson and J. S. Briggs, *Chem. Phys. Lett.*, **43**, 451 (1976).
301. R. B. Walker and R. K. Preston, *J. Chem. Phys.*, **67**, 2017 (1977).
302. (a) D. W. Noid, M. St. Koszykowski, R. A. Marcus, and J. P. McDonald, *Chem. Phys. Lett.*, **51**, 540 (1977). (b) K. D. Hansel, *Chem. Phys. Lett.*, **57**, 619 (1978).
303. C. D. Cantrell and K. Fox, *Opt. Lett.*, **2**, 151 (1978).
304. H. W. Galbraith and J. R. Ackerhalt, *Opt. Lett.*, **3**, 109, 152, (1978).
305. J. R. Ackerhalt and H. W. Galbraith, *J. Chem. Phys.*, **69**, 1200 (1978).
306. I. Schek and J. Jortner, *Chem. Phys. Lett.*, **63**, 5 (1979).
307. D. P. Hodgkins and J. S. Briggs, *J. Phys.*, **10**, 2583 (1977).
308. C. D. Cantrell, W. H. Galbraith, and J. R. Ackerhalt, in *Multiphoton Processes*, J. H. Eberly and P. Lambropoulos, eds., Wiley, New York, 1978, p. 331.
309. (a) J. Stone and M. F. Goodman, *Phys. Rev.*, **A18**, 2618 (1978); (b) J. Stone, M. F.

Goodman, and D. A. Dows, *J. Chem. Phys.*, **71**, 408 (1979).
310. J. G. Black, P. Kolander, M. J. Schultz, E. Yablonovitch, and N. Bloembergen, *Phys. Rev.*, **A19**, 704 (1979).
311. M. Tamir, and R. D. Levine, *Chem. Phys. Lett.*, **46**, 208 (1977).
312. S. Mukamel, *Phys. Rev. Lett.*, **42**, 168 (1979).
313. S. Mukamel, *J. Chem. Phys.*, **70**, 5834 (1979).
314. M. Quack, *J. Chem. Phys.*, **69**, 1282 (1978).
315. J. R. Ackerhalt and J. H. Eberly, *Phys. Rev.*, **A14**, 1705 (1976).
316. J. P. Gordon, C. W. Wang, C. K. N. Patel, R. E. Slusher, and W. J. Tomlinson, *Phys. Rev.*, **179**, 294 (1969).
317. E. Weitz and G. W. Flynn, *Ann. Rev. Phys. Chem.*, **25**, 275 (1974).
318. D. A. Chernoff and S. A. Rice, *J. Chem. Phys.*, **70**, 2521 (1979).
319. A. Amirav, U. Even and J. Jortner, *Chem. Phys. Lett.* **71**, 12 (1980).
320. I. Schek and J. Jortner (unpublished).
321. F. Brunner and D. Proch, *J. Chem. Phys.*, **68**, 4936 (1978).
322. D. S. King and J. C. Stephenson, *Chem. Phys. Lett.*, **66**, 33 (1979).
323. R. Naaman, M. Rossi, J. R. Barker, D. M. Golden, and R. N. Zare, to be published.
324. (a) L. Van Hove, *Physica*, **21**, 517 (1955). (b) R. Zwanzig, *J. Chem. Phys.*, **33**, 1338 (1960). (c) R. Zwanzig, *Lectures in Theoretical Physics III*, Wiley-Interscience, N.Y., 1961, p. 106. (d) R. Zwanzig, *Physica*, **30**, 1109 (1964).
325. A. A. Michelson, *Astrophys. J.* **II**, 251 (1895).
326. R. G. Brewer and A. Z. Genack, *Phys. Rev. Lett.*, **36**, 959 (1976).
327. A. Laubereau and W. Kaiser, *Rev. Mod. Phys.*, **50**, 607 (1978).
328. A. A. Gorokhovskii, R. K. Kaarli, and L. A. Rebane, *Opt. Comm.*, **16**, 282 (1976).
329. H. de Bries and D. A. Wiersma, *Phys. Rev. Lett.*, **36**, 91 (1976).
330. S. Voelker and M. R. MacFarlane, *Chem. Phys. Lett.*, **61**, 421 (1979).
331. W. H. Hesselnik and D. A. Wiersma, *Chem. Phys. Lett.*, **65**, 300 (1979).
332. V. Hizhnyakov and I. Tehver, *Phys. Status Solidi*, **21**, 755 (1967).
333. Y. Toyozawa, *J. Phys. Soc. Japan*, **41**, 400 (1978).
334. A. Lambert and A. H. Zewail, *Chem. Phys. Lett.*, **69**, 270 (1980).
335. F. Legay, in *Chemical and Biochemical Applications of Lasers*, Vol. 2, C. B. Moore, ed., Academic, New York, 1979, Chap. 2.
336. L. E. Brus and V. E. Bondybey, in *Radiationless Transitions*, S. H. Lin, ed., Academic Press, N.Y., 1980.
337. J. Jortner, *Mol. Phys.*, **32**, 379 (1976).
338. R. Rossetti and L. E. Brus, *J. Chem. Phys.*, **71**, 3963 (1979).
339. V. E. Bondybey and L. E. Brus, *J. Chem. Phys.*, **63**, 794 (1975).
340. J. M. Wiesenfeld and C. B. Moore, *J. Chem. Phys.*, **70**, 930 (1979).
341. R. B. Gerber and M. Berkowitz, *Phys. Rev. Lett.*, **39**, 1000 (1977).
342. K. F. Freed and H. Metiu, *Chem. Phys. Lett.*, **48**, 262 (1977).
343. J. Goodman and L. E. Brus, *J. Chem. Phys.*, **65**, 3146 (1976).
344. G. Ewing, private communication.
345. J. P. Maier, A. Seilmeier, A. Laubereau, and W. Kaiser, *Chem. Phys. Lett.*, **46**, 527 (1977).
346. K. Rebane and P. Saari, *J. Luminescence*, **12/13**, 23 (1976).
347. R. M. Hochstrasser and C. A. Nyi, *J. Chem. Phys.*, **70**, 1112 (1979).
348. R. M. Hochstrasser and T. Y. Li, *J. Mol. Spect.*, **41**, 297 (1972).
349. J. Goodman and L. E. Brus, *J. Chem. Phys.*, **67**, 933 (1977).
350. O. Cheshnovsky, B. Raz, and J. Jortner, *J. Chem. Phys.*, **57**, 4628 (1972).
351. W. S. Struve and P. M. Rentzepis, *J. Chem. Phys.*, **60**, 1533 (1974).

352. V. E. Bondybey, *J. Chem. Phys.*, **66**, 995 (1977).
353. R. J. Shaw, J. E. Kent, and M. F. O'Dwyer, *Chem. Phys.* , **18**, 155, 165 (1976).
354. R. M. Hochstrasser and P. Marchetti, *J. Mol. Spect.*, **35**, 335 (1970).
355. R. M. Hochstrasser, H. Lutz, and G. W. Scott, *Chem. Phys. Lett.*, **24**, 162 (1974).
356. M. Kasha, *J. Chem. Phys.*, **20**, 71 (1952).
357. G. W. Robinson, *J. Mol. Spect.*, **6**, 58 (1961).
358. G. W. Robinson and R. P. Frosch, *J. Chem. Phys.*, **38**, 1187 (1963).
359. S. P. McGlynn, T. Azumi, M. Kimoshita, *Molecular Spectroscopy of the Triplet State*, Prentice-Hall, Englewood-Cliffs, N. J. 1969.
360. A. Amirav, U. Even, and J. Jortner, *Chem. Phys. Lett.*, **66**, 9 (1979).
361. P. M. Rentzepis, *Chem. Phys. Lett.*, **2** 177 (1968).
362. D. Huppert, J. Jortner, and P. M. Rentzepis, *Israel J. Chem.*, **16**, 277 (1977).
363. E. P. Ippen, C. V. Shank, R. L. Woerner, *Chem. Phys. Lett.*, **46**, 20 (1977).
364. J. Jortner, *Phil. Mag.*, **B40**, 317 (1979).
365. F. Seitz, *The Modern Theory of Solids*, McGraw-Hill, New York, 1940.
366. K. Huang and A. Rhys, *Proc. Roy. Soc.*, **A204**, 406 (1950).
367. R. Kubo, *Phys. Rev.*, **86**, 929 (1952).
368. R. Kubo and Y. Toyozawa, *Prog. Theor. Phys.*, **13**, 160 (1955).
369. T. Holstein, *Ann. Phys.*, **8**, 343 (1959).
370. T. L. Netzel, P. M. Rentzepis, and J. S. Leigh, *Science*, **18**, 238 (1973).
371. K. J. Kaufman, P. J. Dutton, T. L. Netzel, J. S. Leigh, and P. M. Rentzepis, *Science*, **188**, 1301 (1975).
372. M. G. Rockley, M. W. Windsor, R. J. Cogdell, and W. U. Parson, *Proc. Nat. Acad. Sci. USA*, **72**, 2251 (1975).
373. K. Peters, Ph. Avoires, and P. M. Rentzepis, *Biophys. J.*, **23**, 207 (1978).
374. D. DeVault and B. Chance, *Biophys. J.*, **6**, 825 (1966).
375. J. J. Hopfield, *Proc. Nat. Acad. Sci. USA*, **71**, 3640 (1974).
376. J. Jortner, *J. Chem. Phys.*, **64**, 4860 (1976).
377. R. H. Austin, K. W. Beeson, G. Eisenstein, H. Frauenfelder, and I. C. Garcialus, *Biochemistry*, **14**, 5355 (1975).
378. A. Azuel in *Luminescence of Inorganic Solids*, B. DiBartolo, ed., Plenum Press, New York, 1977, p. 67.
379. J. Ulstrup and J. Jortner, *J. Chem. Phys.*, **63**, 4358 (1975).
380. J. Jortner and E. Buhks (to be published).
381. J. Jortner and J. Ulstrup, *Chem. Phys. Lett.*, **63**, 236 (1979).
382. D. Dexter, C. C. Klick, and G. A. Russell, *Phys. Rev.*, **100**, 603 (1955).
383. R. H. Bartram and A. M. Stoneham, *Solid St. Com.*, **17**, 1593 (1975).
384. L. D. Landau and E. M. Lifschitz, *Quantum Mechanics*, Pergamon Press, Oxford, 1965.
385. V. N. Kernkre, *Phys. Rev.*, **A16**, 766 (1977).
386. J. Jortner: Paper presented at ISOX 3 Conference, Albany, New York (1979).
387. E. Buhks and J. Jortner, *FEBS Lett.*, **109**, 117 (1980).
388. J. J. Hopfield in Ref. 85.
389. J. Jortner, *J. Am. Chem. Soc.* (submitted).
390. T. A. Milne, A. E. Vandergrift, and F. T. Greene, *J. Chem. Phys.*, **52**, 1552 (1970).
391. R. E. Leckenby and E. J. Robins, *Proc. Roy. Soc.* (*London*), **A291**, 389 (1966).
392. D. Golomb, R. E. Good, and R. F. Brown, *J. Chem. Phys.*, **52**, 1545, (1970).
393. J. W. Brady, J. D. Doll, and D. L. Thompson, *J. Chem. Phys.*, **71**, 2467 (1979).
394. A. E. Douglas, *Nature*, **269**, 130 (1977).

Section 1

ASPECTS OF INTRAMOLECULAR DYNAMICS

AN OVERVIEW OF THE DYNAMICS OF INTRAMOLECULAR TRANSFER OF VIBRATIONAL ENERGY

STUART A. RICE

The Department of Chemistry and The James Franck Institute, The University of Chicago, Chicago, Illinois 60637

CONTENTS

I. Introduction . . . 117
II. The Classical Mechanical Description of Coupled Nonlinear Oscillator Systems . . . 119
A. Some Properties of Trajectories . . . 119
B. Numerical Studies of the Dynamics of Nonlinear Oscillator Systems . . . 133
C. Perturbation Theory . . . 147
D. Summary . . . 159
III. Quantum-Mechanical Description of Coupled Nonlinear Oscillator Systems . . . 161
A. The Nordholm–Rice Analysis . . . 162
B. Semiclassical Quantization of Coupled Oscillator Systems . . . 168
C. Conjectures on the Relationship Between Quantum-Mechanical States and Classical Trajectories . . . 176
D. Some Differences Between the Classical and Quantum-Mechanical Description . . . 184
E. Ergodicity and Reaction Rate: Model Considerations . . . 186
F. Summary . . . 195
IV. Closing Remark . . . 195
Note Added in Proof . . . 196
Acknowledgments . . . 198
References . . . 198

I. INTRODUCTION

This chapter is intended to provide an elementary account, and an overview, of the dynamics of intramolecular vibrational energy transfer. This subject has attracted considerable attention recently, in part because of advances in classical mechanics, and in part in response to the need for analysis of the consequences of selective photoexcitation of molecules. The relevant literature is now quite extensive, and it contains contributions representing an enormous range of approaches and goals. Moreover, the techniques of analysis used range from abstract topological description of the general behavior of a class of systems to numerical integration of the equations of motion of a particular system. In the following text I will sketch some of the principle ideas advanced. My presentation draws heavily on the work of others,[1, 2] and I have freely used figures from a variety of sources. The relevant literature is so widely dispersed, and the subject matter so important to advances in photoselective chemistry, that I believe it worthwhile to reproduce these ideas here.

One of the most important features of the classical mechanical dynamics of a typical system of coupled nonlinear oscillators is the existence of a dramatic change in the character of the trajectory of the system at a critical energy, or over a small energy range. Numerical studies show that the trajectory changes from that characteristic of quasiperiodic motion to that characteristic of stochastic motion. Elsewhere in this volume Brumer[3] discusses the several theories for the onset of stochastic behavior in a system of coupled nonlinear oscillators, so these theories will not be included in this article; the presentation that follows gives the background from which the theories that Brumer discusses are developed. I have also omitted an explicit discussion of the set of subjects related to Kolmogorov entropy, although some of its underlying concepts are employed.[4]

A substantial fraction of this article is concerned with the quantum dynamics of coupled nonlinear oscillators, and the relationships between the classical and quantum-mechanical descriptions of the same system.[5] This aspect of the theory of intramolecular vibrational energy exchange is particularly poorly developed, so very few, if any, general results can as yet be reported. Nevertheless, the hypotheses advanced suggest marked differences, as well as marked similarities, between the quantum-mechanical and classical mechanical dynamics of a system of nonlinear oscillators.

Finally, a few comments are made on the possible influence of the dynamics of intramolecular vibrational energy exchange on the rate of a unimolecular reaction.

In an earlier article[6] I summarized the major questions associated with understanding intramolecular energy exchange as follows:

1. Under what conditions, if any, is intramolecular energy exchange slow/rapid relative to other processes, for example, photon emission, isomerization, or fragmentation?
2. How does the intramolecular energy exchange depend on the energy of the molecule and the nature of the initial excitation?
3. If there are situations for which intramolecular energy exchange is slow relative to chemical reaction, why does this behavior occur? Does it derive from special characteristics of the molecular force field? Are

there dynamical or symmetry restrictions on the spectrum of states in these cases? Are these special situations commonly or rarely found?

4. Given the answers to (3), can we devise excitation methods and reaction conditions that permit enhancement of the selectivity of the chemistry that follows?

It is now widely accepted that the vast majority of systems of coupled nonlinear oscillators display quasiperiodic motion at low energy and stochastic motion at high energy. Nevertheless, it is also known that there are exceptions to this generalization. Moreover, both the uniformity of behavior in the quasiperiodic and stochastic domains of the molecular dynamics and the nature of the transition between them are incompletely understood. Prompted by these observations I will augment the preceding list of questions with the following:

5. Are there isolated quasiperiodic trajectories embedded in the stochastic domain of the molecular dynamics?
6. If the answer to (5) is yes, is the existence of such trajectories peculiar to a small set of initial conditions? How does the intramolecular force field influence the existence or nonexistence of these trajectories?
7. If the answer to (5) is no, are there dynamical transients that are quasiperiodic on a time scale that permits interception by competition with other dynamical processes, e.g., collisions?
8. What is the relationship between the eigenstates and the trajectories in the stochastic domain? Which trajectories are "selected" by the quantization conditions?

At present it is not possible to give full answers to any of the questions, but partial answers are suggested by the several analyses and points of view discussed.

II. THE CLASSICAL MECHANICAL DESCRIPTION OF COUPLED NONLINEAR OSCILLATOR SYSTEMS

A. Some Properties of Trajectories

Given a dynamical system with N degrees of freedom described by the Hamiltonian $H(p, q)$ and the equations of motion

$$\dot{q}_s = \frac{\partial H}{\partial p_s} \qquad \dot{p}_s = -\frac{\partial H}{\partial q_s} \tag{2.1}$$

and given the initial values of the coordinates and momenta, q_s°, p_s°, the values of q, p at any other time t

$$q_s = q_s(t, q^\circ, p^\circ) \qquad p_s = p_s(t, q^\circ, p^\circ) \tag{2.2}$$

are unique under very weak conditions. Equations (2.2) can, in principle, be solved for q_s°, p_s°,

$$q_s^\circ = q_s^\circ(t, q, p) \qquad p_s^\circ = p_s^\circ(t, q, p) \tag{2.3}$$

which gives $2N$ functions of the phase-space variables and the time that are constant along any trajectory of the system. Elimination of t between the equations (2.3) leaves $2N-1$ functions of only the phase-space variables; these functions also have the property of being constant along any trajectory. This argument establishes the existence of $2N-1$ functions $C_j(q, p)$, which are integrals of the motion; attributing a set of numerical values to the C_j is equivalent to completely determining the system trajectory in phase space.

Saying that the set of $2N-1$ values of C_j determine the trajectory is one thing; finding the values is quite another! Of course, every constant of the motion must satisfy the Poisson–Bracket relation

$$\{H, C_j\} = 0 \qquad j = 1, 2, \ldots, 2N-1 \tag{2.4}$$

but the only obvious solution is $C_1 = H(q, p)$. Although it is in principle possible to stepwise find $2N-2$ other functions that with C_1 form a complete set of functionally independent integrals of the motion, in practice this is impossible to execute even for very simple mechanical systems.

Note that $C_1 = H$ requires that the trajectory of a conservative system lie on the energy surface $H(q, p) = E$. In general, each of the equations

$$C_j = k_j \qquad j = 1, 2, \ldots, 2N-1 \tag{2.5}$$

for given k_j defines a $2N-1$ dimensional hypersurface in the $2N$-dimensional phase space. The trajectory of the system must lie entirely on each of these surfaces, hence is determined entirely by their hyperdimensional intersection. Put in slightly different words, fixing the value of any C_j restricts the region of phase space in which the trajectory can lie. Specification of all $2N-1$ C_j reduces the allowable dimensionality from $2N$ to 1, which is the trajectory of the system. However, the integrals of the motion C_j are of two types.[7] Some are isolating, in the sense that the domain of phase space to which they restrict the trajectory is compact and readily partitioned from the full phase space—the language used here is loose, but the geometric visualization intended should be clear. The integral $C_1 = H(q, p)$ is of this type. Others, apparently the vast majority, are nonisolating. The regions of phase space to which they restrict the trajectory pass tortuously through the full domain accessible under the isolating integrals of motion. The distinction between these two classes of integrals of the motion is evident even for the simple system of two independent harmonic oscillators whose Hamiltonian is $(m=1)$,

$$H=\tfrac{1}{2}(p_1^2+\omega_1^2 q_1^2)+\tfrac{1}{2}(p_2^2+\omega_2^2 q_2^2) \tag{2.6}$$

which leads to the equations of motion

$$\left.\begin{aligned} p_i \cos\omega_i t+\omega_i q_i \sin\omega_i t&=p_i^\circ \\ \omega_i q_i \cos\omega_i t-p_i \sin\omega_i t&=\omega_i q_i^\circ \end{aligned}\right\} i=1,2 \tag{2.7}$$

Elimination of t, for each value of i, gives

$$\left.\begin{aligned} p_i^2+\omega_i^2 q_i^2=p_i^{\circ 2}+\omega_i^2 q_i^{\circ 2}&=\text{constant} \\ C_i=\tfrac{1}{2}(p_i^2+\omega_i^2 q_i^2)& \end{aligned}\right\} i=1,2 \tag{2.8}$$

Finally, elimination of t between the equations of motion for different $i=1,2$ leads to a third integral of the motion, C_3. The nature of C_3 depends on the ratio ω_2/ω_1, in particular on whether this ratio is rational or irrational. In the case ω_2/ω_1 is rational the projection of the system trajectory on the q_1q_2 plane is a closed curve, in fact a Lissajous figure (Fig. 2.1). In this case $C_3(q, p)$ is a multivalued function with a finite number of branches. On the other hand, if ω_2/ω_1 is irrational, the projection of the system trajectory on the q_1q_2 plane does not generate a closed curve because there can be no rational integer set that leads to matching of the two oscillators. As a result, the motion of the system is not periodic, and the projected trajectory in the q_1q_2 plane passes arbitrarily close to each point lying within the rectangle defined by the maximum amplitudes of the two oscillators (Fig. 2.1). The trajectory thereby densely fills the accessible q_1q_2 space. Although the integral of the motion C_3 exists, it is a pathological function, namely, a multivalued function with an infinite number of branches. Note that when ω_2/ω_1 is rational the projected trajectory is a Lissajous figure that restricts the motion of the representative point to a small portion of the q_1q_2 plane. In this case C_3 is an isolating integral of the motion. But when ω_2/ω_1 is irrational the existence

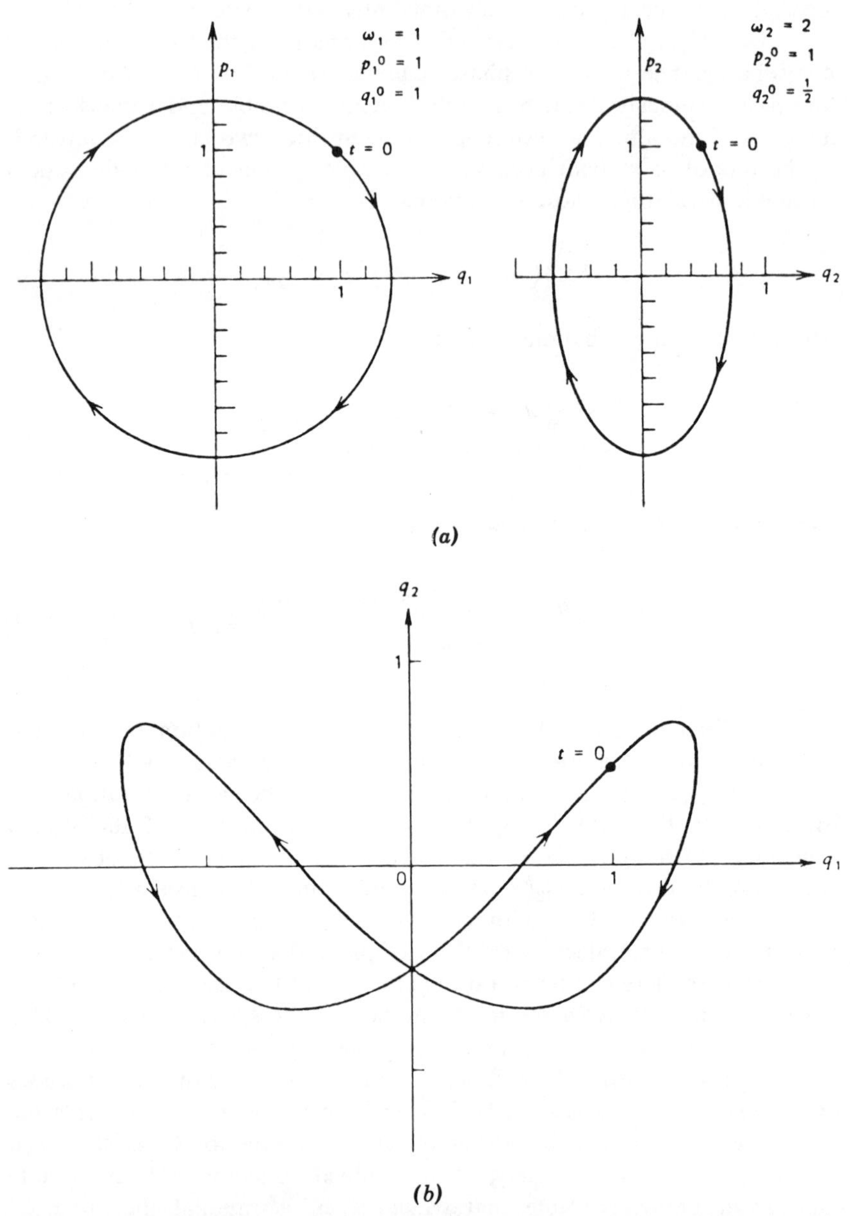

Fig. 2.1. (a) Trajectories of a separable two oscillator system with rational frequency ratio projected on the p_1q_1 and p_2q_2 planes. (b) The trajectory of a separable two oscillator system with rational frequency ratio $\omega_2/\omega_1=2$ projected on the q_1q_2 plane. (c) The trajectory of a separable two oscillator system with irrational frequency ratio, projected on the q_1q_2 plane.

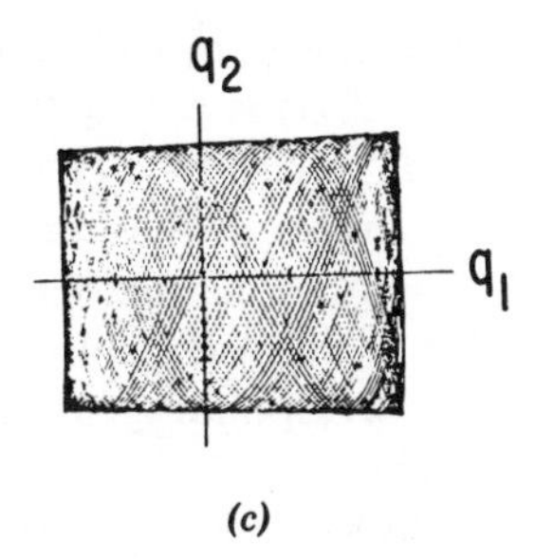

(c)

Fig. 2.1 (*Continued*)

of C_3 does not prevent the projected trajectory from filling the energetically accessible region of the q_1q_2 plane. In this latter case C_3 is a nonisolating integral of the motion.

A different view of dynamics is given by the Hamilton–Jacobi form of mechanics. This representation of the dynamics is based on finding a canonical transformation such that

$$K=H(P) \qquad \dot{Q}_s=\frac{\partial K}{\partial P_s}=\omega_s(P) \qquad \dot{P}_s=-\frac{\partial K}{\partial Q_s}=0 \tag{2.9}$$

The new momenta P_s are integrals of the motion, while the new coordinates Q_s are linear functions of time. Note that only N constants of the motion are determined by the canonical transformation, so the system trajectory is restricted to an $N-1$-dimensional subspace of the full phase space, but not to a smaller space. For a bounded system linear combinations of the P_s define actions J_s, and their conjugate angle variables ϕ_s; the J_s and ϕ_s define the action-angle representation of mechanics. It is only for the case of a system of independent harmonic oscillators that the ω_s are independent of P_s and constant. In more general cases we can, in principle, find $H(J,\phi)$, but the corresponding frequencies will not be independent of J.

Consider the situation in which the system Hamiltonian can be separated into

$$H(J,\phi)=H_0(J)+\epsilon H_1(J,\phi) \tag{2.10}$$

$H_0(J)$ describes an integrable system, the trajectories of which densely cover regions of phase space. As in the example of two harmonic oscillators with ω_2/ω_1 irrational, we expect that when regions of phase space are densely covered the frequencies $\omega_i \equiv (\partial H_0/\partial J_i)$ are not related by a set of rational integers. The term $\epsilon H_1(J, \phi)$ is a "small" perturbation. The traditional view of the influence of anharmonicity on the motion of coupled oscillators suggests that ϵH_1 destroys the topological structure of the trajectories corresponding to $H_0(J)$ no matter how small ϵH_1, if only enough time elapses. The idea is that ϵH_1 causes the trajectory to wander out of densely filled regions corresponding to $H_0(J)$=constant, thereby filling all of accessible phase space.

A remarkable theorem, due to Kolmogoroff, Arnold, and Moser (KAM)[8] implies that the intuitive description of the trajectory just given is incorrect. The theorem says that provided ϵ is "sufficiently small" and $H_1(J, \phi)$ is analytic in J and ϕ in a given domain, the phase space can be separated into two regions of nonvanishing volume. One of these is small, and it shrinks to zero volume as $\epsilon \to 0$. The larger of the two regions has the structure characteristic of $H_0(J)$. Thus, the KAM theorem asserts that for the majority of initial conditions the trajectories of the system have the same character as in the uncoupled oscillator case (Lissajous figures restricted to $N-1$ dimensions). There is a small region (of instability) in which the trajectories are wildly erratic and can depart drastically from the nearby confined trajectories.

To apply the KAM theorem we need to know what is "small enough" with respect to ϵ or, equivalently, for fixed ϵ how the topological behavior of the trajectory changes as the energy of the system increases. At present all of our knowledge concerning this crucial point is derived from numerical solutions of the equations of motion of model systems.[4, 9] Some hypotheses, based on analytical considerations, have been advanced to explain the results of the numerical studies, but these have followed and cannot yet replace the trajectory calculations.[10]

Before examining the results of the numerical solution of the equations of motion of a typical system of coupled nonlinear oscillators, and before describing the analytical considerations advanced to interpret those results, it is desirable to give a qualitative description of the expected behavior of trajectories in such a system. For illustrative purposes consider again a system with two degrees of freedom so that, in angle-action variables,

$$H = H_0(J_1, J_2) + V(J_1, J_2, \phi_1, \phi_2) \tag{2.11}$$

When $V=0$, H_0 generates a motion for which J_1, J_2=constants and $\phi_i = \omega_i(J_1, J_2)t + \phi_i^0$, $\omega_i \equiv \partial H_0/\partial J_i$. The motion of the unperturbed system is conveniently represented on a two-dimensional torus where ϕ_1, ϕ_2 are the angle coordinates and J_1, J_2 the radii (Fig. 2.2). If V is small enough and the Jacobians

$$\frac{\partial(\omega_1, \omega_2)}{\partial(J_1, J_2)} \neq 0 \tag{2.12}$$

KAM show that the most of the unperturbed tori bearing conditionally periodic motion with incommensurate frequencies continue to exist, being only slightly perturbed by V. On the other hand, tori bearing periodic motion or very nearly periodic motion, with commensurate frequencies, or with incommensurate frequencies whose ratio is well approximated by r/s, r, s small integers, are grossly deformed by V and no longer remain close to unperturbed tori. Furthermore, although the unperturbed tori with commensurate frequencies which are destroyed by $V \neq 0$ are everywhere dense, KAM show that the majority (in the sense of measure theory) of initial conditions lead to motion on preserved tori bearing conditionally periodic motion when V is sufficiently small. Thus, KAM theory shows that for small V most initial conditions lead to nonergodic motion.

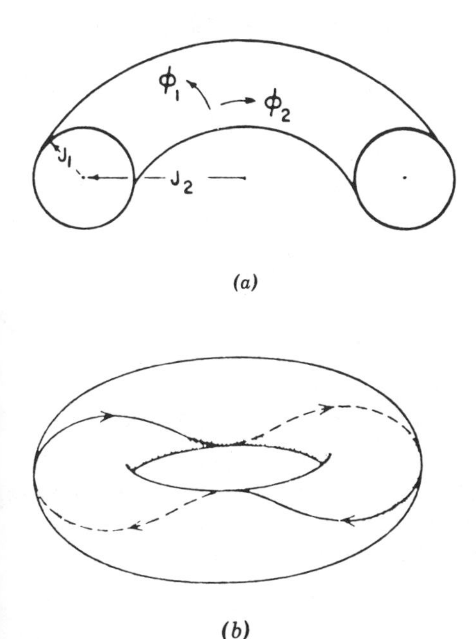

Fig. 2.2. (a) Angle-action variables and the invariant torus for a two-oscillator system. (b) A periodic trajectory on the torus.

What is the character of the motion not on preserved tori? Imagine H expanded in a Fourier series:[11]

$$H = H_0(J_1, J_2) + f_{mn}(J_1, J_2)\cos(m\phi_1 + n\phi_2) + \cdots \tag{2.13}$$

In KAM theory the angle dependent terms are eliminated by successive canonical transformations, each of which is close to the identity transformation. The final Hamiltonian is a function of transformed variables only and is "close" to the original Hamiltonian. If this can be accomplished in some general sense, one finds that the perturbed motion, for the most part, lies on tori close to unperturbed tori.

To illustrate these ideas suppose the only important coupling term in (2.13) is f_{mn} (as displayed). Then to eliminate the term $\cos(m\phi_1 + n\phi_2)$ introduce the canonical transformation

$$F = \mathcal{J}_1\theta_1 + \mathcal{J}_2\theta_2 + B_{mn}(\mathcal{J}_1, \mathcal{J}_2)\sin(m\theta_1 + n\theta_2) \tag{2.14}$$

with $\mathcal{J}, \theta$ the transformed action-angle variables, and B_{mn} to be determined. If $B_{mn} = 0$, then $\mathcal{J}_i = J_i$, $\theta_i = \phi_i$. Applying F to H,

$$H = H_0(\mathcal{J}_1, \mathcal{J}_2) + \{[m\omega_1(\mathcal{J}_1, \mathcal{J}_2) + n\omega_2(\mathcal{J}_1, \mathcal{J}_2)]B_{mn} + f_{mn}(\mathcal{J}_1, \mathcal{J}_2)\}\cos(m\theta_1 + n\theta_2) + \cdots \tag{2.15}$$

$$\omega_i(\mathcal{J}_1, \mathcal{J}_2) = \frac{\partial H_0}{\partial \mathcal{J}_i} \tag{2.16}$$

To lowest order, the angle dependent term is eliminated if

$$B_{mn}(\mathcal{J}_1, \mathcal{J}_2) = -\frac{f_{mn}(\mathcal{J}_1, \mathcal{J}_2)}{m\omega_1(\mathcal{J}_1, \mathcal{J}_2) + n\omega_2(\mathcal{J}_1, \mathcal{J}_2)} \tag{2.17}$$

which requires that $m\omega_1 + n\omega_2 \neq 0$ or be very small compared to f_{mn}. If the denominator is small compared to f_{mn}, B_{mn} is large, and the transformation is not close to the identity transformation, hence the transformed motion is not close to the unperturbed motion. Consequently, if there exists a band of frequencies ω_i for which

$$|m\omega_1(J_1, J_2) + n\omega_2(J_1, J_2)| \ll |f_{mn}(J_1, J_2)| \tag{2.18}$$

then the angle-dependent term grossly distorts an associated zone of unperturbed tori bearing the frequencies satisfying the inequality. In general, if one fourier component satisfies the inequality, there will be additional terms $\cos(m'\phi_1+n'\phi_2)$ in H with ratios m'/n' sufficiently close to m/n that the inequality is also satisfied for them—hence the zone of unperturbed tori distorted by the displayed term will simultaneously be affected by many other angle-dependent terms.

Note that the inequality cited is a kind of resonance relationship which, if satisfied, asserts that $\cos(m\phi_1+n\phi_2)$ couples the oscillators when their frequencies lie in the designated bands. When V is small, hence all f_{mn} small, such resonance zones are narrow and the KAM theorem shows that the totality of all resonant zones is small relative to the measure of the allowed phase space. We expect that as V and f_{mn} increase, or as E increases, the measure of the resonant zones will also increase until most of phase space is filled by them. KAM theory thereby predicts an amplitude instability for conservative nonlinear oscillator systems permitting a transition between predominantly quasiperiodic and ergodic motion.

Walker and Ford have provided a very elegant illustration of the general ideas just sketched.[11] Consider a system described by the Hamiltonian (2.11) with

$$H_0=J_1+J_2-J_1^2-3J_1J_2+J_2^2 \tag{2.19}$$

Set

$$q_i=(2J_i)^{1/2}\cos\phi_i \tag{2.20}$$

$$p_i=-(2J_i)^{1/2}\sin\phi_i \tag{2.21}$$

so that

$$J_i=\tfrac{1}{2}(p_i^2+q_i^2) \tag{2.22}$$

and

$$\omega_1=1-2J_1-3J_2=\frac{\partial H_0}{\partial J_1} \tag{2.23}$$

$$\omega_2=1-3J_1+2J_2=\frac{\partial H_0}{\partial J_2} \tag{2.24}$$

In order that $\omega_1 > 0$, $\omega_2 > 0$ we require E to be in the range $0 < E < 3/13$ and that the J_i lie on the branch that goes to zero with E.

We represent the motion in this system by level curves in the $q_2 p_2$ plane. Note that the unperturbed level curves in the $q_2 p_2$ plane are circles centered on the origin. Points on the level curves in the $q_1 p_1$ plane, defined by $q_2 = 0$, $p_2 > 0$ (or $\phi_2 = 3\pi/2$) also lie on concentric circles. The circular level curves in either plane are enclosed by a bounding level curve representing the intersection of the energy surface with each plane.

Consider the case of a 2–2 resonance described by

$$H = H_0(J_1, J_2) + \alpha J_1 J_2 \cos(2\phi_1 - 2\phi_2) \tag{2.25}$$

which system has the additional constant of the motion

$$\mathsf{C} = J_1 + J_2 \tag{2.26}$$

Now eliminate J_2 from H using C, and set $\phi_2 = 3\pi/2$; we obtain for the level curves in the J_1 plane

$$(3 + \alpha \cos 2\phi_1) J_1^2 - (5\mathsf{C} + \mathsf{C} \cos 2\phi_1) J_1 + \mathsf{C} + \mathsf{C}^2 - E = 0 \tag{2.27}$$

(Recall that $H_0 = J_1 + J_2 - J_1^2 - 3J_1 J_2 + J_2^2$.)

As shown in Fig. 2.3, the unperturbed circular level curves are only slightly distorted except in the 2–2 resonance zone enclosed by the self-intersecting separatrix level curve. The two self-intersection points represent distinct unstable periodic solutions, while the two invariant points at the center of each crescent region represent distinct stable periodic solutions. Since the central point of each crescent represents a distinct periodic orbit the two crescents are not a chain of two islands; the central points of an island chain represent a single periodic orbit. For all four of the mentioned periodic orbits we have

$$\dot{J}_1 = 0, \qquad \dot{J}_2 = 0$$

$$\dot{\phi}_1 - \dot{\phi}_2 = 0 \tag{2.28}$$

For the stable periodic orbits

$$J_1 = \frac{5 + \alpha}{1 + \alpha} J_2 \tag{2.29}$$

$$\phi_1 - \phi_2 = \frac{\pi}{2} \quad \text{or} \quad \frac{3\pi}{2} \tag{2.30}$$

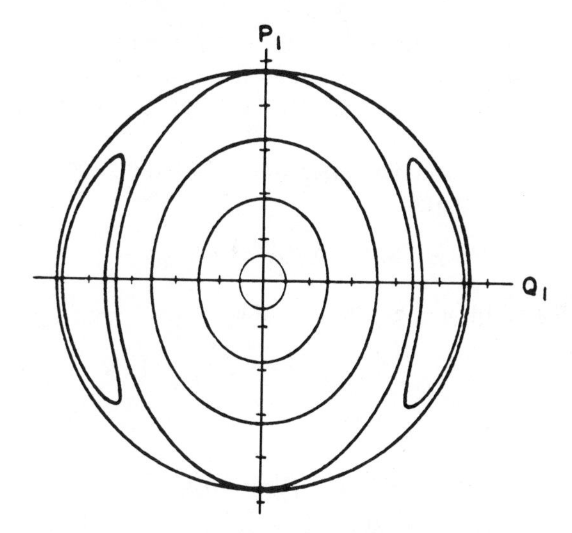

Fig. 2.3. The surface of section for the system with Hamiltonian

$$H = H_0 + \alpha J_1 J_2 \cos(2\phi_1 - 2\phi_2)$$
$$H_0 = J_1 + J_2 - J_1^2 - 3J_1J_2 + J_2^2$$

at an energy corresponding to an isolated 2–2 nonlinear resonance. [From G. H. Walker and J. Ford, *Phys. Rev.*, **188**, 416 (1969).]

while for the unstable periodic orbits

$$J_1 = \frac{5-\alpha}{1-\alpha} J_2 \tag{2.31}$$

$$\phi_1 - \phi_2 = 0 \quad \text{or} \quad \pi \tag{2.32}$$

Note that $2\omega_1 = 2\omega_2$ implies that

$$J_1 = 5J_2 \tag{2.33}$$

Thus, just as expected from the condition stated in (2.18), the 2–2 resonance zone of highly distorted tori occurs in a neighborhood of the unperturbed torus bearing the frequencies $2\omega_1 = 2\omega_2$. Now substituting $J_1 = 5J_2$ into $H_0 = J_1 + J_2 - J_1^2 - 3J_1J_2 + J_2^2$ we find on the unperturbed 2–2 torus that

$$J_1 = \tfrac{5}{13}\left[1 - \left(1 - \tfrac{13}{3}E\right)^{1/2}\right] \tag{2.34}$$

$$J_2 = \tfrac{1}{13}\left[1 - \left(1 - \tfrac{13}{3}E\right)^{1/2}\right] \tag{2.35}$$

Consequently the unperturbed 2–2 torus and the perturbed 2–2 resonance zone exist for all allowed energies $0<E<3/13$. As the energy increases from zero the 2–2 resonance moves out from the origin and increases in width.

The next higher resonance after the 2–2 is the 2–3 or 3–2 resonance. Consider the Hamiltonian

$$H = H_0(J_1, J_2) + \beta J_1^{3/2} J_2 \cos(3\phi_1 - 2\phi_2) \tag{2.36}$$

which has the additional constant of the motion

$$\mathrm{C} = 2J_1 + 3J_2 \tag{2.37}$$

The level curves in the J_1 plane are

$$E = \tfrac{1}{3}\mathrm{C} + \tfrac{1}{9}\mathrm{C}^2 + \left(\tfrac{1}{3} - \tfrac{13}{9}\mathrm{C}\right)J_1 + \tfrac{13}{9}J_1^2 - \tfrac{1}{3}\beta\left(\mathrm{C}J_1^{3/2} - 2J_1^{5/2}\right)\cos 3\phi_1 \tag{2.38}$$

As shown in Fig. 2.4 the 3–2 resonance zone consists of a chain of three islands (points at the centers of each crescent represent a single stable

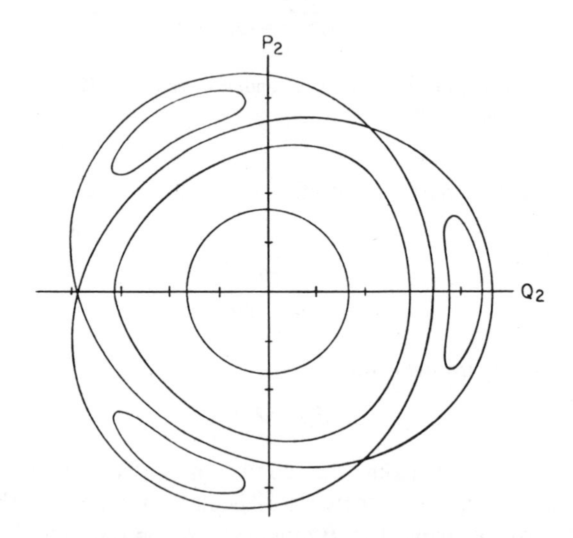

Fig. 2.4. The surface of section for the system with Hamiltonian

$$H = H_0 + \beta J_1^{3/2} J_2 \cos(3\phi_1 - 2\phi_2)$$
$$H_0 = J_1 + J_2 - J_1^2 - 3J_1J_2 + J_2^2$$

at an energy corresponding to an isolated 3–2 nonlinear resonance. [From G. H. Walker and J. Ford, *Phys. Rev.*, **188**, 416 (1969).]

solution). Similarly, the three self-intersecting points on the separatrix represent a single unstable periodic solution. Again setting

$$\dot{J}_1 = 0 \qquad \dot{J}_2 = 0$$
$$3\dot{\phi}_1 - 2\dot{\phi}_2 = 0 \tag{2.39}$$

we find

$$\left.\begin{aligned} J_2 &= \frac{1+2J_1^{3/2}}{13+\frac{9}{2}J_1^{1/2}} \\ 3\phi_1 - 2\phi_2 &= \pi, 3\pi, 5\pi \end{aligned}\right\} \text{stable orbit} \tag{2.40}$$

$$\left.\begin{aligned} J_2 &= \frac{1-2J_1^{3/2}}{13+\frac{9}{2}J_1^{1/2}}, \\ 3\phi_1 - 2\phi_2 &= 0, 2\pi, 4\pi \end{aligned}\right\} \text{unstable orbit} \tag{2.41}$$

As expected, the 3–2 resonance zone lies near to the unperturbed 3–2 zones. Setting $3\omega_1 = \omega_2$, H_0 yields

$$J_1 = \tfrac{5}{13} - \left(\tfrac{3}{13} - E\right)^{1/2} \tag{2.42}$$

$$J_2 = \tfrac{1}{13} \tag{2.43}$$

as the values of J_1, J_2 on the unperturbed torus, to be compared with the perturbed values. Since $J_1 = \frac{1}{2}(p_1^2 + q_1^2) > 0$,

$$E = \tfrac{3}{13} - \left(J_1 - \tfrac{5}{13}\right)^2 \geqslant \tfrac{14}{169} \approx 0.08 \tag{2.44}$$

At $E = 14/169$, $J_1 = 0$, hence the unperturbed 3–2 torus and the 3–2 resonance zone appear abruptly at the origin of the J_1 plane. They appear abruptly in the J_2 plane when the bounding level curve moves out to $J_2 = 1/13$, which is when $E = 14/169$. The corresponding 2–3 resonance is generated from

$$H = H_0(J_1, J_2) + \beta J_1 J_2^{3/2} \cos(2\phi_1 - 3\phi_2) \tag{2.45}$$

which has the additional constant of the motion

$$\mathrm{C} = 3J_1 + 2J_2 \tag{2.46}$$

Level curves in the J_2 plane are found from

$$E=\tfrac{1}{3}\mathrm{C}-\tfrac{1}{9}\mathrm{C}^2+\left(\tfrac{1}{3}-\tfrac{5}{9}\mathrm{C}\right)J_2+\tfrac{23}{9}J_2^2+\beta\left[\tfrac{2}{3}J_2^{5/2}-\tfrac{1}{3}\mathrm{C}J_2^{3/2}\right]\cos 3\phi_2 \quad (2.47)$$

Just as for the 3–2 resonance case, the 2–3 resonance zone appears in the J_2 plane about the unperturbed 2–3 torus, which appears abruptly at $E=0.16$. The level curves are similar to those of the 3–2 case, except that the chain of three islands appears in the J_2 plane.

In general, it is found that:

1. A Hamiltonian of the form

$$H=H_0(J_1, J_2)+f_{mn}(J_1, J_2)\cos(m\phi_1+n\phi_2)$$

has an "extra" well-defined constant of the motion:

$$\mathrm{C}=nJ_1-mJ_2$$

2. An m–n resonance for $m\neq n$ introduces a chain of m islands in the J_1 plane and a chain of n islands in the J_2 plane. (Islands are ovals surrounding points representing stable periodic orbits.)
3. Isolated resonances distort the unperturbed tori by introducing, in pairs, new stable and unstable periodic orbits.
4. An m–n resonance zone appears abruptly, in general at some $E>0$, and is bounded by a separatrix, which passes through the unstable periodic solutions.
5. The m–n resonance zones decrease in size rapidly as m and n increase.

Thus far we have commented only on the consequences of the existence of isolated nonlinear resonances. In general we must expect that a system of coupled nonlinear oscillators will support many different nonlinear resonances of the types described, and that at some energy the different resonant zones will begin to overlap. Walker and Ford and Ford and Lunsford[12] have studied simple models that fully confirm this inference. When the resonant zones corresponding to the several nonlinear resonances overlap there is gross distortion of the level curves on the q_2p_2 plane and higher order nonlinear resonances also begin to strongly influence these curves. The net result is a macroscopic distortion of the system tori, which then implies a macroscopic instability of the trajectory. Clearly, the energy at which there is an onset of overlap of nonlinear resonances is a plausible boundary between quasiperiodic and grossly irregular motion.[13]

B. Numerical Studies of the Dynamics of Nonlinear Oscillator Systems

In a sense, it can be said that numerical solution of the equations of motion of some system is intended to reveal the consequences of the breakdown of integrability and the lack of isolating integrals of the motion other than the energy. Poincaré introduced a representation of the results of trajectory analysis which permits visualization of these consequences.[14] This representation, which is most useful for two-dimensional systems, portrays the motion on a so-called (Poincaré) surface of section. We have already used this representation, without naming it, in the discussion of the level curves in the $q_2 p_2$ plane for a system with nonlinear resonances. Now we generalize our description. Consider, for simplicity, a Hamiltonian of the form $(m=1)$

$$H=\tfrac{1}{2}(p_1^2+p_2^2)+f(q_1,q_2) \tag{2.48}$$

For fixed energy, $H=E$, (2.48) has only three independent variables. One surface of section is defined by the intersection of $H=E$ with $q_1=0$; in that plane the coordinates are p_2 and q_2. To each point in the surface of section there corresponds a unique value of p_2 and q_2, and of E and $q_1=0$. Then p_1 is determined except for sign since, from (2.48),

$$p_1=\pm\left[2E-p_2^2-f(q_2,0)\right]^{1/2} \tag{2.49}$$

A given trajectory of a bound system will repeatedly cross the surface of section, since that trajectory must repeatedly pass through $q_1=0$, half the passages with $p_1>0$ and half with $p_1<0$. We now recognize two possibilities. If there exist isolating integrals of the motion other than the energy, such as C_3 with ω_2/ω_1 rational for the Hamiltonian (2.6), the system trajectory lies on a hypersurface of smaller dimensionality than the energy surface. This hypersurface intersects the surface of section in a smooth closed curve—closed because the motion is periodic. In contrast, if there is not any isolating integral of the motion other than the energy, the intersections of the trajectory with the surface of section will cover that surface. The pattern of intersections will appear random, but is in fact not, since the trajectory satisfies the deterministic equation of motion. The two cases described are schematically sketched in Fig. 2.5.

The surface of section representation has another characteristic that will prove useful in our considerations. Since any point initially on the intersection of a trajectory and the surface of section will repeatedly cross the

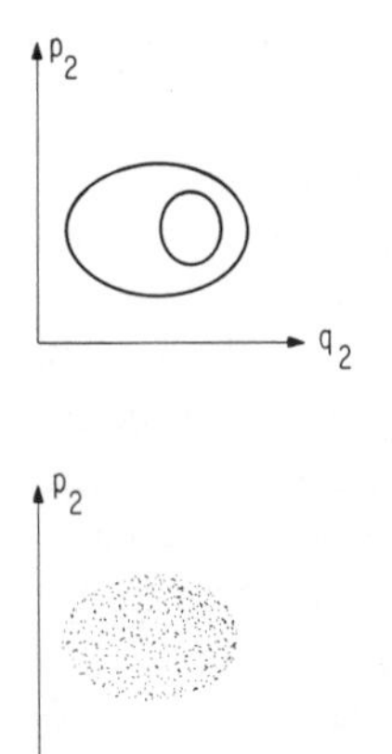

Fig. 2.5. Schematic representations of the surface of section in the quasiperiodic region and the stochastic region.

surface of section at, in general, different points, at each crossing the entire surface of section is mapped into itself. Of course the time taken for the recrossing will be different for each point of the surface of section, but that does not alter the character of the mapping. Given that the dynamics is described by the Hamiltonian equations of motion, this mapping is area preserving (this property is related to Liouville's theorem).

Let T be the operator that maps the point $Z^{(0)}$ on the surface of section into the point $Z^{(1)}$:

$$Z^{(1)} = TZ^{(0)} \tag{2.50}$$

A point $Z^{(0)}$ is called a fixed point when $Z^{(1)} = Z^{(0)}$. It can be shown that when T maps a compact simply connected subset S of the plane onto itself, then S contains a fixed point of T. In general the mappings of surfaces of section for the dynamical systems of interest to us will have many fixed points. The behavior of the trajectories in the vicinities of these fixed points will be shown to determine the overall behavior of the dynamical system.

To begin a description[15] of the nature of the fixed points of the mapping T we note that a set of points U is said to be invariant if $TU = U$, and for an area preserving mapping it is sufficient that TZ belongs to U if Z belongs to U. A fixed point of a mapping is called stable if every neighborhood of this fixed point contains an invariant neighborhood. Suppose, now, that a fixed point of the mapping T is located at $Z \equiv (p, q) = (0, 0)$, which can always be achieved by a translation of the coordinate system. Then in the neighborhood of this fixed point the mapping T can be represented by the linear equations

$$\begin{pmatrix} p^{(1)} \\ q^{(1)} \end{pmatrix} = \begin{pmatrix} a & b \\ c & d \end{pmatrix} \begin{pmatrix} p^{(0)} \\ q^{(0)} \end{pmatrix} \tag{2.51}$$

with

$$\det(T) = \begin{Vmatrix} a & b \\ c & d \end{Vmatrix} = 1 \tag{2.52}$$

since T is an area preserving mapping. Then the eigenvalues of T, denoted as λ_1 and λ_2, satisfy the condition

$$\lambda_1 \lambda_2 = 1 \tag{2.53}$$

If both λ_1 and λ_2 are real, and $\lambda_2 = \lambda_1^{-1} \neq 1$, the fixed point is called hyperbolic; if λ_1 and λ_2 are imaginary conjugates the fixed point is called elliptic; and if $\lambda_1 = \lambda_2 = \pm 1$ the fixed point is called parabolic. The reason for these designations, and the characteristics of each of these kinds of fixed points, will be made apparent below.

Recall that the trajectory representing the motion of a coupled oscillator system lies on a torus, and that in the case when the equations of motion are integrable the intersection of the trajectory with the surface of section generates simple closed curves. We have defined the surface of section in $p_2 q_2$ space. In angle-action space the transverse radius of the torus is, say, J_2, and the corresponding angle is ϕ_2 (see Fig. 2.2); then the invariant curves on the surface of section are circles. A mapping of the system trajectory conserves J_2 but changes ϕ_2. For the case that the time interval between crossings of the surface of section is $\tau = 2\pi/\omega_1$ and the initial angular coordinate is $\phi_1^{(0)}$, the angular coordinate on the next crossing is

$$\phi_1^{(1)} = \phi_1^{(0)} + \omega_2 \tau = \phi_1^{(0)} + 2\pi \frac{\omega_2}{\omega_1} \tag{2.54}$$

where the ratio of frequencies

$$\frac{\omega_2}{\omega_1} = \alpha(J_2) \tag{2.55}$$

is, at constant energy, a function of J_2 only. The topology of successive intersections of the trajectory of an integrable system with the surface of section is, therefore, described by the twist mapping

$$\begin{aligned} J_2^{(1)} &= TJ_2^{(0)} = J_2^{(0)} \\ \phi_2^{(1)} &= T\phi_2^{(0)} = \phi_2^{(0)} + 2\pi\alpha(J_2^{(0)}) \end{aligned} \tag{2.56}$$

The function $\alpha(J_2)$ is called the rotation number. Under the twist mapping, which is area preserving, circles map into circles, and radius vectors into curved arcs passing through the origin. Suppose the mapping of the trajectory is changed to

$$\begin{aligned} J_2^{(1)} &= T_\epsilon J_2^{(0)} = J_2^{(0)} + \epsilon f(J_2^{(0)}, \phi_2^{(0)}) \\ \phi_2^{(1)} &= T_\epsilon \phi_2^{(0)} = \phi_2^{(0)} + 2\pi\alpha(J_2^{(0)}) + \epsilon g(J_2^{(0)}, \phi_2^{(0)}) \end{aligned} \tag{2.57}$$

where the functions f and g have period 2π in $\phi_2^{(0)}$, are zero when $J_2^{(0)}=0$ and are related such that $\det(T_\epsilon)=1$. The condition $f=0$, $g=0$ when $J_2^{(0)}=0$ guarantees that the origin of the coordinates remains a fixed point despite the change in the form of the mapping, and the condition on the determinant of T_ϵ guarantees that it is area preserving. The KAM theorem, applied to the relationship between invariant curves generated by the two mappings, T and T_ϵ, states that most of the points in the surface of section lie on smooth invariant curves which are distortions of the invariant circles generated by the twist mapping T. These smooth invariant curves are sections of tori which are likewise distorted from the original tori. The only possible exceptions to this behavior correspond to points that are near tori on which $\alpha(J_2^{(0)})=\omega_2/\omega_1$ is rational ("commensurable" tori). On commensurable tori the trajectories are closed orbits. It is easy to see that every point on the circle with rotation number

$$\alpha(J_2^{(0)}) = \frac{r}{s} \qquad (r, s \text{ integers}) \tag{2.58}$$

is a fixed point of the twist mapping

$$\begin{aligned} T^s J_2^{(0)} &= J_2^{(0)} \\ T^s \phi_2^{(0)} &= \phi_2^{(0)} + 2\pi s\alpha(J_2^{(0)}) \\ &= \phi_2^{(0)} + 2\pi r\phi_2^{(0)} \\ &= \phi_2^{(0)} \pmod{2\pi} \end{aligned} \tag{2.59}$$

The Poincaré–Birkhoff theorem[16] states that, in general, an even multiple of s (say $2ks$, $k=1,2,\ldots$) fixed points remains when the mapping is changed from T to T_ϵ. This theorem gives supplementary information to that gained from the KAM theorem, which says nothing about the case of transformations of commensurable tori. Note that not all fixed points are preserved under the change $T \to T_\epsilon$; all but a finite number are destroyed.

The nature of a fixed point is determined by the behavior of nearby invariant curves. Consider, first, the case that the eigenvalues of the mapping T are imaginary:

$$\lambda_1 = e^{i\alpha}$$

$$\lambda_2 = \lambda_1^* = e^{-i\alpha} \tag{2.60}$$

In this case T can be written in the form

$$T = \begin{pmatrix} \cos\alpha & \sin\alpha \\ -\sin\alpha & \cos\alpha \end{pmatrix} \tag{2.61}$$

which corresponds to a rotation through the angle α. Thus in this case T is a twist mapping, and the invariant curves are circles. In the more general case that λ_1 and λ_2 are complex, rather than pure imaginary, the invariant curves are ellipses. It is for this reason that a fixed point of this type is designated elliptic. Trajectories in the vicinity of elliptic fixed points are stable, since any trajectory that crosses the surface of section near such a point will still be near it after arbitrarily many iterations of the transformation (Fig. 2.6).

Consider, next, the case that the eigenvalues of T are real numbers, where say $|\lambda_1| > 1$. Then T can be written in the form

$$T = \begin{pmatrix} 1/\lambda & 0 \\ 0 & \lambda \end{pmatrix} \tag{2.62}$$

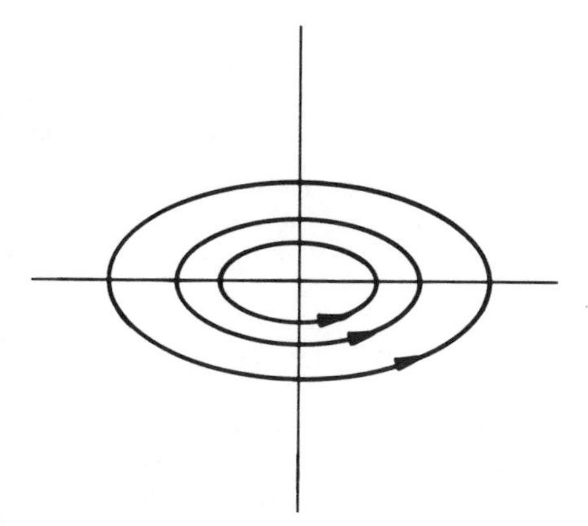

Fig. 2.6. Schematic representation of trajectories near an elliptic fixed point.

In the neighborhood of such a point the invariant curves satisfy the generic equation pq=constant, hence are hyperbolic. When $\lambda_1 > 0$ the fixed point is said to be of the ordinary hyperbolic type. In this case the successive crossings of the surface of section remain on one invariant branch. When $\lambda_1 < 0$ the fixed point is said to be a hyperbolic point with reflection, since successive crossings of the surface of section jump back and forth between opposite branches. In either case trajectories in the vicinity of a hyperbolic fixed point are unstable in the following sense: any point on a trajectory that passes near to, but not coincident with a hyperbolic fixed point, will map far away from the fixed point (Fig. 2.7).

Finally, in the special case when $\lambda_1 = \lambda_2 = \pm 1$, the fixed point is said to be parabolic. When $\lambda_1 = \lambda_2 = +1$, T can be reduced to the form

$$T = \begin{pmatrix} 1 & 0 \\ C & 1 \end{pmatrix} \tag{2.63}$$

with C a constant. The invariant curves are, therefore, straight lines (Fig. 2.8). It is possible for there to be a line of parabolic fixed points. For example, the twist mapping (2.56) generates such a line whenever the curve to which it is applied has a rational rotation number. When the mapping is then changed to that shown in (2.57), what were parabolic fixed points of the mapping (2.56) become a finite set of alternating hyperbolic and elliptic fixed points, since the eigenvalues of the mapping (2.57) are not equal to unity under the addition of the terms in ϵf and ϵg. When ϵ is small these eigenvalues are close to unity, so the hyperbolic fixed points are of the ordinary (not the reflecting) type. Therefore, of the $2ks$ fixed points of

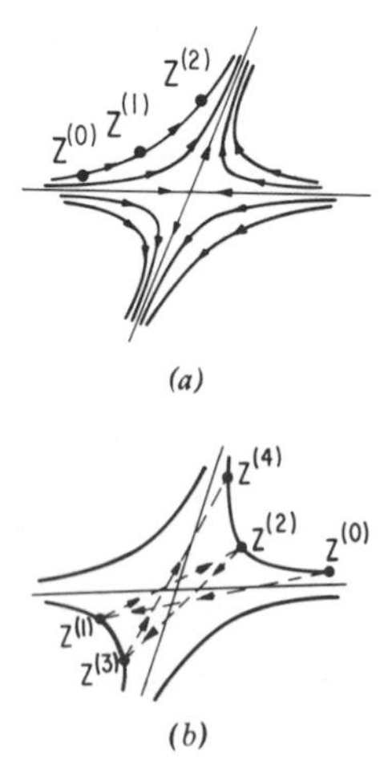

Fig. 2.7. (a) Schematic representation of trajectories near an ordinary hyperbolic fixed point. (b) Schematic representation of trajectories near a hyperbolic fixed point with reflection.

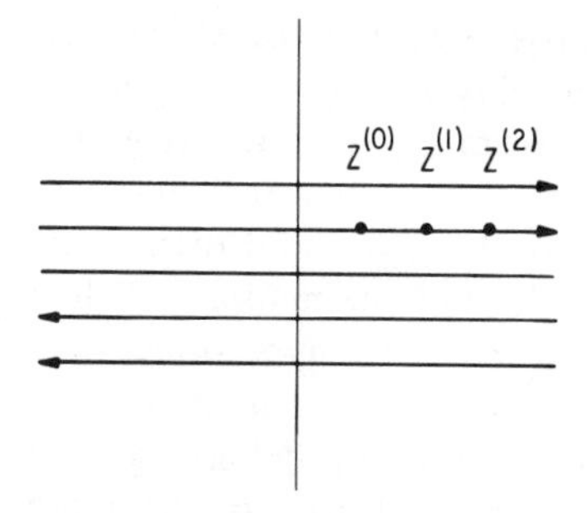

Fig. 2.8. Schematic representation of trajectories near a parabolic fixed point.

T^s that remain after the breakup of the curve with rational rotation number r/s, ks are elliptic and ks are hyperbolic, the two types forming an alternating sequence.

Application of both the Poincaré-Birkhoff and the KAM theorems to the behavior of trajectories in the vicinity of an elliptic fixed point establishes the following: There are closed invariant curves corresponding to motion on tori with irrational frequency ratio and, where there used to be invariant curves of the pure twist mapping (2.56) corresponding to rational rotation number, the use of the twist mapping (2.57) with $\epsilon \neq 0$ generates a new structure of fixed points, half of which are elliptic. In the neighborhoods of these new elliptic fixed points there are invariant curves corresponding to irrational rotation number and, where there used to be invariant curves of the pure twist mapping (2.56) corresponding to rational rotation number,...and so on ad infinitum.

The structure near hyperbolic fixed points is different. At any hyperbolic fixed point P_H four curves meet, two ingoing (H_+) and two outgoing (H_-) (Fig. 2.9). If

$$\lim_{s\to\infty} T^s Z \to P_H \tag{2.64}$$

then Z is on H_+, and if

$$\lim_{s\to\infty} T^{-s} Z \to P_H \tag{2.65}$$

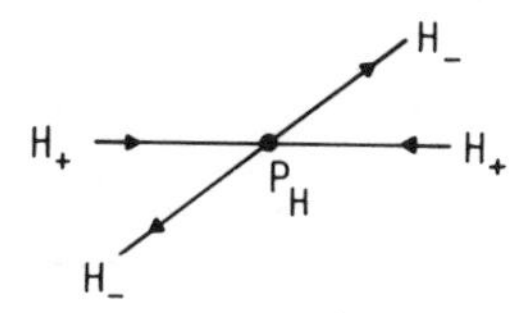

Fig. 2.9. Schematic representation of invariant curves meeting at a hyperbolic fixed point.

then Z is on H_-. Clearly, points on H_+ approach P_H indefinitely slowly, and points on H_- recede from P_H indefinitely slowly, as $s \to \pm\infty$, respectively. Consider following H_+ and H_- away from P_H. For an integrable system the arcs join smoothly, since any point on this invariant curve can be thought of as having started at P_H, mapped along H_- and then along H_+ back to P_H after a double infinity of iterations of the mapping. When P_H is a fixed point of T^s, P_H is one member of a set of s hyperbolic fixed points corresponding to an unstable closed orbit and the outgoing curve H_- from P_H joins smoothly onto the ingoing curve H_+ belonging to a neighboring hyperbolic fixed point. However, this kind of smooth joining is the exceptional behavior. In general, instead, the arcs H_+ and H_- intersect at what is called a homoclinic point (if the arcs belong to the same hyperbolic fixed point or to different fixed points of the same unstable closed orbit) or a heteroclinic fixed point (if the arcs belong to two hyperbolic fixed points not associated with the same closed orbit).

The behavior of the curves H_+ and H_- beyond a homoclinic point is remarkably complicated. Consider one such point Z, and the effect of T^s on trajectories in the neighborhood of Z. Because T is a continuous mapping all the successively generated neighborhoods must be similar near to Z. Therefore the existence of one homoclinic point implies the existence of an infinity ($s \to \infty$) of others. The consequence is that H_+ forms a series of loops intersecting H_-, and vice versa. Indeed, since T is area preserving, every point on H_- between two intersections with H_+ is itself a further intersection. Thus, each point on an "early loop" of, say, H_- maps onto "late loops" that are more and more complicated and wander over more and more of the surface of section, so the point Z will eventually map arbitrarily close to any other point in the surface of section. In this region of phase space neither smooth invariant curves nor tori exist. To sum up, near each rational invariant curve there are hyperbolic fixed points with associated apparently chaotically wandering curves, and elliptic fixed points surrounded with invariant curves which repeat the entire structure ad infinitum (Fig. 2.10). There is, then, an intimate interlacing of regions of integrable and nonintegrable motions. The latter are so chaotic that the motion can be considered stochastic.

We are, finally, ready to examine the results of numerical solution of the equations of motion of a system of coupled nonlinear oscillators. Several such systems have been studied; we choose two to illustrate the range of dynamical behavior possible.

Consider, first, a system described by the Hamiltonian

$$H(q,p)=\tfrac{1}{2}(p_1^2+q_1^2)+\tfrac{1}{2}(p_2^2+q_2^2)+q_1^2q_2-\tfrac{1}{3}q_2^3 \tag{2.66}$$

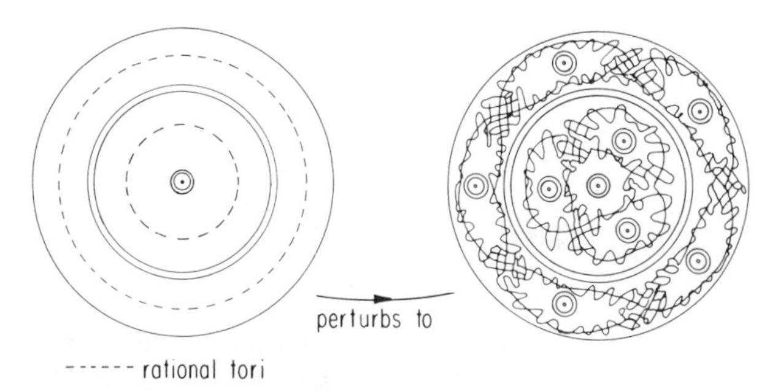

Fig. 2.10. Schematic representation of the change under perturbation of trajectories associated with rational invariant tori. [After M. Berry, in *Topics in Nonlinear Dynamics, A.I.P. Conf. Proc.*, **46**, 16 (1978).]

introduced by Henon and Heiles.[9] This Hamiltonian represents two oscillators coupled by an interaction that is cubic in the amplitude; it describes bounded motion for $E<1/6$, and unbounded motion for $E>1/6$. The trajectories corresponding to (2.66) have been determined by Ford,[4] with the results shown in Figs. 2.11–2.13. In these figures, which represent the surface of section for the system, it is clear that for low energy ($E=1/12$) the motion is periodic, for intermediate energy ($E=1/8$) it is mostly periodic with some nonperiodic regions, and when the energy is close to the dissociation limit ($E=1/6$) the motion is apparently nonperiodic. Therefore for, say, $E<1/8$, there is an isolating integral of the motion other than the energy, but not for all $E<1/6$, which is just the behavior described by the KAM theorem.

It is worthwhile examining the calculations portrayed in Figs. 2.11–2.13 in further detail. Consider again the surface of section representation shown in Fig. 2.11. In this case $E=1/12$. Every trajectory yields a smooth

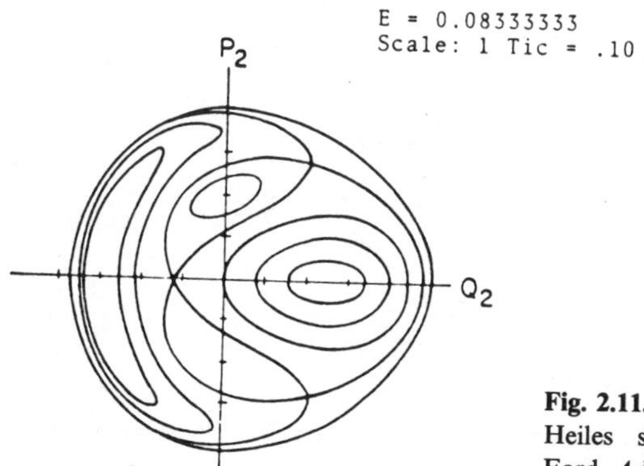

Fig. 2.11. Surface of section for the Henon–Heiles system with $E=1/12$. [From J. Ford, *Adv. Chem. Phys.*, **24**, 155 (1973).]

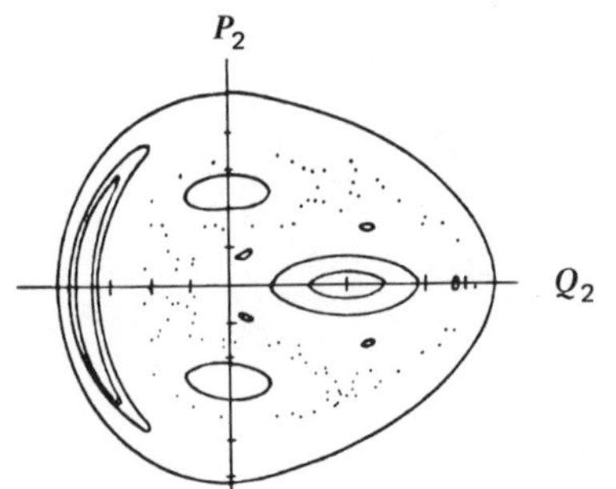

Fig. 2.12. Surface of section for the Henon–Heiles system with $E=1/8$. [From J. Ford, *Adv. Chem. Phys.*, **24**, 155 (1973).]

locus of intersections with the surface of section, indicating that the system is deep within the region of KAM stability. At the center of each region of ovals on the q_2 or p_2 axes is an elliptic fixed point of the mapping generated by a stable periodic orbit, and each oval is an invariant curve generated by those quasiperiodic orbits of H lying on smooth KAM surfaces. The self-intersection points of the self-interacting separatrix curve are hyperbolic fixed points of the mapping generated by unstable periodic orbits. Orbits started elsewhere on this separatrix curve generate mapping iterates which asymptotically approach one or another of the unstable fixed points. Finally, all the allowed trajectories for H at $E=1/12$ intersect the $q_2 p_2$ plane at or within the outermost oval shown in the figure; points outside this bounding curve yield unphysical negative values for p_1^2.

When the system energy is increased to $E=1/8$ large areas of the allowed $q_2 p_2$ plane are still covered with invariant curves generated by trajectories lying on smooth KAM surfaces, but the remaining area is

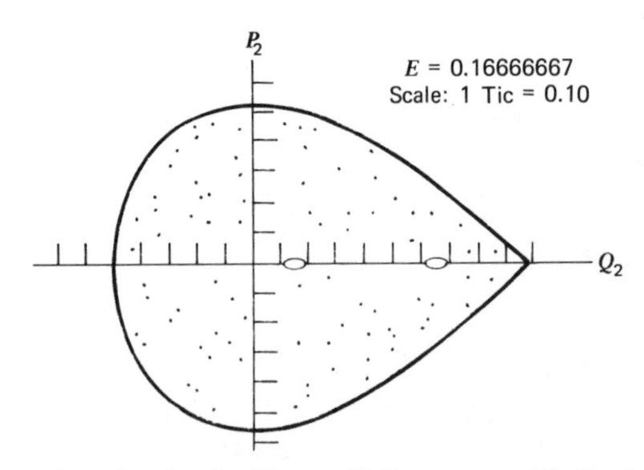

Fig. 2.13. Surface of section for the Henon–Heiles system with $E=1/6$. [From J. Ford, *Adv. Chem. Phys.*, **24**, 155 (1973).]

almost uniformly covered by the intersection points of each (and almost every) orbit initiated in this area (Fig. 2.12). Note that the random-looking splatter of points is generated by a single trajectory; almost any other orbit initiated in the interior of this region would generate a similar pattern of points. It is found that in this region of disintegrated invariant curves, initially close points in the plane map apart exponentially (Fig. 2.14). The computer evidence indicates that a dense set of unstable fixed points exists throughout such regions.

Finally, for the surface of section representation shown in Fig. 2.13 the energy is $E=1/6$. The mapping now reveals a more or less uniform distribution of intersections over the allowed region of the $q_2 p_2$ plane, i.e., the entire allowed region consists of disintegrated invariant curves.

We conclude that even in a mechanical system as simple as that described by the Henon–Heiles Hamiltonian (2.66) there is a smooth transition from nonstatistical to stochastic behavior of the system's trajectories (in this case as E is increased). To connect this behavior with our discussion of the properties of area preserving mappings, consider again Fig. 2.12, especially the region containing ovals surrounding the central fixed point on the positive q_2 axis. Sequential mapping iterates of an initial point (q_2, p_2) lying on an oval rotate about the central fixed point and the average angle of rotation (in radians) is the rotation number α of the invariant oval curve. The value of α associated with each invariant curve varies smoothly as one progresses out from the central fixed point. The

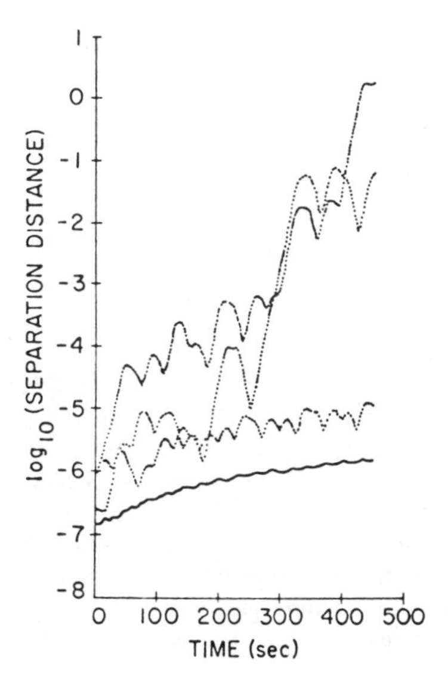

Fig. 2.14. The separation distances vs time for two initially close trajectories of the Henon–Heiles system. $E=1/8$. The lower curves are for trajectory pairs in the quasiperiodic region, the upper curves for trajectory pairs in the stochastic region. [From J. Ford, *Adv. Chem. Phys.*, **24**, 155 (1973).]

KAM theorem leads to the conclusion that invariant curves surround the central fixed point provided, among other things, the associated rotation numbers are irrationals satisfying the inequality

$$\left|\omega - \frac{l}{k}\right| > \frac{\epsilon}{k^{5/2}} \tag{2.67}$$

for all integers l, k, where ϵ is a constant independent of k. Between these nondense invariant curves, where one might expect to find a dense set of invariant curves with rational rotation numbers α, the Poincaré–Birkhoff theorem states that in general only remnants of such α rational invariant curves remain, in the form of interleaved fixed points, half being elliptic and half being hyperbolic. Thus, an extremely accurate numerical calculation of the trajectories in a neighborhood of the central elliptic fixed point on the positive q_2 axis of Fig. 2.12 should reveal a pattern like that in Fig. 2.10. An indication of this structure for the Henon–Heiles system at $E=1/8$ appears in Fig. 2.15, which shows only a few of the hundreds of calculated fixed points found at this energy.

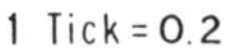

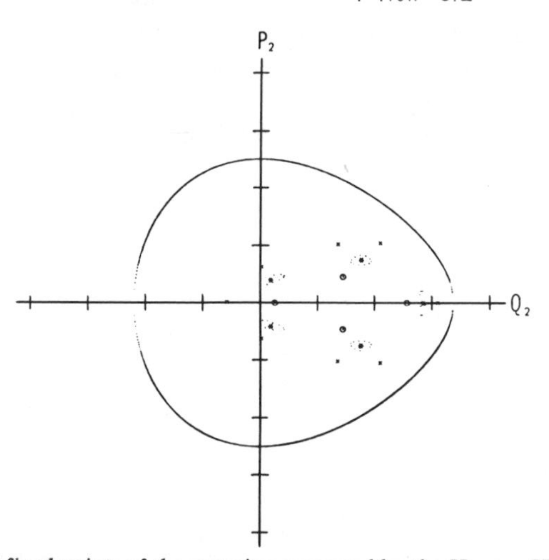

Fig. 2.15. Selected fixed points of the mapping generated by the Henon–Heiles Hamiltonian at energy $E=\frac{1}{8}$. The innermost set of four fixed points located at the dots inside small circles are elliptic members of the T^4 family. The fixed points located at the asterisks are elliptic members of the T^5 family. Surrounding each member of this T^5 family is a set of 11 hyperbolic fixed points belonging to T^{55}. Finally, the eight fixed points located at the $\times$ symbols are hyperbolic fixed points belonging to T^8. With the exception of the T^8 family, each family of fixed points shown represents half an alternating elliptic-hyperbolic set. However, all 16 members of the two T^8 families which lie in the stochastic region are hyperbolic. [From J. Ford, *Adv. Chem. Phys.*, **24**, 155 (1973).]

In summary, it is the ovals surrounding the central fixed point that correspond to the invariant curves generated by quasiperiodic orbits on KAM surfaces. Between these invariant curves on the equivalent KAM surfaces there exists the structure shown in Fig. 2.10, and more! The fixed point families lying between the invariant curves form a dense set; some of these families are alternating hyperbolic–elliptic, but others are alternating hyperbolic–hyperbolic. The separatrices emanating from each hyperbolic point no longer smoothly connect adjacent hyperbolic fixed points as the separatrix in Fig. 2.11 appears to do. Not only do separatrices from the same family intersect each other but they also intersect the separatrices belonging to nearby hyperbolic points. In regions containing these many intersecting separatrices, two close initial points locally map exponentially apart. This exponential behavior is especially marked in those very narrow annular regions containing only hyperbolic fixed points. Thus the phase space for the Henon–Heiles system is always pathologically divided into sets of stochastic and nonstochastic trajectories. At low energies the motion is almost all nonstatistical; as the energy increases the originally microscopic regions containing stochastic trajectories also increase in size, eventually covering most of the allowed phase space. The transition is analytically smooth, although rather abrupt (at about $E=0.11$) on the scale of the dissociation energy for the Henon–Heiles Hamiltonian.

We can also make a connection with the argument that multiple overlap of nonlinear resonances is the source of such behavior in oscillator systems. For the Henon–Heiles system the motion for a specified trajectory involves two frequencies, ω_1 and ω_2, which depend on initial conditions (the system is anharmonic). Since the unperturbed frequencies are equal, we should expect resonant energy exchange between the oscillators in those initial condition regions for which $\omega_1 \approx \omega_2$. Thus, at $\omega_1 = \omega_2$ one should find a periodic orbit surrounded by a region of resonant energy exchange. This is what Fig. 2.11 shows about the central "$\omega_1 = \omega_2$" fixed points. Further, in this nonlinear system there are also resonances of the type $n\omega_1 \approx m\omega_2$ for suitable initial conditions. Again we expect to find a periodic orbit for initial conditions such that $n\omega_1 = m\omega_2$ surrounded by (perhaps) small regions wherein there is energy exchange. This is, essentially, the picture partially verified in Fig. 2.15. As usual, for small E it is expected that all resonances except $\omega_1 = \omega_2$ have (almost) undetectable widths; as E increases the width of each resonance also should increase, thereby destroying the intervening smooth KAM surfaces and, eventually, "overlapping" each other. In such overlap regions, containing many resonances, the final system state is extremely sensitive to the initial state because of the numerous intervening trajectory "collisions" with unstable (and perhaps stable) periodic orbits.

As already mentioned, several other systems of coupled nonlinear oscillators display behavior similar to that of the Henon-Heiles system. It is very important to note, however, that not all such systems behave like the Henon–Heiles system. Consider the Toda Hamiltonian[17]

$$H(q,p)=\tfrac{1}{2}(p_1^2+p_2^2)+\tfrac{1}{24}\Big[\exp\big(2q_2+2\sqrt{3}\,q_1\big)+\exp\big(2q_2-2\sqrt{3}\,q_1\big)+\exp(-4q_2)\Big]-\tfrac{1}{8} \tag{2.68}$$

It is easily shown that expansion of (2.68) to third order in q_1 and q_2 generates the Henon–Heiles Hamiltonian (2.66). The trajectories corresponding to (2.68) were studied by Ford.[18] The results, presented as surfaces of section, are shown in Fig. 2.16. Given the exponential nonlinearity of H, the most plausible guess as to the motion under H is that the only isolating integral is the energy. That expectation is wrong! The surfaces of section for all energies clearly show evidence of periodic behavior and two trajectories started near one another separate only linearly in time (Fig. 2.17). Although this behavior was discovered from numerical integration of the equations of motion, it was later shown by

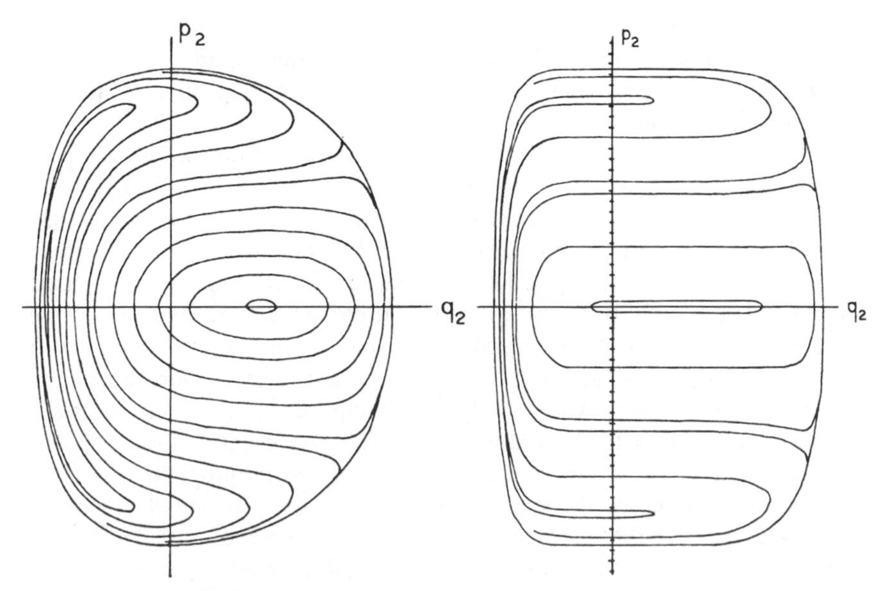

Fig. 2.16. Surface of section for the Toda system. On the left $E=1$, on the right $E=256$. (From J. Ford, *Fundamental Problems in Statistical Mechanics*, Vol. 3, E. D. G. Cohen, ed., North-Holland, Amsterdam, 1975, p. 215).

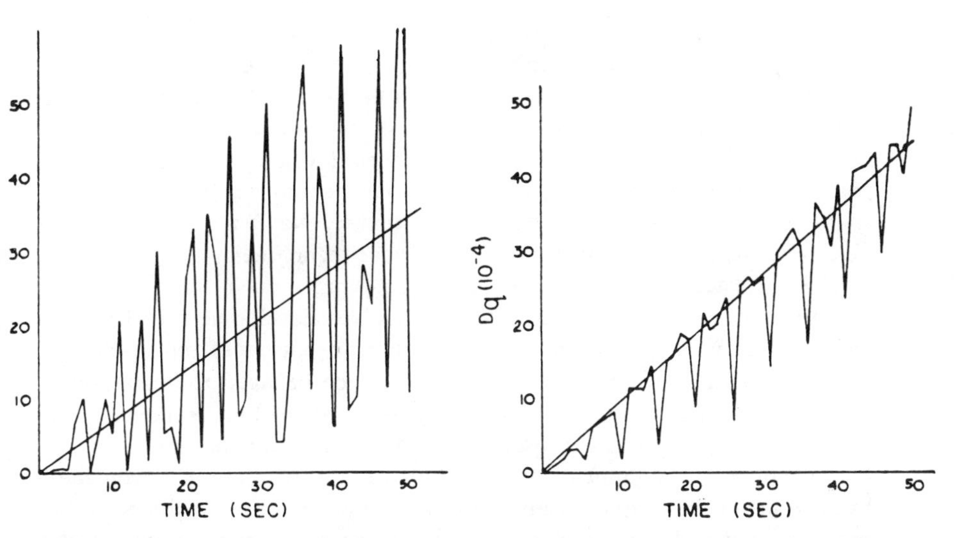

Fig. 2.17. The separation vs time in momentum and position space for two initially close trajectories in the Toda system. (From J. Ford, *Fundamental Problems in Statistical Mechanics*, Vol. 3, E. D. G. Cohen, ed., North-Holland, Amsterdam, 1975, p. 215).

Henon that other isolating integrals of the motion do exist, and that (2.68) corresponds to a completely integrable case, despite the nonlinearity.

C. Perturbation Theory

Given that we wish to study the dynamics of some system of nonlinear oscillators, described by a nonintegrable Hamiltonian, can perturbation theory be used to find invariants of the motion? A vast literature deals with the attempt to answer this question.[19] The straightforward application of perturbation theory to the solution of the classical equation of motion, that is expansion of the solution as a power series in some small parameter about a reference solution for a separable system, followed by successive solution of the equations for the first, second,..., nth order corrections, runs into a serious difficulty. In the nonlinear oscillator system the frequencies depend on the amplitude of motion or, put another way, the nonlinearities alter the frequencies of the system. The existence of this frequency shift is not accounted for in the individual terms in the simplest form of perturbation theory, with the consequence that secular terms arise in the perturbation expansion. A secular term is one that grows linearly with time. Clearly, if any such term appears in the perturbation solution that solution can be valid only for a very limited time. The perturbation theory approach also suffers from the difficulty that successive terms in the perturbation series contain denominators of the form $m_i\omega_i - m_j\omega_j$ which, for ω_i/ω_j irrational, become arbitrarily small for those values of $m_j/m_i = r/s$ that closely approximate the frequency ratio. The nature of this second difficulty has been discussed in the preceding sections. Systematic methods of dealing with the secular problem have been developed by many investigators; an excellent description of the various techniques, and the relationships between them, can be found in the text by Nayfeh.[19] I will mention here only the ideas involved in a few of these techniques.

The Lindstedt–Poincaré method of removing secular terms from the perturbation expansion of the solution involves development of both the system frequencies and amplitudes of motion in powers of the strength of the perturbation, which measures the difference between the equation of motion of interest and that of an integrable system. For example, in the case of a single nonlinear oscillator the perturbation parameter is the coefficient of the nonlinear term in the equation of motion. Given these expansions of the frequency and amplitude, to each order in the perturbation parameter the coefficients are chosen so as to remove the secular terms in the perturbation expansion of the amplitude.

The method of multiple time scales, developed by Storock,[21] Frieman,[22] Nayfeh,[23] and Sandri,[24] is based on the observation that the perturbation expansion of the solution to the equation of motion implies that the system can be thought of as having many different dynamical time scales, which are of the order of $\epsilon t, \epsilon^2 t, \epsilon^3 t, \ldots$. This follows from the way in which the time domain for which a solution is valid is systematically extended in the order-by-order removal of secular terms via the Lindstedt–Poincaré method. Given that the solution of the equation of motion depends explicitly on $t, \epsilon t, \epsilon^2 t, \ldots,$ as well as ϵ itself, the time derivative in the equation of motion is transformed according to

$$\frac{d}{dt} = \frac{\partial}{\partial t} + \epsilon \frac{\partial}{\partial(\epsilon t)} + \epsilon^2 \frac{\partial}{\partial(\epsilon^2 t)} + \cdots \tag{2.69}$$

each term representing a successively slower variation of the system. Substitution of this form of the time derivative, and the expansion of the solution

$$x(t;\epsilon) = \sum_{m=0}^{M-1} \epsilon^m x_m(t, \epsilon t, \ldots, \epsilon^m t) + \mathcal{O}(\epsilon^M t), \tag{2.70}$$

into the equation of motion, followed by equating coefficients of like powers of ϵ, leads to a set of equations for the x_m. The solutions to the equations for the x_m contain arbitrary functions of the m time scales $\epsilon^m t$;

the condition $(x_m/x_{m-1})<\infty$ for all t, $\epsilon t,\ldots,\epsilon^m t$ provides the closure needed to determine the x_m, and it is equivalent to requiring the elimination of the secular terms in the perturbation expansion of the solution to the equation of motion.

Leaving aside the difficulties associated with the appearance of secular terms and small denominators there are two principle categories into which the many proposed perturbation techniques fall. In the so-called method of averaging, developed by Van der Pol,[25] Krylov, Bogolubov, and Mitropolski,[26] and Kruskal,[27] the time variation of the frequencies of the oscillator system is taken to be small relative to the frequencies themselves. When this is the case the drift of the frequencies with time can be separated from the oscillatory motion by transformation to a new set of variables. This transformation consists of a systematic expansion of the variables in powers of a parameter ϵ that describes the deviation of the oscillator system frequencies from the constant values they have in the limit $\epsilon=0$. To lowest order in ϵ the required equations involve an average over the periods of the system frequencies. If the perturbation procedure is carried out as far as terms of order ϵ^n the solution to the original equation which is generated is correct to order ϵ^{n+1}, and that solution is valid for a time of the order of ϵ^{-1}. Thus, this method yields a solution that is asymptotically correct as $\epsilon\to 0$.

In another method, associated with the contributions of Birkhoff,[16] von Zeipel,[28] Hori,[29] Deprit,[30] and Kamel,[31] the Hamiltonian of the system is written in the form

$$H=H_2+\epsilon H_3+\epsilon^2 H_4+\cdots \tag{2.71}$$

where H_2 is purely quadratic in the phase space variables, H_3 is cubic, etc. The procedure used to analyze the motion of the system involves the construction of near identity canonical transformations which reduce (2.71) to a simpler form, for which an invariant can be found. This canonical transformation is carried out step by step, as in the other systematic perturbation theory methods mentioned above.

My purpose in introducing these ideas from perturbation theory is derived from the following questions:

1. Given that an N-oscillator system has quasiperiodic behavior for $E<E_c$ and stochastic behavior for $E>E_c$, are there isolated quasiperiodic solutions embedded in the stochastic domain?
2. If there are not solutions which are asymptotically quasiperiodic, are there solutions which are approximately quasiperiodic for some period of time?
3. If either of the above situations prevails can the corresponding solutions be obtained by the methods of perturbation theory?

These questions are of more than academic interest when one attempts to understand intramolecular vibrational relaxation in isolated molecules.

The results of numerical integration of model equations of motion suggest, but do not establish, that even at the dissociation limit there remain very small regions of quasiperiodic motion embedded in the stochastic domain (see Fig. 2.3). A more satisfying, but not quite complete answer to question 1 is contained in the work of Birkhoff;[16] he proved the existence of an infinite number of periodic orbits near every homoclinic point. Danby showed,[32] by numerical calculation for a model system, that all these orbits originate from stable regular families. That is, a stable periodic orbit (represented by invariant points around a central invariant point) becomes unstable as the perturbation to the integrable system grows, yet at the same time there is generated another stable family, and so on. It is possible that the range of the perturbation over which the new family is stable is smaller than the original, and so on, so that the total range of the perturbation over which all the nested descendants of the original stable family is stable, is finite.

The trajectories just described are regular periodic orbits in the sense that they can be generated from the orbits of the unperturbed system by varying continuously their shape as a function of the perturbation. There is, in addition, another class of trajectories, called irregular orbits; these are independent of the regular orbits and they appear only for a value of the perturbation greater than some threshold. The irregular periodic orbits cannot be continuously transformed into the orbits of the unperturbed system. A pair of irregular families starts at the threshold value of the perturbation, $\epsilon_{\min}$, one of them stable and the other unstable. At some larger value of ϵ the stable family generates, in turn a new set of stable and unstable families, and so on. It is possible, as for the regular families, that above some value of ϵ all nested families of orbits are unstable. Finally, the irregular periodic orbits seem to play a role in promoting ergodicity in that they can join regions of phase space that are associated with different regular periodic orbits.

The following two examples illustrate the character of the quasiperiodic and the periodic solutions embedded in the stochastic domain.

Contoupolos[33] has made an extended study of stable orbits for the system with the Hamiltonian ($m=1$)

$$H=\tfrac{1}{2}(p_1^2+p_2^2)+\tfrac{1}{2}(Aq_1^2+Bq_2^2)-\epsilon q_1 q_2^2 \tag{2.72}$$

This system has also been studied by Barbanis[34] and, like the Henon–Heiles system, has a low energy quasiperiodic domain and a higher energy stochastic domain. The Hamiltonian permits dissociation at the energy

$$E_D = \frac{A^3}{8\epsilon^2} \tag{2.73}$$

for the case $A=B$. The onset of stochastic behavior is a relatively insensitive function of ϵ. It is found that $E_c/E_D \approx 0.76$ for $0.05 < \epsilon < 0.20$ when $A=0.1$. Contoupolos studied the stable orbits of this system as a function of ϵ. He found that stable orbits existed deep inside the region where stochastic motion is the dominant behavior. In somewhat more detail, he concluded that:

"(a) The main families of periodic orbits for very small perturbations ϵ are only: (1) the $\bar{x} \equiv A^{1/2}x$ axis and (2) the "central" periodic orbit near the $\bar{y} \equiv B^{1/2}y$ axis, intersecting the $\bar{x}$ axis perpendicularly at one point, if the ratio $A^{1/2}/B^{1/2} =$ irrational. If $A^{1/2}/B^{1/2}$ is a rational n/m then we also have two resonant periodic orbits (one stable and one unstable) making n and m oscillations along the $\bar{x}$ and $\bar{y}$ axes, respectively. If $A^{1/2}/B^{1/2}$ is "near" a rational n/m then resonant periodic orbits appear as the perturbation increases beyond a minimum value $\epsilon > 0$.

(b) As ϵ increases more and more, families intersecting the $\bar{x}$ axis less than a given number of times appear. If ϵ is not very large most of these families are regular, i.e., they originate at the stable main families above. Families originating at the "central" family or the stable resonant family of periodic orbits have their minimum (in other problems their maximum) ϵ at the point of intersection; i.e.,, the tangent of the characteristic of the new family is, in most cases, perpendicular at the point of intersection.

(c) Every regular periodic orbit makes n and m oscillations along the $\bar{x}$ and $\bar{y}$ axes, respectively, and has a rotation number $n'/m = 2 - n/m$ [more generally $n'/m = -n/m$ (modulo 1)].

(d) From every stable family intersecting the $\bar{x}$ axis m times, new families with the same rotation number are formed, intersecting the $\bar{x}$ axis $2m$ or $3m$, etc., times. These are the families of second "genre" according to the terminology of Poincaré.

(e) Besides the regular families of periodic orbits we have many irregular families not connected with the main families above. Such families are in general, but not always, unstable. In general such irregular families are composed of two branches in quite different parts of the diagram $(\epsilon, \bar{x})$. In many cases, one cannot correspond unambiguously a rotation number to them, but even if we can assign them a rotation number, it is not adjacent to the rotation number of nearby regular families.

(f) Irregular families of orbits intersecting the $\bar{x}$ axis less than a given number m of times appear only for relatively large ϵ, larger than the perturbation at which the corresponding regular families (with the same m)

appear. The appearance of irregular families seems to be connected with the "dissolution" of the invariant curves of nonperiodic orbits.

(g) Beyond the escape perturbation, i.e., for $\epsilon > \epsilon_{esc}$ there are many escape regions, containing escaping orbits. The range of values of $\bar{x}$ for which we have escaping orbits increases as ϵ increases. However, there are left some nonescape regions containing periodic orbits. We may even have stable periodic orbits (and nearby tube orbits of positive measure) for ϵ much larger than the escape perturbation. Further, every orbit escaping after a very large time may be considered nonescaping for practical reasons. We conclude that if, in a given potential field, a moving point has energy greater than the escape energy (but not extremely large), it may not escape for very long times or for all times.

(h) Near the escape regions, a large number of new families of periodic orbits appear. Therefore the transition between orbits escaping too fast (after, say, 1 intersection with the $\bar{x}$ axis, except the initial point) and nonescaping periodic orbits is too abrupt. There are not many intermediate cases of orbits escaping after 2, 3,... intersections, until we reach nonescaping orbits. However, these nonescaping orbits are highly unstable. Therefore small perturbations may make them escaping.

(i) No family of periodic orbits seems to end at an escape orbit or at the family $\bar{y}=0$, or another family. All families seem to extend up to $\epsilon=\infty$. This is the case whenever $A^{1/2}/B^{1/2} > 2/3^{1/2}$. If $A^{1/2}/B^{1/2} > 2/3^{1/2}$, families appearing for small ϵ originate at the family $\bar{y}=0$, reaching their maximum ϵ at or near the "central" family. Even in this case, however, these are families extending to large values of ϵ, and presumably up to $\epsilon=\infty$."

In the system of units used by Contoupolos, the escape perturbation is $\epsilon_{esc}=4.602$. Figures 2.18–2.21 show in increasing detail the nature of the orbits, and the intermingling of stable and unstable behavior deep in the stochastic region. Figure 2.22 shows the behavior of some orbits well above the escape perturbation.

Helleman and Bountis[35] have developed a rapidly convergent variational method for the construction of periodic solutions of the equation of motion, analytic in the time, and with arbitrary period. When applied to the Henon–Heiles system they obtain what appears to be a dense set of one parameter families of periodic solutions wherever the motion is bounded, including the domain where stochastic trajectories dominate the surface of section. The approach to finding periodic solutions is based on the following idea. Consider a system with two degrees of freedom. Instead of specifying an orbit in the usual manner, that is by the initial positions and momenta, one can, for closed orbits, specify the fundamental frequencies and initial phases. When a closed orbit describes the motion in a system with two degrees of freedom the fundamental frequencies are commensurable; the ratio of the frequencies, $m_1/m_2 \equiv \sigma$, is then a rational

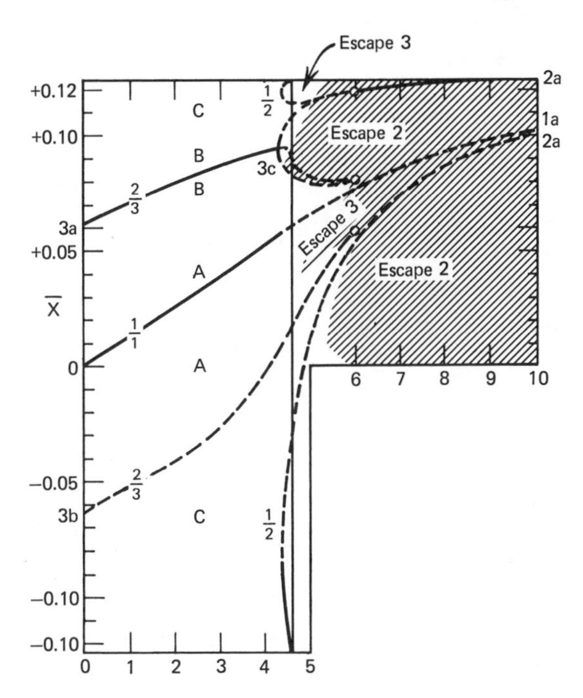

Fig. 2.18. Characteristic curves of the four main families of periodic orbits and the family 1/2: The family 1/1 gives the central periodic orbits intersecting the $\bar{x}$ axis only once perpendicularly. The boundaries $\bar{x} = \pm 0.12363$ represent the family $\bar{y}=0$. The families 2/3 represent resonant periodic orbits with rotation number 2/3. Solid lines represent stable orbits, dashed lines unstable ones. Periodic orbits in the region A, B, C are generated from the families 1/1, the stable family 2/3, and $\bar{x}=0$, respectively. The vertical straight line at $\epsilon = 4.602$ represents the escape perturbation. The dashed regions represent orbits escaping before the 2nd or 3rd intersection with the $\bar{x}$ axis (besides the initial point). [From G. Contoupolos, *Astron. J.*, **75**, 96 (1970).]

fraction and the Poincaré recurrence time of the system is just $T_r = \omega_r^{-1} = m_2\omega_1^{-1} = m_1\omega_2^{-1}$ where ω_r is the greatest common frequency of the system. Helleman and Bountis specify the periodic orbits by the value of T_r and the numbers $2m_1$ and $2m_2$ of zeroes of the velocities within one recurrence period. For given m_1 and m_2 there is an infinite one parameter family of frequency pairs; the parameter is ω_r, i.e., $\omega_1 = m_2\omega_r$ and $\omega_2 = m_1\omega_r$. The solution to the equation of motion is developed from Fourier series of the form

$$
\begin{aligned}
q_1(t) &= \sum_n A_n e^{in\omega_r t} \\
q_2(t) &= \sum_n A'_n e^{in\omega_r t}
\end{aligned}
\tag{2.74}
$$

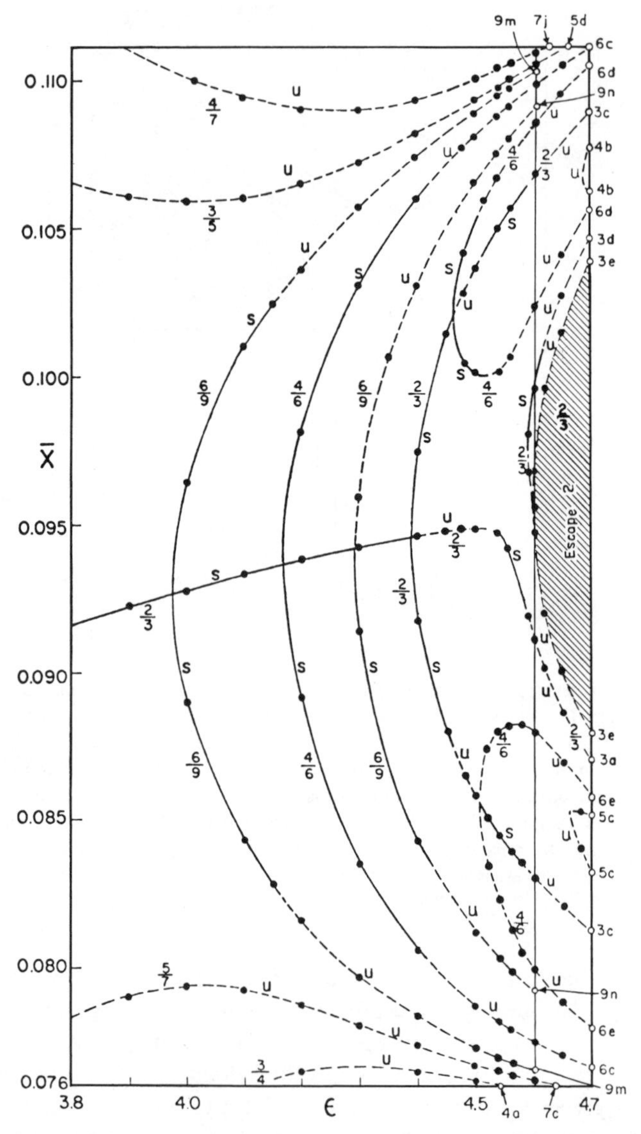

Fig. 2.19. Part of Fig. 2.18 in detail. The vertical line at $\epsilon = 4.602$ represents the escape perturbation. [From G. Contoupolos, *Astron. J.*, **75**, 96 (1970).]

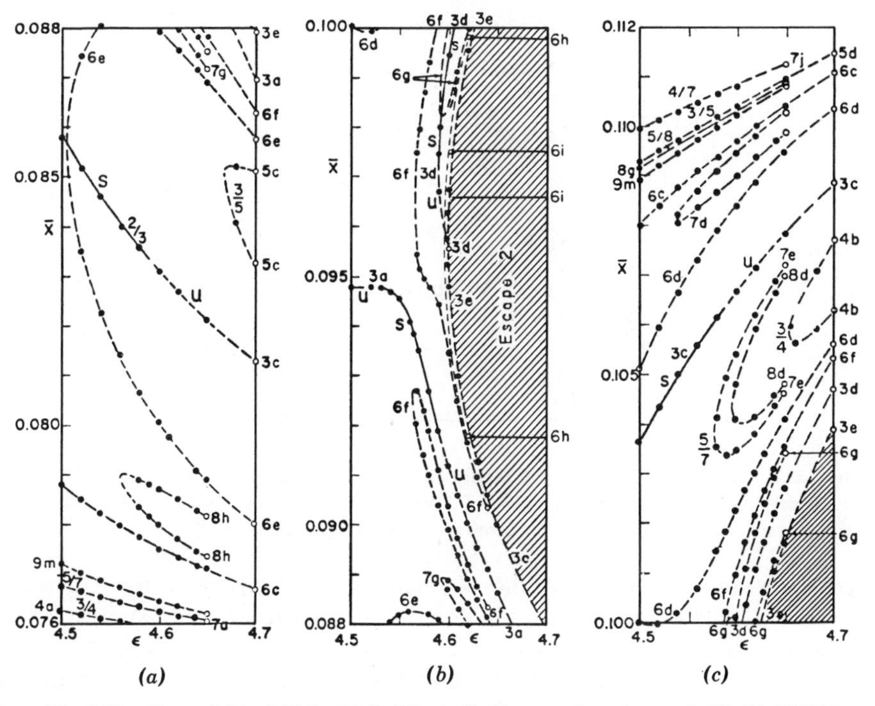

(a) *(b)* *(c)*

Fig. 2.20. Parts of Fig. 2.19 in detail. [From G. Contoupolos, *Astron. J.*, **75**, 96 (1970).]

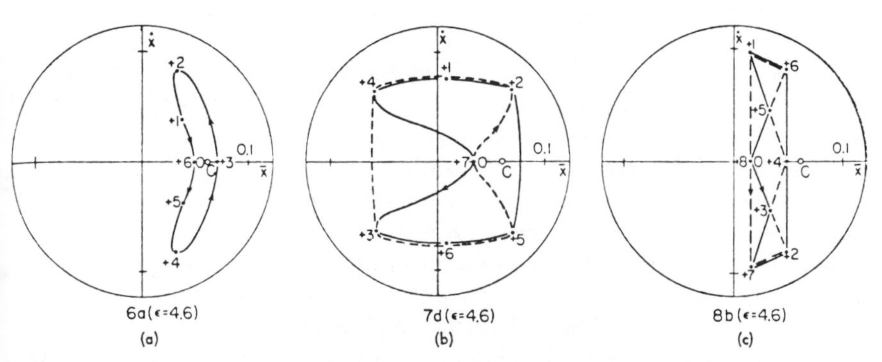

Fig. 2.21. The points of intersection of three periodic orbits of the families 6a, 7d, 8b by the surface of section $\bar{y}=0$ (of the space $\bar{x}\dot{x}\dot{\bar{y}}$) for $\dot{\bar{y}}>0$. The initial point is 0 and successive points are numbered. C represents the "central" periodic orbit. The outermost circle is $\bar{x}^2+\dot{x}^2=2E$. The points of intersection may be joined by one or more smooth curves. [From G. Contoupolos, *Astron. J.*, **75**, 96 (1970).]

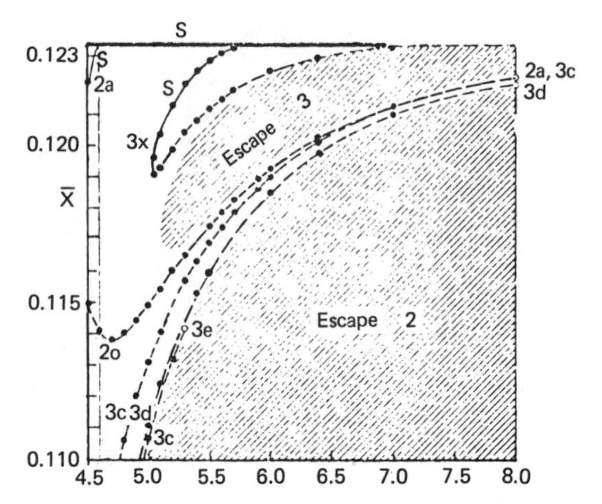

Fig. 2.22. Characteristic curves of the families 2a, 3c, 3d, 3e, and 3x from $\epsilon = 4.5$ up to $\epsilon = 8$: The family $\bar{y} = 0$ is stable up to about $\epsilon = 7.0$, while the family 3x contains a stable branch from its minimum $\epsilon(\epsilon \simeq 5.0)$ up to the termination (starting) point $(\epsilon \simeq 5.8)$.

Helleman and Bountis find the coefficients in (2.74) by a convergent iterative procedure. Of course, the dominant coefficients in (2.74) must be those associated with the fundamental solution $\omega_1 = m_2\omega_r$, $\omega_2 = m_1\omega_r$. From the solutions found one works backwards, for fixed m_2/m_1, to find the initial positions and momenta. Note that, by restricting attention to those families of periodic orbits parameterized by T_r with fixed $m_2/m_1 \equiv \sigma$, the Fourier series representation of the solution is free of problems associated with small divisors.

Figure 2.23 shows some of the results obtained by Helleman and Bountis for the Henon–Heiles system with the small amplitude frequencies $\omega_1 = \omega_2$. The figure shows the loci of periodic solutions for constant σ and varying T_r; along each curve T_r varies continuously, but T_r changes discontinuously as one goes from one value of σ to a neighboring value. The sensitivity of T_r to changes in σ is extreme, the discontinuous jumps being of many orders of magnitude in T_r for miniscule changes in σ. Of course, the curves for $\sigma \neq 1$ represent those periodic solutions where, because the frequencies depend on the amplitudes of motion, $\omega_2 \neq \omega_1$ even though the equality holds in the limit of small amplitude of motion. Note that the curves for $\sigma \neq 1$ bifurcate off that for $\sigma = 1$, and are symmetric with respect to the equipotentials of H (the dotted lines in the figure). Helleman and Bountis find that the curves for different σ cover the $q_1^\circ q_2^\circ$ plane densely; a σ curve can be constructed arbitrarily close to any point of the plane by σ interpolation using only rational values of σ, since the ordering of the nested σ curves varies "continuously" with σ.

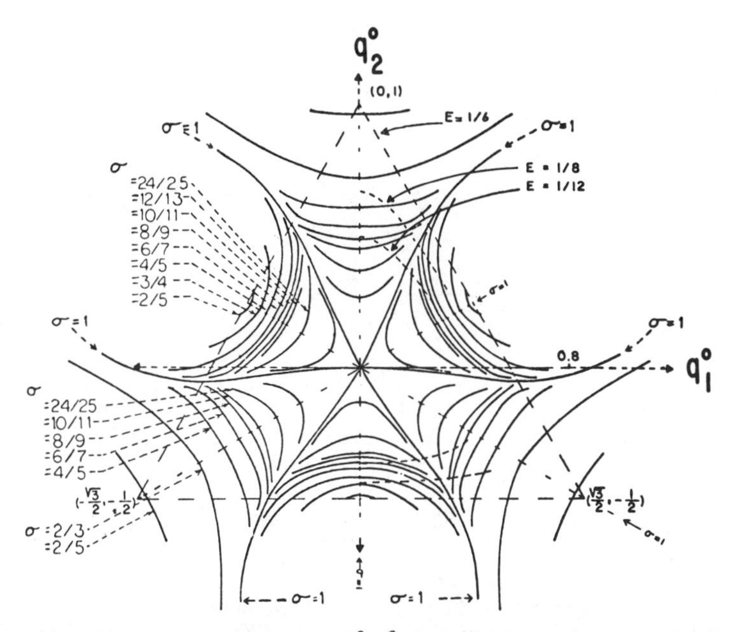

Fig. 2.23. Converged initial displacements q_1^0, q_2^0 of the periodic solutions of the Henon-Heiles system [with $p_1(0)=0=p_2(0)$] at different values of $\sigma \equiv m_2/m_1$; only the "primarily" solutions are used (i.e., m_2, m_1 relative prime). Note the confluence of σ curves (near $\sigma=1$) resulting in a "sensitive dependence on initial conditions" there, yielding "stochastic regions." (From R. Helleman and T. Bountis, *Stochastic Behavior in Classical and Quantum Hamiltonian Systems*, G. Casati and J. Ford, eds., Springer-Verlag, Berlin, 1979, p. 353.)

The preceding example addresses the question of the existence of quasiperiodic trajectories deep in the stochastic domain, but not whether these trajectories can be calculated by perturbation theory. Gustavson[36] has developed an algorithm for constructing formal integrals of a Hamiltonian system with N degrees of freedom, and applied this to the Henon–Heiles system. The method uses a succession of canonical transformations to reduce the Hamiltonian to normal form. For a given system, truncated power series then represent approximate integrals of the motion. Some results are shown in Figs. 2.24–2.26, which display the Poincaré surfaces of section for $E=1/12$, $1/8$ and $1/6$, respectively. These figures should be compared with Figs. 2.11–2.13 in the same order. We note that at low energy, say $E=1/12$, the perturbation method yields results in excellent agreement with direct numerical integration of the equation of motion. At higher energy, say $1/6$, the perturbation method suggests the existence of quasiperiodic trajectories not found in the numerical integration of the equation of motion.

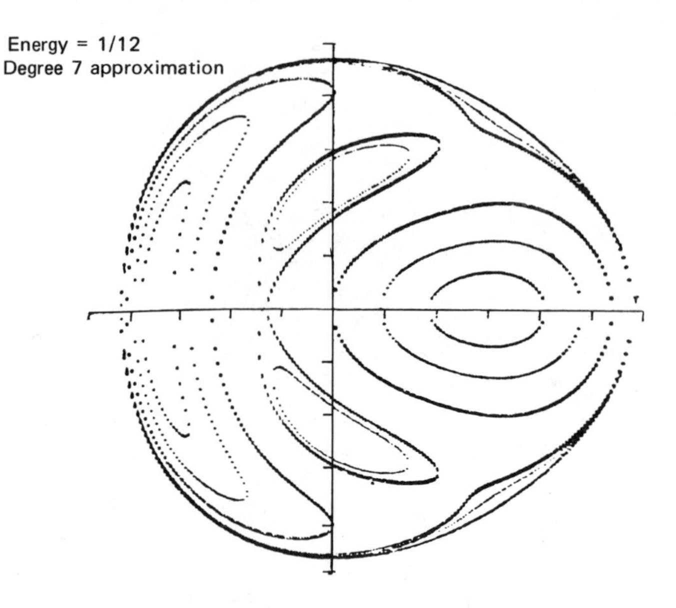

Fig. 2.24. Surface of section for the Henon–Heiles system calculated by successive canonical transformation to Birkhoff normal form. $E = 1/12$. [From F. G. Gustavson, *Astron. J.*, **71**, 670 (1966).]

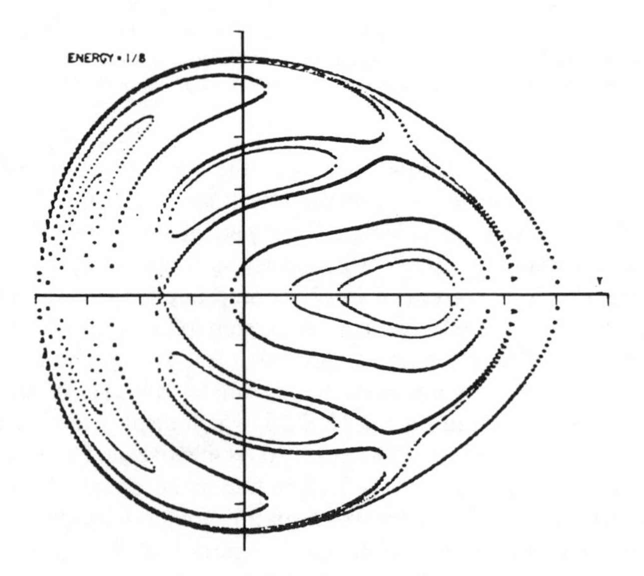

Fig. 2.25. Surface of section for the Henon–Heiles system calculated by successive canonical transformation to Birkhoff normal form. $E = 1/8$. [From F. G. Gustavson, *Astron. J.*, **71**, 670 (1966).]

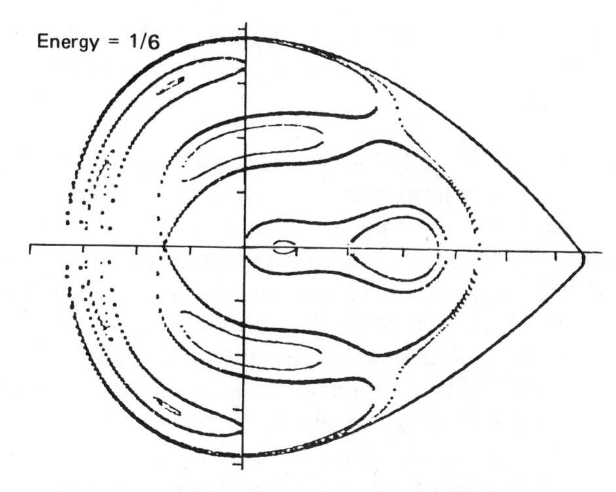

Fig. 2.26. Surface of section for the Henon–Heiles system calculated by successive canonical transformation to Birkhoff normal form. $E = 1/6$. [From F. G. Gustavson, *Astron. J.*, **71**, 670 (1966).]

What interpretation should be attached to the perturbation theory results in the high energy domain? The key point to make is that a truncated power series representation of a Hamiltonian, which is an approximation to some "complete" Hamiltonian, can correspond to an integrable system even when the complete Hamiltonian does not. For example, the KdV equation arises as an approximation to the hydrodynamic description of water waves in a shallow canal, and in other cases as well as an approximation to the exact dynamics of the process. It can be shown that the KdV equation has solitary wave solutions, which are invariant wave form motions in the system, and these are seen in real experiments, hence the approximation is a good representation of the phenomenon described. Similarly, it appears that the truncated Gustavson Hamiltonian, which is an approximation to the complete Henon–Heiles Hamiltonian, describes a system which has quasiperiodic motion where the Henon–Heiles system dynamics is dominantly stochastic. It is tempting to speculate that these quasiperiodic motions are proper solutions to the equations of motion, being part of the small set of quasiperiodic solutions embedded in the stochastic domain. If so, the use of perturbation theory to calculate invariants of the motion is akin to analytic continuation of the invariant solutions. This point of view reappears when the quantum mechanics of motion in the stochastic region is analyzed.

D. Summary

In this section I have discussed the classical mechanics of coupled nonlinear oscillator systems. Although some aspects of the analysis were quite general, the detailed illustrations given focused attention on the two oscillators system, a point to which I will return below.

It has been shown that:

1. A separable N-oscillator system has at least N isolating integrals of the motion, but a nonseparable system will in general have less than N isolating integrals of the motion. Of course, given that the dimensionalities of the separable and nonseparable systems are the same, the total numbers of integrals of the motion of each are the same; the discrepancy is accounted for by nonisolating integrals of the motion.
2. The motion of the separable system is quasiperiodic, the trajectory being restricted to lie on a torus in the phase space of the system. When the system is not separable some of the trajectories are quasiperiodic, but not all. These other trajectories are not restricted to lie on a torus, although they, of course, do lie on the energy surface.
3. In the case that the system is separable the motion can be described simply in terms of action-angle variables and the corresponding Hamilton–Jacobi equation possesses a complete integral that generates the transformation of representation into angle-action coordinates. In contrast, the Hamilton–Jacobi equation for the nonseparable system does not have a complete integral which can generate a transformation to angle-action variables; in this case angle-action variables do not exist.
4. The KAM theorem establishes that most of the quasiperiodic trajectories of some unperturbed system survive under sufficiently small perturbation, but that there is also generated a set of stochastic trajectories. The relative weight of the stochastic trajectories grows as the energy increases, and they eventually fill all of the accessible phase space (the energy surface).
5. The onset of stochastic behavior appears to be relatively sharp, at an energy E_c, but in fact there are some stochastic trajectories even at low energy, and there are periodic and quasiperiodic trajectories embedded in the stochastic domain.
6. Even when periodic trajectories exist in the stochastic domain they do not reflect global invariants of the system because their properties, e.g., periods of the orbits, are extremely sensitive to very small changes in the initial conditions.

There is a difference between systems with two and systems with more than two degrees of freedom. In the former case stochastic trajectories can be isolated from one another in the system phase space by the intervention of a torus supporting a quasiperiodic motion. In the latter case, assuming the energy to be the only isolating integral of the motion, the quasiperiodic trajectories do not isolate the stochastic trajectories from one another. All of the regions of the phase space in which stochastic motion occurs are connected into a network which permeates the entire phase space (called the Arnold web).[8] Elements of the Arnold web are arbitrarily close to every point of the phase space, hence if the initial conditions start the phase point on the Arnold web the motion of the point will be stochastic and will eventually reach every region of the energy surface. The process by which this takes place, namely along a stochastic trajectory inside the Arnold web, is called Arnold diffusion. It is generally believed that the Arnold web covers the energy surface whenever $N>2$, but Contoupolos, Galgani and Giorgilli[37] have shown that in a three oscillator version of the Henon–Heiles system there are noncommunicating stochastic regions of the phase space. One of these stochastic regions is large, the others small. In addition there are regions of the phase space in which the motion is quasiperiodic. Using a perturbation theory construction of integrals of the motion, Contoupolos et al. conclude that there are two isolating integrals of the motion other than the energy in the regions of quasiperiodic trajectories, one extra isolating integral of the motion in the small regions with stochastic trajectories, and no extra isolating integrals of the motion in the large region with stochastic trajectories.

Clearly, there remains an enormous amount to be learned about the classical mechanics of N-oscillator systems.

III. QUANTUM-MECHANICAL DESCRIPTION OF COUPLED NONLINEAR OSCILLATOR SYSTEMS

The results of studies of the classical mechanics of coupled oscillator systems leads, obviously, to three questions about the quantum mechanical description of the same systems. These are:

1. Is there a change in the description of the stationary states of a system of coupled oscillators at some energy E_c analogous to the change in classical trajectory from quasiperiodic to stochastic?
2. Assuming the answer to 1 is yes, what is the nature of the stationary states for $E>E_c$, and how are these related, if at all, to properties of classical trajectories in the stochastic domain?
3. If intramolecular redistribution of energy is regarded as one of several competing processes, how is its rate influenced by the nature of the states when $E>E_c$?

Several approaches to answering these questions are discussed in the following text.

A. The Nordholm–Rice Analysis

The first step in searching for an analog of the KAM transition from quasiperiodic to stochastic behavior in a quantum-mechanical description of a coupled oscillator system is the selection of a criterion to signal that transition. It turns out that the choice of criterion is not trivial, since it quickly drags into the analysis rather esoteric, yet important, properties of quantum states. The existing formulations of quantum ergodic theory provide no assistance for the situations of interest to us, so at present the necessary criterion is postulated on the basis of analogies with the behavior of systems described by classical mechanics. Nordholm and Rice[38] suggest that the important question is how an initially localized excitation spreads over the energy surface of the system, and they propose to classify a state of the system as ergodic if an excitation initially not uniformly distributed over the energy surface becomes uniformly distributed as $t \rightarrow \infty$. This suggestion is deliberately constructed to be in direct correspondence with the usual definition of ergodicity on a classical energy surface.

Implementation of the Nordholm–Rice criterion is based on the principle that the eigenvalues of an arbitrarily complicated Hamiltonian can be computed by use of an expansion in a complete set of basis functions and evaluation of matrix elements, though an actual calculation can be technically difficult and very tedious. Given this methodology, the Nordholm–Rice criterion for the onset of stochastic behavior has the advantage of being easily tested for any given set of basis states. It has the disadvantage that the conclusion of a test for this kind of ergodicity is, in general, basis dependent. Of course, a basis dependence can be connected to reality if there exists an excitation mechanism that prepares the system in one or more of the basis states.

The reader is referred to the paper by Nordholm and Rice[38] for details of the analysis and also for an expanded discussion of the difficulty of selecting a criterion for the transition to stochastic behavior in a quantum-mechanical description of a coupled oscillator system.

It should be noted that the Henon–Heiles system, and several others of those studied, do not have bound states in the quantum-mechanical description. These systems do have resonances in the region of the potential energy surface where there would be classical bound states. In the Nordholm–Rice calculations, and those of Stratt, Handy and Miller,[39] Pomphrey,[40] Jaffe and Reinhardt,[41] and all others that I know of, the distinction between resonances and bound states in these systems has been neglected. The calculations which employ a basis set expansion use functions that span only that part of the space where classical bound states exist, and the several semiclassical quantization schemes use a generalized

Bohr–Sommerfeld–Wilson rule on action integrals defined with respect to periodic or quasiperiodic orbits generated from classically bound orbits of a reference system. Thus, these calculations approximate resonances with stationary state functions in a truncated Hilbert space. It is very unlikely that this procedure introduces any error, since the same kind of behavior is found for systems with bound states, e.g. Morse oscillators.

Nordholm and Rice studied several model systems of coupled nonlinear oscillators, including the Henon–Heiles system so thoroughly described in Section II. Some of the systems have only algebraic nonlinearities; others have exponential nonlinearities such as is characteristic of a Morse potential function. Some of the examples have degenerate small amplitude modes; others have nondegenerate small amplitude modes. The basis states of the representation were, for each system studied, chosen to be the states of the corresponding decoupled set of harmonic oscillators. The calculations lead to the following conclusions:

1. There is nonergodic behavior, in general, below a critical energy (possibly a critical energy region). The nonergodicity is more marked and persistent for nondegenerate systems than for equivalent degenerate systems.
2. For degenerate systems there are occasional global states interspersed in the local states. (A global state is one with amplitude uniformly distributed over the energy surface.)
3. For low energy, typically less than half the dissociation energy E_D, the asymptotic distribution of amplitude over the equienergetic basis states is not very sensitive to the coupling.
4. Initial states with comparable excitation in all oscillators tend to evolve to global states, whereas extremal initial states with excitation mostly localized in one oscillator tend to evolve to local states.
5. For high energy, typically above half the dissociation energy, the asymptotic distribution of amplitude over the basis states is very sensitive to the coupling. In this case the final states achieved tend to have mixed character (e.g., a wide but uneven spread of overlaps).
6. From a very crude analysis of the time evolution of the model systems it is estimated that in these cases it takes of the order of one to ten vibrational periods for an initially localized nonstationary state to achieve its asymptotic form.

Clearly, the quantum dynamics of coupled oscillator systems is analogous to the classical dynamics of these systems. In particular, a KAM-like transition exists. Conclusion (4) is very like the prediction derived from a classical mechanical analysis, namely, that if energy is initially localized in one oscillator it requires a higher total energy to reach the region of overlap of nonlinear resonances than if the same initial energy is spread over several oscillators.[42]

What is the character of the stationary states in the ergodic region? One answer to this question is given by the coefficients in the basis set expansion. A more graphic description is obtained from the nodal patterns of the wave functions. For one-dimensional systems the nodes of the wavefunction and the quantum number of the state have a one-to-one correspondence. What are nodal points in the one-dimensional case become nodal surfaces in higher dimensional cases. Consider, for simplicity, a two-dimensional system, in which case there is some specific pattern of nodal lines which corresponds to a given eigenstate. For example, if the two-dimensional system is separable these nodal lines form a regular rectangular grid (Fig. 3.1a). In contrast, Pechukas[43] showed that if the

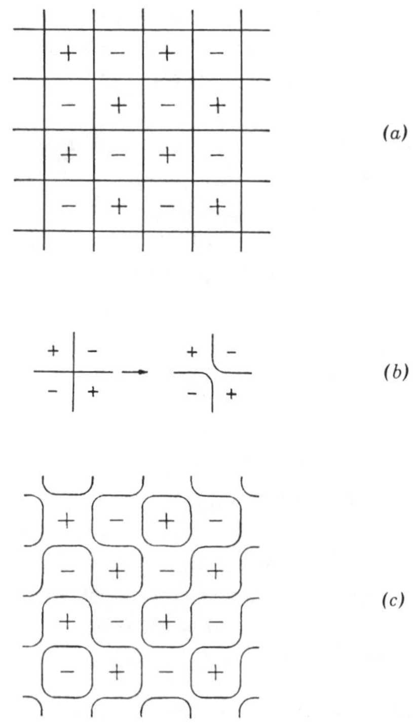

Fig. 3.1. Schematic representation of the nodal lines for a two oscillator system. Case (a) corresponds to separable oscillators, case (b) shows how introduction of interoscillator coupling distorts a nodal crossing, and case (c) is a possible nodal pattern for coupled oscillators.

two-dimensional system is not separable the nodal lines need not, in general, intersect to form a simple grid; instead there can be avoided intersections such as shown in Fig. 3.1b, leading to an overall pattern like that shown in Fig. 3.1c. It is an obvious suggestion that the nodal lines will intersect in the domain where the classical trajectories are quasiperiodic, and will have some complicated, broken-up pattern in the domain where the classical trajectories are stochastic. Stratt, Handy and Miller[39] have examined the nodal structure of the eigenfunctions of the Barbanis Hamiltonian

$$H=\tfrac{1}{2}(p_1^2+p_2^2)+\tfrac{1}{2}(1.6q_1^2+0.9q_2^2)-0.08q_1q_2^2 \tag{3.1}$$

for which the small amplitude frequencies are obviously $(1.6)^{1/2}$ and $(0.9)^{1/2}$. The method of computation was that of Nordholm and Rice, namely expansion in the harmonic oscillator functions of the uncoupled counterpart of (3.1), generated by the first two terms of (3.1). Some of the results are shown in Figs. 3.2 and 3.3. In these diagrams the regions in which the wavefunction is positive are black, and the regions in which the wavefunction is negative are blank. The remaining dotted regions are those in which the wavefunction is very small, hence noise in the numerical calculation prevents assignment of a definite sign. For the pattern shown in Fig. 3.2, which is in the domain where the corresponding classical

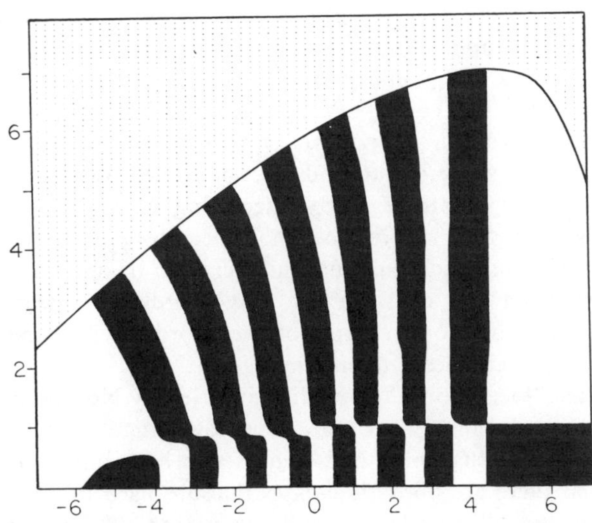

Fig. 3.2. The nodal pattern for an eigenstate of the Hamiltonian (3.1) at an energy below the KAM-like transition.

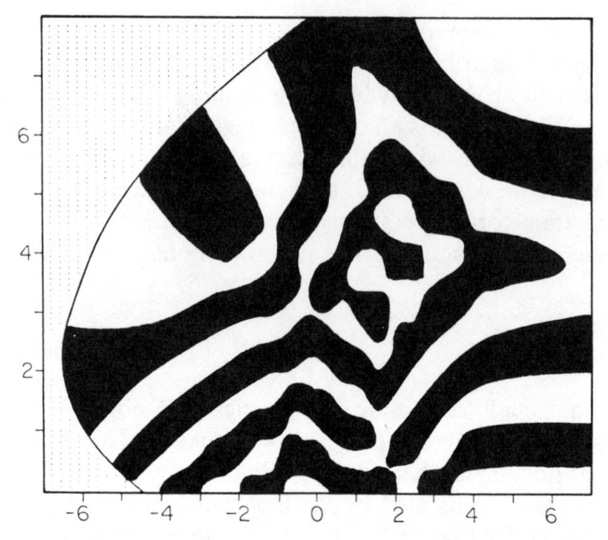

Fig. 3.3. The nodal pattern for an eigenstate of the Hamiltonian (3.1) at an energy above the KAM-like transition. Figures 3.1, 3.2, and 3.3 from R. M. Stratt, N. C. Handy, and W. H. Miller, *J. Chem. Phys.* **71**, 3311 (1979).

trajectory is quasiperiodic, it is easy to visualize two curved orthogonal axes along which black and blank spaces alternate. When this is possible the numbers of nodes corresponding to the separable coordinates (by construction the curved axes are orthogonal) give the quantum numbers of the independent components of the eigenstates. On the other hand, the distribution of black and blank spaces in Fig. 3.3, which is at an energy in the domain where the corresponding classical trajectory is stochastic, seems to involve several nodal patterns superimposed. In this case it is not easy, and possibly impossible, to find a set of curved orthogonal axes that will neatly separate it into regularly alternating sequences of black and blank spaces. But this failure does not mean one cannot find some coordinate system in which quantum numbers can be assigned (see later), only that whatever the character of that coordinate system it has a complicated representation in terms of the coordinates of the harmonic oscillators of the uncoupled reference system.

Just as in the "local-global" categorization used by Nordholm and Rice, Stratt, Handy and Miller find that some low-lying states have greatly distorted nodal patterns, some high-lying states have almost regular patterns, and that there are borderline cases that are hard to assign to either category. The overall correlation with the nature of the corresponding classical trajectory is greatly strengthened by the observation that the expectation values of q_1^2 and q_2^2 are very large in the stochastic region, and have smaller values (comparable to those expected for harmonic oscillators with quantum numbers n_1 and n_2, respectively) in the quasiperiodic region.

Stratt, Handy and Miller suggest that some of the difficulties associated with the basis set dependence of the Nordholm–Rice criterion for stochasticity can be overcome by transforming to "natural orbitals." In a system with two degrees of freedom the wavefunction corresponding to the Nordholm–Rice choice of basis functions is a bilinear combination of harmonic oscillator functions. The natural orbitals for this system are those for which the wavefunction has a diagonal representation; these can be shown to be just the linear combinations of harmonic oscillator functions that give the most rapid convergence of the expansion of the wavefunction. Stratt, Handy and Miller find that in the quasiperiodic domain one of the natural orbitals dominates the wave function, whereas in the stochastic domain many contribute significantly to the wave function. These results confirm completely the Nordholm–Rice analysis, which was based on the bilinear expansion in the harmonic oscillator basis functions, and suggest that basis function sensitivity of the onset of stochasticity may be a less serious problem than originally thought.

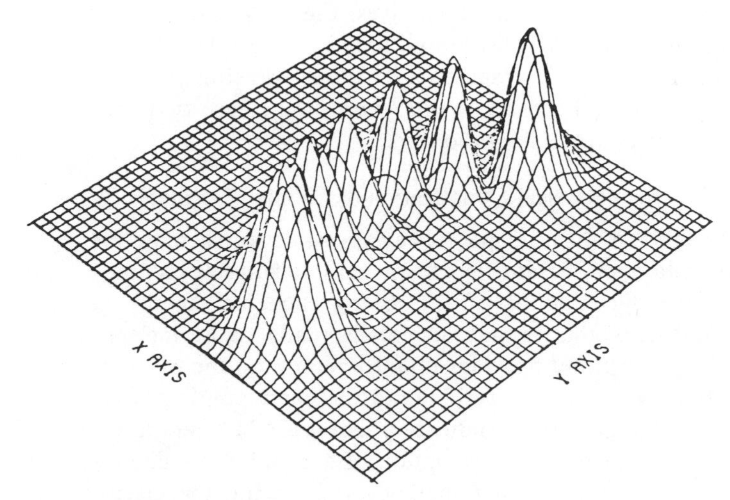

Fig. 3.4. An eigenfunction of the Hamiltonian

$$\hat{H}=\tfrac{1}{2}(p_x^2+p_y^2)+\tfrac{1}{2}(\omega_x^2x^2+\omega_y^2y^2)+\lambda x(y^2+\eta x^2)$$

at an energy where the classical motion is quasiperiodic. In this case $\omega_x=2\omega_y$.

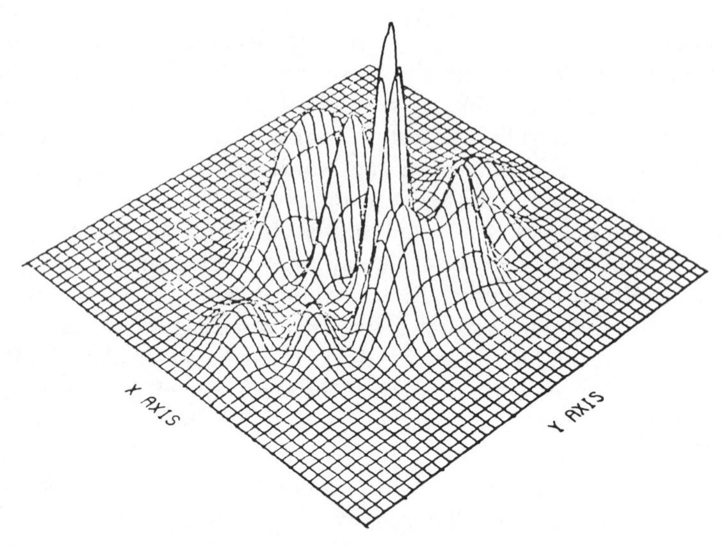

Fig. 3.5. An eigenfunction of the Hamiltonian described in the caption to Fig. 3.4 where the classical motion is stochastic. Figs. 3.4 and 3.5 from R. A. Marcus, D. W. Noid and M. L. Koszykowski in *Stochastic Behavior in Classical and Quantum Hamiltonian Systems*, G. Casati and J. Ford, eds., Springer-Verlag, Berlin, 1979.

As to the wavefunctions themselves, Figs. 3.4 and 3.5 show eigenfunctions of the Henon–Heiles Hamiltonian in the quasiperiodic region and the stochastic region, respectively. These wavefunctions, computed by Marcus, Noid, and Koszykowski,[44] are for the case that the small amplitude frequences are in the ratio $\omega_1/\omega_2 = 2$.

B. Semiclassical Quantization of Coupled Oscillator Systems

Although it is, in principle, always possible to calculate the eigenvalues of a Hamiltonian operator by expansion in a suitable complete set of basis functions and evaluation of the matrix elements thereby generated, for many purposes this is not an intellectually satisfying procedure. For example, this method provides no information on the relationship between the stationary states of a system and the corresponding classical trajectories. Given the qualitative indication that a KAM-type transition occurs in the quantum-mechanical description of a coupled oscillator system, there is reason to examine other methods for finding the eigenvalues of the Hamiltonian operator. The most widely used class of "other methods" is semiclassical quantization. Although the purpose of much of the work on the semiclassical methodology is to calculate eigenvalues as easily as possible, I shall ignore that aspect of the theory. Rather, I shall focus attention on a necessary ingredient of the semiclassical methodology, namely, a scheme for generating, classifying, and selecting periodic and quasiperiodic trajectories of a system. It is just this intermediate part of the semiclassical analysis that can be used to give a vivid connection between the eigenstates of a Hamiltonian operator and the trajectories of the corresponding classical system.

Percival,[45] Berry,[46] Miller,[47] Marcus,[48] Delos,[49] and Reinhardt[41] have advanced the semiclassical theory of bound states. Very important early work by Gutzwiller dealt with periodic orbits.[50] These investigators have adopted the Einstein generalization of the Bohr–Sommerfeld–Wilson quantization condition, namely,[51]

$$J=\frac{1}{2\pi}\oint\sum_{k=1}^{N}p_k\,dq_k=(n+\delta)\hbar \tag{3.2}$$

where the integral over the invariant differential sum $\Sigma p_k\,dq_k$ is along closed curves in coordinate space that need not be classical trajectories. The usual Bohr–Sommerfeld–Wilson condition is defined for each separable coordinate q_k, and the corresponding action integral is taken around one cycle of the motion of the coordinate q_k. The Bohr–Sommerfeld–Wilson condition depends on the choice of coordinates, whereas Einstein's condition, which defines an integration over all N action functions, does not depend on the choice of coordinates. It is readily seen that the Einstein condition can be applied to motion in the region where the trajectory is quasiperiodic, but not above the KAM transition, since in that domain p_k cannot be expressed as a function of the q_k. This has led Percival to classify the spectrum of a system into the categories regular and irregular, the former pertaining to the domain below the KAM transition where the trajectory is quasiperiodic, and the latter to the domain of apparent stochastic behavior of the trajectory. I shall return to the Percival classification, and its connection with other aspects of the dynamics, later.

Percival's development[45] of the semiclassical theory of coupled oscillator systems starts from the time-independent formulation of Hamilton's equations, namely,

$$\boldsymbol{\omega}\cdot\frac{\partial\mathbf{q}}{\partial\boldsymbol{\phi}}=\frac{\partial H(\mathbf{p},\mathbf{q})}{\partial\mathbf{p}} \tag{3.3}$$

$$\boldsymbol{\omega}\cdot\frac{\partial\mathbf{p}}{\partial\boldsymbol{\phi}}=-\frac{\partial H(\mathbf{p},\mathbf{q})}{\partial\mathbf{q}} \tag{3.4}$$

with, in this section, the shorthand notation $\boldsymbol{\omega}\cdot(\partial/\partial\boldsymbol{\phi})=\Sigma_i\omega_i(\partial/\partial\phi_i)$ and similarly $\partial/\partial\mathbf{p}=(\partial/\partial p_1,\ldots,\partial/\partial p_N)$. If the system Hamiltonian is written in the form

$$H=\sum_{n=1}^{N}\frac{1}{2}\left(p_n^2+\lambda_n^2 q_n^2\right)+\epsilon V(\mathbf{q}) \tag{3.5}$$

and the coordinates and momenta are expanded in Fourier series,

$$\mathbf{q}(\mathbf{J},\boldsymbol{\phi})=\sum_{\mathbf{l}}\tilde{\mathbf{q}}_{\mathbf{l}}(\mathbf{J})e^{i\mathbf{l}\cdot\boldsymbol{\phi}} \tag{3.6}$$

$$\mathbf{p}(\mathbf{J},\boldsymbol{\phi})=\sum_{\mathbf{l}}\tilde{\mathbf{p}}_{\mathbf{l}}(\mathbf{J})e^{i\mathbf{l}\cdot\boldsymbol{\phi}} \tag{3.7}$$

then (3.3) and (3.4) become

$$i(\mathbf{l}\cdot\boldsymbol{\omega})\tilde{\mathbf{q}}_{\mathbf{l}}=\tilde{\mathbf{p}}_{\mathbf{l}} \tag{3.8}$$

$$i(\mathbf{l}\cdot\boldsymbol{\omega})\tilde{\mathbf{p}}_{\mathbf{l}}=\tilde{\mathbf{f}}_{\mathbf{l}}^{HO}+\epsilon\tilde{\mathbf{f}}_{\mathbf{l}}^{V} \tag{3.9}$$

where the Fourier components of the force have been separated into a harmonic oscillator term

$$\tilde{\mathbf{f}}_{l}^{HO}=\left(-\lambda_1^2\tilde{q}_{1l},\ldots,-\lambda_N^2\tilde{q}_{Nl}\right) \tag{3.10}$$

and a coupling term

$$\tilde{\mathbf{f}}_{\mathbf{l}}^{V}=\left(\frac{1}{2\pi}\right)^{N}\int_0^{2\pi}d\boldsymbol{\phi}\left(-\frac{\partial V}{\partial\mathbf{q}}\right)e^{i\mathbf{l}\cdot\boldsymbol{\phi}}. \tag{3.11}$$

Equations (3.8) and (3.9) are to be regarded as a set of simultaneous equations for the $\tilde{\mathbf{q}}_{\mathbf{l}}$ and $\tilde{\mathbf{p}}_{\mathbf{l}}$. Elimination of $\tilde{\mathbf{p}}_{\mathbf{l}}$ between them gives

$$\left(\lambda_k^2-(\mathbf{l}\cdot\boldsymbol{\omega})^2\right)\tilde{q}_{k\mathbf{l}}=\epsilon\tilde{f}_{k\mathbf{l}}^{V}\qquad(k=1,\ldots,N) \tag{3.12}$$

The solution to (3.12) is obtained by iteration, using as the zeroth order approximation the solution for $\epsilon=0$, that is, the solution for uncoupled harmonic oscillators. This zero-order solution is used first to calculate the zero-order Fourier coefficient of the coupling force, via (3.11), and that coefficient is used in the $|\mathbf{l}|=1$ term of (3.12) to calculate the first-order frequencies. These are

$$\omega_k^{(1)} = \lambda_k^2 - \epsilon \frac{\tilde{f}_{k\mathbf{l}}^V}{\tilde{q}_{k\mathbf{l}}^{(0)}} \qquad (|\mathbf{l}| > 1, k = 1, \ldots, N) \tag{3.13}$$

The remaining members of (3.12), for $|\mathbf{l}| > 1$, are solved for the first-order Fourier coefficients of the coordinates:

$$\tilde{q}_{k\mathbf{l}}^{(1)} = \frac{\epsilon \tilde{f}_{k\mathbf{l}}^{V(0)}}{\lambda_k^2 - (\mathbf{l} \cdot \omega^{(1)})^2} \tag{3.14}$$

The calculation is closed, at this level, by using specific values of the action to fix $\tilde{q}_{k\mathbf{l}}^{(1)}$. This step imposes the initial conditions on the oscillator system. The entire process is then repeated until the desired accuracy of convergence is achieved.

Chapman, Garrett, and Miller,[47] and Schatz,[52] use the Born periodic generating function approach to define a suitable canonical transformation of a coupled oscillator Hamiltonian to angle-action variables; Jaffe and Reinhardt[41] have presented an improved rapidly converging version of this method. Let the Hamiltonian for the oscillator system be

$$H = H_0(p, q) + \epsilon V(q) \tag{3.15}$$

where $H_0(p, q)$ describes an integrable system. Let $\mathbf{J}^\circ, \boldsymbol{\phi}^\circ$ be the action-angle variables of H_0, and rewrite (3.15) in terms of these variables:

$$H = H_0(\mathbf{J}^\circ) + \epsilon V(\mathbf{J}^\circ, \boldsymbol{\phi}^\circ) \tag{3.16}$$

The equations of motion are then

$$\mu^{(0)}(\mathbf{J}^\circ, \boldsymbol{\phi}^\circ) = \dot{\mathbf{J}} = -\epsilon \frac{\partial V(\mathbf{J}^\circ, \boldsymbol{\phi}^\circ)}{\partial \boldsymbol{\phi}^\circ} \tag{3.17}$$

$$\omega^{(0)}(\mathbf{J}^\circ, \boldsymbol{\phi}^\circ) = \omega^\circ(\mathbf{J}^\circ) + \epsilon \frac{\partial V(\mathbf{J}^\circ, \boldsymbol{\phi}^\circ)}{\partial \mathbf{J}^\circ} \tag{3.18}$$

where $\omega^\circ(\mathbf{J}^\circ) = (\partial H_0(\mathbf{J}^\circ)/\partial \mathbf{J}^\circ)$. Although $\mathbf{J}^\circ$ and $\boldsymbol{\phi}^\circ$ are the proper actions and angles with which to describe the uncoupled oscillators, they are not the action-angle variables of the full Hamiltonian. However, (3.17) and (3.18) show that $\mathbf{J}^\circ$ and $\boldsymbol{\phi}^\circ$ are "almost" the correct action-angle variables, the deviation being of order ϵ. Jaffe and Reinhardt therefore introduce another transformation, this one designed to generate the action-angle variables of H to order ϵ^2, and then another transformation, and so on.

After n such transformations

$$H = H^{(n)}(\mathbf{J}^{(n)}) + \epsilon^{2^n} V^{(n)}(\mathbf{J}^{(n)}, \boldsymbol{\phi}^{(n)}) \tag{3.19}$$

$$\mu^{(n)}(\mathbf{J}^{(n)}, \boldsymbol{\phi}^{(n)}) = \dot{\mathbf{J}}^{(n)} = -\epsilon^{2^n} \frac{\partial V^{(n)}}{\partial \boldsymbol{\phi}^{(n)}} \tag{3.20}$$

$$\omega^{(n)}(\mathbf{J}^{(n)}, \boldsymbol{\phi}^{(n)}) = \dot{\boldsymbol{\phi}}^{(n)} = \omega_0^{(n)}(\mathbf{J}^{(n)}) + \epsilon^{2^n} \frac{\partial V^{(n)}}{\partial \mathbf{J}^{(n)}} \tag{3.21}$$

so the variables generated are the action-angle variables of H correct to order ϵ^{2^n}.

The key step in this analysis is the first transformation from $\mathbf{J}^\circ, \boldsymbol{\phi}^\circ$ to $\mathbf{J}^{(1)}, \boldsymbol{\phi}^{(1)}$. The generating function is

$$F_1(\boldsymbol{\phi}^\circ, \mathbf{J}^{(1)}) = \boldsymbol{\phi}^\circ \cdot \mathbf{J}^{(1)} + \epsilon W^{(1)}(\boldsymbol{\phi}^\circ, \mathbf{J}^{(1)}) \tag{3.22}$$

$$W^{(1)}(\boldsymbol{\phi}^\circ, \mathbf{J}^{(1)}) = \sum_{\mathbf{l} \neq 0} \tilde{W}_{\mathbf{l}}^{(1)}(\mathbf{J}^{(1)}) e^{i\mathbf{l}\cdot\boldsymbol{\phi}^\circ} \tag{3.23}$$

which yields

$$\mathbf{J}^\circ = \mathbf{J}^{(1)} + \epsilon \sum_{\mathbf{l} \neq 0} i\mathbf{l} \tilde{W}_{\mathbf{l}}^{(1)}(\mathbf{J}^{(1)}) e^{i\mathbf{l}\cdot\boldsymbol{\phi}^\circ} \tag{3.24}$$

$$\boldsymbol{\phi}^{(1)} = \boldsymbol{\phi}^\circ + \epsilon \sum_{\mathbf{l} \neq 0} \left(\frac{\partial \tilde{W}_{\mathbf{l}}^{(1)}(\mathbf{J}^{(1)})}{\partial \mathbf{J}^{(1)}} \right) e^{i\mathbf{l}\cdot\boldsymbol{\phi}^\circ} \tag{3.25}$$

and it can be shown that

$$\tilde{W}_{\mathbf{l}}^{(1)}(\mathbf{J}^{(1)}) = -\left(\frac{V_{\mathbf{l}}^{(0)}(\mathbf{J}^\circ)}{i\mathbf{l}\cdot\omega_0^{(0)}(\mathbf{J}^\circ)} \right)_{\mathbf{J}^\circ = \mathbf{J}^{(1)}} \tag{3.26}$$

$$\frac{\partial \tilde{W}_{\mathbf{l}}^{(1)}}{\partial \mathbf{J}^{(1)}} = -\left(\frac{\partial \tilde{V}_{\mathbf{l}}^{(0)}(\mathbf{J}^\circ)/\partial \mathbf{J}^\circ}{i\mathbf{l}\cdot\omega_0^{(0)}(\mathbf{J}^\circ)} \right)_{\mathbf{J}^\circ = \mathbf{J}^{(1)}} \tag{3.27}$$

with

$$V(\mathbf{J}^\circ, \boldsymbol{\phi}^\circ) = \sum_{\mathbf{l} \neq 0} \tilde{V}_{\mathbf{l}}^{(0)}(\mathbf{J}^\circ) e^{i\mathbf{l}\cdot\boldsymbol{\phi}^0} \tag{3.28}$$

The new Hamiltonian is of the same form as the old Hamiltonian, except that the coefficient of the coupling term is now ϵ^2. Therefore the same transformation can be applied again. The generating function of the nth transformation is, in terms of the variables of the $n-1$th Hamiltonian

$$\tilde{W}_{\mathbf{l}}^{(n)}(\mathbf{J}^{(n)}) = -\left(\frac{\tilde{V}_{\mathbf{l}}^{(n-1)}(\mathbf{J}^{(n-1)})}{i\mathbf{l}\cdot\omega_0^{(n-1)}(\mathbf{J}^{(n-1)})}\right)_{\mathbf{J}^{(n-1)}=\mathbf{J}^{(n)}} \tag{3.29}$$

$$\frac{\partial \tilde{W}_{\mathbf{l}}^{(n)}}{\partial \mathbf{J}^{(n)}} = -\left(\frac{\partial \tilde{V}_{\mathbf{l}}^{(n-1)}/\partial \mathbf{J}^{(n-1)}}{i\mathbf{l}\cdot\omega_0^{(n-1)}(\mathbf{J}^{(n-1)})}\right)_{\mathbf{J}^{(n-1)}=\mathbf{J}^{(n)}} \tag{3.30}$$

Jaffe and Reinhardt call their procedure a classical Van Vleck transformation.

For the present purposes the most interesting semiclassical quantization method is that of Swimm and Delos.[49] They have shown that the eigenvalues of the Hamiltonian operator of a coupled oscillator system can be very accurately calculated by transforming the classical Hamiltonian to the Birkhoff normal form, and then quantizing via the Bohr–Sommerfeld–Wilson rule. It was mentioned in Section II.C that the Hamiltonian of a coupled oscillator system could be brought into so called normal form by a succession of canonical transformations. When in normal form the Hamiltonian is a function of the variables combined as $\frac{1}{2}(p_k^2+q_k^2)$, that is, a function of one-dimensional harmonic oscillator Hamiltonians. Clearly, if a Hamiltonian can be brought into normal form, the motion is quasiperiodic.

The succession of transformations required to generate the normal form of the Hamiltonian in general diverges because of small divisors in the coefficients. This divergence arises from the existence of stochastic trajectories which cannot be described by the normal form representation of the Hamiltonian, and which are dense in at least part of the accessible range. Nevertheless, as shown by Gustavson,[36] a truncated series of transformations generates a Hamiltonian which is in normal form up to some well defined order in ϵ, and which is an approximation to the original Hamiltonian. This truncated Hamiltonian describes quasiperiodic trajectories which are good approximations to the quasiperiodic trajectories of the original Hamiltonian.

In brief, the Gustavson procedure involves the following steps, illustrated here for the Henon–Heiles Hamiltonian written in the form

$$\begin{aligned} H &= \tfrac{1}{2}\left(p_1^2+p_2^2+\omega_1^2 q_1^2+\omega_2^2 q_2^2\right)+\epsilon q_2\left(q_1^2+\eta q_2^2\right) \\ &= H^{(2)}+H^{(3)} \end{aligned} \tag{3.31}$$

It is assumed that the ratio ω_2/ω_1 is irrational. The substitutions

$$p_k \to \omega_k^{1/2} p_k' \tag{3.32}$$

$$q_k \to \omega_k^{-1/2} q_k' \tag{3.33}$$

give

$$H^{(2)} = \tfrac{1}{2}(p_1'^2 + q_1'^2)\omega_1 + \tfrac{1}{2}(p_2'^2 + q_2'^2)\omega_2 \tag{3.34}$$

$$H^{(3)} = \frac{\epsilon}{\omega_1 \omega_2^{1/2}} q_1'^2 q_2' + \frac{\epsilon\eta}{\omega_2^{3/2}} q_2'^3 \tag{3.35}$$

The new Hamiltonian is further reduced by the sequential application of canonical transformations defined by a generating function $W^{(s)}$ which has the property

$$Q = q' + \frac{\partial W^{(s)}}{\partial P} \tag{3.36}$$

$$P = p' - \frac{\partial W^{(s)}}{\partial q'} \tag{3.37}$$

$W^{(s)}$ is a homogeneous polynomial of degree s, where $s > 3$. The transformation of variables of degree s leaves unchanged those terms of degree less than s, but changes all other terms. We have, then

$$H(p', q') = \Gamma(P, Q) \tag{3.38}$$

The new Hamiltonian $\Gamma(P, Q)$ is now separated into terms of different degrees and $P + \partial W^{(s)}/\partial q'$ and $q' + \partial W^{(s)}/\partial P$ expanded about P and q', respectively. Collection of terms of equal degree then gives

$$H^{(i)}(P, q') = \Gamma^{(i)}(P, q') \qquad i < s \tag{3.39}$$

$$\hat{D} W^{(s)}(P, q') = \Gamma^{(s)}(P, q') - H^{(s)}(P, q') \qquad i = s \tag{3.40}$$

$$\Gamma^{(i)}(P, q') = H^{(i)}(P, q') + \sum_j \frac{1}{j!}\left[\left(\frac{\partial W^{(s)}}{\partial q'}\right)^j \left(\frac{\partial^{|j|} H^{(l)}}{\partial P^j}\right) - \left(\frac{\partial W^{(s)}}{\partial P}\right)^j \left(\frac{\partial^{|j|} \Gamma^{(l)}}{\partial q'^j}\right)\right] \tag{3.41}$$

with the constraints on the indices $l-|j|+|j|(s-1)=i$, $1<|j|<l<i$, $l>2$, $s>3$ and $j!=j_1!j_2!$. Finally, $\hat{D}$ is the operator

$$\hat{D}=-\omega_1\left(q_1'\frac{\partial}{\partial P_1}-P_1\frac{\partial}{\partial q_1'}\right)-\omega_2\left(q_2'\frac{\partial}{\partial P_2}-P_2\frac{\partial}{\partial q_2'}\right) \tag{3.42}$$

and

$$H^{(i)}(P,q')\equiv\left[H^{(i)}(p,q')\right]_{p=P} \tag{3.43}$$

$$\Gamma^{(i)}(P,q')\equiv\left[\Gamma^{(i)}(P,Q)\right]_{Q=q'} \tag{3.44}$$

To solve (3.40) q' and P are transformed so as to diagonalize $\hat{D}$:

$$P_k=\frac{1}{\sqrt{2}}(\eta_k+i\zeta_k) \tag{3.45}$$

$$q_k'=\frac{1}{\sqrt{2}}(\eta_k-i\zeta_k) \tag{3.46}$$

It is found that $\Phi_{l_1l_2m_1m_2}=\eta_1^{l_1}\eta_1^{l_2}\xi_1^{m_1}\xi_2^{m_2}$ are the eigenfunctions of $\hat{D}$ corresponding to the eigenvalues $i\Sigma_k\omega_k(m_k-l_k)$, and that

$$W^{(s)}=\hat{D}^{-1}\left[\Gamma^{(s)}-H^{(s)}\right] \tag{3.47}$$

$W^{(s)}$ is kept finite by choosing $\Gamma^{(s)}$ to cancel those terms in $H^{(s)}$ which would have vanishing denominator in (3.47) when $\Gamma^{(s)}$ and $H^{(s)}$ are expanded in the basis of the eigenfunctions of $\hat{D}$. There is no difficulty in doing this for ω_2/ω_1 irrational; for ω_2/ω_1 rational a slightly more elaborate procedure is needed. In the case that ω_2/ω_1 is irrational the end result is that the normal form can be written as a sum of products of one-dimensional harmonic oscillator Hamiltonians, and the quantization can be effected to yield analytic formulas. In the case that ω_2/ω_1 is rational additional terms that cannot be written as products of one-dimensional Hamiltonians appear in the normal form, and while one degree of freedom can be quantized analytically the other must be quantized by a one-dimensional (generally numerical) phase integration.

All of the semiclassical quantization schemes described give eigenvalues in excellent agreement with these calculated by direct diagonalization of the Hamiltonian operator. That fact I believe to be an important clue to the interpretation of the relationship between quantum mechanical states and classical trajectories of a system of coupled oscillators. I will return to this point below.

C. Conjectures on the Relationship Between Quantum-Mechanical States and Classical Trajectories

The studies of the quantum states of a system of coupled oscillators described in Section III.B lead to two inferences:

1. There appears to be an analog of the KAM transition in the quantum-mechanical description of a nonintegrable system.
2. The states generated by semiclassical quantization are derived from quasiperiodic trajectories embedded in the stochastic domain of the classical dynamics.

Note that inferences 1 and 2 are not specifically contradictory, but they are not reinforcing. If inferences 1 and 2 are both correct the quantum-mechanical description of the coupled oscillator system must differ from the classical mechanical description more than has been thought to be the case.

The results which lead to inference 1 derive from the definition of quantum ergodicity posed by Nordholm and Rice,[38] or the equivalent posed by Handy, Stratt and Miller.[39] This inference is also supported by calculations of Pomphrey to be described below. Berry[46] has picked up a suggestion made by Nordholm and Rice and developed an analysis of the Wigner function representation of a system of coupled oscillators. For states in the quasiperiodic domain of phase space the classical limit of the Wigner function is a delta function on the corresponding invariant torus. In this case there is a simple connection between quasiperiodic trajectories and stationary states, as might have been expected from the Bohr–Sommerfeld–Wilson quantization condition for separable systems. Berry is then led to the natural conjecture that the Wigner function corresponding to a stochastic trajectory spreads over the stochastic region of the phase space, so that a surface of section for the quantum state ought (?) to show randomly distributed maxima and minima. There is, at present, not even a clue as to what form of randomness (e.g., a Gaussian random distribution) to expect in this surface of section. Nevertheless, if the Wigner function on the surface of section is very irregular, it is reasonable to connect stochastic trajectories and quantum states in a fashion analogous to that used to connect quasiperiodic trajectories and quantum states.

Berry's study of the semiclassical limit of the Wigner function has many interesting features; The reader is referred to the original paper for details.[46] Here we shall consider only the conjecture just described concerning the Wigner function on the surface of section. Hutchinson and Wyatt[54] have constructed the Wigner function for the Henon–Heiles system with Hamiltonian

$$H=\tfrac{1}{2}(p_1^2+p_2^2)+\tfrac{1}{2}(0.49q_1^2+1.69q_2^2)-0.10(q_1q_2^2-q_1^3). \qquad (3.48)$$

The dissociation energy is, in this case, $E_D = 11.46$, and the onset of stochastic behavior is estimated to be $E_c = 5.73$ using the Brumer–Duff–Toda criterion.[55] The method of calculation is the same as that used by Nordholm and Rice. The surface of section Wigner function distribution is defined by

$$\Psi_W(q_1, p_1; q_2=0) = \int_{-\infty}^{\infty} \Psi_W(q_1, q_2=0, p_1, p_2)\, dp_2 \qquad (3.49)$$

with the usual definition

$$\Psi_W \equiv (\pi\hbar)^{-N} \int d\mathbf{X}\, e^{-2i\mathbf{p}\cdot\mathbf{X}/\hbar} \psi(\mathbf{q}+\mathbf{X})\psi^*(\mathbf{q}-\mathbf{X}) \qquad (3.50)$$

Figure 3.6 shows the ground state surface of section ($E = 0.00$), Fig. 3.7 the surface of section for $E = 4.78$, which is close to but below E_c, and Figs. 3.8

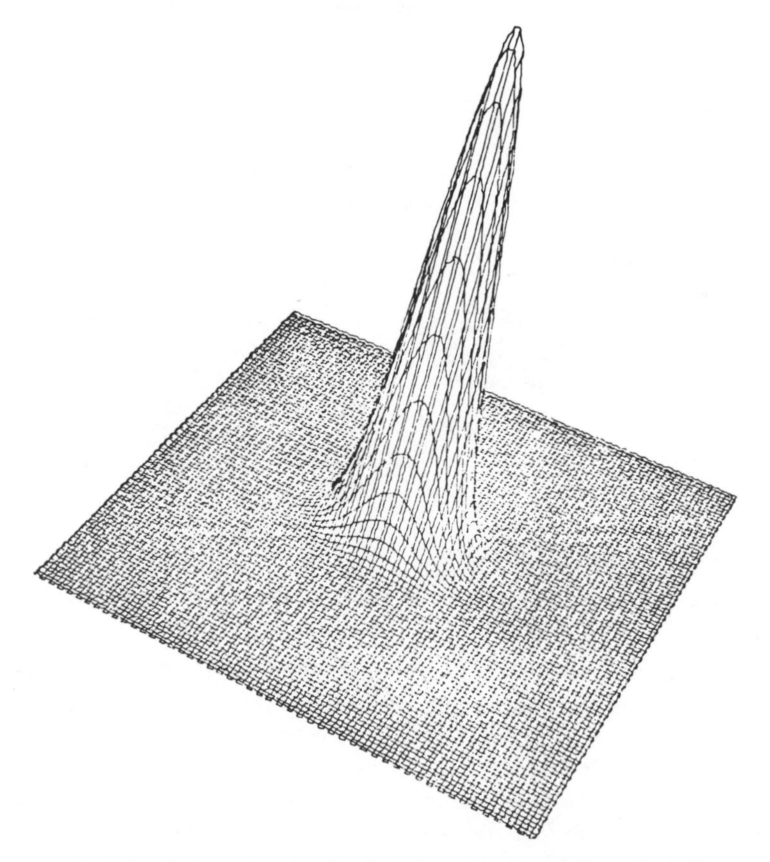

Fig. 3.6. Surface of section for the Wigner function (3.49) at $E = 0.00$.

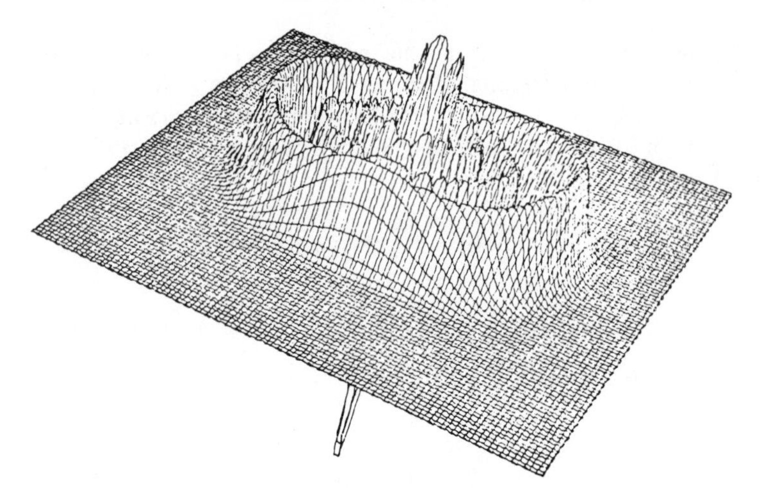

Fig. 3.7. Surface of section for the Wigner function (3.49) at $E = 4.78$.

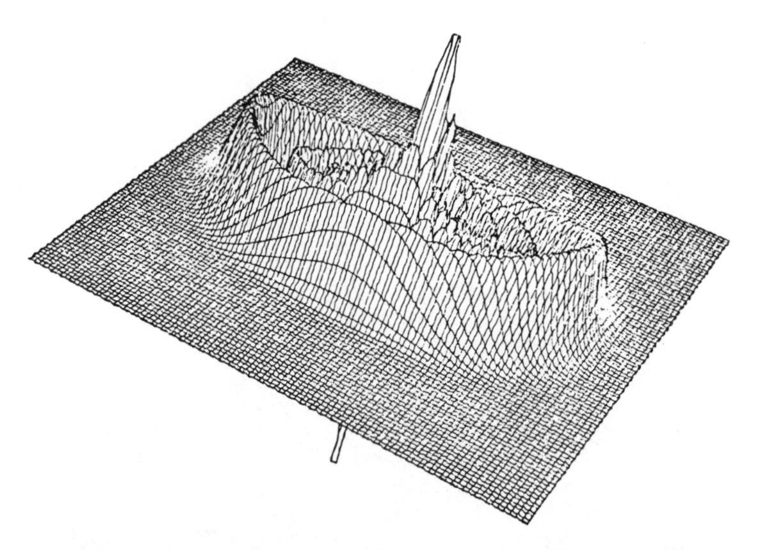

Fig. 3.8. Surface of section for the Wigner function (3.49) at $E = 6.28$.

and 3.9 the surfaces of section for $E = 6.28$ and $E = 9.96$, close to but above E_c and near the dissociation limit, respectively. In contrast to the conjecture made by Berry, these distributions are very regular!

The results of the Wigner function calculation are not the only ones to show regularity in the wavefunction not immediately apparent in the basis set expansion used to infer the analog of the KAM transition. Figure 3.8

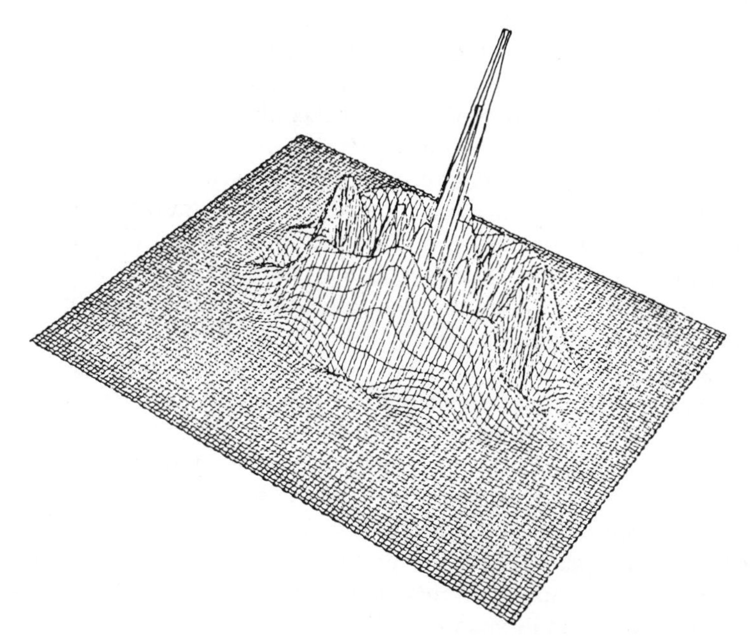

Fig. 3.9. Surface of section for the Wigner function (3.49) at $E = 9.96$. Figures 3.6, 3.7, 3.8, and 3.9 from J. S. Hutchinson and R. E. Wyatt private communication.

shows that the wavefunction of a state in the energy domain where the classical dynamics is dominantly stochastic is rather regular in appearance. Certainly, it is not evident to the eye that the wavefunction in coordinate space has a random component in its amplitude or phase.

Although I originally believed that it was correct, I now think that Berry's conjecture focuses attention on the wrong trajectories. It is correct that the quasiperiodic orbits embedded in the stochastic domain have small measure, but it is arguable that it is only these that are pertinent in the comparison with the eigenstates of the Hamiltonian operator. The argument, which is based on the accuracy of the several semiclassical quantization schemes, is as follows. All of these schemes start from a quasiperiodic trajectory and generate new quasiperiodic trajectories. The calculations of Helleman and Bountis[35] suggest that solutions of this type are dense in the stochastic region, although of very small measure relative to the stochastic solutions. The complete agreement, both in number and value, between the eigenvalues determined by semiclassical quantization and by diagonalization of the Hamiltonian operator then suggests that there is a one-to-one correspondence between a subset of the quasiperiodic trajectories and stationary states of the system. Or, put another way, the intricate substructure of the classical trajectories that leads to stochastic behavior plays no role in determining the stationary states of the system.

The just described point of view is rather close to that stated by Einstein and the recent conjectures of Percival.[45] It remains the case that the Einstein quantization condition cannot be used for the overwhelming majority of classical trajectories above the KAM transition, since in that domain p_k cannot be expressed as a function of the q_k. What has not been adequately emphasized, however, is that there are periodic and quasiperiodic solutions embedded in the stochastic domain, and the Einstein quantization condition does apply to these. As already stated, the numerical evidence suggests but does not prove that these states will exhaust the bound state spectrum of the Hamiltonian operator. As to Percival's perceptive remarks, the Helleman and Bountis results illustrate that the nature of the periodic solutions (and by inference also the quasiperiodic solutions) is inordinately sensitive to the initial conditions, just as inferred by Percival. A further connection between conclusions 3 and 5 of Nordholm and Rice and Percival's notion of regular and irregular spectra has been made in a calculation by Pomphrey.[40] He studied the sensitivity of the parameterized Henon–Heiles Hamiltonian ($m=1$)

$$H=\tfrac{1}{2}(p_1^2+p_2^2)+\tfrac{1}{2}(q_1^2+q_2^2)+\alpha\left(q_1^2q_2-\tfrac{1}{3}q_2^3\right) \tag{3.51}$$

to the value of α. In this case the dissociation energy is $1/6\alpha^2$. Pomphrey computed the eigenvalues of (3.51) for the range $0.090<\alpha<0.086$ and examined the sensitivity of the spectrum as a function of the energy. This sensitivity is measured by the second difference

$$\Delta_i\equiv|\{E_i(\alpha+\Delta\alpha)-E_i(\alpha)\}-\{E_i(\alpha)-E_i(\alpha-\Delta\alpha)\}| \tag{3.52}$$

Perturbation theory yields the result

$$\Delta_i\sim\mathcal{O}(\Delta\alpha^3) \tag{3.53}$$

The calculations show that for $E<16=0.74E_D$ all second differences are very small. This is the regular region of the spectrum, corresponding to localized asymptotic distribution over the basis states, and to quasiperiodic motion in the classical limit. For $E>16$ eigenvalues are found with corresponding Δ_i orders of magnitude larger, that is, the spectrum is very sensitive to small changes in α. This is the irregular region of the spectrum, corresponding to global asymptotic distribution over the basis states, and to apparently stochastic motion in the classical limit (see Fig. 3.10). It is also illuminating to compare the coverage of the surface of section by the

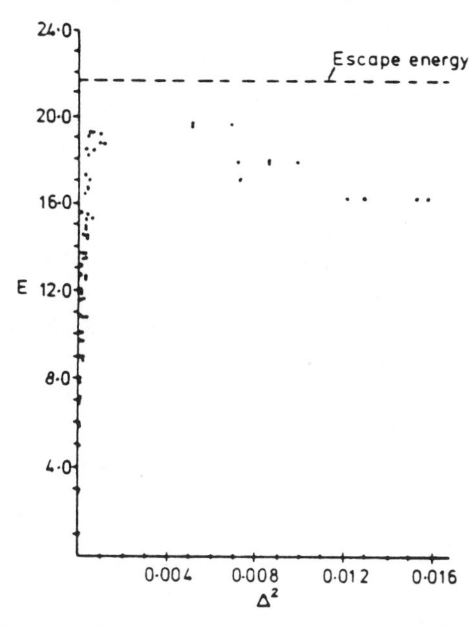

Fig. 3.10. Sensitivity of the eigenvalues of the Henon–Heiles Hamiltonian to small changes in coupling.

apparently stochastic trajectory with the region of Hilbert space wherein the spectrum is very sensitive to the coupling. For the classical Henon–Heiles Hamiltonian the total area covered by unstable trajectories up to energy E is

$$I(E)=\int^{E}\alpha_I(E)\,dE$$

$$\alpha_I(E)=0,\quad E<0.68E_D;\quad \alpha_I(E)=3.125\left(\frac{E}{E_D}\right)-2.125,\quad E>0.68E_D \tag{3.54}$$

where "total area" means the relative area of the surface of section. The quantity is to be compared with

$$S(E)=\frac{1}{E_D}\sum_{i}^{E} n_I(E_i)\langle\Delta E_i\rangle \tag{3.55}$$

corresponding to the part of Hilbert space where the spectrum is very sensitive to a change in α. Here $n_I(E)=1$ if E_i is very sensitive to the value of α, $n_I(E_i)=0$ otherwise. Also,

$$\langle \Delta E_i \rangle = \tfrac{1}{2}(E_{i+1} - E_{i-1}) \tag{3.56}$$

As shown in Fig. 3.11 the quantum-mechanical results follow, qualitatively, the shape of the classical curve. [$I(E)$ and $S(E)$ have different dimensionality, hence cannot agree quantitatively.]

The last topic to be considered in this subsection is the nature of the absorption spectrum below and above the KAM-like transition in the harmonic oscillator basis state representation of the eigenstates of a coupled oscillator system. Stratt, Handy, and Miller[39] have computed the expected infrared spectrum for a system with the Hamiltonian (3.1); typical results are shown in Fig. 3.12. Cases (a) and (b) are for excitation to

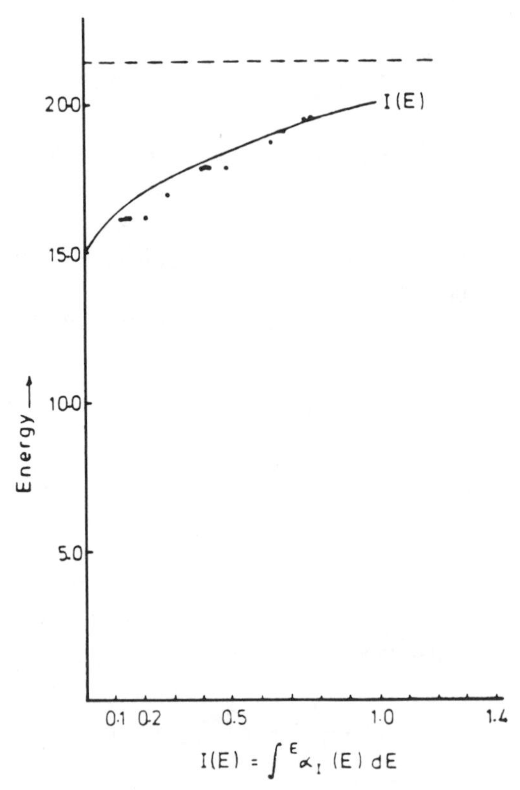

Fig. 3.11. Comparison of the area of surface of section covered by nonperiodic trajectories and the domain of Hilbert space where the eigenvalues are sensitive to small changes in coupling. Figs. 3.10 and 3.11 from N. Pomphrey, *J. Phys. B.*, 7, 1909 (1974).

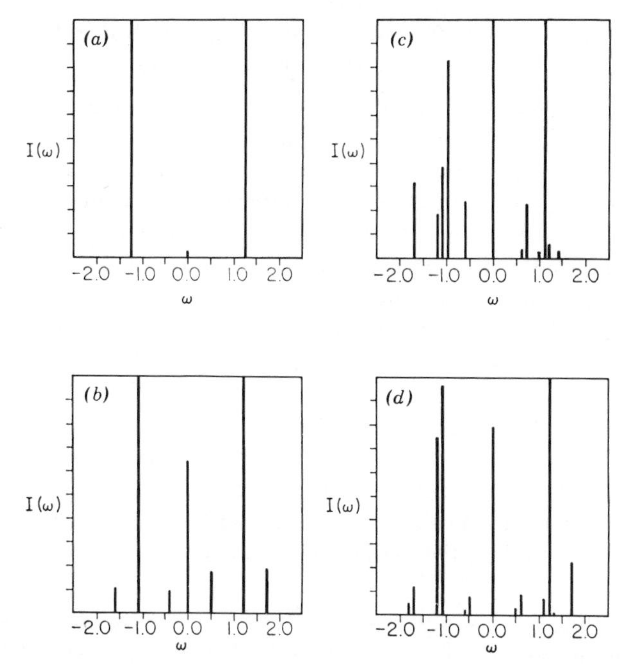

Fig. 3.12. The power spectra of selected eigenstates of the Hamiltonian (3.1). Cases (a) and (b) are for energies below the KAM-like transition, case (c) is borderline, and case (d) is for an energy above the transition. From R. M. Stratt, N. C. Handy and W. H. Miller, *J. Chem. Phys.* **71**, 3311 (1979).

quantum states below the transition, (c) to a borderline state, and (d) to a state above the transition. There is an obvious trend from "regular" to "irregular" behavior just as suggested by Percival.

Given the spectrum it is possible to devise a thought-experiment that creates a pure excited eigenstate of the coupled oscillator system (neglecting radiative damping). What is required is coherent excitation at the transition frequencies shown in, say, Fig. 3.12d. This can, in principle, be achieved by splitting up a single laser beam to drive several tunable sources with phase coherence. The difficulty is, of course, that in a real molecule we do not know which transitions to excite coherently. It is also necessary to enter the caveat that the coupling of vibrational and rotational motions in a molecule could alter the nature of the conclusions drawn from the studies of nonrotating coupled oscillator systems. However, if one could excite high-energy molecular eigenstates (neglecting radiative damping) it is conceivable that the ensuing chemistry is different from that of incoherently excited molecules.

D. Some Differences Between the Classical and Quantum-Mechanical Descriptions

It would be surprising if, for a given system, classical stochastic behavior and quantum stochastic behavior were the same in all details. Heller,[56] in an important contribution, has pointed out the existence of interference effects that lead to a fundamental difference between the quantum-mechanical and classical mechanical behavior of systems of coupled nonlinear oscillators. He showed, using only group theoretical arguments, that nonlinear oscillator systems that have symmetries leading to degenerate quantum states do not transfer energy equivalently to rigorously equivalent phase space locations. In particular, if the initial state is a wavepacket, Heller shows that the time averaged probability of finding the system in the initial state is larger than that of finding the system in states which are symmetrically equivalent. This behavior, which is independent of energy, contradicts the behavior expected when the corresponding classical mechanical trajectory is stochastic, since in the latter case the time-averaged probabilities for finding the system in symmetrically equivalent states are equal.

The interference effects, destructive in some regions of phase space and constructive in other regions, represent only one of the fundamental differences between the quantum-mechanical and classical mechanical descriptions of a system. An equally important difference arises from the nature of the observation process. Kosloff and Rice[57] have pointed out that the means that must be used to prepare the system in an initial state of the type discussed by Heller, or to verify the prediction that the time-averaged probabilities of finding the system in symmetrically equivalent states are not the same, destroy the basis for the lack of equality of time averaged amplitudes. Thus the complete quantum-mechanical description, including the mode of preparation and the nature of the observation, must be used to determine if there is asymptotic behavior which is necessarily different from that of the classical mechanical description.

Heller's argument can be summarized as follows. Let ψ_a be the wavefunction of the initial state. The time averaged probability of finding the system in the state ψ_b is

$$P(a|b)=\lim_{T\to\infty}\frac{1}{T}\int_0^T \mathrm{Tr}(\hat{\rho}_a(t)\hat{\rho}_b)\,dt \tag{3.57}$$

where $\hat{\rho}_a=|a\rangle\langle a|$, and $\hat{\rho}_b=|b\rangle\langle b|$ are the density operators corresponding to ψ_a,ψ_b. Suppose the eigenfunctions of the system Hamiltonian are

ϕ_{ni}, where n labels the energy and i the degeneracy. Then

$$\psi_a = \sum_{n,i} a_{ni}\phi_{ni}$$

$$\psi_b = \sum_{m,i} b_{mi}\phi_{mi} \tag{3.58}$$

so that (3.57) can be rewritten in the form

$$P(a|b) = \sum_n \sum_m \left| \sum_i a_{ni} b_{mi} \right|^2$$

$$= \sum_n |\mathbf{a}_n \cdot \mathbf{b}_n|^2 \tag{3.59}$$

Heller now compares $P(a|a)$ with $P(a|Ra)$, where R is a symmetry operation. Because R is represented by a unitary matrix which preserves length, it is found that

$$|\mathbf{a}_n \cdot \mathbf{b}_n|^2 = |\mathbf{a}_n \cdot \Gamma(R)\mathbf{a}_n|^2 \leqslant |\mathbf{a}_n \cdot \mathbf{a}_n|^2 \tag{3.60}$$

which implies

$$P(a|a) \geqslant P(a|Ra) \tag{3.61}$$

The prediction implied by (3.61) is not meaningful unless the means of preparation of the initial state and of observation are both specified. We now note that in order to differentiate between symmetry-equivalent states of an isolated system it is necessary to break the isolation and the symmetry, that is, both the preparation and measurement processes necessarily introduce into the total Hamiltonian for the system and the subsidiary preparation or measuring apparatus a term that lifts the degeneracy of the states of the isolated-system Hamiltonian. When that degeneracy is lifted Heller's argument ceases to be valid. For simplicity, only the influence of observation will be analyzed below; a parallel argument can be used to describe the preparation process.

Consider, as an example, the Henon–Heiles Hamiltonian. The potential energy surface in this case has threefold symmetry and belongs to the group C_{3v}:

$$v(r,\theta) = \tfrac{1}{2}r^2 + \tfrac{1}{3}\lambda r^3 \sin 3\theta \tag{3.62}$$

We imagine using a two-level system as a measuring apparatus to differentiate the three equivalent states:[58, 59] the spin-up state of the measuring apparatus will be correlated with ψ_a and the spin-down position with ψ_b. The initial state of the measuring apparatus is, then, spin-up. The interaction between the system and the measuring apparatus is taken to be

$$\hat{H}_I = g(t)\sin\left(\frac{\theta}{2}\right)\left(\hat{\sigma}_x - \tfrac{1}{2}\right) \tag{3.63}$$

where $g(t)$ specifies the time dependence of the probing interaction and $\hat{\sigma}_x$ is the appropriate angular momentum operator of the two-level measurement apparatus. A measurement is made as follows: The system is probed, subject to the interaction Hamiltonian (3.63), for a period such that

$$\int_0^t g(t)\,dt = \frac{2\pi}{\sqrt{3}} \tag{3.64}$$

As a result of this probing, the final state of the measuring apparatus records the time averaged probability $P(a|b)$ in a system in which the threefold degeneracy has been lifted. Although the form chosen for $\hat{H}_1$ in (3.63) is specific, the principle implied is generally valid; we conclude that (3.61) describes a situation which is necessarily disturbed by the observation process.

The result of this analysis of the influence of observations on the asymptotic distribution of amplitude in a quantum-mechanical system only shows that the interference effects in a system with degenerate states that would prohibit quasiergodic behavior are perturbed by observation; the result does not show that quasiergodic behavior must be observed, nor does it show that is is impossible to observe the consequences of (3.61). Imagine an ensemble of systems with interaction (3.63) prepared in a coherent state. Suppose the measurement process is carried out, at different times $t_1 < t_2 < \cdots$ on different replicas of the ensemble. Although each measurement on one replica alters the amplitude distribution in that replica, rendering it useless for further measurements, we imagine that in the absence of measurements (3.61) is valid so that measurements on different replicas will yield behavior different from that predicted if (3.61) is not valid. The validity of this interpretation will depend on the nature of the preparation process and of the observation made, and each case must be examined for special characteristics.

E. Ergodicity and Reaction Rate: Model Considerations

I have thus far confined the discussion to the behavior of the bound states of a system of coupled nonlinear oscillators. I now wish to consider the influence, or lack of influence, of the nature of intramolecular energy exchange on the rate of a fragmentation reaction. At first sight it appears that, because the KAM transition typically occurs for $E < E_D$, the rate of fragmentation should be accurately accounted for by a statistical model. A deeper examination reveals that the matter is not so simple. First, the very nature of the irregular spectrum suggests that a decomposition rate might not be a monotone function of the energy. Second, resonances in the localized states of the bond that breaks could conceivably be derived from nonergodic states of the molecule interspersed sparsely in the ergodic region of states. Third, the matrix elements coupling different vibrations of the molecule might vary over such a large range that only a subset of all vibrations is effectively coupled on the time scale of the reaction.

To determine if any of these possibilities is important Nordholm and Rice developed an exact formal theory of fragmentation reactions, and made calculations for a crude model.[60] They showed that if the Hilbert space is partitioned into domains corresponding to bound (B) and free (F) states, then the amplitude of the wavefunction in the bound-state domain evolves according to

$$\frac{\partial}{\partial t}\hat{P}_{\mathrm{B}}\psi(t) = -\frac{i}{\hbar}\hat{H}_{\mathrm{BB}}\hat{P}_{\mathrm{B}}\psi(t) - \frac{1}{\hbar^2}\int_0^t ds\, \hat{H}_{\mathrm{BF}}e^{-is\hat{H}_{\mathrm{FF}}/\hbar}\hat{H}_{\mathrm{FB}}\hat{P}_{\mathrm{B}}\psi(t-s) \tag{3.65}$$

The notation used in (3.65) is as follows: $\hat{P}_{\mathrm{B}}$ is a projection operator which separates the Hilbert space into bound-state and free-motion domains, and its complement is $\hat{P}_{\mathrm{F}} \equiv 1 - \hat{P}_{\mathrm{B}}$. The various projections of the Hamiltonian $\hat{H}$ are defined by $\hat{H}_{\mathrm{BB}} \equiv \hat{P}_{\mathrm{B}}\hat{H}\hat{P}_{\mathrm{B}}$, $\hat{H}_{\mathrm{BF}} = \hat{P}_{\mathrm{B}}\hat{H}\hat{P}_{\mathrm{F}}$, $\hat{H}_{\mathrm{FB}} = \hat{P}_{\mathrm{F}}\hat{H}\hat{P}_{\mathrm{B}}$ and $\hat{H}_{\mathrm{FF}} \equiv \hat{P}_{\mathrm{F}}\hat{H}\hat{P}_{\mathrm{F}}$. Equation 3.65 is easily interpreted: $\hat{H}_{\mathrm{FB}}\hat{P}_{\mathrm{B}}\psi(t-s)$ measures the flow of amplitude from B to F at time $t-s$. Then $\exp(-i\hat{H}_{\mathrm{FF}}s/\hbar)$ propagates this amplitude within the free-motion states F forward in time from $t-s$ to t, and $\hat{H}_{\mathrm{BF}}$ measures how much of it is returned to B at t. The integration sums these effects from $t=0$. The term

$$-\frac{1}{\hbar^2}\int_0^t ds\, \hat{H}_{\mathrm{BF}}e^{-i\hat{H}_{\mathrm{FF}}s/\hbar}\hat{H}_{\mathrm{FB}}\hat{P}_{\mathrm{B}}\psi(t-s)$$

leads to a decrease in the norm of $\hat{P}_{\mathrm{B}}\psi(t)$, hence to dissipation, because the flow out of B will exceed the flow in given that amplitude is propagated away from the boundary of B by the action of $\exp(-i\hat{H}_{\mathrm{FF}}s/\hbar)$, diminishing the net flow back into B. Note that when $\hat{H}_{\mathrm{BF}}$ is set equal to zero the formalism describes exactly the relaxation in the manifold of bound levels.

The advantage offered by the Nordholm–Rice analysis is that at the formal level the equations of motion of $\hat{P}_B\psi(t)$ and $\hat{P}_F\psi(t)$ have been decoupled; the price paid for this decoupling is that the equation of motion of $P_B\psi(t)$ depends on the "memory" of the motion. Nevertheless, there is a net gain in that the structure of the formalism suggests approximations to the N-body dynamics different from those suggested by other formalisms, especially when the analogy with the statistical mechanics of irreversibility is exploited.

Nordholm and Rice have proposed a simple model to reduce (3.65) to useful form. This model has two coupled oscillators, one of which can dissociate. It suffices, here, to state that the model involves both the definition of an effective Hamiltonian for the system, and the evaluation of the coefficient in the imaginary part of that Hamiltonian. For the model described, for which $E_c/E_D=0.48$, it is found that:

1. The imaginary part of the eigenvalues of the effective Hamiltonian, E_1, representing the rate of fragmentation of a system initially prepared in a nonstationary "bound state," varies over a wide range even for $E>E_c$. The range of variation is larger when the oscillators are nondegenerate; the variation is irregular.
2. The states $|0, m_2\rangle$ remain very much localized, even at high energy. (The basis states for the calculation were the harmonic oscillator functions $|m_1, m_2\rangle$).
3. Small values of E_1 are associated with local (nonergodic) states; large values with global (ergodic) states.
4. Coherence effects lead to variation in the rate of decay as a function of t.
5. Initial states deep in the bound region tend to decay in a nonmonotonic fashion; initial states in the transition region decay monotonically. (See Figs. 3.13, 3.14, 3.15 and 3.16.)

The results displayed in Figs. 3.13, 3.14, 3.15 and 3.16 are simply interpreted. An isolated resonant state—one generated by embedding a zero-order discrete state in a zero-order continuum—decays with a constant lifetime (inverse rate). The population decays displayed in these figures, corresponding to different initial states, are monotone, but the rates of decay are not. The variation of the rates of decay is a consequence of competition between energy transfer to the stable oscillator and fragmentation of the unstable oscillator. Clearly, the more effective the localization of energy in the stable oscillator, and the more efficient the energy transfer relative to reaction, the greater should be the variability of the decay rate, just as observed.

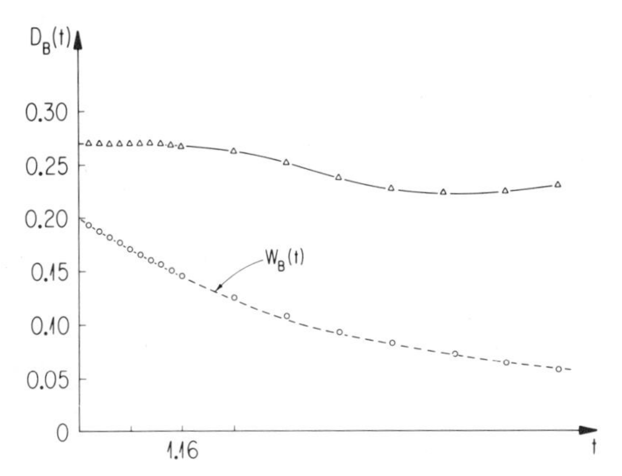

Fig. 3.13. The decay rate $D_B(t)$ and the probability of finding the system to be bound $W_B(t)$, for the zeroth order state $|8,0\rangle$.

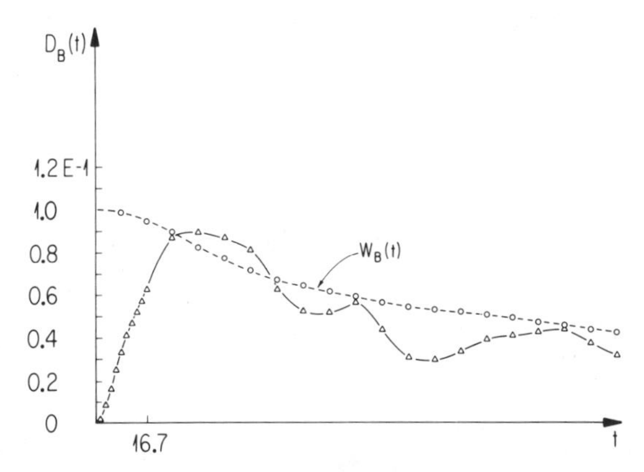

Fig. 3.14. The decay rate $D_B(t)$ and the probability of finding the system to be bound $W_B(t)$, for the zeroth order state $|4,4\rangle$.

Parr and Kupperman have studied, using classical mechanics, the sensitivity to variation of the initial energy of the rate of fragmentation of the model triatomic molecule M_3.[61] They find that with initial energy $E < 1.58E_D$ in a bond of M_3 that does not break, there is delayed fragmentation. The energized M_3 molecule undergoes several (sometimes as many as 20–30) vibrations during which time energy accumulates in the other two bonds, which then break. Moreover, the rate of fragmentation is extremely sensitive to small variations in the initial energy (see Fig. 3.16), and is not a monotone function of the initial energy. Similar results are found for a model of ClNO. And, inclusion of rotational motion, and vibration–

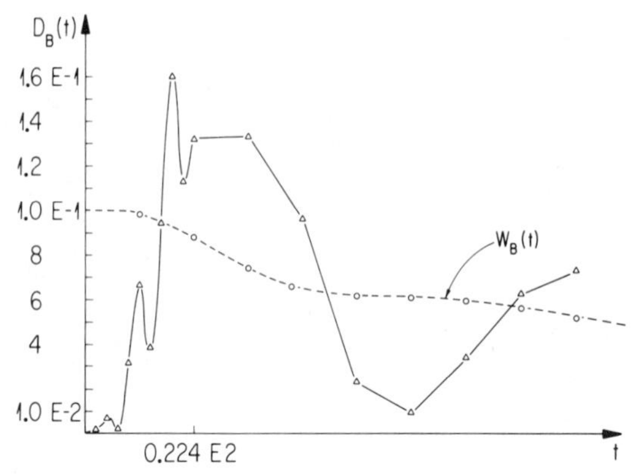

Fig. 3.15. The decay rate $D_B(t)$ and the probability of finding the system to be bound $W_B(t)$, for the initial state $|0,8\rangle$.

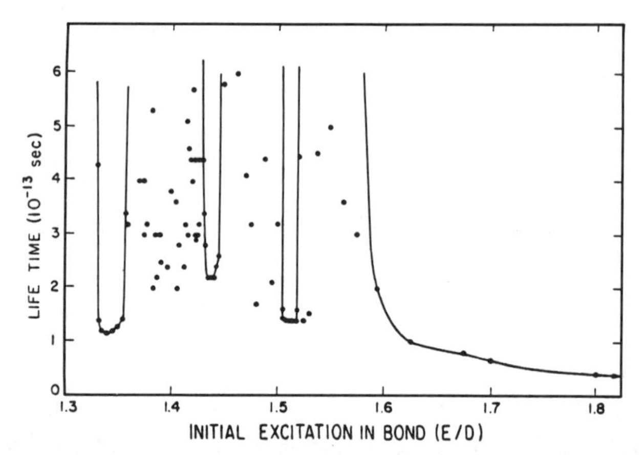

Fig. 3.16. Lifetime of a metastable M_3 molecule as a function of the initial potential energy localized in bond 2. Bonds 1 and 3 break, eventually. From C. A. Parr, Ph.D. dissertation, California Institute of Technology, 1968.

rotation interaction, does not alter the high sensitivity of the lifetime of the energized molecule to small perturbations in the initial energy. Parr and Kupperman conclude that there is not, in general, continuity of classical mechanical molecular lifetimes even on the smallest scale of energy differences in the corresponding quantum-mechanical case. It is not clear to me that the sources of the nonmonotone rates of decomposition in the Nordholm–Rice and Parr–Kupperman models are the same, but the similarity in findings is striking. Both results clearly suggest that, even when intramolecular energy transfer is rapid relative to chemical reaction, the reaction rate may not be adequately described by a statistical model.

A rather different approach to the study of fragmentation dynamics has evolved from the theory of radiationless transitions.[62] In this approach no attempt is made to directly study the molecular dynamics. Rather, a spectrum of zero-order states and their couplings is postulated, and this spectrum is assumed to incorporate all the necessary information about the molecule. The dynamical behavior of the system is then described by following the amplitude of an initially excited zero-order level of the spectrum as a function of time. The calculations can be carried out exactly for several different assumed energy dependences of the matrix elements coupling the different zero-order manifolds. It is amusing to note that what is widely accepted as intuitively plausible with respect to the behavior of these coupling matrix elements can lead to dynamical behavior that is very different from that obtained from integrating the equations of motion under conditions widely accepted as intuitively plausible with respect to energy exchange between oscillators.

Consider the model spectrum of states shown in Fig. 3.17. It is assumed that the system is prepared by excitation of the zero-order state ϕ_s. This state is not directly connected to the fragmentation continuum. Rather, ϕ_s is coupled to the intermediate dense manifold of zero-order levels $\{\phi_l\}$, and these in turn are coupled to the several continua $\{\phi_m^\alpha\}$. This spectrum is designed to model a situation in which the initially prepared state must relax to a different state before reaction occurs. The usual expectation, derived from chemical kinetic arguments, is that the population of ϕ_s will decay sequentially to the continuum via intermediate buildup and decay of population in the manifold of levels $\{\phi_l\}$. The simplest assumption that can be made about the coupling matrix elements of this spectrum is that they are constants independent of the energy. The dynamics of decay of ϕ_s

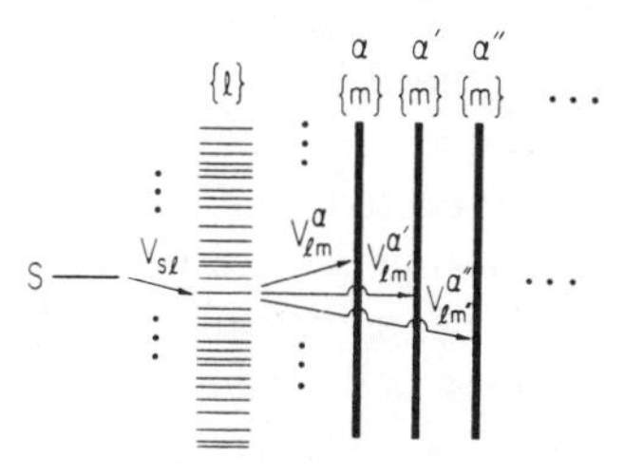

Fig. 3.17. The skeleton spectrum for the discussion of the influence of intramolecular coupling on the fragmentation of a molecule. Each continuum m, α represents a different partitioning between relative translational energy and internal energy of the products.

under this assumption deviate grossly from what is expected from the kinetic arguments cited.[63] Because the matrix elements are constant, amplitude flows coherently from ϕ_s to the $\{\phi_l\}$ and from the $\{\phi_l\}$ to the $\{\phi_m^\alpha\}$ and interference between the coherent amplitude components in the many levels leads to parallel feeding of the intermediate and fragmentation manifolds. There is not, in this case, a buildup and subsequent decay of amplitude in the levels $\{\phi_l\}$, and a buildup of amplitude in the continua $\{\phi_m^\alpha\}$. Even though the assumed constancy of the coupling matrix elements does not appear to be qualitatively incompatible with molecular properties, it leads to decay kinetics which are peculiar.[63]

Since the peculiar coherence effects just described arise from the assumption that the coupling matrix elements are constants, and since no molecule is likely to have this property, the consequences of other assumptions should be studied. Heller and Rice have examined the consequences of assuming that $V_1 \equiv V_{sl}$ and $V_2 \equiv V_l^\alpha$ have random character.[64] The dynamics have been worked out for the case of many coupled continua, with each molecular level coupled to each continuum.

The difference between constant and random coupling models arises as follows: For constant coupling, flux from ϕ_s into $\{\phi_l\}$ appears "near" the V_{lm}^α coupling region but escapes almost as soon as it enters $\{\phi_l\}$, that is, constant coupling leads to a steady-state situation in which the flux into the "V_{lm}^α coupled" region from ϕ_s is equal to the flux out, hence both $\{\phi_l\}$ and $\{\phi_m^\alpha\}$ serve as escaping continua in the sense that there is a flow out of a local interaction region into a noninteractive asymptotic region simultaneously in both manifolds. Thus, constant coupling has the effect of reducing the entire dense manifold, with all its symmetry variations, to the status of a single escaping translational continuum. In the random coupling case sequential behavior is obtained because there is an incoherent flow from ϕ_s into $\{\phi_l\}$, and the flux appears "uniformly in all of $\{\phi_l\}$." Because of the uniformity of the flow into $\{\phi_l\}$ there is a linear buildup in the $\{\phi_l\}$ to $\{\phi_m^\alpha\}$ manifold coupling. For t small the probability of dissociation then builds up as t^2, which is characteristic of sequential flow. Note, for this to be true only $V_{sl}V_{lm}^\alpha$ need be random, not V_{sl}, V_{lm}^α separately. We note in passing that, although the random coupling model predicts a sequential kinetic decay as expected, and a branching of population into the several fragmentation continua, there is no a priori reason why that final branching ratio need be statistical. However, when $\langle (V_{lm}^\alpha)^2 \rangle$ is independent of α the branching is determined by the relative values of the densities of states in the several continua $\{\phi_m^\alpha\}$.

The two examples cited represent extremes in the possible behavior of the matrix elements. Little is known of more general cases. Muthukumar and Rice have studied the consequences of the assumption that there are both systematic functional and random coupling matrix elements; the former were treated as a perturbation of the latter.[65] As particular examples they analyzed mixing of constant coupling and random coupling, and the mixing of Lorentzian coupling and random coupling. As expected, interference effects alter the time dependence of the decay of the system. For the case of mixed constant and random coupling these interference effects can increase or decrease the width of the resonance, that is, the time scale of the decay, depending on subtle variations of the ratio of magnitudes of the matrix elements. Also, the time dependence is no longer simple. For small t one finds a linear combination of t and t^2 terms characteristic of nonsequential and sequential decays, but the general behavior in time is more complex. The case of mixed Lorentzian coupling and random coupling is designed to mimic the situation when a few matrix elements are more important than others, but none can be neglected. That is, there is tight coupling within some subset of levels, which in turn are embedded in a dense manifold of levels. The calculated time evolution of the population of the initially excited level exhibits two time constants for small t, and is rather complex for large t. The shorter time constant corresponds to the redistribution of energy among levels that are nearly resonant with the initially excited level, and the longer time constant corresponds to the relaxation to other levels.

Consideration of the behavior of the coupling matrix elements of a model spectrum such as shown in Fig. 3.17 inevitably leads to the question: When can a Master equation be used to describe intramolecular dynamics? I do not believe we have the answer to this question, although sufficient (but not necessary) conditions for the validity of a Master equation have been established. One analysis, based on the assumption that the off-diagonal elements of the energy matrix (in some suitable representation) are random in sign, has been developed by Gelbart, Rice, and Freed.[66] In the most complete of the attempts to derive a Master equation for a strongly coupled finite system of oscillators, Kay[67] starts with the usual decomposition of the Hamiltonian, $H=H_0+V$. However, unlike the usual decomposition of H, in this case V must not be small. It is necessary that V simultaneously couple many degrees of freedom and lead to multiple (vibrational) quantum exchanges between the states defined by H_0, namely, $H_0|m\rangle=E_m|m\rangle$. Because of the complexity of the system, when the states $|m\rangle$ are ordered according to energy E_m, it is to be expected that the character of $|m-1\rangle$, and of $|m+1\rangle$, will differ considerably from that of $|m\rangle$. This is taken to be a qualitative feature of the many oscillator system.

Kay argues that three important energy ranges characterize the many oscillator system:

1. ΔE: Let the coupling be measured by $V_{nm} \equiv \langle n|V|m\rangle$. It is assumed that V_{nm} and the vibrational density of states are sensibly constant over the energy interval ΔE.
2. Γ: By virtue of nonzero coupling between the levels of H_0 these levels acquire a mean width Γ.
3. $g\epsilon$: Let ϵ be the mean spacing between the levels of H_0, and g a large integer. Kay requires that averages over g adjacent levels be sensibly the same as an average of the same quantity over ΔE.

The three energy ranges 1, 2, and 3 are related as $\Gamma \ll \Delta E$, $g\epsilon \ll \Gamma$. A Master equation describes an irreversible process. The required element of incoherence, leading to irreversible flow of energy on the time scale of interest, is introduced by Kay under the assumption that there is no correlation between coupling matrix elements V_{nm} and $V_{nm'}$ for $|m\rangle$ and $|m'\rangle$ less than ΔE apart:

$$g^{-1} \sum_{\substack{n \\ |n-n_0|<g/2}} V_{nm}=0$$

$$g^{-1} \sum_{\substack{n \\ |n-n_0|<g/2}} V_{mn}V_{nm'}=0, \qquad m \neq m', \qquad |E_m-E_{m'}|<\Delta E \tag{3.66}$$

The two expressions of the assumption of incoherence can be combined to read

$$g^{-1} \sum_{\substack{n \\ |n-n_0|<g/2}} V_{mn}V_{nm'}=v^2\delta_{mn}; \qquad |E_m-E_{m'}|<\Delta E \tag{3.67}$$

Condition (3.67) plays the same role in Kay's analysis as does the condition of diagonal singularity in Van Hove's analysis.[68] There is also a similar condition in the random matrix theory analysis of Gelbart, Rice, and Freed.[66]

As a final condition Kay postulates that there is strong interconnection of states. This means that any pair of states $|m\rangle$, $|m'\rangle$ closer in energy than $g\epsilon$ is connected by a chain of couplings through intermediate states $|n_1\rangle, |n_2\rangle, \ldots, |n_k\rangle$, such that $V_{n_j n_{j+1}} \neq 0$ for $n_0=m$, $n_{k+1}=m'$ and

$$\tfrac{1}{2}(E_m+E_{m'}-g\epsilon) \leqslant E_{n_1}, \ldots, E_{n_k} \leqslant \tfrac{1}{2}(E_m+E_{m'}+g\epsilon)$$

The coupling is characterized as strong interconnection when the intermediate sequences of coupled levels are short enough and numerous enough that all states within $g\epsilon$ are populated on a time scale $t \ll \hbar/\delta$ where $\Gamma \gg \delta \gg g\epsilon$.

With the conditions cited Kay shows that a generalized Master equation describes the evolution of a function related to the occupation probability of a state representing a certain property A, but having zero-order energy specified only to ΔE. Transition probabilities connect any g consecutive levels in ΔE. Because V is not small transitions do occur between states with different energies.

F. Summary

We have seen that if a coupled oscillator system has classical mechanical quasiperiodic trajectories for $E < E_c$ and stochastic trajectories for $E > E_c$, and if the eigenstates of that system are represented as a superposition of harmonic oscillator basis states, then there is a "KAM-like" change in the amplitudes of the contributing basis functions at about E_c. Yet the wavefunction itself, and the corresponding Wigner function, do not display any sign of "randomness," and the evidence from semiclassical quantum theory suggests that only a subset of the quasiperiodic trajectories embedded in the stochastic trajectories correlate with the eigenstates of the system. Thus, despite the fact that the measure of these quasiperiodic trajectories is very small compared to the measure of all trajectories, they are the pertinent ones.

The idea that only the quasiperiodic trajectories correlate with the eigenstates of the system has been proposed many times. The argument (not a proof) I like best is due to Freed, who, from a careful consideration of the semiclassical approximation to the Green's function of the system, shows that interference effects destroy the contributions to the action of all but the quasiperiodic trajectories.[69]

What, then, is the meaning of the KAM-like behavior of the coefficients in the harmonic oscillator basis function expansion of the eigenstate? I do not know, and I think it is important to find out, especially if we wish to learn how to prepare a molecule in a state which will evolve in a way that produces interesting chemistry.

IV. CLOSING REMARK

This chapter is intended to provide the background necessary for the appreciation of the more technical contributions to this volume. Since most of the pertinent qualitative interpretations of the dynamical analyses have already been set out, I will close with the observation that we have much to learn about the dynamics of coupled nonlinear oscillator systems and intramolecular energy transfers; clearly we should expect to find that some of the currently accepted notions concerning the dynamics of such systems will need modification. There is, in my opinion, considerable likelihood that at least some of what we learn about the intramolecular dynamics can be used to control the preparation of reactants and bring into being aspects of photoselective chemistry.

I suggest that well-defined molecular states exist for all energies less than that at which the radiative widths of the levels begin to overlap. Of course, these levels differ dramatically in the character of the nuclear motions, and only a subset are likely to influence in a specific fashion the reactivity of the molecule. If we can learn how to identify and how to excite the molecule into that subset of levels, it is possible that we can selectively modify the rate and selectivity of the ensuing chemistry. In contrast, even narrow-band excitation into the region where the radiative widths of the levels overlap prepares a wavepacket which will disperse. The behavior of a molecule so excited will depend on how the initial wavepacket evolves, but is likely to be statistical. Even if the reaction occurs prior to randomization of the energy, there would likely be less selectivity than by excitation in the discrete levels at the low-energy end of the manifold of states.

Note Added in Proof

Several contributions to the study of chaotic behavior in quantum mechanical systems appeared after this article was submitted for publication.

Noid, Koszykowski, Tabor and Marcus[70] have examined the sensitivity of the eigenvalues of the Henon-Heiles Hamiltonian to variation of the magnitude of the anharmonic coupling. When viewed as a function of the anharmonic coupling parameter, the eigenvalues exhibit "crossings" and "avoided crossings," the latter being responsible for very large second differences between adjacent eigenvalues, as originally found by Pomphrey.[40] An avoided crossing corresponds, in the classical mechanical limit, to a nonlinear resonance. Noid, et al. argue that if a state is simultaneously involved in many avoided crossings its eigenvector has statistical character, which notion is analogous to the Chirikov description of chaos as the consequence of overlapping nonlinear resonances.[42]

Heller[71] has examined the quantum dynamics of wave packet motion and proposed a criterion for stochastic behavior. This criterion is the following: Given an initial nonstationary state ϕ on some potential surface, and the set of eigenfunctions of the potential surface $\{\psi_n\}$, the probabilities $p_n^\phi \equiv |\langle\psi_n|\phi\rangle|^2$ vary smoothly from one to the next for stochastic dynamics and fluctuate from one to the next in the case of nonstochastic dynamics. Since the functions $\langle \psi_n | \phi \rangle$ are the Franck-Condon integrals for the system on that surface, the optical spectrum can be used to diagnose the dynamical behavior. Heller gives a careful discussion of the relationship between the prior knowledge of the nonstationary state, the fluctuations in p_n^{ϕ} and the dependence of the concept of stochastic motion on choice of conditions for an experiment; for details the reader is referred to the original paper.[71] The connection between chaos in a classical mechanical and the corresponding quantum mechanical system is made, in part, by comparison of wave packet motion and classical trajectories.

Kay[72] has examined the dynamics of nonstationary macrostates in a system of two coupled Morse oscillators. He finds, in a number of cases, differences between the classical mechanical and quantum mechanical transition probabilities connecting isoenergetic macrostates, and gives a very interesting discussion of circumstances under which a quantum mechanical system will not display the characteristics of stochastic motion even though the corresponding classical mechanical system does. Kay attributes the difference in behavior, when it occurs, to the fact that if the n-m nonlinear resonance is too narrow to contain many levels it becomes ineffective in influencing the system's behavior. He further conjectures that the source of nonergodic behavior in his model is present in all systems of Morse oscillators and, possibly, in all real molecular systems.

Note that Kay's results suggest that the use of "avoided crossings" as a criterion of quantum mechanical chaos, as suggested by Noid et al,[70] is unlikely to be universally valid.

Miller[73] has examined a two oscillator model of predissociation by tunneling. He finds that the energy dependence of the rate of predissociation appears insensitive to the existence, or nonexistence, of chaos in the corresponding classical mechanical system.

Kosloff and Rice[74] have extended the concept of Kolmogorov entropy to quantum mechanical systems. They show that for a bounded quantum mechanical system the Kolmogorov entropy is zero. Since the Kolmogorov entropy is a measure of chaos in classical mechanics (its value is related to the average over the phase space of the characteristic e-folding time for exponential growth of the distance between initially close trajectories), Kosloff and Rice conclude that there is an essential difference between chaos in classical mechanical and quantum mechanical systems. This conclusion is in one sense trivial, but in another sense profound. It is trivial once it is recognized that a bounded quantum mechanical system always has a discrete spectrum, and a system with a discrete spectrum always has almost periodic behavior—a continuum spectrum is a necessary, but not sufficient, condition for nonzero Kolmogorov entropy and irreversibility in dynamics. The conclusion is profound in that it focuses attention on the interference effects that exist in quantum mechanics but not in classical mechanics. The Kosloff-Rice work leads to the natural examination of state preparation methods that exploit, perhaps by sequential excitation, the interferences inherent to the dynamics. It is at least conceivable that by suitable preparation procedures the effective rate of dephasing of a wave packet can be altered, hence one can attempt to intercept the dispersal of a prepared state with some other process and effect selective chemistry.

Acknowledgments

My research on intramolecular dynamics has been supported by the Air Force Office of Scientific Research and the National Science Foundation.

References

1. J. Ford, *Adv. Chem. Phys.* **24**, 155 (1973). J. Ford, in *Fundamental Problems in Statistical Mechanics*, Vol. 3, E. G. D. Cohen, ed., North-Holland, Amsterdam, 1975, p. 215.
2. M. V. Berry, in *Topics in Nonlinear Analysis*, S. Jorna, ed., A. I. P. Conf. Proc. 46, New York, 1978, p. 16.
3. P. Brumer, *Adv. Chem. Phys.* (this volume).
4. See G. Benettin, C. Froeschle, and J. P. Scheidecker, *Phys. Rev. A.*, **19**, 2454 (1979), and references cited therein.
5. Some aspects of this problem are discussed by M. V. Berry, Ref. 2.
6. S. A. Rice, in *Advances in Laser Chemistry*, A. Zewail, ed., Springer-Verlag, Berlin, 1978, p. 2.
7. A good description of isolating and nonisolating integrals of the motion is given in G. Contoupolos, *Astrophys. J.*, **138**, 1297 (1963).
8. See, for example, V. I. Arnold and A. Avez, *Ergodic Problems of Classical Mechanics*, Benjamin, New York, 1968; Y. M. Treve in Ref. 2, p. 147.
9. M. Henon and C. Heiles, *Astron. J.*, **69**, 73 (1964).
10. See reference 3.
11. G. H. Walker and J. Ford, *Phys. Rev.*, **188**, 416 (1969).
12. J. Ford and G. H. Lunsford, *Phys. Rev. A 1*, **59** (1970).
13. G. M. Zaslavskii and B. V. Chirikov, *Sov. Phys. Usp.*, **14**, 549 (1972). B. V. Chirikov, *Research Concerning the Theory of Nonlinear Resonances and Stochasticity*, Novosibirsk; U.S.S.R., 1969, Unpublished, translated as CERN 71-40 Geneva (1971).
14. H. Poincare, *Les Methodes Nouvelles de la Mecanique Celeste*, Dover, New York, 1957.
15. I have made extensive use of the very clear presentation of M. V. Berry, Ref. 12, in the following paragraphs.
16. G. D. Birkhoff, *Dynamical Systems*, Am. Math. Soc., Providence, R.I., revised edn. (1966).
17. M. Toda, *Prog. Theor. Phys. Suppl.*, **45**, 174 (1970); **59**, 1 (1976). M. Toda, *Phys. Rept. Phys. Lett. C (Netherlands)*, **18**, 1 (175).
18. J. Ford, S. D. Stoddard, and J. S. Turner, *Prog. Theor. Phys.*, **50**, 1547 (1973).
19. Two books that give excellent treatments are: A. H. Nayfeh, *Perturbation Methods*, Wiley, New York, 1973. G. Giacaglia, *Perturbation Methods in Nonlinear Systems*, Springer-Verlag, Berlin, 1972.

20. A. Lindstedt, *Astron. Nach.*, **103**, Col. 211 (1882). H. Poincaré, Ref. 14.
21. P. A. Sturrock, *Plasma Hydromagnetics*, Stanford University Press, Stanford, Calif., 1962.
22. E. A. Frieman, *J. Math. Phys.*, **4**, 410 (1963).
23. A. H. Nayfeh, *J. Math. Phys.*, **44**, 368 (1965).
24. G. Sandri, *Nuovo Cimento*, **B36**, 67 (1967).
25. B. Van der Pol, *Phil. Mag.*, **43**, 700 (1926).
26. N. Krylov and N. N. Bogoliubov, *Introduction to Nonlinear Mechanics*, Princeton University Press, Princeton, N.J., 1947. N. N. Bogoliubov and Y. A. Mitropolski, *Asymptotic Methods in the Theory of Nonlinear Oscillations*, Gordon and Breach, New York, 1961.
27. M. Kruskal, *J. Math. Phys.*, **3**, 806 (1962).
28. H. von Zeipel, *Ark. Mat. Astron. Fysik (Stockholm)*, **11**, No. 1, I; No. 7, I (1916).
29. G. I. Hori, *Publ. Astron. Soc. Japan*, **18**, 287 (1966); **22**, 191 (1970).
30. A. Deprit, *Celest. Mech.*, **1**, 12 (1969).
31. A. A. Kamel, *Celest. Mech.*, **1**, 190 (1969); **3**, 90 (1970); **4**, 397 (1971).
32. J. M. A. Danby, *Celest. Mech.*, **8**, 273 (1973).
33. G. Contoupolos, *Astron. J.*, **75**, 96 (1970).
34. B. Barbanis, *Astron. J.*, **71**, 415 (1966).
35. R. Helleman and T. Bountis in *Stochastic Behavior in Classical and Quantum Hamiltonian Systems*, Springer-Verlag, Berlin, 1979, p. 353.
36. F. G. Gustavson, *Astron. J.*, **71**, 670 (1966).
37. G. Contoupolos, L. Galgani, and A. Giorgilli, *Phys. Rev.*, **A18**, 1183 (1978).
38. K. S. J. Nordholm and S. A. Rice, *J. Chem. Phys.*, **61**, 203 (1975), 61, 768 (1974).
39. R. M. Stratt, N. C. Handy, and W. H. Miller, *J. Chem. Phys.*, **71**, 3311 (1979).
40. N. Pomphrey, *J. Phys.*, **B7**, 1909 (1974).
41. C. Jaffe and W. Reinhardt, *J. Chem. Phys.*, **71**, 1862 (1979).
42. D. W. Oxtoby and S. A. Rice, *J. Chem. Phys.*, **65**, 1676 (1976).
43. P. Pechukas, *J. Chem. Phys.*, **57**, 5577 (1972).
44. R. A. Marcus, D. W. Noid, and M. L. Koszykowski, in *Stochastic Behavior in Classical and Quantum Hamiltonian Systems*, G. Casati and J. Ford, eds., Springer-Verlag, Berlin, 1979, p. 283.
45. For a review see I. C. Percival, *Adv. Chem. Phys.*, **36**, 1 (1977); I. C. Percival, *J. Phys.*, **A7**, 794 (1974). I. C. Percival, *J. Phys.*, **A7**, 794 (1974). I. C. Percival and N. Pomphrey, *J. Phys.*, **B31**, 97 (1976).
46. M. V. Berry, *Phil. Trans. Roy. Soc. (London)*, **A287**, 237 (1977); *J. Phys.*, **A10**, 2083 (1977). M. V. Berry and M. Tabor, *Proc. Roy. Soc. (London)*, **A349**, 101 (1976).
47. S. Chapman, B. C. Garrett, and W. H. Miller, *J. Chem. Phys.*, **64**, 502 (1976). W. H. Miller, *J. Chem. Phys.*, **63**, 936 (1975).
48. D. W. Noid and R. A. Marcus, *J. Chem. Phys.*, **62**, 2119 (1975). W. Eastes and R. A. Marcus, *J. Chem. Phys.*, **61**, 4301 (1974).
49. R. T. Swimm and J. B. Delos, *J. Chem. Phys.*, **00**, 00 (1900); also in *Stochastic Behavior in Classical and Quantum Hamiltonian Systems*, G. Casati and J. Ford, eds., Springer-Verlag, Berlin, 1979, p. 306.
50. M. C. Gutzwiller, *J. Math. Phys.*, **8**, 1979 (1967); **10**, 1004 (1969); **11**, 1971 (1970); **12**, 343 (1971).
51. The quantity δ is usually $\frac{1}{2}$, but can be different. See, for example, Ref. 2.
52. G. C. Schatz and T. Mulloney, *J. Phys. Chem.*, **83**, 989 (1979).
53. M. Born, *The Mechanics of the Atom*, G. Bell, London, 1927.
54. J. S. Hutchinson and R. E. Wyatt, Chem. Phys. Lett. **72**, 378 (1980).

55. P. Brumer and J. W. Duff, *J. Chem. Phys.*, **65**, 3566 (1976). J. W. Duff and P. Brumer, *J. Chem. Phys.*, **67**, 4898 (1977). M. Toda, *Phys. Lett.*, **A48**, 335 (1974).
56. E. J. Heller, *Chem. Phys. Lett.*, **60**, 338 (1979).
57. R. Kosloff and S. A. Rice, Chem. Phys. Lett. **69**, 209 (1980).
58. A. Peress, *Am. J. Phys.*, **42**, 886 (1974).
59. R. Kosloff, *Adv. Chem. Phys.* to be published.
60. K. S. J. Nordholm and S. A. Rice, *J. Chem. Phys.*, **62**, 157 (1975).
61. C. A. Parr and A. Kuperman, from C. A. Parr, Ph.D. thesis, California Institute of Technology, 1968.
62. See, for example, S. A. Rice in *Excited States*, Vol. 11, E. C. Lim, ed., Academic Press, New York, 1975, p. 112.
63. K. G. Kay and S. A. Rice, *J. Chem. Phys.*, **57**, 3041 (1972).
64. E. J. Heller and S. A. Rice, *J. Chem. Phys.*, **61**, 936 (1974).
65. M. Muthukumar and S. A. Rice, *J. Chem. Phys.*, **69**, 1619 (1978).
66. W. M. Gelbart, S. A. Rice, and K. F. Freed, *J. Chem. Phys.*, **57**, 4699 (1972).
67. K. G. Kay, *J. Chem. Phys.*, **61**, 5205 (1974).
68. L. Van Hove, *Physica*, **21**, 517 (1955); **23**, 441 (1957); **25**, 268 (1959).
69. K. F. Freed, *Disc. Faraday Soc.*, **55**, 68 (1973).
70. D. W. Noid, M. L. Koszykowski, M. Tabor and R. A. Marcus, *J. Chem. Phys.* **72**, 6169 (1980).
71. E. J. Heller, *J. Chem. Phys.* **72**, 1337 (1980).
72. K. G. Kay, *J. Chem. Phys.* **72**, 5955 (1980).
73. W. H. Miller, private communication.
74. R. Kosloff and S. A. Rice, *J. Chem. Phys.* in press.

INTRAMOLECULAR ENERGY TRANSFER: THEORIES FOR THE ONSET OF STATISTICAL BEHAVIOR

PAUL BRUMER*

Department of Chemistry, University of Toronto, Toronto, Ontario, Canada

CONTENTS

I. Introduction 201
II. Intramolecular Dynamics: Problems of Interest 203
III. Classical Dynamics and the Stochastic Transition 205
  A. Completely Integrable Systems 206
  B. Mixing Systems 208
  C. Typical Systems 210
  D. K-Entropy 212
IV. Numerical Experiments on Coupled Oscillators 214
V. Models for the Regular to Erratic Transition 220
  A. Mo's Method 220
  B. Variational Equations Approach 221
  C. Critical-Point Analysis 225
  D. Overlapping Resonances 226
  E. Comments 230
VI. Exponentiating Trajectories and Statistical Behavior 231
Appendix: What's New in 1980 233
Acknowledgments 235
References 235

I. INTRODUCTION

Chemists have assumed, for quite some time, the existence of qualitatively different types of intramolecular dynamics in isolated polyatomic molecules. For example, traditional polyatomic vibrational spectroscopy, which samples low-energy behavior, is analyzed in terms of the periodic motion of normal coordinates, whereas behavior at the opposite extreme, that is, statistical intramolecular energy transfer, is assumed to occur in higher energy processes such as unimolecular dissociation. One therefore

*Alfred P. Sloan Foundation Fellow

expects a transition, within a given molecule, between these two characteristically different types of behavior as the total energy increases. A substantial amount of experimental data[1] and trajectory calculations[2] support the view that the detailed dynamics of polyatomics is indeed qualitatively different at high and low energies. Although a transition from regular to erratic behavior with increasing energy is quite naturally assumed there is no known theory that quantitatively accounts for all observations (discussed below) of this transition in model systems. A variety of useful first approximations to a complete theory have, however, been proposed; a review of these theories is presented in this chapter.

The change from regular to erratic behavior as system parameters (e.g., energy or anharmonicity) vary is generally understood to be associated with the nonlinear aspect of the underlying dynamics. Thus it is hardly surprising that this transition is important to a host of nonlinear phenomena other than chemical reaction dynamics. Indeed, the list is almost all-encompassing, including plasma physics,[3] ecology,[4] biology,[5] laser technology,[6] oceanography,[7] pure mathematics,[8] statistical mechanics,[9] celestial mechanics,[10] engineering,[11] global weather prediction,[12] particle physics,[13] and fluid mechanics.[14] As a particular example we cite recent efforts[15] to describe diseases characterized by the breakdown of normal periodic body functions as a manifestation of the erratic regime of the underlying nonlinear equations describing physiological behavior.[16] Modern nonlinear mechanics has therefore seen input from a wide range of subject areas. Some of these developments, selected in accordance with the following limitations, are reviewed below. First, our concern is principally with isolated single molecules and hence with the nonlinear mechanics of conservative Hamiltonian systems. Second, reference to the relevant formal mathematical theory will be qualitative and sketchy. This results from the fact that (1) the formal theory is highly technical in nature,[16] (2) discussions with mathematicians indicate that the formal theory is still at a stage where it can not provide *quantitative* answers to problems of chemical interest, and (3) many of the concerns of mathematicians in the area of dynamical systems and ergodic theory are decidedly different than those of interest in chemistry or physics. Finally, interest in applications of modern nonlinear mechanics has resulted in a variety of excellent recent review articles;[17] overlap of content with these articles will be kept to a minimum although some repetition is unavoidable. Fortunately the principal emphasis of this chapter, on current theories of the regular to erratic transition, has not been the subject of an extensive review.

This chapter is organized as follows. In Section II we discuss some problems of interest in molecular reaction dynamics and the possible utility of modern nonlinear mechanics developments in resolving these problems. In Section III we summarize some features of integrable, mixing, and perturbed integrable systems, which are necessary to understand the characteristics of the transition from regular to erratic behavior in polyatomic molecules. Numerical results on systems displaying this transition are contained in Section IV, whereas Section V contains a discussion of some of the theories proposed to predict the energy of this transition. Finally, in Section VI we discuss several calculations designed to provide insight into the relationship between erratic behavior discussed in nonlinear mechanics and statistical theories prevalent in chemistry.

II. INTRAMOLECULAR DYNAMICS: PROBLEMS OF INTEREST

Renewed interest in the detailed internal dynamics of energized molecules stems from several sources. Of primary importance is the development of laser sources and detection devices which promise the possibility of experimentally preparing, and subsequently probing the dynamics of, coherently excited specifically energized molecules. Indeed interest abounds in the possibility of influencing chemical processes via laser intervention; the efficiency of such processes depends strongly on the relative rate of the process of interest vs. that of competitive processes such as intramolecular energy transfer. At present a variety of experimental and theoretical studies[1] have been interpreted as suggesting that the latter rate may in fact be smaller than previously anticipated and that statistical intramolecular energy transfer is not ubiquitous.

Several of the major chemical questions of interest have been succinctly summarized by Rice in a recent review article[17d]:

1. Under what conditions, if any, is intramolecular energy exchange slow/rapid relative to other processes, for example, photon emission, isomerization or fragmentation?
2. How does the intramolecular energy exchange depend on the energy of the molecule and the nature of the initial excitation?
3. If there are situations for which intramolecular energy exchange is slow relative to chemical reaction, why does this behavior occur? Does it derive from special characteristics of the molecular force fields? Are there dynamical or symmetry restrictions on the spectrum of states in these cases? Are these special situations commonly or rarely found?
4. Given the answers to questions 3 above, can we devise excitation methods and reaction conditions that permit enhancement of selectivity of the chemistry that follows?

Thus our principle concern is with the rate and extent of intramolecular energy transfer in the presence of competitive processes. This focus differs considerably from that of other areas to which modern nonlinear mechanics has been applied, in which competitive processes are unimportant and rates are of little interest.

Additional interest in detailed aspects of intramolecular dynamics stems from efforts[17c] to extend semiclassical theory to bound states of polyatomic molecules. Recent studies[18] in this area have relied heavily on the topological structure of phase space dynamics; this structure is substantially different in energy regimes corresponding to regular and erratic behavior.

Renewed interest in particular problems does not necessarily imply the availability of new theoretical tools with which to tackle these problems. In this case, however, the widespread interest in nonlinear mechanics mentioned above has, in fact, suggested an approach that can be expected to provide new insights into intramolecular dynamics. Efforts based on these developments are now being directed toward obtaining a general understanding of the dynamics, an approach that can be expected to be far more fruitful than pioneering trajectory studies on individual molecular systems. These developments have arisen principally in studies of *classical* dynamical systems, hence the focus of this review.

Studies in semiclassical mechanics[19] have confirmed the utility of classical studies of molecular processes under rather general conditions, that is, where there is substantial averaging over quantum numbers. Nevertheless, intramolecular dynamics is properly described using quantum mechanics. Both computational and conceptual difficulties exist in attempting to treat intramolecular dynamics quantum mechanically. Conceptual problems result from the uncertainty as to how erratic behavior is manifest quantum mechanically. Several possibilities have been discussed. For example, erratic behavior may be[20] reflected in the individual wavefunctions, which are proposed to have irregular nodal structure. Alternatively,[21] wavefunctions expanded in local zeroth order basis sets are expected to display basis set coefficients, which are spread globally over all basis states. Proposals have also been advanced as to the nature of the energy spectrum in the erratic regime. In particular, it is proposed to be characteristically random.[21, 22] The need for a better understanding of quantum intramolecular energy transfer and its relation to classical studies is made amply clear in studies by Heller,[23a] in which the quantum and classical mechanics lead to qualitatively different predictions of system behavior.

The analogy between the current state of the quantum problem and that of the classical problem several decades ago is striking. Developments in modern ergodic theory along with those in nonlinear mechanics have suggested new directions for the study of classical intramolecular dynamics. One principle advance in ergodic theory was the recognition (as discussed below) that classical ergodicity is not necessarily consistent with properties expected of random behavior, leading to the introduction of the stronger concept of mixing. At present, quantum ergodicity has been deemed unrelated[21, 23] to behavior required of random systems, and the proper useful generalization is being sought.

III. CLASSICAL DYNAMICS AND THE STOCHASTIC TRANSITION

Henceforth we regard a polyatomic molecule as a classical N degree of freedom system with Hamiltonian $H(\mathbf{p},\mathbf{q})$. Here $\mathbf{q}$ and $\mathbf{p}$ are the two N-vectors describing the phase space coordinates and conjugate momenta of the system. To discuss the nature of the dynamics of such systems we first discuss two restricted limiting cases: completely integrable systems and mixing systems. In essence these correspond to cases of purely regular motion and highly statistical motion, respectively. These systems are introduced for two reasons. First, they are mathematically well understood, whereas a typical coupled oscillator system is not.[24] Second, a typical polyatomic Hamiltonian system will display regular motion characteristic of integrable systems at low energies and erratic motion characteristic of mixing systems at higher energies. Thus it is useful to emphasize the characteristics of these limiting cases at the outset. The discussion of integrable systems and the influence of small perturbations on such systems (i.e., the KAM theorem), as well as that of highly statistical model systems, is kept deliberately brief. This results from the availability of a recent lucid, hard-to-beat, pedagogically sound review by M. V. Berry,[17e] which is highly recommended.

Numerical studies on simple oscillator systems, discussed below, have relied upon three methods, the Poincaré surface of section (for $N=2$), studies of the rate of divergence of initially adjacent trajectories in phase space and the nature of the power spectrum of the system. We first define these terms and comment later below on their nature in both integrable and mixing systems.

Poincaré Surface of Section: Consider the two degree of freedom system at total energy E with phase-space coordinates and momenta q_1, q_2, p_1, p_2. It proves convenient to study the phase-space dynamics by plotting the (q_1, p_1) coordinates of the trajectory when q_2 assumes a fixed value (say $q_2=0$) and p_2 is of fixed sign. Typical surfaces of section are discussed below and their computational utility made obvious. In essence this technique reduces the dynamics of $N=2$ systems to an area preserving mapping of a plane onto itself.

Divergence of Trajectories: Of importance, as discussed below, is the sensitivity of the dynamics to isoenergetic changes in trajectory initial conditions. This is quantitatively defined in terms of the nature (e.g., linear or exponential in time) and rate of growth of the distance between two trajectories that are initially infinitesimally close to one another in phase space.

Power Spectrum: The power spectrum $I(\omega)$ for any dynamical variable $f(\mathbf{q},\mathbf{p})$ is given by

$$I(\omega)=\frac{1}{2\pi}\int_{-\infty}^{\infty}\langle f[\mathbf{q}(t),\mathbf{p}(t)]f[\mathbf{q}(0),\mathbf{p}(0)]\rangle\exp(-i\omega t)\,dt \qquad (1)$$

Angle brackets $\langle\ \rangle$ denote an ensemble average; unless otherwise specified this notation will refer to a microcanonical average (see, however, Appendix A).

The interest in these three quantities stems from their characteristically different behavior for integrable and mixing systems.

A. Completely Integrable Systems

An integral of the motion is a quantity $F(\mathbf{p},\mathbf{q})$, which remains constant during the time evolution of the system, i.e., $\{F(\mathbf{p},\mathbf{q}), H\}=0$, where $\{\ \}$ denotes a Poisson bracket. A completely integrable N degree of freedom system is one which possesses N independent single-valued integrals of motion $F_k(\mathbf{p},\mathbf{q})$, $k=1,\ldots,N$. Under these circumstances the N functions F_k may be chosen as a new set of generalized momenta; the time dependence of the new coordinates is then particularly simple.

The existence of N independent constants $F_k(\mathbf{p},\mathbf{q})$ ensures that the motion of the system is constrained to an N-dimensional manifold. The topology of the manifold[16] is that of a torus (i.e., an N-dimensional doughnut) and, as a consequence, it proves convenient to linearly recombine the $F_k(\mathbf{p},\mathbf{q})$ to yield a set of N constants of the motion which are the classical actions $\mathbf{I}$ with conjugate angle coordinates $\boldsymbol{\theta}$. Here $\mathbf{I}$ and $\boldsymbol{\theta}$ are N-dimensional vectors, the angles $\boldsymbol{\theta}$ being associated with "circular" motion about the torus; the $N=2$ case is shown in Fig. 1.

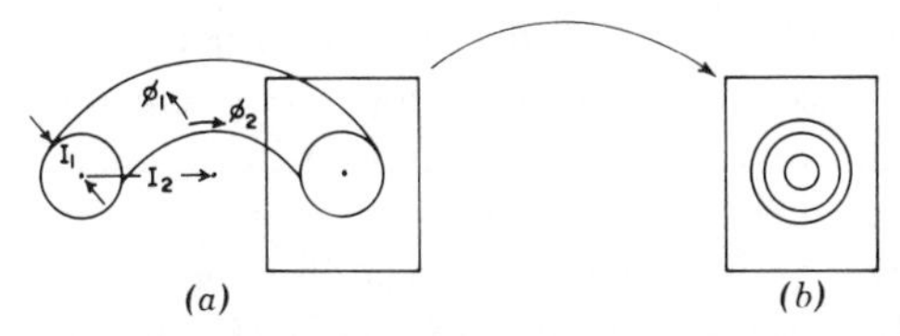

Fig. 1. (a) Cross-section of $N=2$ torus, and (b) corresponding surface of section.

The Hamiltonian is then of the form $H(\mathbf{I})$ and Hamilton's equations are given by

$$\dot{\mathbf{I}}=-\frac{\partial H}{\partial \boldsymbol{\theta}}=0$$

$$\dot{\boldsymbol{\theta}}=\frac{\partial H}{\partial \mathbf{I}}=\boldsymbol{\omega}(\mathbf{I}) \tag{2}$$

where the second equation defines the N frequencies $\boldsymbol{\omega}(\mathbf{I})$. These equations have the simple solution $\boldsymbol{\theta}(t)=\boldsymbol{\omega}(\mathbf{I})t+\boldsymbol{\delta}$ and $\mathbf{I}=$constants; $\boldsymbol{\delta}$ are the initial values of the angles. Elementary arguments show that under these circumstances $\mathbf{q}$ and $\mathbf{p}$ are given by[17e]

$$\mathbf{q}(t)=\sum_{\mathbf{m}} \mathbf{q}_{\mathbf{m}}(\mathbf{I})\exp[\,i(\mathbf{m}\cdot\boldsymbol{\omega}(\mathbf{I})t+\mathbf{m}\cdot\boldsymbol{\delta})]$$

$$\mathbf{p}(t)=\sum_{\mathbf{m}} \mathbf{p}_{\mathbf{m}}(\mathbf{I})\exp[\,i(\mathbf{m}\cdot\boldsymbol{\omega}(\mathbf{I})t+\mathbf{m}\cdot\boldsymbol{\delta})] \tag{3}$$

where the sum is over the integer vector $\mathbf{m}$; the system is said to be quasiperiodic.

A simple example of a quasiperiodic system is a set of N independent harmonic oscillators described by

$$H_{\text{ho}}=\frac{1}{2}\left[\sum_{i=1}^{N}\frac{p_i^2}{m_i}+k_i q_i^2\right]=\sum_i I_i\omega_i$$

with

$$I_i=\frac{1}{2\omega_i}\left[\frac{p_i^2}{m_i}+k_i q_i^2\right], \qquad \omega_i=\frac{1}{2\pi}\left(\frac{k_i}{m_i}\right)^{1/2}$$

which possesses the linear feature that the frequencies are independent of $\mathbf{I}$.

The nature of the Poincaré surface of section for $N=2$ integrable systems is readily discerned from Fig. 1 by cutting the torus with a plane that is nowhere tangent to the torus. As the trajectories travel about the torus with frequencies $\omega_1(\mathbf{I})$, $\omega_2(\mathbf{I})$, they repeatedly cut through the surface, generating a smooth regular curve. Variations in the action I_1 (see Fig. 1*a*) will result in nested tori, which would be reflected, in the simplest case, as nested closed curves in the surface of section (Fig. 1*b*). Two types of tori are distinguished: those with rational $\omega_1(\mathbf{I})/\omega_2(\mathbf{I})$ and those with an irrational ratio.

The response of a quasiperiodic trajectory to changes in initial conditions may be readily ascertained. In particular, introduce the quantities $\delta\mathbf{I}(t)=\mathbf{I}'(t)-\mathbf{I}(t)$ and $\delta\boldsymbol{\theta}(t)=\boldsymbol{\theta}'(t)-\boldsymbol{\theta}(t)$, where the primes denote action and angle variables corresponding to a trajectory initially perturbed relative to the unprimed trajectory. Equations for $\delta\dot{\mathbf{I}}(t)$ and $\delta\dot{\boldsymbol{\theta}}(t)$ may be obtained, as described in Section V, by expanding the primed quantities about the unprimed quantities in a Taylor series expansion. The resultant equations show that $\delta\mathbf{I}(t)$ is zero and that $\delta\boldsymbol{\theta}(t)$ increases linearly in time. Thus the *linear* separation of initially adjacent trajectories is characteristic of completely integrable systems.

Finally, one may readily show[25] that $I(\omega)$ for a completely integrable system is comprised of a series of delta functions positioned at frequencies $\mathbf{m}\cdot\boldsymbol{\omega}(\mathbf{I})$.

Demonstrating that a given system is integrable is tantamount to analytically solving for the dynamics, clearly a difficult task.[26] Nevertheless, recognition of the existence of integrable oscillator systems, with anharmonic $H(\mathbf{I})$ (i.e., ω dependent on $\mathbf{I}$), is important to avoid erroneous conclusions regarding the nature of intramolecular dynamics. For example, the typical normal mode analysis of polyatomics is an effort to approximate the true $H(\mathbf{I})$ of the system (if it exists) by $H_{\text{ho}}(\mathbf{I})$. Intramolecular energy transfer is then often regarded as energy exchange between the zeroth order oscillators as induced by higher order terms in the expansion of H about H_{ho}. This approach can be misleading, as evidenced by considering the completely integrable Toda Hamiltonian:

$$H(\mathbf{p},\mathbf{q})=\tfrac{1}{2}\left(p_1^2+p_2^2\right)+\tfrac{1}{24}\Big[\exp\left(2q_2+2\sqrt{3}\,q_1\right)+\exp\left(2q_2-2\sqrt{3}\,q_1\right)+\exp(-4q_2)\Big]-\tfrac{1}{8} \tag{4}$$

Higher order terms associated with a normal mode analysis of this Hamiltonian will be substantial. Nevertheless, the system possesses two integrals of motion and is quasiperiodic.

B. Mixing Systems

Completely integrable systems lie at one extreme of the spectrum of dynamical behavior. At the other extreme are systems whose motion is highly irregular over all of phase space. Most developments in modern ergodic theory have been directed toward the study of formal systems in which concepts of statisticality and randomness are made increasingly precise. The result is a heirarchy of such systems:[27] ergodic, mixing, K, and Bernoulli, ergodic systems being the least random. We shall briefly discuss ergodic and mixing systems to explain the need for concepts beyond ergodicity.

Qualitative features of ergodic systems are generally familiar, particularly the well-known property that the phase-space average of a dynamical property equals its time average. Ergodic systems do not, however, necessarily possess the property that starting with a nonequilibrium ensemble of systems leads to equilibrium expectation values of dynamical properties as time proceeds. Thus these systems do not necessarily approach an equilibrium state as in the manner expected of statistical systems; mixing systems do, however, relax in this way. Zaslavskii and Chirikov[28] have summarized three principle features of mixing systems:

1. Consider any dynamical property $f[\mathbf{q}(t),\mathbf{p}(t)]$. Then

$$\lim_{t\to\infty} f[\mathbf{q}(t),\mathbf{p}(t)]=\langle f(\mathbf{q},\mathbf{p})\rangle .$$

2. The correlation between any two dynamical properties, i.e.,

$$\langle g[\mathbf{q}(t),\mathbf{p}(t)],f[\mathbf{q}(0),\mathbf{p}(0)]\rangle-\langle g[\mathbf{q}(t),\mathbf{p}(t)]\rangle\langle f[\mathbf{q}(0),\mathbf{p}(0)]\rangle$$

   approaches zero as $t\to\infty$.
3. Subdivide the total phase–space, normalized to unit volume, into regular regions A_i of measure Γ_i, and define $\hat{T}_t A_i$ as the region into which A_i is carried by the system time evolution over time t. In addition, define $w_{ik}(t)$ as the measure of the overlap of $\hat{T}_t A_i$ with region A_k. Then, in a mixing system, $\lim_{t\to\infty} w_{ik}(t)=\Gamma_i\Gamma_k$. That is, in the long time limit an initial volume element A_i is carried into all other regions solely in proportion to the size of the phase-space regions.

Ergodic systems satisfy three analogous conditions, but with the infinite time limit replaced by a time average. Thus mixing systems achieve intuitively desired properties in the long time limit, whereas ergodic systems attain these properties only in a time average sense. A formal proof shows that a mixing system is necessarily ergodic; ergodicity does not, however, imply mixing.

The nature of $I(\omega)$ in mixing systems is the subject of a theorem in ergodic theory,[16, 28] which states that the spectrum is continuous rather than discrete. In addition, property (3) above ensures that over long times the Poincaré surface of section associated with these systems will be filled, presumably randomly. Insight into the expected nature of the divergence of adjacent trajectories in mixing systems comes about via a circuitous route, that is, principally from numerical calculations and behavior in the strongly random C systems introduced by Anosov[16]. Although the definition of a C system is quite formal,[16] the fundamental feature of these systems is that all adjacent phase-space trajectories diverge exponentially from one another in time; this behavior occurs for all initial phase-space points. Although C-system conditions are stronger than mixing (i.e., C systems are mixing but not necessarily vice versa) this property of exponential divergence (termed exponentiating trajectories) is often regarded as fundamental[29] to erratic behavior in realistic systems.

C. Typical Systems

The mixing and integrable systems bracket an entire class of typical systems. The subset of systems of interest are those that display regular quasiperiodic behavior at low energies and mixing-like erratic behavior at higher energies. Although one may still introduce action-angle variables for such systems, they lose their fundamental utility in that the Hamiltonian is no longer independent of $\boldsymbol{\theta}$. That is,

$$H = H(\mathbf{I}, \boldsymbol{\theta}) = H_0(\mathbf{I}) + H'(\mathbf{I}, \boldsymbol{\theta}) \tag{5}$$

where an attempt is often made to select $\mathbf{I}$ to render some major portion of the Hamiltonian angle independent. There are several ambiguities associated with writing H in the form given by (5). First, it presumes that a true set of actions $\mathbf{I}_t$, i.e., a set in which $H = H(\mathbf{I}_t)$, does not exist. Second, it leads to discussions of energy exchange in terms of the zeroth order oscillators H_0 and the coupling H', which are strongly dependent on an arbitrary division of the Hamiltonian into two parts. This dependence demonstrates that a proper description of intramolecular energy exchange even in classical mechanics, is not devoid of conceptual difficulties.

Major formal developments[16] in the theory of systems described by (5) with $H'(\mathbf{I}, \boldsymbol{\theta})$ small are due to Kolmogorov, Arnold, and Moser (KAM) based on work of Poincaré and Birkhoff. These results essentially derive from the development of formal methods to find a true set of action-angle variables and hence tori for perturbed integrable systems. In doing so they discuss[16] the destruction of these tori and hence allow for the possibility of erratic behavior. The KAM theory itself does not, however, provide a useful means of obtaining an estimate of the energy at which system behavior changes. The KAM theory has been described in detail elsewhere,[16, 17e] and we are content to provide a qualitative description of tori breakup for cases with two degrees of freedom. It proves convenient to do so by reference to the nature of the surface of section under the influence of the perturbation $H'(\mathbf{I}, \boldsymbol{\theta})$. As noted above this is equivalent to studying transformations of the type

$$\mathbf{Z}_{n+1} = \mathsf{B}(\mathbf{Z}_n)\mathbf{Z}_n \tag{6}$$

where $\mathbf{Z}_i$ is a 2-vector composed of the coordinates on the surface of section after i iterations of the map defined by the 2×2 matrix B. The area preserving property of the map implies that $\det \mathsf{B} = 1$.

Attention is focused, à la Poincaré, on the fixed points of the map, i.e., points $\mathbf{Z}$ satisfying $\mathbf{Z} = \mathsf{B}^m \mathbf{Z}$, where B^m denotes m applications of B; of principle interest is the case $m = 1$. Two types of fixed points are distinguished by the eigenvalues of B at the point and the nature of the surface of section curves in its immediate vicinity. Elliptic points are those with imaginary eigenvalues; curves in its vicinity are circles, ellipses, etc. Hyperbolic points are characterized by real eigenvalues, and curves in the immediate vicinity form an X with the critical point at its center. Behavior in the region of the hyperbolic point is characteristically unstable. In particular the phase flow in its vicinity approaches the critical point exponentially along opposite arms of the X and diverges exponentially from the point along the other arms.

Consider the case where $H_0(\mathbf{I})$ [Eq. (5)] is characterized by a central elliptic point surrounded by rational and irrational tori [Fig. 2*a*]. A principle result of Poincaré and Birkhoff is that under the perturbation $H'(\mathbf{I}, \boldsymbol{\theta})$ the rational tori break up into a set of alternating hyperbolic and elliptic points. If the result is that shown in Fig. 2*b* then the behavior is, to a large measure, still rather regular. However, if the lines emanating from the hyperbolic points do not join smoothly then behavior in this region becomes horribly complex.[30] The system aimlessly wanders about the region, although still constrained by adjacent irrational tori, which are

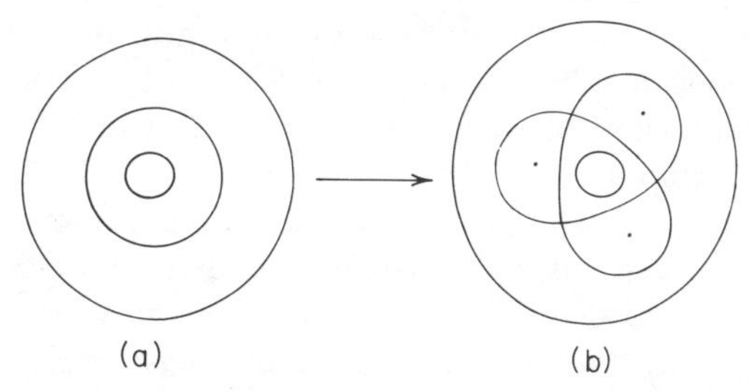

Fig. 2. (a) Schematic surface of section showing central elliptic point surrounded by smooth curves. The middle circle corresponds to a rational torus. (b) Result of applying small perturbation in which the hyperbolic points are assumed to join smoothly.

relatively unaffected by the weak perturbation. Numerical calculations on realistic coupled oscillator systems, discussed below, show that under larger perturbations the regular tori also deteriorate. Thus from this viewpoint coupled oscillator system behavior as a function of energy should show tori at low energies. As the energy increases, $H'(\mathbf{I}, \boldsymbol{\theta})$ becomes more effective, yielding surfaces of section with both tori and erratic regions. At higher energies all tori self-destruct, yielding an erratic pattern on the surface of section.

A detailed description of tori breakup indicates that the structure of the surface of section is complex in that the simultaneous appearance of elliptic points–hyperbolic points and erratic zones is repeated over and over again on smaller and smaller scales. We can note here one distinction between formal mathematics, in which such structure is relevant, and chemical problems in which macroscopic erratic behavior is of prime interest.

The dynamics in phase space after rational tori breakup is somewhat different in systems with $N>2$ and $N=2$. In the latter case the tori are two-dimensional and the energy hypersurface is of dimension three. As a consequence trajectories in the destroyed torus region are still confined by stable irrational tori. For $N>2$, however, the tori are still of dimension N, whereas the energy hypersurface is of dimension $2N-1$. Thus trajectories emanating from the destroyed tori regions can wander about phase space in the space unoccupied by stable tori, a process termed Arnold diffusion.[31]

D. K Entropy

The mathematics of integrable, mixing, and KAM transition systems provides a time-independent viewpoint of the dynamics in being concerned with the topology of the phase flow. However, we also require a useful measure of the rate at which phase space is mixed by the erratic flow. Clearly one could introduce a quantity which measures the time scale based on concepts such as the rate of energy transfer between zeroth order oscillators. Much more satisfying, however, is a measure of the rate for the loss of correlations in phase space—the K entropy of Kolmogorov. At present, the relationship between this quantity and more intuitively obvious decay times is under study.[32]

To define the K entropy,[33] which derives from an information theoretic view of dynamical systems, we consider a subdivision of the unit normalized phase space into disjoint sets $A_1,\dots, A_n$ of measure $\mu(A_i)$. The set $P=\{A_1,\dots, A_n\}$ is said to be a partition of the phase space with associated entropy

$$H(P)=-\sum_{j=1}^{n}\mu(A_j)\ln\mu(A_j) \tag{7}$$

where ln is frequently chosen to base 2. $H(P)$ is just a weighted measure of the number of elements in the partition; for example, if the n partitions are of equal measure $\mu(A_i)=1/n$ then $H(P)=\ln n$.

Under the influence of the dynamics, that is, Hamilton's equations applied over time t, the partition P evolves to a new partition $\hat{T}_t P$. Define then a new partition $P_t = P \vee \hat{T}_t P$ consisting of $\{A_1^{(t)}, A_2^{(t)}, \ldots\}$ the elements of which are the intersection of the elements of P and $\hat{T}_t P$. (For example, consider the partition of the unit square into $\{A_1, A_2\}$ as shown in Fig. 3*a* which evolves to $\hat{T}_t P$ shown in Fig. 3*b*. The resultant partition $P \vee \hat{T}_t P$ is shown in Fig. 3*c* and consists of four elements). If the procedure is repeated m times one obtains the partition P_{mt} with elements $A_j^{(mt)}$ and entropy $H(P_{mt}) = -\Sigma_{j=1}\mu(A_j^{(mt)})\ln \mu(A_j^{(mt)})$. The average entropy accumulated per step in the large m limit is

$$\overline{H}(P) = \lim_{m\to\infty} \frac{1}{m} H(P_{mt}) \tag{8}$$

The Kolmogorov entropy K is defined as the average entropy per step for that initial partition which maximizes $\overline{H}(P)$, i.e.,

$$K = \max_{P} \overline{H}(P) \tag{9}$$

As such it is an invariant of the dynamical system, known to be zero if the system is quasiperiodic and positive if the system is mixing. The strongly random K systems mentioned above are those for which K is positive for any arbitrary nontrivial partition P.

Interest in the K entropy stems from its qualitative interpretation[33] as the inverse of a relaxation time. In particular, a small phase space region of size S_i will, in a mixing system, become distributed uniformly in phase space over a time $-\ln S_i/K$. Although the subject of considerable recent interest,[34] little work has been done to clarify the role of this decay time in intramolecular energy transfer studies.

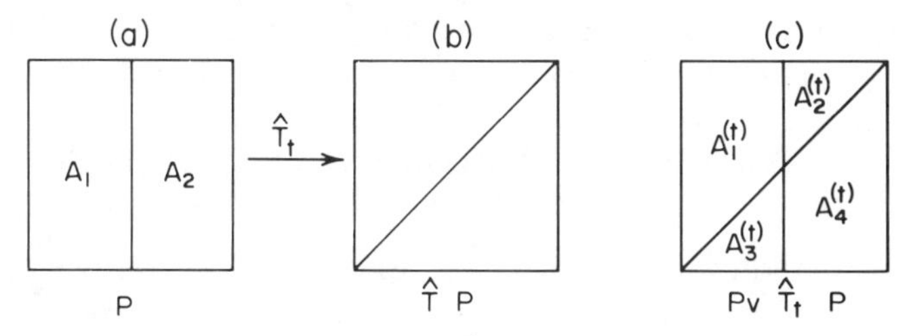

Fig. 3. Simple example for computing (c)$P \vee \hat{T}_t P$ from (a) the initial partition P and (b) the partition $\hat{T}_t P$.

Considerable insight into the dynamic origin of positive K entropy obtains from a study[34] of the relationship between the K entropy and local instabilities, i.e., the rate of growth of the distance $\mathbf{d}(t)$ between adjacent trajectories. In particular, Bennetin et al.[34] introduce a measure of $\mathbf{d}(t)$ growth, $k_n[\tau,\mathbf{x},\mathbf{d}(0)]$, defined by the following computational algorithm. Initiate a trajectory at phase space point $\mathbf{x}=(\mathbf{q},\mathbf{p})$ and a perturbed trajectory at $\mathbf{x}+\mathbf{d}(0)$, with $|\mathbf{d}(0)|$ small. After a time τ the phase-space coordinates of these trajectories are $\mathbf{x}(\tau)$ and $\mathbf{x}(\tau)+\mathbf{d}_1$. The coordinates of the perturbed trajectory are then reset to $\mathbf{x}(\tau)+\mathbf{d}_1|\mathbf{d}(0)|/|\mathbf{d}_1|$ and the process repeated n times to compute

$$k_n[\tau,\mathbf{x},\mathbf{d}(0)]=\frac{1}{n\tau}\sum_{i=1}^{n}\ln\left[\frac{|\mathbf{d}_i|}{|\mathbf{d}(0)|}\right] \tag{10}$$

The quantity $k_n[\tau,\mathbf{x},\mathbf{d}(0)]$ provides an estimate of the maximum Lyapunov characteristic exponent $\lambda_{\max}(\mathbf{q},\mathbf{p})$ which is rigorously related to the K entropy, i.e.,

$$\int_{\Gamma_E}\lambda_{\max}(\mathbf{q},\mathbf{p})\,d\mu_L(\mathbf{q},\mathbf{p})\leqslant K(E)\leqslant(N-1)\int_{\Gamma_E}\lambda_{\max}(\mathbf{q},\mathbf{p})\,d\mu_L(\mathbf{q},\mathbf{p}) \tag{11}$$

Here the energy dependence of the K entropy has been explicitly indicated, the system has N degrees of freedom with a Hamiltonian satisfying $H(\mathbf{q},\mathbf{p})=H(\mathbf{q},-\mathbf{p})$, and μ_L is an invariant measure defined on the energy hypersurface Γ_E with convenient normalization $\mu_L(\Gamma_E)=1$.

Numerical computations[35] on systems with two degrees of freedom indicate that $k_n[\tau,\mathbf{x},\mathbf{d}(0)]$ is essentially zero in phase-space regions occupied by tori and equal to a positive constant $k_c(E)$ in erratic regions. In this case (11) provides the estimate $K(E)=\mu_L(S_E)k_c(E)$, where $\mu_L(S_E)$ is the fraction of Γ_E characterized by erratic behavior at energy E. For cases with $N>2$, $k_n(\tau,\mathbf{x},\mathbf{d}(0))$ can assume different positive values in different erratic regions of phase space.[36] Under these circumstances (11) provides a bound on $K(E)$ composed of a sum of terms, each of which is a product of the characteristic value of k_n for a region of phase space times the measure of that region. This dependence of the K entropy on the size of the erratic region in phase space is important but frequently neglected.

IV. NUMERICAL EXPERIMENTS ON COUPLED OSCILLATORS

Most of the known features of the transition from regular to erratic behavior have been obtained via numerical studies on simple oscillator systems and area preserving mappings.[37] Studies of $N=2$, typical oscillator systems show behavior consistent with that described above. That is, surfaces of section are regular at low energies, show coexistent tori and irregular regions at higher energies, and are almost entirely erratic at yet higher energies. Such studies have been the subject of extensive review articles;[17] we therefore limit our discussion to one example[38] and comments.

Consider three particles of unit mass confined to a line with adjacent pairs coupled by Morse potentials, that is, the coupled Morse system[39, 40]

$$H=\frac{11}{8}p_x^2+\frac{3}{8}p_y^2+\left[1-\exp\left(\frac{-x-y}{2}\right)\right]^2+\left[1-\exp\left(\frac{x-y}{2}\right)\right]^2 \quad (12)$$

where x corresponds to motion along the asymmetric stretch and y to motion along the symmetric stretch. In original studies of this model for unimolecular dissociation Thiele and Wilson[39] detected a transition in behavior at $E\approx 0.6$ for motion confined to the neighborhood of the pure symmetric stretch. Figure 4a shows[38] the (y, p_y) surface of section at

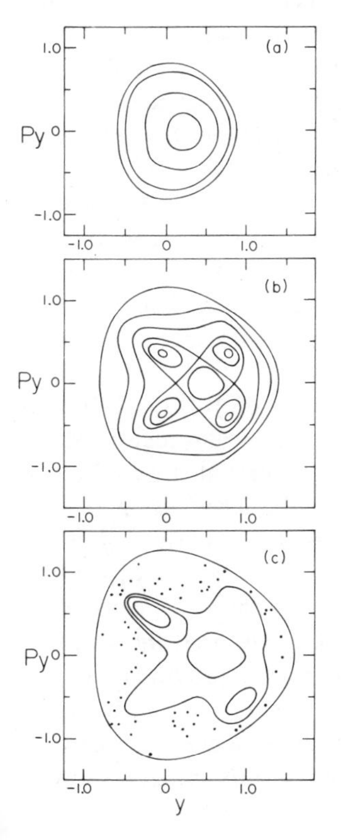

Fig. 4. (y, p_y) surfaces of section for coupled Morse system [Eq. (12)] at (a) $E=0.25$, (b) $E=0.5$, (c) $E=0.6$. In all cases only a small number of trajectories were run; additional structure is expected to arise in more extensive calculations.

$E=0.25$; smooth curves have been drawn through the trajectory (y, p_y) coordinates where possible. At $E=0.5$ (Fig. 4*b*) the system shows somewhat different behavior, although smooth curves can still be drawn. Of particular interest is the appearance of alternating hyperbolic and elliptic points reminescent of the discussion above. Finally, at $E=0.6$ (Fig. 4*c*) the surface of section shows both smooth curves and an erratic region, the splatter of points resulting from a single trajectory. As is typical of other systems studied, two nearby trajectories in the smooth curve region are found to diverge linearly from one another in time whereas those initiated in the erratic region diverge exponentially in time.[38]

The surfaces of section shown in Fig. 4 are also typical of others obtained with perhaps one exception. That is, in most other cases erratic regions of the surface first occur in regions characterized, at lower energies, by hyperbolic points. This is not immediately evident here, a result is discussed in detail elsewhere.[32]

Surface of section studies are of particular use for $N=2$; studies of systems of higher dimensionality have principally characterized phase-space regions as regular or erratic depending on whether adjacent trajectories diverge linearly or exponentially in time. Two techniques for determining the divergence rate have been used. The first is to simply plot the distance $D(t)=|\mathbf{d}(t)|$ between two trajectories providing both the nature of $D(t)$ (i.e., linear or exponential) and the average slope of $D(t)$ vs. t. A result of this kind is shown in Fig. 5*a* for the coupled Morse system; it differs from most studies in that it refers to a collision of an initially free particle A with a bound pair B—C rather than to a bound system. The overall exponential divergence of $D(t)/D(0)$ is clear over most of the trajectory; initial and final regions of nonexponential divergence correspond to portions of the trajectory in the asymptotic regime of noninteracting A and B—C.

The second technique in use is to compute $k_n[\tau, \mathbf{x}, \mathbf{d}(0)]$ [Eq. (10)] as a function of $\mathbf{x}=(\mathbf{p}, \mathbf{q})$; stable regions show zero k_n and erratic regions show positive k_n providing a measure of the exponential growth rate of $D(t)$. The principal advantage of this approach over the $D(t)$ vs. t plot is computational. In particular it eliminates the need for extensive plotting, provides a less ambiguous distinction between linear and exponential growth and is less likely to produce errors resulting from a choice of $D(0)$ which is too large. Numerical studies show,[32, 41] however, that the value of $k_n[\tau, \mathbf{x}, \mathbf{d}(0)]$ is the same as the average slope of $D(t)$ vs. t provided that the choice of $D(0)$ is made sufficiently small.

Studies of the nature of $I(\omega)$ as a function of energy have been carried out on only a few oscillator systems.[25, 35b] The results clearly show a

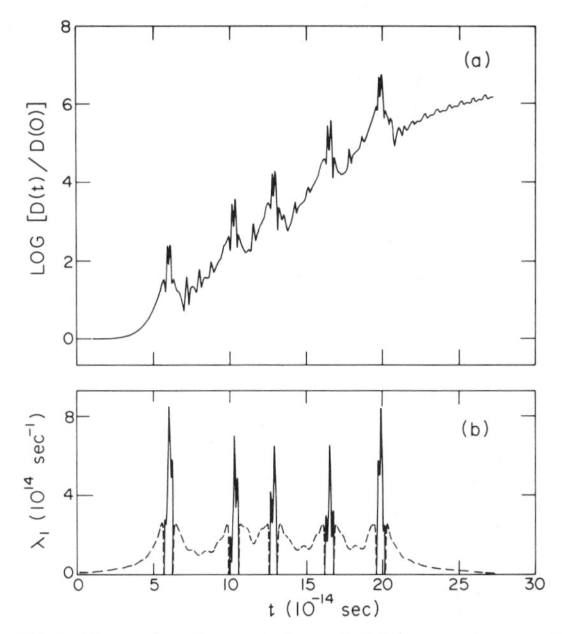

Fig. 5. (a) $\ln D(t)/D(0)$ vs. time for typical coupled Morse trajectory. Initial conditions correspond to the B–C diatom in the (quasiclassical) ground vibrational state and an initial relative translational energy $E_t = 0.1$ eV. (b) λ_1, an eigenvalue of $S(t)$, encountered by the unperturbed trajectory of the pair of trajectories yielding top figure. Dashed curves denote real values of $\lambda_1(t)$ and solid curves denote imaginary values of $\lambda_1(t)$. (From Ref. 40.)

broadening of the spectral lines as the energy increases; possibly a manifestation of the beginnings of a continuous spectrum characteristic of mixing systems.

Thus, Poincaré surfaces of section, divergence of adjacent trajectory studies, and computation of the power spectrum can be used to provide good estimates of the energy E_c, termed the critical energy, at which a transition from regular to erratic behavior occurs. The resultant value is not a precise mathematical quantity in that it depends on the resolution with which one views the surface of section, the extent to which the phase space has been examined for exponentiating trajectories or an ability to computationally identify broadened spectral lines. Nevertheless, of principle interest is the energy at which a reasonable fraction of the phase space displays erratic behavior, a quantity that should be the result obtained with these techniques. As argued above, this operational value of E_c is probably more relevant than a purely formal mathematical value of the transition

TABLE I:
Critical Energies for Model Problems as Determined from Exact Dynamics

Type	$V(x, y)$	m_1	m_2	E_c	E_c (variational equations)
Henon-Heiles	$\frac{1}{2}(x^2+y^2)+x^2y-\frac{1}{3}y^3$	1	1	1/12–1/8 (Ref. 17a)	1/12
Barbanis	$\frac{0.1}{2}(x^2+y^2)-0.1x^2y$	1	1	0.0075 (Ref. 42)	0.00625
Tredgold	$\frac{1}{2}[\frac{3}{2}a(x^2+y^2)-\frac{3}{4}c(x^2+y^2)^2]$	$\frac{3}{2}$	$\frac{3}{2}$	$6a^2/24c$ (Ref. 43)	$5a^2/24c$
Thiele-Wilson	$[1-\exp\{-(x+y)/2\}]^2+$	$\frac{4}{11}$	$\frac{4}{3}$	>0.6 for symmetric (Ref. 39)	>0.5 for symmetric
	$[1-\exp\{-(y-x)/2\}]^2$			—	0.25 for unrestricted
Modified Henon-Heiles	$\frac{1}{2}(x^2+y^2)+x^2y+\frac{1}{3}y^2$	1	1	∞ (Ref. 44)	1/24
Toda	See Eq. (4)	Equal mass		∞ (Ref. 45)	∞
		Unequal mass		Varies with mass (Ref. 45)	∞
$(2-2)+(2-3)$ resonance	See Eq. (26)			~0.2095 (Ref. 46)	—
Henon-Heiles 2	$\frac{1}{2}(x^2+y^2)-\frac{1}{2}x^2y^2$	1	1	>0.5 (Ref. 60)	>0.5

energy, which may indicate the appearance of imperceptibly small regions of erratic behavior in phase space.

Table I provides a list of E_c values determined using these techniques for a variety of $N=2$ oscillator systems of the form

$$H=\frac{1}{2}\left[\frac{p_x^2}{m_1}+\frac{p_y^2}{m_2}\right]+V(x,y)$$

where the value $E_c=\infty$ denotes a completely integrable system.

Additional numerical results of interest have been obtained in studies of systems with $N>2$. Recent results[36] indicate the existence of several erratic regions of phase space characterized by different rates of exponential divergence. This result may prove to be of considerable importance in intramolecular energy transfer studies, indicating that an internal rate depends on the portion of the molecule initially excited vibrationally. The dependence of E_c on the number of particles M in a simple oscillator chain has also been studied,[47] with the result that the ratio E_c/M decreases with increasing M.

Although a substantial amount of numerical work has been done on coupled oscillator systems, augmented by studies[37] of area preserving mappings, much remains unexplored. For example, little is known about the influence of rotation on the onset of statistical intramolecular energy transfer or of the role of Arnold diffusion in simple systems with $N>2$.

Exact numerical studies of the transition between regular and erratic behavior provide little in the way of qualitative insight into the mechanism by which the transition occurs. In particular they have not as yet provided any insight into the connection between system properties (i.e., masses, potentials, etc.) and computed values of E_c and k_n. What is clearly required is a useful almost-analytic theory, which provides accurate estimates of E_c and k_n as well as the sought-after qualitative insight into their relation to system properties. Such a theory should be global in phase space in that it should be able to provide an estimate of the fraction of phase space displaying erratic behavior. In addition it should be global in time so as to recognize that local exponential divergence followed by local compensatory contraction does not signify global instability. The mathematical complexities[16, 24] associated with the transition are such that formal theories provide no hint as to how to construct simplified models. Given this fact, as well as the stringent requirements set out above for a useful model, it comes as no surprise that the models described below are only zeroth order approximations to the desired result.

V. MODELS FOR THE REGULAR TO ERRATIC TRANSITION

A. Mo's Method[48, 49]

Each of the methods described below attempts to predict E_c by focusing attention on a particular system property that undergoes a change in character as E passes through the critical energy. Mo chooses to consider the autocorrelation function $B(t)=\langle \dot{b}(0)\dot{b}(t)\rangle$, where $b(t)=\sum_{i=1}^{N}(p_i^2+q_i^2)$, i.e., the square of the distance of the phase point from the origin. Mo suggests that $B(t)$ is oscillatory at $E<E_c$ and diverges exponentially at $E>E_c$.

$B(z)$, the Laplace transform of $B(t)$, is related to a function $K(z)$ by the expression

$$B(z)=-\langle bb\rangle\frac{K(z)}{(1-K(z)/z)} \tag{13}$$

where

$$K(z)=\frac{1}{\langle bb\rangle}\left\langle biL(1-P_0)\frac{1}{z+(1-P_0)iL}(1-P_0)iLb\right\rangle \tag{14}$$

where L is the Liouville operator and P_0 is a projection operator along b.

Equations 13 and 14 are exact and this approach achieves the status of a model by introducing approximate dynamics in the form of a truncated continued fraction representation of $K(z)$. Truncating the continued fraction to first order gives

$$B(z)=zy_1\frac{z^2+\dfrac{m_2}{m_0}-\dfrac{y_2}{y_1}}{z^4+\dfrac{m_2}{m_0}z^2+\dfrac{m_1}{m_0}} \tag{15}$$

Here $m_0=y_3y_1-y_2^2$, $m_1=y_4y_2-y_3^2$, $m_2=y_4y_1-y_2y_3$ and $y_n=\langle (iL)^n b,(iL)^n b\rangle$. The inverse Laplace transform of (15) gives

$$B(t)=A\cos\omega_1 t+G\cos\omega_2 t \tag{16}$$

where ω_1^2 and ω_2^2 are the roots of the denominator of (15) and A, G are functions of the moments y_n.

Mo's method consists of computing ω_1 and ω_2 as a function of energy. The transition from regular to erratic behavior is then characterized by a

TABLE II:
ω_1, ω_2 [Eq. (16)] of Henon-Heiles System as Computed by Mo[48]

E	ω_1	ω_2	E	ω_1	ω_2
0.08	2.926	0.4483	0.110	2.900	0.1580
0.085	2.922	0.4410	0.114	2.897	0.0638
0.09	2.917	0.3716	0.115	2.896	$0.0326i$
0.095	2.913	0.3291	0.12	2.892	$0.1622i$
0.100	2.909	0.2821	0.15	2.867	$0.4048i$
0.105	2.905	0.2276	0.1666	2.856	$0.4787i$

change from real frequencies to imaginary frequencies, that is $B(t)$ changes from oscillatory to exponential behavior at E_c. The results of Mo's calculations on the Henon-Heiles Hamiltonian (see Table I) are shown in Table II. The frequency ω_2 is seen to be a decreasing function of energy and changes character at $0.114 < E < 0.115$ within the range $1/12 < E_c < 1/8$ determined in surface of section studies. Mo obtains equally good results for the Barbanis system, Henon-Heiles #2, and a completely integrable coupled harmonic oscillator Hamiltonian. Further work would be required to establish the utility of the method for the other systems listed in Table I.

There are several difficulties associated with Mo's method, many of which relate to the use of approximate dynamics and, in particular, the continued fraction representation for $K(z)$. First, like exact dynamics, the method fails to provide useful insight into the relationship between system properties and the critical energy. Second, the utility of the methods relies upon the accuracy of the continued fraction representation of the kernel, an approximation which is poorly understood. Finally, the method postulates, and subsequently demonstrates numerically, exponential behavior of $B(t)$ at $E > E_c$. It is difficult to reconcile this behavior with the expected structure of $I(\omega)$ above E_c, that is, a broadened line spectrum. Further remarks on Mo's method, based upon recent work, are contained in Appendix A.

B. Variational Equations Approach[40, 50, 51]

The transition from linear to exponential divergence of adjacent trajectories has been identified above as one of the principle characteristics of the change from regular to erratic behavior. This method focuses on this feature and adopts the fundamental assumption that the long time exponential instability of trajectories at $E > E_c$ arises from local instabilities. We consider an N degree of freedom system with

$$H = \sum_{i=1}^{N} \frac{p_i^2}{2m_i} + V(q_1, \ldots, q_N) \tag{17}$$

The distance $\mathbf{d}(t)$ between two initially adjacent trajectories is given by

$$\begin{aligned} d_j(t) &= q'_j(t) - q_j(t) \\ d_{j+N}(t) &= p'_j(t) - p_j(t) \qquad j = (1, \ldots, N) \end{aligned} \tag{18}$$

where the primed coordinates are those of a trajectory with initial conditions which are perturbed relative to the unprimed trajectory. The $\mathbf{d}(t)$ evolve in accordance with Hamilton's equations, i.e.,

$$\begin{aligned} \frac{d}{dt}\left[d_j(t) \right] &= \frac{p'_j(t) - p_j(t)}{m_j} \\ \frac{d}{dt}\left[d_{j+r}(t) \right] &= -\left\{ \frac{\partial V}{\partial q'_j} - \frac{\partial V}{\partial q_j} \right\} \qquad j = (1, \ldots, N) \end{aligned} \tag{19}$$

Expanding the right-hand side about the unperturbed trajectory and dropping terms of $O(d^2)$ [i.e., linearizing] yields the first variational equation:

$$\frac{d}{dt} \begin{bmatrix} d_1(t) \\ \vdots \\ d_N(t) \\ d_{N+1}(t) \\ \vdots \\ d_{2N}(t) \end{bmatrix} = \left[\begin{array}{c|ccc} & 1/m_1 & & \\ \mathsf{O} & & \ddots & \\ & & & 1/m_N \\ \hline & & & \\ \mathsf{A}(t) & & \mathsf{O} & \\ & & & \end{array} \right] \begin{bmatrix} d_1(t) \\ \vdots \\ d_N(t) \\ d_{N+1}(t) \\ \vdots \\ d_{2N}(t) \end{bmatrix} \tag{20}$$

Here O is an $N \times N$ matrix of zeroes and $\mathsf{A}(t)$ is an $N \times N$ matrix with elements $-\partial^2 V / \partial q_i \partial q_j$.
We rewrite (20) as

$$\dot{\mathbf{d}}(t) = \mathsf{S}(t)\mathbf{d}(t) \tag{21}$$

and introduce a new set of independent variables $\mathbf{h}(t)$, which are related to $\mathbf{d}(t)$ by the transformation $\mathsf{T}(t)$, i.e.,

$$\mathbf{d}(t) = \mathsf{T}(t)\mathbf{h}(t) \tag{22}$$

Equation 21 then becomes

$$\begin{aligned} \dot{\mathbf{h}}(t) &= \mathsf{C}(t)\mathbf{h}(t) \\ \mathsf{C}(t) &= \mathsf{T}^{-1}(t)\left[\mathsf{S}(t)\mathsf{T}(t) - \dot{\mathsf{T}}(t) \right] \end{aligned} \tag{23}$$

If the principle time dependence of $\mathbf{d}(t)$ resides in $\mathbf{h}(t)$ rather than $\mathsf{T}(t)$, and $\mathsf{C}(t)$ can be obtained in diagonal form, then the behavior of $\mathbf{d}(t)$ is readily ascertained. In particular, if the elements of $\mathsf{C}(t)$ are all imaginary one expects oscillatory $\mathbf{h}(t)$ and $\mathbf{d}(t)$. If, however, one of the elements of $\mathsf{C}(t)$ is real then an exponential time dependence is possible. Unfortunately, obtaining a diagonal $\mathsf{C}(t)$, although possible in principle, is not feasible in practice without prior knowledge of the exact solution to (21). Qualitative arguments have been presented[40] suggesting that if $\mathsf{T}^{-1}(t)\dot{\mathsf{T}}(t)$ is reasonably well behaved then the *nature* of $\mathbf{h}(t)$, that is, whether it exponentially grows or oscillates, is reflected in the nature of $\mathbf{z}(t)$, a function satisfying the auxilary equation

$$\dot{\mathbf{z}}(t)=\left[\mathsf{T}^{-1}(t)\mathsf{S}(t)\mathsf{T}(t)\right]\mathbf{z}(t) \tag{24}$$

Equation 24 immediately suggests that $\mathsf{T}(t)$ be chosen to diagonalize $\mathsf{S}(t)$; the eigenvalues are denoted $\lambda_k(t)$, $k=(1,\ldots,2N)$. If $\mathrm{Re}[\lambda_k(t)]>0$ then $\mathbf{z}(t)$ [and hence $\mathbf{h}(t)$ and hence $\mathbf{d}(t)$] will grow exponentially, whereas if all $\lambda_k(t)$ are imaginary then the distance is expected to oscillate in time. An alternative independent approach due to Toda[50] simply neglects the time dependence of $\mathsf{T}(t)$.

For Hamiltonians of the form given in (17) the eigenvalues $\lambda_k(t)$ are functions of q_k only and are the roots of the determinantal equation

$$\det\left(\mathsf{B}-\lambda_k^2\mathsf{I}\right)=0 \tag{25}$$

where B is an $N\times N$ symmetric matrix with elements $b_{ij}=-(m_im_j)^{-1/2}\partial^2V/\partial q_i\partial q_j$, that is, a generalized force constant matrix evaluated at all q_iq_j. Thus the eigenvalues of S are explicitly determined by the masses m_i and potential V, occur in pairs $(\lambda,-\lambda)$ and, since B is symmetric, are either real or pure imaginary. The system energy and momenta influence the magnitude of λ_k implicitly by determining the coordinate space region encountered during a trajectory, and hence the regions within which (25) is to be solved. This implicit dependence on energy will serve to define the critical energy E_c. In particular, E_c is that energy required to enter the coordinate space region within which the $\mathrm{Re}(\lambda_k)\neq 0$. It is only in this region that exponential separation of nearby trajectories is expected.

The resultant technique for obtaining E_c is then straightforward. First, one determines the eigenvalues λ_k as a function of the coordinates q_i. The lowest energy allowing $\mathrm{Re}(\lambda_k)>0$ is then E_c. In some instances E_c may be determined analytically; most often for interesting chemical systems, however, it must be determined graphically. Results obtained using this eigenvalue criterion for the $N=2$ systems introduced above are included in Table I. All results were obtained analytically (on the "back of an envelope") and the agreement with previous computational estimates of E_c is good for all but the unequal mass Toda case and the integrable systems. Both types of discrepancies are discussed below. In cases where an analytic result is not readily obtained one can (for two degree of freedom systems) plot λ_k as a function of the coordinates and locate the energy above which trajectories can enter the region of real λ_k; the lowest such energy is then E_c. Examples of such contour plots are given in Refs. 40.

The basic assumption that exponential divergence of adjacent trajectories occurs in regions of configuration space characterized by real eigenvalues has been tested in an extensive study[40, 41] of model $A+B-C$ collisions, one of which utilized the coupled Morse potential [Eq. (12)]. Results typical of those obtained are shown in Fig. 5. In particular, Fig. 5*a* discussed above shows the exponential growth of $D(t)=|\mathbf{d}(t)|$ between two initially close [$D(0)=10^{-8}$ a.u.] trajectories. Figure 5*b* shows the corresponding magnitude of one of the eigenvalues encountered by the unperturbed trajectory during the time evolution; real eigenvalues are plotted with solid lines and imaginary eigenvalues by dashed lines. A comparison of Figs. 5*a* and 5*b* shows the expected correlation. That is, the time intervals during which $D(t)/D(0)$ increases exponentially correspond to time intervals during which $\lambda_1(t)$ is real and regions of nonexponential $D(t)/D(0)$ coincide with imaginary $\lambda_1(t)$. Furthermore, the slope of the $\log[D(t)/D(0)]$ vs. t plot, in each of the intervals where $D(t)$ exponentially increases, is well approximated by

$$\langle \lambda \rangle = \int_{t_i}^{t_f} \frac{\lambda_1(t)\, dt}{(t_f - t_i)}$$

where t_i and t_f are the endpoints of the interval during which $\lambda_1(t)$ is real. This provides further evidence that the observed local exponential separation correlates with real eigenvalues of $\mathbf{S}(t)$ and also provides a connection between the rate of exponential growth and details of the potential. Identical correlations have been observed in studies of different model collision systems.[40, 41] The exact correspondence between real eigenvalues and local exponential divergence does not appear to hold, however, for all trajectories in bound state systems.[32, 52] In particular, trajectories predicted to be locally unstable can be stabilized by nearby influential periodic orbits. However, an extensive trajectory study on the Henon-Heiles system due to Cerjan and Reinhardt[52] clearly demonstrates that the onset of erratic behavior is associated with the entrance of trajectories into phase-space regions characterized by real λ_k.

The utility of the eigenvalue criterion for predicting E_c is seen to rely on a variety of assumptions.[53] First, the variational equations must provide a reliable approximation to the true $\mathbf{d}(t)$. Second, both $\mathsf{T}(t)$ and $\mathsf{T}^{-1}(t)\dot{\mathsf{T}}(t)$ must behave in such a way as not to affect the *nature* of $\mathbf{d}(t)$ as predicted from the eigenvalues of $\mathsf{S}(t)$. Third, the local instabilities must determine the global nature of the trajectory. Finally, note that predictions of E_c using this method can depend on the choice of coordinates, a consequence of the neglect of $\mathsf{T}^{-1}(t)\dot{\mathsf{T}}(t)$. Despite the serious limitations imposed by these restrictions the results shown in Table I, as well as studies of individual trajectories (e.g., Fig. 5), suggest that this method identifies an important attribute of the potential affecting the transition from regular to erratic behavior.

C. Critical-Point Analysis[52]

Discrepancies in the E_c predictions of the variational equations, mentioned above, prompted reconsideration of this approach by Cerjan and Reinhardt.[52] In particular they note two approximations inherent in the variational equations approach. First, the time dependence of $\mathsf{T}(t)$ is neglected and second, the global $\mathbf{d}(t)$ behavior is obtained from a *linearized* equation [Eq. (20)]. Although linearization is a common procedure in the qualitative theory of differential equations,[54] it is normally carried out only at the critical points of the differential equation [here $\mathbf{d}(t)$ values where $\dot{\mathbf{d}}(t)=0$]. Cerjan and Reinhardt propose to study (19) by determining the exact critical points after expanding about the unperturbed trajectory but prior to linearization. The nature of these critical points is then studied by linearization about the point and determination of the critical point eigenvalues. As above, stable elliptic points are characterized by imaginary eigenvalues, whereas unstable hyperbolic points are characterized by real eigenvalues. The global behavior of the solutions to the differential equation are then obtained from phase portraits,[54] where the behavior at *all* critical points is important. The fundamental difference between this and typical applications of the qualitative theory of differential equations is that (19), after Taylor series expansion of the right-hand side, is a function of the phase-space coordinate of the unperturbed trajectory. In essence one then has a set of phase portraits, each a function of the unperturbed trajectory coordinates. The basic conclusions of this study are as follows. First, the eigenvalues corresponding to the trivial critical point, that is, at $\mathbf{d}(t)=0$, are precisely those obtained in the variational equations approach outlined above. Second, the behavior of the nontrivial critical points are relevant in those integrable Hamiltonian cases tested (in particular the modified Henon-Heiles) where the variational equations gives incorrect E_c predictions. In particular they find that the phase portraits corresponding to integrable systems display compensating critical points, that is, elliptic and hyperbolic, which switch character as E increases through the variational equations E_c. Thus, although the trivial eigenvalue for the modified Henon-Heiles system changes character from elliptic to hyperbolic at $E=1/24$ the nature of the phase portrait, which is determined by all critical points, does *not* display a change in character. They conclude that such compensating critical points indicate stability, or at worst weak instability. Finally, compensating critical points are not found for those systems tested in which the variational equations prediction of E_c is in good accord with exact computations.

From this viewpoint the failure of the variational equations method for integrable systems stems from the early linearization of (19) and thus the neglect of compensating critical points. Since the property of complete integrability is highly sensitive to changes in the Hamiltonian it is clear that retaining the exact dynamics of $\mathbf{d}(t)$ as far as possible is desirable. Further work is now necessary to demonstrate that systems that are falsely predicted by the variational equations technique to be integrable, of which the unequal mass Toda (see Table I) is the only example I am aware of, are properly described by this critical point analysis.

D. Overlapping Resonances[28, 31]

A well-developed approach to the dynamic origins of the transition is due to Chirikov.[28, 31] A variety of physical problems have been studied by this technique but, to date, there have been few direct applications[55] to coupled oscillatory systems of chemical interest. Chirikov's approach is based upon the concept of overlapping resonances, which is readily explained by example. Consider the model Hamiltonian studied by Ford[46] of the form

$$H=H_0(I_1,I_2)+\alpha I_1 I_2\cos(2\theta_1-2\theta_2)+\beta I_1 I_2^{3/2}\cos(2\theta_1-3\theta_2) \qquad (26)$$

where $H_0(I_1,I_2)$ is a completely integrable Hamiltonian and α,β are constants. Contributions of the form $\cos(n\theta_1-m\theta_2)$ are termed n–m resonance terms and are expected to influence the dynamics most strongly in regions of stationary phase, i.e., where $d/dt(n\theta_1-m\theta_2)=n\omega_1(\mathbf{I})-m\omega_2(\mathbf{I})\approx 0$, termed resonance centers. The Hamiltonian in (26) is constructed to be integrable for either $\alpha=0$ or $\beta=0$; in the former case the second integral of motion is $3I_1+2I_2$, whereas in the latter case it is I_1+I_2. The effect of either the 2–2 or 2–3 resonance term is to distort the $H_0(I_1,I_2)$ tori in the neighborhood of the rational $2\omega_1=2\omega_2$ or $2\omega_1=3\omega_2$ torus.

However, since the $\alpha=0$ or $\beta=0$ perturbed Hamiltonians are also integrable, either distortion is rather gentle. The resultant surfaces of section show alternating stable and unstable fixed points and are qualitatively similar to that shown in Fig. 2*b*. In particular, the surface of section remains composed of smooth curves.

Each resonance center is located at $m\omega_1(\mathbf{I})+n\omega_2(\mathbf{I})=0$, and influences the dynamics over a region in phase space termed the resonance zone, whose size is discussed below. Chirikov's fundamental conjecture is that erratic behavior results when the energy is such that the resonance zones arising from independent resonance terms overlap. This conjecture is beautifully confirmed in Ford's calculations[46] on (26) with both α and β nonzero. In particular, exact calculations for $E<0.2095$ show that the 2–2 and 2–3 resonance zones do not overlap and that the surface of section remains quite regular. At $E>0.2095$ the resonance zones do overlap and the surface of section is highly erratic.

Typical coupled oscillatory systems [Eq. (5)] display resonance term structure when $H'(\mathbf{I},\boldsymbol{\theta})$ is expanded in a Fourier series, yielding

$$H(\mathbf{I},\boldsymbol{\theta})=H_0(\mathbf{I})+\sum_{\mathbf{m}} V_{\mathbf{m}}(\mathbf{I})\exp(i\mathbf{m}\cdot\boldsymbol{\theta}) \tag{27}$$

where $\mathbf{m}$ is an N-dimensional vector of integers. The resonance centers, now multidimensional surfaces, are located at $d(\mathbf{m}\cdot\boldsymbol{\theta})/dt=\mathbf{m}\cdot\boldsymbol{\omega}(\mathbf{I})=0$. Although a large number of resonances now influence the dynamics, the low-order resonances (i.e., small integer $\mathbf{m}$ components) are often regarded as most important for studies of the onset of statistical behavior, since resonance zone widths tend to increase with decreasing order.

The application of the overlapping resonance criterion for determining the onset of ergodicity requires that the Hamiltonian be written in the form of Eq. (27), that the relevant resonance zones be identified and that one can estimate the size of the resonance zones as a function of energy; this allows determination of the critical energy for the onset of resonance overlap. Chirikov has shown, subject to the approximations contained in the description outlined below, that behavior in the neighborhood of a resonance is similar to that of a nonlinear pendulum from which the resonance zone width can be determined.

As a particular example consider the case of $N=2$ with Hamiltonian[31]

$$H(\mathbf{I},\boldsymbol{\theta})=H_0(\mathbf{I})+\sum_{m_1',m_2'} V_{m_1',m_2'}(\mathbf{I})\exp[i(m_1'\theta_1+m_2'\theta_2)] \tag{28}$$

and focus attention on a single resonant term, that is,

$$H(\mathbf{I},\boldsymbol{\theta})\approx H_0(\mathbf{I})+V_{m_1,m_2}(\mathbf{I})\exp[i(m_1\theta_1+m_2\theta_2)] \tag{29}$$

Of particular interest is the behavior in the neighborhood of the resonant action $\mathbf{I}^r$ satisfying $m_1\omega_1(\mathbf{I}^r)+m_2\omega_2(\mathbf{I}^r)=0$. To this end two new independent angles are introduced:

$$\psi_k=\sum_{i=1}^{2}\mu_{ki}\theta_i \qquad (k=1,2) \tag{30}$$

where μ_{ki} are constants such that $\mu_{11}==m_1$ and $\mu_{12}=m_2$, i.e., $\psi_1=m_1\theta_1+m_2\theta_2$, the resonance phase. In addition, we introduce new actions $\mathbf{p}$ via the generating function

$$F(\mathbf{p},\boldsymbol{\theta})=\sum_{i=1}^{2}\left(I_i^r+\sum_{k=1}^{2}p_k\mu_{ki}\right)\theta_i \tag{31}$$

yielding

$$p_k=\sum(I_i-I_i^r)\mu_{ik}^{-1} \tag{32}$$

where $\boldsymbol{\mu}^{-1}$ is the matrix inverse of $\boldsymbol{\mu}$.

The new momenta $\mathbf{p}$ measure deviations of the actions from $\mathbf{I}^r$ and the new angles are such that the resonance is centered at $d\psi_1/dt=0$. To examine the system behavior near the resonance the Hamiltonian is expressed in $\mathbf{p}$ and ψ, $H_0(\mathbf{I})$ is expanded about $\mathbf{I}^r$ [with coefficients $\partial H_0(\mathbf{I})/\partial\mathbf{I}|\mathbf{I}_r=\omega(\mathbf{I}_r)$, etc.], constant terms dropped, and the expansion truncated at $O(p^2)$ to give

$$H_r(\mathbf{p},\psi)=\sum_{i,k}p_k\mu_{ki}\omega_i+\sum_{k,l}\frac{p_kp_l}{2M_{kl}}+V_{m_1,m_2}(\mathbf{I}^r)\cos\psi_1 \tag{33}$$

Here $M_{kl}=\sum_{j,i}^{2}\mu_{ki}(\partial\omega_i/\partial I_j)\mu_{lj}$ is a generalized mass. The Hamiltonian may be simplified by noting that at the resonance $\sum_{i=1}^{2}\mu_{li}\omega_i=0$ and that studies of the ψ_1 resonance $I_2=I_2^r$ imply $p_2=0$. Under these circumstances (33) becomes

$$H_r(\mathbf{p},\psi)=p_1^2/2M_{11}+V_{m_1,m_2}(\mathbf{I}^r)\cos\psi_1 \tag{34}$$

which is the equation of a nonlinear pendulum, here with the variables directly related to the location and deviation from the m_1, m_2 resonance center.

The phase-plane portrait of a pendulum is well known,[28] being composed of repeated cells of elliptic points contained within pairs of hyperbolic points. Thus Chirikov estimates the resonance zone size by the action and frequency width, $(\Delta \mathbf{I}')$ and (Δw^r) of the pendulum cell, here given by[31]

$$(\Delta \mathbf{I})' = 2\mathbf{m}\left(M_{11} V_{m_1, m_2}\right)^{1/2} \qquad (\Delta w^r) = \frac{2}{|\mathbf{m}|}\left(\frac{V_{m_1, m_2}}{M_{11}}\right)^{1/2} \tag{35}$$

where $\mathbf{m}$ has components (m_1, m_2).

Chirikov's method has principally been applied to Hamiltonians of the form

$$H = H_0(\mathbf{I}) + V_{mn}(\mathbf{I})\cos(m\theta - n\tau) \tag{36}$$

where τ is the period of an externally applied oscillatory perturbation. In this case the application of the method is much simplified.

The most chemically relevant application of this approach is due to Oxtoby and Rice[55] in which regions of resonance overlap in phase space were computed for several coupled oscillator systems in order to correlate the extent of intramolecular energy transfer with the fraction of phase space displaying overlapping resonances. To do so it was first necessary to recast the Hamiltonian in the form of (27). This was done by first writing H in bond-angle coordinates and momenta p_i, q_i as

$$H = \frac{1}{2}\sum_{i,j} p_i G_{ij} p_j + \sum_i U_i(q_i) \tag{37}$$

where $U_i(q_i)$ are pairwise potentials and G_{ij} are elements of Wilson's G matrix. With G_{ij} approximated by its equilibrium value G°_{ij}, (38) assumes the form

$$H = \sum_i \left[\frac{1}{2} G^\circ_{ii} p_i^2 + U_i(q_i)\right] + \sum_{i<j} G^\circ_{ij} p_i p_j \tag{38}$$

allowing identification of the first term with $H_0(\mathbf{I})$ and the second with $H'(\mathbf{I}, \boldsymbol{\theta})$.

For the case of $N=2$ the resonance width of the m–n resonance associated with coupled Morse oscillators [each of the form $U_i(q) =$

$D_i(\exp(-2q/a_i)-2\exp(-q/a_i)+1)]$ was determined to be

$$(\Delta E_i)_{mn}=8\sqrt{2}\left[\frac{D_iD_j(G_{ij}^{\circ})^2}{\Omega_i^2\Omega_j^2G_{ii}^{\circ}G_{jj}^{\circ}}\right]^{1/4}\left[\frac{\omega_i^3\omega_j^3}{(\Omega_i^2/D_i)\omega_j^2+(\Omega_j^2/D_j)\omega_i^2}\right]$$
$$\times\left(\frac{1-\omega_i/\Omega_i}{1+\omega_i/\Omega_i}\right)^{m/4}\left(\frac{1-\omega_j/\Omega_j}{1+\omega_j/\Omega_j}\right)^{n/4} \tag{39}$$

where $\Omega_i=(2D_iG_{ii}^{\circ}/a_i^2)^{1/2}$. Here $(\Delta E_i)_{mn}$ is an energy width, closely related to ΔJ about a single oscillator energy E_i.

A graphic study of the resonance overlap regions led to several qualitative conclusions: (1) the fraction of phase space occupied by overlapping resonances increases with energy; (2) the number of resonances decreases as the two frequencies (Ω_i, Ω_j) move apart; (3) large changes in masses and bond angles are necessary to affect the resonances and their overlap; and (4) harmonic bending modes couple strongly to stretching modes only at high energies.

Values of the critical energy for the transition between regular and erratic behavior, as predicted by the overlapping resonances criterion, can be estimated from the graphs presented by Oxtoby and Rice. For example, for a model C—C—H system with C—C dissociation energy D_c and C—H dissociation energy $2D_c$, and frequencies 1000 cm^{-1} and 2900 cm^{-1} the $E_c \sim 0.85D_c$. With the parameters changed to C—C—C with pair bond energies D_c and $1.5D_c$ and frequencies 1000 cm^{-1}, 1300 cm^{-1}, the predicted $E_c \simeq 0.65D_c$. Unfortunately exact calculations of E_c for these systems are unavailable for comparison. The overlapping resonances criterion clearly requires verification for many of the oscillator systems shown in Table I; it is not expected,[31] however, to properly predict $E_c=\infty$ for integrable systems. The accuracy of the method for a variety of other problems has, however, been amply demonstrated.

E. Comments

Space limitations prevent discussion of several other means,[56, 57] both model and computational, for describing the transition. As is clear from the discussion above, however, much remains to be done including applications to $N>2$ systems, before a complete theory is developed. The ideal theory was defined as one which is global in time and in phase space, provided reliable estimates of the critical energy and rate of phase space mixing and gave qualitative insight into the relationship between the transition and properties of the system. No one theory satisfies more than a few of these (admittedly stringent) requirements. In addition, the theories, appear to relate the transition to different system features. In particular, the variational equations approach relies on changes in local system frequencies (note that Mo deals with global system frequencies), whereas Chirikov's approach focuses on overlapping resonances. The success of each of the methods, however limited, suggests that they identify important aspects of the transition phenomena. Perhaps the first step toward a consolidated theory requires an understanding of the relationship between these approaches.

VI. EXPONENTIATING TRAJECTORIES AND STATISTICAL BEHAVIOR

Regions of phase space characterized by exponential divergence of adjacent trajectories are assumed related to behavior that is intuitively regarded as statistical. Work has recently begun on quantifying the relationship between exponentiating trajectories and statistical theories of bimolecular reactions;[40b, 41, 58] studies on unimolecular reactions are in progress.[32]

Consider, for example, the collinear $A+BC$ collision system at fixed initial translational energy E_t, initial A to B—C distance and initial quasiclassical vibrational state n_1. A set of trajectories, used to calculate the product distributions is obtained by randomly selecting q_1, the vibrational phase angle. Our interest is in demonstrating that exponential divergence is a dynamic property which can be used to isolate, from this set of trajectories, a subset which evolves to a statistical product distribution. In this regard we note that a variety of statistical theories have been proposed[59] and interest is in properties which classify the product distribution as being in accord with *any* (rather than a particular) statistical theory. For the case of identical collinear atoms these properties are (1) that the reaction probability $P^R=0.5$, and (2) that the translational energy distribution of the reactive and nonreactive product are identical.

In the initial study reported in Ref. 40b the dynamics was studied on three model potential energy surfaces including the Porter-Karplus $H+H_2$ surface and that described by the coupled Morse Hamiltonian [Eq. (12)]. Qualitative features of the results obtained in each of these systems are similar. For example, the collinear reaction on the Karplus-Porter surface at $E_t=0.6$ eV and $n_1=0$ is clearly nonstatistical as evidenced by substantially different reactive and nonreactive product translational energy distributions. In addition, the reaction probability $P^R=0.77$. Since at this energy E is greater than the E_c obtained from the variational equations criterion (i.e., $E_c=0.255$ eV) and, in addition, *all* of the trajectories exponentially separated from nearby neighbors to some extent during the collision it is clear that these two conditions do not suffice to produce distributions in accord with statistical theories of chemical reactions. It is useful, however, to study subsets of trajectories based on more stringent requirements on the nature of the exponentiating trajectories.

Relying on the property of exponential separation leads to a rather limited choice of possible criteria for statistical behavior. One might require that the trajectories exponentially separate rapidly in time, that is, have a large $\log[D(t)/D(0)]$ vs. t slope. However, as is the case for most $N=2$ systems, the average slope in the Porter-Karplus system is relatively constant for all trajectories (7.5×10^{13} sec^{-1} to 9.6×10^{13} sec^{-1} for both $E_t=0.4$ eV and $E_t=0.6$ eV). An alternative is to study those trajectories that exponentially separate beyond a given minimum value, that is, to require a minimum $D(t_f)/D(0)$, where t_f is the time at the end of the collision. In the Porter-Karplus case ~2000 trajectories were computed and the ratio $D(t_f)/D(0)$ obtained for each. Product distributions were computed for the subsets of trajectories characterized by $D(t_f)/D(0)>10^2$, $D(t_f)/D(0)>10^3$ and $D(t_f)/D(0)>10^4$. The reactive and nonreactive product translational energy distributions were found to converge to those

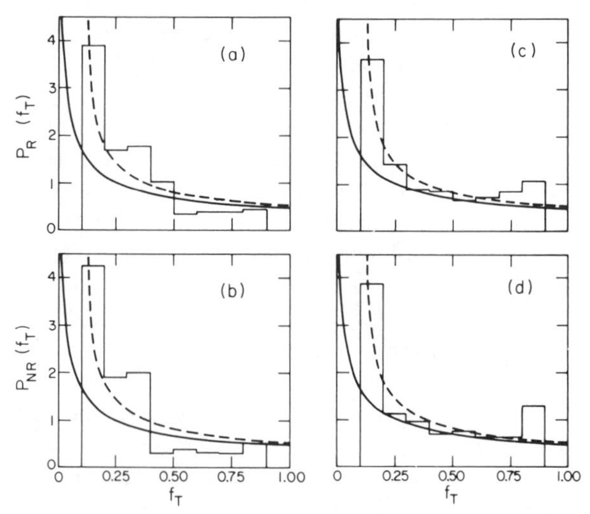

Fig. 6. Reactive (R) and nonreactive (NR) product translational energy in Porter-Karplus system (for conditions see text), where f_t is the fraction of energy in product translation. (a) and (b): subset of trajectories with $D(t_f)/D(0)>10^2$ (histogram); (c) and (d): subset of trajectories with $D(t_f)/D(0)>10^3$ (histogram). Solid + dashed curves are particular statistical theories discussed in Ref. 40b.

shown in Fig. 6, with a reaction probability $P^R = 0.502 \pm 0.015$. The reactive and nonreactive product distributions are clearly in good accord with one another indicating that the subset characterized by $D(t_f)/D(0) > 10^3$ does indeed evolve statistically. Similar results were obtained in more recent studies of three-dimensional $H + ICl$ and $K + NaCl$ collisions[41] as well as in studies[58b] of collinear $F + H_2$. These results clearly establish a quantitative relationship, discussed in further detail elsewhere,[40, 41] between nonlinear mechanics concepts of exponentiating trajectories and behavior required for the validity of statistical theories of bimolecular reactions.

APPENDIX: WHAT'S NEW IN 1980

The main body of this chapter dates from April 1979. This appendix, added in July 1980, provides additional references and remarks on developments over the interim period.

(A1) Additional References

Several useful reviews of theoretical[61, 62] and experimental[63, 64] aspects of intramolecular dynamics are now available or are forthcoming. In addition, we note new experimental work probing the nature of "energy flow" in molecules[65, 66] as well as several recent theoretical papers on the quantum picture of intramolecular dynamics[67–74] in the regular and irregular regimes. Although the focus is perceptively shifting towards an understanding of quantum behavior, several contributions which deal with interesting classical dynamical features in the regular regime have appeared. These include a study of spectra[75], of local vs. normal mode behavior in molecules[76], and of linear divergence in the phase space region dominated by tori[77]. Further insight into the relationship between exponential divergence, K entropies and the Lyapunov characteristic numbers has been provided in several papers[78] which extend results discussed in reference 34.

(A2) Mo's Method and the Variational Equations Approach

We remark on recent developments relating to Mo's method (section VA) and the Variational Equations approach (Sections VB and VC). As discussed above, Mo's method is based on the conjecture that $B(t)$ changes character as E increases through E_c. In particular, $B(t)$ is assumed to diverge exponentially at $E > E_c$, a conjecture which is apparently supported by the appearance of imaginary eigenvalues at these energies (see Table II). Two exact numerical studies have now been performed [62, 79] which clearly show that $B(t)$ decays, rather than diverges at $E > E_c$, casting serious doubt on the validity of the entire approach. Tabor[62], who has considered this method in greater detail, concludes that in defining $B(t)$ via a microcanonical average one necessarily washes out any real changes in dynamical behavior with increasing energy. This argument is supported by a demonstration that the moment y_1 (see below Equation 15) increases smoothly with energy, reflecting increases in the size of phase space and in no way displaying a dramatic change as E passes through E_c.

The variational equations approach has also been the subject of recent scrutiny[32, 77, 80, 81]. In reference 77 Marcus and coworkers present yet another integrable case for which the variational equations approach predicts a finite E_c. (They did not, however, apply the critical point analysis of Cerjan and Reinhardt which has been shown useful for correcting this deficiency.) Casati et al[77] reemphasize earlier concerns[40a, 50], arguing strenuously that the eigenvalues of $\mathsf{S}(t)$ [Equation 21] are sensitive to only highly local behavior (in time) and are unable to identify the important global long time $\underline{d}(t)$ growth. In light of this controversy it is perhaps useful to clearly summarize the method's successes and failures. First, there is little doubt that the method does successfully predict E_c for several $N=2$ problems (see Table I). Integrable cases for which the method incorrectly predicts a finite E_c may be properly treated within the framework of the Cerjan-Reinhardt critical point analysis, but further tests of these approach are required. No satisfactory understanding is available, however, for the method's inability to predict a finite E_c for the unequal mass Toda Hamiltonian.

More relevant, however, are questions relating to the basic conjecture upon which the method is founded, that is, that the local exponential divergence of adjacent trajectories occurs as the system passes through regions of real $\mathsf{S}(t)$ eigenvalues. The validity of this conjecture has been repeatedly confirmed[40, 41, 58a] for simple *scattering* systems, where interest was not in the utility of the variational equations method but rather in questions about the origin of statistical behavior in molecular collisions. The conjecture does not appear to hold[32, 81], however, "where it counts", that is, for bound molecular systems. The differences in behavior between bound and scattering systems suggests the possibility that in the bound system cases the trajectories are strongly influenced by periodic orbits which are not properly treated in the variational equations approach.

Despite the failures of the method it is important to recognize the underlying motivation for approaches of this kind. That is, they seek a useful means of identifying those qualitative system features which induce erratic behavior in molecular systems. In particular, the feasibility of extending the approach to larger "real molecule" systems played a significant role in its development. Quite clearly all of the methods discussed above are only zeroth order in their formulation. What is required now is the development of new theories of the stochastic transition which build upon the successes of the techniques discussed in this article.

Acknowledgments

The support of the Petroleum Research Fund, administered by the American Chemical Society, and support from an internal grant of the University of Toronto and the NSERC is gratefully acknowledged. The section of this paper on Mo's method has benefited greatly from discussions with Professors R. Kapral and J. T. Hynes.

References

1. For a review and references see S. A. Rice in *Excited States*, Vol. 2, E. C. Lim, ed., Academic, New York, 1975.
2. For a review and references see E. Hase in *Dynamics of Molecular Collisions, Part B*, W. H. Miller, ed., Plenum, New York, 1976.
3. J. H. Bartlett, *Classical and Modern Mechanics*, University of Alabama, 1975.
4. For example, R. M. May, *Nature*, **261**, 459 (1976).
5. For example, J. Cronin, *SIAM Rev.*, **19**, 100 (1977).
6. For example, H. Haken, *Synergetics*, Springer-Verlag, New York, 1976.
7. For example, R. Salmon, G. Holloway, and M. C. Hendershott, *J. Fluid Mech.*, **75**, 691 (1976).
8. For example, V. S. Afraimovich, V. V. Bykov, and L. P. Shilnikov, *Dokl. Akad. Nauk. SSSR*, **234**, 336 (1977).
9. For example, *Statistical Mechanics and Statistical Methods*, U. Landman, ed., Plenum, New York, 1977.
10. For example, R. Broucke, *Celestial Mech.*, **16**, 215 (1977).
11. For example, P. J. Holmes and Y. K. Lin, *J. Appl. Mech.*, **45**, 165 (1978).
12. For example, E. N. Lorenz, *J. Atmos. Sci.*, **20**, 130 (1973).
13. For example, R. G. Newton, *J. Math. Phys.*, **19**, 1068 (1978).
14. For example, H. L. Swinney and J. P. Gollub, *Phys. Today*, **31**, 41 (1978).
15. M. C. MacKay and L. Glass, *Science*, **197**, 287 (1977).
16. R. Abraham and J. Marsden, *Foundations of Mechanics*, Benjamin, Mass. 1979; V. I. Arnold, *Mathematical Methods of Classical Mechanics*, Springer-Verlag, New York, 1979; W. Thirring, *Classical Dynamical Systems*, Springer-Verlag, New York, 1979; J. Moser, *Stable and Random Motions in Dynamical Systems*, Princeton University Press, Princeton, N.J., 1973; V. I. Arnold and A. Avez, *Ergodic Problems in Classical Mechanics*, Benjamin, Mass., 1968.
17. (a) J. Ford, *Adv. Chem. Phys.*, **24**, 155 (1973); J. Ford, in *Fundamental Problems in Statistical Mechanics III*, E. D. G. Cohen, ed., North-Holland, Amsterdam, 1975; (b) G. Contopoulos, in *Dynamics of Stellar Systems*, A. Hayli, ed., Reidel, Dordrecht, 1975; (c) I. Percival, *Adv. Chem. Phys.*, **36**, 1 (1977); (d) S. A. Rice in *Advances in Laser Chemistry*, A. Zewail, ed., Springer-Verlag, New York, 1978; (e) M. V. Berry in *Topics in Nonlinear Mechanics*, S. Jorna, ed., American Institute of Physics, New York, 1978; (f) K. J. Wightman, *Rep. Prog. Phys.*, **40**, 1033 (1977); (g) A. S. Wightman, in *Statistical Mechanics at the Turn of the Decade*, E. D. G. Cohen, ed., Marcel Dekker, New York, 1971; (h) R. A. Marcus, *Ber. Bunsenges Phys. Chem.*, **81**, 190 (1977).
18. See, for example, M. V. Berry and M. Tabor, *Proc. R. Soc. Lond.*, **A349**, 101 (1976).

19. W. H. Miller, *Adv. Chem. Phys.*, **25**, 69 (1974).
20. R. A. Marcus, D. W. Noid, and M. L. Koszykowski, in *Advances in Laser Chemistry*, A. Zewail, ed., Springer-Verlag, New York, 1978, M. V. Berry, *J. Phys.* **A10**, 2083 (1977).
21. K. S. J. Nordholm and S. A. Rice, *J. Chem. Phys.*, **61**, 203, 768 (1974); **62**, 157 (1975).
22. M. V. Berry and M. Tabor, *Proc. Roy. Soc. Lond.* **A356**, 375 (1977). G. M. Zaslavsky, *Zh. Eksp. Teor. Fiz.*, **73**, 2089 (1977).
23. (a)E. Heller, *Chem. Phys. Lett.*, **60**, 338 (1979); (b) E. Heller, *J. Chem. Phys.* **72**, 1337 (1980).
24. See, for example, R. C. Churchill, G. Pecelli, and D. L. Rod in *Stochastic Behavior in Classical and Quantum Hamiltonian Systems*, G. Casati and J. Ford, eds., Springer-Verlag, New York, 1978.
25. D. W. Noid, M. L. Koszykowski, and R. A. Marcus, *J. Chem. Phys.*, **67**, 404 (1977).
26. Several techniques for obtaining additional integrals of motion, many of the perturbative type, have been proposed. See the discussion in Ref. 17f and I. C. Percival, *J. Phys.*, **A7**, 794 (1974); *ibid.* **12**, L57 (1979).
27. For an elementary introduction see J. L. Lebowitz and O. Penrose, *Phys. Today*, 23, (1973).
28. G. M. Zaslavskii and B. V. Chirikov, *Sov. Phys. Usp.*, **14**, 549 (1972).
29. J. Ford, *Lectures in Statistical Physics*, W. C. Schieve and J. S. Turner, eds., Springer-Verlag, New York, 1974.
30. This effect is nicely demonstrated, in an area preserving mapping, by Bartlett. See Ref. 3.
31. B. V. Chirikov, *Phys. Repts.*, **52**, 263 (1979).
32. I. Hamilton and P. Brumer (work in progress).
33. We cite two useful qualitative introductions to K-entropy: Wightman's discussion in Ref. 17f and that of Ya. G. Sinai, Acta Physica Austriaca, Suppl. X., 575 (1973).
34. G. Bennetin, L. Galgani, and J. M. Strelcyn, *Phys. Rev.*, **A14**, 2338 (1976).
35. See, for example, (a) M. Casartelli, E. Diana, L. Galgani, and A. Scotti, *Phys. Rev.*, **A13**, 1921 (1976), and (b) K. D. Hansel, *Chem. Phys.*, **33**, 35 (1978).
36. G. Contopoulos, L. Galgani, and A. Giorgilli, *Phys. Rev.*, **A18**, 1183 (1978). A similar effect has been observed in collision systems; see Ref. 41.
37. Many features of the transition from regular to erratic behavior in dynamical systems are seen in studies of area preserving mappings. Such mappings may or may not be related to Hamiltonian Systems. For an introduction see Ref. 3 and J. F. C. Van Velsen, *Phys. Rep.*, **41**, 137 (1978).
38. J. Duff, I. Hamilton, and P. Brumer (unpublished).
39. E. Thiele and D. J. Wilson, *J. Chem. Phys.*, **35**, 1256 (1961).
40. (a) P. Brumer and J. W. Duff, *J. Chem. Phys.*, **65**, 3566 (1976); (b) J. W. Duff and P. Brumer, *ibid.* **67**, 4898 (1977).
41. J. W. Duff and P. Brumer, *J. Chem. Phys.*, **71**, 2693 (1979).
42. B. Barbanis, *Astron. J.*, **71**, 415 (1966).
43. R. H. Tredgold, *Proc. Phys. Soc. Ser.*, **A68**, 920 (1955).
44. Prof. J. Ford called this Hamiltonian to our attention and indicated that its integrability is readily established by transforming to new coordinates $x-y$ and $x+y$.
45. G. Casati and J. Ford, *Phys. Rev.*, **A12**, 1702 (1975).
46. G. H. Walker and J. Ford, *Phys. Rev.*, **188**, 416 (1969).
47. M. Casartelli, G. Casati, E. Diana, L. Galgani, and A. Scotti, *Teor. Mat. Fiz.*, **29**, 205 (1976).
48. K. C. Mo, *Physica*, **57**, 445 (1972).
49. Much of the content of this section results from extensive discussions with Professors R. Kapral and J. T. Hynes.

50. M. Toda, *Phys. Lett.*, **A48**, 335 (1974); see also Ref. 51 for a related discussion of scattering problems.
51. P. Brumer, *J. Comput. Phys.*, **14**, 391 (1973).
52. C. Cerjan and W. P. Reinhardt, *J. Chem. Phys.*, **71**, 1819 (1979). Also, W. P. Reinhardt's short film on Henon–Heiles trajectories.
53. A criticism leveled at the variational equations approach by G. Benettin, R. Brambilla, and L. Galgani, *Physica*, **87A**, 381 (1977) is deserving of comment. They applied the method to a simple periodic anharmonic oscillator and noted that real eigenvalues obtain at energies above the inflection point of the potential, although the oscillator motion is regular. However, in this case (see Ref. 40a, Appendix B) the time dependence of $\mathsf{T}^{-1}(t)\dot{\mathsf{T}}(t)$ is such that application of the eigenvalue criterion is inappropriate.
54. For an elementary discussion see Brauer and Nohel, *Qualitative Theory of Ordinary Differential Equations*, Benjamin, Mass., 1969.
55. D. W. Oxtoby and S. A. Rice, *J. Chem. Phys.*, **65**, 1676 (1976).
56. A. N. Kaufman, *Phys. Rev. Lett.*, **27**, 376 (1971).
57. J. M. Greene, *J. Math. Phys.*, **9**, 760 (1968); for an application of this method to the Henon-Heiles Hamiltonian see G. H. Lunsford and J. Ford, *J. Math. Phys.*, **13**, 700 (1972).
58. See also (a) J. S. Hutchinson and R. E. Wyatt, *J. Chem. Phys.*, **70**, 3509 (1979) and, in relation to this paper (b) J. W. Duff and P. Brumer, *ibid.*, **71**, 3895 (1979).
59. J. C. Light, *Discuss. Faraday Soc.*, **44**, 14 (1968); R. D. Levine and R. B. Bernstein, *Adv. At. Mol. Phys.*, **11**, 216 (1975), Appendix 3; A. F. Wagner and E. K. Parks, *J. Chem. Phys.*, **65**, 4343 (1976).
60. M. Henon and C. Heiles, *Astrophys. J.*, **69**, 73 (1964).
61. Y. M. Treve in *Topics in Nonlinear Dynamics*, S. Jorna, ed., American Institute of Physics, New York, 1978.
62. M. Tabor, "The Onset of Chaotic Motion in Dynamical Systems", *Adv. Chem. Phys.*, (in press).
63. R. G. McDonald, *Ann. Rev. Phys. Chem.*, **30**, 29 (1980); P. A. Schulz, A. S. Sudbo, D. J. Krajnovich, H. S. Kwok, Y. R. Shen and Y. T. Lee, *Ann. Rev. Phys. Chem.*, **30**, 379 (1980).
64. I. Oref and B. S. Rabinovitch, *Accts. Chem. Res.*, **12**, 166 (1979).
65. R. A. Covaleskie, D. A. Dolson and C. S. Parmenter, *J. Chem. Phys.*, **72**, 5774 (1980).
66. J. Chaiken, M. Gurnick and J. D. McDonald, "Average Singlet-Triplet Coupling Properties of Biacetyl and Methylglyoxal Using Quantum Beat Spectroscopy" (preprint), and references therein.
67. R. A. Marcus in *Horizons of Quantum Chemistry*, K. Fukui and B. Pullman, eds., D. Reidel, New York, 1980.
68. R. M. Stratt, N. C. Handy and W. H. Miller, *J. Chem. Phys.*, **71**, 3311 (1979).
69. P. Brumer and M. Shapiro, "Intramolecular Dynamics: Time Evolution of Superposition States in the Regular and Irregular Spectrum" *Chem. Phys.*, Letters (in press).
70. M. J. Davis, E. B. Stechel and E. J. Heller, "Quantum Dynamics in Classically Integrable and Nonintegrable Regions" (preprint).
71. R. Kosloff and S. A. Rice, "The Influence of Quantization on the Onset of Chaos in Hamiltonian Systems: The Kolmogorov Entropy Interpretation", *J. Chem. Phys.*, (in press).
72. K. G. Kay, "Numerical Study of Intramolecular Vibrational Energy Transfer: Quantal, Classical and Statistical Behavior" (preprint).
73. J. S. Hutchinson and R. E. Wyatt, "Quantum Ergodicity and the Wigner Distribution" (preprint).
74. For a discussion of quantization in the irregular spectral regime see C. Jaffé, Ph.D.,

Dissertation, University of Colorado, 1979; Also C. Jaffé and W. P. Reinhardt (to be published).
75. E. J. Heller, E. B. Stechel and M. J. Davis, *J. Chem. Phys.*, **71**, 4759 (1979).
76. R. T. Lawton and M. S. Child, *Mol. Phys.* **37**, 1799 (1979): C. Jaffé and P. Brumer, *J. Chem. Phys.*, (in press).
77. G. Casati, B. V. Chirikov and J. Ford, *Phys.*, Letters **77A**, 91 (1980).
78. G. Benettin, L. Galgani, A. Giorgilli and J-M. Strelcyn, "All Lyapunov Characteristic Exponents are Effectively Computable" (preprint); G. Benettin and L. Galgani in *Intrinsic Stochasticity in Plasmas*, G. Laval and D. Gresillon, eds., Editions de Physique, Orsay, 1979.
79. I. Hamilton and P. Brumer, presented by P. Brumer at the 178th ACS National Meeting, Washington, D.C. (1979).
80. D. W. Noid, M. L. Koszykowski and R. A. Marcus, *J. Chem. Phys.*, **71**, 2864 (1979).
81. D. Carter and P. Brumer (unpublished).

THE INFORMATION THEORETIC APPROACH TO INTRAMOLECULAR DYNAMICS*

R. D. LEVINE

Department of Physical Chemistry and Institute for Advanced Studies, The Hebrew University, Jerusalem, Israel

Abstract

It is shown, both by exact dynamical considerations and by a phenomenological analysis of experimental results, that energy randomization in an activated molecule is as complete as possible subject to constraints. These constraints are constants of the motion and can be identified either by the phenomenological procedure of surprisal analysis or by a dynamical approach. A given energy-rich molecule can demonstrate different dynamical behaviour since the constraints depend not only on the Hamiltonian but also on the activation process. The resulting picture is thus intermediate between the Slater model (complete inhibition of intramolecular energy flow due to ever present constants of the motion) and the RRK model (complete energy randomization).

Surprisal analysis examines the energy-rich molecule only at the terminal stage of its evolution—that is as dissociation products. The central diagnostic conclusion of this chapter is that the surprisal is a constant of the motion and hence serves as a signature of an incomplete (i.e. constrained) energy randomization throughout the time evolution and that the constraints can be identified by surprisal analysis. The chapter thus begins with a survey of the results of surprisal analysis for different energy rich polyatomic molecules excited by all the commonly used techniques. It then offers three technical sections: (1) What is the maximum entropy formalism and why is it required? (2) How to specify the initial state and (3) The exact solution of the dynamics for a given initial state. This final section introduces the concept of the constant of motion as used above and notes how such constants can be identified using sum rules and then can be used in a practical predictive route.

CONTENTS

I. Background 240
II. Surprisal Analysis 244
A. $(C_3H_4O)^{\neq}$ and Other Elimination Reactions 245
B. F+RH 248
C. Intermezzo on Maximum Entropy 250
D. Multiphoton Dissociation 254
E. Collisional Activation and Deactivation 257
F. Distribution of Rotational States 261

*Work supported by the US Air Force Office of Scientific Research, Grant 77-3135.

G. Heavy Ion Transfer Reactions 263
H. Summary 264
III. The Maximum Entropy Formalism 264
A. The Algorithm 265
B. The Initial State—The Unavoidable Need for the Maximum Entropy Principle . 266
C. Variational Approximations 268
D. Phenomenology 269
IV. The Initial State 269
A. The Stationary State 270
B. The Activation Process 271
C. The Most Conservative Inference 272
1. Sufficient Statistics 272
2. The Most Probable State 273
3. The Most Random State 273
4. The Most Conservative Inference 274
V. Intramolecular Dynamics 275
A. Once Is Enough 275
B. Incomplete Resolution of Final States 276
C. Time-Dependent Constants of the Motion 277
D. The Time Evolution 278
E. The Constraints 279
F. Sum Rules 281
G. Approximations 283
H. Summary 284
VI. Outlook 285
Appendix 287
References 288

I. BACKGROUND

The potential applications of photoselective chemistry have reopened the discussion of intramolecular relaxation processes during the time evolution of energy-rich polyatomic molecules. Chemical and collisional activation studies[1] have strongly supported the conclusion that at or above the levels of excitation required for dissociation, intramolecular relaxation rates are faster than the rate of dissociation. Depending on the pressure, the intramolecular relaxation rate is faster or comparable to the rate of collisional deactivation. Hence, on the time scale of interest for either unimolecular or bimolecular chemistry, the highly excited polyatomic molecule is already "equilibrated" by the efficient intramolecular relaxation. In more technical terms, all quantum states with energies in the range E, $E+\delta E$ are equiprobable. Any selective deposition of the initial excitation is thus relaxed on the time scales of chemical interest. The purpose of this chapter is to question this conclusion and to suggest that by proper initial state selection it might be possible to retain selectivity throughout the time evolution.

The kinematic evidence in support of efficient intramolecular relaxation is of two kinds: (1) the magnitude of the dissociation rate constant agrees[2, 3] with that computed using the RRKM theory, where all effectively coupled states (at a given total energy) are given the same weight, and (2) by working at sufficiently high pressures (so that only newly excited molecules have a chance to dissociate) it is just possible to distinguish the initial mode of excitation.[4, 5] These later studies suggest[1] that internal relaxation in energy-rich complex molecules occurs on a time scale of 1–10 psec. Of course, these numbers are quite sensitive to the assumption that deactivating collisions are effective in removing the excess energy. Since the energy content of a molecule formed via chemical activation is often several tens of kcals/mole above the activation barrier for dissociation, this strong collision assumption is subject to question. Indeed, we shall argue that experiments on energy transfer from complex excited molecules are consistent with the assumption that after such a collision the internal energy is effectively randomized.[6, 7] It is possible therefore that the intramolecular energy relaxation is at least partly collision induced.

The results of experiments of multiphoton dissociation of polyatomic molecules have also been interpreted in terms of efficient energy partitioning among available modes.[8, 9] In particular, the velocity distribution of the unimolecular dissociation products was accounted for using such a model.[8] However, and as in previous applications, a quantitative fit often requires the introduction of an "effective number" of degrees of freedom among which the energy is shared.

Detailed studies of unimolecular reactions at the low-pressure regime have also been extensively discussed in terms of strong coupling between participating modes.[10] It was found necessary however to restrict the range of the states which are effectively coupled.[10] Criteria, which reduce to known results as special cases, were developed to decide on which states are included. However, and as in previous studies, states were either strongly coupled (that is, equiprobable) or not coupled at all (and hence excluded). It is this point that distinguishes the present "constrained phase space" approach from previous statistical theories. We allow all states, subject to dynamical constraints. These constraints do not exclude any state but lead to the assignment of different weights to different states. Moreover, the constraints are determined not only by the potential energy surface and the total energy but also by the details of the initial excitation.

Theoretical and computational studies of the classical mechanics of systems below and above the dissociation limit have been carried out.[11–21] At low energies, near the bottom of the well, there is essentially no energy exchange among the normal modes. For a system of n degrees of freedom, the phase space is $2n$ dimensional. Hence, at a given total energy the trajectory could sample a $2n-1$ dimensional space (the so-called "energy shell"). Yet it is found that at low energies the motion of the trajectory is confined to a subspace whose dimension is less than $2n-1$ (e.g., n, for a strictly harmonic molecule). Moreover, at regular intervals, the trajectory returns to its starting point. As the energy is increased, the role of the anharmonic coupling becomes apparent. Two trajectories with rather similar initial conditions may rapidly diverge in their subsequent evolution. Indeed, it has become customary to talk of the onset of ergodic behavior where trajectories appear to sample the entire ($2n-1$ dimensional) energy shell. It remains, however, to be shown that the mathematical definition of ergodicity[17] corresponds to the chemist's idea of intramolecular energy transfer. In particular, what is of concern here is relaxation on the time scale of experimental interest.

It is indeed possible to consider potential energy surfaces, where, depending on the details on the initial excitation, energy randomization does or does not occur. Intermediate examples where energy sharing occurs among only a subset of the available modes have also been documented.[18]

Computational studies[19–21] at energies above the dissociation threshold show that already at fairly low excess energy the randomization is incomplete during the lifetime of the energy-rich molecule. At higher energies the presence of the well becomes less and less pronounced. This transition to a "direct" collision regime is quite evident in, say, the experimental angular distribution of the final products.[22] Even at lower energies, where the energy-rich molecule may survive for at least several rotational periods, its observed mode of dissociation may be determined not only by the total energy but also by the degrees of freedom in which this energy was initially present.[23] Similarly, photoelimination[24–29] at energies above threshold demonstrate that excess energy in the products is not equally shared by all the modes. Chemical activation studies which monitor the energy distribution in the products[30–32] lead to similar conclusions. In the extreme limit of collisions which proceed on a potential energy surface without a deep well, selectivity of energy consumption and specificity of energy disposal are the rule.[23, 33–35]

A particularly instructive example is the study[36] of the

$$O + CN(v) \begin{cases} \to CO(v') + N(^2D) \\ \to CO(v') + N(^4S) \end{cases} \qquad \text{(A)}$$

reaction. The excited, $N(^2D)$ atom is formed via the long living $NCO(X^2\Pi)$ intermediate. Indeed, the corresponding vibrational distribution of CO is essentially a statistical one and is unaffected by changes in the initial vibrational state of CN. (Except for the obvious dependence on the total energy, which does change as the CN vibrational excitation is increased.) The ground, $N(^4S)$, atom is formed by a direct reaction and the accompanying CO vibrational distribution is inverted and changes in a systematic fashion with changes in the initial CN vibrational excitation.

The question discussed in this chapter is whether specificity of energy disposal in the products can serve as an indicator of the evolution of the energy rich molecule and of its mode of preparation. In particular, does it rule out energy randomization. We conclude that it does. Specifically, the picture that emerges is intermediate between the RRK model[37] (free flow of energy among all modes) on the one hand, and the Slater model[38] (strictly harmonic modes, no energy exchange) on the other. It will be shown that the experimental results lead unequivocally to the conclusion that energy randomization is as extensive as possible subject to the dynamic constraints on the system. These constraints are shown to be constants of the motion (just as in the Slater theory). Hence the constraints are determined by the preparation of the initial state and their presence can be discerned by an analysis of the dissociation products.

We are thus faced with a dilemma. Indirect evidence lends considerable support to the concept of efficient intramolecular relaxation in complex energy-rich molecules. Analysis of the distribution of the energy in the dissociation products is often at variance with this simple (RRK) picture. It is possible to remove some of the discord by noting that the presence of constraints is determined also by the refinement of the initial excitation process and that collisional and chemical activation are least selective in this respect. Moreover, surprisal analysis was mostly carried out for dissociation products of comparatively short living species. Experiments on products state analysis for energy-rich molecules with lifetimes well above the picosecond range will thus be particularly welcome.

The RRK and Slater models can be thought of as corresponding to "compound" and "direct" dynamics of bimolecular reactions.[22] As has been noted for many systems, compound behavior is typical when the energy per mode is uniformly low. By preparing the reagents with an excess energy in some mode one observes a transition to direct dynamics. Over the years, the simple RRK model became increasingly more sophisticated (the use of a reduced number of oscillators, RRKM,[39] RRKM plus angular momentum constraints,[40] the adiabatic channel model,[10] etc.). All these refinements served to constrain the range of states that are deemed equiprobable. The recognition that the "most statistical" energy distribution must be subjected to some constraints is therefore not particularly new. What is new in the present discussion is the manner in which such constraints are imposed and three implications. Rather than decide which states are "in" (i.e., equiprobable) and which are excluded, we impose a constraint(s) and determine, using the maximum entropy formalism (Section III), the most statistical state distribution which is consistent with the constraint(s). The result is that each state is assigned its own weight, and this weight can be different for different states. The implications are as follows: (1) The constraints can be identified from experimental (or computational) results by the phenomenological procedure of surprisal analysis (Section II). (2) The nature and magnitude of the constraints is determined by the preparation of the initial state (Section IV). In particular, the less selective is the initial state preparation—the fewer constraints are present during the collision. A given system at a given energy can exhibit considerable variation in its dynamic behavior depending on the initial state. (3) The present "constrained phase space" procedure can be derived on rigorous dynamical grounds (Section V). The use of approximate constraints is then shown equivalent to a variational solution of the equations of motion.

A problem that will not be discussed in this chapter is the computation of the unimolecular rate constant. The connection between specificity of energy disposal (or selectivity of energy consumption), and the magnitude of the rate constant has been discussed elsewhere for both unimolecular[41] and bimolecular[42] processes. Further work along such lines is in progress.

II. SURPRISAL ANALYSIS

Surprisal analysis[33, 35, 43–46] is a phenomenological procedure which centers attention on the deviation of the distribution of the final states from that expected when all such are equally probable. One purpose of this chapter is to argue that such a deviance is a true signature of the lack of randomization during the collision; that even though the surprisal is determined after the event it is a valid indicator of what happened during the time evolution. We do this by showing that the surprisal is a (time-dependent, cf. Section V.C) constant of the motion. Hence its mean value is the same throughout the time evolution and therefore if it is finite at the end it was finite during the entire history of the energy-rich molecule and can be traced back to the specificity built-in in the beginning by the excitation process.

Before turning to the theoretical developments we consider a number of applications of surprisal analysis. The excitation techniques represented by these examples span the available range: chemical activation,[1–5, 24, 26, 30–32, 47–56] single-photon dissociation,[57, 58] multiphoton dissociation,[7–10, 27–29, 62–64, 67–70] photoelimination,[25, 26] and collisional activation. Also included are two examples drawn from computational studies and one example of a nuclear reaction. The final state distributions probed include vibrational, rotational, translational, and electronic.

A. $(C_3H_4O)^{\neq}$ and Other Elimination Reactions.

The vibrational distribution of CO molecules formed in the photodissociation of a methylketene (B.1) has been determined[57, 58] and found to be similar to that found[47] in the $O(^3P)$ + methylacetylene reaction (B.2)

$$CH_3CH{=}CO \xrightarrow{226\ \text{nm}} (CH_3CH{=}CO)^{\neq} \rightarrow CO + CH_3CH \qquad \text{(B.1)}$$

$$O(^3P) + CH_3C{\equiv}CH \rightarrow (CH_3CH{=}CO)^{\neq} \rightarrow CO + CH_3CH \qquad \text{(B.2)}$$

which is thought to proceed via an energy-rich methylketene as the intermediate. Figure 1 shows the CO vibrational distributions for the two alternative modes of preparing the energy-rich $(C_3H_4O)^{\neq}$ intermediate and a schematic energy diagram. The vibrational energy disposal is seen to be essentially independent of the mode of formation of the activated intermediate. As expected under such circumstances the distribution is in

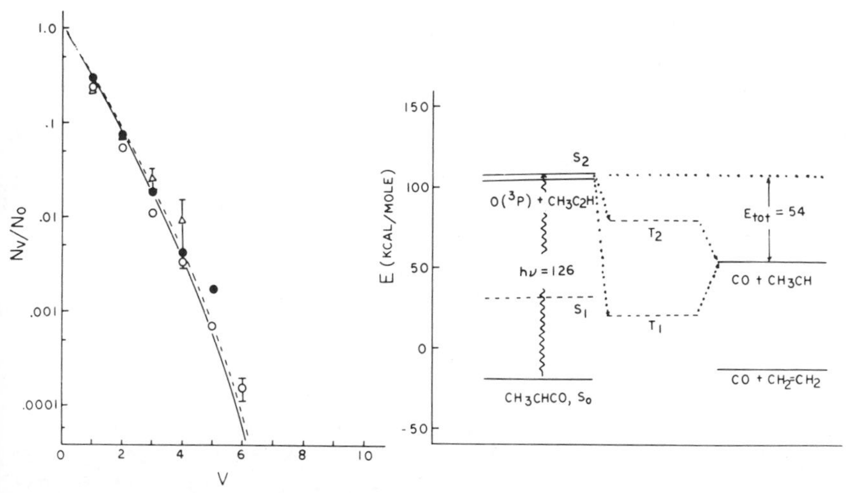

Fig. 1. Energetics and vibrational energy disposal for $(C_3H_4O)^{\neq}$. Left panel: CO vibrational state distribution. Triangles and circles: photodissociation of CH_3CHCO at 226 nm under different conditions. Dots: $O + CH_3C_2H$ reaction. Dashed line: the prior distribution. Solid line: the RRKM prediction. (Figure adapted from Ref. 57.) Right panel: Schematic energy diagram. At 226 nm the photon creates a state at about the same mean total energy as that formed in the $O + CH_3C_2H$ reaction. (Figure adapted from Ref. 57.)

close accord with that expected when all final quantum states are equally probable (the so-called[33, 43–46] "prior" distribution, shown as a dashed line in Fig. 1). Also shown in Fig. 1 is the RRKM prediction (assuming a tight transition state[47]), which accords with the experimental and the prior results.

Not all chemical activation studies using oxygen atoms lead to a prior final state distribution.[48]

Other photoelimination reactions often show considerable specificity of products energy distribution.[26] Indeed, photoelimination has been used to produce excited populations for chemical laser action.[26] Figure 2 compares the HCl vibrational populations for chloroethylene

$$CH_2{=}CHCl \overset{>155\ \text{nm}}{\rightarrow} (CH_2{=}CHCl)^{\neq} \rightarrow HCl + HC{\equiv}CH \qquad \text{(C)}$$

to that expected if all the available energy was randomized among the internal states of HCl, the translational motion, and the internal modes of HC═CH [the so-called prior distribution, denoted $P°(v)$].

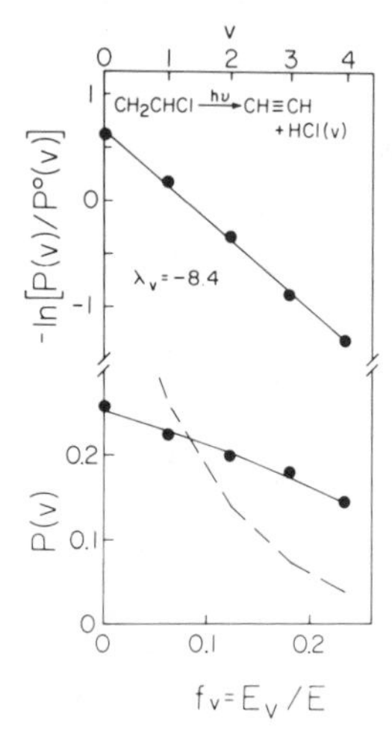

Fig. 2. Disequilibrium in the products of photoelimination. Bottom part: The experimental (dots, Ref. 26) and prior ($P°(v)$, dashed line) HCl vibrational state distribution vs. $f_v = E_v/E$. The vibrational quantum number of HCl is identified in the top scale and f_v is the fraction of the total energy ($E = 133$ kcal/mol) present in the HCl vibration. Top part: the surprisal, $-\ln[P(v)/P°(v)]$ vs. f_v. If one assumes that the surprisal is exactly linear (with a slope of -8.4) one can compute the distribution of vibrational states. The result of this computation [cf. (2.8)] is shown as the solid line in the bottom part. (Adapted from Ref. 25.) The prior distribution used here and in all other figures except 5 is one where all final quantum states are equally probable. This is the prior distribution used also in the theoretical analysis.

Elimination can also result following chemical activation, for example,

$$CH_3 + CF_3 \rightarrow (CH_3CF_3)^{\neq} \rightarrow CH_2{=}CF_2 + HF \tag{D.1}$$

The HF vibrational state distribution[24–26] cannot be accounted for if the energy is statistically distributed (Fig. 3). A second route to $(CH_3CF_3)^{\neq}$ is via multiphoton absorption

$$CH_3CF_3 \xrightarrow{CO_2\ \text{laser}} (CH_3CF_3)^{\neq} \rightarrow CH_2{=}CF_2 + HF. \tag{D.2}$$

Here too[27] the HF vibrational distribution is nonstatistical; moreover, IR emission from vibrationally excited CH_2CF_2 has been observed. Chemical activation can also be used to form the energy-rich adduct in several ways,[49] for example (D.1) as compared to

$$H + CH_2CF_3 \rightarrow (CH_3CF_3)^{\neq} \rightarrow CH_2{=}CF_2 + HF \tag{D.3}$$

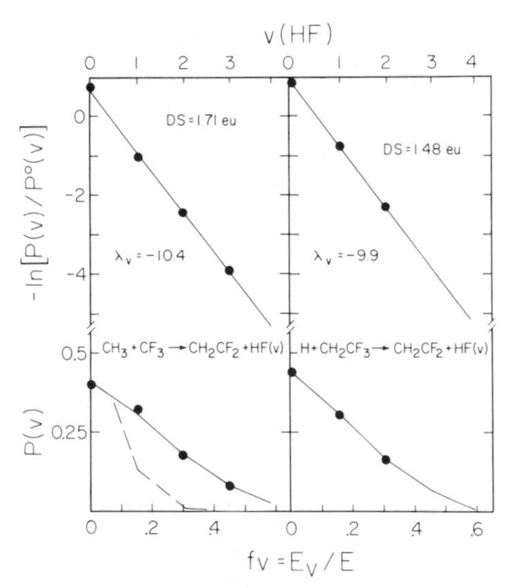

Fig. 3. Surprisal analysis of HF vibrational state distribution in the elimination reactions (D_1) and (D_3). (Adapted from Ref. 25. Experimental results from Ref. 26.) Bottom part: The experimental (dots) and prior (dashed line) distributions vs. $f_v = E_v/E$. Top part: The surprisal $-\ln[P(v)/P^\circ(v)]$ vs. f_v. The virbrational quantum number of HF is shown in the top scale. The distribution (2.8) corresponding exactly to a linear surprisal is shown as a solid line in the bottom part. Within experimental error it is not possible to conclude that the distributions of (D_1) and (D_3) are different. Analysis of earlier data (by Polanyi and co-workers) on reaction (D_1) leads to $\lambda_v = -8.2$. (See Ref. 25). The available information[27] on HF vibrational state distribution following multiphoton absorption of CH_3CF_3 suffices only to show that λ_v, the slope of the surprisal plot, is in the range -9 to -11.

or

$$CH_3 + CH_2F \rightarrow (CH_3CH_2F)^{\neq} \rightarrow CH_2{=}CH_2 + HF \qquad (E.1)$$

compared to

$$H + CH_2CH_2F \rightarrow (CH_3CH_2F)^{\neq} \rightarrow CH_2{=}CH_2 + HF. \qquad (E.2)$$

Both reactions (E) involve the formation of the activated CH_3CH_2F intermediate with nearly the same energy content. The preliminary results[26, 49] are, however, that HF molecules from the reaction (E.1) carry about 20% more vibrational excitation than from (E.2). A memory of the initial mode of excitation, as expected for a reaction where energy partitioning in the products is nonstatistical has not yet been unequivocally demonstrated.

B. F+RH

There are very many studies[50–56] of the vibrational (and, sometimes, rotational) distributions of HF formed in the abstraction of H atoms from a variety of polyatomics. A common characteristic is that the energy disposal, as judged by HF, is highly nonstatistical. Figure 4 is typical. The

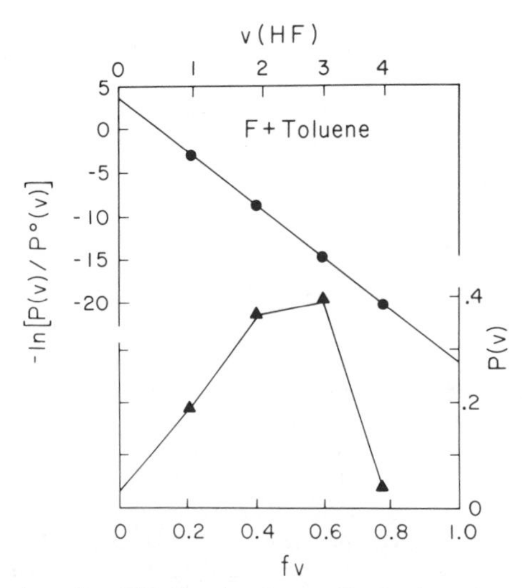

Fig. 4. The observed product HF vibrational state distribution $P(v)$ (triangles, bottom part) and its deviance (circles, upper part) from the prior distribution, $P°(v)$ vs. f_v, for the F+toluene reaction. (Adapted from Ref. 50.)

HF distribution (from Ref. 50; shown in the bottom) is highly deviant from the prior distribution. Reducing the number of degrees of freedom is of no avail. Even in the extreme limit (Fig. 5) where the polyatomic fragment R is treated as a rigid moiety that cannot accept any energy (model I in Fig. 5), so that all the available energy is partitioned between the internal (vibrotational) energy of HF and the relative translational motion of R and HF, one still cannot account for the results, and the deviance (the "surprisal") is quite large. Also shown in Fig. 5 is an "intermediate" case (model II) where only the vibrations of R are frozen. The energy is then partitioned between the vibration and translation of HF, the rotation of R and the relative translation of R and HF.

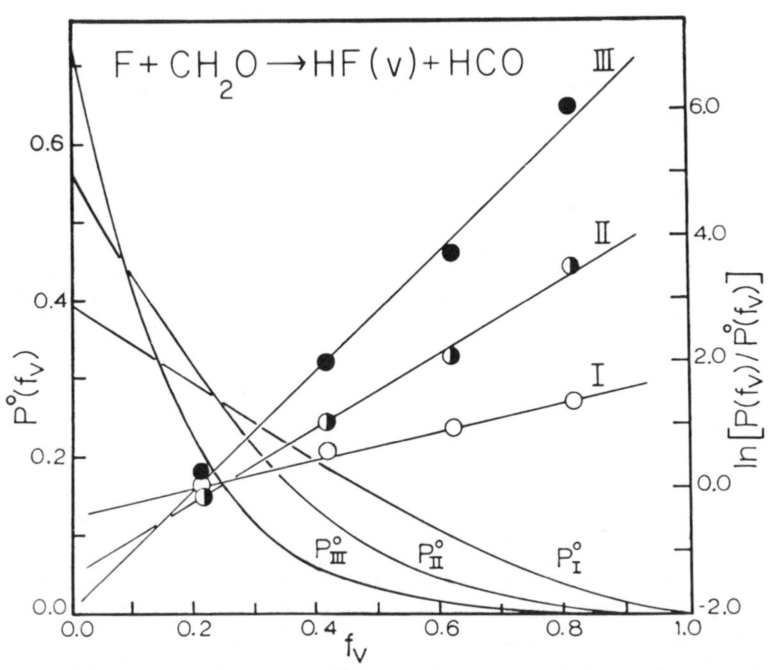

Fig. 5. The deviance of the HF vibrational state distribution in the $F+CH_2O$ reaction from the prior distribution (adapted from Ref. 50). The surprisal line and points labeled III use the proper prior distribution where all final quantum states are equally probable. (Shown as $P^{\circ}{}_{III}$ in the bottom part). In the spirit of the "effective number of oscillators" of the RRK approach one can wonder, however, if by freezing out some degrees of freedom, the deviance between $P(v)$ and $P^{\circ}(v)$ could not be accounted for. The curves labeled II are based on freezing out the vibrations of CHO. The curves labeled I freeze out both the vibrations and the rotations of CHO, regarding it as a rigid particle. The surprisal is indeed diminished as a result, but is not altogether eliminated and remains quite finite even for model I.

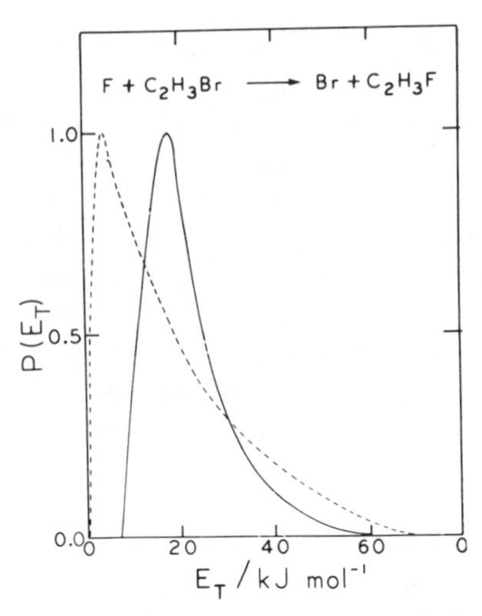

Fig. 6. The observed (solid curve) and the RRKM prediction (dashed line) products translational energy distribution in the $F+C_2H_3Br$ reaction (adapted from Ref. 56). The RRKM theory as employed by Buss et al. is refined in several ways but is unable to account for the observed distribution. The reaction is however highly exoergic and the lifetime of the intermediate adduct is probably below 0.1 psec.

Molecular beam studies verify that chemical activation reactions which appear to proceed via an intermediate that survives for quite a number of vibrational periods, for example,

$$F+C_2H_3Br \rightarrow Br+C_2H_3F \tag{F}$$

yield products' translational energy distributions that do deviate from that expected for complete randomization (Fig. 6).

C. Intermezzo on Maximum Entropy

The picture of intramolecular dynamics proposed in this chapter is one where energy is maximally randomized subject only to such constraints that are present. The experimental support for this picture is provided by surprisal plots such as Figs. 2–5. It is important therefore to see how the proposed picture is implied by the results shown. The technical details will be provided in the later sections. Here we just sketch the argument. The discussion is, at this point, limited to the dissociation products.

To specify a single quantum state of the products R + HF, it is necessary to specify the vibrational state of HF, the rotational angular momentum (and its projection) of HF, the relative momentum of R and HF, and a complete set of (internal) quantum numbers for R. We shall denote the probability of such a quantum state by $P(v,\mathbf{n})$. Here v is the vibrational quantum number of HF and $\mathbf{n}$ is the set of all other quantum numbers which together with v specify a single quantum state of the products. For a distribution over quantum states the entropy is given by[43–46]

$$S[v,\mathbf{n}] = -\sum_{v}\sum_{\mathbf{n}} P(v,\mathbf{n})\ln[P(v,\mathbf{n})]$$

The notation $S[v,\mathbf{n}]$ serves as a reminder of what quantum numbers have been summed over. To obtain the entropy in thermodynamic (e.u.) units, the sum need be multiplied by the gas constant R. This expression for the entropy is further discussed in the Appendix.

There are very many different quantum states, all of which correspond to HF being in a given vibrational level. In these states the energy $E-E_v$, which is not in HF vibrational excitation, is distributed in different ways among the other modes. The probability of HF in a given vibrational level is given by

$$P(v) = \sum_{\mathbf{n}} P(v,\mathbf{n}) \tag{2.1}$$

Summation in (2.1) is over such levels $\mathbf{n}$ that the total energy equals the available energy E.

The most statistical distribution of quantum states (at a given total energy E) is the one where all states are equally probable. This is the distribution whose entropy is maximal, and which is referred to as the prior distribution. Explicitly

$$P^{\circ}(v,\mathbf{n}) = \frac{1}{\sum_{v}\sum_{\mathbf{n}} 1} \tag{2.2}$$

and from (2.1)

$$P^{\circ}(v) = \frac{\sum_{\mathbf{n}} 1}{\sum_{v}\sum_{\mathbf{n}} 1} \tag{2.3}$$

As in (2.1), summations here are restricted to states $(v,\mathbf{n})$ of a given energy E.

Say the observed vibrational distribution, $P(v)$ is not in accord with $P^{\circ}(v)$. One can then seek that distribution $P^{ME}(v,\mathbf{n})$, which is as statistical as possible (i.e., of maximal entropy), but which is chosen not among all possible distributions but only among such distributions where the mean vibrational energy of HF has a given value, $\langle E_{\text{vib}} \rangle$,

$$\sum_v \sum_{\mathbf{n}} E_v P(v,\mathbf{n}) = \langle E_{\text{vib}} \rangle \tag{2.4}$$

The result is[43–46] [cf. (3.3)]

$$P^{ME}(v,\mathbf{n}) = \exp\left[-\lambda_v \frac{E_v}{E} - \lambda_0 \right] \tag{2.5}$$

Here λ_v is a (constant, i.e., v-independent) parameter chosen such that the distribution $P(v,\mathbf{n})$ has the specified value of the mean vibrational energy. Explicitly, the value of λ_v is the solution of the implicit equation

$$\langle E_{\text{vib}} \rangle = \sum_v \sum_{\mathbf{n}} E_v \frac{\exp[-\lambda_v (E_v/E)]}{\sum_v \sum_{\mathbf{n}} \exp[-\lambda_v (E_v/E)]} \tag{2.6}$$

For unharmonic levels, the solution of (2.6) for λ_v is best done numerically and an efficient computer algorithm has been described,[59] λ_0 is a function of λ_v determined by the condition that $P^{\text{ME}}(v,\mathbf{n})$ is normalized,

$$\exp(\lambda_0) = \sum_v \sum_{\mathbf{n}} \exp\left[-\lambda_v \frac{E_v}{E} \right] \tag{2.7}$$

Finally, to the point: The result (2.5) shows explicitly that all final quantum states where HF is in a given vibrational level v are equiprobable. Not so for states where the vibrational quantum number of HF is different. The procedure of maximal entropy subject to the constraint (2.4) led to the most statistical distribution which is consistent with the constraint.

The predicted vibrational distribution is readily computed from (2.5) using (2.1):

$$\begin{aligned} P^{\text{ME}}(v) &= \sum_{\mathbf{n}} P^{\text{ME}}(v,\mathbf{n}) \\ &= \sum_{\mathbf{n}} \exp\left[-\lambda_v \frac{E_v}{E} - \lambda_0 \right] \\ &= \exp\left[-\lambda_v \frac{E_v}{E} - \lambda_0 \right] \sum_{\mathbf{n}} 1 \\ &= \exp\left[-\lambda_v \frac{E_v}{E} \right] \frac{\sum_{\mathbf{n}} 1}{\sum_v \sum_{\mathbf{n}} 1} \left[\sum_v \sum_{\mathbf{n}} \exp(-\lambda_0) \right] \\ &= P^{\circ}(v) \exp\left[-\lambda_v \frac{E_v}{E} - \lambda_0 \right] \end{aligned} \tag{2.8}$$

In the last line in (2.8) we have redefined the value of λ_0, which is now determined by

$$\exp(\lambda_0) = \sum_v P^\circ(v) \exp\left[-\lambda_v \frac{E_v}{E} \right] \tag{2.9}$$

The surprisal of the predicted distribution $P^{ME}(v)$ is given by

$$-\ln\left[\frac{P^{ME}(v)}{P^\circ(v)} \right] = \lambda_v \frac{E_v}{E} + \lambda_0 \tag{2.10}$$

The functional form (2.10) is seen in Figs. 2–5 to be in quite good accord with the data. The data on energy disposal in HF elimination from $(C_7H_8F)^{\neq}$ (and many other[50, 51] $(RHF)^{\neq}$ activated species) can thus be summarized as follows: The distribution of energy among the accessible states is as statistical as possible, subject to a (single and simple) constraint.

The proof that the distribution (2.8) is of maximal entropy among all (normalized) distributions of quantum states having the same mean vibrational energy is immediate. Let $Q(v,\mathbf{n})$ be some such distribution. It follows from the inequality $\ln x \leqslant x-1$ (with equality if and only if $x=1$) that[43–46]

$$-\sum_v \sum_{\mathbf{n}} Q(v,\mathbf{n}) \ln[Q(v,\mathbf{n})] \leqslant -\sum_v \sum_{\mathbf{n}} Q(v,\mathbf{n}) \ln[P^{ME}(v,\mathbf{n})] \tag{2.11}$$

The left-hand side of (2.11) is the entropy of the distribution Q. Since $Q(v,\mathbf{n})$ and $P^{ME}(v,\mathbf{n})$ have the same mean vibrational energy, we have, using (2.5)

$$\begin{aligned} -\sum_v \sum_{\mathbf{n}} Q(v,\mathbf{n}) \ln[P^{ME}(v,\mathbf{n})] &= \sum_v \sum_{\mathbf{n}} Q(v,\mathbf{n}) \left(\lambda_0 + \lambda_v \frac{E_v}{E} \right) \\ &= \sum_v \sum_{\mathbf{n}} P^{ME}(v,\mathbf{n}) \left(\lambda_0 + \lambda_v \frac{E_v}{E} \right) \\ &= -\sum_v \sum_{\mathbf{n}} P^{ME}(v,\mathbf{n}) \ln[P^{ME}(v,\mathbf{n})] \end{aligned} \tag{2.12}$$

The right-hand side of (2.11) is thus the entropy of the distribution of maximal entropy and equality obtains if and only if $Q(v,\mathbf{n}) = P^{ME}(v,\mathbf{n})$ showing that the maximum is unique.

The results in Fig. 4 have also an implication[24, 60] for the role of energy in the reversed, R+HF, collision. For a given HF vibrational level, we have seen that the dissociation reaction produces all quantum states with (about) the same probability. Hence the $R+HF(v)\rightarrow$reaction will be nonselective. Excitation of R will be as effective in promoting the reaction as, say, putting the same amount of energy in the relative translation of R and $HF(v)$. Also, a given excitation of R is as effective irrespective of the distribution of this energy among the modes of R. On the other hand, placing the excitation energy in the vibrational mode of HF will considerably enhance the rate of the addition of HF to R.

The consideration of the previous paragraph points out toward our general conclusion: Selectivity is achieved by exciting an initial state which is subject to (one or more) constraints. Specifically, to enhance the addition of HX molecules to double bonds one should vibrationally excite the HX. Such additions (which are HX elimination in reverse) have been the subject of considerable recent interest.[61]

Specific excitation of a large polyatomic molecule, Q, which is to be used as a reagent in a $Q+A\rightarrow$reaction is of interest only if the reversed, $\rightarrow A+Q$ reaction leads to a population distribution, which is subject to constraint on Q.

In the technical sections below we provide a more detailed discussion of the maximum entropy formalism that led to (2.5) and also follow the surprisal throughout the time evolution.

D. Multiphoton Dissociation

Estimates[62, 63] of the lifetimes of multiphoton excited molecules suggest that there is ample time for energy equilibration. The early results on energy disposal in dissociations that proceed by simple bond rupture,[62] for example,

$$C_2F_5Cl\rightarrow C_2F_5+Cl \tag{G}$$

indicated that the translational energy distribution of the fragments accords with statistical predictions, provided that an effective number of oscillators was employed. Recently, examples of multiphoton eliminations, for example,

$$CH_3CCl_3\rightarrow CH_2CCl_2+HCl \tag{H}$$

were reported.[64] Here it was found inevitable to introduce a deviance between the observed and a statistical distribution in order to fit the data. The functional form that was used to fit the data is of the form

$$P(E_T)=P^\circ(E_T)\exp\left[-\lambda_T\frac{E_T}{E}-\lambda_0\right] \tag{2.13}$$

indicating [cf. (3.3)] the presence of a constraint. We have already noted that other elimination reactions [in particular (D), Fig. 3] also show nonstatistical products-state distributions.

The following practical point should also be noted. Consider such reactions as (H) or (D) where the products are a diatomic molecule and a polyatomic fragment. The prior vibrational distribution (in the RRHO classical limit[25, 46]) of the diatomic molecule is

$$P^{\circ}(E_v) \propto (E - E_v)^{s+3} \tag{2.14}$$

Here s is the number of oscillators of the polyatomic fragment which we assume to be a three-dimensional rotor. E is the total energy available for the products. If, as is often the case, one is interested in the regime $E_v \ll E$, the dependence of $P^{\circ}(E_v)$ on E_v can be approximated by

$$P^{\circ}(E_v) \approx \exp\left[\frac{-E_v}{E/(s+3)}\right] \tag{2.15}$$

In the presence of a constraint on $\langle E_{\text{vib}} \rangle$ [cf. (2.4)], we have that [cf. (2.5)]

$$P(E_v) \approx \exp\left[-\frac{E_v}{E}(s+3+\lambda_v)\right] \tag{2.16}$$

There is some tendency in the literature to conclude from a near linearity of a plot of $\ln P(E_v)$ vs. E_v that the energy has been randomized. As is evident from (2.16), this is by no means necessarily the case. Moreover, since $\lambda_v < 0$, $s_{\text{eff}} = s + \lambda_v < s$!

The concept of "exit state coupling" or "final state interaction" was proposed [65, 66] in order to account for the nonstatistical products state distribution. The picture is that energy is fully equilibrated in the activated molecule. However, past the barrier for dissociation the exit valley coupling among the receding products tends to selectively channel the repulsion energy leading to nonstatistical products distribution. In a technical sense this interpretation is consistent with the point of view adopted in this chapter. For, after all, what the exit state coupling model does is to identify a constraint. One can, if one so wishes, regard the different vibrational states of HCl as different species. Hence if, following MP excitation, HCl in higher vibrational states is preferentially populated we have succeeded in driving the dissociation to preferentially produce the more endothermic products. That, after all, is the name of the game. If you say that is due to the presence of a constraint or that this is due to the presence of a constraint imposed by exit valley repulsions is only a question of how

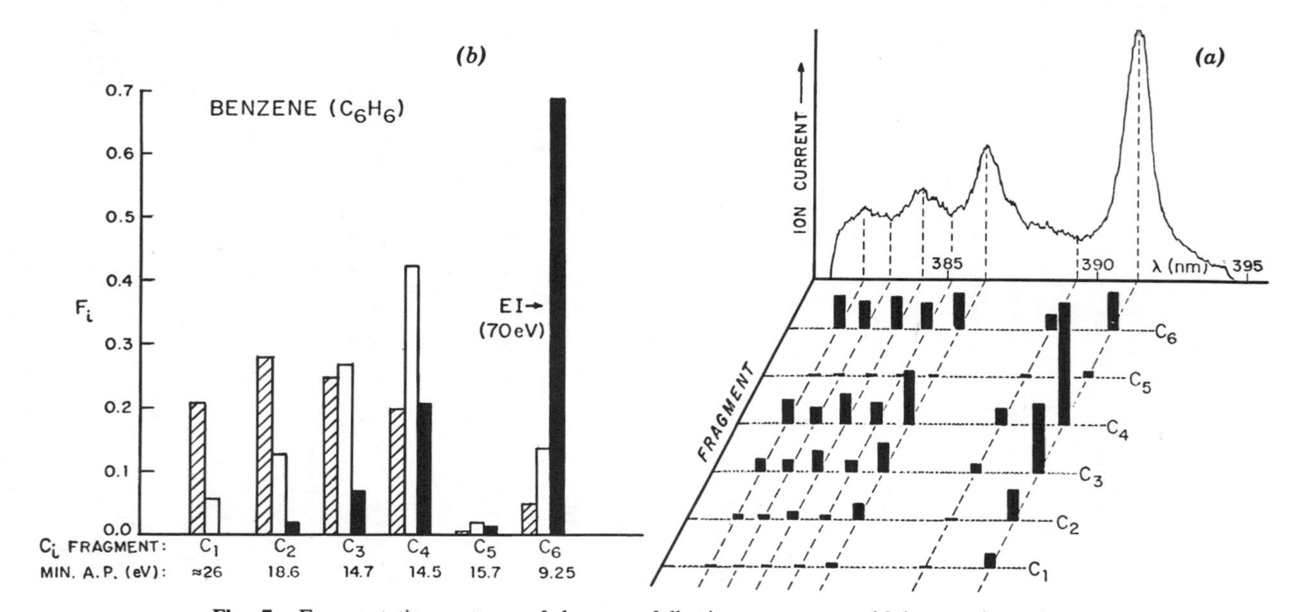

Fig. 7. Fragmentation pattern of benzene following resonant multiphoton absorption (adapted from Ref. 70). (A) Two-dimensional mass spectrum under low power. The wavelength dependence of the total ion production (top) is resolved into the contributions from each of the groups of C_i ions (bars). (B) The fractions of formation of the six possible C_i ions group at 391.4 nm. Open bars: low power. Hatched bars: high power. Solid bars: the results of electron impact spectra, shown for comparison. The minimum appearance potential of any ion of a given C_i group is listed at the bottom.

detailed your mechanism is. Note however that the implication (Section II.C) that HCl vibrational excitation will preferentially enhance the rate of the reversed (association) reaction (H) is independent of the mechanism but is a direct implication of detailed balance.

Recently, multiphoton ionization and fragmentation has been reported.[67–69] The photons used in these experiments are in the visible or the UV and carry considerable energy (e.g., about 3 eV per photon for the results[70] shown in Fig. 7). By increasing the laser power it is possible to drive the fragmentation process of benzene to extreme where the most endoergic product (C^+) is preferentially populated. It has however been found possible to offer a qualitative interpretation of the fragmentation pattern using a simple statistical theory[71].

The observation that by increasing the energy content of the system it is possible to modify its evolution from more statistical to more specific is valid also when the energy is provided by the energy of the reactants when the energy rich intermediate is formed in a collision.[22] A specific example is discussed in Section V below (reaction (O), cf. Fig. 15).

E. Collisional Activation and Deactivation

The effect of increasing collisional energy on the rate of production of electronically excited YO molecules in the $Y+SO_2$ and $Y+CO_2$ reactions[72]

$$Y+SO_2 \rightarrow YO(A^2\Pi)+SO \qquad (I)$$

$$Y+CO_2 \rightarrow YO(A^2\Pi)+CO \qquad (J)$$

is compared in Fig. 8 with that expected if all final quantum states were equally accessible. Similar results were obtained[72–75] for the reactions of Sc and La. Indeed, it is often (but not always, cf. Fig. 12) the case that reactions that lead to products in different electronic states do not show marked deviations from the prior distribution. This is true not only for the distribution among the different electronic states[72–76] but also for the distribution of vibrotational energy for a given electronic state[73–75] (e.g., Fig. 10 of Ref. 73). Contrary again to naive expectations, deviations from the prior are observed for the reactions of metal atoms with larger polyatomics.[74]

Collisional activation and deactivation is, of course, the traditional (Lindemann[39]) mechanism for interpreting unimolecular reactions in the gas phase. More recently, multiphoton excitation experiments at higher pressures have reopened the discussion of the process of collisional deactivation of energy-rich polyatomic molecules. Different points of

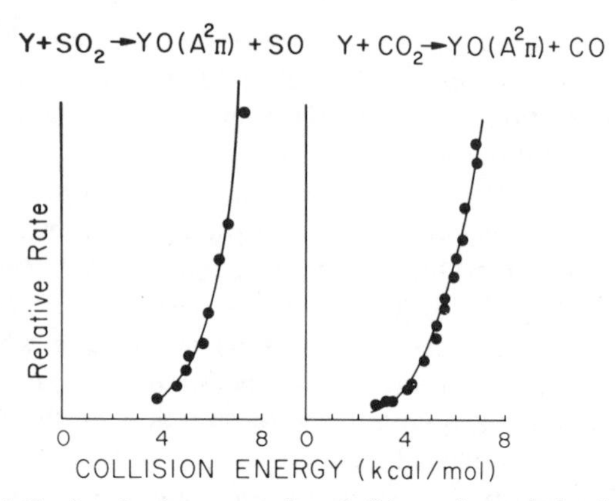

Fig. 8. Observed (dots) and prior expectations (solid curve) translational energy dependence of the rate of $Y+QO_2$ reactions (adapted from Ref. 72). The rotational and vibrational distributions of electronically excited YO molecules also show very little deviance from the corresponding prior distributions, for a variety of oxidation agents.[72–75]

view[6, 7] seem, however, to converge to the same conclusion: Collisional deactivation of an energy-rich polyatomic is highly nonselective. All final quantum states (at the same total energy) are produced at about the same probability. Figure 9 shows the final internal energy distribution of SF_6, for different levels of initial excitation, after collision with a rare gas atom at a translational temperature of 300 K. While the energy removed is substantial, the distribution of the final energy is very skewed and the most probable final energy is not too different from the initial one.

A rough and ready estimate of the fraction of internal energy that is removed out of the polyatomic molecule due to collision with the atom is, in the prior limit, the equipartition estimate $3/(s+3)$. Here s is the number of oscillators of the polyatom (15 for SF_6}, and the estimate is based on equipartitioning the initial excitation between the vibrational modes of the polyatomic and the relative translation.

The results shown in Fig. 9 are the prior ones, that is, in the absence of any constraints on the collisional deactivation process. As such they represent the most random possible disposition of the initial excitation energy. They definately do not conform to the traditional idea that upon collision an energy rich polyatomic will primarily lose energy. In other words, the traditional point of view is that the distribution of final internal energy will be centered at about $\langle \Delta E \rangle$ below the initial excitation energy. The prior expectations shown in Fig. 9 are quite different. Because of the very rapid increase of the number of internal quantum states with excitation energy, an energy-rich polyatomic molecule even stands a chance of getting richer upon collision.

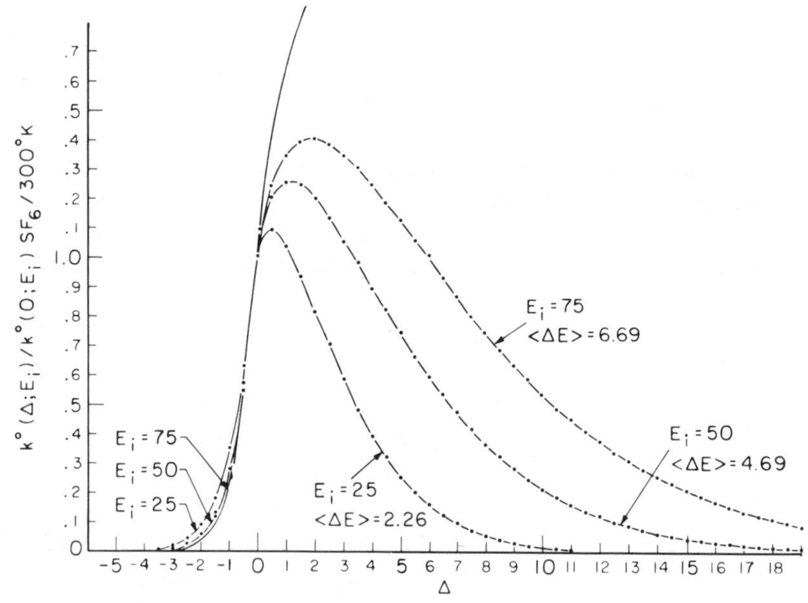

Fig. 9. Prior distribution of SF_6 final internal energy vs. $\Delta = (E_i' - E_i)/2RT$ for three values of the initial excitation, (E_i, quoted in kcal/mole), following collision with a rare gas atom at a translational temperature of 300 K. (Adapted from Ref. 7.) $\langle \Delta E \rangle$ is the mean energy removed by the collision, on prior grounds. Also shown, for comparison, is the prior distribution for collision of a diatomic molecule with an atom (solid line). Note the qualitatively different behavior in the two cases. For the diatomic molecule, deactivation ($\Delta > 0$) is very much favored on prior grounds. Indeed, the most probable final state is one where the diatomic molecule has been completely deactivated. For SF_6 (and other polyatomics) large energy transfers are disfavored on prior grounds. The final vibrational energy distribution is indeed shifted down in energy (i.e., $\langle \Delta E \rangle > 0$), but the most probable final internal energy remains quite near to its initial value. (The most probable ΔE is only $RT = 0.6$ kcal/mole even at $E_i = 75$ kcal/mole).

The prior expectations as shown in Fig. 9 predict the dependence of the mean energy transfer $\langle \Delta E \rangle$ on the initial energy content, which is in accord with experimental results.[7]

The physical picture changes dramatically if we consider collisional vibrational energy transfer to (or from) diatomic molecules. A diatomic molecule in a given vibrational state has, of course, no other vibrational modes. Hence, and in marked contrast to a polyatomic molecule, the prior expectations are (Fig. 9) that a vibrationally excited diatomic molecule will lose a significant fraction of its excitation energy on collision. The "strong deactivating collisions" model used for polyatomics is thus the correct statistical limit for diatomics!

In practice, vibrotational energy transfer in atom-diatomic collisions is often subject to constraints.[77, 78] Figure 10 shows an example for the

$$Cl + Cl_2(v) \to Cl + Cl_2(v') \tag{K}$$

energy transfer process. The considerable deviance between the actual final vibrational state distribution and the prior expectations is evident.

In Section V.F. we introduce the concept of sum rules as a route to the identification of constraints. The right panel in Fig. 10 shows that the energy transfer process (K) is subject to a simple sum rule

$$\langle \Delta E_{vib} \rangle = \beta - \alpha E_v \tag{2.17}$$

α and β are constants related by β/α being the equilibrium mean vibrational energy at the translational temperature T. E_v is the initial vibrational level and $\langle \Delta E_{vib} \rangle$ is the mean vibrational energy transfer

$$\langle \Delta E_{vib} \rangle = \sum_{v'} (E_{v'} - E_v) k(v \to v') \tag{2.18}$$

Here $k(v \to v')$ is the rate constant for the process (K). The sum rule identifies a constraint (see Section V.F. for the reason why) and hence enables us to predict[77] the surprisal (or, equivalently, the final-state distri-

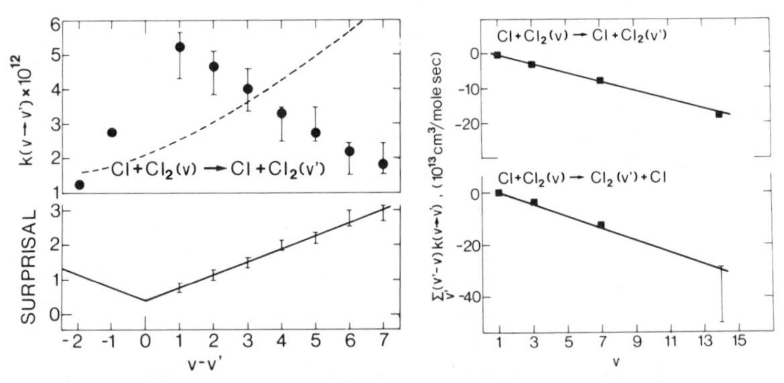

Fig. 10. Surprisal synthesis for the energy transfer process (K). (Adapted from Ref. 77.) Left top panel: Trajectory computations (bars) and prior distribution (dashed line) of the Cl_2 final vibrational state distribution (for $v=7$ and $T=1100$ K. Rate constants in units of $cm^3/mole \cdot sec$). Dots: predicted Cl_2 final vibrational state distribution using the procedure of maximum entropy and the sum rule (2.17). Left bottom panel: The surprisal of the trajectory computed distribution (bars) nd the surprisal predicted using the maximum entropy formalism (solid line). Right panel: The sum rule (2.17) for both the inelastic collision (K) and the atom-exchange collision. Squares (and bar): (2.18) using trajectory computed rates. Line: (2.17). [Note that α and β in (2.17) are not independent parameters.[77]]

bution). The surprisal shown as a solid line in the left panel of Fig. 10 is not a fit to the data but is an independent prediction.[77]

F. Distribution of Rotational States

Due to the large cross-sections for rotational relaxation it is not always the case that the measured distribution of rotational states is indeed the nascent one, that is, that of the products before any subsequent collisions. The first example, Fig. 11, is therefore a computational one,[79] where the distribution of final states is determined by a numerical solution of the Schrödinger equation. It also serves to show that even for systems with a minimal number of accessible states the surprisal, while large, can have a very simple functional form. The reaction is

$$H + H_2(0,0) \rightarrow H_2(0, j) + H \tag{L}$$

at a given total energy. Here v, j are the vibrational and rotational quantum numbers of H_2. The first (left) panel in Fig. 11 shows the deviance of the rotational distribution from the prior limit. The surprisal is seen to be very well represented as a linear function of the rotational

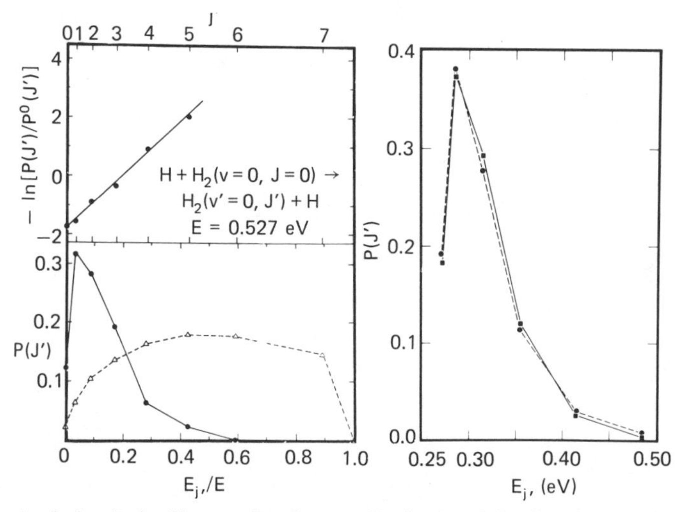

Fig. 11. Analysis of the H_2 rotational state distribution following the $H+H_2$ reactive collision. Left, bottom panel: Computed (dots) and prior (triangles) distribution vs. the fraction of energy in H_2 rotation. Left top panel: The surprisal of the computed distribution. (Adapted from Ref. 79a.) Right: Computed (squares) and distribution of maximum entropy subject to a single constraint (dots, eq. 2.20) for a collision where the total angular momentum is zero. (Adapted from Ref. 80).

energy E_j

$$-\ln\left[\frac{P(j)}{P^{\circ}(j)}\right]=\theta_R\frac{E_j}{E}+\theta_0 \tag{2.19}$$

or

$$P(j)=P^{\circ}(j)\exp\left[-\theta_R\frac{E_j}{E}-\theta_0\right] \tag{2.20}$$

The functional form (2.20) is the distribution of maximal entropy subject to a single constraint

$$\langle E_j\rangle=\sum_j E_j P(E_j) \tag{2.21}$$

The quantal rotational state distribution in the reactive collision (L) is thus as statistical as possible subject only to a single constraint.

The right panel in Fig. 11 examines the final rotational state distribution for such collisions when the H_3 system has a zero total angular momentum. The distribution shown (squares) is a result of a numerical solution of the Schrödinger equation.[80] However, a similar situation is encountered in

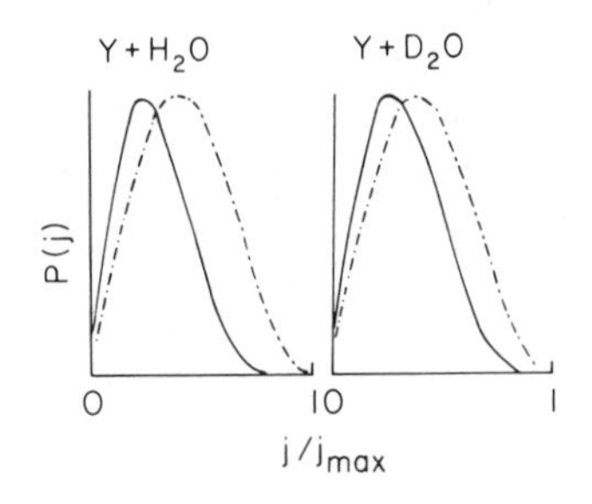

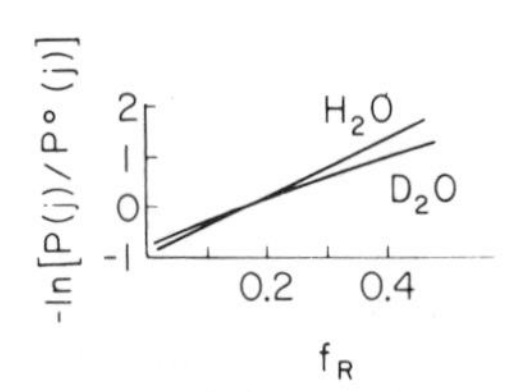

Fig. 12. Top: Observed (solid line) and prior (dashed line) $YO(X^2\Sigma)$ rotational state distributions vs. j/j_{max} where j_{max} is the highest value of j which is energetically accessible. Bottom: The surprisal vs. $f_R=E_j/E$, the fraction of the available energy dumped in the YO rotation (adapted from Ref. 74).

photodissociation when the range of total angular momentum can be quite low.

The right panel in Fig. 11 compares the exactly computed distribution to a distribution (2.20) (dots) which is obtained by the procedure of maximal entropy subject only to $\langle E_j \rangle$ as a constraint. The agreement is seen to be very close.

Experimental results for rotational distributions measured (using laser induced fluorescence) in a molecular beam configuration are compared in Fig. 12 to the prior results for the

$$Y + H_2(D_2)O \rightarrow YO + H_2(D_2) \tag{M}$$

reactions.[74] The deviation is not large, but is quite evident. Indeed, it is seldom the case (Fig. 11 is an exception) that large surprisals are noted for rotational distributions following rearrangement collisions.

G. Heavy Ion Transfer Reactions

Nuclear heavy ion transfer reactions at energies above the Coulomb barrier have several characteristics which indicate that the collisions proceed via a metastable compound state.[81] Despite the nearly macroscopic number of degrees of freedom, and despite the great variety of possible

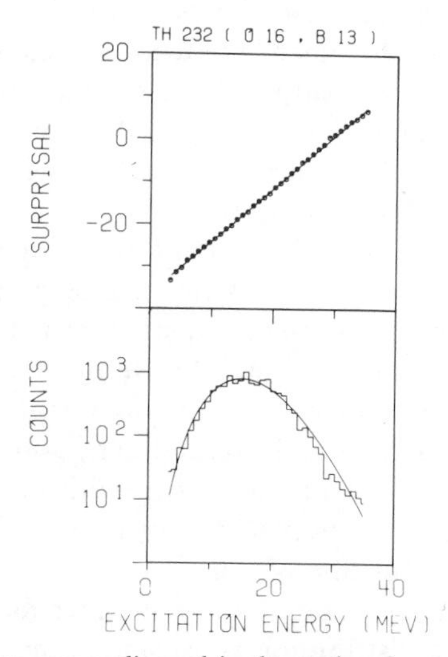

Fig. 13. Internal excitation energy disposal in the nuclear heavy ion transfer reaction (N). (Adapted from Ref. 83.) Bottom panel: Experimental histogram and (continuous curve) the distribution of maximal entropy subject to a single constraint (the mean final excitation energy $\langle E_I \rangle$). Top panel: The surprisal of the experimental results (dots) and the (linear) surprisal predicted by the procedure of maximal entropy subject to the constraint $\langle E_I \rangle$.

rearrangement products (showing that the mixing is quite strong) the distribution of energy among the fragments is distinctly nonstatistical. The results for the Np excitation energy in the

$$^{16}O + ^{232}Th \rightarrow ^{13}B + ^{235}Np \qquad (N)$$

reaction is shown in Fig. 13. The distribution is quite deviant from the prior yet the surprisal is very well approximated as a linear function of the energy.[82, 83]

H. Summary

Surprisal analysis is a particular mode of application of the maximum entropy formalism.[84] It is the phenomenological mode where the experimental results are fitted to a functional form suggested by the theory. There are, however, other modes that do have predictive capabilities (see, e.g., Fig. 10). In fact, we shall argue below that they offer the most economical route, since one only seeks to predict those details that can be observed. The rest of this chapter is an amplification of this cryptic remark. It is not a handbook of surprisal analysis (which is available[45]), nor a review[43–46] of past work, nor a general introduction.[33] Rather it is meant as a statement of principles and outline of potential applications. It is short on equations, since these can be found in abundance elsewhere. Our major purpose is to document the technical backup and the implications of our basic thesis: The surprisal is a constant of the motion. Hence, our two major conclusions: (1) the surprisal is determined by the initial excitation process, and (2) the surprisal is a true signature of the intramolecular dynamics throughout its history.

In this section we have seen that often (but not always) the surprisal of the observed dissociation products is finite. It is a corollary of our basic thesis that a finite surprisal implies a constrained (i.e., not fully relaxed) distribution throughout the collision. On the other hand, the very simple structure of the surprisal implies that the (dominant) constraints are quite simple and hence that a valid and useful description of the intramolecular evolution is: the distribution of states is as statistical as possible (i.e., of maximal entropy) subject, however, to constraints. This picture is intermediate between the imposition of a large number of constraints (equal to the number of vibrational modes) as in the Slater model and the opposite view (RRK) where the distribution is as random as possible, without any constraints.

III. THE MAXIMUM ENTROPY FORMALISM

From a strict technical point of view the procedure of maximal entropy[43–46, 84–90] is an algorithm for specifying the state of the system. In classical mechanics the state is a distribution in phase space, while in quantum mechanics it is a density matrix.[87, 91, 92] This algorithm can be used in several ways and failure to distinguish between distinct modes of applications is responsible for some of the reservations that are sometimes expressed regarding the use of this methodology. We begin by stating the algorithm in purely technical terms and proceed to discuss the three main modes in which it can be applied: (1) specifying the initial state, (2) variational approximations (and exact solutions) for the intramolecular dynamics, and (3) as a basis for the phenomenological procedure of surprisal analysis.

A. The Algorithm

Consider the situation where the information on the state of the system is given as the mean values, $\langle A_r \rangle$, for a set of observables A_r. Typically, these mean values do not suffice to determine the state uniquely, that is, there will be more than one state for which the observables have the given mean values.[93] Even so, these mean values do limit the range of possible states and so the observables are referred to as the constraints.

Among all the states consistent with the values of the constraints the one of maximal entropy subject to R constraints is given by[85–92]

$$\sigma = \exp\left[-\sum_{r=0}^{R} \lambda_r A_r\right] \tag{3.1}$$

Here, $A_0 = \mathrm{I}$ is the normalization constraint. The values of the (Lagrange) parameters, λ_r, in (3.1) are determined by the values of the constraints. Explicitly, if $\langle A_r \rangle$ is the value of the rth constraints, the $R+1$ parameters $(\lambda_0, \lambda_1, \ldots, \lambda_R)$ in (3.1) are determined by the $R+1$ implicit equations

$$\langle A_r \rangle = \mathrm{Tr}\left\{A_r \exp\left[-\sum_{r=0}^{R} \lambda_r A_r\right]\right\} \qquad r = 0, 1, \ldots, R \tag{3.2}$$

In (3.2) we used the notation $\mathrm{Tr}(A_r \sigma)$ for the average value of A_r in the state σ. If the constraints are not contradictory, that is, if there exists a state for which the constraints can have the stated values, than the state of maximal entropy is unique.

In quantum mechanics the observables are operators and σ in (3.1) is the density matrix of the system. In classical mechanics the observables are functions of the coordinates and momenta and σ is the density in phase space. In the quasiclassical limit one retains the quantized nature of bound states but neglects all commutators. Hence, in a basis in which all the operators are diagonal (3.1) reads

$$P(i)=\exp\left[-\sum_{r=0}^{R}\lambda_r A_r(i)\right] \tag{3.3}$$

Here $P(i)$ is the probability of the state i and $A_r(i)$ is the mean value of A_r in the state i. Equations (3.2) now read [cf. (2.6)]

$$\langle A_r\rangle=\sum_i A_r(i)\exp\left[-\sum_{r=0}^{R}\lambda_r A_r(i)\right] \qquad r=0,1,\ldots,R \tag{3.4}$$

In Section II we dealt with cases where the states i could be grouped into levels γ such that $A_r(i)$ had the same value, $A_r(\gamma)$, for all states of a given group γ [cf. eq. (2.1)]. It is then possible to express the probability of the group γ as, $[A_0(\gamma)=1]$,

$$P(\gamma)=\sum_{i\epsilon\gamma}P(i)=P^{\circ}(\gamma)\exp\left[-\sum_{r=0}^{R}\lambda_r A_r(\gamma)\right] \tag{3.5}$$

Here, $P^{\circ}(\gamma)$,

$$P^{\circ}(\gamma)=\frac{\sum_{i\epsilon\gamma}1}{\sum_i 1} \tag{3.6}$$

is the (normalized) number of states in the group γ [and the value of λ_0 has been changed in going from (3.1) to (3.5) by $\ln(\Sigma_i 1)$].

B. The Initial State—The Unavoidable Need for the Maximum Entropy Principle

We begin with the real reason why the maximum entropy formalism (or some alternative prescription) is unavoidable. As will become obvious, the use of the principle is complementary to the use of equations of motion. Both are necessary for an exact description. The maximum entropy formalism is not meant to replace the equations of motion. Rather, what it provides is something that is absent from the equations—namely, the boundary condition.

Classical or quantal mechanics are formulated in terms of differential equations of motion. In any concrete application one must also specify the particular initial state of interest. If it is a single quantum state (or a point in phase space) there is no problem. It is, however, a characteristic of molecular dynamics experiments that one has seldom complete initial state selection. Rather one often starts with a mixture (e.g., a thermal distribution) of states. The maximum entropy principle specifies the weights of the different states in this initial mixture.

At this point one has two options. The first is to separately solve the (linear) equations of motion for each initial quantum state and then average the results over the distribution of states in the initial mixture of interest, using the weights determined by the maximum entropy approach (or some alternative principle). (In one way or another you have, however, to determine these weights in order to perform the required averaging.) There is nothing wrong with this approach except that fine structure in the dynamics of individual states tends to get averaged out. Hence if all one requires is the evolution of the particular initial state of interest, determining the fine structure is a wasted effort. It makes better sense not to have to compute those details that are doomed to be averaged over. Hence the second option and the one that we follow in this chapter: Having determined the initial state via the principle of maximum entropy, solve the equations of motion directly for this state.

By solving the equations of motion for the relevant initial state one does not violate the principle of "conservation of effort." Rather, one eliminates wasted effort. Only such details that are required to describe the experiment at hand are kept in the computation.

It should be stressed that solving the equation of motion for a mixed initial state is a rigorous dynamical procedure. The transition from the Schrödinger equation of motion for a single state to a "Liouville" equation of motion for the density matrix[92, 94] was formulated already by Dirac[95] and von Neuman.[96] The corresponding formulation for classical mechanics was given by Liouville[97] and discussed extensively by Gibbs.[98]

The primary role of the procedure of maximal entropy is thus to specify the initial state. Even if one has solved the equations of motion separately for every initial quantum state one still requires the weights of these different states in the original mix in order to compare with experiment. If, instead, one opts to solve directly for the evolution of the initial state of interest one immediately finds that having specified an initial state of maximal entropy, it will retain this characteristic throughout its time evolution.[89, 99] Moreover, if due to the initial averaging some details get averaged out then there occurs a corresponding simplification in the explicit solution of the equation of motion.

A very special case of the avoidance of irrelevant details is familiar to every chemistry student who is taught that equilibrium is not a static but a dynamic situation. When a gas phase system has come to equilibrium, collisions do not cease but the state of the system is unchanged by collisions. In other words, if one prepares an initial state, which is an equilibrium mixture, then one readily predicts the final, postcollision state. None of the multitude of details of the scattering matrix for the system need be invoked. In formal scattering theory this is sometimes known as the intertwining theorem.[100, 101] Taking for simplicity the case of inelastic collisions the theorem states that any initial state where all initial quantum states (at a given energy) are equiprobable is left invariant by the collision.[100, 102]

What one is after is, in a sense, a generalization of the intertwining theorem. Suppose that there is some (but incomplete) initial state selection. Must we inevitably know the entire S matrix or can one solve only for those details that are relevant at the level of resolution which is of interest. It turns out that one can always do this in principle and that for sufficiently simple situations explicit analytical results can be obtained.

We have left open a central question: Why do we recommend (nay, insist) on using the procedure of maximal entropy to specify the initial state? We return to this question in Section IV.

C. Variational Approximations

The search for a state of maximal entropy is always subject to constraints. For these given constraints (which do not suffice to single out a particular quantum state) one varies the state until the particular one, which is of maximal entropy, is determined. The dynamical approach demonstrates[89, 99] that one can identify a set of constraints, which, when used in the recipe above, specify an exact solution of the equation of motion and its initial conditions.

For the purpose of providing approximations it is advantageous, however, to rephrase the variational procedure.[89, 103] Once we know the relevant constraints, we have the exact solution. Hence what we require is a procedure for incorporating constraints. These may be arrived at by intuitive reasoning, or implied by models or be the result of an incomplete dynamical procedure, etc. We shall return several times below to the art of choosing constraints. For the moment, say one has a set of reasonable constraints. A variational approximation to the exact result is provided by that state which is of maximal entropy subject to the set of assumed constraints. Upon incorporating additional constraints the approximation is guaranteed not to be any worse (that is, it either improves or is unchanged. In the latter case the additional constraints are not informative). In view of the central role of this result it is important to appreciate one technical point. Once one selects a set of constraints, then the explicit functional form of the state of maximal entropy (subject to these constraints) is determined [cf. (3.1)]. This functional form contains, however, (Lagrange) parameters [the λ_r's in (3.1)] whose number equals the number of constraints. One has obtained an explicit variational approximation when the numerical values of these parameters is determined by the condition that the (mean) value of the constraints is the same for the exact and the approximate solution. In practice, one can employ other types of data to explicitly determine the value of the parameters. The point is that only when the numerical values of the parameters are given do we have an explicit solution whose entropy is guaranteed to be above or to equal that of the exact solution. Equality obtains if and only if the "approximate" solution is, in fact, exact. In other words, the variational procedure converges to the exact solution.

Due to the many routes (including sheer guesswork) that are available for selecting constraints, the variational approximation is a versatile tool. We have introduced it here as a convergent approximation, protected by a variational principle, to the exact result. It is, however, possible to argue much more strongly in support of such an approach. Such reasoning is based on the special features of a state of maximal entropy as discussed in Section IV.C. The arguments there are slanted toward the reasoning why the initial state should be specified as one of maximal entropy. However, with suitable modifications several of the arguments apply throughout (and also after) the collision.

D. Phenomenology

The application of the maximal entropy procedure to the analysis of experimental data, which was already discussed in Section II, is essentially a special case of the variational procedure. The required constraints are here identified by the examination of the actual (observed or computed) distribution. The reason why the adaptation of the formalism to phenomenological application is so immediate is that the functional form of a distribution of maximal entropy subject to a given set of constraints can be readily written down. The fit to the experimental results is now obtained by varying the values of the (Lagrange) parameters. Convergence is judged, as in the variational procedure, by the difference between the entropy of the trial distribution and that of the observed distribution.[45] This difference is always nonnegative and vanishes[104] for and only for a perfect fit.

IV. THE INITIAL STATE

This section examines a number of the arguments that have been invoked for specifying the initial state by the procedure of maximal entropy. We offer a variety of arguments since different reasons may appear better suited for different situations. In particular, the discussion of the procedure for inductive inference can equally well be applied directly to, say, the final state after the collision.

A. The Stationary State

A dynamical experiment consists essentially in examining the response of the system to some applied perturbation. The initial state of the system need be stationary for if it is not it will change with time even in the absence of the perturbation. In both classical[98] and quantum[89] mechanics a stationary state is one of maximal entropy. The proof[89] is immediate. Say σ is a stationary state and let ρ be some different normalized state, which is also consistent with our information on the system. In particular, since σ is stationary so is $\ln\sigma$. The mean value of $\ln\sigma$ predicted by σ and by ρ must therefore be the same

$$\mathrm{Tr}(\sigma\ln\sigma)=\mathrm{Tr}(\rho\ln\sigma) \tag{4.1}$$

For any two normalized states[105]

$$\mathrm{Tr}(\rho\ln\rho)\geqslant\mathrm{Tr}(\rho\ln\sigma) \tag{4.2}$$

with equality if and only if $\rho\equiv\sigma$. Since, by assumption, ρ is different from σ, the entropy $-\mathrm{Tr}(\sigma\ln\sigma)$ of σ exceeds that of ρ

$$-\mathrm{Tr}(\sigma\ln\sigma)=-\mathrm{Tr}(\rho\ln\sigma)>-\mathrm{Tr}(\rho\ln\rho) \tag{4.3}$$

Special versions of this conclusion are used routinely; for example, for a stationary system at thermal equilibrium the state of maximum entropy subject to the mean value of the energy is the familiar canonical one:

$$\sigma=\exp(-\beta H_0-\lambda_0) \tag{4.4}$$

Here H_0 is the Hamiltonian and λ_0 is determined by the normalization constraint, $\mathrm{Tr}(\sigma)=1$, so that $\exp(\lambda_0)$ is the partition function,

$$\exp(\lambda_0)=\mathrm{Tr}\{\exp(-\beta H_0)\} \tag{4.5}$$

Since σ is a stationary state, σ and hence $\ln\sigma$ commute with the Hamiltonian of the system. Such quantities are known as constants of the motion. What we have shown by (4.3) is that a stationary state can be represented as one of maximal entropy subject to constraints which are constants of the motion. This result is well known, particularly in classical mechanics.[88] In Section V.D we shall generalize this result to include nonstationary states. Of course, there will be a price tag attached. We shall find it necessary to reexamine the concept of a constant of the motion.

B. The Activation Process

It is often convenient to begin the description of the experiment not with the actual initial state but with an intermediate state that has been obtained via some preparation, for example, optical or collisional excitation. Here too, the procedure of maximal entropy can be applied. We illustrate this by an example, as the general principle will be discussed in connection with the solution of the equation of motion for an initial state of maximal entropy in Section V.D.

Consider a thermal initial state (4.4), which is subjected to a pulse of optical excitation. We write this perturbation as $V(t)$. The state of the system at the time t is the solution of the Liouville equation

$$\frac{i\hbar\,\partial\sigma(t)}{\partial t}=[H(t),\sigma(t)] \tag{4.6}$$

Here the square brackets denote, as usual, the commutator ($[A,B]=AB-BA$) and $H(t)=H_0+V(t)$. The boundary condition is that at $t\to-\infty$, $\sigma(t)$ is of the form (4.4).

The formal solution for $\sigma(t)$ is immediate. Define a time-dependent effective Hamiltonian $\mathcal{H}(t)$ by the first-order differential equation

$$\frac{i\hbar\,\partial\mathcal{H}(t)}{\partial t}=[H(t),\mathcal{H}(t)] \tag{4.7}$$

and the boundary condition that $\mathcal{H}(t)$ reduces to H_0 as $t\to-\infty$. The state of the system at the time t is then given by

$$\sigma(t)=\exp[-\beta\mathcal{H}(t)-\lambda_0] \tag{4.8}$$

To prove this result note that $\sigma(t)$ satisfies both the first-order differential equation (4.6) and the boundary condition [that it reduces to (4.4) as $t\to-\infty$] and hence is the unique solution. By comparing (4.8) to (4.1) it is seen that an alternative route to $\sigma(t)$ is to determine it as the density matrix of maximal entropy subject to the constraint $\langle\mathcal{H}(t)\rangle$.

Before we can compute any matrix elements using the solution (4.8) for $\sigma(t)$ we need first to explicitly solve[106] for $\mathcal{H}(t)$. Hence the solution (4.8) as it stands is purely formal. We have, however, achieved our stated objective. We have determined a constraint, that is, $\mathcal{H}(t)$ such that after (or during) the excitation process the state of the system is one of maximal entropy subject to that constraint. It should also be noted that the two (Lagrange) parameters in (4.8) do not change with time. The reason is that the mean value of $\mathcal{H}(t)$ is independent of time:

$$\frac{d\mathrm{Tr}\{\sigma(t)\mathcal{H}(t)\}}{dt}=0 \tag{4.9}$$

The proof of (4.9) is immediate, using (4.7), (4.6), and the invariance of the trace under cyclic permutations of its arguments.

If the perturbation $V(t)$ is weak and if the change in the initial state is small, then (4.8) reduces to the usual result as employed in linear response theory.[108] However, the exact solution (4.8) is, of course, valid also in this weak coupling limit.

The simplest excitation procedure is one where all quantum states (at the given energy) are equiprobable. This is the state of maximal entropy without any (dynamical) constraints. Such a state is typically assumed when the details of the excitation process are too complex to handle (e.g., Fig. 9). One of the conclusions of the discussion below is that this is indeed the most conservative inference that can be made.

C. The Most Conservative Inference

The concept of the density matrix is introduced when the system may be found in a number of different quantum states. The procedure of maximal entropy is called for when we do not have enough information to deduce the distribution over the possible quantum states. Faced with this apparent lack of precise information one can, of course, simply give up. The reason why one does not is basically that the experimentalists appear oblivious to our dilemma. As far as they are concerned the information that they have provided *is* sufficient to reproduce the experiment.

1. Sufficient Statistics

Assume that the experimentalist can be trusted. This means that the set of mean values $\langle A_r \rangle$, $r=0,1,\ldots,$ is indeed sufficient to fully describe the distribution of states. Since these mean values represent averages over a distribution, they are referred to in the statistical literature as statistics. It follows[109] from the Pitman[110]–Koopman[111] theorem[112] that the state for which the $\langle A_r \rangle$'s are sufficient is the one of maximal entropy subject to the given mean values. The "if" part of this conclusion is obvious. Upon inspection of the functional form (4.1) and the condition (4.2), it is clear that the state of maximal entropy depends only on the values $\langle A_r \rangle$. The "only if" part is harder to prove. We refer to the original literature for both the discrete[112] and the quantal[109] cases.

From the point of view of the concept of sufficient statistics, surprisal analysis is the empirical search for such statistics. Here, one is using the P–K theorem in its reverse version. Having found that the distribution is of the maximal entropy form, for example, (4.1), one concludes that the A_r's are sufficient statistics.

2. The Most Probable State

A simpler version of the previous argument goes back to Boltzmann.[113] The distribution of states observed in an actual experiment is the result of many (N, N large) independent events. Let $\{n_i\}$ be the number of times the ith result was observed. $\Sigma n_i = N$. The number of ways in which a particular set $\{n_i\}$ can be realized is[114] W,

$$W = \exp[NS(\{n_i\})] \tag{4.10}$$

Here $S(\{n_i\})$ is the entropy of the distribution of results

$$S(\{n_i\}) = -\sum_i \frac{n_i}{N} \ln \frac{n_i}{N} \tag{4.11}$$

In determining the distribution of maximal entropy subject to constraints one therefore determines that distribution that is consistent with the constraints and that can be realized in more ways than any other distribution, which is also consistent with the constraints. Because of the factor N in the exponent in (4.10), and since the entropy is independent of N, the distribution of maximal entropy is strongly favored as $N \to \infty$.

3. The Most Random State

A density matrix can always be resolved as a mixture of orthogonal quantum states. (Of course, if the state is "pure" there is only one component in this mixture.) Among all the density matrices consistent with the values of the constraints, the one of maximal entropy assigns the most uniform possible weights to the states in the mixture.

The technical argument is as follows: Let $\{p_i\}$, $i = 1, 2, \ldots$ be the set of weights of the different orthogonal quantum states in the density matrix of maximal entropy $(= -\Sigma p_i \ln p_i)$. We arrange the weights in order of decreasing size the largest one first, etc. Let $\{q_i\}$ be a similar (ordered according to size) set of weights for a different density matrix whose entropy $(-\Sigma q_i \ln q_i)$ is smaller. The result we require is that if for every $m > 0$

$$s_m \equiv \sum_{i=1}^{m} p_i \leqslant \sum_{i=1}^{m} q_i \equiv s'_m \tag{4.12}$$

then

$$-\sum_{i=1} p_i \ln p_i \geqslant -\sum_{i=1} q_i \ln q_i \tag{4.13}$$

with equality in (4.13) if and only if equality holds in (4.12) for every m. The proof is based on the following argument

$$\begin{aligned} -\sum_i p_i \ln p_i &= -\sum_m s_m \ln \frac{p_m}{p_{m+1}} \\ &> -\sum_m s'_m \ln \frac{p_m}{p_{m+1}} \\ &= -\sum_i q_i \ln p_i \\ &\geqslant -\sum_i q_i \ln q_i, \quad \text{Q.E.D.} \end{aligned} \tag{4.14}$$

What the argument shows is that specifying a state of maximal entropy is equivalent to the requirement that for every m, the total probability (s_m) of the m most probable quantum states in the mixture is less than for any other mix. Or, in other words, to reach a given cumulative probability s more states need be included in the maximal entropy mixture than in any other. The maximal entropy procedure is thus equivalent to the most democratic way of incorporating quantum states. This is, of course, obvious if no constraints are required. Then, all states are given the same weight. What the theorem shows is that even when constraints are present, the weights are as uniform as possible while still satisfying the constraints.

4. The Most Conservative Inference

Jaynes[85, 115] and others[116, 117] have strongly advocated the use of the maximum entropy procedure as a general principle of inference. Entropy is introduced (on axiomatic grounds[118, 119]) as a measure of uncertainty. The principle then seeks to determine the least biased distribution (or the broadest, cf. Section VI.C.3), subject to our knowledge on the state of the system. It determines that distribution over quantum states which is consistent with the constraints and is otherwise maximally noncommitted.

Recently the principle has been questioned[120] on the grounds of "why entropy." It is possible to introduce other measures of uncertainty.[121] Indeed, one can argue that any other convex function of the density matrix could equally well be used and will lead to a unique maximum. What distinguishes (in a unique fashion) entropy from all the other convex functions is that it is the only one that satisfies the grouping axiom.[119] We do not propose to go through the argument[103] in detail here.

V. INTRAMOLECULAR DYNAMICS

We have either the initial state, in the form of a distribution of maximal entropy or the final state, in a similar form determined by surprisal analysis. Our problem is to characterize the intervening dynamics, that is, either what will happen or what has happened during the time evolution. The purpose of this section is to show that an initial state of maximal entropy evolves into a state that remains one of maximal entropy throughout the collision.[89, 99] The constraints will be identified as the so called "time-dependent constants of the motion" and methods for their explicit determination will be discussed. The same results obtain, mutis mutandis, for the final state.

A. Once Is Enough

The equations of motion of both classical or quantal mechanics are reversible or the entropy of the state of the system is time independent. The proof is based on the observation[89] that the logarithm of the density matrix satisfies the same equation of motion [cf. (4.6)] as the density matrix itself

$$\frac{i\hbar\partial\ln\sigma(t)}{\partial t}=[H,\ln\sigma(t)] \tag{5.1}$$

Hence, using (5.1), the definition

$$S=-\mathrm{Tr}[\sigma(t)\ln\sigma(t)] \tag{5.2}$$

the normalization condition, $\mathrm{Tr}[\sigma(t)]=1$ and the invariance of the trace under cyclic permutations

$$\frac{dS(t)}{dt}=-\frac{d\mathrm{Tr}\{\sigma(t)\ln\sigma(t)\}}{dt}=0 \tag{5.3}$$

An essential aspect of the proof is that $\sigma(t)$ is the state of the system which evolved from the initial state under the reversible (quantal or classical) Liouville equation (4.6). The result will not necessarily obtain if $\sigma(t)$ provides only a reduced[122] (or "coarse-grained") description of the state of the system.

Consider now the set of admissible initial states, that is, those that are consistent with the given values of the constraints. One particular, unique member of this set is the state σ of maximal entropy. Any other member of the set has a lower entropy. There can be no initial state whose entropy is higher (or equal) to that of σ and which is consistent with the known data (the constraints) on the initial state. Each member of the set of admissible initial states will evolve in time according to the Liouville equation. In this fashion one generates a set of admissible states for any instant of time during the collision. The entropy of each state is, however, not changing. Hence the state $\sigma(t)$, which evolved from the initial state σ, will continue to have a higher entropy than any other state $\rho(t)$, which evolved from some member ρ, $(\rho \not\equiv \sigma)$, of the set of admissible initial states

$$-\mathrm{Tr}\{\sigma(t)\ln\sigma(t)\} = -\mathrm{Tr}\{\sigma\ln\sigma\} > -\mathrm{Tr}\{\rho\ln\rho\} = -\mathrm{Tr}\{\rho(t)\ln\rho(t)\} \tag{5.4}$$

To describe the entire time evolution it is necessary to invoke the principle of maximum entropy only once. Having used it at one instant of time, the reversible equations of motion insure its validity for other times. In particular, it is not necessary to invoke it before the collision. It can equally well be used after the collision. In this fashion one can extrapolate back in time. Indeed, what (5.3) shows is that the mean value of $\ln\sigma$ is constant throughout the collision. Hence if the entropy of the final state is found to be less than its maximal possible value (in the absence of constraints) it follows that constraints were present throughout the collision and that at no time during the collision could all quantum states be equally probable (which is the state of maximal entropy subject to no constraints apart from normalization and other conserved quantities).

Any constraints cause the value of the entropy to drop below its maximal value (in the absence of constraints apart, etc.). Hence surprisal analysis of the final state is a rigorous test for a constrained time evolution. A nonzero surprisal is thus a signature of nonstatistical distribution of states throughout the collision.

B. Incomplete Resolution of Final States

In a typical experiment one is seldom given the complete density matrix of the final state. Typically one only has a rather incomplete data, for example, distribution of vibrational states or distribution of translational energy.[123] This state of affairs does not imply that the considerations of the previous paragraph are not useful. On the contrary. The entropy of the fully detailed distribution of quantum states [i.e., $S(\sigma)$, where σ is the density matrix for the final state] cannot exceed the entropy of that (possibly, partial) distribution that is known to us. The proof is immediate

σ must be consistent with what distribution we do know and possibly must be consistent with additional constraints (which were not revealed by the experimental results, since they govern, for example, the angular distribution which we assume was not observed). These additional constraints will cause a lowering of the entropy of σ with respect to the entropy of the known distribution. Hence if the entropy of the available distribution is found to be below the global maximal value, that is, the observed distribution is subject to constraints, then all the more so for the entropy of the fully detailed description.

The conclusion that the state of the system must be at least as constrained (and typically more so) than any partial distribution of states that is revealed by a less than fully state-analyzed experiment is of prime importance for the phenomenological application. It means that any finite surprisal of the products is a direct proof of incomplete randomization during the time evolution. We therefore offer, in the appendix, an explicit comparison of the entropy of a complete description to that estimated from a lower resolution of the final states.

There is however one essential difference between the fully detailed and the reduced description. The entropy of the latter need not stay constant during the interaction.[122] It may even decrease,[124] but, more typically, it will increase, and the lower the resolution the faster is the increase. Hence surprisal analysis of low resolution (e.g., vibrational distribution in one particular mode as in Figs. 1 or 3 or 10), final state distribution may fail to detect any constraints, particularly so for longer living intermediates. That, however, does not necessarily imply that no constraints are present. It only calls for higher resolution distributions. (In particular,[125] joint distributions such as vibrotational or in two vibrational modes or velocity-angle,[22] etc.). Angular momentum disposal has been discussed[126] in detail from this point of view.

C. Time-Dependent Constants of the Motion

A constant of the motion is an observable whose mean value is not changing with time. Often one considers only such observables that are not explicit functions of time. We shall refer to these as conserved variables. Neither in classical[122, 127] nor in quantum[128] mechanics need one limit the discussion to time independent constants of the motion. Hence, in general, we shall admit also time-dependent constants. Say I is such a constant. Then for any density matrix $\rho(t)$ its mean value is time independent:

$$\frac{d\langle I\rangle}{dt}=\frac{d\mathrm{Tr}\{I\rho(t)\}}{dt}=0 \tag{5.5}$$

Using the Liouville equation of motion (4.6) and the cyclic invariance of the trace, (5.5) implies that for any $\rho(t)$

$$\mathrm{Tr}\left\{\rho(t)\left([I,H]-\frac{i\hbar\,\partial I}{\partial t}\right)\right\}=0 \tag{5.6}$$

If I is not an explicit function of time, then (5.6) implies that it need commute with H. However, I can be taken to depend on time. In order for its mean value to be time independent, I need satisfy the equation of motion [cf. (5.6)]

$$\frac{i\hbar\,\partial I}{\partial t}=[H,I] \tag{5.7}$$

The Heisenberg equation of motion[129] for any observable O is

$$\frac{i\hbar\,dO}{dt}=[O,H]+\frac{i\hbar\,\partial O}{\partial t} \tag{5.8}$$

Hence a constant of the motion can equally well be defined by

$$\frac{dI}{dt}=0 \tag{5.9}$$

whether I is or is not an explicit function of time.

On comparing the equation of motion (4.6) for the density matrix to (5.7) it is seen that the celebrated Liouville theorem[94] that the density matrix is a constant of the motion is a particular case of the general result (5.7).

It follows from the definition (5.9) that functions of constants of the motion are themselves constants of the motion. A particular example is as follows: Let $\rho(t)$ be a solution of the Liouville equation ($d\rho/dt=0$), and let I be a constant of the motion ($dI/dt=0$), then $I\rho(t)$ is also a solution of the Liouville equation ($dI\rho/dt=0$). In other words, constants of the motion transform solutions of the Liouville equation into solutions of the Liouville equation. This result is a generalization of the familiar conclusion that symmetry operations transform eigenfunctions of the Schrödinger equation into eigenfunctions of the same equation.

D. The Time Evolution

Given an initial state of maximal entropy subject to the constraints $\langle A_r\rangle$, $r=0,1,\ldots$

$$\sigma=\exp\left[-\sum_{r=0}\lambda_r A_r\right] \tag{5.10}$$

we are now in a position to write down the formal solution of the Liouville equation. Let $I_r(t)$ be a (possibly, time-dependent) constant of the motion whose initial value is A_r. The differential equation (5.7) and the boundary condition define I_r uniquely. The normalization constraint (A_0, the identity) leads to I_0 = the identity, which is time independent. Other constraints, if present in the initial state, may well lead to much more complex constants of the motion and the problem of explicit solution for the I_r's is discussed in the next section.

In terms of the set of constants of the motion $I_r(t)$, $r=0,1,\ldots$, the solution of the Liouville equation subject to the initial condition (5.10) is

$$\sigma(t)=\exp\left[-\sum_{r=0}\lambda_r I_r(t)\right] \tag{5.11}$$

The proof consists of noting that, since (5.11) is a function of constants of motion, it is a solution of the Liouville equation. Since, by construction of the I_r's, $\sigma(t)$ satisfies the initial condition (5.10), it is the required solution. An alternative characterization of $\sigma(t)$ as given by (5.11) is that among all those states $\rho(t)$ which yield the given values of $\langle I_r(t)\rangle$,

$$\langle I_r(t)\rangle=\langle A_r\rangle, \qquad r=0,1,\ldots \tag{5.12}$$

the state (5.11) is the one of maximal entropy.

We have thus shown that the solution of the Liouville equation throughout the time evolution is given by a state of maximal entropy subject to constraints which are (possibly, time-dependent) constants of the motion. Both the identity of the constraints and their mean values [cf. (5.12)] are uniquely specified by the initial state.

The reader will correctly point out that once we have specified the initial state to be of maximal entropy [i.e., to be of the form (5.10)] then all the rest follows inevitably from the Liouville equation of motion. Indeed, this is as it should be. The procedure of maximal entropy does not profess to replace quantal (or classical) dynamics. All that it offers is to specify the initial state and then lets the dynamics take over. Hence, the only criticism that can be offered is regarding the choice of initial state. It is for this reason that we spent so many words on this topic in Section IV.

E. The Constraints

An explicit solution for the time dependent constants of the motion may be possible, for example, by the method of separation of variables.[99, 122] This last method implies that $I_s(t)$ can be expressed as a linear combination of time-independent operators with time-dependent coefficients

$$I_s(t)=\sum_r G_{rs}(t)A_r \tag{5.13}$$

The initial condition $I_s(t) \to A_s$ is assured by requiring that initially the coefficient matrix G is diagonal.

To see whether (5.13) can be a solution we substitute it in the equation of motion (5.7) for $I_s(t)$. This leads to

$$\sum_r A_r \frac{\partial G_{rs}(t)}{\partial t} = \sum_q G_{qs}(i\hbar)^{-1}[H, A_q] \tag{5.14}$$

Since the operators A_r are linearly independent, (5.14) can obtain if the coefficient of each and every operator is the same on both sides of the equation. It follows that a sufficient condition for (5.13) to be a solution is that the commutator $[H, A_q]$ is itself a linear combination of the same type as $I_s(t)$ itself

$$(i\hbar)^{-1}[H, A_q] = \sum_r A_r g_{rq} \tag{5.15}$$

Substituting (5.15) in (5.14) one equates coefficients of A_r to obtain an equation of motion for the G matrix

$$\frac{\partial G_{rs}}{\partial t} = \sum_q g_{rq} G_{qs} \tag{5.16}$$

If one can find a set of operators $\{A_q\}$ such that (1) every operator which acts as a constraint for the initial state is a member of the set (this is necessary to satisfy the boundary condition on the I_r's; the set may, however, include additional operators); and (2) the set is closed under the operation of commutation with the Hamiltonian [cf. (5.15)] then, an explicit solution for $\sigma(t)$ is possible: if there are $N+1$ operators A_s in the closed set

$$\begin{aligned}\sigma(t) &= \exp\left[-\sum_{r=0} \lambda_r I_r(t)\right] \\ &= \exp\left[-\sum_{s=0}^{N} \lambda_s(t) A_s\right]\end{aligned} \tag{5.17}$$

where, using (5.13),

$$\lambda_s(t) = \sum_{s=0}^{N} G_{sr}(t)\lambda_r \tag{5.18}$$

The initial condition for (5.18) (introduced via the **G** matrix) is that those Lagrange parameters which are conjugate to operators that do not act as constraints for the initial state have the initial value zero.

We have brought the solution of the equation of motion for $\sigma(t)$ to the point [cf. (5.16)] where all that is required is a numerical solution of a set of coupled linear and first-order differential equations. That may or may not be a formidable numerical problem but is no longer a problem of principle. Note also that even without a determination of the values of the elements of **G** we have an explicit functional form (5.17) for the density matrix. The $n+1$ parameters $\lambda_r(t)$ are available not only by solving for **G** and using (5.18). An equally valid route, which will give identical results, is to determine the Lagrange parameters from the mean values of the $N+1$ time independent constraints A_s. This is the procedure that will be followed when variational approximations are introduced in Section V.G.

How do we know that sets of operators as required by (5.15) exist? One can show[122] that every system on n degrees of freedom has at least one set, containing $2n$ operators. The proof in Ref. 122 is for classical mechanics but goes unchanged for the quantal case since the only difference is that $(i\hbar)^{-1}\cdot$commutator of the quantal problem (cf. (5.7) and (5.15)) is replaced in classical mechanics by the Poisson bracket. Of course, the set just mentioned may fail to include some or all the operators required to specify the initial state of interest [cf. condition (1) above]. Other sets have however been reported[89, 99, 130] and a general technique for the construction of such sets has been discussed.[131]

The functional form (5.17) is that of a state of maximal entropy subject to the mean values of $N+1$ time-independent and linearly independent operators. If there are m possible quantum states of the system it takes the mean values of m^2 linearly independent operators to specify a unique state.[92] Hence if the dimension, $N+1$, of the closed set is below m^2 we have the situation discussed in Section III.B. There will, however, be Hamiltonians and initial states for which the present line of reasoning would not be advantageous. The use of approximations, based on the variational character of the maximum entropy formalism (Appendix A of Ref. 89 and Section V.G) is then unavoidable.

F. Sum Rules

Since $\langle I_r(t)\rangle$ is time independent, it follows from (5.12) and (5.13) that

$$\langle A_s\rangle^{\text{in}} = \sum_{r=0}^{N+1} G_{rs}(t)\langle A_r\rangle(t) \tag{5.19}$$

where $\langle A_s\rangle^{\text{in}}$ is the initial mean value of A_s, while $\langle A_s\rangle(t)$ is its mean value at the time t. In particular, it follows that the initial values of the constraints are a linear combination of their final values and vice versa,

$$\langle A_r \rangle^{\text{out}} = \sum_{s=0}^{N+1} (\mathbf{G}^{\text{out}})^{-1}_{sr} \langle A_s \rangle^{\text{in}} \tag{5.20}$$

Here $\mathbf{G}^{\text{out}}$ is the $\mathbf{G}$ matrix for $t \to \infty$, determined by solving the equation of motion (5.16).

Since the $\mathbf{G}$ matrix is independent of the values of the constraints, (5.20) can be used to identify the members of the set $\{A_r\}$ in a phenomenological fashion. An explicit example is shown in Fig. 14 for the reaction[132]

$$C^+(^2P) + H_2 \to CH^+(A^1\Pi) + H \tag{O}$$

which proceeds via a CH_2^+ intermediate[132, 133] (which at low energies is long living). The results of surprisal analysis[134] for the $CH^+(A^1\Pi)$ vibrational energy distribution at different collision energies (Fig. 15) show that at very low energies (just above the threshold which is at 3.4 eV), the distribution is essentially unconstrained [i.e., λ_v in (2.10) is about zero]. As the collision energy is increased, the CH^+ vibrational distribution becomes more and more deviant from prior expectations.

The results shown in Figs. 7 and 15 (and other examples) suggest that one can pump molecules in a manner which insures a constrained time evolution. The empirical rule of thumb suggested by these results is that one degree of freedom should be pumped to excess. In other words, it should carry much more energy than the other modes. While this rule is obvious on intuitive grounds, it is pleasing to note that it is also implied by

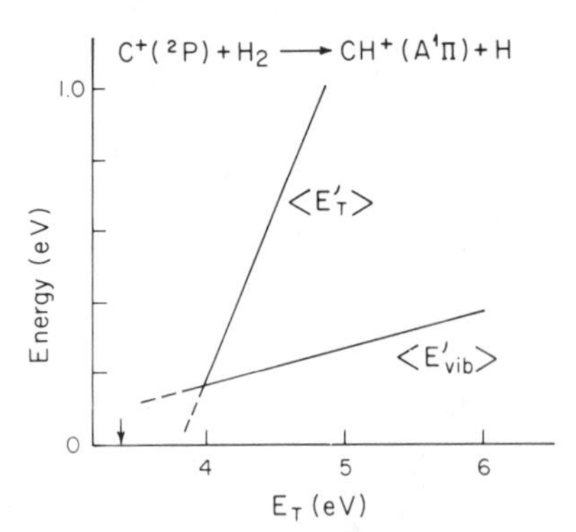

Fig. 14. Observed[132] mean final vibrational, $\langle E'_{\text{vib}} \rangle$ and translational $\langle E'_T \rangle$ energies for the reaction (O) as a function of the initial translational energy. The arrow indicates the threshold energy for the reaction.

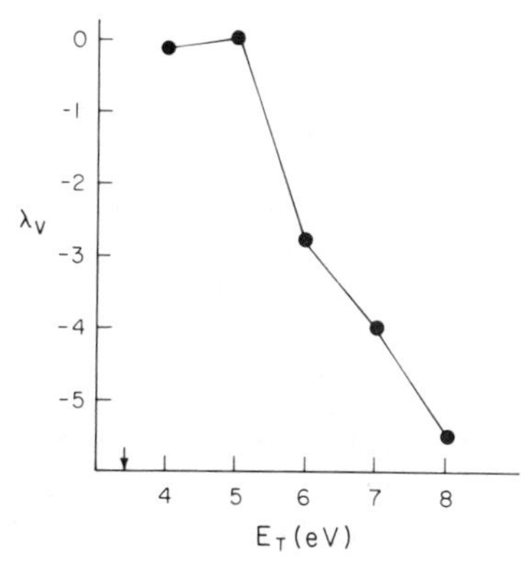

Fig. 15. Vibrational surprisal parameter [cf. (2.10)] for the $CD^+(A^1\Pi)$ vibrational state distribution vs. the initial translational energy. The solid arrow indicates the threshold energy for reaction. The reaction, (O), proceeds via the CD_2^+ intermediate which at low energies is long-living. (Adapted from Ref. 134.)
Chapter 5

the momentum transfer constraint[135, 136] and that it is confirmed by the available experimental results.

Sum rules offer a particularly useful route to identifying constraints because of the different ways that one can arrive at a relation of the type (5.20). The most direct one is empirical, as in Figs. 14 or 10. Then, models can often be stated in the form of a sum rule. Some of the most common models in molecular[135] and nuclear[83] physics are indeed of this form and have been used to suggest constraints. Sum rules can also be related to the macroscopic relaxation of the system.[77]

G. Approximations

Variational approximations to the state of the system whether during or after the collision are obtained by determining a state $\rho^R(t)$ of maximal entropy subject to a set of any constraints B_s, $s=0,1,\ldots$ provided that the mean value of the constraints is that determined by the exact state,

$$\mathrm{Tr}\{B_s\rho^R(t)\}=\mathrm{Tr}\{B_s\sigma(t)\} \tag{5.21}$$

$\rho^R(t)$ is thus of the form

$$\rho^R(t)=\exp\left[-\sum_{s=0}\gamma_s(t)B_s\right] \tag{5.22}$$

where the values of the Lagrange parameters $\gamma_s(t)$ are those implied by the values of the constraints via (5.21).

It follows from (5.22) and (5.21) that

$$\mathrm{Tr}\{\sigma(t)\ln\rho^R(t)\}=\mathrm{Tr}\{\rho^R(t)\ln\rho^R(t)\} \tag{5.23}$$

and hence that the entropy of $\rho^R(t)$ is above or equal to that of $\sigma(t)$

$$-\mathrm{Tr}\{\sigma(t)\ln\sigma(t)\}\leqslant-\mathrm{Tr}\{\sigma(t)\ln\rho^R(t)\}=-\mathrm{Tr}\{\rho^R(t)\ln\rho^R(t)\} \tag{5.24}$$

with equality if and only if $\rho^R\equiv\sigma$. One readily shows that to linear order in the difference between $\rho^R(t)$ and $\sigma(t)$ their respective entropies are equal. It is only terms in the second (or higher) order in $\rho^R(t)-\sigma(t)$ that cause the difference in (5.24).

On adding constraints the entropy of the new state is either equal to or below that of $\rho^R(t)$ [the proof goes as in (5.21)–(5.24)], but is above or equal to that of $\sigma(t)$. The method is therefore convergent.

Constraints for use in the approximation are, e.g., those operators present already as constraints in the initial state or those operators which have been identified via empirical sum rules or via models. The derivation (5.21)–(5.24) makes it clear that any operator can be used. At worst it will be found that its Lagrange parameter is zero, that is, it does not constrain the state.

An illustration of the use of (5.21) is shown in Fig. 11. Using just one constraint (the rotational energy) beyond normalization a rather accurate reproduction of the rotational state distribution was obtained. As expected, the same constraint was found[80] to work well not only after but also during the collision.

The example in Fig. 11 and others like it (e.g., Fig. 3) raise the following open question: It is usually observed that very few (one, two) constraints are dominant, in the sense that they suffice to very accurately approximate the distribution. It remains to develop a procedure in order to identify those members of the closed set (5.15) that are the dominant ones.

H. Summary

As a summary of the discussion in this section we return to the point that by solving the equations of motion directly for an initial state of maximal entropy one avoids the computation of details which are averaged out when the final state of interest is determined. Let ρ_α be an initial state which is a pure quantum state. It evolves into $\rho_\alpha(t)$. Clearly one can always put

$$\rho_\alpha(t)=\rho^R(t)+\delta\rho_\alpha(t) \tag{5.25}$$

where $\rho^R(t)$ is a density matrix of maximal entropy subject to some given set of constraints. $\delta\rho_\alpha(t)$ is for our present purpose defined by (5.25), although one can derive an explicit expression for it.[122, 137] What is clear already is that, since $\rho^R(t)$ is normalized, the difference term $\delta\rho_\alpha(t)$ averages out to zero, $\mathrm{Tr}[\delta\rho_\alpha(t)]=0$.

The initial state of interest σ is some linear combination of pure states

$$\sigma=\sum_\alpha p_\alpha \rho_\alpha \tag{5.26}$$

and evolves into the state $\sigma(t)$. Since the Liouville equation is linear

$$\sigma(t)=\rho^R(t)+\sum_\alpha p_\alpha\,\delta\rho_\alpha(t) \tag{5.27}$$

The procedure of maximal entropy is a way of assigning the weights p_α in the initial state (5.26). Say, the mean values $\langle A_r\rangle$

$$\langle A_r\rangle=\sum_\alpha p_\alpha \mathrm{Tr}(A_r\rho_\alpha) \qquad r=0,1,\ldots \tag{5.28}$$

are used as the constraints. Then we have shown that (1) If the constraints on $\rho^R(t)$ are chosen as indicated in Section V.E, then the second term in (5.27) is identically zero; in other words, the terms $\delta\rho_\alpha(t)$ average out exactly; and (2) if the constraints on $\rho^R(t)$ are any set $\{B_r\}$, whose average value for $\sigma(t)$ and $\rho^R(t)$ coincide, then for all the average values $\langle B_r\rangle$,

$$\delta\langle B_r\rangle\equiv\mathrm{Tr}\left\{B_r\sum_\alpha p_\alpha\,\delta\rho_\alpha(t)\right\}=0 \tag{5.29}$$

vanish, by construction, for any time t. An observer who is only interested in the average values $\langle B_r\rangle$ is thus unable (or uninterested) to detect the second term in (5.27). It is also important to note that $\rho^R(t)$ will predict correctly not only the mean values of the B_r's but of all of those other operators which act as noninformative constraints for $\rho^R(t)$.

Molecular dynamics experiments typically carry out only a partial resolution of the final states. Hence the second situation described above is not without merit. The formalism is used not to predict the state in all of its detail but only such details that can be measured in the experiment under consideration.[122]

VI. OUTLOOK

The time evolution of activated molecules shows the characteristics of both direct and statistical dynamics. This chapter has argued that the maximum entropy approach offers a particularly suitable method for the analysis and interpretation of such behavior. The concept of the state, which is maximally statistical subject to constraints, was shown to account for the available results and to have a rigorous dynamical foundation.

More work on both the phenomenological and th : theoretical levels is, however, required in order to bring the ideas explored in this chapter to maturity. Surprisal analyses of dissociation products for longer living intermediates and for processes with very selected initial states would be instructive. It would help clarify two important issues: (1) Does the number of dominant constraints (cf. Section V.E) decline as the lifetime increases? The theoretical position (Section V.B and Ref. 122) is that it does, and the less detailed is the distribution examined, the faster does it relax during the time evolution. Hence it would be particularly useful to determine joint products distributions, that is the distributions in two or more degrees of freedom.[125] (2) The theoretical position is that by using state selected reactants more constraints are introduced. This has been demonstrated by experiments (e.g., Fig. 15 and the discussion therein), where the internal energy of the reagents is held constant and their kinetic energy is increased (so that the mean internal and mean kinetic energies are very different). Experiments and trajectory computations have also verified this for direct collisions,[138] for example,

$$OH(v)+Cl \rightarrow O+HCl(v') \tag{P}$$

where[139] the HCl vibrational excitation is highly correlated with the OH vibrational state. The energy is thus effectively locked into the vibration.[136] It remains however to map out such behavior for laser pumped polyatomics. The results shown in Figs. 6 or 7 are, we believe, in the shape of things to come.

The formalism discussed in Section V is primarily algebraic in character. It resembles spectroscopy much more than kinetics. The traditional approach to delineating the main dynamic effects is through the major topographical features of the potential energy surface. It is desirable to forge better links between the two points of view. In particular, how are the dominant constraints related to the main passes in the potential energy surface. (What is already clear is that beside the surface, the masses and the total energy are also relevant.) Another question which, as chemists, we should address, is how do the constraints vary as we make systematic changes in the structure of the reagents. Work along such lines has started (e.g., for collinear reactive collisions;[89] preliminary studies on structure–reactivity correlations have also been carried out[140]), but most of the work is yet to be done. One is also not yet in a position to specify, for polyatomic molecules, how to excite an initial state in such a manner that it remain constrained and hence locked onto the direction of some desirable end product.

An important topic, not discussed in this chapter, is the surprisal analysis and the prediction of the absolute (as opposed to the relative) rate constant. Much of our current activity is in this direction.

APPENDIX: THE DECLINE OF ENTROPY UPON IMPROVED EXPERIMENTAL RESOLUTION

Does increased resolution of the products state distribution necessarily lower the entropy? This is how it should be if entropy is a measure of uncertainty. It is therefore worthwhile to prove that this is indeed the case. The proof below is for the quasiclassical case. The proof in the quantal case is roughly along the same lines.

Let $P(i)$ be the detailed distribution and $P(\gamma)$ be a less detailed one,

$$P(\gamma) = \sum_{i \in \gamma} P(i) \tag{A.1}$$

where the states i have been grouped together [cf. (2.1)]. Define $P(i|\gamma)$, the probability of the state i within the group γ,

$$P(i|\gamma)P(\gamma) = P(i) \tag{A.2}$$

and, from (A.1)

$$\sum_{i \in \gamma} P(i|\gamma) = 1 \tag{A.3}$$

The entropy of the detailed distribution, $P(i)$, can now be written as

$$\begin{aligned} S[i] &= -\sum_{i} P(i)\ln[P(i)] \\ &= -\sum_{\gamma}\sum_{i \in \gamma} P(i|\gamma)P(\gamma)\ln[P(\gamma)] \\ &\quad -\sum_{\gamma}\sum_{i \in \gamma} P(i|\gamma)P(\gamma)\ln[P(i|\gamma)] \\ &= -\sum_{\gamma} P(\gamma)\ln[P(\gamma)] + \sum_{\gamma} P(\gamma)\left\{-\sum_{i \in \gamma} P(i|\gamma)\ln[P(i|\gamma)]\right\} \end{aligned} \tag{A.4}$$

Information theory defines the first term in (A.4) as the entropy of the distribution $P(\gamma)$. Since $P(i|\gamma) \leqslant 1$, $\ln[P(i|\gamma)] \leqslant 0$, and so the second term in (A.4) is nonnegative. The result is that it proves the opposite of what we want. The resolution of the paradox is that the first term in (A.4) is not necessarily the entropy of the distribution $P(\gamma)$. The reason is that each group γ may contain more than one, say g_γ, quantum states, $g_\gamma \geqslant 1$. The entropy is then[43–46, 94]

$$S[\gamma] = -\sum_\gamma P(\gamma)\ln\frac{P(\gamma)}{g_\gamma}$$

$$= -\sum_\gamma P(\gamma)\ln P(\gamma) + \sum_\gamma P(\gamma)\ln g_\gamma \tag{A.5}$$

But (see below)

$$-\sum_{i\in\gamma} P(i|\gamma)\ln[P(i|\gamma)] \leqslant \ln g_\gamma \tag{A.6}$$

and hence, comparing (A.4) and (A.5)

$$S[i] \leqslant S[\gamma] \tag{A.7}$$

To prove the lemma (A.6) recall that there are g_γ states in the group γ. The left-hand side of (A.6) is the entropy of the distribution of states i within the group γ. It is maximal when all states within the group are equiprobable, that is, when (and only when) $P(i|\gamma) = 1/g_\gamma$, Q.E.D.

The entropy of the detailed state distribution, $S[i]$, is thus less than or equal to the entropy, $S[\gamma]$ of the distribution with a lower resolution. Equality obtains if and only if all states i within each group γ are equiprobable, that is, when no additional information is provided by the refinement.

References

1. I. Oref and B. S. Rabinovitch, *Act. Chem. Res.*, **12**, 166 (1979).
2. E. A. Hardwidge, B. S. Rabinovitch, and R. C. Ireton, *J. Chem. Phys.*, **58**, 340 (1973). F. H. Dorer and B. S. Rabinovitch, *J. Phys. Chem.*, **69**, 1952 (1974).
3. B. E. Holmes and D. W. Setser, *J. Phys. Chem.*, **79**, 1320 (1975); *ibid.*, **82**, 2461 (1978).
4. J. N. Butler and G. B. Kistiakowski, *J. Am. Chem. Soc.*, **82**, 759 (1960).
5. J. D. Rynbrandt and B. S. Rabinovitch, *J. Phys. Chem.*, **75**, 2164 (1971). A. N. Ko, B. S. Rabinovitch, and K. J. Chao, *J. Chem. Phys.*, **66**, 1374 (1977). A. N. Ko and B. S. Rabinovitch, *Chem. Phys.*, **30**, 361 (1978).
6. I. Oref and B. S. Rabinovitch, *Chem. Phys.*, **26**, 385 (1977).
7. C. C. Jensen, J. I. Steinfeld, and R. D. Levine, *J. Chem. Phys.*, **69**, 1432 (1978).
8. M. J. Coggiola, P. A. Schulz, Y. T. Lee, and Y. R. Shen, *Phys. Rev. Lett.*, **38**, 17 (1977). E. R. Grant, M. J. Coggiola, Y. T. Lee, P. A. Schulz, and Y. R. Shen, *Chem. Phys. Lett.*, **52**, 595 (1977).

9. J. G. Black, E. Yablonovitch, N. Bloembergen, and S. Mukamel, *Phys. Rev. Lett.*, **38**, 1131 (1977).
10. M. Quack and J. Troe, in *Gas Kinetics and Energy Transfer*, Vol. II, Chemical Society, London, 1977, p. 175; *Ber. Bunsenges. Phys. Chem.*, **78**, 240 (1974); *ibid.*, **79**, 170 (1975). J. Troe, *J. Chem. Phys.*, **66**, 4745, 4758 (1977).
11. D. L. Bunker, *J. Chem. Phys.*, **37**, 393 (1962); *ibid.*, **40**, 1946 (1964); *Ber. Bunsenges. Phys. Chem.*, **81**, 155 (1977).
12. D. W. Oxtoby and S. A. Rice, *J. Chem. Phys.*, **65**, 1676 (1976). S. A. Rice in *Advances in Laser Chemistry*, A. Zewail, ed., Springer, Berlin, 1978.
13. J. D. McDonald and R. A. Marcus, *J. Chem. Phys.*, **65**, 2180 (1976). D. L. Bunker, K. R. Wright, W. L. Hase, and F. A. Houle, *J. Phys. Chem.*, **83**, 933 (1979).
14. E. R. Grant and D. L. Bunker, *J. Chem. Phys.*, **68**, 628 (1978).
15. W. L. Hase, in *Dynamics of Molecular Collisions*, Part B, W. H. Miller ed., Plenum, New York, 1976, p. 121.
16. K. D. Hansel, *J. Chem. Phys.*, **70**, 1830 (1979).
17. See, for example, G. Benettin, L. Galgani, and J. M. Strelcyn, *Phys. Rev.*, **A14**, 2339 (1976).
18. R. J. Harter, E. B. Alterman, and D. J. Wilson, *J. Chem. Phys.*, **40**, 2137 (1964). C. A. Parr, A. Kuppermann, and R. N. Porter, *J. Chem. Phys.*, **66**, 2914 (1977). P. J. Nagy and W. L. Hase, *Chem. Phys. Lett.*, **54**, 73 (1978).
19. E. Pollåk, *J. Chem. Phys.*, **68**, 534 (1978). P. Brumer and J. W. Duff, *J. Chem. Phys.*, **65**, 3566 (1976).
20. C. Rebick, R. D. Levine, and R. B. Bernstein, *J. Chem. Phys.*, **60**, 4977 (1974).
21. A. B. Lees and G. H. Kwei, *J. Chem. Phys.*, **58**, 1710 (1973). J. C. Whitehead, *Mol. Phys.*, **29**, 177 (1975).
22. R. D. Levine and R. B. Bernstein, *Molecular Reaction Dynamics*, Clarendon Press, Oxford, 1974.
23. L. Zandee and R. B. Bernstein, *J. Chem. Phys.*, **68**, 3760 (1978).
24. M. J. Berry and G. C. Pimentel, *J. Chem. Phys.*, **49**, 5190 (1968).
25. E. Zamir and R. D. Levine, *Chem. Phys. Lett.*, **67**, 237 (1979). *Chem. Phys.* (1980).
26. M. J. Berry in *Molecular Energy Transfer*, R. D. Levine and J. Jortner, eds., Wiley, New York, 1976, p. 114.
27. G. A. West, R. E. Weston, Jr. and G. W. Flynn, *Chem. Phys. Lett.*, **35**, 275 (1978).
28. C. R. Quick and C. Wittig, *Chem. Phys.*, **32**, 75 (1978); *J. Opt. Soc. Am.*, **68**, 693 (1978).
29. D. S. King and J. C. Stephenson, *Chem. Phys. Lett.*, **51**, 48 (1977). J. C. Stephenson and D. S. King, *J. Chem. Phys.*, **69**, 1485 (1978).
30. D. W. Setser, in *MTP Int. Rev. Sci.–Phys. Chem.*, **9**, 1 (1972).
31. B. E. Holmes and D. W. Setser, in *Physical Chemistry of Fast Reactions*, Vol. 3, I. W. Smith, ed., Wiley, New York, 1980.
32. J. F. Durana and J. D. McDonald, *J. Chem. Phys.*, **64**, 2518 (1976); J. T. Gleaves and J. D. McDonald; *ibid.*, **62**, 1582 (1975).
33. R. D. Levine and R. B. Bernstein, *Acct. Chem. Res.*, **7**, 393 (1974).
34. S. H. Bauer, *Chem. Rev.*, **78**, 147 (1978).
35. R. D. Levine, *Ann. Rev. Phys. Chem.*, **29**, 59 (1978).
36. K. J. Schmatjko and J. Wolfrum, *Ber. Bunsenges. Phys. Chem.*, **82**, 419 (1978).
37. O. K. Rice and H. C. Ramsperger, *J. Am. Chem. Soc.*, **49**, 1617 (1927); *ibid.*, **50**, 617 (1928). L. S. Kassel, *J. Phys. Chem.*, **32**, 225 (1928). *Kinetics of Homogeneous Gas Reactions*, Chemical Catalog, New York, 1932.
38. N. B. Slater, *Proc. Camb. Phil. Soc.*, **35**, 56 (1939); *Theory of Unimolecular Reactions*, Cornell University Press, New York, 1959).

39. R. A. Marcus, *J. Chem. Phys.*, **20**, 359 (1952); *ibid.*, **43**, 2658 (1965).
40. S. A. Safron, N. D. Weinstein, D. R. Herschbach, and J. C. Tully, *Chem. Phys. Lett.*, **12**, 564 (1972). J. C. Light, in *Atom Molecule Collision Theory—A Guide for the Experimentalist*, R. B. Bernstein, ed., Plenum, New York, 1979).
41. R. D. Levine, *Ber. Bunsenges. Phys. Chem.*, **78**, Section VIII, 111 (1974).
42. F. Kaufman and R. D. Levine, *Chem. Phys. Lett.*, **54**, 407 (1978). E. Pollak and R. D. Levine, *J. Chem. Phys.*, **72**, 2484 (1980).
43. R. D. Levine and R. B. Bernstein, in *Dynamics of Molecular Collisions*, Part B, W. H. Miller, ed., Plenum, New York, 1976, p. 323.
44. R. D. Levine and A. Ben-Shaul, in *Chemical and Biochemical Applications of Lasers*, Vol. II, C. B. Moore, ed., Academic Press, New York, 1977, p. 145.
45. R. D. Levine and J. I. Kinsey, in *Atom–Molecule Collision Theory—A Guide for the Experimentalist*, R. B. Bernstein, ed., Plenum, New York, 1979, Ch. 22.
46. A. Ben-Shaul, Y. Haas, K. L. Kompa, and R. D. Levine, *Lasers and Chemical Change*, Springer, Berlin, 1980.
47. M. C. Lin, R. G. Shortridge, and M. E. Umstead, *Chem. Phys. Lett.*, **37**, 279 (1976); M. E. Umstead, R. G. Shortridge, and M. C. Lin, *Chem. Phys.*, **20**, 271 (1977).
48. D. J. Bogan, *J. Phys. Chem.*, **81**, 2509 (1977).
49. E. R. Sirkin and M. J. Berry, *IEEE J. Quant. Elec.*, **QE-10**, 701 (1974).
50. D. J. Bogan and D. W. Setser, *J. Chem. Phys.*, **64**, 586 (1976).
51. D. Bogan and D. W. Setser, in *Kinetics and Dynamics of Fluorine Reactions*, J. W. Root, ed., ACS Symposium Series 66, Washington, D.C., 1977.
52. M. J. Berry, *J. Chem. Phys.*, **59**, 6229 (1973).
53. D. S. Perry and J. C. Polanyi, *Chem. Phys.*, **12**, 419 (1976).
54. J. G. Moehlmann and J. D. McDonald, *J. Chem. Phys.*, **62**, 3061 (1975).
55. J. M. Farrar and Y. T. Lee, *J. Chem. Phys.*, **65**, 1414 (1976).
56. R. J. Buss, M. J. Coggiola, and Y. T. Lee, *J. Chem. Soc. Faraday Discuss.*, **67**, 162 (1979).
57. M. E. Umstead, R. G. Shortridge, and M. C. Lin, *J. Phys. Chem.*, **82**, 1455 (1978).
58. A. Baronavski, M. E. Umstead, and M. C. Lin, this volume.
59. Y. Alhassid, N. Agmon, and R. D. Levine, *Chem. Phys. Lett.*, **53**, 22 (1978); *J. Comput. Phys.*, **30**, 250 (1979).
60. H. Kaplan, R. D. Levine, and J. Manz, *Chem. Phys.*, **12**, 447 (1976).
61. M. J. Berry, private communication. See, however, I. P. Herman and J. B. Marling, *J. Chem. Phys.*, **71**, 643 (1979); E. Zamir, Y. Haas and R. D. Levine, *J. Chem. Phys.*, **73**, 2680 (1980).
62. Aa. S. Sudbø, P. A. Schulz, Y. T. Lee, and Y. R. Shen, *J. Chem. Phys.*, **68**, 1306 (1978).
63. P. A. Schulz, Aa. S. Sudbø, D. J. Karjnovich, H. S. Kwok, Y. R. Shen, and Y. T. Lee, *Ann. Rev. Phys. Chem.*, **30**, 379 (1979).
64. Aa. S. Sudbø, P. A. Schulz, Y. R. Chen and Y. T. Lee, *J. Chem. Phys.*, **69**, 2312 (1978).
65. R. A. Marcus, *J. Chem. Phys.*, **62**, 1372 (1975). G. Worry and R. A. Marcus, *J. Chem. Phys.*, **67**, 1636 (1977).
66. D. J. Zvijac and J. C. Light, *Chem. Phys.*, **21**, 411, 433 (1977).
67. S. Rockwood, J. P. Reilly, K. Hohla, and K. L. Kompa, *Opt. Comm.*, **28**, 175 (1979).
68. L. Zandee, R. B. Bernstein, and D. A. Lichtin, *J. Chem. Phys.*, **69**, 3427 (1978).
69. V. S. Antonov, I. N. Knyazev, V. S. Letokhov, V. M. Matiuk, V. G. Movshev, and V. K. Potapov, *Opt. Lett.*, **3**, 37 (1978).
70. L. Zandee and R. B. Bernstein, *J. Chem. Phys.*, **71**, 1359 (1979).
71. J. Silberstein and R. D. Levine, *Chem. Phys. Lett.*, **74**, 6 (1980).
72. D. M. Manos and J. M. Parson, *J. Chem. Phys.*, **69**, 231 (1978).
73. D. M. Manos and J. M. Parson, *J. Chem. Phys.*, **63**, 3575 (1975).

74. K. Liu and J. M. Parson, *J. Chem. Phys.*, **68**, 1794 (1978).
75. C. L. Chalek and J. L. Gole, *Chem. Phys.*, **19**, 59 (1977).
76. M. B. Faist and R. D. Levine, *Chem. Phys. Lett.*, **47**, 5 (1977).
77. I. Procaccia and R. D. Levine, *J. Chem. Phys.*, **63**, 4261 (1975).
78. I. Procaccia and R. D. Levine, *J. Chem. Phys.*, **64**, 808 (1976).
79. R. E. Wyatt, *Chem. Phys. Lett.*, **34**, 167 (1975). See also G. C. Schatz and A. Kuppermann, *J. Chem. Phys.*, **65**, 4668 (1976).
80. D. C. Clary and R. K. Nesbet, *Chem. Phys. Lett.*, **59**, 437 (1978).
81. P. E. Hodgson, *Nuclear Heavy Ion Reactions*, Clarendon Press, Oxford, 1978. V. V. Volkov, *Phys. Rep.*, **44**, 93 (1978).
82. R. D. Levine, S. G. Steadman, J. S. Karp, and Y. Alhassid, *Phys. Rev. Lett.*, **41**, 1537 (1978).
83. Y. Alhassid, R. D. Levine, J. S. Karp, and S. G. Steadman, *Phys. Rev. C*, **20**, 1789 (1979).
84. R. D. Levine and M. Tribus, eds., *The Maximum Entropy Formalism*, MIT Press, Cambridge, Mass., 1979.
85. E. T. Jaynes, *Phys. Rev.*, **106**, 620 (1957).
86. M. Tribus, *Thermostatics and Thermodynamics*, Van Nostrand, N.J., 1961.
87. E. Wichman, *J. Math. Phys.*, **4**, 884 (1963).
88. A. Katz, *Principles of Statistical Mechanics—The Information Theory Approach*, Freeman, San Francisco, Calif., 1967.
89. Y. Alhassid and R. D. Levine, *J. Chem. Phys.*, **67**, 4321 (1977).
90. R. D. Levine, *Phys. Rep.*, in press.
91. E. T. Jaynes, *Phys. Rev.*, **108**, 171 (1957).
92. U. Fano, *Rev. Mod. Phys.*, **29**, 83 (1957).
93. For a system with n accessible quantum states it takes n^2 linearly independent operators to uniquely specify the state. See, e.g., Ref. 92.
94. R. C. Tolman, *The Principles of Statistical Mechanics*, Clarendon Press, Oxford, 1938.
95. P. A. M. Dirac, *Proc. Camb. Phil. Soc.*, **25**, 62 (1928); *ibid.*, **27**, 240 (1930).
96. J. von Neumann, *Gottingen. Nach.*, 245 (1927).
97. L. Liouville, *J. Math.*, **3**, 349 (1838).
98. W. J. Gibbs, *Elementary Principles in Statistical Mechanics*, Yale University Press, New Haven, Conn., 1902.
99. Y. Alhassid and R. D. Levine, *Phys. Rev.*, **A18**, 89 (1978).
100. R. D. Levine, *Quantum Mechanics of Molecular Rate Processes*, Clarendon Press, Oxford, 1969.
101. J. R. Taylor, *Scattering Theory*, Wiley, New York, 1972.
102. R. D. Levine in Ref. 84.
103. R. D. Levine, *J. Chem. Phys.*, **65**, 3302 (1976).
104. Due to inevitable experimental scatter (or computational round-off) one cannot and should not expect perfect convergence when comparing to experimental or computational results. The stopping criterion for such cases is discussed by J. L. Kinsey and R. D. Levine, *Chem. Phys. Lett.*, **65**, 413 (1979).
105. See, for example, G. Lindbland, *Comm. Math. Phys.*, **33**, 305 (1973).
106. It may be possible to explicitly compute $\mathcal{H}(t)$. One very simple case is that of a harmonic oscillator excited by a monochromatic light source. Then $H_0 = a^+a$ and the solution of (4.7) is[99, 102, 107] $\mathcal{H}(t) = [a^+ - c^*(t)]\cdot[a - c(t)]$ where $c(t)$ is an explicit function of the time determined by the frequency of the field. See Refs. 99, 102, or 107 for more details.
107. W. H. Louisell, *Quantum Statistical Properties of Radiation*, Wiley, New York, 1973.
108. See, for example, R. Kubo, *Lect. Theoret. Phys.*, **1**, 127 (1959).

109. R. D. Levine, *Phys. Rev. A*, in press.
110. E. J. G. Pitman, *Proc. Camb. Phil. Soc.*, **32**, 567 (1936).
111. B. O. Koopman, *Trans. Am. Math. Soc.*, **39**, 399 (1936).
112. E. B. Dynkin, *Selected Translations Math. Stat. Prob.* (*Am. Math. Soc.*), **2**, 23 (1961). H. Jeffreys, *Theory of Probability*, 3rd ed., Cambridge University Press, 1962, p. 168.
113. L. Boltzmann, *Wein. Ber.*, **63**, 397 (1871); *ibid.*, **76**, 373 (1877).
114. R. B. Bernstein and R. D. Levine, *Adv. Atom. Mol. Phys.*, **11**, 215 (1975).
115. E. T. Jaynes in Ref. 84.
116. M. Tribus in Ref. 84.
117. I. J. Good, *Ann. Math. Stat.*, **34**, 911 (1963).
118. C. E. Shannon, *Bell Syst. Tech. J.*, **27**, 397 (1948).
119. R. Ash, *Information Theory*, Wiley-Interscience, New York, 1965.
120. See, for example, J. S. Rowlinson, *Nature*, **225**, 1196 (1970).
121. J. Aczel, B. Forte, and C. T. Naj, *Adv. Appl. Prob.*, **6**, 131 (1974). For the quantal case see W. Ochs, *Rep. Math. Phys.*, **8**, 109 (1975).
122. Y. Alhassid and R. D. Levine, *Phys. Rev. C*, **20**, 1775 (1979).
123. J. L. Kinsey, *J. Chem. Phys.*, **54**, 1206 (1971).
124. G. L. Hofacker and R. D. Levine, *Chem. Phys. Lett.*, **33**, 404 (1975). Numerical results for the entropy of a reduced distribution throughout the collision have been reported by J. N. L. Connor, W. Jakubetz, and J. Manz, *Chem. Phys. Lett.*, **44**, 516 (1976) (vibrational). R. K. Nesbet, *ibid.*, **42**, 197 (1976) and D. C. Clary and R. K. Nesbet, *ibid.*, **59**, 437 (1978) (rotational); and J. S. Hutchinson and R. E. Wyatt, *J. Chem. Phys.*, **70**, 3509 (1979).
125. D. A. Case and D. R. Herschbach, *J. Chem. Phys.*, **69**, 150 (1978).
126. G. M. McClelland and D. R. Herschbach, in press.
127. E. T. Whittaker, *A Treatise on the Analytical Dynamics of Particles and Rigid Bodies*, Cambridge University Press, 1904.
128. Y. Dothan, *Phys. Rev.*, **D2**, 2944 (1970). E. B. Aronson, I. A. Malkin, and V. I. Man'ko, *Sov. J. Particles Nucl.*, **5**, 47 (1974).
129. See, for example, A. Messiah, *Quantum Mechanics*, Wiley, New York, 1961.
130. R. D. Levine and C. E. Wulfman, *Chem. Phys. Lett.*, **60**, 372 (1979).
131. C. E. Wulfman and R. D. Levine, to be published.
132. I. Kusunoki and Ch. Ottinger, *J. Chem. Phys.*, **71**, 4227 (1979).
133. C. A. Jones, K. L. Wendell, J. J. Kaufman, and W. S. Koski, *J. Chem. Phys.*, **65**, 2345 (1976).
134. E. Zamir, R. D. Levine, and R. B. Bernstein, *Chem. Phys.*, (1980).
135. A. Kafri, E. Pollak, R. Kosloff, and R. D. Levine, *Chem. Phys. Lett.*, **33**, 201 (1975).
136. This is as expected for the momentum transfer constraint.[135] See also E. Pollak, and R. D. Levine, *Chem. Phys. Lett.*, **39**, 199 (1976); E. Pollak, *Chem. Phys.*, **22**, 151 (1977); and Ref. 77.
137. I. Oppenheim and R. D. Levine, *Physica*, **99A**, 383 (1979).
138. R. D. Levine and J. Manz, *J. Chem. Phys.*, **63**, 4280 (1975). R. D. Levine, in *The New World of Quantum Chemistry*, B. Pullman and R. Parr, eds., Reidel, Dortrecht-Holland, 1976, p. 103.
139. B. A. Blackwell, J. C. Polanyi, and J. J. Sloan, *Chem. Phys.*, **24**, 25 (1977). For other examples see, e.g., Z. Karny and R. N. Zare, *J. Chem. Phys.*, **68**, 3360 (1978) and Ref. 36. For the role of electronic excitation, see Z. Karny, R. C. Estler, and R. N. Zare, *J. Chem. Phys.*, **69**, 5199 (1978).
140. R. D. Levine, *J. Phys. Chem.*, **83**, 159 (1979). N. Agmon and R. D. Levine, *Is. J. Chem.*, **19**, 330 (1980).

BOND MODES

MARTIN L. SAGE

Department of Chemistry, Syracuse University, Syracuse, New York, 13210

and

JOSHUA JORTNER

Department of Chemistry, Tel-Aviv University, Tel Aviv, Israel

CONTENTS

I. Introduction 293
II. Beyond Normal Coordinates 295
III. Energetics 298
IV. Intensities 302
V. Intramolecular Effects on Energetics 306
VI. Cooperative Excitations of Bond Modes 309
VII. Intramolecular Dynamics Involving X—H Bonds 313
VIII. Absorption Lineshapes 315
References 321

I. INTRODUCTION

The proper description of highly excited bound vibrational states of polyatomic molecules in their ground electronic configuration is essential for the understanding of intramolecular dynamics.[1] Experimental information regarding the energetics and the dynamics of these high-energy vibrationally excited states originates from the following sources:

1. Absorption spectroscopy of high vibrational overtones.[2–17]
2. Visible emission spectra of mixed crystals.[18]
3. Infrared luminescence in molecular beams.[19]
4. Picosecond optical spectroscopy.[20]
5. Multiphoton optical double resonance.[21]
6. High-order multiphoton molecular photofragmentation.[22]

The outstanding experimental and theoretical questions in this area are:

1. What is the proper description of high-energy vibrational excitations in the range where the conventional normal mode description breaks down?
2. What is the proper description of intramolecular vibrational energy transfer and energy redistribution in large molecules?
3. Can a specific vibrational mode of a large molecule be selectively excited, and are chemical consequences of such mode-selective excitation amenable to experimental observation?

We shall be concerned with the description of the energetics and dynamics of high vibrational molecular excitations that can be described in terms of anharmonic quasilocal bond modes and that couple weakly to other vibrational modes of the same molecule. Organic chemists have routinely used the empirical concept of group vibrational frequencies for the interpretation of infrared spectra of polyatomic molecules. On the other hand, physical chemists have adopted the rigorous approach based on the well-established theory of normal modes for the description of the low-lying vibrational excitations. The gap between the empirical and the rigorous approach was partially bridged by the introduction of the notion of bond modes[2–17, 23–25] for the description of the energetics and dynamics of X—H (X≡C, O, N, etc.) molecular bonds. The concept of X—H bond modes has been established as an adequate (zero order) picture of the spectroscopic data, that is, energy levels and intensities, as well as a proper starting point for the description of intramolecular dynamics of molecules containing light atoms, such as hydrogen or deuterium. Currently available experimental evidence, which supports the concept of X—H bond modes, stems from the following sources:

1. Energetics: The energy levels of high C—H vibrational overtones ($v=2$–6) of aliphatic and aromatic hydrocarbons can be adequately described in terms of the one-dimensional Morse oscillator.[3, 5–17, 24]
2. Relative intensities: The intensity of the transition to the 0–6 C—H overtone of C_6H_6 is roughly six times as large as the intensity in C_6D_5H.[5]
3. Absolute intensities: The intensity of high C—H vibrational overtones was found to be determined by additive equal contributions from all the local C—H bond modes.[26]
4. Cooperative excitations: Additional experimental support for the X—H bond modes model was obtained from the observation of a simultaneous excitation of two distinct X—H bond modes by one photon.[14,

[15, 26–28] Because of anharmonicity defects this cooperative excitation of a combination of two bond modes is located at a higher energy than the corresponding excited local-mode overtone.

5. Two-photon absorption: Two-photon spectroscopy of high vibrational overtones is expected to result in the simultaneous excitation of two distinct bond modes, as in the case of the one-photon cooperative excitation.[29] This difficult experiment has not yet been performed.

This experimental information pertains essentially to spectroscopic data. This review will survey the current state of the theoretical understanding of this spectroscopic information and its relevance for the elucidation of intramolecular dynamics in polyatomic molecules containing X—H bonds.

II. BEYOND NORMAL COORDINATES

The complete treatment of vibration and rotation in a polyatomic molecule involves the solution of a $3N-3$-dimensional Schrödinger equation in which the kinetic energy can be written in terms of $3N-3$ Cartesian momenta and the potential energy depends on $3N-6$ internal coordinates ($3N-5$ for the linear molecule). Usually the kinetic energy is expressed in terms of the momenta conjugate to the internal coordinates and three suitably chosen angular momenta and contains pure vibration, pure rotation, and vibration–rotation terms. Furthermore if true or "curvilinear" coordinates[30] are used the G-matrix elements in the kinetic energy depend on coordinates and quantization of the classical Hamiltonian is difficult.[31, 32] On the other hand, the potential is more complicated if "rectilinear" coordinates are used.

High resolution spectroscopy, which interrogates individual rotation–vibration levels, requires a detailed treatment of the Hamiltonian. For low amplitude vibrations, the potential can be expanded in a Taylor series about the equilibrium configuration where only the quadratic terms are retained. Likewise, the G-matrix elements and moments of inertia can be expanded about equilibrium. Rotation–vibration interaction corrections to the G-matrix elements, and higher order terms in the potential are treated as perturbations. The observed energy levels are, to a first approximation, those of a rigid rotator with $3N-6$ normal vibrations with corrections for Coriolis coupling, centrifugal distortion, and anharmonicity. For large amplitude vibrations complicated procedures must be followed, particularly when the molecule has a great deal of symmetry.[33, 34]

We are mainly interested in the general features of the vibrational states and will neglect rotational structure from now on. For an N-atom molecule we shall use $3N-6$ internal valence coordinates to describe the nuclear motion. These will be separated into two classes:

1. s coordinates that behave like independent oscillators, which will be denoted by $\mathbf{R}=(R_1, R_2,\ldots, R_s)$ with the conjugate momenta $\mathbf{P}=(P_1, P_2,\ldots, P_s)$ and
2. $3N-6-s$ other valence coordinates $\mathbf{q}=(q_{s+1}, q_{s+2},\ldots, q_{3N-6})$ with their conjugate momenta $p=(p_{s+1}, p_{s+2},\ldots, p_{3N-6})$.

The potential energy $V(R,q)$ will be represented using a generalized valence force field. For many molecules a simple harmonic valence force field provides a good approximation to the harmonic potential. Large interaction constants between different valence coordinates prevail only for molecules in which resonance structures or delocalization lead to obvious couplings between bond stretches. For example, the interaction constants are no more than a few percent in H_2O, H_2S, H_2Se, SO_2, and SeO_2, while they are considerably larger in CO_2, CS_2, COS, N_2O, NO_2, and O_3.[35, 36] Furthermore, cubic and higher order diagonal terms in the potential are large compared to corresponding off-diagonal terms.[25, 36, 37] Thus we are led to a separable potential that consists of a number of uncoupled, anharmonic oscillators. We note in passing that a similar approach is widely used for molecules with a single large amplitude vibration, where the special vibration, such as inversion in ammonia, pseudorotation in cyclopentane, or internal rotation in ethane is sorted out and described in terms of a potential with several minima, while all other coordinates are treated by a quadratic potential.

Following Sage and Jortner[38] we shall represent the potential energy in terms of the anharmonic simple valence force model.[35, 36] We shall further assert that the bond stretches, which correspond to coordinates of class 1, can be described in terms of diatomic-type potential functions $V_j(R_j)$ $(j=1,\ldots,s)$, such as Morse functions. The potential energy of all other valence coordinates (class 2) will be taken in a nonseparable form and will be denoted as $u(q)$. Such a representation of the potential energy is sufficient for a reasonable description of the energy levels but not for intramolecular dynamics which require the incorporation of (small) coupling terms $w(\mathbf{R},\mathbf{q})$ between the X—H bond modes and all the other valence modes. The potential energy is

$$V(\mathbf{R},\mathbf{q})=\sum_{j=1}^{s} V_j(R_j)+u(\mathbf{q})+w(\mathbf{R},\mathbf{q}) \tag{1}$$

The kinetic energy is expressed in the form

$$T(\mathbf{P},\mathbf{p})=T(\mathbf{P})+T(\mathbf{p})+t_1(\mathbf{P})+t(\mathbf{p},\mathbf{P}) \tag{2}$$

where

$$T(\mathbf{P}) = \sum_{i=1}^{s} g_{ii} P_i^2 \tag{3a}$$

$$t_1(\mathbf{P}) = \sum_{i=1}^{s} \sum_{j=1}^{s} g_{ij} P_i P_j \tag{3b}$$

$$T(\mathbf{p}) = \sum_{i=s+1}^{3N-6} \sum_{j=s+1}^{3N-6} g_{ij} p_i p_j \tag{3c}$$

$$t(P, p) = \sum_{i=s+1}^{3N-6} \sum_{j=1}^{s} g_{ij} p_i P_j \tag{3d}$$

In general, the coefficients g_{ij} are dependent on internal coordinates. The total Hamiltonian can be rewritten as

$$H = H_0 + W \tag{4}$$

where

$$H_0 = H_0^L + H_0^S \tag{5}$$

with

$$H_0^L = \sum_{j=1}^{s} \left[g_{jj} P_j^2 + V_j(R_j) \right] \tag{6}$$

$$H_0^S = T(\mathbf{p}) + u(\mathbf{q}) \tag{7}$$

and

$$W = w(\mathbf{R}, \mathbf{q}) + t(\mathbf{P}, \mathbf{p}) + t_1(\mathbf{P}) \tag{8}$$

The total Hamiltonian (4) has the conventional appearance of a zero-order part and a perturbation term. The zero-order Hamiltonian, H_0 (5), consists of two separable contributions H_0^L and H_0^S, describing local modes and skeletal modes, respectively. Furthermore, the part describing local modes (6) corresponds to separable contributions from individual bond modes. The perturbation term W contains a kinetic energy term, $t_1(\mathbf{P})$, which couples the bond modes. The other two contributions to W involve intermode potential energy coupling as well as kinetic energy and potential energy coupling between the bond and skeletal modes.

The spectrum of the zero-order Hamiltonian is expected to provide a reasonable description of the molecular eigenstates because of two reasons. First, the diagonal coefficients $g_{ii}(i=1,2,\ldots,s)$ in the kinetic energy expansion (3a) for bonds containing light atoms, such as X—H bonds, are inversely proportional to the reduced mass of the bond, so that the diagonal kinetic energy term in (6) considerably exceeds the off-diagonal kinetic energy contributions $t(\mathbf{p},\mathbf{P})$ and $t_1(\mathbf{P})$ in (8). Second, the adequacy of the valence force model for the potential energy insures that the perturbation contribution $W(\mathbf{R},\mathbf{q})$ is small. These considerations provide proper rationalization for the applicability of the bond-mode picture for X—H bonds.

III. ENERGETICS

The use of local-mode states in the analysis of the energies of high overtone spectra in molecules containing X—H bonds is now well established. These modes were introduced by Henry and Siebrand,[9, 10] who showed that anharmonicity coupled C—H stretching vibrations in benzene were more simply described using local rather than normal modes, particularly for high vibrational states. Wallace gave a simple picture of the vibrations in water and benzene,[23] which showed that the local-mode picture was an appropriate and accurate description of X—H vibrational levels. Henry and co-workers,[9–17, 24] Albrecht and co-workers,[2–5, 26] and others,[6–8] have made numerous studies of C—H, N—H, and O—H stretching modes which are adequately described in terms of local modes.

The X—H bond modes are described in terms of one-dimensional Morse oscillators, each of which being characterized by a reduced mass μ and the potential

$$V(R)=D\{1-\exp[-a(R-R_{\mathrm{eq}})]\}^2 \tag{9}$$

where R is the bond coordinate, R_{eq} represents the equilibrium configuration, D corresponds to the bond dissociation energy, while the characteristic inverse length is $a=(\mu/2D)^{1/2}\omega_e/\hbar$. The energy level E_v of the Morse potential, which is characterized by the vibrational quantum number v, is given by

$$E_v=\omega_e\left(v+\tfrac{1}{2}\right)-x\left(v+\tfrac{1}{2}\right)^2 \tag{10}$$

where the characteristic frequency, ω_e, and the anharmonicity constant x are related to D by $D=\omega_e^2/4x$ and are explicitly given in terms of the relations

$$\omega_e = \hbar a \left(\frac{2D}{\mu} \right)^{1/2} \tag{11a}$$

$$x = \frac{\hbar^2 a^2}{2\mu} \tag{11b}$$

while the number of bound states in the potential well is given by the integer closest from below to $[(\omega_e/2x)+\frac{1}{2}]$. Obviously μ, the potential parameters D, a and R_{eq}, and other features of the bond depend on the specific X—H bond.

The experimental data for the energies defined as the peak energies of the broad absorption bands, corresponding to the 0→3 through to 0→7 vibrational overtones of the C—H vibrations in a variety of aromatic and aliphatic hydrocarbons are consistent with (10), as is evident from Figs. 1 and 2. In Table I we present a compilation of the energetic data for a variety of C—H modes. There is a slight spread of the energetic parameters, but the overall picture provides strong support to the notion that local C—H bond modes constitute a reasonable description of the energy levels. Further energetic information is obtained from the analysis of the deuterium isotope effect on the C—H local modes. The results for the potential parameters for benzene and for perdeuterobenzene, as calculated from (11), are shown in Table II. The experimental deuterium isotope shift

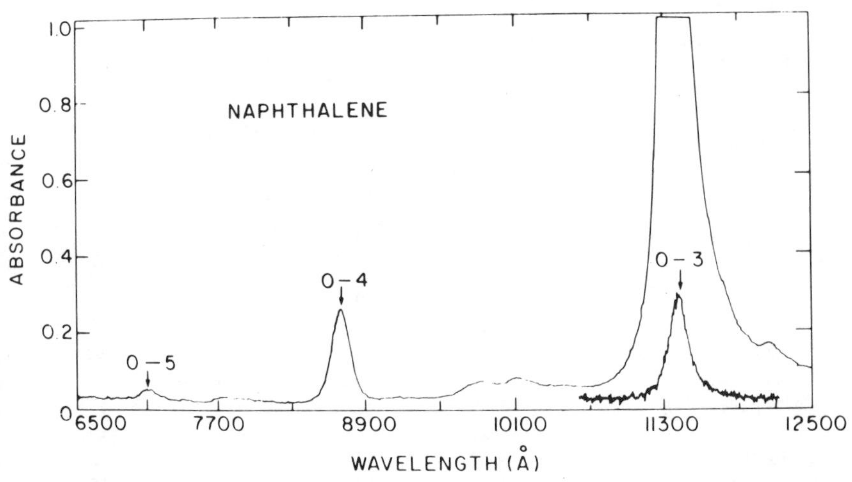

Fig. 1. The absorption spectra of high C—H overtones of crystalline naphthalene obtained by Perry and Zewail reproduced from Ref. 8.

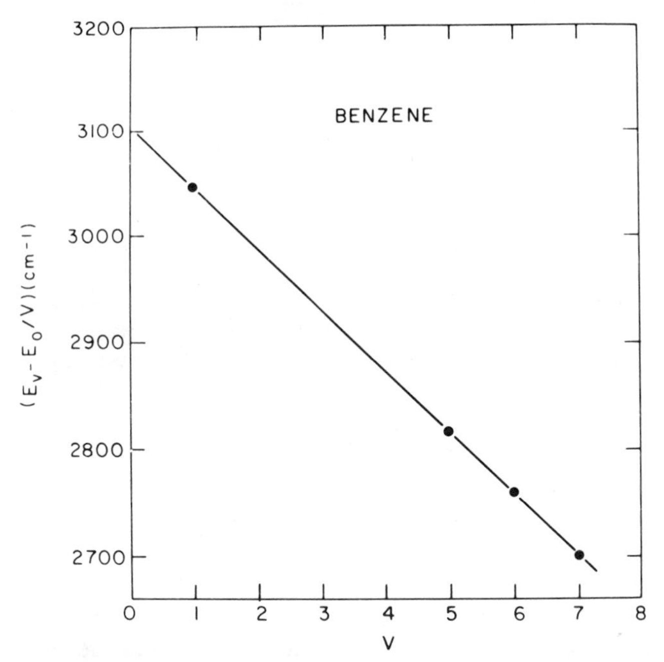

Fig. 2. Analysis of the peak energies of the C—H overtones of gaseous benzene according to relation (10). Experimental data from Bray and Berry (Ref. 7).

TABLE I
Energetics of C—H Bond Modes

Molecule	ω_e (cm^{-1})	x (cm^{-1})	C—H type[a]	Ref.
Benzene	3152	58	ar	26
	3150	58	ar	13
	3162	57.7	ar (gas)	7
Toluene	3142	57	ar	26
	3155	60	ar	13
	3030	58	al	26
	3018	55	al	13
p-Xylene	3139	59	ar	26
	3127	58	ar	13
	3028	59	al	26
	3024	55	al	13
1,2,4-Trimethyl benzene	3132	59	ar	26
Naphthalene	3142	55.8	ar (crystal)	8
Propane	3049	61	al-p	15
	3027	64	al-s	15
n-Butane	3053	62	al-p	15
	3035	68	al-s	15

TABLE I *(continued)*

Molecule	ω_e (cm^{-1})	x (cm^{-1})	C—H type[a]	Ref.
n-Pentane	3040	60	al-p	15
	3007	64	al-s	15
n-Hexane	3045	61	al-p	15
	3001	62	al-s	15
n-Heptane	3046	62	al-p	15
	3003	63	al-s	15
Cyclopentane	3061	63	al	17
	3049	67	al	17
Cyclohexane	3009	60	al	17
	2987	62	al	17
Cycloheptane	3003	62	al	17
Cyclooctane	3027	66	al	17
Cyclopentene	3057	67	al	17
	3163	59	ol	17
Cyclohexene	2996	60	al	17
	3111	59	ol	17
Cycloheptene	3016	64	al	17
	3121	63	ol	17

[a]ar, aromatic C—H; al, aliphatic C—H; ol, olefinic C—H; p, primary H; s, secondary H. All are neat liquid unless otherwise indicated.

TABLE II
Experimental Morse Parameters for C_6H_6 and C_6D_6 (Ref. 3)[a]

	Experimental		Predicted
	C_6H_6	C_6D_6	C_6D_6
ω_e	3153 cm^{-1}	2337	2314
x	58.4 cm^{-1}	31.7	31.4

[a]Predicted parameters for C_6D_6 assuming same Morse potential. Reduced mass equal to that of diatomic C—H and C—D.

for ω_e is somewhat smaller than predicted, indicating that the C—D local-modes couple quite effectively to the remaining vibrational modes. The effects of deviation of the C—D vibrations from the local bond-mode picture is exhibited more dramatically in the failure of a single bond-mode dipole function to predict the deuterium isotope effect on the intensity of vibrational overtones, a problem that we shall discuss in Section IV.

Recently Lawton and Child[39] carried out a classical trajectory calculation for the stretching vibrations of the water molecule using a realistic potential model.[40] They find local-mode behavior for most trajectories with energy greater than or equal to the energy of the second excited vibrational state. This semiclassical calculation provides further support for the local-mode picture in highly excited states of molecules containing X—H bonds.

IV. INTENSITIES

Albrecht and co-workers[26] have recently presented a detailed experimental study of the intensity of C—H stretching local mode overtone spectra for a variety of hydrocarbons. They found that, for a given overtone, all C—H local modes contributed equally to the intensity. The intensity per C—H bond (in units of liters mol^{-1} cm^{-2}) is 2.2 ± 0.4 for the 0–3 overtone, 0.18 ± 0.04 for the 0–4 overtone, $(2.3 \pm 0.3) \times 10^{-2}$ for the 0–5 overtone and $(3.4 \pm 0.6) \times 10^{-3}$ for the 0–6 overtone. Data for individual compounds is shown in Table III. Although there are some minor

TABLE III
Intensity[a] per Hydrogen (Ref. 26a)

	0→3		0→4	
	Aromatic	Aliphatic	Aromatic	Aliphatic
Benzene	2.2		0.18	
Toluene	2.0	2.0	0.17	0.16
o-Xylene	2.5	1.8	0.15	0.25
m-Xylene	2.4	1.9	0.17	0.16
p-Xylene	2.2	1.9	0.18	0.16
1,2,4-Trimethylbenzene	3.1	1.8	0.21	0.15
Isooctane[b]		1.8		0.17
n-Hexane		2.3		0.21
Cyclohexane		2.7		0.26
Average	2.4 ± 0.4	2.0 ± 0.3	0.18 ± 0.02	0.19 ± 0.04
	0→5		0→6	
Benzene	2.3×10^{-2}		3.5×10^{-3}	
Toluene	2.4	2.4×10^{-2}	3.0	3.1×10^{-3}
o-Xylene	2.0	2.0	—[c]	—[c]
m-Xylene	2.1	1.9	3.6	3.6
p-Xylene	2.0	2.0	4.0	3.5
1,2,4-Trimethylbenzene	2.1	2.1	3.7	1.8
Isooctane[b]		2.1		2.9
n-Hexane		2.8		3.7
Cyclohexane		3.0		3.6
Average	2.1 ± 0.2	2.3 ± 0.4	3.6 ± 0.43	3.2 ± 0.6

[a] Units are liters $mole^{-1}$ cm^{-2}.
[b] Only the peak from the primary hydrogen is included.
[c] Data unavailable.

differences between aromatic and aliphatic hydrogens, the intensities are remarkably constant. This pattern is in marked contrast to the situation for the fundamental C—H transition in which aliphatic hydrogens contribute four times the intensity of olefinic or aromatic hydrogens.[41] These results provide compelling evidence for the validity of the local bond-mode picture.

The calculations of the intensities requires the eigenstate $|v\rangle$ of the Morse oscillator, which is given in terms of the associated Laguerre polynomial

$$|v\rangle = N_v Y^{C-v-1/2} L_v^{2(C-v)-1}(Y) \exp\left(-\frac{Y}{2}\right) \tag{12}$$

with

$$Y = 2C \exp\left[-a(R - R_{\text{eq}})\right] \tag{12a}$$

$$C = \left(\frac{D}{x}\right)^{1/2} = \frac{\omega_e}{2x} \tag{12b}$$

$$N_v = \left[\frac{v!(2C-2v-1)}{(2C-v-1)!}\right]^{1/2} \tag{12c}$$

The intensity I_{0-v} of the 0–v transition, apart from irrelevant proportionality constants, is given by

$$I_{0-v} = (E_v - E_0)|\langle 0|\mu(R)|v\rangle|^2 \tag{13}$$

where $\mu(R)$ is the electric dipole moment operator that depends parametrically on the nuclear C—H bond coordinate. In what follows we shall utilize the dipole length representation rather than the dipole velocity representation for the calculation of the vibrational intensities, following the recipe of Mead and Moskowitz.[42] To proceed with the analysis of the intensities of the local C—H bond modes, one needs information concerning the functional form of $\mu(R)$. On the basis of general arguments it is apparent that $\mu(R)\to 0$ when $R\to 0$ and when $R\to\infty$. It is interesting to inquire whether one can obtain a universal dipole moment function for X—H bond modes. Burberry and Albrecht[43] and Schek et al.[44] addressed the problem of the dependence of the dipole moment on the internuclear distance for C—H bond modes. Burberry and Albrecht[43] used a simple quadratic dipole function

$$\mu(R) = M_1(R - R_{\text{eq}}) + M_2(R - R_{\text{eq}})^2 \tag{15}$$

which just corresponds to a Taylor expansion of $\mu(R)$ around the equilibrium configuration. They find the parameters $M_1 = -0.81$ debye Å^{-1} and $M_2 = 0.20$ debye Å^{-2} for the Morse oscillator wavefunctions, or alternatively, $M_1 = -0.59$ debye Å^{-1} and $M_2 = -1.16$ debye Å^{-2} for the quartic oscillator wavefunctions, to fit their experimental intensity data for a single C—H bond. Drastically different M_1 and M_2 parameters are required to predict the intensities of C—D local modes. This is not surprising since the C—D local modes are more strongly coupled to the skeletal states. Schek et al.[44] used the exponential dipole function:

$$\mu(R) = KR\exp\left(-\frac{R}{R^*}\right) \tag{16}$$

where K is a constant while R^* marks the maximum of $\mu(R)$. The dipole function (16) exhibits the proper asymptotic behavior at $R=0$ and at

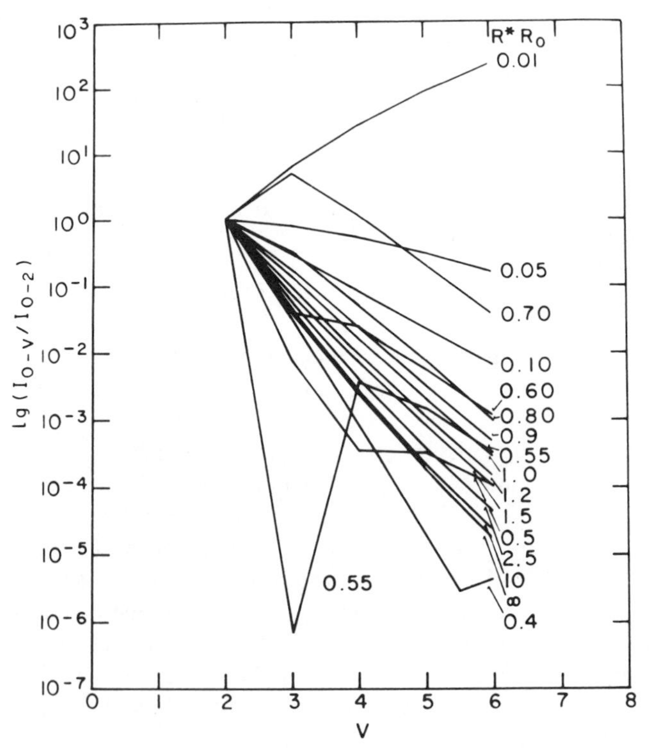

Fig. 3. Model calculations of the relative intensities normalized to the 0–2 transition for a C—H bond mode. The dipole function, eq. (15), is characterized by various values of R^*/R_{eq} as indicated. The linear approximation corresponds to $R^*/R_{eq} = \infty$.

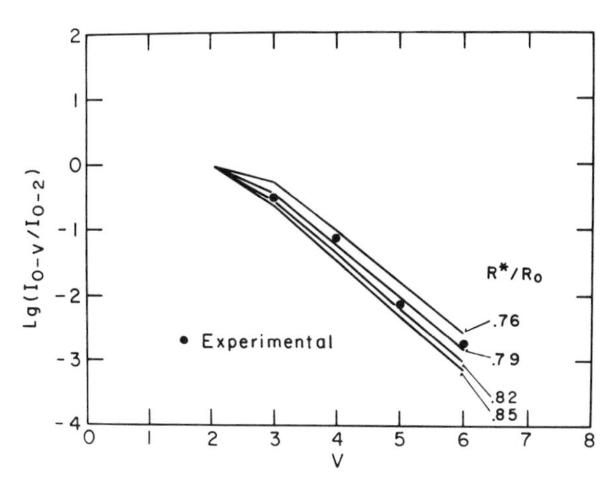

Fig. 4. Fit of the experimental intensity data for C—H bond modes in crystalline naphthalene (Ref. 8) with the results of model calculations based on the exponential dipole function, eq. (15), with $R^*/R_{eq}=0.75-0.85$. Best fit is obtained for $R^*/R_{eq}=0.79$.

$R=\infty$. Analytic matrix elements of this dipole function are known in a Morse oscillator basis set.[45] The relative intensities of the various overtone transitions is a strong function of R^*/R_{eq}. Figures 3 and 4 portray dependence of the intensity of the $0\to v$ overtone relative to the $0\to 2$ transition as a function of R^*/R_{eq} for Morse parameters corresponding to the C—H stretch in an aromatic hydrocarbon. A value of $R^*/R_{eq}=0.79$ is required to reproduce the experimental intensity data for crystalline naphthalene. The same exponential dipole function with $R/R^*=0.75$ can be used to reproduce the results of the theoretical calculations of Lie, Hinze, and Liu[46] for the dipole function of the CH radical in its ground $(X^2\pi)$ state for $0.4<R/R_{eq}<1.15$, which is in good agreement with the value of this parameter found from the analysis of the intensity data for C—H bond modes. The universal intensities of Albrecht et al.[26] fall off somewhat more rapidly with increasing v than the experimental data of Perry and Zewail,[8] as is apparent from Table IV. The reason for this discrepancy between the liquid data for a variety of aromatic hydrocarbons and the crystal data for naphthalene is not understood. It is quite unlikely that vibrational exciton effects and crystal field effects will affect the intensities of crystalline naphthalene overtones, and further work is required to resolve this inconsistency.

From the foregoing discussion of the spectroscopy of the high C—H vibrational overtones in terms of local modes, it is apparent that the simple model Hamiltonian, H_0 (5), is adequate for a reasonable description of the energy levels for these high-energy C—H excitations. These individual

TABLE IV
Relative Intensities

	Experimental naphthalene (Ref. 8)	Universal intensities for aromatic and aliphatic hydrocarbons (Ref. 26a)	Calculated from the exponential dipole function (15) with $R^*/R_{eq}=0.79$ (Ref. 44)
$\log \frac{I_{0\to 3}}{I_{0\to 4}}$	0.64	1.09	0.79
$\log \frac{I_{0\to 4}}{I_{0\to 5}}$	1.00	0.89	0.82
$\log \frac{I_{0\to 5}}{I_{0\to 6}}$	0.58	0.83	0.79

local mode states are only weakly coupled and the dipole moment consists of a superposition of separate contributions for each bond-dipole term. This concept of additivity of contributions from individual X—H bond modes provides a reasonable and internally consistent zero-order description of the C—H excitations, but not of C—D overtones. In the latter case, the effects of kinetic energy scrambling begin to become noticeable, as expected. The simple, but quite adequate, description of the spectroscopy of C—H bond modes, which rests on the Hamiltonian (5) will be extended in two directions. First, we shall consider in Section V the weak but systematic dependence of the energy of C—H stretches on the intramolecular environment, which lead to the description of "dressed" Morse oscillator for these bond modes. Subsequently, in Section VI we shall discuss the energetics and intensities of cooperative excitations, which may provide interesting information regarding the (weak) coupling between individual bond modes.

V. INTRAMOLECULAR EFFECTS ON ENERGETICS

Local mode energy levels for C—H stretches depend to a limited extent on the intramolecular environment of the C—H bond. For the series benzene, toluene, *p*-xylene and 1,2,4-trimethyl benzene,[26] there is a monatonic decrease in ω_e from 3153 to 3132 cm^{-1} with increasing methylation, as is apparent from Table I. On the other hand, x remains constant at 58.6 ± 0.6 cm^{-1}. A slight decrease in C—H bond strength, the parameter D for the Morse oscillator, with increasing numbers of methyl groups accounts for the observed changes in ω_e and still leads to negligible changes in anharmonicity. The methyl hydrogens in molecules with non-neighboring methyl groups are characterized by $\omega_e = 3036 \pm 15$ cm^{-1} and

$x = 59.0 \pm 2.0\ \text{cm}^{-1}$. The energetic parameters for methyl hydrogen stretches in liquid alkanes are virtually identical with those of aromatic methyls.[17, 26] In molecules containing neighboring methyl groups intermethyl interactions result in a 200 cm^{-1} splitting of the absorption band for the $0 \to 5$ transitions.[26] In crystalline naphthalene the $0 \to 5$ transition consists of two overlapping Lorentzians split by about 100 cm^{-1}, originating from the α and β hydrogens.[8] The C—H local bond parameters for naphthalene are similar to those of benzene (see Table I). A similar effect is exhibited by the energetic splitting of the excitation between primary and secondary aliphatic hydrogens.

Recent experimental studies addressed the central problem of the applicability of the bond-mode picture. Henry and Hung[14] studied the effect of mass on the validity of local mode description of high overtone transitions in dihalomethanes. They find that for dichloro-, dibromo-, and diodomethane the local mode description is adequate for the $0 \to 3$ through to $0 \to 6$ overtones. In difluoromethane the $0 \to 3$ transition is correctly described in terms of normal coordinates, the $0 \to 5$ agreeing with the local-mode picture, while the $0 \to 4$ is somewhere in between. This unique behavior of difluoromethane is rationalized by proposing that the smaller mass of fluorine allows for a stronger kinetic energy coupling of the C—H bond modes in CH_2F_2 than in the other dihalomethanes. Similarly, Lawton and Child[39] find in their classical trajectory calculation that the stretching vibrations are less well described in terms of local modes in D_2O than they are in H_2O. Finally, it is worthwhile to point out that Greenlay and Henry[15] observed some transitions which were assigned by them to the combination of local-mode and skeletal-mode excitations.

Sage[47] has investigated the effects of other motions of the C—H bond on the observed local-mode transition energies. He considered the in-plane bends and the out-of-plane bends β and α, and the C—H stretch, s, in C_6D_5H. The diagonal G-matrix elements are

$$G_{ss} = \mu_H + \mu_C \tag{16a}$$

$$G_{\beta\beta} = \frac{\mu_H + \mu_C}{s^2} + \frac{\mu_C}{st} + \frac{3\mu_C}{4t^2} \tag{16b}$$

$$G_{\gamma\gamma} = \frac{\mu_H + \mu_C}{s^2} + \frac{4\mu_C}{st} + \frac{6\mu_C}{t^2} \tag{16c}$$

while the only nonvanishing off-diagonal element is

$$G_{s\beta} = \frac{3^{1/2}\mu_C}{4t} \tag{17}$$

where μ_H and μ_C are the reciprocals of the mass of the H and C atoms, respectively, and t is the equilibrium length of the C—C bond. For the high overtone spectrum of the C—H stretching mode all other vibrations can be treated in terms of small amplitude motions.

The full vibrational Hamiltonian for the C—H bond is now

$$H_{CH}=\frac{G_{\beta\beta}p_\beta^2}{2}+\frac{G_{\gamma\gamma}p_\gamma^2}{2}+\frac{G_{ss}p_s^2}{2}+G_{\beta s}p_\beta p_s+\frac{k_\beta\beta^2}{2}+\frac{k_\gamma\gamma^2}{2}$$
$$+D\{\exp[-2a(s-s_{eq})]-2\exp[-a(s-s_{eq})]\} \qquad (18)$$

The in-plane and out-of-plane bending force constants are k_β and k_γ, respectively, while D and a are parameters of the Morse potential describing the C—H bond and s_{eq} is the equilibrium value of the C—H bond length. The Hamiltonian (18) is not separable due to the s dependence of the G-matrix elements for the bends and the term involving $G_{s\beta}$ (17). Nevertheless, an approximate solution was represented in terms of a product of three vibrational wavefunctions. Since the bending vibrations are characterized by a frequency that is low compared to the stretching frequency, an adiabatic type approximation for the wavefunction for the stretching motion was obtained ignoring the bending motion. These wavefunctions were then used to find effective G-matrix elements for the bends. These matrix elements decrease with increasing local-mode state v and are 30 and 24% smaller than their equilibrium values at $v=10$ for the in-plane and for the out-of-plane bends. The local-mode transition occurs at an energy that includes the direct energy difference for the C—H stretching states in addition to the changes in zero-point energy for the two bends. The frequency of the in-plane and out-of-plane bends shift from 1202 cm^{-1} and 1025 cm^{-1} to 1018 cm^{-1} and 900 cm^{-1}, as v increases from 0 to 10, resulting in zero-point energy shifts of up to 154 cm^{-1} for the C—H motion. The observed energy levels for the C—H stretching (Table V) can

TABLE V
Morse Parameters for C—H Stretch:
"Dressed" Parameters for C_6H_6 and C_6D_5H
Parameters for Isolated C—H in C_6D_5H

	"Dressed" parameter		Isolated C—H
	C_6H_6[a]	C_6D_5H[a]	C_6D_5H[a]
x cm^{-1}	57.7	58.7	58.5
D cm^{-1}	4.33×10^4	4.31×10^4	4.33×10^4
a A^{-1}	1.86	1.87	1.87

[a] The equilibrium bond length for C—H and C—C was assumed to be 1.07 and 1.40 Å, respectively.

be described by a "dressed" Morse oscillator, which is slightly more anharmonic than the oscillator used to describe the "isolated" C—H bond.

VI. COOPERATIVE EXCITATIONS OF BOND MODES

The additivity concept for individual C—H bond modes leads to a propensity rule for optical excitation of a molecule containing several C—H bonds, implying that the vibrational excitation will be exhibited within a single C—H bond. For a given degree of vibrational excitation to a vibrational state $|v_1 v_2 v_3 \ldots\rangle$, where v_j represents the excitation of the jth C—H local mode, the optically accessible state is $|v,00\ldots\rangle$ with $v=\Sigma_j v_j$, being characterized by the greatest anharmonicity defect and by the lowest energy. For example, consider the $v=6$ C—H excitations in C_6H_6. There are 462 distinct vibrational states spanning a broad energy range from 16,550 cm^{-1} for the $|6,0^5\rangle$ state to 18,280 cm^{-1} for the $|1^6\rangle$ state. The simple uncoupled model described by the Hamiltonian (5) predicts optical excitation from the $|0^6\rangle$ ground state to a single $|6,0^5\rangle$, E_{1u} type, vibrationally excited state. Indeed, the $|0^6\rangle \rightarrow |6,0^5\rangle$ transition is the dominant spectral feature in that energy range. Recently, additional weak transitions, in which the vibrational excitations v_1 and v_2 of two bond modes are simultaneously excited, have been experimentally observed. In isooctane, where the $0 \rightarrow 6$ transition of the C—H bond mode of the primary hydrogen peaks occurs at 15,700 cm^{-1}, a weak absorption band at 16,300 cm^{-1} has been assigned to the $|0^3\rangle \rightarrow |5,1,0\rangle$ transition,[26] while another weak absorption band at 16900 cm^{-1} was tentatively assigned to a $|0^3\rangle \rightarrow |3^2,0\rangle$ transition[26] (Fig. 5). The origin of these combination bands, which correspond to one-photon cooperative excitation of two bond modes, is of considerable interest as they provide experimental evidence for the breakdown of the additivity concept for excitation of C—H bonds. The understanding of the nature of these cooperative excitation has to surpass the additivity concept and may result in interesting information regarding the intramolecular coupling between bond modes.

We shall now proceed to consider the theory of cooperative excitations $|0,0\rangle \rightarrow |v_1, v_2\rangle$ in a system containing two equivalent C—H (or other X—H) bonds characterized by the nuclear coordinates R_1 and R_2, respectively. The cardinal question pertains to the origin of the intensity of such cooperative excitation. Two mechanisms will be considered:

1. Contribution of cross-terms in the dipole moment operator. Burberry and Albrecht[26] suggested that the separability assumption for the dipole moment has to be extended to include cross-terms. Thus, the

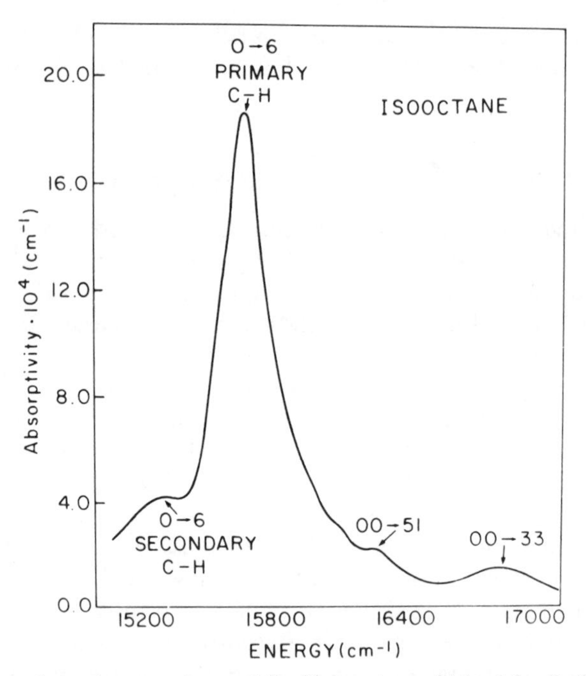

Fig. 5. The $v=6$ overtone spectrum of liquid isooctane obtained by Burberry, Morrel, Albrecht, and Swofford reproduced from Ref. 26. The positions of the $|0\rangle \rightarrow |3,3\rangle$ and $|0\rangle \rightarrow |5,1\rangle$ cooperative excitations are marked by arrows.

dipole moment function $\mu(R_1, R_2)$ for the system is

$$\mu(R_1 R_2) = \mu(R_1) + \mu(R_2) + \sum_{n,m=1} M_{mn}(R_1 - R_{1\,\mathrm{eq}})^m (R_2 - R_{2\,\mathrm{eq}})^n \tag{19}$$

where $\mu(R_j)$; $j = 1,2$ are the dipole moment functions of the individual bonds, while the double sum with constant coefficients M_{nm} represent the appropriate cross-terms. The intensity of the $|00\rangle \rightarrow |v_1 v_2\rangle$ transition is

$$I_{00 \rightarrow v_1 v_2} = (E_{v_1 v_2} - E_0) \cdot \left| \sum_{nm} M_{mn} \langle 0 | R_1 - R_{1\,\mathrm{eq}})^m | v_1 \rangle \langle 0 | (R - R_{\mathrm{eq}})^n | v_2 \rangle \right|^2 \tag{20}$$

which is determined by the cross-terms in the dipole expansion, which are at present unknown.

2. Simultaneous interband and radiative coupling resulting in an intramolecular cooperative excitation. Sage and Jortner proposed[28] an alternative mechanism for the origin of the intensity of the cooperative excitation, which essentially rests on the notion of (weak) coupling between the local bond modes, so that the coupled system is subjected to radiative interactions. Such cooperative excitations, where a single photon simultaneously excites two weakly coupled systems, provide the molecular analog of the phenomenon of bremsstralung radiation, where an electron is simultaneously subjected to coulomb and radiative interactions. Intermolecular cooperative excitations are known in the area of electronic spectroscopy, for example, the simultaneous excitation $O_2(^3\Sigma_g)O_2(^3\Sigma_g) \rightarrow O_2(^1\Delta_g)O_2(^1\Delta_g)$ of a pair of oxygen molecules[48] resulting in the blue color of liquid oxygen. In solid-state spectroscopy, cooperative electronic excitations of a pair of rare-earth ions in a crystal were recorded.[49] In the field of infrared spectroscopy the cooperative vibrational excitation HCl ($v=0$) HCl ($v=0$)$\rightarrow$HCl ($v=1$) HCl ($v=1$) of a pair of molecules in solid HCl was studied.[50, 51] The $|0^3\rangle \rightarrow |5,1,0\rangle$ and $|0^3\rangle \rightarrow |3^2,0\rangle$ weak transitions in isooctane[26] (Fig. 5) are assigned by us to a one-photon intramolecular cooperative excitation of two bond modes. These zero-order anharmonic bond modes, correspond to the eigenstates of H_0^L (6), with the intramolecular intermode coupling being exerted by the perturbation W (8). The intermode coupling, which can be described in terms of a Fermi resonance between anharmonic bond modes provides the intramolecular interaction, which induces the cooperative excitation. A simplified energy level scheme exhibiting the simultaneous effects of intramolecular interaction and radiative coupling is portrayed in Fig. 6. The transition probability $T_{00 \rightarrow v_1 v_2}$ for the cooperative excitation $|00\rangle \rightarrow |v_1 v_2\rangle$ is obtained within the framework of second-order perturbation theory. The dominant intermode coupling is expected to prevail between the states $|v0\rangle \rightarrow |v_1 v_2\rangle$ and $|0v\rangle \rightarrow |v_1 v_2\rangle$. These large intramolecular terms combine with relatively small radiative coupling $\mu\epsilon$ terms, where ϵ is the strength of the electromagnetic field. On the other hand, lower intermediate states $|0v_i\rangle$ with $v_i < v$ are characterized by a weaker intermode coupling but by a stronger radiative interaction. The transition probability to second-order is

$$T_{00 \rightarrow v_1 v_2} = 2 \sum_{v_j=1}^{v} \frac{\langle 00|\mu\epsilon|v_j 0\rangle \langle v_j 0|W|v_1 v_2\rangle}{\Delta E_j} \tag{21}$$

with the energy gaps

$$\Delta E_j = E_{v_1 v_2} - E_{v_j} \tag{22}$$

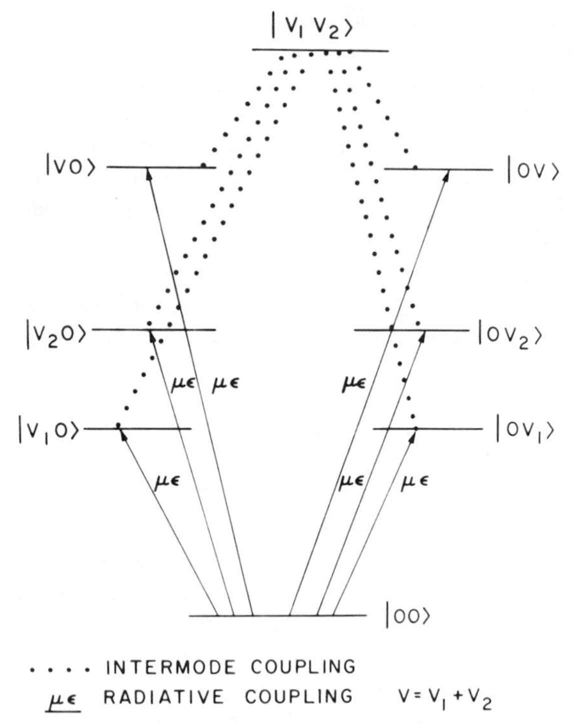

Fig. 6. Energy scheme for the simultaneous effects of intramolecular interbond coupling and radiative interaction.

the intensity for the cooperative transition being

$$I_{00\to v_1v_2}=(E_{v_1v_2}-E_0)|T_{00\to v_1v_2}|^2 \tag{23}$$

As a rough estimate we shall pick from (21) the contribution of the leading intermode term, and (23) yields

$$I_{00\to v_1v_2}\simeq 4(E_{v_1v_2}-E_0)I_{0\to v}\frac{(\langle W\rangle/\Delta E)^2}{(E_v-E_0)} \tag{24}$$

with $\langle W\rangle=\langle v0|W|v_1v_2\rangle$ and $\Delta E=2v_1v_2x$. The relative intensity of the cooperative excitation relative to the conventional one-mode excitation is $r_{v_1v_2}=I_{00\to v_1v_2}/I_{0\to v}\simeq 4(\langle W\rangle/\Delta E)^2$. For $r_{v_1v_2}=0.01$ we estimate $(\langle W\rangle/\Delta E)^2 \sim 2\times 10^{-3}$. Taking $v_1=v-1$ and $v_2=1$ with $\Delta E=2v_1x$, then for $v=6$ one gets $\langle W\rangle\simeq 5\ \text{cm}^{-1}$, which is a reasonable value for intermode coupling due to kinetic energy and anharmonicity effects. Thus, moderate intermode coupling can induce the cooperative excitation. The rough estimate of $\langle W \rangle$ obtained from this analysis will be relevant for the understanding of the spectral lineshape of high vibrational C—H overtones in "isolated" molecules, which will be the subject matter of Section VIII.

Finally, we would like to emphasize that there is no dichotomy between the two mechanisms which induce the intensity of cooperative transition, that involving cross terms in the dipole moment operator[26] and that which corresponds to simultaneous intermode and radiative coupling.[28] In fact, both mechanisms may peacefully coexist, contributing to the observation of simultaneous excitations of pairs of C—H bond modes.

VII. INTRAMOLECULAR DYNAMICS OF MOLECULES INVOLVING X—H BONDS

The zero-order description of C—H bond modes appears to provide a useful and general conceptual framework for spectroscopic purposes. A theory of intrastate, intramolecular, nuclear dynamics of a polyatomic molecule within the framework of quantum-mechanical relaxation theory cannot get away with just a description of zero-order spectroscopic states and one has to consider the following additional features:

1. The specification of near-resonant and off-resonance intrastate coupling, which provides the small residual interactions responsible for intramolecular time evolution of nonstationary metastable states.
2. The nature of the "preparation" of initial nuclear metastable states of the system by optical or other means of excitation.

We shall first address the problem of intrastate coupling. The partition of the vibrational Hamiltonian given above in (4) allows for zero-order wavefunctions $|\mathbf{v}, \xi\rangle = |\mathbf{v}\rangle |\xi\rangle$, where $|\mathbf{v}\rangle$ is a local mode eigenfunction and $|\xi\rangle$ is a skeletal mode state. Figure 7 presents a schematic energy level diagram for the zero-order states of a hydrocarbon. There are three relevant types of coupling terms, $\langle \mathbf{v}, \xi | W | \mathbf{v}', \xi' \rangle$,

a. Off-resonance interactions between $|\mathbf{v}, \xi\rangle$ and $|\mathbf{v}', \xi\rangle$ corresponding to different local-mode states and identical skeletal states.
b. Off-resonance interactions between $|\mathbf{v}, \xi\rangle$ and $|\mathbf{v}, \xi'\rangle$ corresponding to the same local-mode states and different skeletal states.
c. Resonance (and nonresonance) interactions between $|\mathbf{v}, \xi\rangle$ and $|\mathbf{v}', \xi'\rangle$ corresponding to completely different states.

Two theories were advanced[38, 52] to describe intrastate coupling, which differ in their emphasis. Heller and Mukamel[52, 53] focus on type (a)

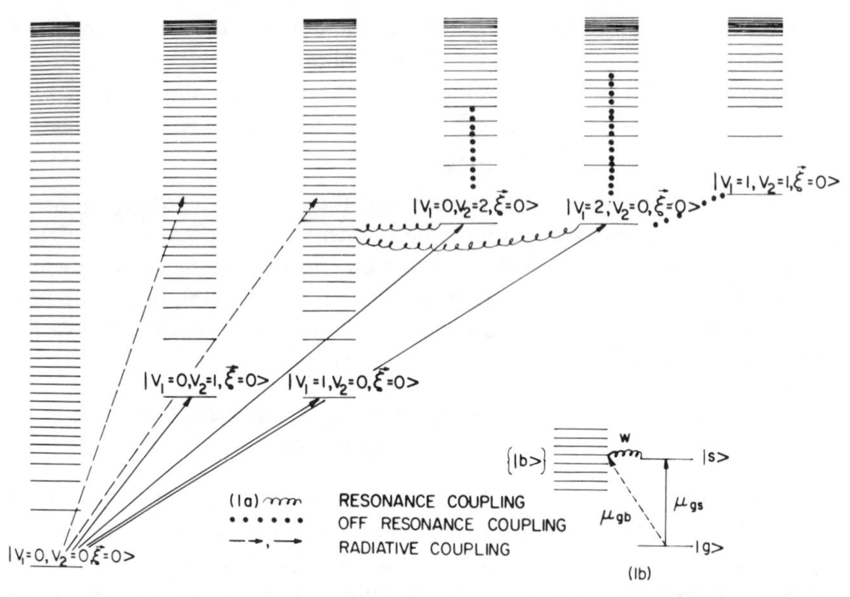

Fig. 7. Vibrational level structure on the ground state potential surface of a large molecule. Figure 7a provides an artist's view of the zero-order states of a molecule containing two X—H bonds. Figure 7b presents the level scheme for the Fano problem, with the radiative coupling and the intramolecular resonance coupling W designating random coupling terms.

off-resonance coupling, which Sage and Jortner neglect. Sage and Jortner[38] assume that type (c) interactions lead to the main linewidth contributions.

Sage and Jortner[38] dismiss type (b) off-resonance coupling by prediagonalizing each local-mode manifold. In any event, skeletal modes for each local-mode manifold may be slightly different, since skeletal G-matrix elements depend on local mode coordinates.[47] Intermanifold couplings (a) will result in level shifts and, more importantly, lead to some additional line broadening. If the states coupled by mechanism (a) are well separated in energy, the contribution of these coupling effects relative to the contribution of near-resonance interactions (c) will be of the order $|\langle \mathbf{v}, \boldsymbol{\xi}|W|\mathbf{v}', \boldsymbol{\xi}\rangle/\Delta E$, where ΔE is a typical energy gap. This contribution will be small for high excitations of the C—H bonds. For highly excited nuclear states of molecules like benzene, in which there may be many close-lying local-mode manifolds, the effects cannot be large, since deuteration drastically affects the number of local-mode states.

The zero-order level structure (Fig. 7) consists of a set of nuclear manifolds $|\mathbf{v}, \boldsymbol{\xi}\rangle$. The lowest-lying state in each manifold is $|\mathbf{v}, \boldsymbol{\xi}=0\rangle$, which is well separated in energy (by the energy corresponding to the lowest characteristic molecular frequency) from the other states in this manifold. Each $|s\rangle \equiv |\mathbf{v}, \xi=0\rangle$ state, where $\mathbf{v}$ is sufficiently high, is quasi-degenerate with skeleton excited states of the manifolds $|b\rangle = |\mathbf{v}', \xi\rangle$, where all the origins $|\mathbf{v}', \xi=0\rangle$ of these manifolds are located below $|s\rangle$. The resonance intrastate coupling between nuclear levels on the ground state potential surface, which is induced by kinetic energy and potential anharmonic interactions, bears a close analogy to nonadiabatic interstate coupling between two electronic configurations in polyatomic molecules.[54, 55] Using the conventional approach[54, 55] to the problem of intrastate coupling, we subdivide the features of nuclear dynamics in molecules containing C—H bonds into two categories:

1. The statistical limit: The intrastate s–b coupling W_{sb} is strong, $W_{sb}\rho_b \gg 1$, where ρ_b is the (average) density of background states. These background states are not resolvable relative to their widths, γ_b, which originate for infrared decay so that $\gamma_b \rho_b > 1$. As $\gamma_b \sim 10^{-8}$ cm^{-1}, one expects the statistical limit to be exhibited for $\rho_b > 10^8$ cm.
2. The small molecule limit: Now strong intrastate coupling prevails, i.e., $W_{sb}\rho_b \gg 1$, but the background states are well resolved in energy relative to their widths, i.e., $\gamma_b \rho_b < 1$. The resulting mixed molecular eigenstates will be individually active in absorption.

It is important to realize that the small molecule limit and the statistical limit for intrastate coupling can be realized within the same molecule at different energy regions. At low energies the level structure is sparse and C—H bond modes may exhibit Fermi resonances resulting in effective intrastate scrambling. At higher energies for "large" polyatomic molecules the statistical limit is realized. Intrastate scrambling is exhibited both in the small molecule case and in the statistical limit, while intramolecular relaxation of a C—H bond mode state to the skeleton can be considered, in principle, only in the statistical limit. This concept of intramolecular intrastate relaxation is meaningful provided that a C—H bond state can be selectively excited and dynamic observables can be interrogated either by "short-time" and by "long-time" experiments. Important information regarding this problem stems from the absorption lineshapes of C—H overtones at low pressure, which we shall now discuss.

VIII. ABSORPTION LINESHAPES

The experiments of Bray and Berry[6, 7] (Fig. 8) on the absorption high C—H overtones of low-pressure gaseous benzene reveal the following features relevant to intramolecular dynamics:

1. Nearly Lorentzian lineshapes.
2. The Lorentzian widths for the overtone quantum numbers $v=5$, 6, and

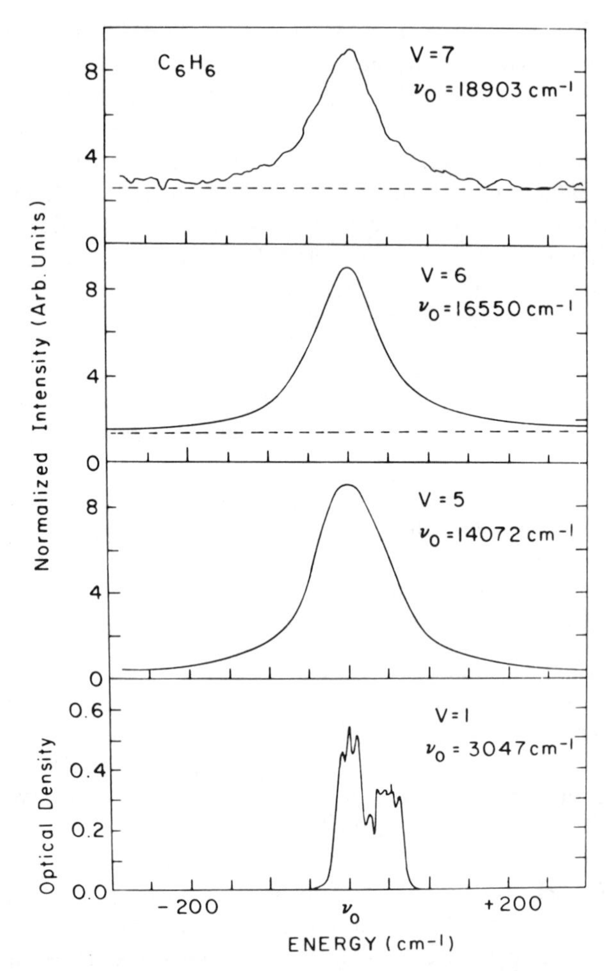

Fig. 8. Optical absorption lineshapes for the $v = 5$–7 vibrational overtones of gaseous benzene taken from the work of Bray and Berry (Ref. 7).

7 are 90–110 cm^{-1}, which considerably exceed the widths due to rotational structure.

3. The C—H stretching overtone bands are superimposed on a continuum of background absorption, whose intensity relative to the Lorentzian intensity increases with increasing v. Current measurements of the intensity of the continuum absorption are still inaccurate.[6, 7, 56] On the basis of a comparison of the benzene overtone spectra with other systems, characterized by a sparse density of states, Berry[56] concludes that the continuum absorption is real.

We shall now outline the theory of absorption lineshapes advanced by Sage and Jortner.[38] The notion of intramolecular relaxation of a C—H bond mode state to the skeleton background states constitutes a meaningful concept provided that the $|s\rangle$ state, which is well isolated in energy, carries oscillator strength from the ground state $|g\rangle = |\mathrm{v}=0, \xi=0\rangle$, while the background $|b\rangle$ states, which are quasidegenerate with $|s\rangle$ are inactive. One can argue that the intensity of higher overtones decreases exponentially with increasing vibrational quantum number. Then, in view of the high frequency of the C—H bond modes relative to the skeleton modes, one expects that $\mu_{gs} \gg \mu_{gb}$, where μ is the transition-moment operator. Nevertheless, the contribution of the μ_{gb} terms may be amenable to the experimental detection. We have examined the lineshape for the overtone spectra where both $|s\rangle$ as well as the background states $\{|b\rangle\}$ carry oscillator strength from $|g\rangle$. The Fano problem[57] for interference between resonance absorption and background absorption is well known and can be applied to the problem at hand.

The lineshape at zero temperature is

$$L(\omega) = -\frac{1}{\pi}\mathrm{Im}\langle g|\mu G(\omega)\mu|g\rangle \tag{25}$$

$$G(\omega) = (\hbar\omega - H + i\eta)^{-1} \qquad \eta \to 0^+ \tag{26}$$

For the system described in Fig. 7, one finds

$$L(\omega) = -\frac{1}{\pi}\mathrm{Im}\{G_{ss}[|\mu_{sg}|^2 + A_{sg}\mu_{gs} + \tilde{A}_{gs}\mu_{g} + A_{sg}\tilde{A}_{gs}]\} + \sum_b |\mu_{gb}|^2\delta(\hbar\omega - E_0) \tag{27}$$

where

$$A_{gs}(\omega) = \sum_b \frac{\mu_{gb}W_{bs}}{\hbar\omega - E_b + i\eta} \tag{28}$$

$$\tilde{A}_{sg}(\omega) = \sum_b \frac{W_{sb}\mu_{bg}}{\hbar\omega - E_b + i\eta} \tag{29}$$

In the present case of intrastate coupling of C—H and skeleton modes, random coupling effects with the background quasicontinuum will preclude the observation of any interference effects.[38] One expects the intramolecular near-resonance coupling terms W_{sb} and the radiative coupling terms μ_{gb}, to exhibit a wide and wild variation with respect to their magnitudes and signs. This particularly random variation of the intramolecular s–b and of the radiative g–b coupling terms is a general feature of the intramolecular vibrational quasicontinuum in large molecules. The corresponding configurational averages, $\langle\ \rangle$, would vanish, i.e., $\langle \mu_{gb} \rangle = 0$ and $\langle W_{sb} \rangle = 0$. Introducing the general correlation factor

$$f = \frac{\langle \mu_{gb} W_{bs} \rangle}{\left[\langle \mu_{gb}^2 \rangle \langle W_{bs}^2 \rangle \right]^{1/2}} \tag{30}$$

and performing the configurational averages reduces the interference factor (29) to

$$A_{gs}(\omega) = \sum_b \frac{\left[\langle \mu_{gb}^2 \rangle \langle W_{bs}^2 \rangle \right]^{1/2}}{\hbar\omega - E_b + i\eta} f \tag{31}$$

where the correlation factor is assumed to be independent of E_b.

Taking the radiative and the intramolecular matrix elements to be real and neglecting the principle part contribution yields the following expression for the lineshape function normalized to background absorption strength

$$L(\epsilon) = \frac{(1 - f^2) + q_0^2 + 2q_0 f\epsilon + \epsilon^2}{1 + \epsilon^2} \tag{32}$$

Here

$$\epsilon = \frac{2(\hbar\omega - E_s)}{\Gamma_s} \tag{33}$$

E_s is the energy of $|s\rangle$, while

$$\Gamma_s = 2\pi \langle W_{sb}^2 \rangle \rho_b \tag{34}$$

is the width of the zero-order state and Fano's interference factor is given by

$$q_0 = \frac{\mu_{gs}}{\pi \left[\langle \mu_{gb}^2 \rangle \langle W_{bs}^2 \rangle \right]^{1/2} \rho_b} \tag{35}$$

Under the conditions of completely random coupling $f=0$ and the lineshape becomes

$$L(\epsilon)=1+\frac{q_0^2}{1+\epsilon^2} \tag{36}$$

This result represents the superposition of background absorption and a Lorentzian due to the resonance. Completely random coupling erodes all interference effects, and there is peaceful coexistence between background absorption and resonance absorption. The ratio r between the intensity of the background absorption and the peak intensity of the Lorentzian (at $\epsilon=0$) is given by $r=q_0^2$. Thus a careful measurement of r will provide pertinent information regarding the product of the ratio of the transition moments $|\mu_{gs}/\langle\mu_{gb}^2\rangle^{1/2}\rho_b^{1/2}|$ and the normalized coupling strength $\langle W_{bs}^2\rangle^{1/2}\rho_b^{1/2}$. An independent measurement of the Lorentzian halfwidth, Γ_s, correcting for rotational broadening in the gas phase, results in $\langle W_{bs}^2\rangle\rho_b$. Thus $r\Gamma_s=\frac{2}{\pi}|\mu_{gs}^2/\langle\mu_{gb}^2\rangle\rho_b|$ yielding the ratio of the transition moments. Accurate values of r are not available at present.[56] Such data will be of considerable interest for the determination of the ratio of the transition moments, which will lead to pertinent information regarding the radiative coupling of $|g\rangle$ with the background quasicontinuum.

The width, Γ_s, of the Lorentzian, corrected for rotational broadening effects, represents the decay probability of a C—H bond mode, provided that in a hypothetical "Gedanken experiment" such a bond mode can be initially excited. Interference effects will be washed out by the effects of random coupling. Thus in a large molecule whose intrastate coupling corresponds to the statistical limit the decay of C—H (and other X—H) modes is characterized by a single time scale $\tau\sim\hbar/\Gamma_s$. This conclusion differs from that of Heller and Mukamel,[52, 53] who essentially considered two time scales for the problem. They have assumed[52] that radiative interactions essentially couple the system through local-mode states. In terms of the effects of off-resonance coupling they have attempted to account for the small (up to 20%) decrease in the Lorentzian absorption linewidth Γ_s for the C—H overtones of benzene and some deuterated benzenes when one passes from $v=5$ to $v=7$. We would like to emphasize that the truly surprising feature of the experimental data is not that small decrease of Γ_s but rather the weak variation of this linewidth with increasing v. Some semiquantitative estimates of off-resonance coupling terms are required in order to subject the proposal of Heller and Mukamel to a crucial test. Relevant information regarding the strength of intermode coupling may be obtained from the intensity of cooperative excitations utilizing the theory of Section VI. Unfortunately, such data are not yet available for the benzene molecule.

Sage and Jortner[29] have extended their model for two-photon excitation of different local-mode states. They find the excitation can transfer from the discrete local mode that is absorbing the radiation into the quasicontinuum. This is a natural way of looking at the crossover from the sparse molecular region I with coherent multiphoton coupling effects[58] into the dense[22b, 59] region II which is characterized by incoherent single photon pumping. Consider two-photon transitions from the ground state $|g\rangle$ to a final local-mode state $|f\rangle$ through an intermediate $|s\rangle$. The states $|s\rangle$ and $|f\rangle$ are coupled to quasicontinuum states $|b\rangle$ and $|c\rangle$, respectively, via type (c) resonance coupling. The analysis results in a complicated expression with many interference terms. Under the assumption of random coupling the two-photon lineshape reduces to[29]

$$L_2(\omega) = L_2^{LL}(\omega) + L_2^{LS}(\omega) + L_2^{SS}(\omega) \tag{37}$$

where the first term involves 2-photon transitions between local mode states, the second term involves the 1-photon coupling between local-mode states followed by 1-photon interaction between quasicontinuum skeletal mode states, while the third term involves 2-photon excitation from the ground state directly into the quasicontinuum.

What are the implication of the theory of absorption lineshapes for the problem of intramolecular vibrational redistribution in large molecules containing C—H bonds? Three comments should be made concerning the "initial" selective excitation of C—H bond modes in a real-life experiment. First, detailed experimental studies of the background absorption, together with the results of the present analysis, will provide evidence for the notion of random intramolecular and radiative coupling which was recently introduced[59–62] in the theory of multiphoton excitation of polyatomic molecules. Second, from the point of view of general theory, one-photon optical excitation from $|g\rangle$ will result in the simultaneous excitation of the C—H overtone resonance as well as background quasicontinuum states. The subsequent time evolution of the $|s\rangle$ state and of the $\{|b\rangle\}$ manifold is uncorrelated because of random coupling. However, strictly speaking, the "Gedanken experiment" of selective $|g\rangle \to |s\rangle$ excitation results in a contaminated excited manifold. In high-order multiphoton experiments where radiative coupling prevails between groups of states in the quasicontinuum, which are separated by laser frequency, the role of the background states $\{|b\rangle\}$ may be even more important. Third, from the practical point of view we note that as $\Gamma_b \sim 100\ \text{cm}^{-1}$ for high C—H overtones of benzene an initial excitation on the time scale shorter than $<10^{-14}$ sec is required for initial excitation of the (contaminated) C—H bond mode state. Currently available light sources can result only in the excitation of a small fraction of the energetic spread of the Lorentzian. A proper approach towards a meaningful physical description of the realistic photon induced excitation should be based on the nuclear molecular eigenstates (NMES) which constitute a superposition of the $|s\rangle$ and of the $\{|b\rangle\}$ states. These NMES will provide a useful theoretical tool for the understanding of the dynamic implications of one-photon excitation in the overtone range as well as of high-order multiphoton excitation of large molecules.

References

1. S. A. Rice, in *Excited States*, Vol. II, edited by E. C. Lim, Academic press, New York, 1975, p. 111.
2. M. E. Long, R. L. Swofford, and A. C. Albrecht, *Science*, **191**, 183 (1976).
3. R. L. Swofford, M. E. Long, and A. C. Albrecht, *J. Chem. Phys.*, **65**, 179 (1976).
4. R. L. Swofford, M. E. Long, M. S. Burberry, and A. C. Albrecht, *J. Chem. Phys.*, **66**, 664 (1977).
5. R. L. Swofford, M. S. Burberry, J. A. Morrell, and A. C. Albrecht, *J. Chem. Phys.*, **66**, 5245 (1977).
6. K. V. Reddy, R. G. Bray, and M. J. Berry in *Advances in Laser Chemistry*, edited by A. H. Zewail, Springer-Verlag, Berlin, 1978, p. 48.
7. R. G. Bray and M. J. Berry, *J. Chem. Phys.*, **71**, 4909 (1979).
8. J. W. Perry and A. H. Zewail, *J. Chem. Phys.*, **70**, 582 (1979).
9. B. R. Henry and W. Siebrand, *J. Chem. Phys.*, **49**, 5369 (1968).
10. R. J. Hayward, B. R. Henry, and W. Siebrand, *J. Mol. Spect.*, **46**, 207 (1973).
11. R. J. Hayward and B. R. Henry, *J. Mol. Spect.*, **50**, 58 (1974).
12. R. J. Hayward and B. R. Henry, *J. Mol. Spect.*, **57**, 221 (1975).
13. R. J. Hayward and B. R. Henry, *Chem. Phys.*, **12**, 387 (1976).
14. B. R. Henry and I. F. Hung, *Chem. Phys.*, **29**, 465 (1978).
15. W. R. A. Greenlay and B. R. Henry, *J. Chem. Phys.*, **69**, 82 (1978).
16. B. R. Henry and R. J. D. Miller, *Chem. Phys. Lett.*, **60**, 81 (1978).
17. B. R. Henry, I. F. Hung, R. A. MacPhail, and H. T. Strauss. To be published.
18. D. D. Smith and A. H. Zewail, *J. Chem. Phys.*, **71**, 540 (1979).
19. J. G. Moehlam, J. T. Gleaves, J. W. Hudgens, and J. D. McDonald, *J. Chem. Phys.*, **60**, 4790 (1974).
20. J. P. Maier, S. Selmeier, A. Laubereau, and W. Kaiser, *Chem. Phys. Lett.*, **46**, 527 (1977).
21. R. V. Ambartzumian, Yu A. Gorokhov, V. S. Letokhov, G. N. Makarov, *Zh. ETF Pis. Red.* (*Sov.*), **22**, 96 (1975).
22. See, for example (a) R. V. Ambartzumian and V. S. Letokhov, *Acct. Chem. Res.*, **10**, 61 (1977); (b) C. D. Cantrell, S. M. Freund, and J. L. Lyman, *Laser Handbook*, Vol. III, North Holland, Amsterdam, 1978; (c) N. Blumbergen and E. Yablonovitch, *Phys. Today*, **31**(5), 23 (1978).
23. R. Wallace, *Chem. Phys.*, **11**, 189 (1975).
24. B. R. Henry, *Acct. Chem. Res.*, **10**, 207 (1977).
25. M. L. Elert, P. R. Stannard, and W. M. Gelbart, *J. Chem. Phys.*, **67**, 5395 (1977).

26. (a) M. S. Burberry, J. A. Morell, A. C. Albrecht, and R. L. Swofford, *J. Chem. Phys.*, **70**, 5522 (1979); (b) M. S. Burberry and A. C. Albrecht, *J. Chem. Phys.*, **71**, 4768 (1979).
27. M. S. Burberry and A. C. Albrecht, *J. Chem. Phys.*, **71**, 4631 (1979).
28. M. L. Sage and J. Jortner, to be published.
29. M. L. Sage and J. Jortner, to be published.
30. A. R. Hoy, J. M. Stone and J. K. G. Watson, *J. Mol. Spectr.*, **42**, 393 (1972).
31. R. McWeeny, *Quantum Mechanics: Principles and Formalism*, Pergamon, Oxford, 1972.
32. E. C. Kemble, *Fundamental Principles of Quantum Mechanics*, McGraw-Hill, New York, 1937.
33. G. D. Carney, L. L. Sprandel, and C. W. Kern, in *Advances in Chemical Physics*, Vol. 31, edited by I. Prigogine and S. A. Rice, Wiley, New York, 1978.
34. W. G. Harter, C. W. Patterson, and F. J. da Paixano, *Rev. Mod. Phys.*, **50**, 37 (1978).
35. I. M. Mills, in *Critical Evaluation of Chemical and Physical Structural Information*, edited by D. R. Lide and M. A. Paul, National Academy of Sciences, Washington, D.C., 1974, p. 269.
36. J. Pliva, *ibid.*, p. 289.
37. A. B. Anderson, *J. Chem. Phys.*, **66**, 4709 (1977).
38. M. L. Sage and J. Jortner, *Chem. Phys. Lett.*, **62**, 451 (1979).
39. R. T. Lawton and M. S. Child, *Mol. Phys.*, **37**, 1799 (1979).
40. K. S. Sorbie and J. N. Murrell, *Mol. Phys.*, **29**, 1387 (1975).
41. A. S. Wexler, *Spectrochim. Acta*, **21**, 1725 (1965).
42. A. Mead and A. Moskowitz, *Int. J. Quantum Chem.*, **1**, 243 (1967).
43. M. S. Burberry and A. C. Albrecht, *J. Chem. Phys.*, **70**, 147 (1979).
44. I. Schek, J. Jortner and M. L. Sage, *Chem. Phys. Lett.*, **64**, 209 (1979).
45. M. L. Sage, *Chem. Phys.*, **35**, 375 (1978).
46. (a) G. C. Lie, J. Hinze and B. Liu, *J. Chem. Phys.*, **59**, 1872 (1973); (b) G. C. Lie, J. Hinze and B. Liu, *J. Chem. Phys.*, **59**, 1887 (1973).
47. M. L. Sage, *J. Phys. Chem.*, **83**, 1455 (1979).
48. (a) H. Salow and W. Steiner, *Z. Phys.*, **99**, 137 (1936); (b) G. W. Robinson, *J. Chem. Phys.*, **46**, 572 (1967).
49. F. Varsanyi and G. H. Dieke, *Phys. Rev. Lett.*, **7**, 442 (1961).
50. A. Ron and D. F. Hornig, *J. Chem. Phys.*, **39**, 1129 (1963).
51. J. Jortner and S. A. Rice, *J. Chem. Phys.*, **44**, 3364 (1966).
52. D. F. Heller and S. Mukamel, *J. Chem. Phys.*, **70**, 463 (1979).
53. D. F. Heller, *Chem. Phys. Lett.*, **61**, 583 (1979).
54. M. Bixon and J. Jortner, *J. Chem. Phys.*, **50**, 3284 (1969).
55. J. Jortner and S. Mukamel, *In Molecular Energy Transfer*, edited by R. D. Levine and J. Jortner, Wiley, New York, 1974, p. 178.
56. M. J. Berry, private communication.
57. U. Fano, *Phys. Rev.*, **124**, 1866 (1961).
58. S. Mukamel and J. Jortner, *J. Chem. Phys.*, **65**, 5204 (1976).
59. J. Jortner, *SPIE*, **113**, 88 (1977).
60. I. Schek and J. Jortner, *J. Chem. Phys.*, **70**, 3016 (1979).
61. S. Mukamel, *J. Chem. Phys.*, **70**, 5834 (1979).
62. B. Carmeli and A. Nitzan, *Chem. Phys. Lett.*, **62**, 457 (1979).

VAN DER WAALS MOLECULES

DONALD H. LEVY

The James Franck Institute and The Department of Chemistry, The University of Chicago, Chicago, Illinois 60637

CONTENTS

I. Introduction . . . 323
II. Chemical Synthesis and Experimental Probes . . . 325
III. Spectroscopy and Structure . . . 330
  A. Diatomic Species . . . 331
  B. Polyatomic Species: Infrared, Microwave, and Radiofrequency Spectra . . . 334
  C. Polyatomic Species: Visible and Ultraviolet Spectra . . . 337
IV. Photoselective Chemistry of van der Waals Molecules . . . 343
  A. Lifetimes . . . 345
  B. Competing Processes . . . 348
  C. Binding Energies . . . 349
  D. Product-State Distributions . . . 352
    1. Complexes of Iodine and Helium . . . 353
    2. Complexes of Iodine and Neon . . . 355
    3. Complexes of Iodine and Argon . . . 358
Acknowledgments . . . 360
References . . . 360

I. INTRODUCTION

Chemists are accustomed to studying molecules which are held together by chemical bonds with dissociation energies on the order of tens of thousands of cm^{-1}. Since bond energies are very much larger than kT at room temperature, chemically bound molecules are usually stable with respect to binary collisions with inert partners. van der Waals forces are roughly 10^2-10^3 times weaker than chemical forces, and at room temperature produce bonds with dissociation energies comparable to or weaker than kT. Therefore binary collisions with inert room temperature partners are likely to lead to the breaking of van der Waals bonds, and if van der Waals molecules form at all under ordinary laboratory conditions, their existence will be transitory and the study of the properties of individual van der Waals molecules will be difficult.

In spite of the experimental difficulty there has been a longstanding interest in the study of van der Waals forces, inasmuch as these forces have a profound effect on the bulk properties of matter even at temperatures where bound van der Waals molecules are unstable. The traditional approach has been to attempt to infer the nature of the microscopic interactions from measurements of bulk properties such as transport coefficients, which are effected by van der Waals forces.[1] Much of our knowledge of intermolecular forces comes from measurements of this type, but there is a limit to the information contained in measurements of bulk properties. If the nature of the microscopic force is known in detail it is often possible to do the appropriate statistical averaging to obtain an accurate calculated value of a bulk property. However, the process of unaveraging a bulk measurement to calculate a microscopic interaction is much more difficult, and there is always a question about whether or not the derived microscopic interaction is unique. Bulk properties are frequently most heavily influenced by either the long-range attractive part of the intermolecular potential or the short-range repulsive part, and therefore even in those cases where microscopic information can be inferred from measurements of bulk properties, this information does not provide a very good description of the intermediate well region of the potential.

In the last decade there has been an increasing interest in attacking these problems by the study of van der Waals molecules under conditions where the bound molecule is stable for an appreciable period of time. This review will concern itself with studies of van der Waals molecules, which have been prepared and examined in the gas phase under conditions where they are more or less free from environmental perturbation and where the goal of the study is to examine the intrinsic properties of the individual molecule. In the spirit of this series, this review is not intended to be comprehensive but will be heavily slanted toward those topics of particular interest to the author. Fortunately there already exist a number of excellent reviews[2] that, taken together, provide a quite comprehensive survey of the field.

There are three topics that this chapter will discuss. As has already been mentioned, free stable van der Waals molecules do not usually exist under ordinary laboratory conditions and, as is frequently the case in chemical research, there is a substantial problem in chemical synthesis of the sample, which must be solved before any study of the properties of van der Waals molecules can be undertaken. In recent years there have been significant advances made in the synthesis of van der Waals molecules, and in Section II the synthesis problem is discussed and various techniques used to study van der Waals molecules are described. The question of the nature of van der Waals forces themselves is of great importance, and the study of the spectra and structure of individual van der Waals molecules can produce the same detailed information about van der Waals forces that the spectroscopy of chemically bound molecules have provided about chemical forces. In Section III the spectra and structure of van der Waals molecules is discussed. The third area of interest involves the dynamics and photochemistry of van der Waals molecules, an area of particular interest in my own laboratory. In Section IV there is a discussion of the dissociation of van der Waals molecules induced by laser radiation.

II. CHEMICAL SYNTHESIS AND EXPERIMENTAL PROBES

The problem in preparing a suitable sample of van der Waals molecules arises from the fact that the van der Waals bond is weak and under ordinary conditions is unstable with respect to dissociation by binary collisions. There have been three approaches used to solve the synthesis problem:

1. Study molecules with unusually large binding energies so that even at room temperature and at low density there is a large enough equilibrium concentration of van der Waals molecules to detect.
2. Lower the temperature so that even at low density there is a large enough equilibrium concentration of van der Waals molecules to detect.
3. Work on a system that is not in thermodynamic equilibrium.

The first approach has been used to study hydrogen-bonded dimers of HF and HCl in the gas phase.[3] The question of how strong a van der Waals bond is necessary to allow study at room temperature is in part a question of how sensitive a probe is being used in the study. At any temperature and density there is some finite concentration of van der Waals molecules, which could in principle be observed if one had enough sensitivity. The study of HCl and HF dimers was possible partly because of the relatively strong bond between the monomers and partly because the probe technique, infrared absorption spectroscopy, was unusually sensitive due to the large infrared transition probability of the monomers.

It is also true that at high enough density a reasonable concentration of even weakly bound van der Waals can be produced at room temperature, but high densities inferfere with the measurements which one wants to make. To date our best source of information on van der Waals molecules has been the analysis of radiofrequency, infrared, and optical spectra, and the pressure broadening of spectral lines limits the density at which useful information can be obtained.* Roughly speaking, the pressure-broadened width must be less than the typical spacing between spectral lines. At room

*Of course the study of pressure broadening can itself yield information about van der Waals forces. This approach is useful but more in the spirit of a bulk property measurement and therefore outside the scope of this article.

temperature this problem is aggravated by the fact that van der Waals molecules tend to have small rotational constants and low-frequency vibrational modes, and the large number of rovibronic states populated at room temperature leads to a very dense spectrum. In a limited number of favorable cases, room-temperature measurements can yield important information, but it seems unlikely that this approach will be generally useful.

The second approach, studies of low-temperature static gases, has also been useful but seems to be limited to a small number of favorable cases. In most systems condensation to the liquid and solid phase occurs before the temperature can be lowered enough to produce an adequate concentration of van der Waals molecules. Only in cases where the substituent gases retain a high vapor pressure down to very low temperature is this technique likely to be successful. An outstanding example of this method is the studies of Welsh and co-workers[4] on the spectroscopy of van der Waals complexes of molecular hydrogen. The potential surface of the van der Waals molecule H_2Ar was determined by the analysis of the spectra obtained in this work, and this is probably the most complete and detailed van der Waals surface that we know.

The third possibility is to prepare the sample in a system not in thermodynamic equilibrium where the constraints described above do not apply. Much recent work has taken this approach and used a supersonic expansion of an appropriate gas mixture to generate the sample.[5] A supersonic expansion involves expanding a gas a pressure P_0 and temperature T_0 through a nozzle of diameter D to a final lower pressure P_1 and temperature T_1 under conditions where random thermal enthalpy in the static gas is converted into directed mass flow in the downstream region. Under conditions that are easily achieved the expansion can be isentropic, and in this case the expansion produces an increase in the flow velocity and a decrease in the translational temperature. The translational temperature is determined by the width of the velocity distribution about the most probable velocity. In the static gas the most probable velocity in the direction of flow is zero and the width is determined by T_0. After the expansion the velocity distribution both shifts and narrows so that the most probable velocity is the flow velocity, and the width is determined by the new lower temperature T_1. The width of the velocity distribution is the quantity of interest in determining the stability of van der Waals molecules because it determines the distribution of kinetic energies available in bimolecular collisions between the van der Waals molecules and the surrounding gas. The flow velocity and its associated kinetic energy is irrelevant until the flowing gas encounters an obstruction.

If the gas in the postnozzle flow were allowed to come into thermodynamic equilibrium with the cold translational bath, the effect would be the same as if the prenozzle gas were refrigerated, and nothing of interest would have been accomplished in the supersonic expansion. However, as one proceeds downstream in the postnozzle expansion the density is decreasing along with the temperature, and therefore there is only a finite distance over which there is sufficient density to provide collisions between molecules. Once a molecule enters the collision-free region it can no longer exchange energy with its surroundings, and therefore the state distribution is frozen in and does not change from this point on. Consequently, the state distribution of molecules in the collision-free region is determined more by kinetics than by thermodynamics. Those degrees of freedom that equilibrate rapidly with translations will be at or near equilibrium and those degrees of freedom that equilibrate more slowly will be far from thermodynamic equilibrium.

In our laboratory[6] typical operating conditions would be an upstream pressure of 100 atm and a nozzle diameter of 0.025 mm. Using a gas mixture which is primarily helium, the downstream translational temperature in the collision-free region is 0.05 K. It is clear that even fairly far upstream from the collision-free region the kinetic energy of binary collisions is insufficient to dissociate even the weakest van der Waals molecules, and it is also clear that if the system came to thermodynamic equilibrium at the translational temperature everything would condense out of the gas phase. Since molecular rotational motion equilibrates fairly rapidly with the translational bath, the downstream rotational state distribution is greatly cooled, and even the molecular vibrations are fairly efficiently cooled before the sample enters the collision-free region. Condensation, however, is a relatively slow process, and by choosing the appropriate conditions the experimenter in many instances may achieve the synthesis of a given van der Waals molecule without producing complete condensation.

The problem of synthesis of van der Waals molecules in a supersonic expansion is surprisingly similar to the more familiar problem of chemical synthesis of stable compounds. As in the case of chemical synthesis the major problem is not just to produce the species of interest but to stop the reaction before the species of interest is consumed by subsequent reactions. The variables which we have at our disposal are the solvent (carrier gas), the concentrations of various reactants, the total pressure, and the downstream temperature. The downstream temperature is affected by both the initial pressure and the nozzle temperature, and therefore one must consider these two variables together. The overall composition of the gas mixture is determined by the initial conditions, but the relative concentration, total pressure, and temperature change as a function of position downstream of the nozzle. The conditions at a particular point downstream will be the result of integrating over the concentration–temperature–pressure profile between the nozzle and the point of interest, and, since different processes can take place at different regions of the expansion, the kinetics leading up to the conditions at a given point downstream can be complicated. For instance, van der Waals molecule formation only occurs in regions of the jet where the temperature is low enough to make the van der Waals molecule stable with respect to two-body collisions. On the other hand, the production of van der Waals molecules may require many-body collisions and therefore only occurs in regions where the density is high enough to produce these collisions. It is usually the case that the region just downstream of the nozzle is too hot and the region very far downstream of the nozzle is too rarified to lead to the formation of van der Waals molecules, and that formation takes places in some limited intermediate region.

As a general rule, van der Waals molecule formation is favored by low nozzle temperature, high stagnation pressure, and high concentrations of the constituent species. However, taking any of these variables to their extreme limit frequently produces side reactions or sequential reactions that destroy the species of interest, and some intermediate value of these variables is usually used to maximize the concentration of a given species. For example, it was found that in the production of the van der Waals molecule NaAr, the optimum argon concentration was not 100% but ~5% argon in 95% helium.[7] In this case the exothermic sequential reaction

$$Ar + NaAr \rightarrow Na + Ar_2$$

requires the use of helium as a solvent.

In the production of iodine–neon complexes of the form I_2Ne_a, as might be expected, it was found that higher concentrations of neon favored the formation of larger complexes, and the production of I_2Ne and I_2Ne_2 was optimized by a mixture of a few percent in helium, while the production of a large cluster such as I_2Ne_6 required a mixture richer in neon. The production of mixed clusters of the form $I_2Ne_aHe_b$ required a careful control of concentration, and in general any given mixed cluster could only be produced over a rather narrow range of concentration.[8]

Variation of nozzle temperature can also be used to regulate the synthesis of van der Waals molecules. In the production of the van der Waals clusters of I_2, the ratio of complexed iodine to free iodine is decreased by raising the nozzle temperature. Nonetheless, a mild heating of the nozzle to ~100°C was found to increase the absolute concentration of complex, since the reduction in mole fraction was more than compensated by the very large increase in iodine vapor pressure. At some point one would expect further heating to decrease both the relative and absolute concentration of van der Waals molecules. While this point was never reached in our iodine experiments, the use of a 500°C nozzle seems to completely eliminate van der Waals formation when free-base phthalocyanine is used as a reactant.[9]

Although some thought and much experimentation may be necessary to produce a given new species, the supersonic expansion has been such a powerful synthetic tool that the problem of chemical synthesis no longer seems to be the principal obstacle in the study of van der Waals molecules. It is likely that almost any binary van der Waals molecule can be prepared in sufficient concentration for study, and the preparation of rather complicated larger complexes is possible in many cases. Until fairly recently the preparation and identification of a new van der Waals species was a matter of some interest, but there are now enough species known so that synthesis in the absence of any other information is rather less exciting than it once was. The production of free uncomplexed molecules cooled in a supersonic expansion may now be the more difficult problem.

There are a number of techniques that have been used to probe the properties of van der Waals molecules. If we exclude techniques which measure bulk properties, all of the methods that have been used thus far are spectroscopic measurements of one sort or another. These can be divided into two broad classes depending on the method of detection. In the first class are those techniques that infer the occurence of a spectroscopic transition from a change in the electromagnetic radiation incident on or emitted or scattered by the sample. These are the usual range of spectroscopic techniques used on static samples, and in principle any form of spectroscopy with sufficient sensitivity can be used to study van der Waals molecules. Molecular beams, even supersonic molecular beams, tend to be of low density and short path length, but the cooling of the internal degrees of freedom of the sample leads to small partition functions, and therefore the sensitivity problem is not as severe as might at first be imagined. Any spectroscopic technique that can be used on a low-pressure static gas sample should at least be considered for the study of van der Waals molecules. Thus far infrared absorption in static samples and visible and ultraviolet laser-induced fluorescence in supersonic beam samples have actually been used, but it is likely that in the near future a wide variety of spectroscopic techniques will be applied to the study of van der Waals molecules.

The second class of detection techniques are used only with molecular beams and infer the occurrence of a spectroscopic transition from a change in the nature of the molecular beam itself. A very powerful method has been the use of electric resonance spectroscopy in which the trajectory of the molecular beam is altered by absorption of photons by the molecules in the beam. Since the magnitude of the change is not necessarily correlated with the energy of the photon, this method has long been useful in doing microwave or radiofrequency spectroscopy on molecular beams and it has recently been used in the infrared. A second method that has recently been developed uses bolometer detection to measure energy deposition in the sample.[10] A spectroscopic absorption increases the internal energy of the absorbing molecule, and if the molecular beam impinges on a bolometer detector, this leads to an increased response by the detector. Alternatively, the spectroscopic absorption might alter the trajectory of the beam so as to cause some molecules to miss the detector, and this leads to a decreased response. Inasmuch as well-designed bolometers appear to be very sensitive, this technique appears very promising.

III. SPECTROSCOPY AND STRUCTURE

The development of our understanding of the structure of van der Waals molecules parallels that of our understanding of chemically bound molecules, but it is happening some fifty years later. The principal technique used to provide structural information is an analysis of the molecule's spectrum. It takes some effort to analyze the spectrum of a diatomic van der Waals molecule, but the interactions that must be considered are similar to those one must consider in analyzing the spectra of chemically bound diatomics. Although there are not a lot of examples of analyzed diatomic van der Waals spectra, it appears that there is no fundamental barrier to future progress. As with the case of chemically bound diatomics, the prospects for successfully analyzing a measured diatomic van der Waals spectrum are quite good.

Our understanding of polyatomic van der Waals spectra is not nearly as good, and the analysis is full of special problems that are not usually encountered in chemically bound molecules. These arise from the weakness of van der Waals forces, which produce large amplitude vibrational motion in a polyatomic van der Waals molecule. The goal of a structural study would be a complete and detailed description of the potential surface of the molecule. Various studies have made some progress in this direction, but our knowledge of polyatomic van der Waals potential surfaces is still meager, and this is largely an unsolved problem. To be successful future work will require both experimental methods of probing the potential surface more completely and theoretical methods for attacking the problem of strongly coupled degrees of freedom. Studies that have been carried out to date suggest that many of our simple assumptions about van der Waals potentials are wrong, at least in the binding region, but a correct comprehensive description is a subject for future research.

A. Diatomic Species

Examples of diatomic van der Waals spectra that have been observed are NaAr,[7, 11] NaNe,[12], XeF,[13], and XeCl.[14] The molecules NaAr, NaNe, and XeF have been studied in supersonic expansions. The xenon halides are chemically bound in their excited states and the bond strength of the ground state of XeF is intermediate between that of a chemical bond and a van der Waals bond, and therefore these species have also been studied under static gas conditions.

Both the fluorescence excitation spectrum and the dispersed fluorescence spectrum of NaAr have been observed and analyzed, and the results of these studies indicate the information potentially available from diatomic van der Waals spectra. The ground state is held together by a weak van der Waals bond and has a well depth of 40.4 cm^{-1}. The $A^2\Pi$ excited state has a much deeper well depth, and as shown in Fig. 1 the observed fluorescence excitation transitions were from the lowest two vibrational levels of the ground electronic state to those several excited vibrational levels of the $A^2\Pi$ electronic state which had the largest Franck–Condon factors. The spectral resolution available in the fluorescence excitation spectra was sufficient to resolve rotational and nuclear hyperfine structure, and a typical vibronic band is shown in Fig. 2.

The fluorescence excitation spectrum could be analyzed in the same way that one would analyze a high resolution spectrum of a $^2\Sigma \rightarrow {}^2\Pi$ transition of a chemically bound molecule, and the analysis yielded the spectroscopic constants shown in Table I. As is usually the case in supersonic molecular beam spectroscopy, the fluorescence excitation spectrum sampled only the lowest vibrational levels of the ground electronic states, and therefore the information provided about the excited electronic state was much more extensive than that provided about the ground state, since the spectrum sampled a larger number of excited state vibrational levels. Because of the low temperature of the jet and the restrictive rotational selection rules, the

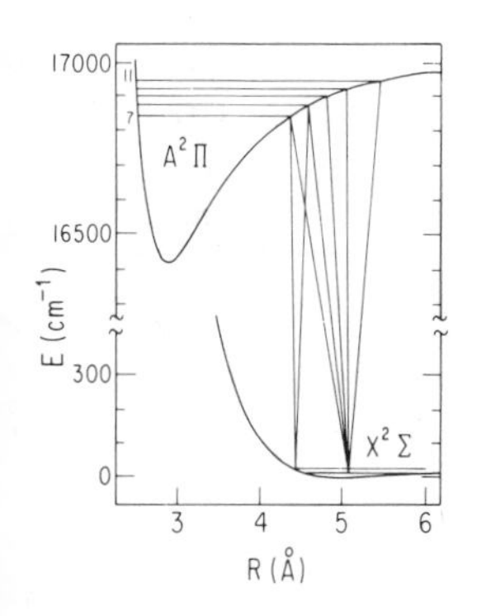

Fig. 1. Potential curves of $A^2\Pi$ and $X^2\Sigma$ electronic states of NaAr showing the transitions observed in the fluorescence excitation spectrum.

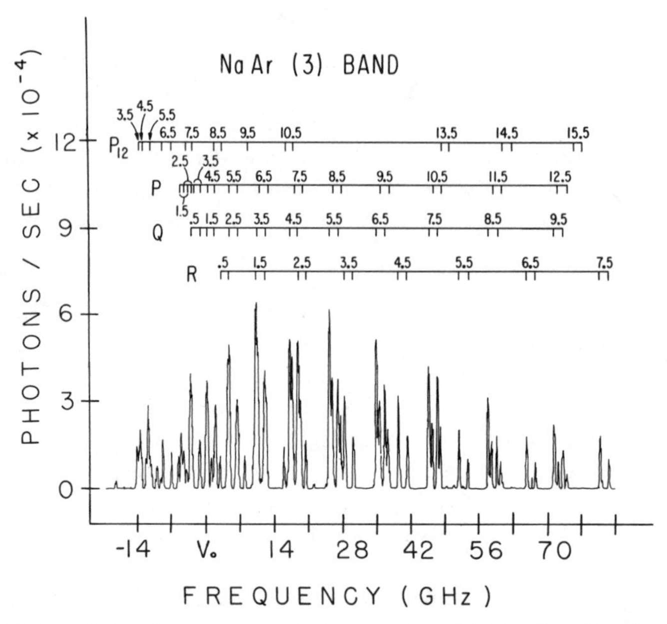

Fig. 2. High resolution fluorescence excitation spectrum of NaAr. The vibronic transition is $A^2\Pi_{1/2}(v''=10) \leftarrow X^3\Sigma^+_{1/2}(v''=0)$ and the laser resolution was 300 MHz FWHM. Rotational and hyperfine assignments are indicated.

TABLE I

Spectroscopic Constants of NaAr Determined from the Fluorescence Excitation and Dispersed Fluorescence Spectra

Constant	$X^2\Sigma^+$	$A^2\Pi_{1/2}$	$A^2\Pi_{3/2}$	Unit
T_e	0	16428.37	16455.20	cm^{-1}
ω_e	13.557 ±0.065	83.906	80.720	cm^{-1}
$\omega_e x_e$	1.155 ±0.018	4.1400	3.8585	cm^{-1}
$\omega_e y_e$		0.07005	0.0606	cm^{-1}
$\mathfrak{D}_e$	40.4 ±1.0	568.2	558.6	cm^{-1}
B_e	1.3902±0.0012	4.09684		GHz
α_e	0.1156±0.0020	0.2015		GHz
D_e	$0.74\times10^{-4} \pm 0.12\times10^{-4}$			GHz
r_e	4.991 ±0.002	2.907		Å
A		12±1		cm^{-1}

spectrum only probed low rotational levels of either electronic state, and therefore the spectrum contains little information about higher order vibration–rotation interactions that become important at high values of J. On the other hand, the absence of these higher order interactions makes the spectrum much easier to interpret.

The dispersed fluorescence spectrum obtained when NaAr is excited to the $A^2\Pi(v'=7)$ state is shown in Fig. 3. Since the A state potential curve minimum is shifted to much shorter internuclear distances than the ground state curve, most of the fluorescence is bound-free emission to the repulsive wall of the ground-state potential. In Fig. 3 we see a structured continuum in the long wavelength region of the spectrum leading up to a few sharp bound-bound features at wavelengths only slightly to the red of the exciting wavelength. Although the resolution of the dispersed fluorescence spectrum is determined by the dispersing monochromator and is therefore much lower than the laser-limited resolution of the fluorescence excitation spectrum, the dispersed spectrum samples a much larger portion of the ground-state potential, and therefore contains valuable information.

The results from both the fluorescence excitation and dispersed fluorescence experiments have been used to generate potential curves for both the

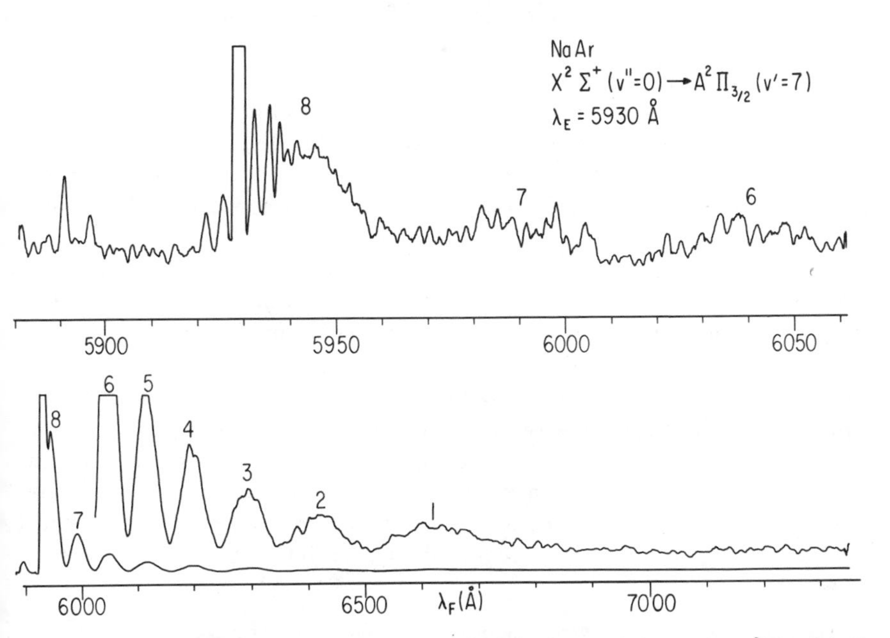

Fig. 3. Dispersed fluorescence spectrum of NaAr following excitation to the $A^2\Pi_{3/2}(v'=7)$ state. The lower trace shows the structured bound-free continuum. The upper trace was taken at higher resolution and also shows the sharper bound–bound transitions.

ground and excited electronic states.[11] The spectra sampled the entire well region of the ground electronic state as well as the repulsive wall up to a few thousand cm^{-1} above the dissociation energy. Therefore in this region, our knowledge of the ground-state curve is quite good. The spectra sampled a more limited region of the excited-state curve, but the data are very precise. Our knowledge of the curve in the sampled region is very good, and extrapolations over the entire well region are probably fairly reliable. While quantitative improvement is certainly possible, the potential curves of NaAr are now reasonably well known, and the prospects for obtaining more or less complete structural information on diatomic van der Waals molecules seem very good.

B. Polyatomic Species: Infrared, Microwave, and Radiofrequency Spectra

The problem of determining the van der Waals potential of a polyatomic molecule from an analysis of its spectrum is very much more difficult than that presented by a diatomic molecule. The presence of at least two additional degrees of freedom* greatly complicates the problem in that it is hard to obtain spectral data that probe the full potential surface, and it is hard to deduce a unique potential even if the data are available. The best determination[15] of a triatomic van der Waals surface was done for the hydrogen rare-gas complexes, using the infrared spectra of Welsh and co-workers.[4] The general problem of reducing spectral data to a potential surface as well as the specific problem of the hydrogen rare-gas complexes was recently reviewed by LeRoy and Carley[16] in an excellent article in this series, and there is nothing that the present author can add to their discussion.

Infrared spectra of the triatomic van der Waals molecules N_2-Ar[17] and O_2-Ar[18] have been observed by Ewing and co-workers along with the larger molecules $(O_2)_2$[19] and $(N_2)_2$.[20] In principle the triatomics could be subject to the same detailed analysis given to H_2Ar, but in practice this was not possible. The van der Waals potentials of N_2-Ar and O_2-Ar are much more anisotropic than that of H_2-Ar, and the rotational constants are much smaller. These factors lead to a decreased spectral resolution and an increased number of bound van der Waals states, and this severely limits the amount of information that can be extracted from the spectra. Henderson and Ewing[17, 18] did conclude that both O_2-Ar and N_2-Ar had

*In a triatomic van der Waals molecule consisting of a chemically bound diatomic fragment van der Waals bound to an atom, the van der Waals potential is a function of three coordinates, for instance the chemical bond length, the van der Waals bond length, and the angle between the chemical bond and the van der Waals bond.

a T-shaped equilibrium structure and that the barrier to internal rotation of the argon about the diatomic were 30 cm^{-1} and 20 cm^{-1}, respectively.

The first substantial body of spectroscopic data to be produced using a supersonic expansion is the work of Klemperer[21] and his colleagues who studied the electric resonance spectra of van der Waals molecules of rare-gas atoms bound to a stable molecule. In particular they have studied the series ArHCl,[22] ArHF,[23] ArClF,[24] and KrClF,[25] and, since there is some similarity in the analysis of the spectra of all of these molecules, we discuss them together.

The observed spectra consist of microwave transitions between various rotational levels ($\Delta J=1$) of the molecule. In addition to these transitions, which were observed at zero field, the application of a DC electric field causes a Stark splitting of the various M sublevels of a given J state, and radiofrequency transitions between the M levels ($\Delta J=0, \Delta M=1$) were observed. In all cases only transitions in the ground vibrational state were observed. Analysis of the microwave transitions provides a value of the rotational constant, $\bar{B}\equiv(B+C)/2$, and in those cases where several transitions are observed, the centrifugal distortion constant D. Moreover, in most cases this information could be obtained for several isotopic species.

In addition to the rotational structure, in those molecules with quadrupolar nuclei, each rotational level was split by nuclear quadrupole coupling. Analysis of this hyperfine splitting provided a measure of the quantity $(eqQ)_a$, the nuclear quadrupole coupling constant along the A inertial axis. Finally, the radiofrequency transitions between the Stark split M levels provided a measure of μ_a, the component of the electric dipole moment along the A inertial axis.

In considering the structure of these molecules, it is important to remember that the average structure and the equilibrium structure are not necessarily the same. Of course this is always the case for any molecule, but in most chemically bound molecules in their lowest vibrational states, the amplitude of the zero-point vibrational motion is sufficiently small that the difference between the average and equilibrium structure is not so significant. The weak forces responsible for the van der Waals bonds allow very large zero-point motions, and the average and equilibrium structures may be quite different.

All of the spectroscopic constants give direct information only about the average structure. The rotational constants give average values of the reciprocal moments of inertia. Assuming that the magnitude of the dipole moment and quadrupole coupling constant are identical in the complex and in the free molecule, and that any observed change in Stark interaction or quadrupole splitting is due to a shift in the direction of the inertial axes, the quantities μ_a and $(eqQ)_a$ provide a measure of $\langle\cos\theta\rangle$ and

$\langle \cos^2 \theta \rangle$, where θ is the angle between the diatomic molecule (HCl, HF, or ClF) bond axis and the A inertial axis of the complex. In a diatomic molecule the centrifugal distortion constant is given by $D = 4B^3/\omega_e^3$ where ω_e is the vibrational frequency. If one assumes that this relationship is true in the van der Waals complex, the measured values of D and B may be used to provide a value for the vibrational frequency.

To obtain information on the equilibrium geometry of the complex, one must assume a model potential to describe the van der Waals bond, calculate the various average quantities predicted by the model, and vary the parameters in the model to produce the best fit to the observed average values for all observed isotopic species. By following this procedure it was concluded that all four of the molecules ArHF, ArHCl, ArClF, and KrClF have linear equilibrium geometries, and that the equilibrium distance between the rare gas atom and its nearest neighbor was always somewhat smaller than that which would be predicted from van der Waals radii. The suggested linear equilibrium geometry is on somewhat firmer ground for the nonhydrides where even the average geometry is not far from linear, the average bond angle being 168.9° and 169.9° for ArClF and KrClF, respectively. In the hydrides the very light hydrogen undergoes a large zero-point motion and it requires a rather larger extrapolation to get from the bent average geometries (bond angles of 122.9° and 118.5° for ArHCl and ArHF) to the linear equilibrium geometries.

If the anisotropy in the molecule is dominated by pure dispersion forces, a linear geometry would be expected inasmuch as in the diatomic molecule the polarizability parallel to the bond is usually larger than the polarizability perpendicular to the bond. However, in the region of the equilibrium distance between the rare gas atom and the diatomic molecule, one would expect shorter range forces to be important. For example, one common model assumes that the anisotropy is dominated by short-range repulsions, which can be represented as a sum of pair potentials between the rare gas atom and the various atoms in the molecule. This model would produce a close-packed structure that would never be linear. As a demonstration that all van der Waals molecules are not linear, the supersonic molecular beam electric resonance spectra of ArOCS[26] and $ArCO_2$ [27] have been observed, and it was determined that these molecules have T-shaped nonlinear structures rather like that which would be predicted by a close-packing model.

To explain the geometries of these van der Waals molecules, Harris et al.[26] put forth the interesting idea that the geometry was determined by a short-range attractive force not unlike that produced by Mulliken's model of donor–acceptor complexes. In this picture the molecule acts as the acceptor, and the rare gas atom donor chooses a geometry which allows maximum overlap between its highest filled orbital and the lowest unoccupied orbital of the acceptor. Of course the resulting bond is much weaker than that of the more usual donor–acceptor complex, but is still strong enough to dominate the other anisotropic forces and determine the geometry. This model has the virtue that it correctly predicts the geometry of many known van der Waals molecules, that it explains why the bending frequencies of the linear molecules are similar to the stretching frequencies, and that it explains why the equilibrium van der Waals bond length is somewhat shorter than the sum of the van der Waals radii. It would be useful to have more evidence to establish under what conditions this model is correct.

Finally, it should be noted that in addition to the study of van der Waals molecules, supersonic molecular beam electric resonance has been used in the study of hydrogen bonded complexes[28] where the binding forces are stronger than the van der Waals forces discussed above, but still much weaker than those of ordinary chemical bonds.

C. Polyatomic Species: Visible and Ultraviolet Spectra

The visible and ultraviolet fluorescence excitation spectra of a number of van der Waals molecules have now been observed. In all cases the van der Waals spectrum appears as a satellite spectrum slightly displaced from an electronic transition of the uncomplexed parent molecule. For instance, the large number of van der Waals complexes of iodine that we have observed have their spectra slightly blue shifted from the related $B^3\Pi_{0_u^+}(v')\leftarrow X^1\Sigma_{0_g^+}(v'')$ transition of uncomplexed iodine.

The spectra of the van der Waals band carry information about the van der Waals potentials of both the ground and excited electronic states of the molecule. As is usually the case in electronic spectroscopy, the spectra provide better information about the differences between ground and excited state properties than they do about the properties of a single electronic state. For example, the spectral shift between the van der Waals band origin and the band origin of the uncomplexed molecule is given by

$$\Delta\nu = \nu_C - \nu_U = D_0'' - D_0'$$

(see Fig. 4), and therefore gives a measure of the difference between the zero-point van der Waals dissociation energies of the two states. It does not provide a measure of the van der Waals binding energy in either the ground or excited electronic state.

One might expect that the van der Waals potential would be a sensitive function of the electronic state of the molecule, and if this were the case one would expect to see large spectral shifts between the bands of the

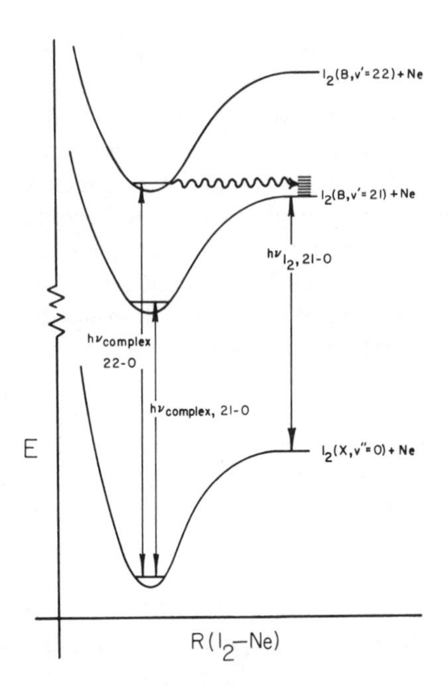

Fig. 4. Potential energy of the van der Waals molecule I_2Ne as a function of the I_2—Ne bond length. Vibrational predissociation from the bound $v'=22$ state to the continuum $v'=21$ state is indicated by wavy line. The fluorescence excitation transition of the complex and of uncomplexed I_2 are indicated by solid arrows.

complex and those of the uncomplexed molecule. This certainly was the case with the diatomic species NaAr where the excited state van der Waals well depth was more than an order of magnitude greater than that of the ground-state molecule. Contrary to this expectation, NaAr seems to be an anomaly rather than a representative species. In Table II there are listed the spectral shifts of a number of van der Waals molecules whose electronic spectra have been observed. As may be seen from the table, with the exception of NaAr, the spectral shift of the complex is always small probably amounting to no more than 20% of the binding energy, and the direction of the shift indicates that more often than not the ground-state potential is deeper than that of the excited electronic state.

Another interesting and unexpected feature of the electronic spectra was the fact that for larger clusters of rare-gas atoms around a given chemically bound molecule, the spectral shift per atom was additive. If one forms the species CA_aB_b where C is a chemically bound molecule, A and B are different rare gases, and a and b are the number of atoms of A and B in the complex, then the spectral shift is given by

$$\Delta\nu = \nu_{\mathrm{COM}} - \nu_{\mathrm{UNCOM}} = a\nu_{\mathrm{A}} + b\nu_{\mathrm{B}}$$

TABLE II
Spectral Shifts Between Fluorescence Excitation Bands of Various van der Waals Complexes and the Corresponding Bands of the Uncomplexed Substrate

Molecule	Spectral shift (cm^{-1})	Direction
I_2He	3.5	Blue
I_2Ne	6.6	Blue
I_2Ar	13.3	Blue
Tetrazine–He	1.4	Red
Tetrazine–H_2	5.6	Red
Tetrazine–Ar	22.4	Red
Benzene–He[a]	2.3	Blue
NO_2He	1.5	Blue
CrO_2Cl_2He	4.2	Blue
NaAr	450.	Red

[a]Ref. 29.

In this expression ν_A and ν_B are constants characteristic of the type of rare gas atom and chemically bound substrate, but independent of the number of atoms in the complex. This band-shift rule is illustrated in Table III where we have listed the spectral shifts between two complexes of a given series differing by one rare gas atom. Although several of the series only extend for a few members, the band-shift rule is seen to hold for the $I_2Ne_aHe_b$ series where complexes as large as I_2Ne_6 have been observed. In Fig. 5 we show the fluorescence excitation spectrum of a number of van der Waals molecules of the series I_2Ne_a along with mixed complexes of the form $I_2Ne_aHe_b$. The regularity of the band spacing is obvious. The physical basis for the band-shift rule is not well understood, but the fact that it seems to be a general rule has greatly aided the assignment of spectra in systems where several van der Waals species are present.

In a number of cases rotational structure has been observed in the electronic spectra of van der Waals molecules, and analysis of this structure has provided information about the geometry of the complex. A word of caution should be inserted at this point. In no case have the data from the electronic spectrum been extensive enough to allow a complete structural determination involving the full potential surface and all the degrees of freedom. Therefore it has been necessary to make approximations in analyzing the data, and the quality of the inferred geometric information depends both on the quality of the spectroscopic data and on the validity of the assumption used in the analysis.

TABLE III
Spectral Shifts Between van der Waals Clusters Differing by One Rare-Gas Atom

Species	Spectral Shift (cm^{-1})
I_2-I_2He	3.78
$I_2He-I_2He_2$	3.76
$I_2He_2-I_2He_3$	3.76
I_2Ne-I_2NeHe	3.73
$I_2NeHe-I_2NeHe_2$	3.70
$I_2Ne_2-I_2Ne_2He$	3.75
$I_2Ne_3-I_2Ne_3He$	3.74
I_2Ar-I_2ArHe	3.70
$I_2Ar_2-I_2Ar_2He$	3.65
I_2-I_2Ne	6.65
$I_2Ne-I_2Ne_2$	6.69
$I_2Ne_2-I_2Ne_3$	6.65
$I_2Ne_3-I_2Ne_4$	6.76
$I_2Ne_4-I_2Ne_5$	6.74
$I_2Ne_5-I_2Ne_6$	6.87
I_2-I_2Ar	13.26
$I_2Ar-I_2Ar_2$	13.44
$I_2Ar_2-I_2Ar_3$	13.50

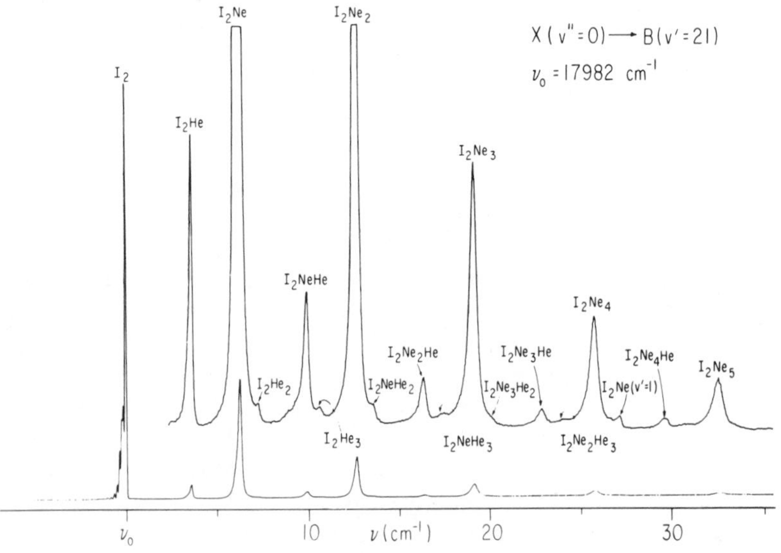

Fig. 5. The fluorescence excitation spectrum of a supersonic expansion of iodine in a mixture of neon and helium. Assignment of the various van der Waals species formed in the expansion is shown.

The rotational structure of van der Waals complexes of the ring molecules benzene[29] and *s*-tetrazine[30] have been analyzed. In all cases the complex was found to consist of either a single rare-gas atom or hydrogen molecule located above the plane of the ring on the inertial axis, or two rare-gas atoms or hydrogen molecules located symmetrically above and below the plane of the ring on the inertial axis. When the highest upstream pressures were used behind the supersonic nozzle, the major species was the doubly complexed van der Waals molecule. Nonetheless a triply complexed molecule was never observed, and it was concluded that the equivalent out of plane sites were by far the preferential binding sites. The fact that the two sites were geometrically equivalent helps to explain the constant spectral shift for each rare-gas atom.

Of the many complexes of molecular iodine with various rare gas and diatomic partners that have been observed, only the rotational structure of the species I_2He has been analyzed.[31] Splittings in the P branch were interpreted as the K structure of a near-prolate asymmetric top, and from this it was inferred that this complex was nonlinear. The analysis provided reasonably precise values of the distance from the helium atom to the center of the I_2 bond, but gave only qualitative information about the angle of the van der Waals bond with respect to the I_2 chemical bond. The geometric information obtained from electronic spectrum is summarized in Table IV.

Vibrational structure due to vibrational progressions in the van der Waals modes of the complex have been observed in most of those species where visible spectra have been measured. Since van der Waals bonds are so weak, to a very good approximation the vibrational modes of a van der Waals complex are the set of high frequency vibrational modes of the chemically bound substrate plus a set of new low-frequency modes associated with the van der Waals bond. That is to say that in a van der Waals complex the vibrational modes can be separated into a set of chemical

TABLE IV
Measured Geometric and Spectroscopic Constants of I_2He

Constant	$X(v''=0)$	$B(v'=10)$
A	0.214 ± 0.012 cm^{-1}	0.187 ± 0.016 cm^{-1}
B	0.0367 ± 0.0010 cm^{-1}	0.0267 ± 0.0012 cm^{-1}
C	0.03177 ± 0.00028 cm^{-1}	0.02381 ± 0.00028 cm^{-1}
R^a	4.47 ± 0.13 Å	4.79 ± 0.22 Å
$\nu_0{}^b$	-0.0824 ± 0.0045 cm^{-1}	

[a]Perpendicular distance from He atom to I—I bond.
[b]Measured relative to R branch head.

modes and a set of van der Waals modes with very little mixing between them due to the large frequency mismatch.

If a van der Waals complex is formed between a single rare-gas atom and a polyatomic molecule, the three translational degrees of freedom of the free atom are converted into three new van der Waals vibrational modes. If the chemical substrate is a linear molecule, and if the van der Waals complex is nonlinear, then the three translational degrees of freedom of the free atom are converted into one rotational degree of freedom of the complex plus two new van der Waals vibrational modes. It is sometimes useful to think of the van der Waals modes as consisting of one stretching mode involving extension of the van der Waals bond between the atom and the center of mass of the substrate, and one or two bending modes. Of course there is no guarantee that the normal modes of the molecule will be pure stretch or pure bend, and the extent of mixing depends on the details of the van der Waals potential. Moreover, if the anisotropic van der Waals forces are weak enough, or if the bending vibration is excited, the bending vibration is better thought of as hindered internal rotation.

If the van der Waals molecule has any symmetry, and we believe that this is the case in all of the complexes which we have studied, then symmetry arguments can be used to identify and describe the van der Waals vibrations. As an example let us consider the series of van der Waals molecules I_2X where X is a rare gas atom. As already mentioned, we believe that I_2He is T-shaped* with point group symmetry C_{2v}. In this case the van der Waals stretching vibration preserves the symmetry and transforms according to the totally symmetric representation A_1. The van der Waals bending vibration is non-totally symmetric transforming like the representation B_2 (using the usual convention[32] that the molecular plane is the YZ plane and the C_2 axis is the Z axis).† This means that the stretching and bending motions do not mix and that this convenient picture does in fact describe the normal modes.

*In fact, the spectroscopic evidence implies that it is nonlinear. If the structure is not T-shaped, for instance if it is L-shaped, there is no symmetry and the following arguments do not apply. If, contrary to our interpretation of the spectroscopic data, it is linear, then the molecule is again symmetric and the following arguments are still valid.

†It should be noted that the language used in describing the vibrational modes of a C_{2v} triatomic van der Waals molecule is somewhat different from that used in describing a chemically bound C_{2v} triatomic such as water. In the case of water the two totally symmetric modes are usually referred to as the symmetric stretch and the bend, while the asymmetric B_2 vibration is referred to as the asymmetric stretch. It should be noted that the symmetric stretch in water correlates with the van der Waals stretch, the asymmetric stretch in water correlates with the van der Waals bend, and the bend in water correlates with the stretch of the chemically bound diatomic substrate in the case of the van der Waals molecule.

TABLE V
Vibrational Spacing Between the Zero-Point Level and First Excited Level of the van der Waals Stretching Mode of Various van der Waals Complexes

Molecule	Vibrational spacing (cm^{-1})
I_2He	6
I_2Ne	21.1
I_2Ar	24.2
Tetrazine–He	38
Tetrazine–H_2	95
Tetrazine–Ar	110

These symmetry arguments can be used to identify the vibrational mode involved in the progression that we observe in the fluorescence excitation spectra of van der Waals molecules. Because the van der Waals bending mode is non-totally symmetric, electronic transitions between states where Δv is odd are forbidden. Moreover, transition between states where Δv is even are expected to be extremely weak. The intensity of these transitions depends on the difference in vibrational frequency between ground and excited state. It has been shown[33] that even if the frequencies differ by a factor of 2, the $\Delta v = 0$ transition still has 94% of the total intensity. As has already been described, the spectral shifts between the complex and the uncomplexed species are always very small and from this we infer that the ground and excited state potentials are very similar. It follows that the vibrational frequencies are similar, a factor of two difference being difficult to imagine. Therefore, it seems likely that the observed vibrational progression involves the van der Waals stretching mode where the observed intensity can be accounted for by a slight shift of the equilibrium point of the vibration. In Table V we have summarized the vibrational spacings which have been observed in the excited electronic states of various van der Waals molecules.

IV. PHOTOSELECTIVE CHEMISTRY OF VAN DER WAALS MOLECULES

Most studies of van der Waals complexes are concerned with the structure and spectroscopy of these molecules. Recently we have begun to investigate the photochemistry of van der Waals molecules, and this section is a discussion of these experiments. The motivation for studying the photochemistry of van der Waals molecules is the hope that principles that govern more general photochemical reactivity will be revealed in these weakly interacting systems which are amenable to more detailed experimental and theoretical study.

In a photochemical reaction involving the breaking of a chemical bond, the molecule must be excited to energy levels far above the zero-point level. The addition of sufficient energy to break a bond means that the relevant portion of the molecular potential surface will be far above the minimum, and usually relatively little is known about this region of the surface. Moreover, vibrational modes are no longer normal and there is considerable question as to how one goes about describing highly excited states with large coupling between the modes.

In a van der Waal molecule, the reactive bond is weak and far less excitation leads to reaction. For this reason the necessary energy may be a reasonably small perturbation on the unexcited molecule, and therefore it is rather easier to study the reacting system. The relevant portion of the potential surface involves those states of the chemically bound modes that are described by the usual spectra of the uncomplexed molecule plus the van der Waals modes. Since the van der Waals well supports relatively few bound states, the complete description of the potential surface is somewhat easier. Moreover, even in a reacting van der Waals molecule, the chemically bound vibrational modes need not be highly excited, and they therefore may be only weakly coupled to each other and to the van der Waals modes. In many cases excitation to a single known quantum state is experimentally possible, and this allows very detailed study of the photochemical reaction.

All of our experiments are directed toward describing the details of a process referred to by Herzberg[34] by the three equivalent terms case II predissociation, vibrational predissociation, and unimolecular decomposition. This involves exciting the van der Waals molecule from its zero-point level to a given quantum state in which one or more chemically bound vibrational modes are excited. This is followed by transfer of one or more vibrational quanta from the chemical modes to the van der Waals mode producing rupture of the van der Waals bond and production of free fragments with nonzero relative kinetic energy. In this process energy may be transferred to other, initially unexcited, chemical vibrational modes, and energy may also be transferred to rotational energy of the fragments. If in the initial excitation step the complex is electronically excited as well as vibrationally excited, the product fragments may themselves be electronically excited and the resulting product emission spectrum can provide an experimentally convenient way of studying the details of the photochemical reaction. An example of the type of reactions that we have studied is the photodissociation of complexes of molecular iodine with rare gas atoms as shown in Fig. 4. A typical reaction sequence would be (X=rare gas atom)

$$I_2+X \rightarrow I_2X(v''=0) \qquad \text{formation}$$

$$I_2X(v''=0) \rightarrow I_2X^*(v' \neq 0) \qquad \text{electronic excitation}$$

$$I_2X^*(v') \rightarrow I_2^*(v'-z)+X \qquad \text{photodissociation}$$

$$I_2^*(v'-z) \rightarrow I_2(v'')+h\nu \qquad \text{fluorescence of product}$$

The questions that we are interested in answering are: (1) What is the time for energy redistribution and photodissociation? (2) What is the final distribution of the energy initially stored in a single mode, i.e., what is the product state distribution? (3) What other competing processes, if any, are present? and (4) How do the reaction times, product state distributions, and competing processes depend on the initially excited state and on the composition of the complex.

A. Lifetimes

The first observed optical spectrum of a van der Waals molecule was the laser-induced fluorescence excitation spectrum of $HeNO_2$.[35] This spectrum is shown in Fig. 6, and one of the distinguishing characteristics of van der Waals spectral bands is immediately obvious from this figure; the linewidth of the van der Waals band is much larger than that of the uncomplexed parent molecule. The spectral origin of the visible–near-infrared spectrum of NO_2 is a matter of some controversy, but it is certainly several thousand cm^{-1} to the low-frequency side of the exciting frequencies used in Fig. 6, and therefore the van der Waals molecules that are excited by the laser contain several thousand cm^{-1} of vibrational energy in their chemically bound vibrational modes. After some finite lifetime some of this energy transfers to the van der Waals mode, breaks the van der Waals bond, terminates the lifetime of the quantum state, and broadens the spectral line. Measurements of the linewidth of the fluorescence excitation spectrum indicate that the photodissociation lifetime of $HeNO_2$ is $\sim$10 psec. The radiative lifetime of uncomplexed NO_2 is also a matter of controversy, but it is certainly of the order of microseconds. The radiative lifetime of the complex must be similar to that of uncomplexed NO_2, and this means that the photodissociation lifetime is orders of magnitude shorter than the radiative lifetime. One of the consequences of this is that essentially all of the fluorescence detected in the fluorescence excitation spectrum is produced by the product NO_2 fragment and not by the complex itself.

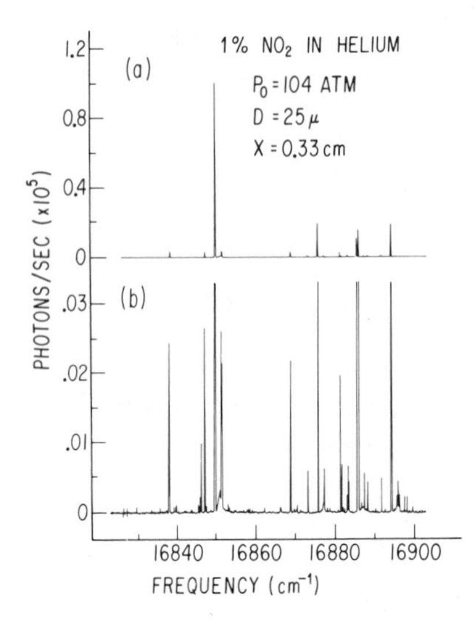

Fig. 6. The fluorescence excitation spectrum of NO_2 and $HeNO_2$ shown at low (a) and high (b) sensitivity. The broad features slightly to the blue of the largest NO_2 features are assigned to $HeNO_2$.

One might ask how the photodissociation lifetime of the complex varies with the degree of vibrational excitation in the chemically bound part of the molecule. This is a difficult question to answer in the case of NO_2 complexes because so little is known about the vibrational states of the parent molecule, but the question has been studied in detail in the case of I_2He where the vibrational structure of the parent molecule is very well known. By measuring the line-broadening of the fluorescence excitation spectrum of the complex I_2He, the photodissociation lifetime was measured as a function of the vibrational state of the I_2 stretching vibration originally excited.[36] The photodissociation lifetime was found to decrease nearly quadratically with v', varying from 221 psec at $v'=12$ to 38 psec at $v'=26$. The effect of the shortening of the lifetime with increasing v' may be seen in Fig. 7 which shows the fluorescence excitation spectrum of I_2He excited at $v'=7$ and $v'=27$. The results of the photodissociation lifetime measurements are shown in Fig. 8. Beswick et al.[37] have developed a theory of vibrational predissociation in van der Waals molecules and have used it to calculate expected lifetimes in I_2He. The details of this theory are (presumably) discussed in a separate article in this volume. The results of the theory are sensitive to the assumed van der Waals potential between the helium atom and the iodine, but using reasonable estimates of the potential parameters, Beswick et al. are able to get good agreement between their calculated lifetimes and the lifetimes measured in the fluorescence excitation spectra.

Lifetimes of other iodine van der Waals complexes have been estimated from spectral linewidths but have not been studied in the same detail as

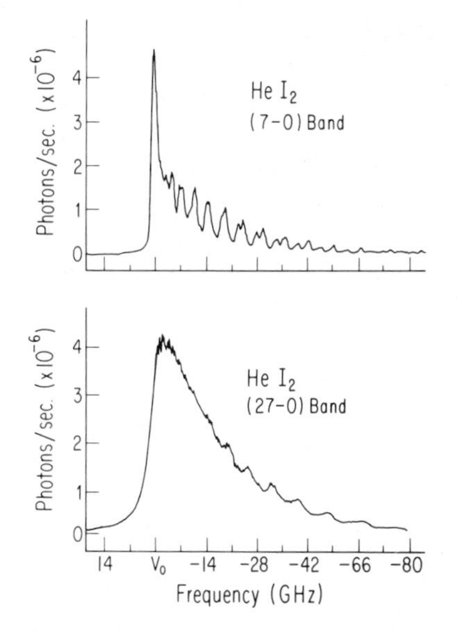

Fig. 7. The fluorescence excitation spectrum of I_2He excited in the $B(v'=7) \leftarrow X(v''=0)$ and $B(v'=27) \leftarrow X(v''=0)$ bands.

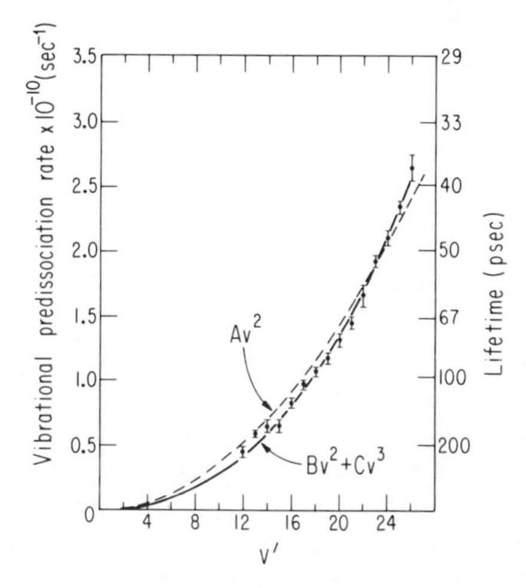

Fig. 8. Vibrational predissociation rate (and lifetime) of I_2He as a function of the vibrational state of the I_2 stretching mode that was excited. Points are experimental measurements and dashed and solid curves are the best least squares fits of functional form Av^2 and $Bv^2 + Cv^3$.

has I_2He. One of the experimental problems that one faces with larger or heavier complexes is the fact that the increased moment of inertia causes the rotational structure to collapse, and it is difficult to resolve a single rotational line. If one can only observe a profile, part of the spectral linewidth is due to homogeneous lifetime broadening but part is due to the heterogeneous broadening caused by spectral overlap, and it is frequently difficult to separate these two effects. In I_2Ne individual rotational features could be at least partially resolved, and the inferred lifetimes of I_2Ne appear to be slightly shorter than those of I_2He but certainly of the same order of magnitude.[38] In the spectrum of I_2Ar one observes only a rotational contour, and therefore one can only estimate the lower limit of the lifetime.[39] The limit obtained for the $v' = 15$ state of I_2Ar is 30 psec. The lifetime for this state in I_2He is 153 psec, and this indicates that the I_2Ar lifetime cannot be much shorter than that of I_2He and might even be larger.

B. Competing Processes

Vibrational predissociation and fluorescence are not the only decay channels that are open to photoexcited van der Waals molecules. If one looks upon a van der Waals molecule as a particularly sticky collision, then the usual collection of collisional fluorescence quenching processes must be considered. For example, in the case of I_2 collisionally induced electronic predissociation is known to be an important quenching process, and this channel does compete with vibrational predissociation in iodine van der Waals molecules. The two processes are

$$I_2X^*(v') \rightarrow X + I_2^*(v' - z) \rightarrow X + I_2 + h\nu$$
$$I_2X^*(v') \rightarrow X + 2I$$

The vibrational predissociation channel leads to fluorescent products, while the electronic predissociation channel does not. A dramatic consequence of this competition between light emitting and dark decay channels was observed in the complex I_2Ar where the intensity of the fluorescence excitation spectrum is a strong function of the vibration state that is originally excited.[39] As may be seen in Fig. 9, fluorescence is not observed when I_2Ar is excited to low vibrational states where electronic predissociation is expected to be the dominant process, and at higher vibrational states the fluorescence intensity oscillates as a function of v' presumably due to oscillations in the electronic predissociation cross section.

One general consequence of competing decay processes which quench the fluorescence is that it may be difficult to observe fluorescence excitation spectra of van der Waals complexes excited to the zero-point level of the excited electronic state. In this case the chemically bound vibrational modes do not contain any excess energy and the vibrational predissociation channel is closed. Therefore any quenching process which is fast compared to fluorescence will become important even if it was unimportant when compared to the much faster vibrational predissociation.

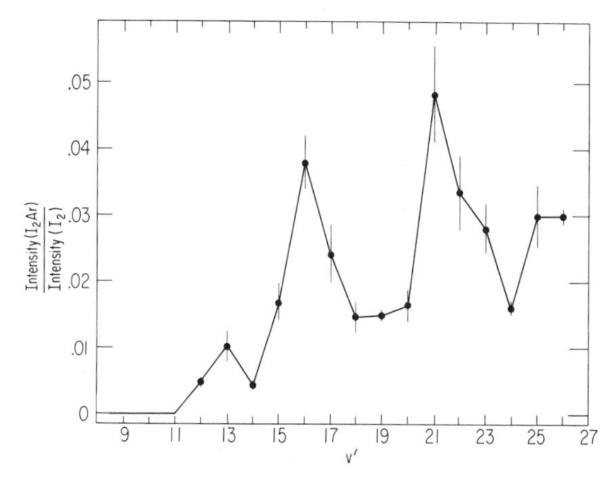

Fig. 9. Relative intensity of the fluorescence excitation spectrum of I_2Ar as a function of the vibrational state of the I_2 stretch that was originally excited.

Electronic predissociation was responsible for our failure to observe the fluorescence excitation spectrum of the $v'=0$ level of I_2He,[40] and the fluorescence excitation spectrum of the origin band of CrO_2Cl_2He[41] was only seen with difficulty.* In general we depend on vibrational predissociation to shake off the perturbing partners before they can quench the fluorescence from the parent molecule.

C. Binding Energies

It has already been suggested that the photochemical behavior of van der Waals molecules is strongly affected by the nature of the van der Waals potential. A particularly critical potential parameter is the binding energy, and most theories will predict a wide range of dynamical behavior if one is free to pick any value for binding energy. The theory of Beswick and Jortner[37] stresses the importance of an energy-gap law in determining photochemical properties.

Unfortunately van der Waals binding energies are difficult to measure experimentally, and the theoretical models that have been used to estimate binding energies are of questionable validity. Many of these models try to determine properties of the well region of the potential such as the binding energy by extrapolating measurements which are sensitive to the long-range

*At the time our paper[41] on van der Waals molecules of chromyl chloride was written we had not observed the fluorescence excitation spectrum of the origin band. Since that time such a spectrum has been observed.[42] The origin spectrum is both weaker and narrower than that from excited vibrational states.

r^{-6} part of the potential, and recent work has cast doubt on this approach.

We have recently[43] been able to measure van der Waals binding energies of iodine–rare-gas complexes by taking advantage of the rapid vibrational predissociation that takes place in these systems, and this approach may prove to be of some general use. If the rate of vibrational predissociation is fast compared to the rate of fluorescence, then most of the emitted light comes from the dissociation products and not from the complex itself. If collisions between product molecules can be eliminated between the time of dissociation and the time of emission, then the dispersed emission spectrum of the product fragments contains information about the product state distribution of the dissociation reaction. Since an iodine complex can be excited to a single known vibronic state, the product state distribution may reveal among other things the minimum amount of energy used in the dissociation reaction. Since this minumum amount of energy must be at least as large as the binding energy, measurement of the product state distribution sets a limit on the binding energy.

This can be illustrated by considering the complex I_2Ar. In Fig. 10 we see the dispersed fluorescence spectrum of the I_2^* produced by the dissociation of the complex I_2Ar initially excited to the $v'=21$ state. The first observed fluorescence band is the $v'=18 \rightarrow v''=0$ band, the product of a three-quantum dissociation, and the absence of products from one or two-quantum dissociation channels indicates that at least three iodine

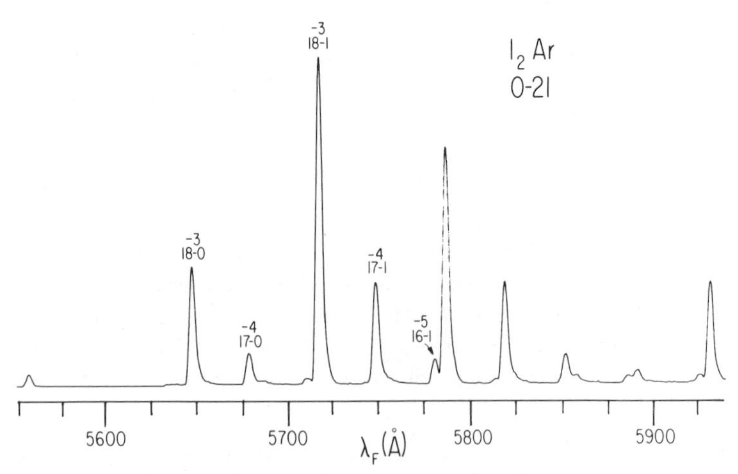

Fig. 10. Dispersed fluorescence spectrum of the product I_2^* produced upon dissociation of the van der Waals complex I_2Ar. The complex was originally excited to the B($v'=21$) state, and the feature near 5560 Å is scattered light from the exciting laser.

vibrational quanta are necessary to break the van der Waals bond. The known vibrational spacings[44] of iodine thus set upper and lower limits on the binding energy.

A more accurate estimate of the binding energy can be obtained by tuning the energy in a vibrational quantum using the anharmonicity of the iodine stretching vibration. As one excites to higher vibrational levels of the excited electronic state, the energy contained in three vibrational quanta gets smaller, and eventually one would expect the three-quantum dissociation channel to close when three vibrational quanta contain less than the binding energy. This is in fact what happens as seen in Fig. 11, which shows the dispersed fluorescence spectra from the products of the dissociation of I_2He when the complex is excited to $v'=29$ and $v'=30$. In the case of $v'=29$ excitation the three-quantum channel is open, whereas in the case of $v''=30$ excitation it is closed. This puts rather narrow limits on the binding energy and requires the zero-point dissociation energy, D_0, to be in the range 220–226 cm^{-1}.

A similar set of experiments has been done using the complex I_2Ne, and from them we have determined that D_0 is in the range 65–67 cm^{-1}. In the case of I_2He the same procedure was attempted. Dispersed fluorescence spectra were observed after excitation to various vibrational states up to

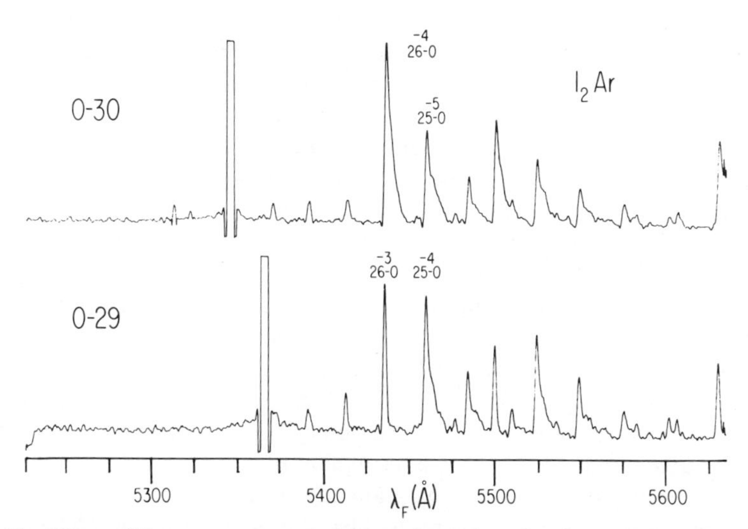

Fig. 11. Dispersed fluorescence spectra of the product I_2* produced upon dissociation of the van der Waals complex I_2Ar. In the upper spectrum the complex was originally excited to the B($v'=30$) state, and in the lower spectrum the complex was excited to the B($v'=29$) state. The off-scale features to the short wavelength side of each spectrum are from scattered laser light.

$v' = 58$. Even at $v' = 58$ the one-quantum channel was still open, which allows us to place a limit on the dissociation energy of $D_0 \leqslant 18.5\ \text{cm}^{-1}$. At the time of writing we have been unable to observe an adequate spectrum with excitation to $v' > 58$, but experiments are still in progress.

It should be noted that in all cases we are measuring the dissociation energy from the zero-point level. To get the well depth, D_e, the zero-point energy must be added to our measured values of D_0. In the cases of I_2Ne and I_2Ar, the zero-point energy is reasonably small, and D_e is only slightly larger than D_0. In the case of I_2He the zero-point energy is probably a substantial fraction of the total well depth and D_e could differ from D_0 by as much as a factor of 2.

D. Product-State Distributions

The dynamics of the photodissociation process can be probed by measuring the product-state distribution in the chemically bound fragment that is produced in the dissociation reaction. Because van der Waals bonds are weak, it is possible to photoexcite the chemically bound modes of the substrate such that these modes contain far more energy than is needed to break all of the van der Waals bonds in the complex. This internal energy, initially contained in a known vibrational state, will be used in the dissociation reaction to break the van der Waals bond, and the excess will be partitioned between relative kinetic energy of the fragments, rotational energy of the chemically bound fragment, and vibrational energy of the chemically bound fragment. If, as is the case of iodine van der Waals molecules, the chemically bound fragment is electronically excited, analysis of its emission spectrum will give the product-state distribution at the time of emission. If there is little or no energy relaxation between the time of dissociation and the time of emission, the product-state distribution measured from the emission spectrum will be the product-state distribution of the photodissociation reaction.

There are a number of requirements which the system of interest must satisfy in order for the experiment to be feasible. As we will see, iodine is a suitable chemical substrate, and the experiment which will be discussed all involve van der Waals molecules built around an I_2 core. In the first place, both inter- and intramolecular relaxation between the time of formation and the time of emission must be negligible. Iodine does have an intramolecular relaxation channel via spontaneous electronic predissociation, but in the energy range of interest this will be of little importance.[45] The extent of intermolecular relaxation via collisions is more a function of conditions in the jet than the nature of the molecule. Because the cross-section for collisional vibrational relaxation is greatly increased for low-energy collisions,[46] even the low densities of our translationally cold jet are sufficient to produce measurable collisional effects. Nonetheless these are small and are themselves measurable, and the product-state distribution inferred from the emission spectrum can be corrected to provide a measure of the initial distribution.

A second requirement for a useful substrate is that its spectroscopy be thoroughly understood. This is necessary so that spectral line positions can be calculated and that a particular quantum state of the emitter can be associated with a given observed spectral line. It is also necessary that Franck–Condon factors be calculable so that a measured spectral intensity can be used to infer the population of the quantum state involved in a particular transition.[47] Very extensive spectroscopic work has been done on iodine, and the appeal of this species from this point of view should be obvious.

In the rest of this chapter we will discuss a number of iodine van der Waals molecules whose product-state distributions have been measured. In conjunction with our experimental work, a number of theoretical treatments of such processes have been published.[48] Since the authors of much of this theoretical work are also contributing to these volumes, we will not discuss the theory in detail in this chapter. In the following sections we will discuss the experimental observations and make some qualitative generalizations about the dissociation process. In other chapters in this volume the state of our theoretical understanding of van der Waals photodissociation are summarized. The data show some striking general trends, many of them not intuitively obvious, and the theory has been able to give us an understanding of some but not all of these trends. During the relatively short period in which van der Waals photodissociation has been studied there has been a very fruitful interchange between theory and experiment, and there is every indication that this will continue in the foreseeable future.

1. Complexes of Iodine and Helium

Three complexes of iodine and helium have been studied: I_2He, I_2He_2, and I_2He_3.[49] In Fig. 12 we show the emission spectrum produced by the product I_2^* that is formed following excitation of the complexes I_2He and I_2He_2 to the $v'=22$ vibrational level of the $B\ 0_u^+$ electronic state. The quantum number v' refers to the chemically bound iodine stretching mode of the complex. Measurement of the product state distribution requires correction of the emission intensity for collisional relaxation and Franck–Condon factors, but these corrections are small and a qualitative picture of the product state distribution may be obtained from the raw spectral intensities.

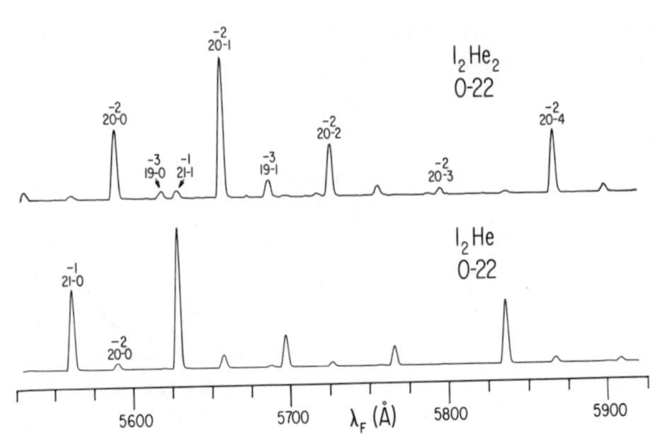

Fig. 12. Dispersed fluorescence spectra of the product I_2^* formed upon dissociation of the van der Waals complexes I_2He and I_2He_2. In both cases the complex was originally excited to the $B(v'=22)$ state.

As may be seen in Fig. 12, the dominant decay channel for the complex I_2He involves the loss of one quantum of vibrational energy from the iodine stretching mode. Some portion of this quantum is used to break the I_2–He van der Waals bond, and the remainder goes into relative kinetic energy and rotational energy of the I_2 fragment. The two-quantum channel (product emission from $v'=20$ following excitation of the complex to $v'=22$) is observable, but its relative cross-section is very small, only a few percent of that of the one-quantum channel. It is clear that I_2He dissociates with a very strong $\Delta v=-1$ propensity.

The larger complex I_2He_2 produces a different product-state distribution. The first observable channel is the two-quantum ($\Delta v=-2$) channel and a small three-quantum channel is also observed. If the complex dissociates via a one-quantum process, its relative cross-section is unobservably small. The binding energy of the complex I_2He is known[43] to be less than 18.5 cm^{-1}. If the binding energy of I_2He_2 is roughly twice that of I_2He, a single vibrational quantum of the I_2 stretch in the vicinity of $v'=22$ has more than enough energy ($\sim$90 cm^{-1}) to completely dissociate I_2He_2 and even I_2He_3. Thus the one-quantum channel is expected to be energetically open, and the fact that we do not observe this process is due to a dynamic and not an energetic restriction. Apparently in I_2He_2, the two rare-gas atoms are not able to share a single vibrational quantum. This suggests that the dissociation of the two helium atoms is a sequential rather than a concerted process. In the case of I_2He_3, the first observed channel is the three-quantum process ($\Delta v=-3$), and this is further evidence of the sequential dissociation.

The relative cross-section for the higher quantum channels ($\Delta v = -2$ for I_2He and $\Delta v = -3$ for I_2He_2) are small and difficult to measure quantitatively, but insofar as we can measure them, the relative cross section for the three-quantum process in I_2He_2 is twice that of the two-quantum process in I_2He. This would be consistent with an independent particle model where the product-state distributions for the two steps were identical. That is to say that the vibrational product-state distribution was insensitive to the nature of the larger fragment, and the reactions

$$I_2He_2 \rightarrow I_2He + He$$

and

$$I_2He \rightarrow I_2 + He$$

produce the same distribution.

2. *Complexes of Iodine and Neon*

Rather large clusters of iodine and neon have been identified.[38] Product-state distributions have been measured in clusters as large as I_2Ne_6 as well as in a number of mixed clusters of iodine, neon, and helium. Figure 13 shows the emission spectrum of the I_2 fragment produced when I_2Ne, I_2Ne_2, and I_2Ne_3 are photodissociated following excitation of the complex to the $v' = 22$ level of the $B\ 0_g^+$ state.

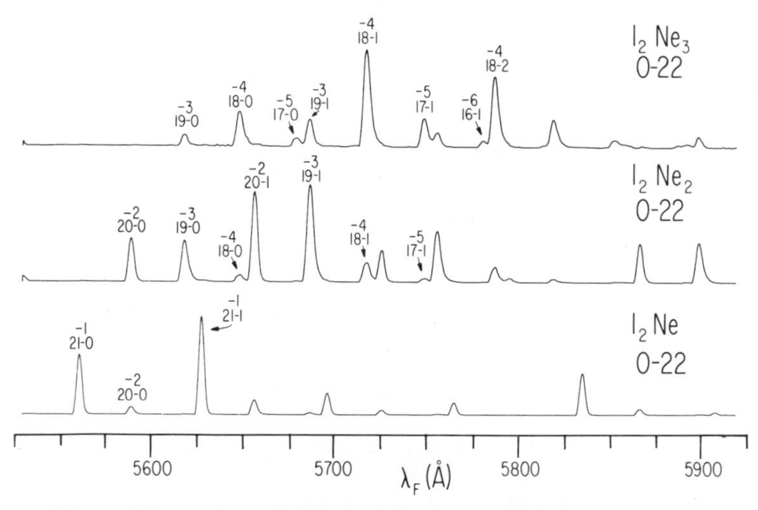

Fig. 13. Dispersed fluorescence spectra of the product I_2^* formed upon dissociation of the van der Waals complexes I_2Ne, I_2Ne_2, and I_2Ne_3. In all cases the complex was originally excited to the $B(v' = 22)$ state.

The product-state distribution of I_2 produced when I_2Ne dissociates is strikingly similar to that produced when I_2He dissociates. The cross-section for the one-quantum process is far larger than that of any other channel, the two-quantum channel having a relative cross-section of only 7%. As in the case of I_2He_2, the first observed dissociation channel for I_2Ne_2 is the two-quantum channel. Because of the stronger binding energy of neon to iodine, the one-quantum channel is expected to be energetically closed, and our failure to observe it tells us nothing about the dynamics of the reaction.

The case of I_2Ne_2 does show some striking differences from that of I_2He_2 when one examines the higher quantum channels. In I_2Ne_2 relative cross-sections for the three-quantum channel is much larger than it was for I_2He_2. This trend continues for the larger complexes as shown in Table VI where we have listed the ratio of the cross-sections for the first observed channel (the z-quantum channel) and for the next higher channel (the $z+1$-quantum channel) for a number of van der Waals complexes. As may

TABLE VI
State Distribution of Product $I_2^*(v')$ Produced by Photodissociation of Rare Gas Iodine van der Waals Complexes[a]

Species	z	k_{z+1}/k_z
I_2He	1	0.04
I_2Ne	1	0.07
I_2Ar	3	0.21
I_2He_2	2	0.05
I_2HeNe	2	
I_2Ne_2	2	0.75
I_2HeAr	3	1.69
I_2Ar_2	6	1.73
I_2Ne_2He	3	0.54
I_2Ne_3	3	3.73
I_2Ne_3He	4	1.16
I_2Ne_4	4	4.8
I_2Ne_5	6	0.74
I_2Ne_6	7	1.16

[a]The first observed channel $[XI_2(v_i) \rightarrow X+I_2(v_i-z)]$ involves the loss of z quanta from the iodine stretching vibration. The quantity k_{z+1}/k_z is the branching ratio between the second and first observed channels.

be seen from this table the higher quantum channels become relatively more important as the complex gets larger.

There are two qualitative conclusions that may be drawn from these observations. First, contrary to the case of helium, the product-state distributions of the individual dissociation steps (assuming that the overall process involves sequential dissociation of individual atoms) depends on the nature of the larger fragment. For example, one cannot reproduce the final product-state distribution for I_2Ne_2 dissociation by assuming that both individual steps

$$I_2Ne_2 \rightarrow I_2Ne + Ne$$
$$I_2Ne \rightarrow I_2 + Ne$$

have the same product-state distribution as observed when I_2Ne dissociates.

The second qualitative conclusion is that the photodissociation process gets less efficient as the complex gets larger, that is, more vibrational quanta are required per rare-gas atom. In the smaller complexes the one quantum per particle channel is always important and even dominant, while in I_2Ne_6 the one quantum per particle channel is not observed.

Both of these observations may be understood if we consider those theoretical models of van der Waals photodissociation that have been developed thus far. These models deal only with single rare gas atom complexes, but one can understand in a qualitative way the effect of the more complex vibrational structure of the larger complexes.

The theory of Beswick and Jortner[37] emphasizes the importance of the energy gap in determining the vibrational predissociation rate of a van der Waals complex. The energy gap is the mismatch between the vibrational frequency of the chemical mode where the energy is originally stored and the vibrational frequency of the dissociating van der Waals mode that most accept energy before the complex can dissociate. The theory predicts that the smaller the energy gap the larger the dissociation rate, and small changes in energy gap can have a large effect on the rate.

In a single-atom van der Waals molecule, there are only two van der Waals modes, the stretch and the bend. The stretching frequency is likely to be higher than the bending frequency, and the energy-gap law would predict that energy would preferentially flow from the iodine stretch into the van der Waals stretch. The van der Waals stretching mode is a dissociating mode, while the van der Waals bend is not, and therefore the energy-gap law would predict an efficient transfer of energy from the storage mode to the dissociating mode, bypassing the nondissociating mode.

As one adds additional atoms to the complex the number of van der Waals vibrational modes rapidly increases. While the energy-gap law will still produce a stronger coupling between the storage mode and the dissociating stretching modes, the number of more weakly coupled nondissociating bending and torsional modes will increase. Since these are lower frequency modes, the density of nondissociating states will rapidly increase and eventually the density of states will overwhelm the stronger coupling of the dissociating modes. Another way of saying this is that in smaller complexes the energy flow is dominated by the strength of the couplings, and energy can flow along specific favorable paths. However, as the complex becomes larger, the energy flow becomes more statistical, and therefore energy originally in the storage mode must be used to heat the nondissociating modes as well as the dissociative modes. This would lead to a less efficient overall dissociation process.

3. *Complexes of Iodine and Argon*

The complexes of I_2Ar, I_2Ar_2, and I_2Ar_3 have been observed along with various mixed argon–helium complexes.[49] It seems unlikely that there would be any intrinsic difficulty in forming higher argon complexes, and our failure to observe higher complexes is probably due to reduced fluorescence quantum yield rather than insufficient concentration. Even in the single atom complex I_2Ar, quenching due to electronic predissociation induced by the single argon atom is known to compete with vibrational predissociation, and it may be that in the higher complexes this nonradiative channel dominates the decay of the excited state.

To the extent that one can distinguish a trend from measurements on just two species, the product-state distributions in the argon complexes seem to have the same qualitative features as those of the neon complexes. The higher quantum channels are relatively more important in I_2Ar_2 than they are in I_2Ar, and the overall product-state distribution of the I_2Ar_2 reaction cannot be generated by convoluting the individual steps with a single I_2Ar product-state distribution.

One new feature does appear when one considers the photodissociation of the mixed complex I_2ArHe. As already mentioned, the first observed dissociation channel in I_2He is the one-quantum process and the first observed channel in I_2Ar is the three-quantum channel, presumably due to the fact that the one- and two-quantum channels are energetically closed. If the total van der Waals binding energy of I_2ArHe is the same or greater than the sum of van der Waals binding energies of I_2He and I_2Ar, and if the dissociation of the mixed complex occurred sequentially, one would expect that the first observed channel in I_2ArHe would be the four-quantum channel. In fact, as seen in Table VI, while the four-quantum cross-section is largest, the three-quantum channel is observed. As opposed to the pure helium complexes, the two rare gas atoms in I_2ArHe seem to be able to share a vibrational quantum.

Consideration of the energy-gap law may provide some insight into this apparent quantum sharing. The vibrational frequency of the iodine rare-gas van der Waals stretch is known to be larger in I_2Ar than in I_2He. The energy-gap law would then predict that in the mixed complex energy would flow preferentially from the iodine storage mode to the iodine–argon van der Waals stretch. However, unlike the case of I_2He, transfer of one vibrational quantum to the argon vibration still produces a bound state. It is possible that the argon van der Waals modes themselves act as secondary storage modes for transfer of energy to the helium stretching mode. If this is the case, less than a full iodine stretching quantum can be transferred to the helium, the remainder being available for dissociation of the argon.

In the discussion thus far, the product states of interest have been the vibrational states of the fragment iodine. To examine the rotational state distribution one in principle only needs higher spectral resolution. In practice the measurement of the rotational-state distribution of the product has been a difficult task, because the requirement of increased spectral resolution carries with it the price of reduced sensitivity. Because iodine has such a large moment of inertia, transfer of a large number of quanta of angular momentum results in only very little rotational energy deposition in the fragment.

The requirement of overall conservation of angular momentum means that every quantum of rotational angular momentum left in the iodine fragment must be balanced by one quantum of orbital angular momentum in the recoiling fragments. In the classical limit the orbital angular momentum is given by $\mu v b$ where μ is the reduced mass, v is the relative recoil velocity, and b is the impact parameter in the exit channel. As an indication of the restriction of angular momentum conservation, an I_2Ar complex dissociating with a recoil velocity equivalent to 90 cm^{-1} of kinetic energy (roughly one iodine stretching vibrational quantum in the excitation region of interest) and an impact parameter of 3 Å could transfer 44 $\hbar$ of angular momentum.

The qualitative effect of rotational excitation of the I_2 fragment may be seen in Fig. 11 which shows the emission spectrum of the fragment I_2 produced when the I_2Ar complex is excited to $v'=29$. As previously discussed, the $v'=29$ level is the last level where the three-quantum dissociation channel is still energetically open, and therefore in the three-quantum channel essentially all of the energy initially stored in the iodine stretching mode is required to break the van der Waals bond. This means that there is very little energy left to go into rotational excitation. In fact the real limit on rotational excitation is due to the small amount of energy that can go into relative kinetic energy, which limits the amount of orbital angular momentum that can be produced in the recoiling atom. Conservation of angular momentum therefore limits the amount of angular momentum that can be left in the rotating molecular fragment. The effect of this may be seen by comparing the 26–0 (three-quantum channel) emission band with the 25–0 (four-quantum channel) emission band in Fig. 11. Even at the limited instrumental resolution of this spectrum, the 25–0 band is distinctly broader due to the unresolved rotational envelope. The four-quantum channel has an additional quantum of energy to be distributed between kinetic energy and rotational energy, and this leads to a distinctly broader rotational distribution. At the time of writing, analysis of the rotational distribution of product fragments of the dissociation of van der Waals molecules was still in progress.

Acknowledgements

The research discussed in this chapter that was carried out in my own laboratory has greatly benefited from collaboration with a number of co-workers. The original work on van der Waals molecules and supersonic jet spectroscopy was done in collaboration with Rick Smalley and Lennard Wharton, and the extent of this collaboration will be evident by examination of the references. Their impact on the work has been enormous, but I at least have the satisfaction of acknowledging their contribution by proper references to the literature. The more recent work discussed in this article has been done in collaboration with my present co-workers: Joe Blazy, Ben DeKoven, Ken Johnson, John Kenny, Tim Russell, and Wayne Sharfin. Because this work is new, reference to it has been in the form "in press" or "to be published", and these literature references neither inform the reader nor adequately describe the very substantial contributions of my present collaborators. The reader should be warned that the efforts of my collaborators will have made this article out-of-date before it is published.

This material is based upon work supported by the National Science Foundation under Grant CHE78-25555, by the Louis Block Fund of the University of Chicago, and by the donors of The Petroleum Research Fund, administered by the American Chemical Society.

References

1. J. O. Hirschfelder, C. F. Curtiss, and R. B. Bird, *Molecular Theory of Gases and Liquids*, Wiley, New York, 1964.
2. G. E. Ewing, *Angew. Chem. Int. Ed.*, **11**, 486 (1972); W. Klemperer, *Ber. Bunsenges. Phys. Chem.*, **78**, 128 (1974); G. E. Ewing, *Acct. Chem. Res.*, **8**, 185 (1975); G. E. Ewing, *Can. J. Phys.*, **54**, 487 (1976); B. L. Blaney and G. E. Ewing, *Ann. Rev. Phys. Chem.*, **27**, 553 (1976).
3. J. L. Himes and T. A. Wiggins, *J. Mol. Spectry.*, **40**, 418 (1971); M. Larvor, J.-P. Houdeau, and C. Haeusler, *Can. J. Phys.*, **56**, 334 (1978).
4. A. Kudian, H. L. Welsh, and A. Watanabe, *J. Chem. Phys.*, **47**, 1553 (1967); A. K. Kudian and H. L. Welsh, *Can. J. Phys.*, **49**, 230 (1971); A. R. W. McKellar and H. L. Welsh, *J. Chem. Phys.*, **55**, 595 (1971); *ibid., Can. J. Phys.*, **50**, 1458 (1972); **52**, 1082

(1974); A. R. W. McKellar, *J. Chem. Phys.*, **61**, 4636 (1974).
5. R. E. Smalley, L. Wharton, and D. H. Levy, *Acct. Chem. Res.*, **10**, 139 (1977); D. H. Levy, L. Wharton, and R. E. Smalley, in *Chemical and Biochemical Applications of Lasers*, Vol. II, Academic Press, New York, 1977, Chap. 1; R. Campargue, *J. Chem. Phys.*, **52**, 1795 (1970). See also references cited in these reviews.
6. R. E. Smalley, D. H. Levy, and L. Wharton, *J. Chem. Phys.*, **64**, 3266 (1976).
7. R. E. Smalley, D. A. Auerbach, P. S. H. Fitch, D. H. Levy, and L. Wharton, *J. Chem. Phys.*, **66**, 3778 (1977).
8. J. E. Kenney, K. E. Johnson, W. F. Sharfin, and D. H. Levy, ed. Bradley Moore, *J. Chem. Phys.*, **72**, 1109 (1980).
9. P. S. H. Fitch, L. Wharton, and D. H. Levy, *J. Chem. Phys.*, **69**, 3424 (1978); **70**, 2018 (1979).
10. T. E. Gough, R. E. Miller, and G. Scoles, *J. Chem. Phys.*, **69**, 1588 (1978).
11. J. Tellinghuisen, A. Ragone, M. S. Kim, D. J. Auerbach, R. E. Smalley, L. Wharton, and D. H. Levy, *J. Chem. Phys.*, **71**, 1283 (1979).
12. R. Ahmad-Bitar, W. P. Lapatovich, D. E. Pritchard, and I. Renhorn, *Phys. Rev. Lett.*, **39**, 1657 (1977).
13. D. L. Monts, L. M. Ziurys, S. M. Beck, M. G. Liverman, and R. E. Smalley, *J. Chem. Phys.*, **71**, 4057 (1979); J. Tellinghuisen, P. C. Tellinghuisen, G. C. Tisone, J. M. Hoffman, and A. K. Hays, *J. Chem. Phys.*, **68**, 5177 (1978); P. C. Tellinghuisen, J. Tellinghuisen, J. A. Coxon, J. E. Velazco, and D. W. Setser, *J. Chem. Phys.*, **68**, 5187 (1978); J. Tellinghuisen, G. C. Tisone, J. M. Hoffman, and A. K. Hays, *J. Chem. Phys.*, **64**, 4796 (1976); A. L. Smith and P. C. Kobrinsky, *J. Mol. Spectry.*, **69**, 1 (1978).
14. A. Sur, A. K. Hui, and J. Tellinghuisen, *J. Mol. Spectry.*, **74**, 465 (1979); J. Tellinghuisen, J. M. Hoffman, G. C. Tisone, and A. K. Hays, *J. Chem. Phys.*, **64**, 2484 (1976).
15. R. J. Le Roy and J. Van Kranendonk, *J. Chem. Phys.*, **61**, 4570 (1974); R. J. Le Roy, J. S. Carley, and J. E. Grabenstetter, *Faraday Disc. Chem. Soc.*, **62**, 169 (1977); J. S. Carley, *Faraday Disc. Chem. Soc.*, **62**, 303 (1977).
16. R. J. Le Roy and J. Scott Carley, *Adv. Chem. Phys.*, **42** 353 (1980).
17. G. Henderson and G. E. Ewing, *Mol. Phys.*, **27**, 903 (1974).
18. G. Henderson and G. E. Ewing, *J. Chem. Phys.*, **59**, 2280 (1973).
19. C. A. Long and G. E. Ewing, *J. Chem. Phys.*, **58**, 4824 (1973).
20. C. A. Long, G. Henderson, and G. E. Ewing, *Chem. Phem. Phys.*, **2**, 485 (1973).
21. W. Klemperer, *Ber. Bunsenges. Phys. Chem.*, **78**, 128 (1974).
22. S. C. Holmgren, M. Waldman, and W. Klemperer, *J. Chem. Phys.* **69**, 1661 (1978); S. E. Novick, K. C. Janda, S. L. Holmgren, M. Waldman, and W. Klemperer, *J. Chem. Phys.*, **65**, 1114 (1976); S. E. Novick, P. Davies, S. J. Harris, and W. Klemperer, *J. Chem. Phys.*, **59**, 2273 (1973).
23. S. J. Harris, S. E. Novick, and W. Klemperer, *J. Chem. Phys.*, **60**, 3208 (1974).
24. S. J. Harris, S. E. Novick, W. Klemperer, and W. E. Falconer, *J. Chem. Phys.*, **61**, 193 (1974).
25. S. E. Novick, S. J. Harris, K. C. Janda, and W. Klemperer, *Can. J. Phys.*, **53**, 2007 (1975).
26. S. J. Harris, K. C. Janda, S. E. Novick, and W. Klemperer, *J. Chem. Phys.*, **63**, 881 (1975).
27. J. M. Steed, T. A. Dixon, and W. Klemperer, *J. Chem. Phys.*, **70**, 4095 (1979).
28. T. R. Dyke and J. S. Muenter, *J. Chem. Phys.*, **57**, 5011 (1972); ibid. **60**, 2929 (1974); T. R. Dyke, K. M. Mack, and J. S. Muenter, *J. Chem. Phys.*, **66**, 498 (1977).
29. S. M. Beck, M. G. Liverman, D. L. Monts, and R. E. Smalley, *J. Chem. Phys.*, **70**, 232 (1979).

30. R. E. Smalley, L. Wharton, D. H. Levy, and D. W. Chandler, *J. Mol. Spectry.*, **66**, 375 (1977); R. E. Smalley, L. Wharton, D. H. Levy, and D. W. Chandler, *J. Chem. Phys.*, **68**, 2476 (1978).
31. R. E. Smalley, L. Wharton, and D. H. Levy, *J. Chem. Phys.*, **68**, 671 (1978).
32. R. S. Mulliken, *J. Chem. Phys.*, **23**, 1997 (1955).
33. G. Herzberg, *Molecular Spectra and Molecular Structure*, Vol. 3 Van Nostrand, New York, 1966, p. 150.
34. Ref. 33, p. 469.
35. R. E. Smalley, L. Wharton, and D. H. Levy, *J. Chem. Phys.*, **66**, 2750 (1977).
36. K. E. Johnson, L. Wharton, and D. H. Levy, *J. Chem. Phys.*, **69**, 2719 (1978).
37. J. A. Beswick and J. Jortner, *Chem. Phys. Lett.*, **49**, 13 (1977); *J. Chem. Phys.*, **68**, 2277, 2525 (1977); J. A. Beswick, G. Delgado-Barrio, and J. Jortner, *J. Chem. Phys.*, **70**, 3895 (1979).
38. J. E. Kenny, K. E. Johnson, W. F. Sharfin, and D. H. Levy, *J. Chem. Phys.*, **72**, 1109 (1980).
39. G. Kubiak, P. S. H. Fitch, L. Wharton, and D. H. Levy, *J. Chem. Phys.*, **68**, 4477 (1978).
40. R. E. Smalley, L. Wharton, and D. H. Levy, *Chem. Phys. Lett.*, **51**, 392 (1977).
41. J. A. Blazy and D. H. Levy, *Chem. Phys. Lett.*, **51**, 395 (1977).
42. J. A. Blazy and D. H. Levy, *J. Chem. Phys.*, **69**, 2901 (1978).
43. J. A. Blazy, B. M. de Koven, T. Russell, and D. H. Levy, to be published.
44. J. Wei and J. Tellinghuisen, *J. Mol. Spectry.*, **50**, 317 (1974).
45. J. Tellinghuisen, *J. Chem. Phys.*, **57**, 2397 (1972).
46. W. Sharfin, K. E. Johnson, L. Wharton, and D. H. Levy, *J. Chem. Phys.*, **71**, 1292 (1979); J. Tusa, M. Sulkes, and S. A. Rice, *J. Chem. Phys.*, **70**, 3136 (1979).
47. J. Tellinghuisen, *J. Quant. Spec. Rad. Trans.*, **19**, 149 (1978).
48. G. Ewing, *Chem. Phys.*, **29**, 253 (1978); J. A. Beswick and J. Jortner, *J. Chem. Phys.*, **68**, 2277, 2525 (1977).
49. W. Sharfin, K. E. Johnson, L. Wharton, and D. H. Levy, *J. Chem. Phys.*, **71**, 1292 (1979).

INTRAMOLECULAR DYNAMICS OF VAN DER WAALS MOLECULES

J. A. BESWICK*

*Laboratoire de Photophysique Moléculaire*** Université de Paris Sud, 91405 Orsay, France*

and

JOSHUA JORTNER

Department of Chemistry, University of Tel-Aviv, Tel Aviv, Israel

CONTENTS

I Introduction 364
II Models for Vibrational Predissociation of Triatomic van der Waals Molecules 370
II.A Some Facts 370
II.B Collinear Analytical Models 372
1. Frozen diatomic model and the simplest golden-rule expressions 372
2. Rosen's relative coordinate treatment 388
3. The diabatic distorted wave treatment 390
4. General formalism for VP decay incorporating final-state interactions . . . 395
5. Some matrix elements 407
II.C Numerical Methods 413
1. Scattering formalism 413
2. Comparison between numerical and distorted wave calculations 417
3. Application to perpendicular vibrational predissociation 420
III Intramolecular Dynamics of van der Waals Dimers 427
III.A Some Facts and Correlations 427
III.B VP Dynamics of Heterodimers 428
III.C The V→T Process in Heterodimers 431
III.D The V→V+T Process in Heterodimers 435
III.E Intramolecular Dynamics and VP of Homodimers 439
IV Vibrational Predissociation of Polyatomic van der Waals Molecules 453
IV.A Background Information 453
IV.B Model Calculations of VP of Polyatomic Complexes 454
IV.C A Comment on VP of Electronically Excited Polyatomic Complexes 463

*Work supported in part by Université Pierre et Marie Curie, UER 52, Paris, France
**Laboratoire du Central National de la Recherche Scientifique

V Concluding Remarks 464
VI Recent Progress in VP Dynamics 466

I. INTRODUCTION

According to Pauling[1] there is a chemical bond whenever it is convenient for the chemist to consider a molecular species, resulting from interatomic or intermolecular interactions, to be distinct and independent. This subjective definition encompasses both traditional chemical bonds, characterized by dissociation energies of a few eV, as well as weakly bound interatomic and intermolecular aggregates stabilized by interaction energies of 10^{-2}–10^{-3} eV. van der Waals (vdW) molecules[2,3] are weakly bound molecular complexes held together at their equilibrium configuration by attractive intermolecular interactions between closed-shell atoms or molecules. The term "vdW molecule" is not restricted to the stabilization of an aggregate by dispersive, London-type, forces but pertains to a broad class of intermolecular clustering due to electrostatic forces, dispersive forces, charge transfer interactions, hydrogen bonding, etc. The primary characteristics of the vdW molecules are their low (10 cm^{-1}–500 cm^{-1}) dissociation energies, the large length of the vdW bond, as well as the retention of the individual properties of the molecular constituents within the vdW aggregate.

The elucidation of the structure and energetics of vdW molecules pertains to the basic understanding of intermolecular interactions in chemistry.[4,5] The implications of vdW binding in determining a wide variety of gas-phase and interface properties were invoked in many areas of chemical physics. These include equations of state and transport properties of gases,[3] vdW binding in three-body recombination reactions,[6] the influence of vdW molecules on gas-surface absorption and on the primary steps of nucleation phenomena,[6] as well as the role of vdW molecules in vibrational relaxation[7] and relaxation of electron spin in low-pressure optical pumping experiments.[8] Van der Waals molecules exist at very low concentration in any gas phase system at room temperature. For example, gaseous Ar contains about 1% of the Ar_2 molecule at 100 K.[6] Conventional spectroscopic studies, which are expected to provide useful information concerning the structure and the energetics of vdW molecules, are severely hampered by the low concentration of the species, by the overlap of their spectra with those of the constituents, as well as by the conventional effects of rotational broadening and of the appearance of vibrational sequence congestion exhibited at room temperature. To overcome the problem of spectral overlap, which was found to be severe in the case of the infrared spectra of Ar–HCl and other hydrogen halide complexes,[9] Welsh et al.[10] and Ewing et al.[11] conducted extensive studies of the infrared spectra of vdW molecules, where the constituents are infrared inactive. A powerful way to overcome the limitations inherent in high-temperature spectroscopic studies of vdW molecules rests on the use of isentropic nozzle beam expansions.[12] At the low translational and rotational temperatures achieved in supersonic jets, the binding energy of the vdW molecules considerably exceeds kT, the problems of spectral overlap are reduced, while the effects of thermal rotational broadening and of sequence congestion are practically eliminated. Thus in supersonic jets the complexes are stable and their spectra are well resolved. Supersonic-free expansion has been utilized to prepare a wide variety of weakly bound vdW molecules. Extensive and exciting body of information is emerging from magnetic and electrical resonance studies pioneered by Klemperer and colleagues,[13] who have explored the structure and the electrical dipole moments of HF dimers and its isotopes, Ar–HCl, Ar–HF, Ar–ClF, Kr–ClF, Ar–COS, as well as large complexes such as the benzene dimer.

These modern spectroscopic experiments do not only probe the structure and the energetics of the complexes but also provide central information regarding the intramolecular nuclear motion in a variety of vdW molecules. Of considerable interest is the nature of the intramolecular dynamics of vdW molecules in vibrationally excited states of the ground electronic configuration and in electronically excited configurations.[14] A central and most interesting feature of excited-state intramolecular relaxation processes in such systems involves the breaking of weak chemical bonds, which are characterized by bond dissociation energies of 10–500 cm^{-1}. This new class of photochemical photofragmentation via vibrational or electronic-vibrational excitation of vdW molecules provides an unambiguous example for vibrational predissociation (VP) of a polyatomic molecules. In the investigation and interpretation of excited-state nonradiative decay processes of medium-sized molecules it is extremely difficult to provide an unambiguous distinction between intrastate VP and interstate electronic relaxation, so that experimental evidence for VP is sparse. The intramolecular dynamics in excited states of vdW molecules provides the first demonstration of the experimental manifestations of VP. This new class of VP of vdW molecules in supersonic beams is of considerable experimental interest as the vibrational excitation energy is relatively low and photoselective chemistry following excitation of individual vibrational (and even vibrational-rotational) levels can be conducted. Recently, Smalley, Levy, Wharton, and their collaborators[15] have provided a pioneering study of the photodissociation dynamics of RI_2 (R=He, Ne, and Ar) vdW molecules; the central results of these important investigations are surveyed by Levy in this volume. Other experimental studies of the dynamics of VP of polyatomic vdW complexes were performed recently. These involve the study by Dixon and Herschbach on collisional excitation of the $(Cl_2)_2$ dimer,[16] the investigation of the electronically-vibrationally excited $HeNO_2$ complex,[17] as well as infrared studies of the $(N_2O)_2$,[18] $(NH_3)_2$,[19] and ethylene dimers.[20]

It is the purpose of this chapter to summarize the recent theoretical developments in the area of intramolecular dynamics of vdW molecules, exploring the chemical and photophysical consequences of the reactive processes occurring on a single nuclear potential surface of a weakly-bound vdW complex. These investigations provide a unified theoretical framework for the understanding of interstate, nonradiative, VP relaxation phenomena. Until 1976 the experimental information concerning molecular VP processes was fragmentary, incomplete, and inclusive. It is not surprising that theory did not precede experiment, and that the interesting area of VP processes was not theoretically explored from the microscopic point of view. Two important classical theoretical contributions to this field were advanced a long time ago. The remarkable 1933 paper of Rosen[21] on the mechanism of decomposition of metastable molecules produced by collisions provides a pioneering contribution to the field. From the point of view of general methodology, this work essentially considered the role of Franck–Condon nuclear vibrational overlap integrals as a dominating ingredient for the proper description of intramolecular dynamics, a long time before this concept became so popular in the fields of reactive collision and photofragmentation. From the technical point of view, Rosen's work considered adequately the effects of anharmonicities of molecular bonds on intramolecular dynamics. An entirely different, but extremely relevant point of view, was advanced by Stepanov[22] in 1946 when he suggested that vibrational predissociation may constitute an important relaxation mechanism in excited states of hydrogen bonded systems. Accordingly, he proposed that the smooth contours of the infrared absorption broad bands experimentally found in $XH\cdots Y$ complexes originate from the superposition of a large number of transitions corresponding to the combination of a $\nu(XH)$ frequency and the low-energy hydrogen bond frequencies $\nu(HX\cdots Y)$, each line being broadened by vibrational predissociation. Thirty years since the publication of Stepanov's paper elapsed before hectic activity started in the area of theory of VP of molecular complexes. Four independent lines of attack on these problems were recently advanced.

A. VP of Hydrogen-Bonded Systems

Coulson and Robertson[23] have conducted detailed calculations on the dynamics of VP of hydrogen-bonded systems. The calculated VP lifetimes were found to be extremely long, so it was concluded that the diffuse spectra in hydrogen-bonded systems cannot be accounted for in terms of Stepanov's mechanism. Coulson and Robertson used a linear triatomic model for the hydrogen-bonded system X-H$\cdots$Y, and the potential energy surface was expressed in terms of two pairwise additive contributions, that is, a harmonic potential for the X-H bond and a Morse potential for the H$\cdots$Y interaction. The calculation of the VP lifetimes were performed by first-order perturbation theory using distorted wavefunctions and golden-rule expressions. Subsequently, Ewing[24] had considered the role of coupling between bending and stretching modes, as well as the effect of interbond vibrational energy exchange on the VP of hydrogen-bonded systems.

B. Vibrational Relaxation in Gases

Ewing[25] addressed himself to the problem of the effect of vdW molecules on the rate of vibrational relaxation in gases at moderately low temperatures. The VP process was treated within the framework of a linear triatomic model using golden-rule rate expressions.

C. VP of ArHCl

Ashton and Child[26] have studied the VP of ArHCl, taking into account the effect of coupling between vibrational and rotational motion.

D. Dynamics of VP and vdW molecules

Beswick and Jortner[27] were inspired by the experimental work of the Chicago group[15] and attempted to provide a comprehensive theoretical framework for the molecular VP process. They advanced a theoretical model for vibrational predissociation of linear triatomic vdW molecules consisting of a rare-gas atom X bound to a diatomic BC. Their analytical treatment rests on the decay of a bound state into a manifold of coupled translational continua. They explored the dependence of the vibrational predissociation rates on the molecular parameters of the complex and have investigated the role of the anharmonicity of the diatomic fragment, the effects of the intercontinuum coupling (final states interaction) on the dynamics of the vibrational predissociation process and on the final state vibrational distributions. The zero-order states for the internal motion were chosen as products of a vibrational wavefunction for the BC bond and a bound (or unbound) nuclear wavefunction representing the motion of X relative to the center of mass of the BC molecule which has been frozen at its equilibrium configuration. The residual interaction representing the deviation between the total interaction potential of X with the vibrating BC molecule and the interaction of X with the frozen diatomic induced discrete–continuum and continuum–continuum couplings. On the basis of the analysis of these coupling terms they have asserted that the zero-order resonance widths are usually small relative to their spacings and, furthermore, that continuum–continuum coupling prevails essentially only between adjacent continua. The dynamics of VP was then reduced to the problem of the decay of a single discrete state into a manifold of adjacently coupled continua. Closed analytical expressions for the rate of VP of the complex and for the vibrational distribution of the BC product were derived by incorporating the effects of discrete–continuum and continuum –continuum coupling. Beswick and Jortner have also considered[28] the dynamics of a T-shaped vdW molecule consisting of a rare-gas atom X constrained to move on a line perpendicular to the interatomic axis of the BC molecule. In contrast to the colinear case, this problem does not result in an analytical solution and numerical methods have been applied. One of these methods involves the use of conventional scattering solutions of the Schrödinger equation resting on the numerical solutions of the resulting close-coupling equations. From the study of the probabilities for inelastic transitions in the collision of X with BC, it is possible to extract information about the quasibound states of the complex. These states (also called resonances, collision complexes, compound states, metastable states) manifest themselves as dips or peaks at energies which approximately correspond to their zero-order energies. The widths of these resonances can then be related to the rate for vibrational predissociation. This method was applied for the investiation of the VP of HeI_2.[29] A by-product of this study involves the application of this numerical technique to the case of colinear systems and in this way to provide a test for the accuracy of the approximate analytical results. Good correspondence between exact numerical and approximate analytical results for the linear systems XI_2 was found.[28] Subsequently, Beswick and Jortner extended their formalism to treat intramolecular dynamics in more complicated vdW complexes.[30] In an attempt to understand the surprising stability of the $(Cl_2)_2$ dimer with respect to VP, discovered by Dixon and Herschbach,[16] Beswick and Jortner[30] have considered VP of linear AB—CD dimers, focusing attention on the competition between intramolecular interchange of energy between the molecular bond modes and the energy flow from a bond mode to the vdW bond. The interesting consequences of interband vibrational energy transfer, which result in a dramatic enhancement of the efficiency of the VP process, were also explored. Finally, further studies have been conducted[31] on VP of polyatomic vdW molecules, exploring both intramolecular as well as intermolecular energy exchange in these interesting systems.

Before alluding to a review of the interesting results of all these theoretical studies, we would like to emphasize that the problem of VP of vdW complexes is intimately related to a broad class of processes involving intramolecular and intermolecular vibrational energy exchange in molecular systems and in condensed phases, which pertain to some basic problems of reactive processes in chemistry. The elucidation of these dynamic processes in vdW complexes is of interest for the following reasons:

1. The fragmentation of the vdW bond provides a unique example for VP on a single electronic potential surface. A complete understanding of the role of vibrational energy exchange in VP will elucidate the nature of a basic molecular relaxation process.
2. The understanding of VP of polyatomic vdW molecules is relevant for establishing the general features of bond breaking processes in chemical systems.
3. These processes will determine the VP mechanism and the internal energy content of the polyatomic fragments A+B resulting from the fragmentation of vibrationally excited vdW complexes, as well as conventional polyatomic molecules.
4. The process of intramolecular vibrational energy flow from a conventional chemical bond to a vdW bond is a problem of current interest in the area of intramolecular dynamics. In particular, studies of intramolecular vibrational energy flow in vdW complexes will be helpful in assessing the range of applicability of statistical theories for vibrational energy redistribution in molecular systems. This physical situation is relevant for the understanding of possible nonstatistical vibrational energy redistribution in large molecules, which pertain to basic chemical processes, such as unimolecular reactions and high-order multiphoton molecular excitation.
5. Vibrational energy exchange in vdW complexes is central for the understanding of the mechanism of collision-induced vibrational relaxation particularly at low translational temperatures. Recent studies[32] of such atom–diatom vibrational relaxation should be extended to low-energy collision of polyatomic molecules.
6. The problem of vibrational relaxation of guest polyatomic molecules in host matrices may also benefit from the understanding of vibrational energy energy exchange. Specific chemical interactions of the vdW type have a profound effect on the vibrational relaxation of some diatomic radicals in mixed rare-gas solids.[33] The role of intermolecular vibrational energy exchange in the VP of vdW complexes may be important for the elucidation of vibrational relaxation of polyatomic molecules in a monoatomic or polyatomic host matrix.
7. The VP dynamics of hydrogen-bonded complexes is of considerable interest. Several workers have considered[22–24] the V$\rightarrow$T decay channel,

concluding that the inefficiency of the latter process precludes any appreciable line broadening. Nevertheless, intramolecular and intermolecular vibrational energy exchange may play a central role in the intramolecular dynamics of hydrogen-bonded complexes and may be important for the broadening of their IR absorption bands.[24]

We have just stated the general goals of the theory of intramolecular dynamics of vdW molecules. We shall now proceed to discuss the theoretical concepts, as well as the implications and applications of the theory of basic VP processes. The exposition will be organized not in a historical order but rather according to the increasing complexity of the molecular constituents of the vdW complex.

II. MODELS FOR VP OF TRIATOMIC vdW MOLECULES

A. Some Facts

Wharton, Levy and their collaborators[15] have conducted extensive experimental studies of the VP of triatomic $X—I_2(B)$, X=He, Ne, and Ar vdW molecules formed between a rare-gas atom and an iodine molecule. A comprehensive survey of this pioneering work is provided by Levy[15f] in this volume. In what follows we shall briefly review some of these experimental results, which will subsequently be confronted with our theory. Utilizing the techniques of laser spectroscopy in supersonic beams, Levy, Wharton et al. have provided extensive and important data concerning the lifetimes for VP and the identifications of the decay channels of these triatomic vdW molecules, where the I_2 component is electronically-vibrationally excited to its $B^3\Pi$ electronic configuration and to a v vibrational state. The first relevant question one has to consider is whether the VP channel

$$X—I_2(B, v) \rightarrow X + I_2(B, v-n) \qquad n=1,2,3\ldots \tag{2.A.1}$$

is dominating or whether a second competing nonradiative decay channel, which involves electronic predissociation, induced by the presence of the rare-gas atom

$$X—I_2(B, v) \rightarrow X + 2I \tag{2.A.2}$$

also prevails. For the case of $HeI_2(B)$ the contribution of the electronic predissociation channel is negligible.[15b] On the other hand, for ArI_2 the rates of the electronic predissociation and of VP are comparable, and one has to separate the contribution of the VP channel.[15d] The following experimental data for VP of X—I_2(B) triatomic vdW molecules were reported:

1. He—I_2 exhibits efficient VP, the rates being 4.5×10^9 sec^{-1} for $v=12$, increasing to 2.6×10^{10} sec^{-1} for $v=26$.[15a]
2. The VP rates $\Gamma_{diss}(v)$ for the He–I_2(B) follow a superlinear dependence on the vibrational quantum number v of the I—I bond, which can be fit by the form[15c]

$$\Gamma_{diss}(v)=5.55\times 10^{-5}v^2+1.74\times 10^{-6}v^3(cm^{-1}) \qquad (2.A.3)$$

   This superlinear relationship clearly reflects the effects of the anharmonicity of the molecular bond on the VP dynamics.
3. For Ar–I_2 the VP rate, obtained after separating out the contribution of the electronic predissociation channel, exhibits a steep vibrational dependence, which is roughly proportional to v^5.[34]
4. The propensity rule $\Delta v = -1$ for the VP of HeI_2 was established[15b] for a wide range of vibrational levels from $v=1$ up to $v=43$. This propensity rule holds as long as level spacing between adjacent vibrational states exceeds the energy of the vdW bond.
5. In the VP of Ne–I_2 and of Ar–I_2 two or three quantum jumps in the vibrational ladder of the I—I bond are exhibited.[34] These results exhibit the features of stronger vdW bonding in these complexes, as compared to the case of He–I_2

These experimental results are of considerable interest as they demonstrate the efficiency of VP of the X—I_2(B) complexes, clearly indicating the vibrational energy of the I—I bond, which in the $B^3\Pi$ state is characterized by the frequency of 124 cm^{-1} and anharmonicity of 0.83 cm^{-1}, can be effectively transferred to rupture the vdW bond. A VP process induced by the flow of one vibrational quantum of $\leqslant 124$ cm^{-1} constitutes an efficient process in He–I_2(B), where the energy of the vdW bond, D, being low, $D \leqslant 18$ cm^{-1}.[34] For Ar–I_2(B), where the vdW bond energy is higher, $D \cong 130$–160 cm^{-1}, one-quantum jumps are impossible, while 2 and 3 quantum jumps constitute an efficient intramolecular VP process.[34]

We shall now proceed to outline the theory of VP for colinear triatomic molecules. The advantage of such colinear models is that closed analytical solutions can be derived for the VP rate and for the vibrational distribution of the diatomic fragment. This simple model provides considerable insight into the dynamics of VP and yields some predictions which will be useful in designing new experiments.

B. Colinear Analytical Models

1. Frozen Diatomic Model and the Simplest Golden-Rule Expression

In the colinear models the triatomic vdW molecule is restricted to one-dimensional motion. In order to describe the nuclear dynamics of a triatomic system X$\cdots$BC in one dimension, a possible set of coordinates particularly well adapted to the asymptotic region, with X very far from the BC molecule, involves the interatomic distance R_{BC} of the conventional molecule BC, and the distance $R_{X,BC}$ between the atom X and the center of mass of the molecule BC

$$R_{X,BC} = R_{XB} + \gamma R_{BC}$$
$$\gamma = \frac{m_C}{m_B + m_C} \tag{2.B.1}$$

The Hamiltonian for nuclear motion, obtained after the separation of the center of mass motion of the entire system, assumes the form

$$H = -(\hbar^2/2\mu_{X,BC})\frac{\partial^2}{\partial R^2_{X,BC}} - (\hbar^2/2\mu_{BC})\frac{\partial^2}{\partial R^2_{BC}}$$
$$+ V_{BC}(R_{BC}) + U(R_{X,BC}, R_{BC}) \tag{2.B.2}$$

where μ_{BC} and $\mu_{X,BC}$ are the reduced masses:

$$\mu_{BC} = \frac{m_B m_C}{m_B + m_C}$$
$$\mu_{X,BC} = \frac{m_X(m_B + m_C)}{m_X + m_B + m_C} \tag{2.B.3}$$

In (2.B.2), $V_{BC}(R_{BC})$ is the potential energy for the free BC molecule, so that $U(R_{X,BC}, R_{BC})$ is the vdW interaction which should vanish when R_{XB} goes to infinity. Notice that there is no loss of generality in writing the

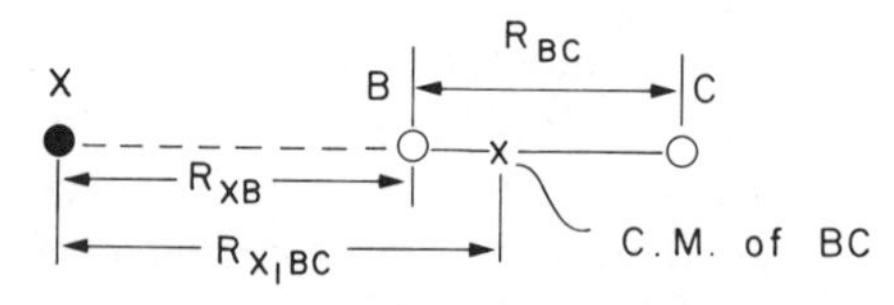

Fig. 1. Coordinate system used to describe the one-dimensional motion of the linear triatomic vdW molecule X—AB, where X is a rare-gas atom.

total Hamiltonian (2.B.2) in this form. A zero-order basis set may be obtained now by segregating the Hamiltonian in the following manner:

$$H = H_0 + V \tag{2.B.4}$$

the zero-order Hamiltonian H_0 being[23c, 27]

$$H_0 = H_{BC} + H_{X,BC} \tag{2.B.5}$$

$$H_{BC} = -(\hbar^2/2\mu_{BC})\frac{\partial^2}{\partial R_{BC}^2} + V_{BC}(R_{BC}) \tag{2.B.5a}$$

$$H_{X,BC} = -(\hbar^2/2\mu_{X,BC})\frac{\partial^2}{\partial R_{X,BC}^2} + U(R_{X,BC}, \bar{R}_{BC}) \tag{2.B.5b}$$

where $\bar{R}_{BC}$ is an arbitrary fixed distance between atoms B and C, while the residual perturbation term is

$$V = U(R_{X,BC}, R_{BC}) - U(R_{X,BC}, \bar{R}_{BC}) \tag{2.B.6}$$

The zero-order Hamiltonian (2.B.5) corresponds to separable contributions from a vibrating BC bond and from the motion of X relative to the center of mass of BC, which is frozen at distance $\bar{R}_{BC}$ (usually the equilibrium interatomic distance for the free BC molecule). The zero-order "nuclear diabatic" solutions for the zero-order Hamiltonian (2.B.5) consist of the following:

1. Discrete, bound, vibrational states of the vdW molecule

$$\langle R_{BC}, R_{X,BC}|vl\rangle = \chi_v(R_{BC})\phi_l(R_{X,BC}) \tag{2.B.7}$$

   where v denotes the discrete vibrational quantum number of the BC molecule bond, while l corresponds to the vibrational quantum number of the vdW bond. These bound states are characterized by the energies $E_{vl} = W_{BC}(v) + \epsilon_l$, where $W_{BC}(v)$ and ϵ_l correspond to the energies of the discrete levels χ_v and ϕ_l, respectively. The χ_v and ϕ_l wavefunctions satisfy the eigenvalue equations:

$$H_{BC}\chi_{BC}(R_{BC}) = W_{BC}(v)\chi_v(R_{BC}) \tag{2.B.8a}$$

$$H_{X,BC}\phi_l(R_{X,BC}) = \epsilon_l\phi_l(R_{X,BC}) \tag{2.B.8b}$$

   with H_{BC} and $H_{X,BC}$ defined by (2.B.5),

2. Continuum states of the fragments X+BC

$$\langle R_{\mathrm{Bc}}, R_{\mathrm{X,BC}} | v\epsilon \rangle = \chi_v(R_{\mathrm{BC}})\phi_\epsilon(R_{\mathrm{X,BC}}) \tag{2.B.9}$$

where v denotes again the vibrational quantum number for the BC molecular bond and ϵ designates the relative kinetic energy between X and BC. The continuum wavefunctions are assumed to be energy normalized, being characterized by the energies $E_{v\epsilon} = W_{\mathrm{BC}}(v) + \epsilon$. The zero-order nuclear diabatic states are coupled by V (2.B.6), the relevant discrete–discrete (d–d), discrete–continuum (d–c), and continuum–continuum (c–c) coupling terms are

$$\begin{aligned} V^{\mathrm{d\text{-}d}}_{vl,v'l'} &\equiv \langle vl | V | v'l' \rangle \\ V^{\mathrm{d\text{-}c}}_{vl,v'\epsilon'} &\equiv \langle vl | V | v'\epsilon' \rangle \\ V^{\mathrm{c\text{-}c}}_{v\epsilon,v'\epsilon'} &\equiv \langle v\epsilon | V | v'\epsilon' \rangle \end{aligned} \tag{2.B.10}$$

The simplest models[23c] for VP will now invoke the following simplifying assumptions:

1. Discrete–discrete interactions are neglected.
2. Continuum–continuum coupling interactions are also neglected.
3. The square of the modulus of the discrete–continuum coupling terms $|V^{\mathrm{d\text{-}c}}_{vl,v'\epsilon'}|^2$ (which are given in units of energy), are assumed to be appreciably smaller than the energy spacing between the discrete levels, that is,

$$|V^{\mathrm{d\text{-}c}}_{vl,v'\epsilon'}|^2 \ll |E_{vl} - E_{v''l''}| \tag{2.B.11}$$

for all v'' and l''. Thus, in this treatment all interference effects between resonances are disregarded.

These points pertain to the description of the energy levels. An additional assumption has to be invoked to specify the "preparation" of the vibrationally excited state or the electronically-vibrationally excited state of the vdW molecule which undergoes VP:

4. Only the discrete zero-order excited states carry oscillator strength from the ground state $|g; v=0, l=0\rangle$ where g refers to the ground state electronic wavefunction. The optical excitations of the vdW molecule are thus described in terms of the radiative coupling $|g; v=0, l=0\rangle \rightarrow |g; v', l'\rangle$ between zero-order discrete levels for infrared excitation or by $|g; v=0, l=0\rangle \rightarrow |s; v'l'\rangle$ for optical excitation, where

s represents a higher electronic configuration. The transition moments to the continuum states that correspond to the radiative couplings $|g; v=0, l=0\rangle \to |g; v'\epsilon'\rangle$ or $|g; v=0, l=0\rangle \to |s; v'\epsilon'\rangle$ are assumed to be negligibly small. For example, in the case of electronic–vibrational excitation it is expected that the transition moment in the Condon approximation, $|\mu_{gs}|^2 |\langle 0|v'\rangle|^2 |\langle 0|l'\rangle|^2$, where $\mu_{gs} \equiv \langle s|\mu|g\rangle$ corresponds to the electronic matrix element of the dipole moment operator, will be largest for $l'=0$ and that the nuclear vibrational overlap $|\langle l=0|\epsilon\rangle|^2$ will be very small. This expectation is borne out by the spectroscopic data of Wharton, Levy, and co-workers[15] on HeI_2 which indicate that the electronic–vibrational transition moment from $|X'\Sigma; v=0, l=0\rangle$ to the $|B^3\Pi; v', l=0\rangle$ state is 1–2 orders of magnitude larger than the transition moment to the $|B^3\Pi; v', l=1\rangle$ state and no evidence is exhibited for Fano type resonances,[35] which indicates that the (zero-order) continuum states do not carry appreciable oscillator strength.

In Fig. 2 the level scheme and coupling terms, according to assumptions 1 to 4, are represented. It should be emphasized that these simplifying assumptions can be relaxed, and in the following sections we shall consider more general situations. In particular, we shall present in Section 2.B.4 the formalism to incorporate the role of c–c coupling. We turn now to the description of the dynamics of the photofragmentation process. Two limiting experimental conditions are usually envisaged to describe photon excitation processes:

1. Energy-resolved (ER) experiments, characterized by infinite spectral resolution and corresponding to a time-independent type of experiment. The observables in this case are the cross sections for all possible scattering events, as well as branching ratios and quantum yields.
2. Time-resolved (TR) experiments, characterized by infinite time resolution and instantaneous excitation. The observables are in this case lifetimes and quantum yields.

Let consider the case of ER experiments. The molecular states $|g; v, l\rangle$, $|g; v, \epsilon\rangle$, $|s; v, l\rangle$, $|s; v, \epsilon\rangle$ correspond to the product of an electronic part ($|g\rangle$ for the ground electronic state and $|s\rangle$ for the excited configuration) and vibrational eigenstates of the zero-order molecular Hamiltonian H_0 defined in (2.B.5). The total Hamiltonian $\mathcal{H}$ for the molecular system and the radiation field involves four terms $\mathcal{H} = H_0 + V + H_{\text{rad}} + H_{\text{int}}$. The molecular terms H_0 and V have been defined in (2.B.4)–(2.B.6). The H_{rad} term is the free radiation field Hamiltonian while H_{int} is the radiation field-matter interaction term. For weak photon fields, the relevant eigenstates of H_{rad} are the zero-photon state $|\text{vac}\rangle$ and the one photon states $|\mathbf{k}, \mathbf{e}\rangle$, where $\mathbf{k}$ is

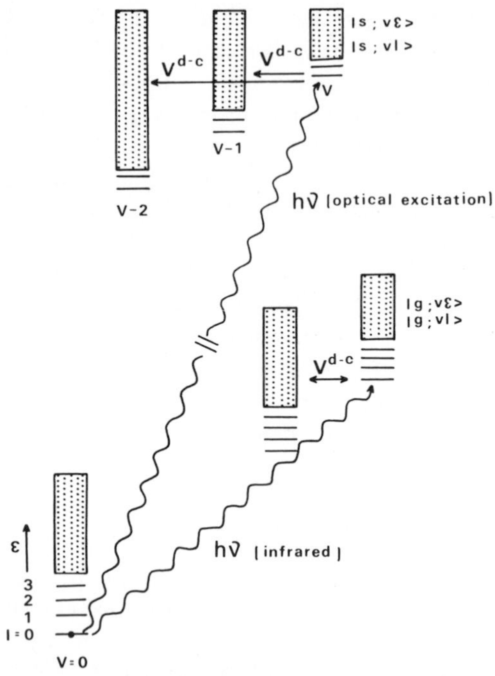

Fig. 2. Spectrum of zero-order Hamiltonian and the relevant coupling terms appropriate for the description of VP of van der Waals molecule X—AB, where X is a rare-gas atom and BC is a conventional diatomic molecule. The system can be initially excited by infrared excitation to the vibrationally excited state $|g; vl\rangle$ and by excitation in the optical region to the electronically–vibrationally excited state $|s; vl\rangle$. Here v is the quantum number associated with the vibration of the BC molecule, while l is the quantum number associated with the bound motion of X with respect to BC.

the photon wavevector and $\mathbf{e}$ its polarization vector. In the case of optical excitation the relevant zero-order states are $|g; v_0, l_0; k_0, e_0\rangle$, which correspond to the initial state (usually the ground state with $v_0=0$ and $l_0=0$) and an incident photon of energy $\hbar k_0 c$, the doorway state $|s; v, l; \text{vac}\rangle$, which is in resonance or quasiresonance with the exciting field, that is, $E_{svl}-E_{gv_0l_0}\sim\hbar k_0 c$, and the final continuum states $|s; v', l'; \text{vac}\rangle$ and $|g; v'', l''; \mathbf{k}, \mathbf{e}\rangle$. The initial state $|g; v_0, l_0; \mathbf{k}_0, \mathbf{e}_0\rangle$ is coupled by H_{int} to $|s; v, l; \text{vac}\rangle$, which in turn is coupled by V to the continua $|s; v', l'; \text{vac}\rangle$ and by H_{int} to the $|g; v'', l''; \mathbf{k}, \mathbf{e}\rangle$ states. These coupling terms correspond to the processes of absorption of light, dissociation, and reemission of light, respectively.

Scattering theory furnishes a general expression for the cross section of this process at total energy $E=E_{gv_0l_0}+\hbar k_0c=E_{sv'\epsilon'}$:

$$\sigma(E)^{gv_0l_0 \overset{\mathbf{k}_0,\mathbf{e}_0}{\rightarrow} sv'}=\frac{2\pi}{\hbar c}|\langle s;v',\epsilon';\text{vac}|T(E)|g;v_0,l_0;\mathbf{k}_0,\mathbf{e}_0\rangle|^2 \tag{2.B.12}$$

Where $T(E)$ is the transition operator

$$T(E)=(H_{\text{int}}+V)\left[1+G^+(E)(H_{\text{int}}+V)\right] \tag{2.B.13}$$

$$G^+(E)=(E^+-\mathcal{H})^{-1} \tag{2.B.14}$$

and E^+ stands for $E+i\eta$ with $\eta\rightarrow 0^+$.

The Hilbert space is segregated into two parts,

$$P\equiv|s;v,l;\text{vac}\rangle\langle s;v,l;\text{vac}|$$

$$Q\equiv\sum_{v'}\int d\epsilon'|s;v',\epsilon';\text{vac}\rangle\langle s;v',\epsilon';\text{vac}|$$

$$+\sum_{v''}\sum_{l''}\sum_{\mathbf{e}}\int d^3\mathbf{k}|g;v'',l'';\mathbf{k},\mathbf{e}\rangle\langle g;v'',l'';\mathbf{k},\mathbf{e}| \tag{2.B.15}$$

where the subspace spanned by P contain only the discrete doorway state, while the projection operator Q spans all the continuum channels. With these definitions, it is noticed that the right-hand side of (2.B.12) is given in terms of the operator $QT(E)Q$. Making use of assumptions 1 to 4, one obtains from (2.B.13),

$$QT(E)Q=Q(H_{\text{int}}+V)PG^+(E)P(H_{\text{int}}+V)Q \tag{2.B.16}$$

The operator $PG^+(E)P$ can be expressed in terms of the level shift operator $R(E)$,

$$PG^+(E)P=(E^+-E_{svl}-PR(E)P)^{-1}P \tag{2.B.17}$$

with

$$R(E)=(H_{\text{int}}+V)\left[1+(E^+-Q\mathcal{H}Q)^{-1}(H_{\text{int}}+V)\right] \tag{2.B.18}$$

Explicitly, from (2.B.18) and (2.B.15),

$$PR(E)P=P[\Delta(E)-i\Gamma(E)] \tag{2.B.19}$$

with

$$\Delta(E)=\sum_{v''l''\mathbf{e}}\mathcal{P}\int d^3\mathbf{k}\frac{|\langle s;\,vl;\mathrm{vac}|\,H_{\mathrm{int}}|g;\,v''l'',\mathbf{ke}\rangle|^2}{E-E_{gv''l''}-\hbar kc}$$
$$+\sum_{v'}\mathcal{P}\int d\epsilon'\frac{|\langle s;\,vl;\mathrm{vac}|V|s;\,v'\epsilon';\mathrm{vac}\rangle|^2}{E-E_{sv'\epsilon'}} \tag{2.B.20a}$$

where $\mathcal{P}$ denotes the Cauchi principal part, and

$$\Gamma(E)=\pi\sum_{v''l''\mathbf{e}}\int d\Omega_k k^2|\langle s;\,vl;\mathrm{vac}|\,H_{\mathrm{int}}|g;\,v''l'';\mathbf{ke}\rangle|^2\Big|_{k=\frac{(E-E_{gv''l''})}{\hbar c}}$$
$$+\pi\sum_{v'}|\langle s;\,vl;\mathrm{vac}|V|s;\,v'\epsilon';\mathrm{vac}\rangle|^2\Big|_{\epsilon'=E-E_{sv'}} \tag{2.B.20b}$$

Substituting (2.B.19) into (2.B.17) and the result into (2.B.16) and (2.B.12),

$$\sigma(E)^{gv_0l_0\overset{\mathbf{k}_0\mathbf{e}_0}{\to}sv'}=\frac{2\pi}{\hbar c}|\langle s;\,v',\epsilon';\mathrm{vac}|V|s;\,vl;\mathrm{vac}\rangle|^2$$
$$\times\frac{|\langle s;\,vl;\mathrm{vac}|\,H_{\mathrm{int}}|g;\,v_0l_0;\mathbf{k}_0\mathbf{e}_0\rangle|^2}{(E-E_{svl}-\Delta(E))^2+\Gamma^2(E)} \tag{2.B.21}$$

Expression (2.B.21) provides the final formal expression for the cross-section for dissociation into channel s, v' when the system is excited with monochromatic light in the vecinity of the transition $(gv''l'')\to(svl)$. If all the coupling matrix elements are slowly varying functions of the energy, (2.B.21) gives the usual Lorentzian predissociation profile centered at $\tilde{E}_{svl}=E_{svl}+\Delta(E_{svl})$ with half-width equal to $\Gamma(E_{svl})$. Notice that Γ given by (2.B.20) is equal to the sum of the radiative widths and the dissociation widths. Usually the former are small compared with the latter.

All these cross-sections are obtained for a well-defined photon energy. In an actual experiment, the exciting light source is not necessarily monochromatic. Excitation of the system by a photon wavepacket $|\psi_0\rangle=\Sigma_{\mathbf{e}_0}\int d^3\mathbf{k}_0A_{\mathbf{ke}_0}|\mathbf{k}_0,\mathbf{e}_0\rangle$, will result in a cross-section that can be expressed in terms of the integral of σ over the power spectrum of the pulse

$$\sigma^{gv_0l_0\to sv'}=\sum_{\mathbf{e}_0}\int d^3\mathbf{k}_0|A_{\mathbf{k}_0,\mathbf{e}_0}|^2\sigma(E)^{gv_0l_0\overset{\mathbf{k}_0\mathbf{e}_0}{\to}sv'},\;E=E_{gv_0l_0}+\hbar k_0c \tag{2.B.22a}$$

In the limit of a very broad pulse, i.e., when the width of the photon distribution is much larger than Γ, one obtains (neglecting the energy dependence of the coupling matrix elements and thresholds effects),

$$\sigma^{gv_0l_0\to sv'} \propto |\langle s;\, v'\epsilon';\mathrm{vac}|V|s;\, vl;\mathrm{vac}\rangle|^2 \times \frac{|\langle s;\, vl;\mathrm{vac}|\, H_{\mathrm{int}}|g;\, v_0 l,\mathbf{ke}\rangle|^2}{\Gamma(\tilde{E}_{svl})} \qquad (2.B.22b)$$

where all the matrix elements are evaluated for $\hbar kc = \tilde{E}_{svl} - E_{gv_0l_0}$ and for $\epsilon' = \tilde{E}_{svl} - E_{sv'}$. The total cross-section for dissociation is in this case, $\sigma^{gv_0l_0\to s} = \sum_{v'}\sigma^{gv_0l_0\to sv'}$, and the final relative vibrational distribution will be given by $P_{s,v'} = \sigma^{gv_0l_0\to sv'}/\sigma^{gv_0l_0\to s}$, which, according to (2.B.22b), will be

$$P_{s,v'} = \pi|\langle s;\, v'\epsilon';\mathrm{vac}|V|s;\, vl;\mathrm{vac}\rangle|^2/\Gamma_{\mathrm{diss}}\big|_{\epsilon'=\tilde{E}_{svl}-E_{sv'}} \qquad (2.B.23)$$

where Γ_{diss} is the dissociative contribution to the total width Γ, i.e.,

$$\Gamma_{\mathrm{diss}} = \pi\sum_{v'}|\langle s;\, v'\epsilon';\mathrm{vac}|V|s;\, cl;\mathrm{vac}\rangle|^2\Big|_{E'=\tilde{E}_{svl}-E_{sv'}} \qquad (2.B.24)$$

Let us now consider the TR experimental situation. In this case, after the excitation has taken place, the initial condition $|\psi(t=0)\rangle = |s;\, vl;\mathrm{vac}\rangle$ is invoked and the probability for dissociation at time t is given by,

$$p_{s,v'}(t) = \int d\epsilon'|\langle s;\, v'\epsilon';\mathrm{vac}|U(t,0)|s;\, vl;\mathrm{vac}\rangle|^2 \qquad (2.B.25)$$

where $U(t,0)$ is the time evolution operator. The final (relative) probability distribution of the fragments among the various vibrational channels is $P_{s,v'} \equiv p_{s,v'}(\infty)/[\sum_{v'}p_{s,v'}(\infty)]$ while the decay probability of the initial state is

$$p_{svl}(t) = |\langle s;\, vl;\mathrm{vac}|U(t,0)|s;\, vl;\mathrm{vac}\rangle|^2 \qquad (2.B.26)$$

Using the resolvent operator methods,

$$p_{sv'}(t) = (4\pi^2)^{-1}\int d\epsilon'\left|\int_{-\infty}^{\infty} dE\langle s;\, v'\epsilon';\mathrm{vac}|G^+(E)|s;\, vl;\mathrm{vac}\rangle\exp\left(\frac{-iEt}{\hbar}\right)\right|^2 \qquad (2.B.27)$$

where $G^+(E)$ has been already defined in (2.B.14). In terms of the projection operators P and Q defined in (2.B.15), the evaluation of (2.B.27) requires the calculation of the operator $QG^+(E)P$, which can be shown to be of the form,

$$QG^+(E)P=(E^+-Q\mathcal{H}Q)^{-1}Q\mathcal{H}PPG^+(E)P \tag{2.B.28}$$

where $PG^+(E)P$ is given by (2.B.17). Using (2.B.28) with (2.B.17) and (2.B.19),

$$p_{sv'}(t)=(4\pi^2)^{-1}\int d\epsilon'\left|\int_{-\infty}^{\infty} dE\frac{\langle s;\, v'\epsilon';\text{vac}|V|s;\, vl;\text{vac}\rangle e^{-iEt/\hbar}}{(E^+-E_{sv'\epsilon'})(E^+-E_{svl}-\Delta(E)+i\Gamma(E))}\right|^2 \tag{2.B.29}$$

The integration in (2.B.29) may now be performed by invoking the usual assumptions regarding the weak dependence of $\Delta(E)$ and $\Gamma(E)$ on the energy and the negligible effects of the thresholds. The probability distribution is then given by

$$p_{sv'}(t)=\frac{\pi}{\Gamma}|\langle s;\, v'\epsilon';\text{vac}|V|s;\, vl;\text{vac}\rangle|^2\left[1-\exp\left(\frac{-2\Gamma t}{\hbar}\right)\right] \tag{2.B.30}$$

where again all the matrix elements are evaluated for $E_{sv'}+\epsilon'=\tilde{E}_{svl}\equiv E_{svl}+\Delta(E_{svl})$. For $t\to\infty$

$$p_{sv'}(\infty)=\frac{\pi}{\Gamma}|\langle s;\, v'\epsilon';\text{vac}|V|s;\, vl;\text{vac}\rangle|^2 \tag{2.B.31}$$

and the quantum yield for dissociation will be

$$Y_{\text{diss}}\equiv\sum_{v'}p_{sv'}(\infty)=\frac{\Gamma_{\text{diss}}}{\Gamma} \tag{2.B.32}$$

where Γ_{diss} is the dissociation rate defined by (2.B.24). The relative vibrational distribution of the diatomic fragments among the different vibrational open channels will be

$$P_{sv'}=\frac{\pi|\langle s;\, v'\epsilon';\text{vac}|V|s;\, vl;\text{vac}\rangle|^2}{\Gamma_{\text{diss}}} \tag{2.B.33}$$

which is identical to the result obtained for the broad pulse excitation distribution given by (2.B.23). Using the same techniques, it is possible to show that

$$p_{svl}(t)=\exp\left(\frac{-2\Gamma t}{\hbar}\right) \tag{2.B.34}$$

which show the initial bound state $|s;vl\rangle$ decaying with the lifetime $\tau=\hbar/2\Gamma$. Finally, the fluorescence quantum yield may be obtained by the conservation of the probability,

$$Y_{fl}=1-Y_{\mathrm{diss}}=\frac{\Gamma-\Gamma_{\mathrm{diss}}}{\Gamma} \tag{2.B.35}$$

Usually, in the case of fast vibrational predissociation, the fluorescence quantum yield Y_{fl} is much smaller than 1. In that case $\Gamma\sim\Gamma_{\mathrm{diss}}$ and all the relevant quantities needed to describe VP processes under the assumptions (1) to (4) advanced above, are the "Golden Rule" individual rates

$$\Gamma_{svl\to sv'}=\pi|\langle s;vl|V|s;v'\epsilon'\rangle|^2 \tag{2.B.36}$$

for VP from the initial state $|svl\rangle$ to the final channel $|sv'\epsilon'\rangle$. If the usually small energy shift $\Delta(E_{svl})$ are neglected, the matrix elements in (2.B.36) should be evaluated at the energy E_{svl} of the discrete level $|svl\rangle$. In what follows we shall concentrate on the calculation of these matrix elements.

Before alluding to more detailed calculations it is instructive to advance a simple perturbation type argument, which will elucidate the gross features of the dependence of the VP rate on the excess vibrational energy of the BC bond, and of the final vibrational distribution of the fragments. From (2.B.6) we note that the molecular coupling term V inducing the dissociation depends on the two distances R_{BC} and $R_{\mathrm{X,BC}}$ (see Fig. 1). Let expand this potential in a Taylor series around the equilibrium position $\bar{R}_{\mathrm{BC}}$ of the diatomic molecule BC, keeping only the linear term in the intramolecular displacement of the BC bond,

$$V=\left(\frac{\partial U}{\partial R_{\mathrm{BC}}}\right)_{\bar{R}_{\mathrm{BC}}}\left(R_{\mathrm{BC}}-\bar{R}_{\mathrm{BC}}\right) \tag{2.B.37}$$

Introducing (2.B.37) into (2.B.36) and using (2.B.7) and (2.B.9), we obtain

$$\Gamma_{svl\to sv'}=\pi|\langle\phi_l|(\partial U/\partial R_{\mathrm{BC}})_{\bar{R}_{\mathrm{BC}}}|\phi_{\epsilon'}\rangle|^2$$

$$\otimes|\langle x_v|\left(R_{\mathrm{BC}}-\bar{R}_{\mathrm{BC}}\right)|x_{v'}\rangle|^2 \tag{2.B.38}$$

so that the individual VP rate $\Gamma_{svl \to sv'}$ is approximated by the product of an intermolecular term, depending on the parameters of the van der Waals interaction potential and on the final relative kinetic energy of recoiling fragments ϵ' (determined by conservation of the energy), and an intramolecular factor which depends only on intrinsic properties, (i.e., frequency and anharmonicity), of the normal BC bond. It should be noted that the intramolecular term in (2.B.38) will give a strong propensity rule $\Delta v = -1$ for VP, as even for an anharmonic BC potential the matrix elements $|\langle x_v|(R_{\mathrm{BC}} - \bar{R}_{\mathrm{BC}})|x_{v'}\rangle|^2$ will strongly favor the transition with $v' = v-1$. Thus within the first order approximation $\Gamma_{\mathrm{diss}} \cong \Gamma_{svl \to s(v-1)}$. This result is consistent with the experimental data of Kim, Smalley, Wharton, and Levy[15] for the $\mathrm{HeI_2(B)}$ VP. From the analysis of the fluorescence of the $\mathrm{I_2(B)}$ fragments they concluded that more than 98% of the dissociation proceeds via the $(v-1)$ channel.[15b] For $\mathrm{NeI_2(B)}$ and $\mathrm{ArI_2(B)}$, two or more channels are significantly populated indicating that higher order terms in the Taylor expansion of V should be included.[34]

For the case of $\mathrm{HeI_2(B)}$, it is possible to utilize (2.B.38) to determine the relative contributions of the intramolecular term and of the intermolecular term to the dependence of the VP rate on the excess vibrational energy of the normal bond. When a harmonic model is used for the BC potential, the intermolecular factor is independent of the vibrational excitation, as the relative kinetic energy of the recoiling fragment is constant, and the intramolecular factor is solely responsible for the v dependence of $\Gamma_{\mathrm{diss}} \propto v$. For an anharmonic BC bond the linear v dependence of the intramolecular contribution is not significantly modified for sufficient low values of v. For a Morse type BC potential the intramolecular contribution is

$$|\langle x_v|(R_{\mathrm{BC}} - \bar{R}_{\mathrm{BC}})|x_{v-1}\rangle|^2 = \frac{(2K_{\mathrm{BC}} - 2v + 1)(2K_{\mathrm{BC}} - 2v - 1)}{(2K_{\mathrm{BC}} - v)(K_{\mathrm{BC}} - v)^2} v \otimes (\hbar K_{\mathrm{BC}} / 4\mu_{\mathrm{BC}}\omega_{\mathrm{BC}}) \qquad (2.\mathrm{B}.39)$$

where $K_{\mathrm{BC}} = \omega_{\mathrm{BC}} / 2(\omega x_e)_{\mathrm{BC}}$, with ω_{BC} being the frequency of the BC bond and $(\omega x_e)_{\mathrm{BC}}$ its anharmonicity. For $\mathrm{I_2(B)}$ we get $K_{\mathrm{BC}} = 76.7$. Thus for sufficiently low values of $v < K_{\mathrm{BC}}$ we expect that the intramolecular contribution (2.B.39) for an anharmonic BC bond is still proportional to v. This is apparent from a numerical evaluation of (2.B.39) presented in Fig. 3, where for comparison we have also plotted the experimental data of Johnson, Wharton, and Levy[15c] obtained for $\mathrm{HeI_2(B)}$. They found that the dependence of Γ_{diss} on the vibrational quantum number v of $\mathrm{I_2(B)}$ were fit by the superlinear relationship (2.A.3). When the experimental result is confronted with the intramolecular contribution (2.B.39), it is evident that

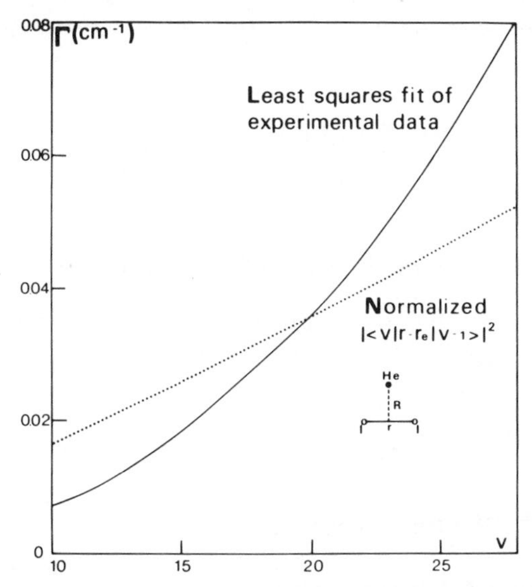

Fig. 3. Dependence of the VP linewidth Γ for VP of HeI_2 on the vibrational quantum number v of I_2. Solid line represents the experimental results of Johnson, Wharton, and Levy,[15(c)] while the dashed line corresponds to the intramolecular contribution.

an additional contribution to the v dependence of the VP rate for an anharmonic BC potential should originate from the intermolecular contribution to (2.B.38). This intermolecular contribution to the v dependence originates from a dynamic effect which is determined by the relative kinetic energy of the fragments. As v increases the relative velocity of the fragments decreases, because of the anharmonicity of the BC potential, and accordingly Γ_{diss} is enhanced. To proceed further, and to prove this assertion an actual calculation of the matrix elements $\langle\phi_l|(\partial U/\partial R_{\text{BC}})_{\bar{R}_{\text{BC}}}|\phi_{\epsilon'}\rangle$ is necessary. For this purpose the vdW interaction U needs to be specified. The simplest choice leading to analytical expressions, is to consider a Morse potential interaction between the two neighboring atoms X and B (see Fig. 1),

$$U = D_{\text{XB}}\left\{\exp\left[-2\alpha_{\text{XB}}\left(R_{\text{XB}}-\bar{R}_{\text{XB}}\right)\right]-2\exp\left[-\alpha_{\text{XB}}\left(R_{\text{XB}}-\bar{R}_{\text{XB}}\right)\right]\right\} \tag{2.B.40}$$

where D_{XB} and $\bar{R}_{\text{XB}}$ are the minimum energy and the equilibrium distance, respectively. It is convenient to define an effective frequency ω_{XB} for the van der Waals bond $\omega_{\text{XB}} = [(\partial^2 U/\partial R_{\text{XB}}^2)/\mu_{\text{X,BC}}]^{1/2}$, where $\mu_{\text{X,BC}} = m_{\text{X}}(m_{\text{B}}+m_{\text{C}})/(m_{\text{X}}+m_{\text{B}}+m_{\text{C}})$ is the reduced mass for the X$\cdots$BC motion. The characteristic inverse length for the vdW bond can then be expressed as

$$\alpha_{XB} = \omega_{XB}(\mu_{X,BC}/2D_{XB})^{1/2} \tag{2.B.41}$$

Defining

$$K_{XB} = 2D_{XB}/\hbar\omega_{XB} = (2\mu_{X,BC}D_{XB})^{1/2}/\alpha_{XB}\hbar \tag{2.B.42}$$

it follows that the number of bound levels supported by a Morse potential is given by N=integer $(K_{XB}+1/2)$ and that the energy of the levels is given by

$$\begin{aligned}\epsilon_l &= -D_{XB} + \hbar\omega_{XB}\left(l+\tfrac{1}{2}\right)\left[1-(2K_{XB})^{-1}\left(l+\tfrac{1}{2}\right)\right] \\ &= -\left(\hbar^2\alpha_{XB}^2/2\mu_{X,BC}\right)\left(K_{XB}-l-\tfrac{1}{2}\right)^2\end{aligned} \tag{2.B.43}$$

Using (2.B.41) and (2.B.1) we have

$$\langle\phi_l|(\partial U/\partial R_{BC})_{\bar{R}_{BC}}|\phi_{\epsilon'}\rangle = -2\alpha_{XB}\left(B_{l\epsilon'}^{(2)} - B_{l\epsilon'}^{(1)}\right) \tag{2.B.44}$$

where

$$B_{\alpha\beta}^{(j)} = D_{XB}\int \partial R_{X,BC}\phi_\alpha(R_{X,BC})\exp\left[-j\alpha_{XB}\left(R_{X,BC}-\bar{R}_{X,BC}\right)\right]$$
$$\otimes\phi_\beta(R_{X,BC}) \quad j=1,2 \quad \alpha,\beta\equiv l \text{ or } \epsilon' \tag{2.B.45}$$

In order to perform the integration in (2.B.45) it is necessary to find the discrete and continuum wavefunctions $\phi_\alpha(R_{X,BC})$ of the $H_{X,BC}$ Hamiltonian defined in (2.B.5b). The discrete normalized eigenfunctions are given by[21]

$$\phi_l(R_{X,BC}) = \left[\frac{(2K_{XB}-2l-1)\alpha_{XB}}{l!\Gamma(2K_{XB}-l)}\right]^{1/2} z^{-1/2}W_{K_{XB},K_{XB}-l-1/2}(z) \tag{2.B.46}$$

where

$$Z = 2K_{XB}\exp\left[-\alpha_{XB}\left(R_{X,BC}-\bar{R}_{X,BC}\right)\right] \tag{2.B.47}$$

with $W_{k,\mu}(Z)$ being the Whitaker's function.[36] The energy-normalized continuum eigenfunctions are similarly given by

$$\phi_{\epsilon'}(R_{\mathrm{X,BC}})=(\pi\hbar)^{-1}\left[\left(\frac{\mu_{\mathrm{X,BC}}}{\alpha_{\mathrm{XB}}}\right)\sinh(2\pi\theta_{\epsilon'})\right]^{1/2} \otimes|\Gamma(\tfrac{1}{2}-K_{\mathrm{XB}}-i\theta_{\epsilon'})|Z^{-1/2}W_{K_{\mathrm{XB}},i\theta_{\epsilon}}(Z) \quad (2.B.48)$$

with

$$\theta_{\epsilon'}=(\hbar\alpha_{\mathrm{XB}})^{-1}(2\mu_{\mathrm{X,BC}}\epsilon')^{1/2}\equiv\frac{2(D_{\mathrm{XB}}\epsilon')^{1/2}}{\hbar\omega_{\mathrm{XB}}} \quad (2.B.49)$$

Using the integrals[21]

$$\int_0^{\infty}W_{k,i\theta}(Z)W_{k,k-l-1/2}(Z)\,dZ=|\Gamma(\tfrac{1}{2}+K-i\theta)|^2\left[(K-l-\tfrac{1}{2})^2+\theta^2+2K\right] \quad (2.B.50)$$

and

$$\int_0^{\infty}W_{K,i\theta}(Z)W_{K,K-l-1/2}(Z)(\partial Z/Z)=|\Gamma(\tfrac{1}{2}+K-i\theta)|^2 \quad (2.B.51)$$

it is obtained

$$B_{l\epsilon'}^{(1)}=\frac{1}{2}\left[\frac{D_{\mathrm{XB}}}{2}\sinh(2\pi\theta_{\epsilon'})\frac{(2K_{\mathrm{XB}}-2l-1)}{l!\Gamma(2K_{\mathrm{XB}}-l)}\right]^{1/2} \otimes\frac{|\Gamma(\tfrac{1}{2}+K_{\mathrm{XB}}-i\theta_{\epsilon'})|}{\left[\cos^2(\pi K_{\mathrm{XB}})+\sinh^2(\pi\theta_{\epsilon'})\right]^{1/2}} \quad (2.B.52a)$$

$$B_{l\epsilon'}^{(2)}=(B_{l\epsilon'}^{(1)}/2K_{\mathrm{XB}})\left[(K_{\mathrm{XB}}-l-\tfrac{1}{2})^2+\theta_{\epsilon'}^2+2K_{\mathrm{XB}}\right] \quad (2.B.52b)$$

Introducing (2.B.52) into (2.B.44) and the result together with (2.B.39) into (2.B.38) it is obtained

$$\Gamma_{\mathrm{diss}}\approx(\pi/8)\hbar\omega_{\mathrm{BC}}(\mu_{\mathrm{X,BC}}/\mu_{\mathrm{BC}})v\frac{(2K_{\mathrm{BC}}-2v+1)(2K_{\mathrm{BC}}-2v-1)}{2K_{\mathrm{BC}}(2K_{\mathrm{BC}}-v)} \times\left\{\frac{\sinh(2\pi\theta_{\epsilon'})}{\cos^2\pi K_{\mathrm{XB}}+\sinh^2(\pi\theta_{\epsilon'})}\right\}|\Gamma(K_{\mathrm{XB}}+\tfrac{1}{2}-i\theta_{\epsilon'})|^2 \times\left[\frac{2K_{\mathrm{XB}}-2l-1}{l!\Gamma(2K_{\mathrm{XB}}-l)}\right] \quad (2.B.53)$$

where $\theta_{\epsilon'}$ has been defined in (2.B.49). Usually the factor $\pi\theta_{\epsilon'}$ is much larger than 1. Utilizing the expansion of the gamma function[36]

$$|\Gamma(N-i\theta)|^2=\left\{\prod_{m=0}^{N-1}\left[(N-m)^2+\theta^2\right]\right\}2\pi\theta\exp(-\pi\theta) \quad (2.B.54)$$

then for $\theta \gg N$, (2.B.53) result in

$$\Gamma_{\text{diss}} \approx \pi^2\hbar\omega_{BC}(\mu_{X,BC}/\mu_{BC})v\frac{(2K_{BC}-2v+1)(2K_{BC}-2v-1)}{2K_{BC}(2K_{BC}-v)}$$

$$\otimes\left[\frac{(N-l-1)}{l!(2N-l-1)!}\right]\theta_{\epsilon'}^{2N-1}\exp(-\pi\theta_{\epsilon'}) \quad (2.B.55)$$

where N is the number of bound levels supported by the vdW bond ($N \sim K_{XB}+1/2$). Equation (2.B.55) together with the definition of $\theta_{\epsilon'}$ given in (2.B.49) provides a useful semiquantitative description of the dependence of the VP dynamics on the molecular parameters of the vdW molecule, such as the relative kinetic energy of the fragments ϵ', the molecular frequency ω_{BC}, the dissociation energy of the vdW molecule D_{XB}, and other parameters, such as $\mu_{X,BC}$, ω_{XB}, or α_{XB}. This result also accounts for the dependence of the VP rate on the quantum number v of the molecular bond as well as on the quantum number l of the vdW bond. The following comments are now in order.

Energy-Gap Law. For large values of $\pi\theta_{\epsilon'}$, Γ_{diss} given by (2.B.55) is dominated by the factor $\exp(-\pi\theta_{\epsilon'})$. We can therefore write, using (2.B.49),

$$\ln\Gamma_{\text{diss}}=-\frac{\pi(2\mu_{X,BC}\epsilon')^{1/2}}{\hbar\alpha_{XB}}+\text{const} \quad (2.B.56)$$

From (2.B.56), the VP rate is expected to decrease fast with increasing the excess energy ϵ', which is determined by the conservation of the energy,

$$\epsilon'=\hbar\omega_{BC}\left(\frac{1-v}{2K_{BC}}\right)-\frac{\hbar^2\alpha_{XB}^2}{2\mu_{X,BC}}\left(K_{XB}-l-\tfrac{1}{2}\right)^2 \quad (2.B.57)$$

The VP rate is expected to decrease fast with increasing the molecular frequency ω_{BC}. We thus expect that the VP rate will be severely retarded when there is an appreciable mismatch between the molecular frequency, which breaks the molecular complex, and the dissociation energy of the vdW bond. This result advanced by us has been referred to as the "energy gap law" for VP.[27a] It has also been noted that the dimensionless parameter $\theta_{\epsilon'}$ contains $p=(2\mu_{X,BC}\epsilon')^{1/2}$, i.e., the relative translational momentum of the departing fragments, so that it may be appropriate to refer to relation (2.B.56) as the "momentum-gap law."[37] Apart from semantic matters, we would like to point out that it is easy to rationalize the exponential dependence of the VP rates on the translational momentum in terms of the semiclassical approximation in quantum mechanics. The final wavefunction will be a rapid oscillating function if p is large. The rate for VP, on the other hand, is essentially determined by the overlap between the bound initial state and this rapid oscillating wavefunction. The nuclear overlap integral is just expressed in the form $\exp(-bp)$, where b is some constant. Thus, the VP rate will decrease fast with increasing p.

Dependence on the Quantum Number v. We have already shown that the intramolecular contribution fo the VP rate follows an almost linear dependence on v. The situation is rather different for the intermolecular term, since due to the exponential factor $\exp(-\pi\theta_{\epsilon'})$ even a slight dependence on v of $\theta_{\epsilon'}$ may produce significant changes on the VP rate. From (2.B.57) we see that as v increases, ϵ' decreases by the anharmonicity and therefore Γ_{diss} will increase. For $v \ll K_{BC}$, we get

$$\Gamma_{diss} \propto v \exp[\beta v] \tag{2.B.58}$$

with

$$\beta = \frac{\pi(D_{XB}\hbar\omega_{BC})^{1/2}}{2K_{BC}\hbar\omega_{XB}} \tag{2.B.59}$$

To proceed further it is necessary to evaluate the factor β. For $HeI_2(B)$ this is of the order of 0.1–0.2 and therefore it is expected from (2.B.58) a linear dependence on v for $v \ll 10$ and a superlinear dependence for higher vibrational excitation. This is in agreement with the experimental result (2.A.3) of Johnson, Wharton, and Levy.[15c]

Dependence on the Parameters of the vdW bond. From (2.B.56) it is immediately seen that an increasing of the characteristic inverse length α_{XB} should increase the VP rates. Essentially, α_{XB} governs the slope of the potential for small distances, so that we conclude that as the slope becomes steep the rates increase. Another possible way to write (2.B.56) is by using (2.B.41),

$$\ln \Gamma_{diss} = -\frac{2\pi(D_{XB}\epsilon')^{1/2}}{\hbar\omega_{XB}} + \text{const} \tag{2.B.60}$$

so that increasing D_{XB} while keeping ω_{XB} constant should decrease rates, the opposite occurring when ω_{XB} is increased while D_{XB} is kept constant.

Mass Effect. To explore the dependence of the VP rate on the masses we consider again (2.B.56), which, using the definition (2.B.3) of the reduced mass $\mu_{X,BC}$, we get

$$\ln \Gamma_{\text{diss}} = -a\left[\frac{m_X(m_B+m_C)}{m_X+m_B+m_C}\right]^{1/2} \tag{2.B.61}$$

Thus it is expected that for similar molecular parameters, the VP rates become larger in the case of a light mass atom forming a vdW molecule with diatomic of any mass or a heavy atom with a light diatomic molecule. However, the vibrational frequency of the BC molecule decreases when the masses of the constituent atoms increase and, therefore, the energy gap ϵ' will generally be larger for light diatomic molecules. We conclude that the best combination for a large VP rate is a light atom bound to a heavy diatomic molecule.

2. *Rosen's Relative Coordinates Treatment*

An alternative description of vibrational predissociation for a linear vdW molecule can be obtained in terms of Rosen's relative coordinates treatment.[21] This amounts to the choice of the two interatomic distances R_{BC} and R_{XB} as independent coordinates. The nuclear Hamiltonian, after the separation of the center of mass motion, becomes

$$H = -\frac{\hbar^2}{2\mu_{BC}}\frac{\partial^2}{\partial R_{BC}^2} - \frac{\hbar^2}{2\mu_{XB}}\frac{\partial^2}{\partial R_{XB}^2} + \frac{\hbar^2}{m_B}\frac{\partial^2}{\partial R_{BC}\,\partial R_{XB}} + V_{BC}(R_{BC}) + V_{XB}(R_{XB}) \tag{2.B.62}$$

where $\mu_{BC} = m_B m_C/(m_B+m_C)$ and $\mu_{XB} = m_X m_B/(m_X+m_B)$ are the reduced masses for diatomics BC and XB, respectively. The Hamiltonian is segregated in the following manner:

$$H = H_0 + V \tag{2.B.63a}$$

with

$$H_0 = -\frac{\hbar^2}{2\mu_{BC}}\frac{\partial^2}{\partial R_{BC}^2} - \frac{\hbar^2}{2\mu_{XB}}\frac{\partial^2}{\partial R_{XB}^2} + V_{BC}(R_{BC}) + V_{XB}(R_{XB}) \tag{2.B.63b}$$

and

$$V = \frac{\hbar^2}{m_B} \frac{\partial^2}{\partial R_{BC}\, \partial R_{XB}} \tag{2.B.63c}$$

The zero-order Hamiltonian (2.B.63b) is now separable in the two coordinates, and its spectrum consists again of discrete and continuum wavefunctions of the form

$$\langle R_{BC}, R_{XB} | vl \rangle = x_v(R_{BC}) \phi_l(R_{XB}) \tag{2.B.64a}$$

$$\langle R_{BC}, R_{XB} | v'\epsilon' \rangle = x_{v'}(R_{BC}) \phi_{\epsilon'}(R_{XB}) \tag{2.B.64b}$$

where v denotes the discrete vibrational quantum number of the BC bond, l is the discrete vibrational quantum number of the XB bond, and ϵ' designates the relative kinetic energy between X and B. This eigenfunctions are essentially the same as those considered in the last section the difference being for the ϕ_α wavefunctions given in (2.B.46) to (2.B.48), where $\mu_{X,BC}$ should be replaced by μ_{XB} and $R_{X,BC}$ and $\bar{R}_{X,BC}$ by R_{XB} and $\bar{R}_{XB}$, respectively. The coupling between discrete and continuum "zero-order" states is now

$$V^{d-c}_{vl,v'\epsilon'} = \frac{\hbar^2}{m_B} \left\langle v \left| \frac{\partial}{\partial R_{BC}} \right| v' \right\rangle \left\langle l \left| \frac{\partial}{\partial R_{XB}} \right| \epsilon' \right\rangle \tag{2.B.65}$$

Using the relationships

$$\left\langle v \left| \frac{\partial}{\partial R_{BC}} \right| v' \right\rangle = \left\langle v \left| \frac{\partial V_{BC}}{\partial R_{BC}} \right| v' \right\rangle \Big/ [W_{BC}(v) - W_{BC}(v')] \tag{2.B.66a}$$

$$\left\langle l \left| \frac{\partial}{\partial R_{XB}} \right| \epsilon' \right\rangle = \left\langle l \left| \frac{\partial V_{XB}}{\partial R_{XB}} \right| \epsilon' \right\rangle \Big/ [\epsilon_l - \epsilon] \tag{2.B.66b}$$

and the integrals (2.B.50) and (2.B.51) together with (2.B.39) we obtain for Morse-type potentials of the form (2.B.40) for the two bonds BC and XB, the result (on the energy shell)

$$\begin{aligned} \underset{(v'=v-1)}{V^{d-c}_{vl,v'\epsilon'}} &= \frac{\hbar \omega_{BC} \alpha_{XB} \bar{K}_{XB}}{2 m_B} \left(\frac{\hbar \mu_{BC}}{2 \omega_{BC}} \right)^{1/2} v^{1/2} \left(\frac{K_{BC}}{2} \right)^{1/2} \\ &\otimes \left[\frac{(2K_{BC} - 2v + 1)(2K_{BC} - 2v - 1)}{(2K_{BC} - v)(K_{BC} - v)^2} \right]^{1/2} \left[\left(\frac{1}{2} D_{XB} \right) \sinh(2\pi \bar{\theta}_{\epsilon'}) \right. \\ &\times \left. \frac{(2\bar{K}_{XB} - 2l - 1)}{l!\, \Gamma(2\bar{K}_{XB} - l)} \right]^{1/2} \frac{|\Gamma(\frac{1}{2} + \bar{K}_{XB} - i\bar{\theta}_{\epsilon'})|}{[\cos^2(\pi \bar{K}_{XB}) + \sinh^2(\pi \bar{\theta}_{\epsilon'})]^{1/2}} \end{aligned} \tag{2.B.67}$$

with the definitions

$$\bar{K}_{XB}=(\hbar\alpha_{XB})^{-1}(2\mu_{XB}D_{XB})^{1/2} \tag{2.B.68a}$$

and

$$\bar{\theta}_{\epsilon'}=(\hbar\alpha_{XB})^{-1}(2\mu_{XB}\epsilon')^{1/2} \tag{2.B.68b}$$

The quantities $\bar{K}_{XB}$ and $\bar{\theta}_{\epsilon'}$ differ from the corresponding ones K_{XB} and $\theta_{\epsilon'}$ defined in (2.B.42) and (2.B.49) by the appearance of the reduced mass $\mu_{XB}=m_X m_B/(m_X+m_B)$ instead of $\mu_{X,BC}=m_X(m_B+m_C)/(m_X+m_B+m_C)$. The decay width is given by

$$\Gamma_{diss}=\frac{\pi}{8}\hbar\omega_{BC}\frac{\mu_{BC}\mu_{XB}}{m_B^2}v\frac{(2K_{BC}-2v+1)(2K_{BC}-2v-1)}{2K_{BC}(2K_{BC}-v)}$$
$$\otimes\left[\sinh\frac{(2\pi\bar{\theta}_{\epsilon})}{\cos^2\pi\bar{K}_{XB}+\sinh^2(\pi\bar{\theta}_{\epsilon'})}\right]\left|\Gamma\left(\bar{K}_{XB}+\frac{1}{2}-i\bar{\theta}\epsilon'\right)\right|^2$$
$$\times\left[\frac{2\bar{K}_{XB}-2l-1}{l!\Gamma(2\bar{K}_{XB}-l)}\right] \tag{2.B.69}$$

This result can be compared with (2.B.53) obtained with the reaction coordinates R_{BC} and $R_{X,BC}$. The two formulas look very similar. We note that they become identical if we replace (m_B+m_C) by m_B everywhere. The validity of the relative coordinates treatment has already been discussed in the literature. In particular, it has been shown that the asymptotic behavior of the continuum wavefunctions in this representation is not correct and a renormalization procedure should be applied.[42] From the analysis of this section we can conclude that the relative coordinates result (2.B.69) and the "frozen" distorted wave expression (2.B.53) will give similar results only in the limit $m_B \gg m_C$. This conclusion is easy to understand since in this case of a very massive middle atom, $R_{X,BC}$ and R_{XB} are almost the same and the center of mass of the diatomic molecule BC coincides with the position of the atom B.

3. *The Diabatic Distorted Wave Treatment Without Linearization*

We now proceed to the calculation of the total dissociation rate Γ_{diss} (2.B.24) and the individual rates $\Gamma_{svl\to sv'}$ (2.B.36) without the linearization approximation (2.B.37). Using the Morse potential (2.B.40) and the wavefunctions (2.B.46) and (2.B.48) the d–c coupling terms (2.B.10) take the form

$$V^{\text{d-c}}_{vl,\,v'\epsilon'} = A^{(2)}_{vv'} B^{(2)}_{l\epsilon'} - 2A^{(1)}_{vv'} B^{(1)}_{l\epsilon'} \tag{2.B.70}$$

where

$$A^{(j)}_{vv'} = \int x_v(R_{\text{BC}})\left\{\exp\left[\,j\alpha_{\text{XB}}\gamma\left(R_{\text{BC}} - \bar{R}_{\text{BC}}\right)\right] - 1\right\} x_{v'}(R_{\text{BC}})\,dR_{\text{BC}}$$
$$j = 1, 2 \tag{2.B.71}$$

The $B^{(j)}_{l\epsilon'}$, $j=1,2$ coefficients are the integrals (2.B.45) and the result has been given in (2.B.52a) and (2.B.52b). The integrals $A^{(2)}_{vv'}$ and $A^{(1)}_{vv'}$ for the harmonic bond were given by Rapp and Sharp[38] as

$$A^{(j)}_{vv'} = \left(\frac{v!}{v'!}\right)^{1/2} \beta_j^{v'-v} \exp\left(\frac{\beta_j^2}{2}\right) L_v^{v'-v}\left(-\beta_j^2\right) \qquad (j=1,2) \tag{2.B.72}$$

where $\beta_j = j\hbar\alpha_{\text{XB}}\gamma/(2\mu_{\text{BC}}\hbar\omega_{\text{BC}})^{1/2}$ and $L_v^{v'-v}$ is the generalized Laguerre polynomial, while for the Morse potential these are given by[39]

$$A^{(j)}_{vv'} = -\delta_{vv'} + \left[\frac{(v')!(2K_{\text{BC}} - 2v' - 1)(2K_{\text{BC}} - 2v - 1)}{v!\,\Gamma(2K_{\text{BC}} - v)}\Gamma(2K_{\text{BC}} - v')\right]^{1/2}$$
$$\otimes (2K_{\text{BC}})^{j\alpha_{\text{XB}}\gamma/\alpha_{\text{BC}}} \sum_{m=0}^{v'} (-1)^m$$
$$\times \frac{\Gamma(v - j\alpha_{\text{XB}}\gamma/\alpha_{\text{BC}} - m)\Gamma(-v - j\alpha_{\text{XB}}\gamma/\alpha_{\text{BC}} - m + 2K_{\text{BC}} - 1)}{m!(v'-m)!\Gamma(2K_{\text{BC}} - v' - m)\Gamma(-j\alpha_{\text{XB}}\gamma/\alpha_{\text{BC}} - m)} \tag{2.B.73}$$

for $j=1,2$ and $v \geqslant v'$. From the discussion of Section II.B.2 it follows that the harmonic model is accurate only for very low vibrational excitation ($v \sim 1,5$) and that for higher excitation the anharmonicity effects are important.

Equations (2.B.70), (2.B.72), and (2.B.73) together with the (2.B.52a) and (2.B.52b) provide the final results needed to calculate the individual rates $\Gamma_{svl \to sv'}$ (2.B.36), as well as the total dissociation rate Γ_{diss} (2.B.24). In what follows we shall attempt to account for some features of the VP of linear rare-gas-diatomic vdW molecules using these formulas. The basic input

TABLE I
Compilation of Morse Potential Parameters for some van der Waals Bonds Between Rare-Gas Atoms and Diatomic Molecules

Molecule	D_{XB} (cm^{-1})	$\bar{R}_{XB}$ (Å)	α_{XB} ($Å^{-1}$)	ω_{XB} (cm^{-1})	Number of bound states	ΔE_{01} (cm^{-1})	Ref.	Molecular configuration
HeI_2	52.0	4.24	1.42	42.4	2	25.1	40	
	13.5	~4.0	1.18	18.0	2	6.0	15	T-shaped
NeI_2	97.4	4.36	1.38	26.0	7	22.5	40	
ArI_2	181.5	4.71	1.27	24.0	15	22.4	40	
ArHCl	130.0	4.4	1.36	30.0	9	26.5	40	
	160.0	3.9	1.54	36.7	9	32.5	41	Linear

Useful Formulas for Morse Potentials
$V(R)=D\{\exp[-2\alpha(R-\bar{R})]-2\exp[-\alpha(R-\bar{R})]\}$; $H=-(\hbar^2/2\mu)\,d^2/\partial R^2+V(R)$
$\omega=\mu^{-1}(d^2V/dR^2)_{\bar{R}}=\alpha(2D/\mu)^{1/2}$
Number of bound states $=$ Integer$(\kappa+\frac{1}{2})$ with $\kappa=2D/\hbar\omega$
$\omega_l=-(\hbar\omega/2\kappa)(\kappa-l-\frac{1}{2})^2$; $l=0,1,\ldots,$ Integer$(\kappa-\frac{1}{2})$
$\epsilon_l=-D+\hbar\omega\{(l+\frac{1}{2})-\chi(l+\frac{1}{2})^2\}$; $\chi=(2\kappa)^{-1}$
$\Delta E_{01}\equiv\epsilon_1-\epsilon_0=\hbar\omega(1-1/\kappa)$

data involve the well-known spectroscopic parameters for the molecular BC bond and the potential parameters for the vdW bond. For the latter we have utilized the currently available information summarized in Table I.

a) Energy-Gap Law for VP. To explore the gross features of the dependence of the VP rate on the energetic parameters of the vdW molecules we portray in Fig. 4 a sample of numerical results of model calculations for a number of such molecules. The BC bond was taken to be harmonic and for the sake of a simplified representation of the VP rate we have taken the same vdW parameters D_{XB} and ω_{XB} for all the molecules. These data for a series of linear vdW molecules, where the normal BC bond frequency varies in the range 128–3000 cm^{-1}, exhibit the energy gap law for VP, the rate decreasing linearly with increasing $\omega_{BC}^{1/2}$ which is in accord with the prediction of (2.B.56) and (2.B.57). The variation of the VP lifetimes for X–BC linear triatomic molecules is exhibited over 20 orders of magnitude. The VP lifetimes vary from the picosecond range up to times which are comparable to the age of the universe. The results of Fig. 4 should be considered to provide guidelines concerning general trends, relations and correlations, rather than quantitative estimates of VP rates. The pedantic reader will argue that the VP rates are rather sensitive to the details of the potential parameters of the vdW bond, which we did not consider carefully. More important, our energy gap law considers only the

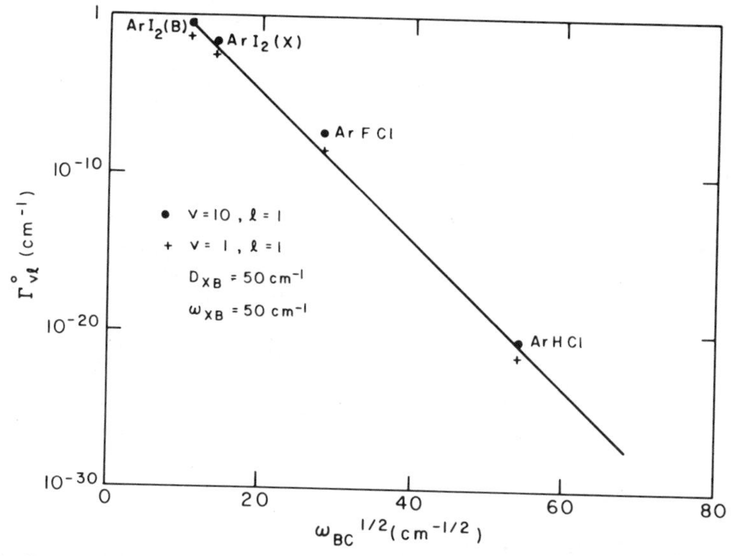

Fig. 4. Dependence of the VP rate on the intramolecular frequency ω_{BC}. The calculations were performed for a harmonic BC potential, neglecting continuum–continuum couplings. The points correspond to the VP from the level $v=10$ and $l=1$, while the crosses correspond to $v=1$ and $l=1$. The parameters for the van der Waals interaction are marked on the figure.

interconversion of vibrational energy to translational energy, while the effects of rotations are disregarded. Rotational effects will provide additional channels for interconversion of vibrational energy to rotational energy enhancing the VP rate. To be specific, we note that our simple model predicts that the VP rate of Ar–HCl is exceedingly small $\Gamma \sim 10^{-8}$ $\sec^{-1}$. We do not want to take the numerical result seriously, but rather conclude that the Ar–HCl vdW molecule is stable with respect to VP. This conclusion concurs with the results of Ashton and Child,[26] who incorporated rotational effects in the study of the VP process of Ar–HCl, finding that the rate is surprisingly low, $\Gamma \sim 10^{2}-10^{-1}$ sec.

b) Anharmonicity Effects. To provide a visual demonstration of the dependence of the VP rate on the vibrational quantum number v of the BC bond we present in Fig. 5a–c the results of model calculations of Γ_{diss} for $l=\bar{l}$, where the VP rate reaches its maximum value for a given v. For the harmonic BC potential Γ_{diss} exhibits a linear dependence on v. When an anharmonic bond potential is utilized the increase of the VP rate with v is superlinear and can be quite adequately fit by the relation $\Gamma_{\text{diss}} = v\exp(\beta v)$ (2.B.58), where β is a constant for a given vdW molecule. As is evident

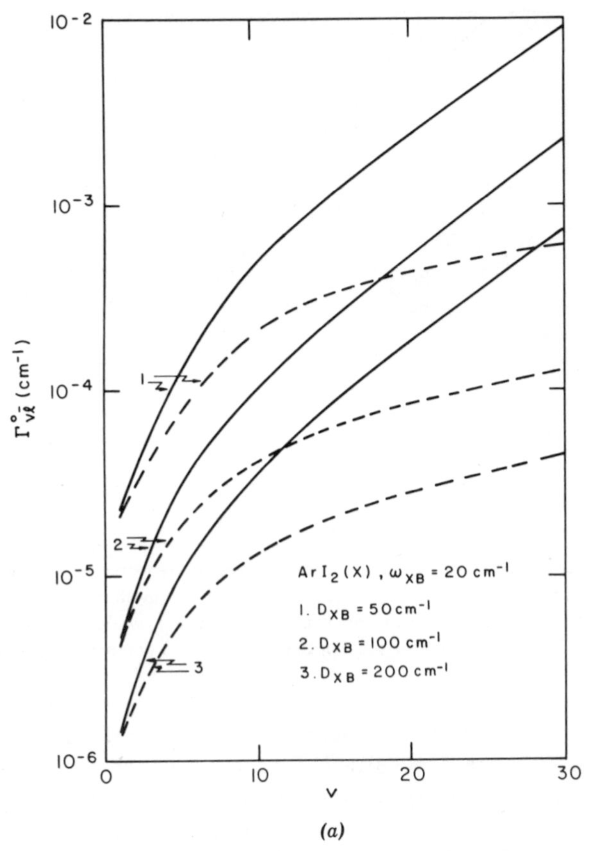

(a)

Fig. 5. (a) Model calculations of the dependence of the VP rate for a linear $ArI_2(X)$ van der Waals molecule, neglecting intercontinuum couplings, on the vibrational quantum number v of the $I_2(X)$ bond. Dashed lines represent the values of the VP rate $\Gamma(0/vl)$ calculated in the harmonic approximation for the I_2 molecule, while solid lines correspond to an anharmonic I—I bond characterized by a Morse potential. The parameters of the van der Waals interaction are marked on the Figure. (b) Same as Fig. 5(a) for linear $NeI_2(B)$. (c) Same as Fig. 5(a) for linear $HeI_2(B)$.

from Fig. 5a–c anharmonicity effects have a minor influence on the VP rate for low value of v, as expected, however, for high v values the anharmonicity of the BC bond considerably enhances the VP rate. As it has been noted before, this result can easily be rationalized by noting that anharmonicity decreases the effective energy gap between the levels sv and $s(v-1)$, resulting in a somewhat better matching between the vibrational energy of the BC bond which is transferred to the vdW bond and the dissociation energy of the latter, thus resulting in a considerable enhancement of the VP rate at higher values of v.

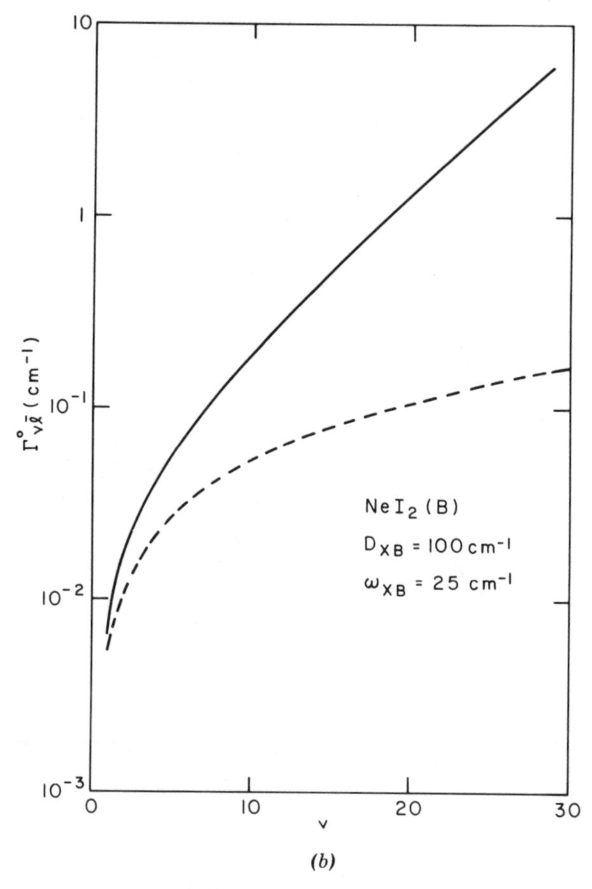

(*b*)

Fig. 5. (*continued*)

c) Dependence of VP Rate on *l*. To assess the dependence of the VP rate on the vibrational quantum number l of the bound states of the vdW bond a series of calculations are presented in Fig. 6*a–d*. From these figures it is apparent that for vdW molecules, where the weak bond supports a small number of bound states, the VP rate at constant value of v decreases when increasing l, while when the number of bound states is large Γ_{diss} increases with increasing l, reaching a maximum at some value $l=\bar{l}$ and then subsequently decreases with further increase of l.

4. *General Formalism for VP Decay Incorporating Final-State Interactions*

The golden-rule perturbative result of the VP rate, which we have studied in some detail, disregards the effects of c–c interactions. The role

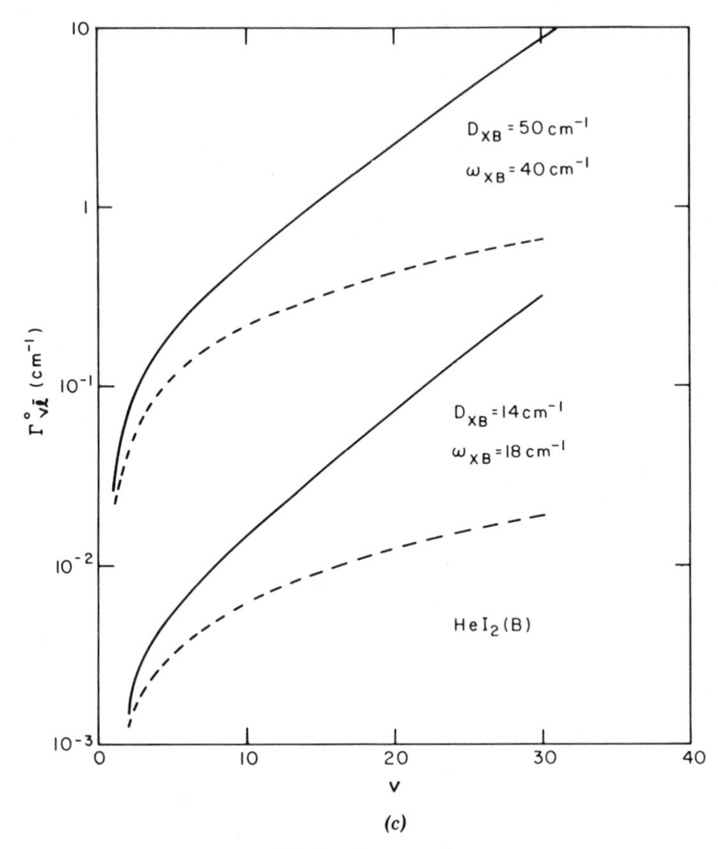

(c)

Fig. 5. (*continued*)

of c–c coupling effects, sometimes referred to as final-state interactions, is crucial for a quantitative description of the VP dynamics. Analogous effects of c–c coupling were encountered in the theory of photofragmentation of conventional molecules[42] and were considered semiclassically in terms of a "half-collision" model.[43] We shall now proceed to provide a general treatment of the VP of a linear vdW molecule starting with the general expression for the dissociation probability, which is given by (2.B.27). The dissociation probability can be expressed in terms of the matrix elements of the operator $QG^{+}(E)P$ defined in (2.B.28). An alternative way to express this operator is[27]

$$QG^{+}(E)P=(E^{+}-Q\mathcal{H}_0Q)^{-1}QR(E)P(E^{+}-\mathcal{H}_0-PR(E)P)^{-1} \tag{2.B.74}$$

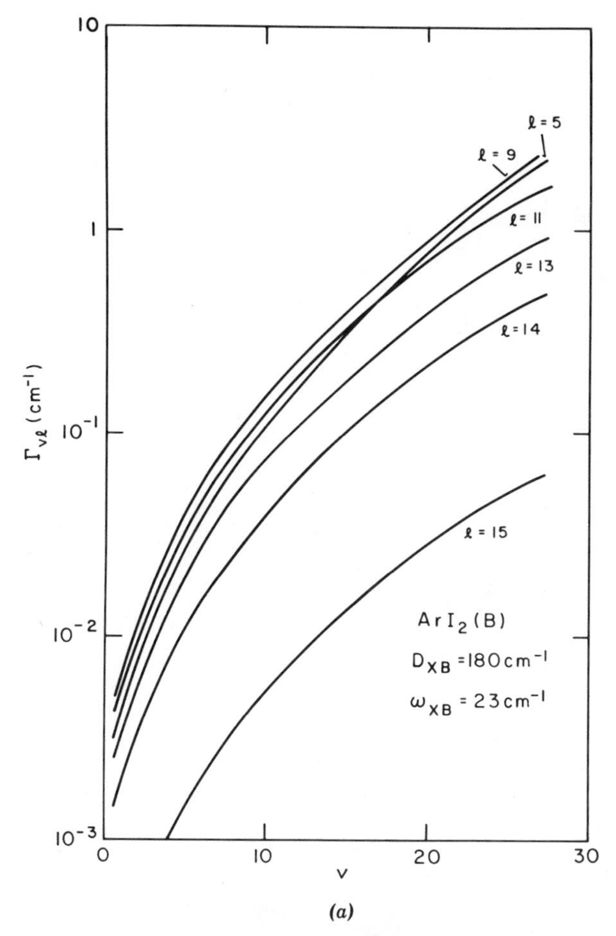

(a)

Fig. 6. Model calculations of the dependence of the VP rate of X—I_2 complexes on the vibrational quantum number v and on the vibrational quantum number l of the van der Waals bond. The I—I bond was anharmonic characterized by a Morse interaction, while the parameters of the van der Waals bond are marked on the figure. (a) ArI_2(B) dependence of Γ on v for several values of l. (b) HeI_2(B) dependence of Γ on v for several values of l. (c) HeI_2(B) dependence of Γ on l. (d) ArI_2(B) dependence of Γ on l.

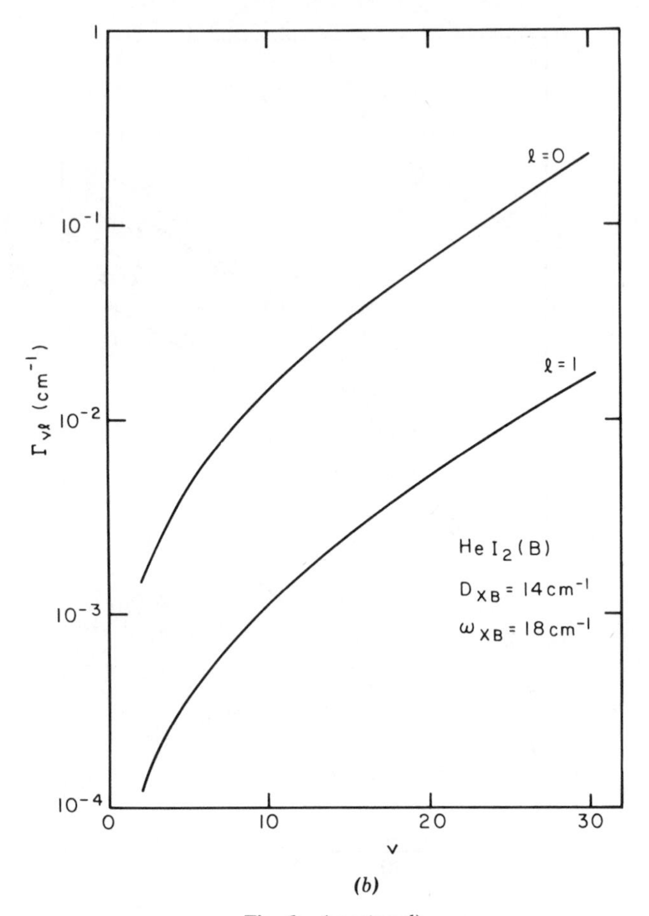

(b)

Fig. 6. *(continued)*

where R is the level-shift operator defined in (2.B.18). Using (2.B.74) we obtain for the relevant matrix elements of $G^{+}(E)$

$$\langle s;\, v'\epsilon';\mathrm{vac}|G^{+}(E)|s;\, vl;\mathrm{vac}\rangle = \frac{\langle s;\, v'\epsilon';\mathrm{vac}|QRP|s;\, vl;\mathrm{vac}\rangle}{(E^{+}-E_{sv'\epsilon'})(E^{+}-E_{svl}-\Delta_{svl}(E)+(\Gamma_{svl}(E))} \tag{2.B.75}$$

where $\Delta(E)$ and $\Gamma(E)$ are, respectively, the real and imaginary parts of the matrix element $\langle s;\, vl;\mathrm{vac}|\, R\,|s;\, vl;\mathrm{vac}\rangle$. We may perform the integration

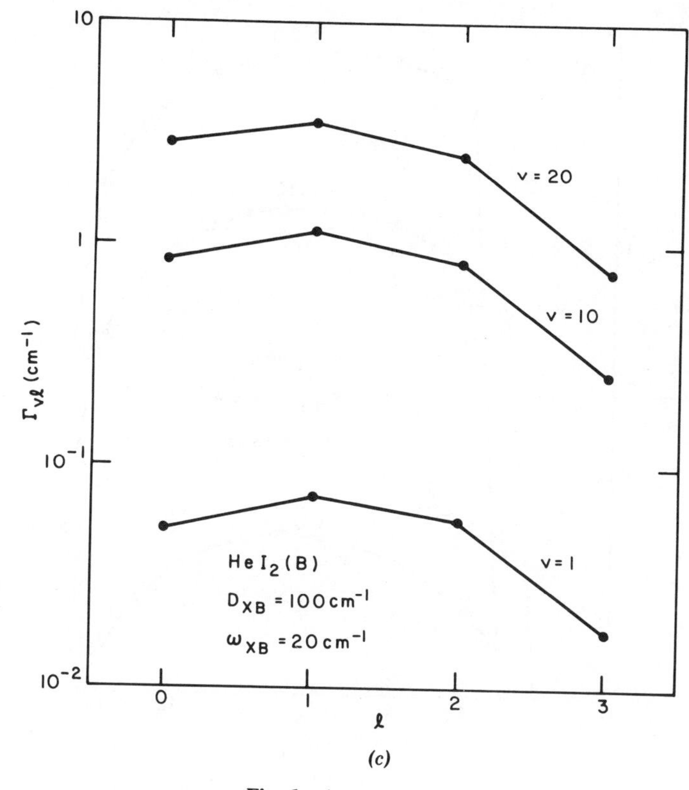

(c)

Fig. 6. (*continued*)

in (2.B.27) by invoking the usual assumptions regarding the weak dependence of R on the energy and the negligible effects of the thresholds. The probability distribution is then given by

$$P_{sv'}(t)=\frac{\pi}{\Gamma}|\langle s;\, v'\epsilon';\mathrm{vac}|QRP|s;\, vl;\mathrm{vac}\rangle|^2[1-\exp(-2\Gamma t/\hbar)] \tag{2.B.76}$$

and for $t\to\infty$

$$P_{sv'}(t)=\frac{\pi}{\Gamma}|\langle s;\, v'\epsilon';\mathrm{vac}|QRP|s;\, vl;\mathrm{vac}\rangle|^2 \tag{2.B.77}$$

It is possible to separate formally the contribution of the c–c interaction

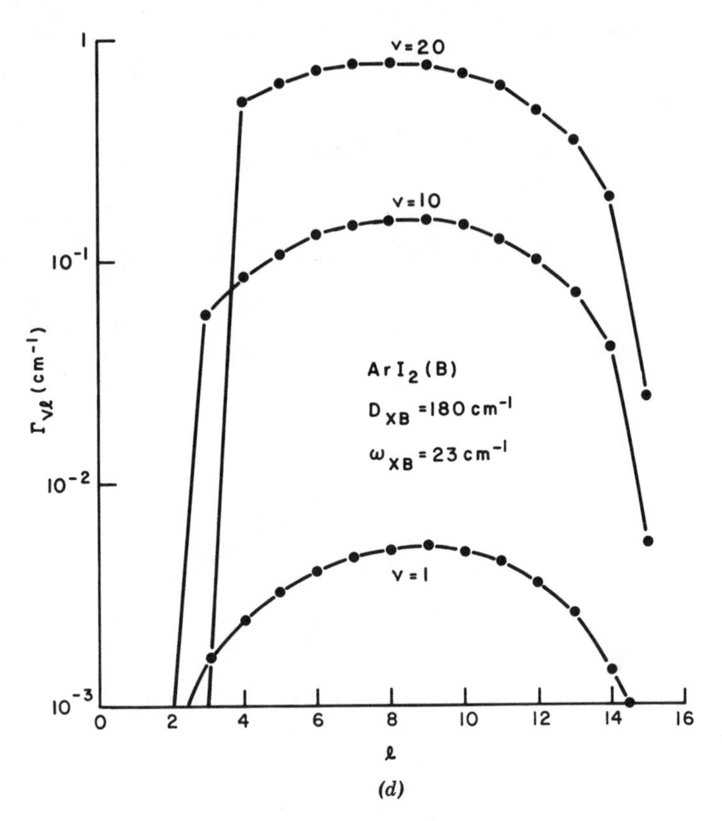

(d)

Fig. 6. *(continued)*

by defining a transition $T^{\text{c-c}}$ operator which acts only in the Q subspace

$$T^{\text{c-c}} = V^{\text{c-c}} + V^{\text{c-c}}(E^+ - QHQ)^{-1}V^{\text{c-c}} \equiv QRQ \qquad (2.\text{B}.78)$$

The operator QRP takes the form

$$QRP = V^{\text{c-d}} + V^{\text{c-c}}(E^+ - QHQ)^{-1}V^{\text{c-d}} \qquad (2.\text{B}.79)$$

and using the relation

$$(E^+ - QHQ)^{-1} = (E^+ - QH_0Q)^{-1} + (E^+ - QHQ)^{-1}V^{\text{c-c}}(E^+ - QH_0Q)^{-1} \qquad (2.\text{B}.80)$$

we obtain

$$QRP = \left[1 + T^{\text{c-c}}(E^+ - QH_0Q)^{-1}\right]V^{\text{c-d}} \qquad (2.\text{B}.81)$$

The operator $(E^+ - QH_0Q)^{-1}$ is formally written as

$$(E^+ - QH_0Q)^{-1} = PP\left[(E - QH_0Q)^{-1}\right] - i\pi\delta(E - QH_0Q) \tag{2.B.82}$$

where PP stands for the principle part distribution. Invoking the first-order K matrix approximation we neglect the principal part on the right-hand side of (2.B.82), which amounts to the assumption that the coupling varies slowly with energy. Setting $(E^+ - QH_0Q)^{-1} = -i\pi\delta(E - QH_0Q)$ in (2.B.81) results in the following final expression for the probability distribution:

$$P_{sv'} = \frac{\pi}{\Gamma}\left|\sum_{v''}(\delta_{v'v''} - i\pi T^{\text{c-c}}_{v'v''})V^{\text{d-d}}_{v'',vl}\right|^2 \tag{2.B.83}$$

This result can be recast in the form

$$P_{sv'} = \frac{\pi}{\Gamma}\left|\sum_{v''}\tilde{S}_{v'v''}V^{\text{c-d}}_{v'',vl}\right|^2$$

$$\tilde{S} = 1 - i\pi T^{\text{c-c}} \tag{2.B.84}$$

where $\tilde{S}$ is the scattering matrix for a half-collision. Similarly we have

$$PRP = V^{\text{d-c}}(E^+ - QHQ)^{-1}V^{\text{c-d}} \tag{2.B.85}$$

To calculate $\Gamma = -\text{Im}\, PRP$ we utilize the relation

$$(E^+ - QHQ)^{-1} = (E^+ - QH_0Q)^{-1} + (E^+ - QH_0Q)^{-1}T^{\text{c-c}}(E^+ - QH_0Q)^{-1} \tag{2.B.86}$$

which results in

$$PRP = V^{\text{d-c}}(E^+ - QH_0Q)^{-1}\left[1 + T^{\text{c-c}}(E^+ - QH_0Q)^{-1}\right]V^{\text{c-d}} \tag{2.B.87}$$

Thus, Γ is given by

$$\Gamma = \pi\,\text{Re}\left[\sum_{v'v''}V^{\text{d-c}}_{vl,v'}(\delta_{v'v''} - i\pi T^{\text{c-c}}_{v'v''})V^{\text{c-d}}_{vl,v''}\right] \tag{2.B.88}$$

in deriving this result we have used again the relation $(E^+ - QH_0Q)^{-1} = -i\pi\delta(E - QH_0Q)$. Defining the wave operator

$$F^{\text{c-c}} = 1 - i\pi T^{\text{c-c}} \tag{2.B.89}$$

we note from the definition of $T^{\text{c-c}}$ that provided we neglect again the principal part integrals,

$$T^{\text{c-c}} = V^{\text{c-c}} F^{\text{c-c}} \tag{2.B.90}$$

and so the wave operator assumes the form

$$F^{\text{c-c}} = 1 - i\pi V^{\text{c-c}} F^{\text{c-c}} \tag{2.B.91}$$

The sum of the final probabilities is then

$$\sum_{v'} P_{v'} = \frac{(F^{\text{c-c}} V^{\text{c-d}})^+ (F^{\text{c-c}} V^{\text{c-d}})}{\text{Re}(V^{\text{d-c}} F^{\text{c-c}} V^{\text{c-d}})} \tag{2.B.92}$$

Multiplying now (2.B.91) by $F^{\text{c-c}\dagger}$ from the left we have

$$F^{\text{c-c}\dagger} F^{\text{c-c}} = \text{Re}\, F^{\text{c-c}} \tag{2.B.93}$$

and using the complex conjugate of (2.B.91) on the right-hand side of (2.B.93) we obtain

$$F^{\text{c-c}\dagger} F^{\text{c-c}} = \text{Re}\, F^{\text{c-c}} \tag{2.B.94}$$

Conservation rules hold as by substituting (2.B.94) into (2.B.92) we get $\Sigma_{v'} P_{v'} = 1$, so that our treatment is self-consistent. In terms of the wave operator $F^{\text{c-c}}$ defined in (2.B.89), the decay half-width Γ is given by

$$\Gamma = \pi\, \text{Re}\left[\sum_{v'} \sum_{v''} V^{\text{d-c}}_{vl,v'} F(v', v'') V^{\text{d-c}}_{v'',vl} \right] \tag{2.B.95}$$

where $V^{\text{d-c}}_{vl,v'}$ denotes the coupling between $|vl\rangle$ and the continuum state $|v'\epsilon'\rangle$ on the energy shell, that is, $W_{\text{BC}}(v) + \epsilon_l = W_{\text{BC}}(v') + \epsilon'$. The quantities $F(v', v'')$ are the matrix elements $\langle v'\epsilon' | F | v''\epsilon'' \rangle$ of the wave operator F on the energy shell, that is, $W_{\text{BC}}(v) + \epsilon_l = W_{\text{BC}}(v') + \epsilon' = W_{\text{BC}}(v'') + \epsilon''$. The time dependence of a population of a given dissociative channel (2.B.76) is

$$P_{sv'}(t) = \frac{\pi}{\Gamma} \left| \sum_{v''} F(v', v'') V^{\text{d-c}}_{v'',vl} \right|^2 \left[1 - \exp\left(-\frac{2\Gamma t}{\hbar} \right) \right] \tag{2.B.96}$$

while the final branching ratio among vibrational channels (2.B.77) is

$$P_{sv'}(t)=\frac{\pi}{\Gamma}\left|\sum_{v''}F(v',v'')V^{\mathrm{d-c}}_{v'',vl}\right|^2 \tag{2.B.97}$$

Thus, the dynamics of VP is now determined by the d–c coupling terms and the wave operator, which is determined by the c–c couplings. From (2.B.91) we have

$$F=(1+i\pi V^{\mathrm{c-c}})^{-1} \tag{2.B.98}$$

When the coupling prevails only between adjacent continua the F matrix can be expressed in the form[42]

$$F(v',v'')=\frac{Q_\alpha \mathring{Q}_\beta}{Q_n}\prod_{j=\alpha}^{\beta-1}\left(-i\pi V^{\mathrm{c-c}}_{j,j+1}\right) \tag{2.B.99}$$

where, as before, $V^{\mathrm{c-c}}_{j,j+1}$ denotes the coupling between two adjacent continua on the energy shell of the initial discrete state $|vl\rangle$, $\alpha=\min(v',v'')$, $\beta=\max(v',v'')$ and Q_α and $\bar{Q}_\beta$ are polynomials determined by the recurrence relations

$$\begin{aligned}&Q_0=Q_1=1\\&Q_{j+1}=Q_j+\pi^2|V^{\mathrm{c-c}}_{j-1,j}|^2Q_{j-1}\end{aligned} \tag{2.B.100a}$$

and

$$\begin{aligned}&\bar{Q}_{v-1}=\bar{Q}_{v-2}=1\\&\bar{Q}_{j-1}=\bar{Q}_j+\pi^2|V^{\mathrm{c-c}}_{j,j+1}|^2\bar{Q}_{j+1}\end{aligned} \tag{2.B.100b}$$

where v is the vibrational quantum number of the BC bond in the initial discrete state $|vl\rangle$. Equations (2.B.96) and (2.B.97) together with (2.B.95), (2.B.99)–(2.B.100) and the explicit expressions for the coupling matrix elements presented above, provide a theory of VP on a single electronic potential surface, which incorporates both the effects of d–c coupling as well as the effects of c–c coupling. This description of the VP process involves basically feeding of the continuum states, induced by d–c coupling and a "half-collision" process within the dissociative states on the single potential surface which originates from c–c coupling. A simplified treatment which disregards the effects of intercontinuum coupling will result in the conventional description of a metastable resonance $|vl\rangle$ decaying into a manifold of uncoupled continua, presented in Section II.B.1. Under these circumstances (2.B.98) gives $F(v', v'') = \delta_{v'v''}$ and the resonance half-width (2.B.95) reduces to the golden-rule result (2.B.24), while the vibrational distribution (2.B.97) yields the simplified result of (2.B.31). Using the results of the general treatment, together with the calculation of the c–c coupling terms outlined in Section II.5, it is possible to study the effects of the incorporation of c–c coupling in both the

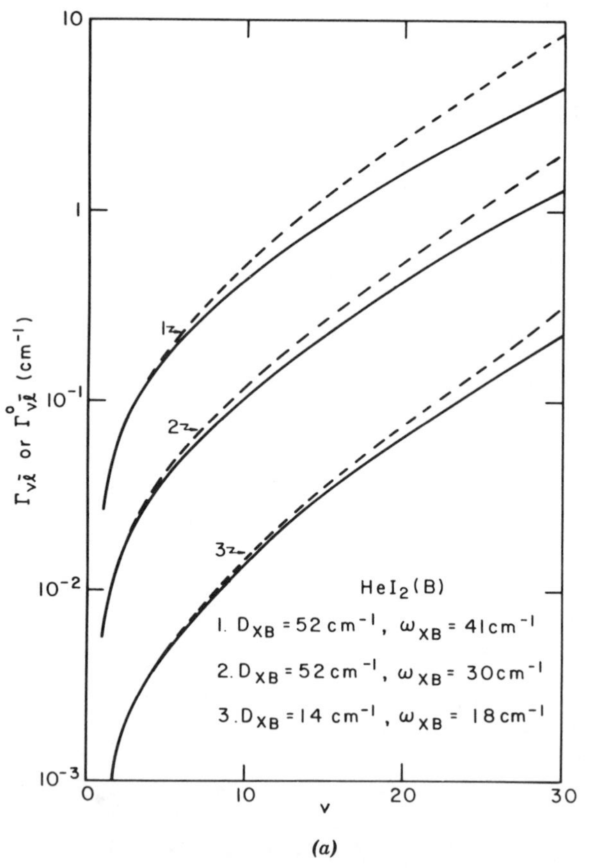

(a)

Fig. 7. (a) Model calculations for the VP rate of linear HeI_2(B) in function of the vibrational quantum number v of the initial discrete level $|v, \bar{l}\rangle$. Dashed lines are the results obtained by neglecting continuum–continuum interaction ($\Gamma^0_{v\bar{l}}$), while solid lines correspond to the results for $\Gamma_{v\bar{l}}$). The I—I bond is characterized by a Morse potential. The parameters of the van der Waals interaction are marked on the Figure. (b) Same as Fig. 7(a) for NeI_2(B). (c) Same as Fig. 7(a) for ArI_2(B).

half-width Γ and on the final vibrational distribution $P_{sv'}$. As is evident from Fig. 7a–c the effects of intercontinuum couplings are rather small for low values of v. However, at higher values of v the VP rate is retarded by intercontinuum coupling effects and $\Gamma \sim (0.3-0.5)\Gamma^{\circ}$. Such a retardation effect due to coupling between "smooth" continua, where the c–c interaction is weakly varying with energy, is well known in the theory of relaxation phenomena.

Another interesting physical consequence of c–c coupling involves the final vibrational distribution of the diatomic product in the VP process. When intercontinuum coupling effects are disregarded (2.B.31) demonstrates that the branching ratios for the population of different vibrational states $v' < v$ of the diatomic fragment is determined by the squares of the

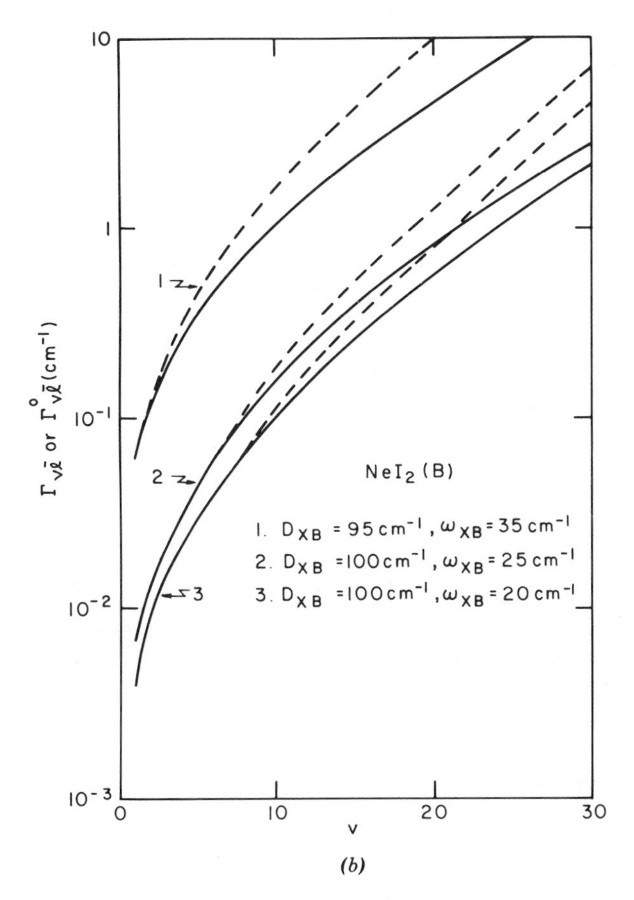

(b)

Fig. 7. (*continued*)

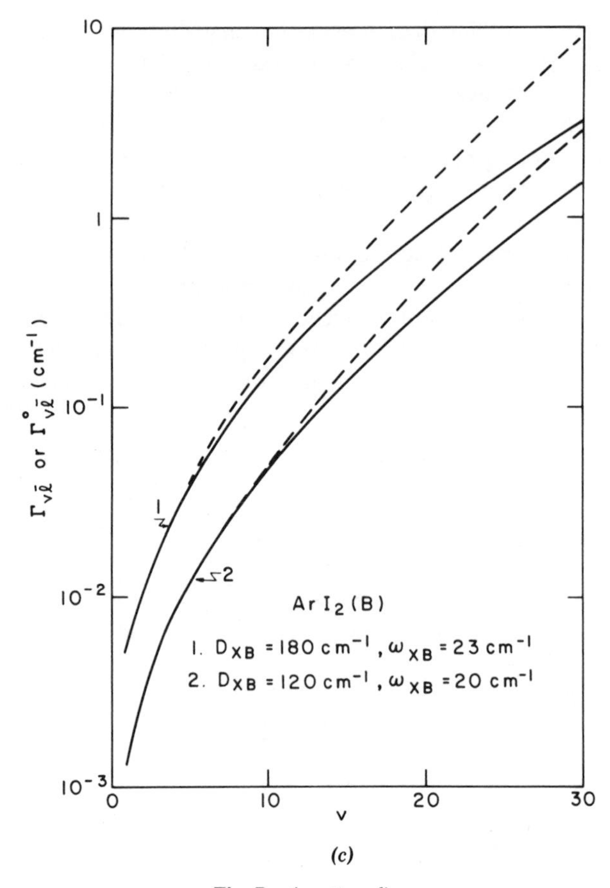

(c)

Fig. 7. (*continued*)

d–c coupling terms. Following the discussion of Section II.B.1 we recall that the $V^{\mathrm{d-c}}_{vl, v'\epsilon'}$ $(v'<v)$ terms evaluated on the energy shell are dominated by the term $v'=v-1$ and are negligible small for $v'<(v-1)$. Without alluding to any further numerical calculations we can assert that provided c–c coupling effects are disregarded the dominant decay channel of the $|vl\rangle$ initial state will involve the population of the $(v-1)$ state of the fragment. Intercontinuum coupling effects may result in relaxation of the propensity rule $\Delta v=-1$. As is evident from Fig. 8a–c we note that $P_{v-1}\simeq 0.6-0.7$ and lower vibrational states may be populated for $ArI_2(B)$, but $P_{v-1}\simeq 0.9-1$ for $HeI_2(B)$.

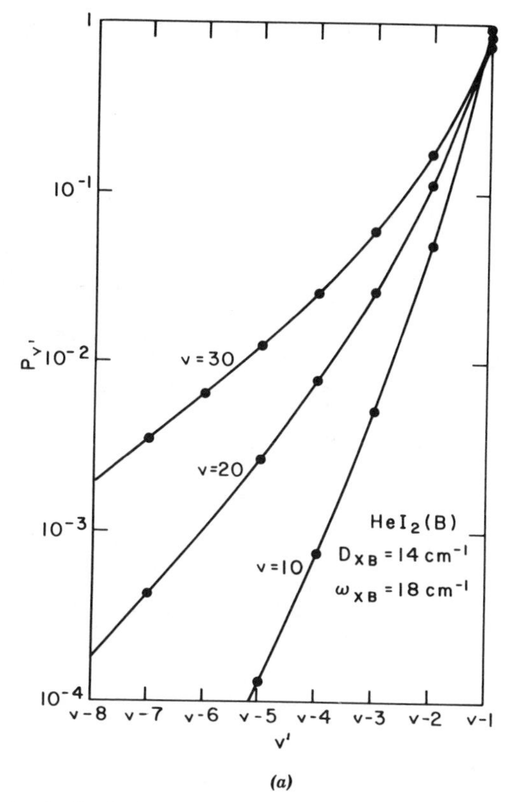

(a)

Fig. 8. (a) Vibrational distribution of the I_2 molecule resulting from an initial state of $HeI_2(B)$ characterized by quantum number v and $\bar{l}$. The I_2 bond is described by a Morse potential. The parameters of the van der Waals interaction are marked on the Figure. (b) Same as Fig. 8(a) for $NeI_2(B)$. (c) Same as Fig. 8(a) for $ArI_2(B)$.

5. *Some Matrix Elements*

It is customary that every technical calculation in the area of theoretical chemistry boils down finally to evaluation of the relevant matrix elements. This is also the case for the present theory of VP dynamics, where the relevant coupling terms for d–d, d–c, and c–c interactions are defined in (2.B.10). The d–c terms were evaluated in Section II.B.3. Here we outline the evaluation of the d–d and the c–c interactions. The d–d terms have to be evaluated in order to demonstrate that interference effects are negligible in the VP dynamics. The evaluation of the c–c terms is crucial for the incorporation of final-state interactions. Using the Morse potential (2.B.41)

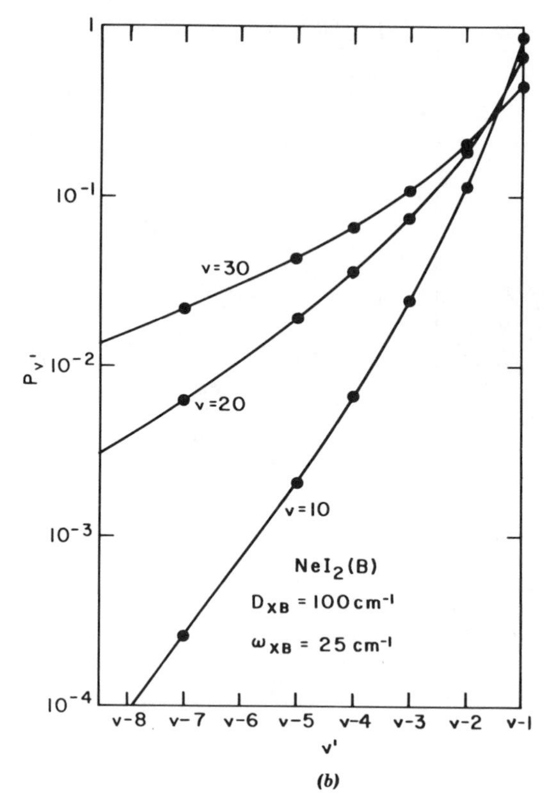

(b)

Fig. 8. (*continued*)

they can be written, in general, as

$$V^{\text{a-b}}_{v\alpha, v'\beta} = \left(A^{(2)}_{vv'} B^{(2)}_{\alpha\beta} - 2 A^{(1)}_{vv'} B^{(1)}_{\alpha\beta} \right) \tag{2.B.101}$$

where a, b ≡ d or c (discrete or continuum); $\alpha, \beta \equiv l$ or ϵ;

$$B^{(j)}_{\alpha\beta} = D_{\text{XB}} \int dR_{\text{X, BC}} \phi_\alpha(R_{\text{X, BC}}) \exp\left[-j\alpha_{\text{XB}}\left(R_{\text{X, BC}} - \bar{R}_{\text{X, BC}} \right) \right] \phi_\beta(R_{\text{X, BC}}) \tag{2.B.102}$$

and $A^{(j)}_{vv'}$, $j = 1, 2$ are given in (2.B.72) and (2.B.73). Using the wavefunc-

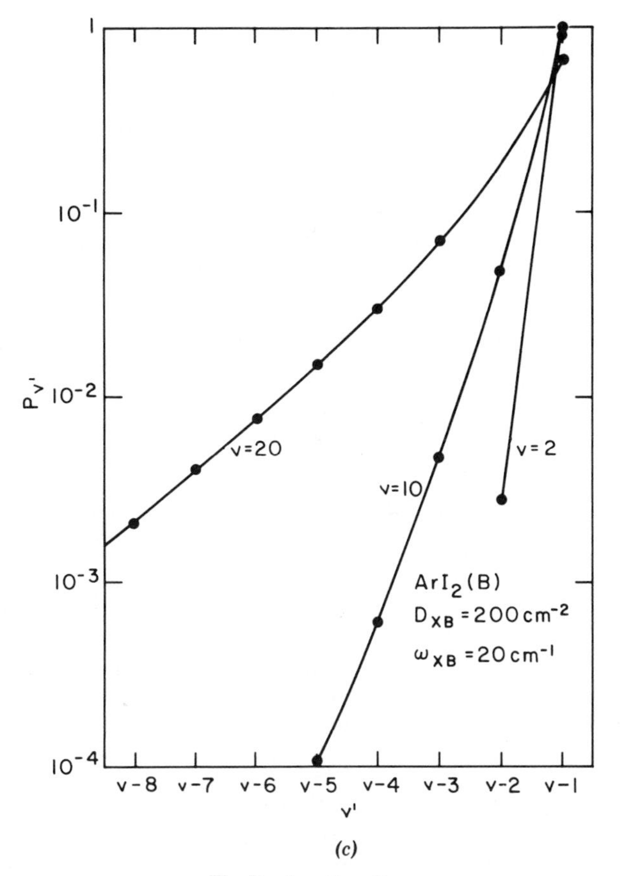

(c)

Fig. 8. (*continued*)

tions (2.B.46) and the integrals[21]

$$\int_0^{\infty} W_{K,\,K-l'-1/2}(z)\,W_{K,\,K-l-1/2}(z)\,dz$$
$$=\Gamma(l'+1)\Gamma(2K-l')\left[\,l'(2K-l'-1)-l(2K-l-1)+2K\right] \tag{2.B.103a}$$

$$\int_0^{\infty} W_{K,\,K-l'-1/2}(z)\,W_{K,\,K-l-1/2}(z)\,\frac{dz}{z}=\Gamma(l'+1)\Gamma(2K-l') \tag{2.B.103b}$$

valid for $l'>l-1$, it is obtained

$$B_{l'l}^{(1)}=(D_{XB}/2K_{XB})\left[\frac{(2K_{XB}-2l-1)(2K_{XB}-2l'-1)}{l!l'!\Gamma(2K_{XB}-l)\Gamma(2K_{XB}-l')}\right]^{1/2}$$

$$\times\Gamma(l'+1)\Gamma(2K_{XB}-l'),\qquad l'>l-1 \tag{2.B.104a}$$

$$B_{l'l}^{(2)}=(B_{l'l}^{(1)}/2K_{XB})[l'(2K_{XB}-l'-1)-l(2K_{XB}-l-1)+2K_{XB}],$$
$$l'>l-1 \tag{2.B.104b}$$

Equations (2.B.104a) and (2.B.104b), together with (2.B.70), (2.B.72), and

TABLE II
Discrete–Discrete Coupling Terms Between Neighboring Levels Calculated for Model Linear VDWM. The Molecular Bond BC is Characterized by the Following Parameters: $D_{BC}=4911\ cm^{-1}$, $\omega_{BC}=128\ cm^{-1}$ for $I_2(B^3\Pi)$

Molecule	Molecular parameters	$(v,l)\to(v'l')$	$E_{v'l'}-E_{vl}$ (cm^{-1})	$V_{vl,v'l'}^{d\text{-}d}$ (cm^{-1})
HeI_2	$D_{XB}=14\ cm^{-1}$	(0,0)→(0,1)	6.4	0.53 10^{-2}
		(5,0)→(5,1)	6.4	0.66 10^{-1}
	$\omega_{XB}=18\ cm^{-1}$	(10,0)→(10,1)	6.4	0.14
		(20,0)→(20,1)	6.4	0.37
NeI_2	$D_{XB}=100\ cm^{-1}$	(0,0)→(0,1)	22.6	0.21
		(0,2)→(0,3)	15.9	0.20
		(0,4)→(0,5)	9.1	0.1
		(0,6)→(0,7)	2.3	0.11 10^{-1}
	$\omega_{XB}=26\ cm^{-1}$	(10,0)→(10,1)	22.62	5.58
		(10,2)→(10,3)	15.9	5.42
		(10,4)→(10,5)	9.1	2.85
		(10,6)→(10,7)	2.34	0.33
NeI_2		(20,0)→(20,1)	22.6	14.16
		(20,2)→(20,3)	15.9	14.0
		(20,4)→(20,5)	9.1	7.6
		(20,6)→(20,7)	2.3	0.96
ArI_2	$D_{XB}=180\ cm^{-1}$	(0,0)→(0,1)	22.4	0.28
		(0,4)→(0,5)	16.0	0.36
		(0,8)→(0,9)	9.6	0.21
		(0,12)→(0,13)	3.2	0.42 10^{-1}
	$\omega_{XB}=24\ cm^{-1}$	(10,0)→(10,1)	22.4	7.38
		(10,4)→(10,5)	16.0	9.54
		(10,8)→(10,9)	9.6	5.6
		(10,12)→(10,13)	3.2	1.23
		(20,0)→(20,1)	22.4	18.4
		(20,4)→(20,5)	16.0	24.2
		(20,8)→(20,9)	9.6	14.6
		(20,12)→(20,13)	3.2	3.5

(2.B.73), give the final analytical expression for the d–d coupling $V^{\text{d-d}}_{vl,v'l'}$. For the $X-I_2$ systems (X≡He, Ne, Ar) these coupling terms are usually small relative to the energy spacing between the discrete levels. In Table II some numerical results which demonstrate this point, are presented. From these numerical data it is apparent that

$$|E_{vl}-E_{v'l'}|\gg|V^{\text{d-d}}_{vl,v'l'}| \tag{2.B.105}$$

for most of the cases. This result clearly demonstrates that at least for low v the shifts of the discrete levels due to the d–d coupling are small.

The c–c couplings $V^{\text{c-c}}_{v\epsilon,v'\epsilon'}$, are obtained by replacing the wavefunctions (2.B.48) into (2.B.102) and by making use of the integrals[44]

$$\int_0^\infty W_{K,i\theta}(z)W_{K,i\theta'}(z)\,dz=2\pi^2(\cosh 2\pi\theta-\cosh 2\pi\theta')^{-1} \times\left[\frac{\theta'^2-\theta^2+2K}{|\Gamma(\frac{1}{2}-K+i\theta)|^2}+\frac{\theta'^2-\theta^2-2K}{|\Gamma(\frac{1}{2}-K+i\theta')|^2}\right] \tag{2.B.106a}$$

and

$$\int_0^\infty W_{K,i\theta}(z)W_{K,i\theta'}(z)(\partial z/z)=2\pi^2(\cosh 2\pi\theta-\cosh 2\pi\theta')^{-1} \times\left[|\Gamma(\tfrac{1}{2}-K+i\theta)|^{-2}-|\Gamma(\tfrac{1}{2}-K+i\theta')|^{-2}\right] \tag{2.B.106b}$$

The result is

$$B^{(1)}_{\epsilon'\epsilon}=(K_{\text{XB}}/2)|\Gamma(\tfrac{1}{2}-K_{\text{XB}}-i\theta_\epsilon)\Gamma(\tfrac{1}{2}-K_{\text{XB}}-i\theta_{\epsilon'})| \times\frac{[\sinh(2\pi\theta_\epsilon)\sinh(2\pi\theta_{\epsilon'})]^{1/2}}{\cosh(2\pi\theta_\epsilon)-\cosh(2\pi\theta_{\epsilon'})}\left[|\Gamma(\tfrac{1}{2}-K_{\text{XB}}-i\theta_\epsilon)|^{-2} -|\Gamma(\tfrac{1}{2}-K_{\text{XB}}-i\theta_{\epsilon'})|^{-2}\right] \tag{2.B.107a}$$

$$B^{(2)}_{\epsilon'\epsilon}=(\tfrac{1}{4})|\Gamma(\tfrac{1}{2}-K_{\text{XB}}-i\theta_\epsilon)\Gamma(\tfrac{1}{2}-K_{\text{XB}}-i\theta_{\epsilon'})| \times\frac{[\sinh(2\pi\theta_\epsilon)\sinh(2\pi\theta_{\epsilon'})]^{1/2}}{\cosh(2\pi\theta_\epsilon)-\cosh(2\pi\theta_{\epsilon'})}\left[\frac{\theta_\epsilon^2-\theta_{\epsilon'}^2+2K_{\text{XB}}}{|\Gamma(\frac{1}{2}-K_{\text{XB}}-i\theta_\epsilon)|^2} +\frac{\theta_\epsilon^2-\theta_{\epsilon'}^2-2K_{\text{XB}}}{|\Gamma(\frac{1}{2}-K_{\text{XB}}-i\theta_{\epsilon'})|^2}\right] \tag{2.B.107b}$$

where θ_ϵ and $\theta_{\epsilon'}$ are defined in terms of (2.B.49). This expression is closely related to Devonshire's results.[44] From extensive calculations of the c–c couplings it follows that the dominant contributions to the c–c coupling originate from those terms where $\Delta v = \pm 1$. For the harmonic description of the BC bond the $V^{\mathrm{c-c}}_{v\epsilon, v'\epsilon'}$ terms identically vanish for $\Delta v \neq \pm 1$ if the exponential in the Morse interaction (2.B.41) is expanded in α_{XB} and only the linear term is retained. For the case of an anharmonic BC molecule numerical studies of the c–c coupling have been conducted, a sample of the results is portrayed in Fig. 9 for one-the-energy-shell c–c coupling terms, at the energy of a discrete level (v, l). We denote by $V^{\mathrm{c-c}}_{v', v''}$ the matrix element $V^{\mathrm{c-c}}_{v'\epsilon', v''\epsilon''}$ with the condition that $W_{\mathrm{BC}}(v') + \epsilon' = W_{\mathrm{BC}}(v'') + \epsilon'' = W_{\mathrm{BC}}(v) + \epsilon_l$. From this figure it is apparent that $V^{\mathrm{c-c}}_{v', v''}$ decreases very fast with decreasing v'' and that $V^{\mathrm{c-c}}_{v', v'-1}$ is the dominant c–c coupling term. We may accordingly set

$$V^{\mathrm{c-c}}_{v', v''} = V^{\mathrm{c-c}}_{v', v'-1} \delta_{v', v''+1} \qquad v' > v'' \tag{2.B.108}$$

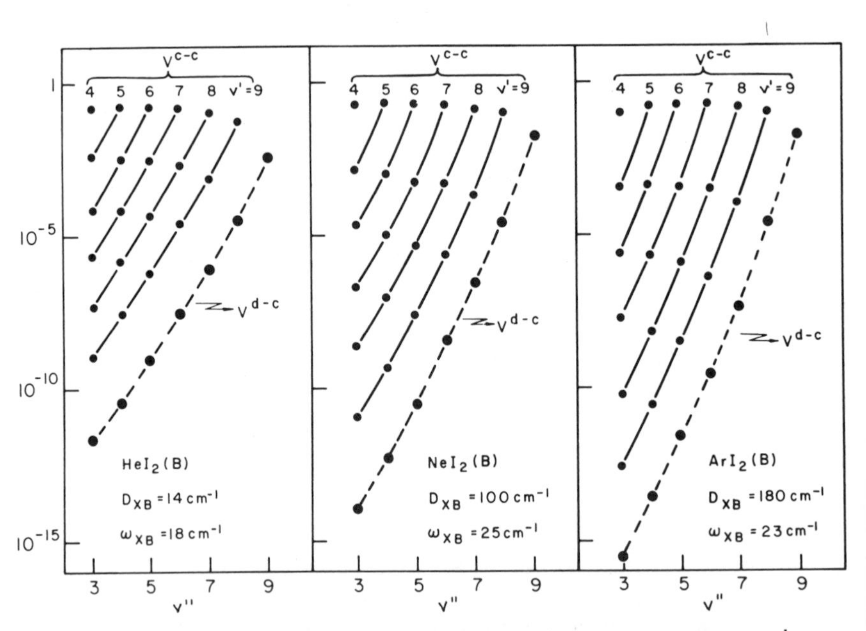

Fig. 9. On-the-energy-shell continuum–continuum $V^{\mathrm{c-c}}$ and discrete–continuum $V^{\mathrm{d-c}}$ couplings involved in the predissociation of some model $\mathrm{XI_2(B)}$, X≡He, Ne, Ar, from level $v = 10$ [v being the quantum number of the $\mathrm{I_2(B)}$ molecule]. The points connected by solid lines correspond to the continuum–continuum couplings $V^{\mathrm{c-c}}_{v'v''}$, where v' is specified on top of each line and v'' is marked on the abscissa. The points connected by dashed lines correspond to the discrete–continuum coupling $|V^{\mathrm{d-c}}_{10, n''}|/(\hbar\omega_{\mathrm{I_2}})^{1/2}$. Note that with these definitions all the couplings are adimensional. The parameters for the van der Waals interaction between X and the $\mathrm{I_2(B)}$ molecule are marked on the figure (see also Table I).

disregarding c–c coupling for $\Delta v \neq \pm 1$. The incorporation of the c–c couplings (final-states interaction) into the general VP problem is possible using relation (2.B.108) and which was discussed in the preceding section.

C. Numerical Methods

1. Scattering formalism

We were able[28] to provide closed analytic solutions of the quantum dynamics of the VP process of a linear X···B—C triatomic molecule. Two reasons motivated the implementation of these results by numerical techniques to study VP processes. First, we wanted to test the accuracy of the approximate results obtained in the distorted wave approaches of Section II.B. Second, and more important, we wanted to be able to handle situations where analytical solutions are no longer possible. These include:

1. The consideration of other types of vdW interaction potentials. The choice of the Morse function to mimic the vdW interaction is convenient but inadequate at least at large distances. Thus, the use of alternative representations of the weak vdW potential is desirable.
2. The study of VP processes in molecules with different equilibrium geometries, such as T-shaped vdW molecules, is of considerable interest. In particular, the HeI_2 molecule is believed to be T-shaped[34] and we were tempted to explore its VP dynamics.

Several numerical methods have been proposed in the literature to handle half-collision problems. One of these involves the use of conventional scattering solutions of the resulting close-coupling equations. From the study of the probabilities of inelastic transitions in the collision of X with BC, it is possible to extract information about the quasibound states of the van der Waals molecule X···BC. These states (also called resonances, collision complexes, compound states, metastable states) manifest themselves as dips or peaks at energies which approximatively correspond to their zero-order energies. The widths of these resonances can then be related to the rate of VP.

Let us consider again the triatomic system X···BC, where X is a rare-gas atom and BC a conventional diatomic molecule, in a fixed geometrical configuration but not necessarily colinear. If we denote by R_{BC} and $R_{X,BC}$ the interatomic distance between atoms B and C and the distance between atom X and the center of mass of BC, respectively, then (2.B.1) will not hold any more but the nuclear Hamiltonian is still given by (2.B.2). In the potential coupling scheme the nuclear wavefunction at total energy E is expanded in terms of the eigenstates $x_v(R_{BC})$ of the Hamiltonian for the free diatomic fragment BC, H_{BC}, which is given by (2.B.5a).

Explicitly, by,

$$\psi_E(R_{BC}, R_{X,BC}) = \sum_v x_v(R_{BC})\phi_{vE}(R_{X,BC}) \tag{2.C.1}$$

After substitution of this expansion in the Schrodinger equation, we obtain the set of the familiar close-coupled equations

$$\left\{-\frac{\hbar^2}{2\mu_{X,BC}}\frac{\partial^2}{\partial R_{X,BC}^2} + U_{vv}(R_{X,BC}) + W_{BC}(v) - E\right\}\phi_{vE}(R_{X,BC})$$

$$= -\sum_{v' \neq v} U_{vv'}(R_{X,BC})\phi_{v'E}(R_{X,BC}) \tag{2.C.2}$$

where

$$U_{vv'}(R_{X,BC}) = \int x_v(R_{BC})U(R_{BC}, R_{X,BC})x_{v'}(R_{BC})\, dR_{BC} \tag{2.C.3}$$

These coupled differential equations (2.C.2) may be solved by any of the currently available numerical integration methods. These algorithms usually give the elements of the S matrix. The absolute value squared of the elements of the S matrix, namely, $|S_{vv'}(E)|^2$, for a collision between atom X and diatomic BC at total energy E, gives the probability for BC, which initially was at the distant past in the state v (or v'), to be after the collision in the state v' (or v). The coupled equations (2.C.2) involve, for a given total energy E, open $E - W_{BC}(v) > 0$ and closed $E - W_{BC}(v) < 0$ channels. Provided that a diagonal potential term U_{vv} has a minimum then it can be shown[46] that the absolute value squared of the S matrix exhibits resonances located at approximatively the position of the bound states of these U_{vv} potentials. In the case of vdW complexes X$\cdots$(BC) where BC is a normal diatomic and X a rare-gas atom, the widths of the resonances specify the VP rate, experimentally determined for the VP of the vdW molecule. A simple parametrization of these resonances can be obtained by using partitioning techniques and the Feshbach formalism. Neglecting for a moment the couplings between closed and open channels, a system of coupled equations for the closed channels is obtained, which is solved exactly results in a set of (zero-order) discrete eigenfunctions ψ_i^0 corresponding to the bound states of the complex. The projection operator P is then defined by

$$P=\sum_{i}|\psi_i^0\rangle\langle\psi_i^0| \tag{2.C.4}$$

For the open channels the zero-order ingoing plane waves solutions of the asymptotic Hamiltonian

$$\left(-\frac{\hbar^2}{2\mu_{\mathrm{X,BC}}}\frac{\partial^2}{\partial R_{\mathrm{X,BC}}^2}+W_{\mathrm{BC}}(v)-E\right)\phi_{vE}^0(R_{\mathrm{X,BC}})=0 \tag{2.C.5}$$

will define the complement of P,

$$Q=\sum_{v}\int\partial E|\psi_{vE}^0\rangle\langle\psi_{vE}^0| \tag{2.C.6}$$

with

$$\langle R_{\mathrm{X,BC}},R_{\mathrm{BC}}|\psi_{vE}^0\rangle=x_v(R_{\mathrm{BC}})\phi_{vE}^0(R_{\mathrm{X,BC}}) \tag{2.C.7}$$

In terms of this complete set, the Hamiltonian assumes the form

$$H=H_0+V \tag{2.C.8}$$

with

$$\begin{aligned}H_0&=\sum_{i}E_i|\psi_i^0\rangle\langle\psi_i^0|+\sum_{v}\int\partial E\times E\times|\psi_{vE}^0\rangle\langle\psi_{vE}^0|\\&=PH_0P+QH_0Q\end{aligned} \tag{2.C.9}$$

and

$$V=PVQ+QVQ+QVP \tag{2.C.10}$$

where there is no term PVP as P has been chosen to diagonalize PHP. The matrix elements of V are

$$\begin{aligned}\langle\psi_i^0|V|\psi_{vE}^0\rangle=\int\partial R_{\mathrm{BC}}\,\partial R_{\mathrm{X,BC}}\langle\psi_i^0|R_{\mathrm{BC}},R_{\mathrm{XBC}}\rangle\\\times U(R_{\mathrm{BC}},R_{\mathrm{X,BC}})x_v(R_{\mathrm{BC}})\phi_{vE}^0(R_{\mathrm{X,BC}})\end{aligned} \tag{2.C.11a}$$

for bound-continuum coupling, and

$$\langle \psi_{vE}^{0} | v | \psi_{v'E'}^{0} \rangle = \int \partial R_{BC} \, \partial R_{X,BC} \, x_v(R_{BC}) \phi_{vE}^{0*}(R_{X,BC}) \times U(R_{BC}, R_{X,BC}) x_{v'}(R_{BC}) \phi_{v'E'}^{0}(R_{X,BC}) \tag{2.C.11b}$$

for c–c coupling. The S matrix is defined in the usual way[46]

$$S_{vv'}(E) = \delta_{vv'} - 2i\pi T_{vv'}(E) \tag{2.C.12}$$

where $T_{vv'}(E)$ are on-the-energy-shell matrix elements $\langle \psi_{\epsilon}^{0} | T(E) | \psi_{v'E}^{0} \rangle$ of the transition operator defined in (2.B.13). In what follows the interaction of the molecule with the electromagnetic field is neglected. It can be shown, that the projection $QT(E)Q$ can be written in the form

$$QT(E)Q = QR(E)Q + QR(E)P(E^{+} - H_0 - PR(E)P)^{-1} PR(E)Q \tag{2.C.13}$$

where the level shift operator $R(E)$ has been defined in (2.B.18). Notice that this operator only involves c–c interactions and therefore is expected to vary slowly with energy. The first term in (2.C.13) is equal to the transition operator T^{c-c} for a collision process in which the bound states are neglected [see (2.B.78)]. This operator represents direct transitions between open channels. On the other hand, the second term in (2.C.13) exhibits peaks due to the presence of the energy dependent denominator. Defining Δ and Γ as the Hermitian and negative-value of the anti-Hermitian part of PRP, respectively, namely,

$$\Delta = PVQ\mathcal{P}(E - QHQ)^{-1} QVP \tag{2.C.14}$$

and

$$\Gamma = PVQ\partial(E - QHQ)QVP \tag{2.C.15}$$

where $\mathcal{P}$ denotes the "Cauchy principal part distribution," we can write

$$QT(E)Q = T^{c-c} + QRP[E - (H_0 + \Delta) + i\Gamma]^{-1} PRQ \tag{2.C.16}$$

where the weak dependence on energy of the operator R has been omitted. If the resonances do not overlap, that is,

$$\Gamma_i, \Gamma_j \ll |E_i + \Delta_i - (E_j + \Delta_j)| \tag{2.C.17}$$

for all i and j, for energies E in the vicinity of E_i, the probability for a transition from v to v' channel with $v \neq v'$ will be given approximatively by

$$P_{vv'}(E) = |S_{vv'}(E)|^2 \simeq 4\pi^2 \left| T_{vv'}^{\text{c-c}} + \frac{R_{vi} R_{iv'}}{E - (E_i + \Delta_i) + i\Gamma_i} \right|^2 \quad (2.C.18)$$

where

$$T_{vv'}^{\text{c-c}} \equiv \langle \psi_{vE}^0 | T^{\text{c-c}} | \psi_{v'E}^0 \rangle \quad (2.C.19a)$$

$$R_{vi} \equiv \langle \psi_{vE}^0 | R | \psi_i^0 \rangle \quad (2.C.19b)$$

$$\Delta_i - \Gamma_i \equiv \langle \psi_i^0 | R | \psi_i^0 \rangle \quad (2.C.19c)$$

The final expression (2.C.18) which is valid for an isolated resonance results in the general dispersion formula

$$P_{vv'}(E) = C + \frac{\Gamma_i B + (E - E_i - \Delta_i) A}{(E - E_i - \Delta_i)^2 + \Gamma_i^2} \quad (2.C.20)$$

with

$$C = 4\pi^2 |T_{vv'}^{\text{c-c}}|^2$$
$$B = 4\pi^2 \left[|R_{vi} R_{iv'}|^2 \Gamma_i^{-1} - 2\,\text{Im}(T_{vv'}^{\text{c-c}} R_{vi}^* R_{iv'}^*) \right]$$
$$A = 8\pi^2 \text{Re}(T_{vv'}^{\text{c-c}} R_{vi}^* R_{iv'}^*) \quad (2.C.21)$$

A fit of the energy dependence of the probabilities $P_{vv'}(E)$ will yield the coefficients A, B, and C as well as Γ_i, which can be considered as the exact half-width of the level $|\psi_i\rangle$, and the exact position $(E_i + \Delta_i)$ of this level. The results can be contrasted in the case of colinear geometry with the approximate results of Section II.

2. *Comparison Between Numerical and Distorted Wave Calculations*

We now proceed to consider the application of the numerical methods to the colinear VP of XBC molecules, and comparisons with the approximate distorted wave results shown in Section II.B. In the colinear case, the coordinate $R_{\text{X,BC}}$ describing the distance between the atom X and the center of mass of the diatomic molecule BC is given by (2.B.1). The vdW bond is characterized by the Morse potential defined in (2.B.40). The $U_{vv'}(R_{\text{X,BC}})$ terms in the closed-coupled equations (2.C.2) are then given by

$$U_{vv'}(R_{\mathrm{X,BC}}) = D_{\mathrm{XB}}\left\{A_{vv'}^{(2)} \exp\left[-2\alpha_{\mathrm{XB}}\left(R_{\mathrm{X,BC}} - \bar{R}_{\mathrm{X,BC}}\right)\right]\right.$$
$$\left. - 2A_{vv'}^{(1)} \exp\left[-\alpha_{\mathrm{XB}}\left(R_{\mathrm{X,BC}} - \bar{R}_{\mathrm{X,BC}}\right)\right]\right\} \quad (2.C.22)$$

where

$$\bar{R}_{\mathrm{X,BC}} = \bar{R}_{\mathrm{XB}} + \gamma \bar{R}_{\mathrm{BC}} \quad (2.C.23)$$

The coefficients $A_{vv'}^{(1)}$ and $A_{vv'}^{(2)}$ have been defined already in (2.B.71) and their explicit expressions are given in (2.B.72) and (2.B.73) for harmonic and for anharmonic BC potentials, respectively.

The closed-coupled equations (2.C.2) are solved numerically, subjected to the usual scattering boundary conditions.[28] A typical plot of the probability P_{01} for a transition between states $v=0$ and $v=1$ in the energy region of a resonance corresponding to a quasibound state of the closed channel $v=2$ is depicted in Fig. 10. The energy scale for the probability $P_{01}(E)$ is represented relative to the approximate position of the resonance, as given

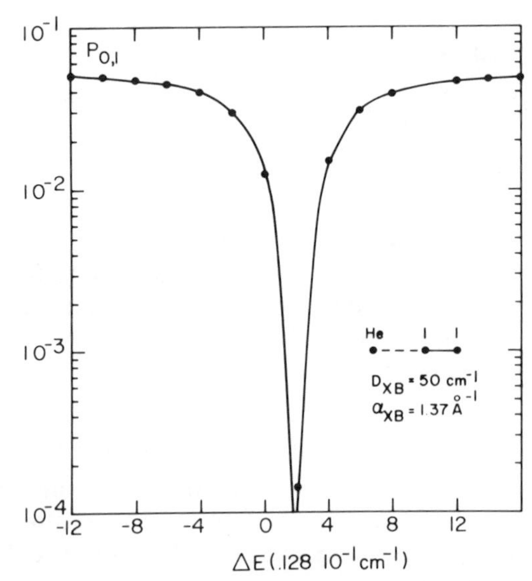

Fig. 10. Transition probability between vibrational states of $v=1$ and $v=0$ of I_2 for the linear collision of He with I_2(B) in the energy region of a resonance corresponding to the closed channel $v=2$. The total energy is measured for the center of the resonance.

by a variational calculation, namely,

$$E_{vl} = \hbar\omega_{BC}\left(v+\tfrac{1}{2}\right) - \frac{\hbar^2\alpha_{BX}^2}{2\mu_{X,BC}}\left(K'_{XB} - l - \tfrac{1}{2}\right)^2 \tag{2.C.24}$$

where

$$K'_{XB} = \frac{K_{XB}A_{vv}^{(1)}}{\left(A_{vv}^{(2)}\right)^{1/2}} \qquad K_{XB} = \left(2D_{XB}\mu_{X,BC}\right)^{1/2}/\hbar\alpha_{XB} \tag{2.C.25}$$

The $P_{01}(E)$ curve is fitted to (2.C.20) and the half-width Γ_{vl}^{num} is obtained. A series of calculations for VP of linear vdW molecules, $XI_2(B)$ (X= Ar, Ne, He) have been conducted.[28] The results are presented in Table III.

TABLE III
Comparison Between Numerical and Distorted Wave Half-widths for Some Model Colinear van der Waals Molecules

System	D_{XB} (cm^{-1})	α_{XB} (A^{-1})	ω_{XB} (cm^{-1})	$E_{v,1}-E_{v,0}$ (cm^{-1})	N	v	l	$\Delta E_{v,l}$ (cm^{-1})	$\Gamma_{v,l}^{num}$ (cm^{-1})	$\Gamma_{v,l}^{DW}$ (cm^{-1})
$ArI_2(B)$	200	2.02	40	36	10	2	2	$0.30\ 10^{-1}$	0.26	0.32
							3	$-0.55\ 10^{-1}$	0.34	0.41
							4	-0.17	0.38	0.45
							5	-0.12	0.39	0.45
							6	-0.24	0.35	0.40
							7	-0.24	0.23	0.32
$NeI_2(B)$	100	1.31	25	22	8	2	0	$0.78\ 10^{-2}$	$0.13\ 10^{-2}$	$0.15\ 10^{-2}$
							1	$-0.26\ 10^{-2}$	$0.37\ 10^{-2}$	$0.46\ 10^{-2}$
							2	$-0.90\ 10^{-2}$	$0.67\ 10^{-2}$	$0.78\ 10^{-2}$
							3	$-0.12\ 10^{-1}$	$0.87\ 10^{-2}$	$0.10\ 10^{-1}$
							4	$-0.12\ 10^{-1}$	$0.93\ 10^{-2}$	$0.11\ 10^{-1}$
$HeI_2(B)$	50	1.37	40	24	2	2	0	$0.22\ 10^{-1}$	$0.42\ 10^{-1}$	$0.44\ 10^{-1}$
							1	$0.11\ 10^{-1}$	$0.38\ 10^{-1}$	$0.40\ 10^{-1}$
						10	0	0.18	0.16	0.20
							1	$0.96\ 10^{-1}$	0.16	0.18
						20	0	0.29	0.35	0.37
							1	0.20	0.30	0.33
	14	1.18	18	6.4	2	2	0	$0.17\ 10^{-2}$	$0.12\ 10^{-2}$	$0.12\ 10^{-2}$
							1	$0.12\ 10^{-3}$	$0.81\ 10^{-4}$	$0.10\ 10^{-3}$
						10	0	$0.12\ 10^{-1}$	$0.57\ 10^{-2}$	$0.60\ 10^{-2}$
							1	$0.72\ 10^{-3}$	$0.36\ 10^{-3}$	$0.49\ 10^{-3}$
						20	0	$0.24\ 10^{-1}$	$0.11\ 10^{-1}$	$0.12\ 10^{-1}$
							1	$0.11\ 10^{-2}$	$0.49\ 10^{-3}$	$0.95\ 10^{-3}$
						30	0	$0.35\ 10^{-1}$	$0.15\ 10^{-1}$	$0.17\ 10^{-1}$
							1	$0.93\ 10^{-3}$	$0.44\ 10^{-3}$	$0.14\ 10^{-2}$

N is the number of bound states. ΔE_{vl} gives the numerical result for the position of the resonance with respect to the approximate result (2.C.24). Finally, Γ^{num} and Γ^{dw} represent the numerical value and the analytical distorted wave value of the half-width of the resonance, respectively. The total rate for VP is then given by $2\Gamma/\hbar$. From the results of Table III, we may conclude that the distorted wave treatment provides a very good estimate of the VP rates.

3. *Application to Perpendicular Vibrational Predissociation*

We now proceed to present the results of the numerical methods for the perpendicular VP of a T-shaped vdW molecule

$$\begin{matrix} & \text{B} \\ \text{X}\cdots\cdot & | \\ & \text{C} \end{matrix}$$

(with B$\equiv$C) consisting of a rare-gas atom X bound to a homonuclear diatomic BC and where X moves on a line perpendicular to the BC axis. The coordinate $R_{\text{X,BC}}$ describing the distance between the atom X and the center of mass of BC, is now

$$R_{\text{X,BC}} = \left(R_{\text{XB}}^2 - \tfrac{1}{4}R_{\text{BC}}^2\right)^{1/2} \tag{2.C.26}$$

Thus the nuclear Hamiltonian, takes the form

$$H = -\frac{\hbar^2}{2\mu_{\text{X,BC}}}\frac{\partial^2}{\partial R_{\text{X,BC}}^2} - \frac{\hbar^2}{2\mu_{\text{BC}}}\frac{\partial^2}{\partial R_{\text{BC}}^2} + V_{\text{BC}}(R_{\text{BC}})$$
$$+ 2V_{\text{XB}}\left[\left(R_{\text{X,BC}}^2 + \tfrac{1}{4}R_{\text{BC}}^2\right)^{1/2}\right] \tag{2.C.27}$$

where the factor 2 in the last term of (2.C.27) originates from the sum of two identical interactions between X and atoms B and C. In order to avoid time-consuming numerical evaluation of the integrals (2.C.3), the V_{XB} term in (2.C.27) is expanded in a Taylor series around the equilibrium position $\overline{R}_{\text{BC}}$ of the diatomic molecule BC. In the calculation on the HeI_2(B) system only the first two terms in the expansion have been kept,

$$V(R_{\text{BC}}, R_{\text{X,BC}}) = V(\overline{R}_{\text{BC}}, R_{\text{X,BC}}) + \left.\frac{\partial V}{\partial R_{\text{BC}}}\right|_{\overline{R}_{\text{BC}}}(R_{\text{BC}} - \overline{R}_{\text{BC}}) \tag{2.C.28}$$

This linearization approximation has been tested for that system and it was concluded that it gives reasonably good accuracy.[29] In one set of calculations the interaction V_{XB} in (2.C.27) has been specified by a Morse-type potential of the form (2.B.41). The V_{BC} interatomic potential for the conventional BC molecule is represented by an anharmonic potential described by a Morse potential with parameters

$$D_{BC}=\frac{\omega_{BC}^2}{4\omega x_{BC}} \qquad \alpha_{BC}=\frac{(2\mu_{BC}\omega x_{BC})^{1/2}}{\hbar} \tag{2.C.29}$$

where ω_{BC} and ωx_{BC} are the frequency and the anharmonicity of the BC molecule, obtained from spectroscopic data.

The approximate position of the bound states can be obtained by making use of the approximation

$$\begin{aligned}\left(R_{X,BC}^2+\tfrac{1}{4}R_{BC}^2\right)^{1/2} &\simeq \left(\bar{R}_{X,BC}^2+\tfrac{1}{4}\bar{R}_{BC}^2\right)^{1/2}\\ &\quad+\left(\bar{R}_{X,BC}^2+\tfrac{1}{4}\bar{R}_{BC}^2\right)^{-1/2}\bar{R}_{X,BC}\left(R_{X,BC}-\bar{R}_{X,BC}\right)\end{aligned} \tag{2.C.30}$$

in (2.C.27). In analogy to (2.C.24) the approximate position for the resonance which defines the origin of the energy scale will be given by

$$\begin{aligned}E_{vl} &\simeq W_{BC}(v)-\frac{\hbar^2\alpha_{XB}'^2}{2\mu_{X,BC}}\left(K''-l-\tfrac{1}{2}\right)^2\\ \alpha_{XB}' &= \alpha_{XB}\left(\bar{R}_{X,BC}^2+\tfrac{1}{4}\bar{R}_{BC}^2\right)^{-1/2}\bar{R}_{X,BC}\\ K'' &= \left(\frac{4D_{XB}\mu_{X,BC}}{\hbar^2\alpha_{XB}^2}\right)^{1/2}\frac{B_{vv}^{(1)}}{\left(B_{vv}^{(2)}\right)^{1/2}}\\ B_{vv}^{(j)} &= \int x_v^2(R_{BC})\exp\left\{j\alpha_{XB}\left(\bar{R}_{X,BC}^2+\tfrac{1}{4}\bar{R}_{BC}^2\right)^{-1/2}\right.\\ &\quad\left.\times\bar{R}_{BC}\left(R_{BC}-\bar{R}_{BC}\right)\right\}dR_{BC}\end{aligned} \tag{2.C.31}$$

The results for the VP rates Γ_{vl} ($l=0$) obtained by numerical integration of the close-coupling equations using the molecular parameters[47]

$$\begin{aligned}\omega_{BC} &= 128\ \text{cm}^{-1}\\ \omega x_{BC} &= 0.834\ \text{cm}^{-1}\end{aligned} \tag{2.C.32a}$$

for the $I_2(B)$ molecule, and the parameters

$$\begin{aligned}\alpha &= 1.25\ \text{Å}^{-1}\\ \bar{R}_{XB} &= 4\ \text{Å}\\ D_{XB} &= 7\ \text{cm}^{-1}\end{aligned} \tag{2.C.32b}$$

for the vdW bond, are presented in Fig. 11. These values were advocated in previous work[27, 28] on the basis of the spectroscopic data of Smalley, Levy, and Wharton.[15] The v dependence of the calculated VP rates is superlinear being well accounted by the empirical relation $\Gamma = Av^2$, in the range $v = 10$–30. This v dependence of the VP rates is in excellent qualitative agreement with the experimental data of Johnson, Wharton, and Levy.[15c] However, it should be noted (see Fig. 11) that the absolute values of the VP rates calculated for the potential parameters (2.C.32) are lower

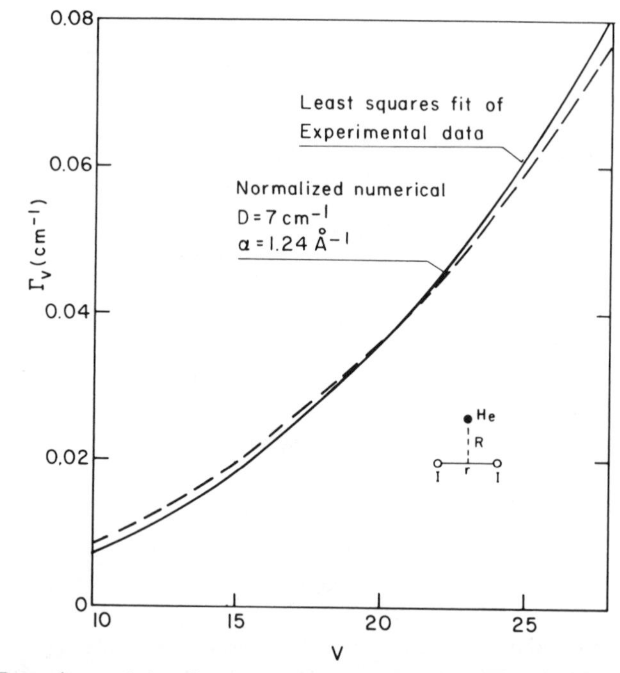

Fig. 11. Dependence of the vibrational predissociation linewidths $\Gamma_{v,l}(l=0)$ on the vibrational quantum number v of the $I_2(B)$ diatomic molecule in the T-shaped van der Waals complex HeI_2. The full line corresponds to the experimental data of Johnson, Wharton, and Levy, while the dashed line is the result of the numerical integration of the close-coupling equations using the molecular parameters given in the text.

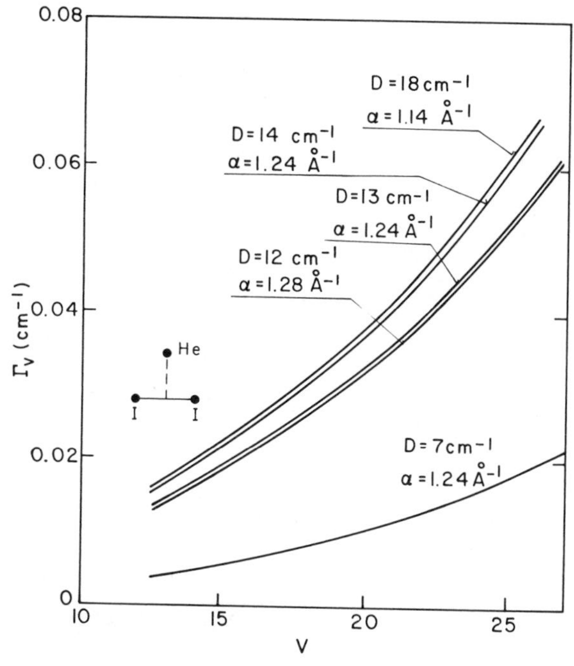

Fig. 12. Numerical calculations of the vibrational predissociation rates for the T-shaped HeI_2 complex as a function of the vibrational quantum number v of the I_2(B) diatomic molecule. The parameters D and α correspond to the Morse potential with $\bar{R}_{XB} = 4$ Å.

by a numerical factor of ~ 3 than the experimental values. Numerical calculations of the VP rates for $l=0$ for several other values of the potential parameters (2.C.32b) characterizing the vdW interaction are presented in Fig. 12. It is seen that increasing the parameter D_{XB} (notice that $2D_{XB}$ gives the minimum of the potential surface) results in an enhancement of the VP rate. Also other calculations are presented where α_{XB} has been slightly changed. These calculations give an idea of the sensitivity of the vibrational predissociation rate to changes in the potential parameters of the vdW bond. In Fig. 13 two possible theoretical fits of the experimental data are displayed. For α_{XB} in the range 1.1–1.2 Å^{-1}, D_{XB} is in the range 13–18 cm^{-1}. The dissociation energy of the vdW complex (being equal in this model to $2D_{XB}$ minus the zero point energy of the vdW bond) will be in the range 20–30 cm^{-1}, while ΔE_{01} (the energy difference between the two levels $l=1$ and $l=0$ supported by the vdW bond) is ~ 13 cm^{-1}. Recent experimental data indicate that the dissociation energy of the HeI_2(B) vdW complex is < 18.5 cm^{-1}, while $\Delta E_{01} = 6$

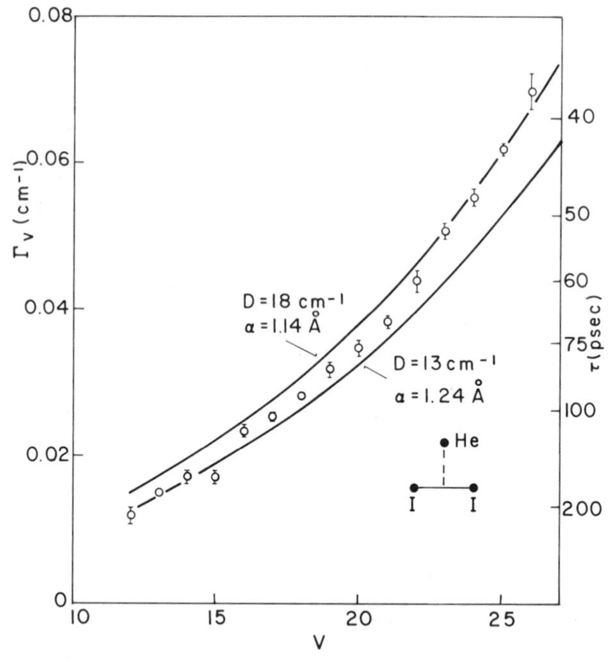

Fig. 13. Two possible numerical fits of the experimental data of Johnson, Wharton, and Levy for vibrational predissociation linewidths of the T-shaped $HeI_2(B)$ van der Waals complex. The parameters D and α correspond to the Morse potential with $\bar{R}_{XB} = 4$ Å.

cm^{-1} as reported by Smalley et al.[15a] We are not seriously concerned about the discrepancy between the theoretical value of ΔE_{01} and the experimental energy splitting as the vibrational excitation of the HeI_2 molecule observed by Smalley et al.[15a] may correspond to some other vibrational mode of the vdW complex. However, the dissociation energy $2D = 20-30$ cm^{-1} required to fit the VP data is definitely higher than the experimental values.[34] We shall examine now the possible reasons for this discrepancy. The numerical calculations for the VP rates of $HeI_2(B)$ rest on the following assumptions:

1. A dumbbell model potential is used, the total vdW potential being represented by the sum of two atom–atom interactions.
2. The atom–atom interactions are represented by Morse potentials.
3. The system is restricted to vibrational motion in a fixed T-shaped configuration. Rotational effects are disregarded.
4. The potential is expanded in powers of the displacement from the equilibrium interatomic distance of I_2, up to the linear term.

Furthermore while confronting the results with the experimental data we have assumed that:

5. The measured linewidth correspond to pure vibrational predissociation occurring on a single adiabatic potential surface.
6. The photodissociation is assumed to proceed through the metastable state: only the metastable state carry oscillator strength from the ground state and the transition to the continuum states are assumed to be negligibly small.

Assumption 1 is likely to be a good approximation for a vdW molecule like HeI_2. An interesting question arises concerning the validity of assumption 2. A Morse potential is a reasonable description at small and at intermediate distances but certainly very poor for long-range interactions, where the usual R^{-6}, R^{-8}, dependence due to dispersion forces is expected. However, the VP dynamics is essentially determined by bound–continuum interaction, so that one can hope that the VP rates will be essentially determined by the details of the potential at internuclear distances corresponding to the minimum of the potential surface. In order to investigate this cardinal point numerical calculations have been performed using the modified Buckingham potential

$$V_{XB}(R_{XB})=A\exp(-\beta R_{XB})-CR_{XB}^{-6} \tag{2.C.33}$$

for the vdW bond. The parameters A, C, and β have been chosen in such a way so that the potential function and its second derivative at the equilibrium position will be equal to the corresponding values for the Morse potential. This prescription results in the relations

$$A=\frac{6D_{XB}e^{\gamma}}{\gamma-6}$$

$$C=\frac{D_{XB}\bar{R}_{XB}^{+6}\gamma}{\gamma-6}$$

$$\gamma=\frac{7}{2}+\frac{\gamma'^2}{6}+\left[\left(\frac{7}{2}+\frac{\gamma'^2}{6}\right)^2-2\gamma'^2\right]^{1/2}$$

$$\gamma=\beta\bar{R}_{XB} \qquad \gamma'=\alpha_{XB}\bar{R}_{XB} \tag{2.C.34}$$

In Table IV we present the results of the numerical integration of close-coupling equations using the Buckingham potentials (2.C.33) and the Morse potential for two different sets of parameters α_{XB} and D_{XB} ($\bar{R}_{XB}=4$

TABLE IV
Comparison Between Vibrational Predissociation Rates Obtained with Morse and Buckingham Potentials

Parameters of the Morse potential[a]	v	Morse Γ_v (cm^{-1})	Buckingham Γ_v (cm^{-1})
$D=13$ cm^{-1}, $\alpha=1.24$ Å^{-1},	15	0.19 10^{-1}	0.20 10^{-1}
$\bar{R}_{XB}=4$ Å	20	0.33 10^{-1}	0.34 10^{-1}
	25	0.53 10^{-1}	0.55 10^{-1}
$D=7$ cm^{-1}, $\alpha=1.24$ Å^{-1},	15	0.61 10^{-2}	0.64 10^{-2}
$\bar{R}_{XB}=4$ Å	20	0.11 10^{-1}	0.11 10^{-1}
	25	0.18 10^{-1}	0.19 10^{-1}

[a]The parameters of the Buckingham potentials have been determined following the procedure described in the text, Eq. (2.C.34).

Å). The calculated VP rates are insensitive to the form of the potential at large distances. This is to be expected, since the rates for VP are essentially determined by the overlap between a continuum wavefunction and a bound state, so that the only nonvanishing contribution originates from the region of the potential well. We conclude that for the determination of VP rates the Morse potential provides a reasonable description of the vdW interaction.

Assumption 3 should also be critically examined. It can be related to the rotational infinite order sudden approximation[48] (RIOSA) used in collision problems, in which the vibrational motion is considered to be much faster than rotation. For a system like HeI_2 this will be particularly acceptable as the vibrational frequencies for the I—I stretch and for the $He\cdots I_2$ motion are 128 cm^{-1} and ~6 cm^{-1}, respectively, while the rotational constants associated to the I_2 and the complex are 0.037 cm^{-1} and 0.27 cm^{-1}, respectively. The rotational motion can be considered in that case as adiabatic and θ the angle between the interatomic axis of I_2 and the line between the He atom and the center of mass of I_2 ($\theta=\pi/2$ for T=shaped configuration) can be taken as a parameter. Then a vibrational predissociation amplitude that depends parametrically on θ will be obtained. The final result is obtained after an average over rotational wavefunctions. Further work in this direction is under progress (see section VI). We have already shown above that the linearization approximation 4 introduces an error of <20%. For the purpose of approximate fits of the VP rates this is an acceptable error. The calculation can be however easily improved by incorporating additional terms in the expansion. Assumption 5 seems to be borne out by the experimental results of Smalley et al.[15a] If a second decay channel of induced electronic predissociation of the type (2.A.2) will significantly contribute to the linewidth, a strong irregular vibrational dependence of the fluorescence intensity is expected to be exhibited. The situation is completely different for the ArI_2 vdW molecule where electronic relaxation competes with VP.[15d] Finally, assumption 6 has already been discussed in Section II.B and seems to be justified in view of the spectroscopic data of Smalley et al.[15a] We may conclude at this stage that the discrepancy between the dissociation energy, inferred from the analysis of the VP data and the corresponding experimental value of the energy, will be settled after a further examination of rotational effects on the energetics and dynamics of vdW molecules (see section VI).

III. INTRAMOLECULAR DYNAMICS OF vdW DIMERS

A. Some Facts and Correlations

Dixon and Herschbach[16] have observed remarkably long lifetimes of the VP of a vibrationally excited Cl_2—Cl_2 vdW dimer. They were able to set a lower limit of the VP lifetime, τ, by observing that the dimer in a supersonic beam survived the transit time from the point of collisional vibrational excitation up to the point of detection without undergoing fragmentation. Thus, for the Cl_2—Cl_2 dimer $\tau \geqslant 10^{-4}$ sec and this vdW molecule undergoes $\sim 10^8 - 10^9$ vibrations of the Cl_2 subunit, whose vibrational energy exceeds the dissociation energy of the vdW bond before bond breaking occurs. The ineffective process of intramolecular vibrational energy flow from a conventional chemical bond to a vdW bond is a problem of considerable current interest in the area of intramolecular dynamics. Dixon, Herschbach, and Klemperer[49] have suggested that the VP lifetime τ $(=\tau_0 z_{01})$ can be phenomenologically represented[16] by a product of the translational-vibrational energy transfer probability per collision, z_{01}, and $\tau_0 \cong 10^{-12\pm 1}$ sec, which corresponds to the duration of a collision. This approach for VP of a vdW bond bears a close analogy to the phenomenological "half-collision" concept,[43] which was invoked to describe photofragmentation of ordinary chemical bonds. On the basis of the energy gap law $\ln(1/\tau) \propto \omega_{BC}^{1/2}$ for the VP process, we can assert that the mismatch between the molecular frequency ω_{BC} and the frequency of the vdW bond severly inhibits the VP process. The stability of the Cl_2—Cl_2 dimer with respect to VP can be qualitatively rationalized in terms of our energy gap law.[27]

We shall proceed to outline the theory of intramolecular dynamics and VP of linear vdW dimers AB—CD, where AB and CD are normal molecules. There are two novel features of the problem of linear dimer, as compared to the simpler case of a linear X—AB triatomic vdW molecule. First, while in the case of linear X—AB the only decay channel involves the interconversion of vibrational energy to translational energy, in the case of linear vdW dimers an additional VP decay channel, a new VP decay channel, is exhibited involving partial conversion of vibrational energy of one bond to vibrational energy of the second bond, i.e., inter-bond vibrational energy transfer. Second, the general problem of inter-bond, intramolecular vibrational energy flow in vdW dimers is of considerable interest.

B. VP Dynamics of Heterodimers

The VP of linear vdW heteronuclear dimers (heterodimers) AB $\cdots$ CD, where AB and CD are two normal molecules which are characterized by the vibrational frequencies ω_{AB} and ω_{CD}, respectively, involves two decay channels:

1. Conversion of vibrational energy of AB (or CD) to translational energy i.e., $V \rightarrow T$ process.
2. Conversion of vibrational energy of AB to the vibrational energy of CD (provided that $\omega_{AB} > \omega_{CD}$), while the energy balance is made up by translational energy. This is a $V \rightarrow V + T$ process.

When the mismatch between the molecular frequency (ω_{AB} or ω_{CD}) and the dissociation energy of the vdW heterodimer is large, the $V \rightarrow T$ process is severely inhibited, as implied by the energy gap law.[27] On the other hand, an alternative VP channel of the dimer may involve the conversion of the vibrational energy of AB to the vibrational energy of CD (provided that $\omega_{AB} > \omega_{CD}$), while the energy balance is made up by translational energy. The relative kinetic energy of the two molecular fragments is consequently lowered for the $V \rightarrow V + T$ process relative to the case of $V \rightarrow T$ process, whereupon the VP rate for the former mechanism may be considerably enhanced. The new VP mechanism, which involve a $V \rightarrow V + T$ process is important for establishing the VP mechanism and the internal energy content of the diatomic molecules resulting from fragmentation of the vibrationally excited dimer.

We consider the one-dimensional motion of the AB—CD dimer, which is represented by the Hamiltonian

$$H = H_0 + V \tag{3.B.1}$$

$$H_0 = -\frac{\hbar^2}{2\mu_{AB}}\frac{\partial^2}{\partial X_{AB}^2} - \frac{\hbar^2}{2\mu_{CD}}\frac{\partial^2}{\partial X_{CD}^2}$$
$$-\frac{\hbar^2}{2\mu_{AB,CD}}\frac{\partial^2}{\partial X_{AB,CD}^2} + V_{AB}(X_{AB}) + V_{CD}(X_{CD})$$
$$+ V_{BC}\left(X_{AB,CD} - \gamma_{AB}\bar{X}_{AB} - \gamma_{CD}\bar{X}_{CD}\right) \tag{3.B.1a}$$

$$V = V_{AB}(X_{AB,CD} - \gamma_{CD}X_{CD}) - V_{AB}\left(X_{AB,CD} - \gamma_{AB}\bar{X}_{AB} - \gamma_{CD}\bar{X}_{CD}\right) \tag{3.B.1b}$$

where X_{AB} and X_{CD} represent the internal interatomic distances of the two diatomic molecules, respectively, $\bar{X}_{AB}$ and $\bar{X}_{CD}$ represent the corresponding equilibrium distances for the two molecules, while $X_{AB,CD}$ corresponds to the distance between the centers of mass of AB and CD. μ_{AB} and μ_{CD} are the reduced masses of the two diatomic molecules, respectively. $\mu_{AB,CD}$ denotes the reduced mass of the pair of the diatomic molecules AB and CD, while the mass ratios are defined by $\gamma_{AB} = m_A(m_A + m_B)^{-1}$ and $\gamma_{CD} = m_D(m_C + m_D)^{-1}$. The interatomic potentials $V_{AB}(X_{AB})$ and $V_{CD}(X_{CD})$ for the AB and CD chemical bonds will be represented in terms of a Morse potential. The vdW interactions V_{BC} was specified in terms of a Morse potential[30]

$$V_{BC}(X_{BC}) = D_{BC}\left\{\exp\left[-2\alpha_{BC}\left(X_{BC} - \bar{X}_{BC}\right)\right] - 2\exp\left[-\alpha_{BC}\left(X_{BC} - \bar{X}_{BC}\right)\right]\right\} \tag{3.B.2}$$

where D_{BC} and $\bar{X}_{BC}$ are the dissociative energy and the equilibrium distance, respectively. The effective frequency for the vdW bond is $\omega_{BC} = (\mu_{AB,CD})^{-1}(\partial^2 V_{BC}/\partial X_{AB,CD}^2)$, while the characteristic inverse length for the vdW bond is $\alpha_{BC} = \omega_{BC}(\mu_{AB,CD}/2D_{BC})$. Finally, the number of bound states in the vdWs bond in $N = (K_{BC} + \frac{1}{2})$ where $K_{BC} = (2D_{BC}/\hbar\omega_{BC})$. A compilation of the Lennard-Jones (LJ) potential parameters for some VDW AB$\cdots$AB dimers is available.[3] The potential parameters for the dimers AB$\cdots$CD were guessed by invoking the combination rules for the (LJ) potential parameters. The Morse potential parameters were subsequently evaluated from the LJ parameters using the recipe described in

Section II.4. The use of the Morse potential in the present context is quite adequate as our model calculations[29] have demonstrated that the dynamics of the VP process is dominated by the short-range part of the potential, being insensitive to the long-range form of the interaction.

Turning now to the VP dynamics of a heteronuclear dimer (heterodimer) AB$\cdots$CD, we note that the zero-order Hamiltonian, (3.B.1a) is separable and its spectrum consists of discrete and continuum eigenvalues corresponding to the eigenfunctions $|v_{AB}, v_{CD}, l\rangle$ and $|v_{AB}, v_{CD}, \epsilon\rangle$, respectively, where v_{AB} and v_{CD} are quantum numbers associated with the internal vibrational motion of molecules AB and of CD, respectively; l is a discrete quantum number characterizing the bound states of the vdWs bond, while ϵ is the relative kinetic energy of the diatomic fragments AB and CD. The VP dynamics on the ground-state potential surface can be considered by defining a preparation process where only the discrete zero-order states $|v_{AB}, v_{BC}, l\rangle$ are amenable to infrared or collisional excitation, while the continuum states $|v'_{AB}, v'_{BC}, \epsilon\rangle$ are inactive. We have demonstrated that the VP rate of the heterodimer is dominated by discrete–continuum (d–c) resonance interactions. The decay rate of the metastable state $|v_{AB}, v_{CD}, l\rangle$ is

$$\Gamma_{v_{AB}v_{CD}l} = \pi \sum_{v'_{AB}} \sum_{v'_{CD}} \left| V^{\text{d–c}}_{v_{AB}v_{CD}l,\, v'_{AB}v'_{CD}\epsilon} \right|^2 \tag{3.B.3}$$

The d–c coupling terms can be expressed in the form[30]

$$\begin{aligned} V^{\text{d–c}}_{v_{AB}v_{CD}l,\, v'_{AB}v'_{CD}\epsilon} &= A^{(2)}_{v_{AB}v_{CD},\, v'_{AB}v'_{CD}} C^{(2)}_{l\epsilon} \\ &\quad - 2A^{(1)}_{v_{AB}v_{CD},\, v'_{AB}v'_{CD}} C^{(1)}_{l\epsilon} \end{aligned} \tag{3.B.4}$$

where the coefficients $A^{(p)}_{v_{AB}v_{CD},\, v'_{AB}v'_{CD}}$ $(p=1,2)$ are

$$\begin{aligned} A^{(p)}_{v_{AB}v_{CD},\, v'_{AB}v'_{CD}} &= \langle v_{AB}v_{CD} | \{ \exp(p\alpha_{BC}[\gamma_{AB}(X_{AB} - \overline{X}_{AB}) \\ &\quad + \gamma_{CD}(X_{CD} - \overline{X}_{CD})]) - 1 \} | v'_{AB}v'_{CD} \rangle \qquad p=1,2 \end{aligned} \tag{3.B.5}$$

and the coefficients $C^{(p)}_{l\epsilon}$ $(p=1,2)$ are given by

$$\begin{aligned} C^{(1)}_{l\epsilon} &= \frac{1}{2} \left[\frac{D_{BC}}{2} \sinh(2\pi\theta_\epsilon) \frac{(2K_{BC} - 2l - 1)}{l!\Gamma(2K_{BC} - l)} \right]^{1/2} \\ &\quad \times \frac{|\Gamma(\frac{1}{2} + K_{BC} - i\theta_\epsilon)|}{[\cos^2(\pi K_{BC}) + \sinh^2(\pi\theta_\epsilon)]^{1/2}} \end{aligned} \tag{3.B.6a}$$

and

$$C_{l\epsilon}^{(2)}=\frac{C_{l\epsilon}^{(1)}}{2K_{\mathrm{BC}}}\left[\left(K_{\mathrm{BC}}-l-\tfrac{1}{2}\right)^{2}+\theta_{\epsilon}^{2}+2K_{\mathrm{BC}}\right] \tag{3.B.6b}$$

where $\Gamma(Z)$ stands for the Gamma function of the complex variable Z and where we have introduced the following definitions

$$K_{\mathrm{BC}}=(\hbar\alpha_{\mathrm{BC}})^{-1}\left[2\mu_{\mathrm{AB,CD}}D_{\mathrm{BC}}\right]^{1/2}=2D_{\mathrm{BC}}/\hbar\omega_{\mathrm{BC}} \tag{3.B.7a}$$

$$\omega_{\mathrm{BC}}=\alpha_{\mathrm{BC}}\left(\frac{2D_{\mathrm{BC}}}{\mu_{\mathrm{AB,CD}}}\right)^{1/2} \tag{3.B.7b}$$

$$\theta_{\epsilon}=(\hbar\alpha_{\mathrm{BC}})^{-1}\left[2\mu_{\mathrm{AB,CD}}\epsilon\right]^{1/2}=\frac{2\sqrt{D_{\mathrm{BC}}\epsilon}}{\hbar\omega_{\mathrm{BC}}} \tag{3.B.7c}$$

ϵ being the final relative kinetic energy between fragments AB and CD, taken on the energy shell

$$\begin{aligned}\epsilon=&W_{\mathrm{AB}}(v_{\mathrm{AB}})-W_{\mathrm{AB}}(v'_{\mathrm{AB}})+W_{\mathrm{CD}}(v_{\mathrm{CD}})-W_{\mathrm{CD}}(v'_{\mathrm{CD}})\\&-\frac{\hbar\omega_{\mathrm{BC}}}{2K_{\mathrm{BC}}}\left(K_{\mathrm{BC}}-l-\tfrac{1}{2}\right)^{2}\end{aligned} \tag{3.B.8}$$

where $W_{\mathrm{IJ}}(n_{\mathrm{IJ}})$, with $\mathrm{IJ}\equiv\mathrm{AB}$ or CD and $v_{\mathrm{IJ}}\equiv v_{\mathrm{AB}}, v_{\mathrm{CD}}, v'_{\mathrm{AB}}$ or v'_{CD} is the internal energy of the corresponding chemical bond in the vibrational state n_{IJ}.

Equations (3.B.3)–(3.B.8) constitute a comprehensive and quite complete theory of VP of a linear vdW heterodimer. The VP rate, (3.B.3) and (3.b.4), is expressed in terms of products of intermolecular contributions $C_{l\epsilon}^{(p)}$, (3.B.6a) and (3.B.6b), and intramolecular contributions $A^{(p)}_{v_{\mathrm{AB}}v_{\mathrm{CD}},\,v'_{\mathrm{AB}}v'_{\mathrm{CD}}}$ (3.B.5).

C. The $V\to T$ Process in Heterodimers

The first-order approximation to the intramolecular contribution to the VP rate involves a linearization approximation, where $A^{(p)}$ is expanded up to linear terms in the intramolecular displacements $(X_{\mathrm{AB}}-\overline{X}_{\mathrm{AB}})$ and $(X_{\mathrm{CD}}-\overline{X}_{\mathrm{CD}})$. The first-order intramolecular contribution is

$$\begin{aligned}A^{(p)}_{v_{\mathrm{AB}}v_{\mathrm{CD}},\,v'_{\mathrm{AB}}v'_{\mathrm{CD}}}\cong p\alpha_{\mathrm{BC}}\Bigg\{&\gamma_{\mathrm{AB}}\left(\frac{\hbar}{2\mu_{\mathrm{AB}}\omega_{\mathrm{AB}}}\right)^{1/2}\\&\times\left(v'^{1/2}_{\mathrm{AB}}\delta_{v_{\mathrm{AB}},\,v'_{\mathrm{AB}}+1}+(v_{\mathrm{AB}}+1)^{1/2}\delta_{v_{\mathrm{AB}}+1,\,v'_{\mathrm{AB}}}\right)\delta_{v_{\mathrm{CD}}v'_{\mathrm{CD}}}\\&+\gamma_{\mathrm{CD}}\left(\frac{\hbar}{2\mu_{\mathrm{CD}}\omega_{\mathrm{CD}}}\right)^{1/2}\left(v'^{1/2}_{\mathrm{CD}}\delta_{v_{\mathrm{CD}},\,v'_{\mathrm{CD}}}+1+(v_{\mathrm{CD}}+1)^{1/2}\delta_{v_{\mathrm{CD}}+1}\right)\\&\delta_{v_{\mathrm{AB}}v'_{\mathrm{AB}}}\Bigg\}\qquad p=1,2\end{aligned} \tag{3.C.1}$$

Thus, in first-order the nonvanishing d–c coupling originates from two combinations $\Delta v_{AB}=0$, $\Delta v_{CD}=+1$ and $\Delta v_{CD}=0$, $\Delta v_{AB}=+1$, which correspond to the $V\to T$ mechanism. The decay problem reduces just to the simple case of parallel decay of a simple discrete state into two continua. From (3.B.3), together with (3.B.6), we obtain the total width for VP of the heterodimer

$$\Gamma_{v_{AB}v_{CD}l}=\Gamma^{(1)}_{v_{AB}v_{CD}l}+\Gamma^{(2)}_{v_{AB}v_{CD}l} \tag{3.C.2}$$

where the partial widths for the channels specified by the propensity rules (3.C.1) is given by

$$\begin{aligned}\Gamma^{(1)}_{v_{AB}v_{CD}l}&\equiv\pi\left|V^{\text{d-c}}_{v_{AB}v_{CD}l,(v_{AB}-1)v_{CD}\epsilon}\right|^2\\&=\frac{\pi}{8}\hbar\omega_{AB}v_{AB}m_1\left[\frac{(2K_{BC}-2l-1)}{l!\Gamma(2K_{BC}-l)}\right]\\&\quad\times\left[\frac{\sinh(2\pi y_1)}{(\cos^2(\pi K_{BC})+\sinh^2(\pi y_1))}\right]\left|\Gamma\left(K_{BC}+\tfrac{1}{2}-iy_1\right)\right|^2\end{aligned} \tag{3.C.3}$$

where

$$m_1=\frac{(m_C+m_D)m_A}{m_B M} \tag{3.C.4a}$$

$$y_1=\left[\beta_1-\left(K_{BC}-l-\tfrac{1}{2}\right)^2\right]^{1/2} \tag{3.C.4b}$$

$$\beta_1=\frac{4D_{BC}\omega_{AB}}{\hbar\omega_{BC}^2}=\frac{2\omega_{AB}\mu_{AB,CD}}{\hbar\alpha_{BC}^2} \tag{3.C.4c}$$

if $y_1^2>0$, otherwise $\Gamma^{(1)}_{v_{AB}v_{CD}l}=0$, and

$$\begin{aligned}\Gamma^{(2)}_{v_{AB}v_{CD}l}&\equiv\pi\left|V^{\text{d-c}}_{v_{AB}v_{CD}l,\,v_{AB}(v_{CD}-1)\epsilon}\right|^2\\&=\frac{\pi}{8}\hbar\omega_{CD}v_{CD}m_2\left[\frac{2K_{BC}-2l-1}{l!\Gamma(2K_{BC}-l)}\right]\\&\quad\times\left[\frac{\sinh(2\pi y_2)}{(\cos^2(\pi K_{BC})+\sinh^2(\pi y_2))}\right]\left|\Gamma\left(K_{BC}+\tfrac{1}{2}-iy_2\right)\right|^2\end{aligned} \tag{3.C.5}$$

with

$$m_2 = \frac{(m_A + m_B)m_D}{m_C M} \tag{3.C.6a}$$

$$y_2 = \left[\beta_2 - \left(K_{BC} - l - \tfrac{1}{2}\right)^2\right]^{1/2} \tag{3.C.6b}$$

$$\beta_2 = \frac{4D_{BC}\omega_{CD}}{\hbar\omega_{BC}^2} = 2\omega_{CD}\mu_{AB,CD}/\hbar\alpha_{BC}^2 \tag{3.C.6c}$$

if $y_2^2 \geqslant 0$, while otherwise $\Gamma^{(2)}_{v_{AB}v_{CD}} = 0$.

Equations (3.C.2)–(3.C.6) constitute our final result for the VP of a heterodimer. To explore the characteristics of the VP dynamics of this system we shall be interested in the situation $y_1 \gg 1$ and $y_2 \gg 1$, as for the vdW dimer $\omega_{AB}/\omega_{BC} \gg 1$, as well as $\omega_{CD}/\omega_{BC} \gg 1$, so that according to (3.C.4c) and (3.C.6c) $\beta_1, \beta_2 \gg 1$. Utilizing the expansion of the gamma function, the partial widths $\Gamma^{(j)}_{v_{AB}v_{CD}l}$ $(j=1,2)$ can be recast in the form[25, 27, 30]

$$\Gamma^{(1)}_{v_{AB}v_{CD}l} \simeq \pi^2 \hbar\omega_{AB} v_{AB} m_1 \left[\frac{N-l-1}{l!(2N-l-1)!}\right] \times y_1^{2N-1} \exp(-\pi y_1) \tag{3.C.7a}$$

and

$$\Gamma^{(2)}_{v_{AB}v_{CD}l} \simeq \pi^2 \hbar\omega_{CD} v_{CD} m_2 \left[\frac{N-l-1}{l!(2N-l-1)!}\right] \times y_2^{2N-1} \exp(-\pi y_2) \tag{3.C.7b}$$

The VP rate for the $V \to T$ process is given by

$$\Gamma_{v_{AB}v_{CD}l} \cong \pi^2 \left[\frac{N-l-1}{l!(2N-l-1)!}\right] \times \left\{\hbar\omega_{AB} v_{AB} m_1 y_1^{2N-1} \exp(-\pi y_1) + \hbar\omega_{CD} v_{CD} m_1 y_2^{2N-1} \exp(-\pi y_2)\right\} \tag{3.C.8}$$

The theory of VP of linear heterodimers AB$\cdots$CD constitutes an extension of the theory of the nuclear dynamics of linear triatomic vdW molecules. The characteristics of the VP process of the heterodimer bears a close analogy to that of the triatomic vdW molecule, except that in the present case two parallel effective decay channels for each metastable state are involved. Usually, a single term in (3.C.2) dominates, resulting in the approximate relation for the $V \rightarrow T$ mechanism

$$\Gamma_{v_{AB}v_{CD}l} \propto \alpha_{BC}\gamma\left(\frac{\hbar}{2\mu\omega}\right)^{1/2} |\Gamma(K_{BC}+\tfrac{1}{2}-i\theta_\epsilon)|^2$$

$$\simeq \alpha_{BC}\gamma\left(\frac{\hbar}{2\mu\omega}\right)^{1/2} \exp(-\pi\theta_\epsilon)$$

$$= \alpha_{BC}\gamma\left(\frac{\hbar}{2\mu\omega}\right)^{1/2} \exp\left[-\pi(2\mu_{AB,CD}\epsilon)^{1/2}/\hbar\alpha_{BC}\right] \quad (3.C.9)$$

where $\gamma(\equiv\gamma_{AB}$ or $\gamma_{CD})$, $\mu(\equiv\mu_{AB}$ or $\mu_{CD})$ and $\omega(\equiv\omega_{AB}$ or $\omega_{CD})$ represent typical intramolecular parameters, while the relative kinetic energy is now

$$\epsilon = \hbar\omega_{AB} - \frac{\hbar\omega_{BC}}{2K_{BC}}\left(K_{BC}-\tfrac{1}{2}-l\right)^2 \qquad \omega_{AB}>\omega_{CD} \qquad (3.C.10a)$$

or

$$\epsilon = \hbar\omega_{CD} - \frac{\hbar\omega_{BC}}{2K_{BC}}\left(K_{BC}-\tfrac{1}{2}-l\right)^2 \qquad \omega_{CD}>\omega_{AB} \qquad (3.C.10b)$$

Equation (3.C.9) exhibits the energy gap law for VP of heterodimers, as well as for homodimers. The exponential dependence of the VP rate for the

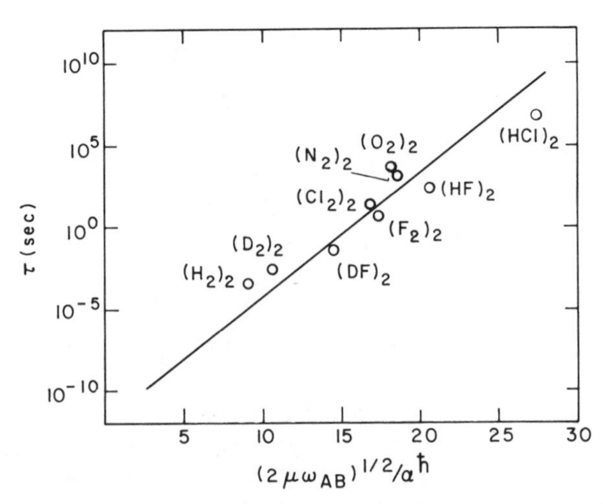

Fig. 14. Energy gap law for VP of van der Waals dimers. The lifetimes calculated from (3.B.3) and (3.B.4) are plotted vs the momentum $(2\mu\omega_{AB})^{1/2}/\alpha\hbar$ according to (3.B.17).

$V \rightarrow T$ process on the momentum $(\mu_{AB,CD}\epsilon)^{1/2}$ is borne out by model calculations for this VP channel in a series of dimers. In Fig. 14 we present a compilation of such VP lifetimes calculated by Ewing and by us, which were plotted by Ewing.[37] The plot of VP lifetime τ vs $q \equiv (2\mu\omega_{AB})^{1/2}/\alpha\hbar$, over a broad range of 10 orders of magnitude variation of τ, exhibits a slope $-d\ln\tau/dq \simeq 2.2$, which is quite close to the value $-d\ln\tau/dq = \pi$ inferred from our theoretical treatment. Slight improvements of our simple energy gap law (3.C.9) can be incorporated. However, we feel that there is little point in improving on this semiquantitative and useful guideline.

D. The $V \rightarrow V + T$ Process in Heterodimers

The $V \rightarrow T$ channel dominates the VP dynamics provided that the linearization approximation inherent in (3.C.9) is valid. To assess the validity of the linearization approximation we first examine the intramolecular contribution (3.B.5), whose first-order expansion is justified provided that $Z \equiv \alpha_{BC}\langle R^2\rangle^{1/2} \ll 1$, where $\langle R^2\rangle^{1/2} = (\hbar/\mu\omega)^{1/2}$ represents a characteristic root-mean-square displacement of a zero-point intramolecular vibration. For typical values of $\alpha_{BC} \sim 2$ Å^{-1} and $\langle R^2\rangle^{1/2} \sim 0.1$ Å, $Z \simeq 0.05$ and the linearization approximation for $A^{(p)}$ seems to be justified. It should, however, be noted that the intermolecular contribution to the VP rate falls off exponentially with $(\epsilon)^{1/2}$, as is evident from (3.C.9), whereupon for large values of ϵ, (3.C.10), the $V \rightarrow T$ mechanism is severely retarded. Under these circumstances, high-order terms in the expansion of the intramolecular contribution $A^{(p)}$ may result in a significant contribution to the VP rate. For example, the second-order term in nuclear displacements, $(X_{AB} - \overline{X}_{AB})$ and $(X_{CD} - \overline{X}_{CD})$ yields

$$A^{(p)}_{v_{AB}, v_{CD}; v'_{AB}, v'_{CD}} \simeq \left(\frac{p^2\alpha_{BC}^2}{2}\right)\gamma_{AB}\gamma_{CD}\left(\frac{\hbar^2}{4\mu_{AB}\omega_{AB}\mu_{CD}\omega_{CD}}\right)^{1/2}$$

$$\times\left(v'^{1/2}_{AB}\delta_{v_{AB}, v'_{AB}+1} + (v_{AB}+1)^{1/2}\delta_{v_{AB}+1, v'_{AB}}\right)$$

$$\left(v'^{1/2}_{CD}\delta_{v_{CD}, v'_{CD}+1} + (v_{CD}+1)^{1/2}\delta_{v_{CD}+1, v'_{CD}}\right)$$

$$\text{for } v'_{AB} = v_{AB} \pm 1;\ v'_{CD} = v_{CD} \mp 1 \qquad (3.D.1)$$

Thus, the nonvanishing second-order contribution corresponds to two-quantum jumps $\Delta v_{AB} \pm 1$ and $\Delta v_{CD} \mp 1$, which result in the $V \rightarrow V + T$ VP mechanism. Although the second-order intramolecular $V \rightarrow V + T$ term, eq. (12), is small $(\sim Z)$ relative to the first-order $V \sim T$ term, eq. (5), the intermolecular contribution to the $V \rightarrow V + T$ rate may now be considerably higher than for the case of $V \rightarrow T$ transfer. For the $V \rightarrow V + T$ mechanism

(3.B.3), (3.B.4), (3.B.6), and (3.D.1) result in the following relation

$$\Gamma_{v_{AB}v_{CD}l} \propto \alpha_{BC}^2 \gamma^2 \left(\frac{\hbar}{\mu\omega}\right) \exp\left[\frac{-\pi(2\mu_{AB,CD}\epsilon)^{1/2}}{\hbar\alpha_{BC}}\right] \tag{3.D.2}$$

for the $V \to V+T$ mechanism, while the relative kinetic ($\epsilon > 0$) energy for the $V \to V+T$ process is

$$\epsilon = \pm\hbar\omega_{AB} \mp \hbar\omega_{CD} - \frac{\hbar\omega_{BC}}{2K_{BC}}\left(K_{BC} - \frac{1}{2} - l\right)^2 \tag{3.D.3}$$

As the kinetic energy (3.D.3) for $V \to V+T$ mechanism may now be considerably lower than the kinetic energy (3.C.10) for the $V \to T$ process, the intermolecular contribution to the VP rate for the $V \to V+T$ mechanism (3.D.2) may be greatly enhanced in spite of the relatively small contribution to the preexponential factor. We thus conclude that for systems where $\hbar|\omega_{AB} - \omega_{CD}|$ is relatively low the new $V \to V+T$ mechanism can be efficient.

To obtain numerical estimates and to provide theoretical predictions for the VP mechanism involving $V \to V+T$, we have performed model calculations for some linear heterodimers making use of (3.B.3)–(3.B.8). The molecular bonds AB and CD were characterized by anharmonic Morse potentials specified by the frequencies $(\omega_e)_{AB}$ and $(\omega_e)_{CD}$ and by the anharmonicity constants $(x_e\omega_e)_{AB}$ and $(x_e\omega_e)_{CD}$. The intramolecular contributions to the d–c coupling were evaluated numerically. To explore the gross features of the VP dynamics we present in Table V some typical

TABLE V
Model Calculations for V→T and for V→V+T VP Rates of Linear vdW Heterodimers

System	v_{AB}	v_{CD}	v'_{AB}	v'_{CD}	l	τ (sec)	VP mechani
Br_2—Cl_2; $\alpha_{BC} = 1.395$ Å^{-1}; $D_{BC} = 300$ cm^{-1};	0	1	0	0	0	0.14×10^4	V→T
$N = 22$					2	0.34×10^2	
					4	0.39×10	
					6	0.52	
$(\omega_e)_{Cl_2} = 564.9$ cm^{-1}; $(x_e\omega_e)_{Cl_2} = 4$ cm^{-1}	0	1	1	0	2	0.2×10^{-3}	V→V+
$(\omega_e)_{Br_2} = 325.2$ cm^{-1}; $(x_e\omega_e)_{Br_2} = 1$ cm^{-1}					4	0.47×10^{-4}	
					6	0.19×10^{-4}	
					8	0.10×10^{-4}	
					10	0.70×10^{-5}	
					12	0.57×10^{-5}	
					14	0.55×10^{-5}	
					16	0.65×10^{-5}	

TABLE V (*continued*)

tem	v_{AB}	v_{CD}	v'_{AB}	v'_{CD}	l	τ (sec)	VP mechanism
—NO	1	0	0	0	0	0.17×10^{10}	V→T
$=1.714$ cm^{-1}; $D_{BC}=80$ cm^{-1}; $N=5$					2	0.21×10^{9}	
					4	0.61×10^{9}	
$)_{CO}=2170.2$ cm^{-1}; $(x_e\omega_e)_{CO}=13.46$ cm^{-1};	1	0	0	1	0	0.86×10^{-5}	V→V+T
$)_{NO}=1904.0$ cm^{-1}; $(x_e\omega_e)_{CO}=13.97$ cm^{-1};					2	0.26×10^{-5}	
					4	0.11×10^{-4}	
—N_2	0	1	0	0	0	0.1×10^{10}	V→T
$=1.62$ Å^{-1}	0	1	1	0	0	0.78×10^{-6}	V→V+T
100 cm^{-1}					1	0.33×10^{-6}	
6					2	0.24×10^{-6}	
$)_{CO}=2170.2$ cm^{-1}					3	0.25×10^{-6}	
$\omega_e)_{CO}=13.46$ cm^{-1}					4	0.39×10^{-6}	
$)_{N_2}=2359.6$ cm^{-1}					5	0.34×10^{-5}	
$\omega_e)_{N_2}=14.46$ cm^{-1}							
—H_2	0	1	0	0	0	0.22×10^{-5}	V→T
$=2.05$ Å^{-1}							
$_C=25.7$ cm^{-1}							
1							
$)_{H_2}=4395.2$ cm^{-1}							
$\omega_e)_{H_2}=117.9$ cm^{-1}							
—HD							
$=2.05$ Å^{-1}	1	0	0	0	0	0.76×10^{-5}	V→T
$_C=25.7$ cm^{-1}							
1							
$)_{H_2}=4395.2$ cm^{-1}	1	0	0	1	0	0.12×10^{-8}	V→V+T
$\omega_e)_{H_2}=117.9$ cm^{-1}							
$)_{HD}=3817.09$ cm^{-1}							
$\omega_e)_{HD}=93.96$ cm^{-1}							
—O_2	1	0	0	0	0	0.84×10^{-3}	V→T
$=1.85$ Å^{-1}							
$_C=60$ cm^{-1}	1	0	0	1	0	0.40×10^{-3}	V→V+T
1							
$)_{H_2}=4395.2$ cm^{-1}	1	0	0	2	0	0.12×10^{-2}	V→2V+T
$\omega_e)_{H_2}=117.9$ cm^{-1}							
$)_{O_2}=1580.36$ cm^{-1}							
$\omega_e)_{H_2}=12.07$ cm^{-1}							
—NO	1	0	0	0	0	0.42×10^{-1}	V→T
$=1.85$ Å^{-1}							
$_C=60$ cm^{-1}					1	0.13	
2							
$)_{HD}=3817.09$ cm^{-1}	1	0	0	1	0	0.4×10^{-3}	V→V+T
$\omega_e)_{HD}=94.96$ cm^{-1}							
$)_{NO}=1904.0$ cm^{-1}					1	0.13×10^{-2}	
$\omega_e)_{NO}=13.97$ cm^{-1}							

input data, α_{BC} and D_{BC} and N, for the vdW Morse potential, as well as the computed results for the VP lifetimes $\tau=\hbar/\Gamma_{v_{AB}v_{CD}l}$ for the $V\to T$, as well as for the $V\to V+T$ VP mechanism. From these VP relaxation data the following conclusions emerge:

1. For vdW dimers, which are characterized by large intramolecular frequencies and where the reduced mass $\mu_{AB,CD}$ is high, the VP rates for the $V\to T$ process are exceedingly slow. Typical examples are Br_2—Cl_2, CO—NO, and CO—N_2, where the calculated VP rates for $V\to T$ in our model system considerably exceed the infrared radiative decay times, whereupon the vdW heterodimers are stable with respect to the $V\to T$ VP decay. The astronomically low rates for the $V\to T$ process in these cases should not be considered as quantitative estimates as rotational effects will increase these rates. However, it is safe to assert that these vdW heterodimers are stable with respect to $V\to T$, VP.
2. For some vdW heterodimers, which are characterized by large intramolecular frequencies as well as by moderately high values of the reduced mass and where the kinetic energy (3.D.3) for the second-order $V\to V+T$ mechanism is relatively low, a dramatic contribution of the $V\to V+T$ decay channel is exhibited. The rate for the $V\to V+T$ decay process is huge relative to the rate for the $V\to T$ VP. Thus, the enhancement η of the VP rates for $V\to V+T$ relative to those for $V\to T$ is $\eta\sim10^7$ for Cl_2—Br_2, $\eta\sim10^{15}$ for CO—NO and $\eta\sim10^{16}$ for CO—N_2.
3. The $V\to V+T$ lifetimes for the Br_2—Cl_2, CO—NO, and CO—N_2 model systems are comparable to or shorter than the lifetimes for spontaneous infrared emission of these complexes. We thus conclude that the $V\to V+T$ channel will provide an efficient VP mechanism for some heterodimers, which is amenable to experimental observation. The experimental demonstration of the $V\to V+T$ VP process has to rest on the analysis of the relative kinetic energy as well as the internal energy content of the diatomic molecules resulting from the VP process.
4. For vdW dimers, which are characterized by a light reduced mass, for example, H_2—H_2 or H_2—HD, the VP process via $V\to T$ is quite efficient, in accord with Ewing's data.[37] This result can easily be rationalized in terms of the mass effect on the VP rate which depends exponentially on $(\mu_{AB,CD})^{1/2}$, as is evident from (3.C.9). In this case, the second-order $V\to V+T$ process can be somewhat enhanced relative to the $V+T$ channel, as is the case for the H_2—HD system where $\eta\sim10^3$, and for the HD—NO dimer where $\eta\sim10^2$, while for the H_2—O_2 dimer $\eta\sim2$. In these systems containing H_2 or HD the

efficiency of the $V \to T$ mechanism may make the observation of the new $V \to V+T$ channel somewhat difficult. Modern methods for probing the internal vibrational energy content of the fragments, such as CARS techniques, will be extremely useful in this context.

5. High-order intermolecular vibrational energy exchange processes accompanying VP of a heterodimer are of some interest. In Table V we have considered the process $|v_{H_2}=1,\ v_{O_2}=0, l\rangle \to |v_{H_2}=0,\ v_{O_2}=2, \epsilon\rangle$, which involves a two-quantum jump, being labeled in Table I as $V \to 2V+T$. In the H_2—O_2 system the branching ratio, b, for the $V \to 2V+T$ high-order process is $b \sim 0.3$, whereupon the process may be amenable to experimental observation.

We conclude that intermolecular vibrational energy exchange may considerably enhance the VP rate of some vdW heterodimers and that the $V \to V+T$ process may be amenable to experimental observation.

E. Intramolecular Dynamics of VP of Homodimers

The VP dynamics of a heterodimer involves essentially the decay of a single, isolated, discrete zero-order state into several continua. In this case the d–d coupling terms are negligible whereupon the VP process can be adequately characterized in terms of a simple single exponential decay law. In the case of a homodimer a new physical effect should be considered, which stems from the degeneracy of some of the zero-order state. Consider, for example, the local-mode description of a homodimer whose bond zero-order states will be denoted by $|v_1 v_2 l\rangle$, where v_1 and v_2 represent the vibrational quantum numbers of the two diatomic fragments. The pairs of zero-order states $|v_1 v_2 l\rangle$ and $|v_2 v_1 l\rangle$ are degenerate. Under these circumstances even a small d–d coupling term between zero-order degenerate states can go a far way in determining the intramolecular dynamics of the homodimer. Thus, in contrast to the case of VP of triatomic vdW molecules and of vdW heterodimers, where interference effects are negligible, the role of d–d coupling may be crucial in determining the intramolecular dynamics of homodimers. Consequently, the VP process of a homodimer may not be characterized by a simple exponential decay law specified just by a VP width, but rather by an oscillatory time evolution. To consider the intramolecular dynamics and VP of homodimers, we shall proceed in three steps. First, we shall discuss some zero-order basis sets which can be used for the study of the VP of the homodimer. These basis set will be introduced without alluding to any real physical situation. For each basis set we shall obtain different values of the resonance half width and of the d–d coupling term. However, now the resonance half-width does not represent (an exponential) decay probability. Different basis sets will result in different energetic parameters and the characteristics of the VP dynamics depend on the initially excited metastable state. Second, we shall define excitation processes which result in physically meaningful initial conditions for the dynamic problem. These initial conditions will determine the subsequent time evolution of the system. Third, we shall develop a complete theory of intramolecular dynamics and apply it to the problem at hand.

We shall first consider three sets of zero-order states which are appropriate for the description of the intramolecular dynamics of the homodimer. For each basis set we shall provide explicit expressions for the d–d coupling terms and for the half-width of the resonance.

1. The Local Mode Basis

This basis set consisting of discrete states $|v_1 v_2 l\rangle$ and of continuum states $|v_1 v_2 \epsilon\rangle$ is displayed in Fig. 15a. When a harmonic approximation is adopted for the potential energy surfaces of the two AB subunits, all the states with the same l, and for which $v_1 + v_2 = M$, are degenerate. Part of this degeneracy is accidental and can be lifted by describing the molecular bond modes in terms of more realistic anharmonic potential. Now the (permutation) symmetry determined that degeneracy prevails between the

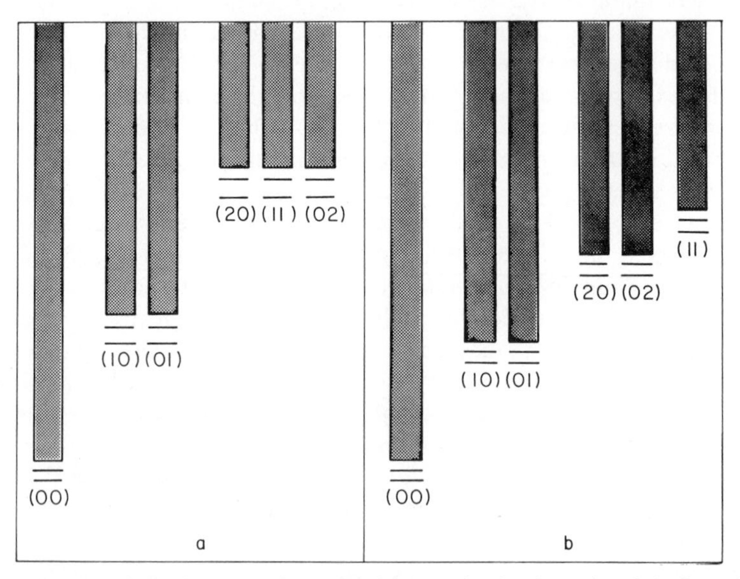

Fig. 15. Local-mode basis sets for the study of intramolecular dynamics of the homodimer AB···AB. (a) Harmonic approximation. (b) Anharmonic molecules: The labels $(v_{AB} v'_{AB})$ specify different vibrational levels of the individual subunits AB.

pairs of discrete states $|v_1 v_2 l\rangle$ and $|v_2 v_1 l\rangle$, and similarly for continuum states (see Fig. 15b). As anharmonicity defects for the intramolecular AB modes are expected to be of the order of a few cm^{-1} the accidental degeneracy is practically completely lifted and we can disregard d–d coupling between states which were accidentally degenerate in the harmonic description.

Consider now the half-widths of the resonances for a homodimer consisting of two identical AB diatomics. These half-widths of the resonances will be presented in the harmonic approximation for the bond modes, which is adequate for low values of v_1 and of v_2. For the linear arrangement AB$\cdots$BA which is characterized by a centre of inversion symmetry we obtain[30]

$$\Gamma_{v_1 v_2 l} = \frac{\pi}{8}\hbar\omega_{AB} m_1 \left[\frac{2K_{BB}-2l-1}{l!\Gamma(2K_{BB}-l)}\right] \times \left[\frac{\sinh(2\pi y_1)}{(\cos^2(\pi K_{BB})+\sinh^2(\pi y_1))}\right] \times |\Gamma(K_{BB}+\tfrac{1}{2}-iy_1)|^2 (v_1+v_2) \tag{3.E.1}$$

with $m_1 = m_A/2m_B$. For this homodimer $\Gamma_{v_1 v_2 l} = \Gamma_{v_2 v_1 l}$ as the two molecular subunits are equivalent. For the configuration $AB(n_1)\cdots AB(n_2)$, the half-width is given by

$$\Gamma_{v_1 v_2 l} = (\pi/8)\hbar\omega_{AB}\left[\frac{2K_{AB}-2l-1}{l!\Gamma(2K_{AB}-l)}\right] \times \left[\sinh(2\pi y_1)/(\cos^2(\pi K_{AB})+\sinh^2(\pi y_1))\right] \times |\Gamma(K_{AB}+\tfrac{1}{2}-iy_1)|^2 (v_1 m_1 + v_2 m_2) \tag{3.E.2}$$

where

$$m_1 = \frac{m_A}{2m_B} \qquad m_2 = \frac{m_B}{2m_A} \tag{3.E.3}$$

The resonance half-widths given in terms of (3.E.1) and (3.E.2) for the homodimer exhibit a characteristic energy gap law, a typical dependence on the parameters of the vdW bond and of the molecular bonds, which is similar to that obtained for the heterodimer.

To predict the time evolution of the degenerate discrete levels such as $|v_1 v_2 l\rangle$ and $|v_2 v_1 l\rangle$ it is necessary, to study the appropriate d–d couplings which could be no longer negligible. Obviously we shall have to consider only coupling term of the type $V^{\rm d-d}_{v_1 v_2 l, v_2 v_1 l}$ with the same value of l, all the other d–d couplings are still negligible. It may be shown[30] that for couplings with the same value of l, utilization of the linearization approximation, i.e., expanding $A^{(p)}$ (3.B.5) to first-order in the intramolecular displacements results in $V^{\rm d-d}_{v_1 v_2 l, v_2 v_1 l}=0$ The linearization approximation results in the vanishing of the d–d coupling between degenerate states, which are characterized by the same value of l; this effect will be exhibited irrespective of the specific form of the potential for the bond modes; it will happen both for harmonic and anharmonic description of the AB potential. We conclude that the coupling between zero-order discrete degenerate levels originates from high-order terms in the intramolecular displacements which have to be explicitly incorporated in the calculations of the VP dynamics of homodimers.

2. *The Miniexciton Basis*

A traditional way to treat degenerate zero-order excited states involving vibrational or electronic excitations in dimers and in molecular aggregates rests on the construction of symmetry adopted wavefunctions with the interaction lifting the degeneracy. We shall adopt a similar approach here, which bears a close analogy to exciton theory applied to the homodimer, that is, a miniexciton. We shall consider the subset of discrete zero-order states $|v_1 v_2 l\rangle$ and in this subspace of the Hilbert space we shall construct the symmetry adapted functions from each pair of degenerate bond-mode states

$$|(+)v_1 v_2 l\rangle = 2^{-1/2}(|v_1 v_2 l\rangle + |v_2 v_1 l\rangle)$$
$$|(-)v_1 v_2 l\rangle = 2^{-1/2}(|v_1 v_2 l\rangle - |v_2 v_1 l\rangle) \tag{3.E.4}$$

This basis will be referred to as the miniexciton basis set. As we shall be interested in the homodimers of the halogen dimers A_2—A_2 ($A \equiv$ F, Cl, Br, I) we shall avoid unnecessary complication and from now on consider the linear AB$\cdots$BA dimer only. Each pair of the $|(+)\rangle$ and $|(-)\rangle$ states (3.E.4) is now uncoupled with respect to d–d interaction

$$V^{\rm d-d}_{(+)v_1 v_2 l, (-)v_2 v_1 l} = 0 \tag{3.E.5}$$

The degeneracy is split and the (zero-order) energies of the discrete states are $E(+)_{v_1 v_2 l} = -V^{\rm d-d}_{v_1 v_2 l, v_2 v_1 l}$ and $E(-)_{v_1 v_2 l} = V^{\rm d-d}_{v_1 v_2 l, v_2 v_1 l}$. The half-widths of the resonances are readily obtained from (3.E.1) and (3.E.2) in the form

$$\Gamma_{(+)v_1 v_2 l} = 2\Gamma_{v_1 v_2 l}$$
$$\Gamma_{(-)v_1 v_2 l} = 0 \tag{3.E.6}$$

It is interesting to note that each $|(-)\rangle$ discrete state is stable with respect to direct decay into the continuum due to destructive interference. As these $|(-)\rangle$ states are not coupled to $|(+)\rangle$ states in view of (3.E.5), we expect the $|(-)\rangle$ levels to constitute truly "isolated" states which are stable with respect to intramolecular decay.

3. *The Normal Mode Basis*

This basis is obtained by adopting the conventional procedure for construction normal modes for the intramolecular motion of the homodimer, which rests on the formation of symmetric and antisymmetric combination of the displacements in the bond modes. This approach rests on the harmonic model for the AB bond modes. Thus in this description both accidental and symmetry determined degeneracies will be exhibited for the homodimer. This intrinsic limitation of the normal mode basis prevents us from using it in our general treatment of the VP dynamics.

From the foregoing discussion we conclude that basis sets, the bond-mode basis and the miniexciton basis, will be useful for the subsequent treatment of intramolecular dynamics of the linear homodimer on its ground-state potential surface. The physical preparation of the "initially excited" state is of cardinal importance, as this will determine the intramolecular dynamics. For example, we have already noted that the $|(-)v_1v_2l\rangle$ miniexciton states are stable with respect to time evolution. Thus if such $|(-)v_1v_2l\rangle$ state could be initially prepared it will not exhibit intramolecular time evolution. This is not surprising, as other examples of states that are stable with respect to subsequent decay are well known in related areas of molecular and of solid-state physics. To provide a simple analogy, let us recall that in the simple didactic theory of Frenkel[50] exciton states, all the states with $\mathbf{k}\neq 0$ are stable with respect to radiative decay to the ground state. The cardinal question in relation to our problem is whether such a stable state can be excited in a heterodimer? The answer usually is negative. To be more specific let us consider two limiting excitation modes of the homodimer, which result in reasonably well defined "initial" metastable states.

Optical Excitation. We shall consider optical excitation of the homodimer by infrared radiation as we are concerned with vibrational excitation on the ground state potential surface. To consider properly a "short-time" excitation process we have to specify which of the zero-order states carries oscillator strength from the ground state. We invoke the reasonable assumption that only the discrete zero-order states carry oscillator strength from the ground state $|v_1=0,\ v_2=0,\ l=0\rangle$. Accordingly we have to consider radiative coupling with the discrete subspace of the Hilbert space where the miniexciton basis provides a diagonal representation of the

Hamiltonian. Thus two optical excitation processes should be considered for each nearly degenerate pair

$$|v_1=0, v_2=0, l=0\rangle \to |(+)v_1v_2l\rangle \tag{3.E.7a}$$

$$|v_1=0, v_2=0, l=0\rangle \to |(-)v_1v_2l\rangle \tag{3.E.7b}$$

For the linear homodimer only the optical transition (3.E.7a) where the transition moments of the two subunits are in phase is allowed, while the transition (3.E.7b) where the transition moments of the time subunits cancel is forbidden. This state of affairs bears a close analogy to exciton theory. On the other hand, when the optical excitation of a nondegenerate state is considered, such as $|v_1=v, v_2=v, l\rangle$ this discrete zero-order state is initially prepared by optical excitation. We conclude that optical excitation results in the initial selection of the $|(+)v_1v_2l\rangle$ state from each degenerate pairs and in the initial excitation of a discrete nondegenerate state. These metastable states will exhibit subsequent time evolution.

Finally, it is worthwhile to comment on the nature of the transition moments of the two subunits (TMTS) whose superposition determines the total transition moment for infrared excitation. For a homodimer consisting of two heteronuclear diatomics AB each of these TMTS corresponds to the infrared transition moment of the AB molecule. For a homodimer $A_2 \cdots A_2$ consisting of a pair of homonuclear diatomics the infrared transition moment of each "isolated" A_2 molecule vanishes. In this case charge transfer vibronic type mixing[51] will lead to finite TMTS.

Collisional Excitation. Excitation of the homodimer by collision will result in an incoherent superposition of zero-order degenerate states, which can be expressed as

$$|a\rangle = A|v_1v_2l\rangle + B|v_2v_1l\rangle \tag{3.E.8}$$

where the constant coefficients determined by the experimental collisional excitation A nd B are uncorrelated, in contrast to the case of optical excitation. We can immediately construct the complementary state to (3.E.8) which cannot be initially excited in that particular excitation experiment

$$|b\rangle = C|v_1v_2l\rangle + D|v_2v_1l\rangle \tag{3.E.9}$$

where the constant coefficients C and D are related to the "experimental" coefficients A and B via the conventional orthonormality relations

$$|C|^2 + |D|^2 = 1$$

$$AC^* + BD^* = 0 \tag{3.E.10}$$

The relevant energetic parameter which determine the intramolecular dynamics of the homodimer are

$$\mathcal{V}_{ba}^{\text{d-d}} = V_{v_1v_2l,\, v_2v_1l}^{\text{d-d}}(AD^* + BC^*) \tag{3.E.11}$$

$$\Gamma_a = (A^2 + B^2)\Gamma_{v_1v_2l}; \qquad \Gamma_b = (C^2 + D^2)\Gamma_{v_1v_2l} \tag{3.E.12}$$

As the coefficients A and B do not bear any phase relationship it is apparent from (3.E.11) and (3.E.12) that a reasonable description of the intramolecular dynamics can be based on a bond mode state as an initial state. This is justified as now for incoherent excitation $(A^2 + B^2) \sim 1$ and $(AD^* + BC^*) \sim 1$, so that the energetic parameters calculated for an initially excited bond mode and thus the subsequent time evolution of the system will faithfully reproduce the VP dynamics of the collisionally excited homodimer.

We are now in a position to discuss the time evolution of the quasi-degenerate states. From the foregoing discussion we conclude that a reasonable description of a physically meaningful excitation condition of the homodimer in an energy range containing a degenerate (in the bond-mode basis) or nearly degenerate (in the miniexciton bond) pair of zero-order discrete levels can be specified by the initial condition

$$\Psi(0) = |(+)v_1v_2l\rangle \tag{3.E.13}$$

for a coherent optical excitation, and

$$\Psi(0) = |v_1v_2l\rangle \tag{3.E.14}$$

for collisional incoherent excitation. In defining these initial conditions we have asserted that the duration of the "short-time" optical or collisional excitation is short relative to the energy spread of the two zero-order levels. Such separation between excitation and subsequent time evolution is acceptable in the modern theory of relaxation phenomena. The problem we are facing is essentially that of the dynamics of two coupled levels, which will be denoted by $|\alpha\rangle$ and $|\beta\rangle$ and which are coupled to a common continuum $\{|\epsilon\rangle\}$. The d–c coupling terms are $\mathcal{V}_{\alpha\epsilon}$ and $\mathcal{V}_{\beta\epsilon}$ and the corresponding half-widths, which specify the d–c couplings, are given in terms of the elements of the off-diagonal decay matrix. The diagonal terms are

$$\Gamma_{\alpha\alpha} = \pi|\mathcal{V}_{\alpha\epsilon}|^2 \tag{3.E.15}$$

$$\Gamma_{\beta\beta} = \pi|\mathcal{V}_{\beta\epsilon}|^2 \tag{3.E.16}$$

while the off-diagonal terms are given by

$$\Gamma_{\alpha\beta}=\Gamma^{*}_{\beta\alpha}=\pi\mathcal{V}_{\alpha\epsilon}\mathcal{V}_{\epsilon\beta} \tag{3.E.17}$$

These off-diagonal terms (Eq. 3.E.17) are related to the diagonal terms by

$$\Gamma_{\alpha\beta}\Gamma_{\beta\alpha}=\frac{\Gamma_{\alpha\alpha}\Gamma_{\beta\beta}}{\pi} \tag{3.E.18}$$

The d–d coupling among the states $|\alpha\rangle$ and $|\beta\rangle$ is denoted by $\mathcal{V}_{\alpha\beta}$. The initial condition is $\Psi(0)=|\alpha\rangle$, where $|\alpha\rangle$ is given in terms of either (3.E.13) or (3.E.14). The probability $p_{\alpha}^{(\alpha)}(t)$ for the system to remain in the initial state $|\alpha\rangle$ at time t is

$$P_{\alpha}^{(\alpha)}(t)=|\langle\alpha|e^{-iHt/\hbar}|\alpha\rangle|^{2} \tag{3.E.19}$$

while the probability to populate the $|\beta\rangle$ state at time t is

$$P_{\beta}^{(\alpha)}(t)=|\langle\beta|e^{-iHt/\hbar}|\alpha\rangle|^{2} \tag{3.E.20}$$

The VP probability $P_{\mathrm{VP}}^{(\alpha)}(t)$ is given in terms of the population probability of all the continuum states which in view of basic conservation relations is given by

$$P_{\mathrm{VP}}^{(\alpha)}(t)=1-P_{\alpha}^{(\alpha)}(t)-P_{\beta}^{(\beta)}(t) \tag{3.E.21}$$

The time evolution of the two-level system is well known.[52] In what follows we shall just consider two limiting cases which are of interest for the elucidation of the VP dynamics of homodimers.

Case (A): Two quasi-degenerate levels coupled to a common continuum

$$\Gamma_{\alpha},\Gamma_{\beta}\gg\mathcal{V}_{\alpha\beta} \tag{3.E.22}$$

so that the resonances widths considerably exceed the d–d coupling. Under these conditions the decay pattern of the system is exponential and no oscillatory terms are exhibited. In this case

$$P_{\alpha}^{(\alpha)}(t)=(\Gamma_{\alpha\alpha}+\Gamma_{\beta\beta})^{-2}\left[\Gamma_{\beta\beta}+\Gamma_{\alpha\alpha}\exp\left(\frac{-(\Gamma_{\alpha\alpha}+\Gamma_{\beta\beta})t}{\hbar}\right)\right]^{2}$$

$$P_{\beta}^{(\alpha)}(t)=(\Gamma_{\alpha\alpha}+\Gamma_{\beta\beta})^{-2}\Gamma_{\alpha\alpha}\Gamma_{\beta\beta}\left[1-\exp\left(\frac{-(\Gamma_{\alpha\alpha}+\Gamma_{\beta\beta})t}{\hbar}\right)\right]^{2} \tag{3.E.23}$$

The VP probability is

$$P_{\mathrm{VP}}^{(\alpha)}(t)=1-(\Gamma_{\alpha\alpha}+\Gamma_{\beta\beta})^{-1}[\Gamma_{\beta\beta}+\Gamma_{\alpha\alpha}\exp(-2\gamma t)] \qquad \gamma=\frac{\Gamma_{\alpha\alpha}+\Gamma_{\beta\beta}}{\hbar} \tag{3.E.24}$$

While the VP yield at $t=\infty$ is given by

$$P_{\mathrm{VP}}^{(\alpha)}(t=\infty)=\frac{\Gamma_{\alpha\alpha}}{\Gamma_{\alpha\alpha}+\Gamma_{\beta\beta}} \tag{3.E.25}$$

Four comments are now in order. First, the time evolution of a two-level system where the resonance widths exceed the d–d coupling exhibits a nontrivial decay pattern, not just a simple exponential decay. Second, in the limit $\Gamma_{\beta}=0$ the simple exponential decay law

$$\begin{aligned} P_{\alpha}^{(\alpha)}(t)&=\exp\left[\frac{-2\Gamma_{\alpha\alpha}t}{\hbar}\right] \\ P_{\beta}^{(\alpha)}(t)&=0 \\ P_{\mathrm{VP}}^{(\alpha)}(t)&=1-\exp\left[\frac{-2\Gamma_{\alpha\alpha}t}{\hbar}\right] \end{aligned} \tag{3.E.26}$$

is regained while the $|\beta\rangle$ state is not populated. Third, in general, when $\Gamma_{\beta}\neq 0$ the $|\beta\rangle$ state is populated not by direct coupling between $|\alpha\rangle$ and $|\beta\rangle$ but by indirect coupling between these discrete states via the continuum. Fourth, in the general case the VP yield is smaller than unity.

Case (B): Two effectively coupled levels weakly coupled to a common continuum

$$\Gamma_{\alpha\alpha},\ \Gamma_{\beta\beta}\ll\mathcal{V}_{\alpha\beta} \tag{3.E.27}$$

Thus the d–d coupling considerably exceeds the decay widths. For short times $t\ll\hbar\Gamma_{\alpha\alpha}^{-1},\ \hbar\Gamma_{\beta\beta}^{-1}$ the time evolution of the system is oscillatory

$$\begin{aligned} P_{\alpha}^{(\alpha)}(t)&=\cos^{2}\left(\frac{\mathcal{V}_{\alpha\beta}t}{\hbar}\right) \\ P_{\beta}^{(\alpha)}(t)&=\sin^{2}\left(\frac{\mathcal{V}_{\alpha\beta}t}{\hbar}\right) \end{aligned} \tag{3.E.28}$$

$\mathcal{V}_{\alpha\beta}$ determining the oscillation frequency. The intramolecular dynamics will now involve quasiperiodic energy exchange between the zero-order states $|\alpha\rangle$ and $|\beta\rangle$. Only for sufficiently long times of the order of the decay widths Γ_α and Γ_β effective damping will be exhibited. The decay is so slow relative to the period of the oscillations that in this limit the oscillatory behavior cannot be observed by monitoring the VP decay as the fast oscillations will be smeared out.

On the basis of the foregoing discussion, we can now elucidate the distinct dynamic consequences of optical excitation and of collisional excitation of homodimers. Optical excitation of a metastable state of the homodimer corresponds to the initial state $|\alpha\rangle = 1(+)v_1v_2l\rangle$, the complementary state β is just $|(-)v_1v_2l\rangle$, whereupon $\mathcal{V}_{\alpha\beta}=0$ and $\Gamma_\beta=0$. As the d–d coupling vanishes identically it is apparent that the time evolution following optical excitation of the homodimer corresponds to case A of a pair of quasidegenerate levels coupled to a common quasicontinuum. Furthermore, as $\Gamma_\beta=0$ the decay pattern of the optically excited initial state is described in terms of a simple exponential decay the VP rate being given by $2\Gamma_\alpha/\hbar=4\Gamma_{v_1v_2l}/\hbar$. On the other hand, collisional excitation of a metastable state of the homodimer can be described as resulting in the initial state $|\alpha\rangle=|v_1v_2l\rangle$, the complementary state being $|\beta\rangle=|v_2v_1l\rangle$. Now $\mathcal{V}_{\alpha\beta}^{\mathrm{d-d}}=V_{v_1v_2l,\,v_2v_1l}^{\mathrm{d-d}}$ is finite and $\Gamma_\alpha=\Gamma_\beta=\Gamma_{v_1v_2l}$. The nuclear dynamics can correspond either to case A or to case B or even to an intermediate situation of oscillatory decay. In order to confront the theory with the experimental results of Dixon and Herschbach[16] where a collisional excitation of the Cl_2—Cl_2 vdW homodimer was performed it will be interesting to obtain information concerning the energetic parameters which determine the VP dynamics of this and similar systems.

TABLE VI
Energetic Parameters for Halogen Dimers

	Parameters of the diatomic fragments[a]		Parameters of the van der Waals bond[b]			
Dimer	ω_{AB} (cm^{-1})	ωX_{AB} (cm^{-1})	ω_{BC} (cm^{-1})	D_{BC} (cm^{-1})	α_{BC} (Å^{-1})	Number of bond levels
$(F_2)_2$	802.	4	27.3	77.8	1.64	6
$(Cl_2)_2$	564.9	4	33.0	334.0	1.3	20
$(Br_2)_2$	325.2	1	24.7	361.4	1.41	29
$(I_2)_2$	214.6	6	17.2	382.0	1.2	45

[a]The frequencies and anharmonic factors for the diatomic fragments have been taken from Ref. 47.
[b]The parameters for the vdW bond for $(F_2)_2$, $(Br_2)_2$, and $(I_2)_2$ have been obtained from viscosity data reported in Ref. 3. For $(Cl_2)_2$ we have used the estimate given in Ref. 16.

Finally, we shall proceed to model calculations of VP of homodimers. We shall now consider the VP of halogen dimers A_2—A_2 ($A \equiv F, Cl, Br, I$) on the ground state potential surface, adopting our simple model for nuclear dynamics of homodimer excited by collisions. The energetic parameters which determine the VP dynamics were evaluated numerically, $V^{d-d}_{v_1v_2l, v_2v_2l}$ was calculated without invoking the linearization approximation, while $\Gamma_{n_1n_2l}$ was calculated from (3.B.3). The spectroscopic and structural input data are summarized in Table VI. To explore the gross features of the nuclear dynamics we present in Table VII some numerical results for the d–d coupling terms and for the half-widths of the resonance

TABLE VII
Energetic Data for Intramolecular Dynamics and VP of Halogen Dimers

Molecule	Levels					Energy difference (cm^{-1})	d–d couplings (cm^{-1})	$\Gamma^{(cm^{-1})}_{v_{AB}v_{CD}l}$
	v_{AB}	v_{CD}	v'_{AB}	v'_{CD}	l			
$(F_2)_2$	1	0	0	1	0	0	$0.15\ 10^{-4}$	$0.62\ 10^{-12}$
					1	22.5	$0.43\ 10^{-4}$	
					1	0	$0.25\ 10^{-3}$	$0.26\ 10^{-11}$
					2	17.7	$0.48\ 10^{-3}$	
					2	0	$0.14\ 10^{-2}$	$0.56\ 10^{-11}$
					3	12.9	$0.19\ 10^{-2}$	
					3	0	$0.35\ 10^{-2}$	$0.74\ 10^{-11}$
					4	8.1	$0.35\ 10^{-2}$	
					4	0	$0.44\ 10^{-2}$	$0.59\ 10^{-11}$
					5	3.4	$0.20\ 10^{-2}$	
					5	0	$0.11\ 10^{-2}$	$0.11\ 10^{-11}$
$(Cl_2)_2$	1	0	0	1	0	0	$0.48\ 10^{-44}$	$0.18\ 10^{-12}$
					1	31.4	$0.30\ 10^{-43}$	
					10	0	$0.20\ 10^{-34}$	$0.80\ 10^{-9}$
					11	15.1	$0.30\ 10^{-34}$	
					19	7.6	$0.49\ 10^{-33}$	
					19	0	$0.36\ 10^{-33}$	$0.28\ 10^{-9}$
$(Br_2)_2$	1	0	0	1	0	0	$0.19\ 10^{-74}$	0
					1	23.7	$0.14\ 10^{-73}$	
	1	0	0	1	10	0	$0.66\ 10^{-63}$	$0.2\ 10^{-7}$
					11	15.4	$0.13\ 10^{-62}$	
	1	0	0	1	20	0	$0.19\ 10^{-58}$	$0.1\ 10^{-6}$
					21	7.	$0.24\ 10^{-58}$	
$(I_2)_2$	1	0	0	1	11			$0.46\ 10^{-8}$
					16			$0.22\ 10^{-7}$
					20		$< 10^{-78}$	$0.50\ 10^{-7}$
					24			$0.88\ 10^{-7}$
					28			$0.12\ 10^{-6}$
					32			$0.13\ 10^{-6}$
					36			$0.11\ 10^{-6}$

for VP of halogen dimers in the low-lying vibrational states of the bond modes. From these results it is apparent that off-resonance interactions between zero-order states characterized by different values of l are negligible relative to the energy spacing between the zero-order levels. Thus we have to consider d–d coupling only between degenerate zero-order levels.

From a cursory examination of the energetic data for degenerate zero-order states for the halogen dimers it is immediately apparent that they fall into two categories:

1. The $(Cl_2)_2$, $(Br_2)_2$ and $(I_2)_2$ homodimers are characterized by negligible small d–d coupling terms. For these dimers condition (3.E.22) is well satisfied and the nuclear dynamics of these systems following collisional excitation is characterized by case (A). Thus the VP process of these dimers will exhibit a nontrivial exponential decay determined by (3.E.24) and (3.E.25). In this case no oscillatory energy exchange between the bond modes will be exhibited as the d–d coupling is too weak. The only exchange mechanism between the bond modes pertains to high-order coupling via the common dissociative channel. The widths $\Gamma_{n_1 n_2 l}$ provide a proper rule of the thumb characterization of the VP decay rates.
2. The $(F_2)_2$ homodimer is characterized by discrete-discrete coupling terms which overwhelm the widths of the zero-order degenerate states, whereupon condition (3.E.27) is strictly obeyed and this system corresponds to case (B). In this case collisional excitation will result in direct energy exchange between the bond modes as the VP decay is inefficient. The dimer collisionally excited into one bond mode will play a game of musical chairs, oscillating 10^8–10^9 times between the time bond modes before VP occurs. In this case the oscillations are too fast to be interrogated by following the VP decay, which will occur on the time scale $\sim\hbar/\Gamma_{n_1 n_2 l}$.

To gain some insight into the time scale that characterizes the VP process in linear halogen dimers some further numerical model calculations were performed. In Fig. 16 we portray the l dependence of the half-width of the resonances, of the $(Cl_2)_2$ homodimer for several vibrational states. These results qualitatively demonstrate the enhancement of the VP rate with increasing the excess vibrational energy of the dimer. The dependence of the VP rate on the vibrational quantum number l of the bound states in the vdW bond reveals an increase of the VP rate with decreasing l at high l values until a maximum is reached at $l=\bar{l}$. This pattern is analogous to that previously obtained by us for VP of triatomic

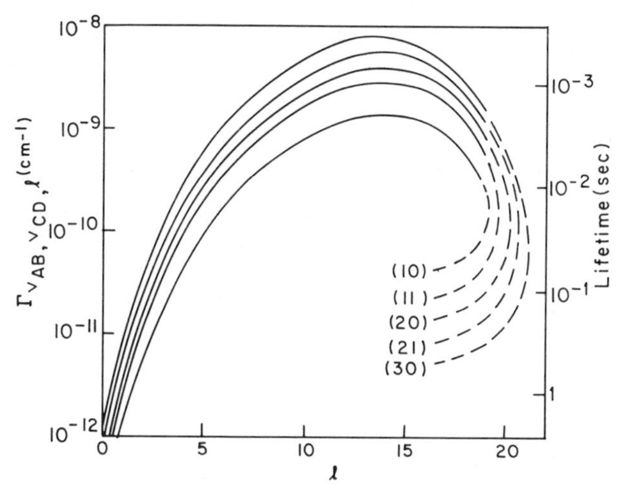

Fig. 16. Dependence of the half-width of the resonances of the $(Cl_2)_2$ homodimer for several vibrational states on the quantum number l, corresponding to bound motion along the van der Waals coordinate.

linear vdW molecules.[27] To gain some insight into the semiquantitative aspects of the VP dynamics we present in Fig. 17 the reciprocal decay widths for the VP of the $|n_1=1, n_2=0, l\rangle$ zero-order state at $l=\bar{l}$ for the halogen dimers. For the $(Cl_2)_2$, $(Br_2)_2$, and $(I_2)_2$ dimers which correspond to class A these reciprocal decay widths indeed represent the lifetimes with respect to VP, while for the $(F_2)_2$ dimer, which belongs to class B, again the reciprocal width marks the time scale for VP, after averaging over the fast energy-exchange oscillation. The effective lifetimes for VP of linear $(A_2)_2$ halogen dimer exhibit the following features:

1. The lifetimes are remarkably long, being in the range of 10^{-1} sec. for $(F_2)_2$ to 10^{-5} sec. for $(I_2)_2$.
2. For the $(Cl_2)_2$, $(Br_2)_2$, and $(I_2)_2$ dimers efficient intramolecular energy flow between the bond modes in not exhibited on the remarkably long time scale of 10^{-3}–10^{-5} sec.
3. The low values of the lifetimes and their dependence on the chemical composition of the dimer can be adequately rationalized in terms of our energy gap law for VP, $\ln \tau \propto \omega^{1/2}$.
4. The lifetime of 10^{-3} sec evaluated for the $(Cl_2)_2$ dimer is in fortuitous excellent agreement with the experimental value $\tau \geqslant 10^{-4}$ sec of Dixon and Herschbach.[16] This numerical agreement should not be taken too seriously because the goal of the present theory is not to provide

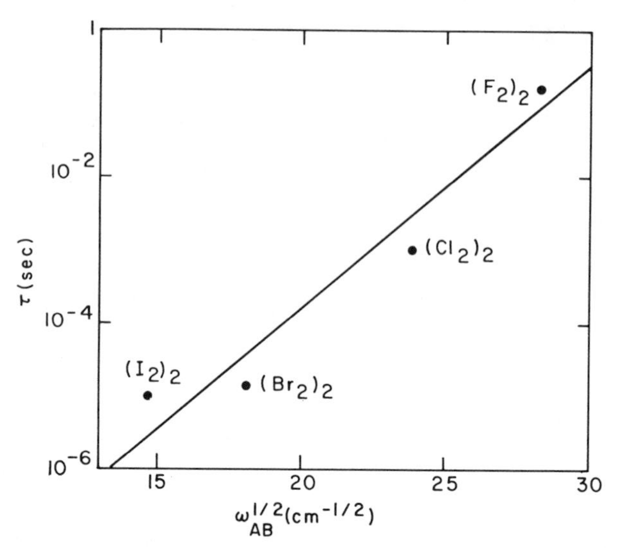

Fig. 17. Reciprocal decay widths and lifetimes for vibrational predissociation of the $|n_1 = 1, n_2 = 0, l\rangle$ zero-order state at $l = \bar{l}$ for the halogen dimers.

numerical results, as it is too primitive for that purpose, but rather to elucidate the characteristics of the intramolecular dynamics of a new and interesting class of chemical systems.

We have been concerned with a detailed quantum mechanical study of the VP dynamics of vdW dimers. A relevant question which may be raised at this stage by a skeptical chemist is whether all this effort is indeed justified and, in particular, whether reasonable estimates of the VP rate can be derived using the conventional statistical RRKM theory of unimolecular reactions.[53] It might be argued that a four-atom vdW dimer may be a sufficiently large molecular system to warrant the application of statistical theories. According to the celebrated RRKM theory the VP rate is

$$\frac{1}{\tau} \simeq \tilde{\nu}\left[\frac{E-E_t}{E}\right]^{s-1}$$

where $\tilde{\nu} \sim 10^{12}\ \text{sec}^{-1}$ is a characteristic frequency factor, E_t is the threshold energy, which in our case corresponds to the vdW bond dissociation energy, E is the excess energy which we shall identify with the bond vibrational energy ω_{AB}, while s is the number of vibrational degrees of freedom of the complex. Taking the parameters for the Cl_2—Cl_2 dimer,

$E_t = 330$ cm^{-1}, $E = 564$ cm^{-1}, and $s = 6$, we get $1/\tau \simeq 10^{10}$ sec^{-1} for the RRKM result for the VP rate. The VP rate estimated on the basis of the RRKM theory overestimates by about 7 orders of magnitude both our quantum-mechanical result and the experimental lower limit for this rate. This huge discrepency between the RRKM estimate and the quantum mechanical result clearly demonstrates the failure of the basic assumption of statistical intramolecular vibrational energy redistribution inherent in the RRKM theory. The elucidation of the VP dynamics of vdW dimers is not only interesting in relation to these isoteric and interesting systems, but is also relevant for the understanding of the general basic problem of nonstatistical vibrational energy redistribution in polyatomic molecules.

IV. VIBRATIONAL PREDISSOCIATION OF POLYATOMIC MOLECULES

A. Background Information

Very little experimental information is currently available concerning the intramolecular dynamics and VP of polyatomic vdW molecules A—B, consisting of a polyatomic-molecule–rare-gas-atom complex or corresponding to a complex between two polyatomic molecules. This sparse information falls into two categories:

1. VP in an electronically-vibrationally excited state. The polyatomic complex He—NO_2 in the ^2B electronically excited state of NO_2 exhibits an efficient VP process, which is characterized by lifetime $\tau \sim 10$ psec.[17]
2. The $(N_2O)_2$ dimer excited to the ν_1 vibration of the N_2O component exhibits VP on the ground electronic state potential surface.[18] The VP lifetime, τ, for this complex was estimated[18] to be in the range 10^{-12} sec $\leqslant \tau \leqslant 10^{-4}$ sec; the lower limit, which seems unrealistic, was obtained by assigning the entire width of the infrared absorption band to lifetime broadening, while the upper limit was obtained from a time of flight experiment.

Some other polyatomic vdW dimers, such as $(NH_3)_2$[19] and $(C_2H_4)_2$,[20] have been identified in supersonic beams by infrared spectroscopy, but dynamic studies have not yet been performed (see however Casassa et al. in the bibliography).

It is apparent from the present embryonic stage of development of this interesting area that theory will be extremely useful in providing general guidelines for the understanding of the VP dynamics of such complexes.

Two novel features of the intramolecular dynamics of vdW complexes containing polyatomic molecules should be considered. These pertain to dynamic process on the ground electronic potential surface and in electronically excited states, which will now be discussed separately. Consider first the dynamics on the ground state potential surface. Here the conventional $V \rightarrow T$ channel and the intermolecular $V \rightarrow V+T$ decay channel, already encountered for the VP of vdW dimers, will prevail. We can consider an additional decay process involving intramolecular energy exchange within the initially excited polyatomic component A of the complex, resulting in the degradation of a vibrational quantum of A into a vibration of lower frequency of the same molecule, that is, an intramolecular $V \rightarrow V+T$ process. Next, we shall focus attention on the VP dynamics in an electronically excited state of a polyatomic component. Here the role of interstate nonadiabatic coupling between zero-order vibronic levels, corresponding to two electronic configurations, will be crucial in determining the molecular level structure and the VP dynamics.

B. Model Calculations of VP of Polyatomic Complexes

We shall proceed to discuss the intramolecular dynamics and the VP process on the ground-state potential surface of a vdW complex consisting of two polyatomic molecules. Consider a vdW molecule A··· B, where A and B represent two polyatomic molecules. We shall denote by R the distance between the centers of mass of the two molecules. (See Fig. 18). The total Hamiltonian of the system, after separation of the motion of the center of mass of the entire system is

$$H = -\frac{\hbar^2}{2\mu}\nabla_{\mathbf{R}}^2 + H_{\mathrm{A}} + H_{\mathrm{B}} + U_{\mathrm{AB}} \tag{4.B.1}$$

where H_{A} and H_{B} are the internal Hamiltonians for the "free" molecules A

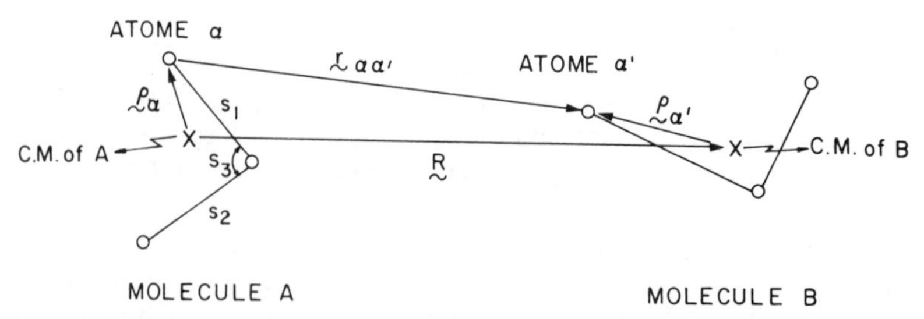

Fig. 18. Coordinate system used to describe a van der Waals complex consisting of two triatomic molecules.

and B, respectively, while U_{AB} is the intermolecular interaction. Finally, $\mu = M_A M_B/(M_A + M_B)$ is the reduced mass for the relative motion of A and B. We now invoke the dumbbell model for the intermolecular interaction, representing U_{AB} in terms of a sum of atom–atom interactions

$$U_{AB} = \sum_{\alpha, \alpha'} U_{\alpha\alpha'}(|\mathbf{r}_{\alpha\alpha'}|) \tag{4.B.2}$$

where $\mathbf{r}_{\alpha\alpha'}$ is the distance between atom α on molecule A and atom α' on molecule B. In general we have

$$\mathbf{r}_{\alpha\alpha'} = \mathbf{R} - (\mathcal{P}_\alpha - \mathcal{P}_{\alpha'}) \tag{4.B.3}$$

where $\mathcal{P}_\alpha$ and $\mathcal{P}_{\alpha'}$ are distances of the atoms α and α' from the centers of mass of the molecules A and B, respectively (see Fig. 1). Denoting by $\overline{\mathcal{P}}_\alpha$ and $\overline{\mathcal{P}}_{\alpha'}$ the corresponding equilibrium distances, we have

$$|\mathbf{r}_{\alpha\alpha'}| = \Big[|\mathbf{R} - (\overline{\mathcal{P}}_\alpha - \overline{\mathcal{P}}_{\alpha'})|^2 + |\Delta\mathcal{P}_\alpha - \Delta\mathcal{P}_{\alpha'}|^2 - 2\big[\mathbf{R} - (\overline{\mathcal{P}}_\alpha - \overline{\mathcal{P}}_{\alpha'})\big] \cdot (\Delta\mathcal{P}_\alpha - \Delta\mathcal{P}_{\alpha'}) \Big]^{1/2} \tag{4.B.4}$$

where $\Delta\mathcal{P}_\alpha$ and $\Delta\mathcal{P}_{\alpha'}$ are infinitesimal displacements. Expanding (4.B.4), and retaining only the linear terms in $\Delta\mathcal{P}_\alpha$ and $\Delta\mathcal{P}_{\alpha'}$ we obtain

$$|\mathbf{r}_{\alpha\alpha'}| \simeq |\mathbf{R} + \overline{\mathcal{P}}_{\alpha'} - \overline{\mathcal{P}}_\alpha| - \hat{n}_{\alpha\alpha'} \cdot (\Delta\mathcal{P}_\alpha - \Delta\mathcal{P}_{\alpha'}) \tag{4.b.5}$$

where we have defined the unit vector

$$\hat{n}_{\alpha\alpha'} = \frac{\mathbf{R} + \overline{\mathcal{P}}_{\alpha'} - \overline{\mathcal{P}}_\alpha}{|\mathbf{R} + \overline{\mathcal{P}}_{\alpha'} - \overline{\mathcal{P}}_\alpha|} \tag{4.B.6}$$

In general, we can express the displacements $\Delta\mathcal{P}_\alpha$ and $\Delta\mathcal{P}_{\alpha'}$ in terms of two sets of internal coordinates[55]) $\{s_t\}$ and $\{s'_{t'}\}$ for molecules A and B, respectively,

$$\Delta\mathcal{P}_\alpha = \sum_t (a_{\alpha t}\hat{x} + b_{\alpha t}\hat{y} + c_{\alpha t}\hat{z}) s_t$$
$$\Delta\mathcal{P}_{\alpha'} = \sum_{t'} (a'_{\alpha' t'}\hat{x}' + b'_{\alpha' t'}\hat{y}' + c'_{\alpha' t'}\hat{z}') s'_{t'} \tag{4.B.7}$$

where $a_{\alpha t}$, $b_{\alpha t}$, and $c_{\alpha t}$ (and $a'_{\alpha' t'}$, $b'_{\alpha' t'}$, $c'_{\alpha' t'}$) represent the elements of the appropriate transformations, while $\{\hat{x}, \hat{y}, \hat{z}\}$ and $\{\hat{x}', \hat{y}', \hat{z}'\}$ are unit Cartesian vectors with respect to the molecular axis of molecules A and B, respectively. Equation (4.B.2), together with (4.B.5) and (4.B.7), specify the intermolecular interactions in terms of the intermolecular distance **R** and the internal coordinates $\{s_t\}$ and $\{s'_{t'}\}$.

To proceed with the description of the intramolecular dynamics we have to define an appropriate zero-order Hamiltonian. To accomplish this goal we use the distorted wave treatment where the zero-order Hamiltonian is

$$H_0 = -(\hbar^2/2\mu)\nabla_{\mathbf{R}}^2 + H_A + H_B + U_{AB}(|\mathbf{R}|) \tag{4.B.8}$$

where the intermolecular interaction U_{AB} corresponds to the two molecules frozen at their equilibrium nuclear configurations. The residual interaction

$$V = H - H_0 = U_{AB}(\{\mathbf{r}_{\alpha\alpha'}\}) - U_{AB}(|\mathbf{R}|) \tag{4.B.9}$$

induces the nonradiative intramolecular VP transitions. The zero-order Hamiltonian (4.B.8) is separable and its eigenstates correspond to the discrete states

$$\psi_{v_A, v_B, l}(\{s_t\}, \{s'_{t'}\}, |\mathbf{R}|) = x_{v_A}(\{s_t\}) x_{v_B}(\{s'_{t'}\}) \phi_l(|\mathbf{R}|) \tag{4.B.10}$$

and to the continuum states

$$\psi_{v_A, v_B, \epsilon}(\{s_t\}, \{s'_{t'}\}, |\mathbf{R}|) = x_{v_A}(\{s_t\}) x_{v_B}(\{s'_{t'}\}) \phi_\epsilon(|\mathbf{R}|) \tag{4.B.11}$$

Here the rotational degrees of freedom are not considered so that (4.B.10) and (4.B.11) represent vibrational wavefunctions for a fixed angular configuration. The functions $x_{v_A}(\{s_t\})$ and $x_{v_B}(\{s'_{t'}\})$ represent general vibrational eigenstates of the molecules A and B, respectively. These nuclear eigenstates are characterized by the collection of nuclear vibrational quantum numbers v_A and v_B, and by the energies E_{v_A} and E_{v_B}, so that

$$\begin{aligned} (H_A - E_{v_A})|x_{v_A}\rangle &= 0 \\ (H_B - E_{v_B})|x_{v_B}\rangle &= 0 \end{aligned} \tag{4.B.12}$$

The eigenstates $\phi_l(|\mathbf{R}|)$ and $\phi_\epsilon(|\mathbf{R}|)$ correspond to a bound state and to a continuum state of the vdW bond, respectively, l is a discrete quantum number characterizing bound states of the vdW bond with energies E_l while ϵ is the relative kinetic energy of A and B. The VP dynamics on the ground state potential surface can be adequately described by defining a "preparation" process where a single $\psi_{v_A, v_B, l}$ discrete state is amenable to

IR or collisional excitation. The VP rate is essentially dominated by the d–c resonance interaction, while the effect of c–c coupling terms is relatively small. Accordingly, the decay rate $\Gamma_{v_A v_B l}$ of the metastable state $\psi_{v_A, v_B, l}$ is given by the golden rule

$$\Gamma_{v_A v_B l} = \pi \sum_{v'_B} \sum_{v'_B} |\langle \psi_{v_A, v_B, l} | V | \psi_{v'_A, v'_B, \epsilon} \rangle|^2 \tag{4.B.13}$$

where the relative kinetic energy is now taken on the energy shell

$$\epsilon = E_{v_A} + E_{v_B} + E_l - E_{v_{A'}} - E_{v_{B'}} \tag{4.B.14}$$

while the interaction V (4.B.9), which induces the VP process can be expressed with the help of (4.B.2) in the form

$$V = \sum_{\alpha, \alpha'} \left[U_{\alpha\alpha'}(|\mathbf{r}_{\alpha\alpha'}|) - U_{\alpha\alpha'}(|\mathbf{R}|) \right] \tag{4.B.15}$$

Equations (4.B.13)–(4.B.15), together with (4.B.5), constitute a comprehensive and quite complete theory of VP of polyatomic vdW complexes. What remains to be done is to specify explicitly the atom–atom interaction potentials appearing in (4.B.15). These will be represented in terms of Morse potentials

$$\begin{aligned} U_{\alpha\alpha'}(|\mathbf{r}_{\alpha\alpha'}|) = D_{\alpha\alpha'} \{ &\exp[-2\beta_{\alpha\alpha'}(|\mathbf{r}_{\alpha\alpha'}| - |\bar{\mathbf{r}}_{\alpha\alpha'}|) \\ &- 2\exp[-\beta_{\alpha\alpha'}(|\mathbf{r}_{\alpha\alpha'}| - |\bar{\mathbf{r}}_{\alpha\alpha'}|)] \} \end{aligned} \tag{4.B.16}$$

where $D_{\alpha\alpha'}$ is an effective dissociation energy, while $\beta_{\alpha\alpha'}$ denotes a characteristic inverse length. The use of a superposition of Morse potentials (4.B.16), to represent the intermolecular interaction is quite adequate as the VP dynamics are dominated by the short-range part of the intermolecular interaction potential. Using the fact that d–c coupling terms are essentially determined by the details of the interaction potential in the region of the minimum, $\mathbf{R} \simeq \bar{\mathbf{R}}$, we may write

$$|\mathbf{R} + \bar{\mathcal{P}}_{\alpha'} - \bar{\mathcal{P}}_{\alpha}| = |\bar{\mathbf{R}} + \bar{\mathcal{P}}_{\alpha'} - \bar{\mathcal{P}}_{\alpha}| + \hat{n}_{\alpha\alpha'} \cdot \Delta\mathbf{R} \tag{4.B.17}$$

where $\Delta\mathbf{R}$ is the displacement from the minimum, and $\hat{n}_{\alpha\alpha'}$ is the unit vector [defined in (4.B.6)] evaluated at $\mathbf{R} = \bar{\mathbf{R}}$. Using the Morse-type interaction and (4.B.5)–(4.B.7) and (4.B.10), together with (4.B.15)–(4.B.17), the d–c coupling can be written in terms of products of intramolecular contributions and of intermolecular factors,

$$\langle \psi_{v_A, v_B, l} | V | \psi_{v'_A, v'_B, \epsilon} \rangle = \sum_{a, \alpha'} D_{\alpha\alpha'} \Big[A^{(2)}_{v_A v_B, v'_A v'_B}(\alpha, \alpha') B^{(2)}_{l\epsilon}(\alpha, \alpha') - 2 A^{(1)}_{v_A v_B, v'_A v'_B}(\alpha, \alpha') B^{(1)}_{l\epsilon}(\alpha, \alpha') \Big] \quad (4.B.18)$$

where the intramolecular contributions are

$$A^{(p)}_{v_A v_B, v'_A v'_B}(\alpha, \alpha') = \langle x_{v_A} x_{v_B} | \exp(p\beta_{\alpha\alpha}' \hat{n}_{\alpha\alpha'} \cdot (\Delta \mathscr{P}_\alpha - \Delta \mathscr{P}_{\alpha'}) - 1 | x_{v'_A} x_{v'_B} \rangle \qquad p = 1, 2 \quad (4.B.19)$$

with $\Delta\mathscr{P}_\alpha$ and $\Delta\mathscr{P}_{\alpha'}$ given by (4.B.7), while the intermolecular contributions are given by

$$B^{(p)}_{l\epsilon}(\alpha, \alpha') = \exp\Big[-p\beta_{\alpha\alpha'}\big(|\bar{\mathbf{R}} + \bar{\mathscr{P}}_{\alpha'} - \bar{\mathscr{P}}_\alpha|\big) - |\bar{\mathbf{r}}_{\alpha\alpha'}|\big) \Big] \cdot \otimes \langle \phi_l | \exp\big[-p\beta_{\alpha\alpha'} \hat{n}_{\alpha\alpha'} \cdot \Delta \mathbf{R} \big] | \phi_\epsilon \rangle \qquad p = 1, 2 \quad (4.B.20)$$

The eigenfunction $\phi_l(|\mathbf{R}|)$ and $\phi_\epsilon(|\mathbf{R}|)$ are obtained by fitting the potential $U_{AB}(|\mathbf{R}|)$ in (2.B.8), that is,

$$U_{AB}(|\mathbf{R}|) = \sum_{\alpha, \alpha'} D_{\alpha\alpha'} \big\{ \exp\big[-2\beta_{\alpha\alpha'}(|\mathbf{R} + \mathscr{P}_{\alpha'} - \mathscr{P}_\alpha| - |\bar{\mathbf{r}}_{\alpha\alpha'}|) \big] - 2\exp\big[-\beta_{\alpha\alpha'}(|\mathbf{R} + \mathscr{P}_{\alpha'} - \mathscr{P}_\alpha| - |\bar{\mathbf{r}}_{\alpha\alpha'}|) \big] \big\} \quad (4.B.21)$$

to the single Morse function

$$U_{AB}(|\mathbf{R}|) = D\Big\{ \exp\Big[-2\beta\big(|\mathbf{R}| - \bar{\mathbf{R}}|\Big] - 2\exp\Big[-\beta\big(|\mathbf{R}| - |\bar{\mathbf{R}}|\big) \Big] \Big\} \quad (4.B.22)$$

The explicit expressions for the intermolecular matrix elements $\langle \phi_l | \exp[-\eta(|\mathbf{R}| - |\bar{\mathbf{R}}|)] | \phi_\epsilon \rangle$ needed in (4.B.20) are well known

$$\langle \phi_l | \exp\Big[-\eta\big(|\mathbf{R}| - |\bar{\mathbf{R}}|\big) \Big] | \phi_\epsilon \rangle = (\beta\pi\hbar)^{-1} \otimes \left[\frac{l!(2K - 2l - 1)!}{\Gamma(2K - l)} \right]^{-1} (-1)^l [\mu \sinh(2\pi\theta)]^{1/2} \otimes |\Gamma(\tfrac{1}{2} - K - i\theta)| (2K)^{-\eta/\beta} \otimes \sum_{n=0}^{l} \frac{(-1)^{n-l}\Gamma(2K - l)|\Gamma(K + i\theta - n + \eta/\beta - \frac{1}{2})|^2}{n!(l - n)!\Gamma(2K - l - n)\Gamma(-n + \eta/\beta)} \quad (4.B.23a)$$

with the definitions

$$K=(\hbar p)^{-1}(2\mu D)^{1/2}=\frac{2D}{\hbar\omega} \tag{4.B.23b}$$

$$\theta=(\hbar\beta)^{-1}(2\mu\epsilon)^{1/2}=\frac{2(D\epsilon)^{1/2}}{\hbar\omega} \tag{4.B.23c}$$

Equations (4.B.13) and (4.B.18) constitute a comprehensive, quite complete and practical theory of VP of polyatomic vdW molecules. It should be emphasized at this point that all the approximations introduced up to the present stage essentially pertain only to the description of the intermolecular interaction, while the intramolecular vibrational states of the individual molecules are retained in a general form without alluding to any specific approximation. An approximate description of the intermolecular potential in terms of a harmonic potential will considerably simplify the analysis for specific model systems, and this will now be introduced.

We have considered[31] four model systems corresponding to the colinear and perpendicular VP of polyatomic vdW molecules, consisting of an atom-linear-triatomic-molecule complex and of a complex between two linear triatomics. The intermolecular interaction (4.B.2) has been represented in our model calculations by a single atom-atom interaction between the nearest atoms of the two molecules. This is sufficient for the understanding of the general qualitative behavior of the VP dynamics, in view of our poor knowledge of the details of the interaction potential. The internal coordinates for the linear triatomic molec- between atoms 1 and 2 and 2 and 3, respectively, and s_3 which corresponds to the variation of the bending angle. The coefficients $a_{\alpha t}, b_{\alpha t}, c_{\alpha t}$ of (4.B.7) relating the internal coordinates to the displacements $\Delta\mathcal{P}_\alpha$ from the center of mass are given by (choosing the z-axis along the molecular axis)

$$b_{13}=a_{13}=-\frac{m_2 m_3 l_2}{N}$$

$$b_{23}=a_{23}=\frac{m_1 m_3(l_1+l_2)}{N}$$

$$b_{33}=a_{33}=-\frac{m_1 m_2 l_1}{N}$$

$$c_{11}=-\frac{m_2+m_3}{M}$$

$$c_{12}=c_{22}=-\frac{m_3}{M}$$

$$c_{21}=c_{31}=\frac{m_1}{M}$$

$$c_{32}=\frac{m_1+m_2}{M}$$

where m_1, m_2, and m_3 are the masses of the atoms, l_1 and l_2 are the equilibrium distances of the two bonds, M is the total mass of the triatomic molecule, and

$$N=\frac{m_1m_2l^2+m_1m_3(l_1+l_2)^2+m_2m_3l_2^2}{l_1l_2} \tag{4.B.25}$$

All other coefficients are zero. We now have to specify the vibrational eigenstates of the fragments. In the model calculations presented in this section we have adopted the harmonic approximation. The eigenstates are thus products of normal modes wavefunctions, that is,

$$|x_{v_A}\rangle=|v_A^{(1)}\rangle_{Q_1}|x_{v_A}^{(2)}\rangle_{Q_2}\cdots|x_{v_A}^{(k)}\rangle_{Q_k}\cdots \tag{4.B.26}$$

where $Q_1, Q_2, \ldots$ and so on are the normal coordinates. The relation between the internal coordinates s_t and the normal coordinates Q_k is given by the linear transformation

$$s_t=\sum_k L_{tk}Q_k \tag{4.B.27}$$

For the case of linear triatomic molecules, expressing the intramolecular potential as

$$W=\tfrac{1}{2}(f_{11}s_1^2+f_{12}s_1s_2+f_{22}s_2^2+f_{33}s_3^2) \tag{4.B.28}$$

results in the normal frequencies ω_1, ω_2, and ω_3 with

$$\begin{aligned}\omega_1^2+\omega_3^2&=\left(\frac{f_{11}}{\mu_{12}}\right)+\left(\frac{f_{22}}{\mu_{23}}\right)-2\left(\frac{f_{12}}{m_2}\right)\\ \omega_1^2\omega_3^2&=\frac{(f_{11}f_{22}-f_{12}^2)M}{m_1m_2m_3}\\ \omega_2^2&=\frac{f_{33}N}{l_1l_2m_1m_2m_3}\end{aligned} \tag{4.B.29}$$

where $\mu_{12}=m_1m_2/(m_1+m_2)$ and $\mu_{23}=m_2m_3/(m_2+m_3)$ are reduced masses. The coefficients L_{tk} for the transformation (4.B.27) between normal modes and internal coordinates are given by standard methods.[55]

We have considered the linear molecule N_2O with force constants and equilibrium distances taken from the literature.[56, 57] We shall now discuss the colinear He–N_2O complex. This system is of interest as it exhibits the effect of the intramolecular V→V+T process. In this case, we have from (4.B.18)

$$A^{(P)}_{v_A v_B, v'_A v'_B} = \langle x_{v_A} x_{v_B} | \{ \exp[p\beta(g_1 Q_1 + g_3 Q_3)] - 1 | x_{v'_B} x_{v'_B} \rangle \tag{4.B.30}$$

with

$$\begin{aligned} g_1 &= c_{31} L_{11} + c_{32} L_{21} \\ g_3 &= c_{31} L_{13} + c_{32} L_{23} \end{aligned} \tag{4.B.31}$$

In the VP of the linear He–N_2O complex the two decay channels involve V→T and intramolecular V→V+T processes. The contribution of the conventional V→T channel is obtained from the linearization approximation to the intramolecular term $A^{(P)}$ (4.B.30), which is expanded up to the linear term in the intramolecular normal modes Q_1 and Q_2. The contribution of the intramolecular $V \rightarrow V+T$ channel is obtained from the expansion of $A^{(P)}$ (4.B.30), up to second-order in the normal modes. In Table VIII we summarize the results of model calculations of the VP of linear He–N_2O, where the triatomic molecule is initially excited to the $\nu_1 = 2325.6$ cm^{-1} vibrational state. The V–T channel is exceedingly slow, as is evident from the energy gap law, which makes the rupture of a vdW bond of dissociation energy $D = 50$ cm^{-1} by dissipation of a ν_1 quantum very inefficient. The intramolecular V→V+T mechanism for the process $\nu_1 \rightarrow \nu_2 + T$ with $\nu_1 = 2325$ cm^{-1} and $\nu_2 = 1261$ cm^{-1} is more efficient, the VP rate

TABLE VIII
VP of the He—N_2O Complex

vdW Bond Parameters: $\alpha = 1.5$ Å^{-1}
$D = 50$ cm^{-1}
$h\omega = 45.5$ cm^{-1}
$K = 2.2$
$E_0 = 29.6$ cm^{-1}
Excited State: $\nu_1 = 2325$ cm^{-1}
Final States: V→T: $\epsilon = 2325$ cm^{-1}
V→V+T: $\nu_1 \rightarrow \nu_2 + T$; $\nu_2 = 1261$ cm^{-1}

Molecule	V→T τ(sec)	V→V+T τ(sec)
N—N—O··· He	33	0.19
0—N—N··· He	7	0.05

being enhanced by about 2 orders of magnitude relative to the V→T channel. Similar model calculations performed by us[31] for the VP of the linear and of the T-shaped $(N_2O)_2$ dimer excited to ν_1, demonstrated the enhancement of the intramolecular V→V+T and the intermolecular V→V +T processes, relative to the *V*→T channel. However, the VP lifetimes are long $\tau \sim 10^{-3}-10^{-5}$ sec. The present calculations rest on a harmonic approximation for the triatomic molecule. Incorporation of intramolecular anharmonicity effects will tend to enhance the VP process. Nevertheless, we are in a position to draw some conclusions from the simplified analysis of the VP dynamics presented herein. First, we have demonstrated the effectiveness of intramolecular and intermolecular V→V+T processes in the VP of vdW complexes consisting of polyatomic molecules. Second, when the translational kinetic energy $\epsilon \simeq \nu_1 - \nu_2 - D$ is high, the two V→V +T processes are still rather inefficient and VP is expected to be slow. This state of affairs provides an interesting example for inefficient nonresonant exchange of vibrational energy in a polyatomic molecular complex. Third, we expect that when the kinetic energy for the second-order V→V+T processes is relatively low, a dramatic contribution of those two channels will make the VP process accompanied by intramolecular and intermolecular vibrational energy exchange amenable to experimental observation. From the practical point of view, two conclusions emerge from this analysis. First, for the specific case of VP of the $(N_2O)_2$ dimer experimentally studied in molecular beams, our model calculations indicate that the VP process is slow. Although our numerical estimate $\tau \sim 10^{-3} - 10^{-5}$ sec for the VP of the dimer,[31] originating mainly from intramolecular and intermolecular V→V+T, should not be taken seriously, we can consider the theoretical estimate as a rough guide for the upper limit of the lifetime. We can assert that the lifetime for VP of $(N_2O)_2$ should be close to the experimental upper limit of $\tau \sim 10^{-4}$ sec deduced from time of flight experiments,[18] and that the experimental lower limit of $\tau \gtrsim 10^{-12}$ sec inferred from the assignment of the entire line broadening to lifetime broadening is unrealistic. Second, some conclusions concerning line broadening in hydrogen-bonded molecular complexes emerge from the present study. Coulson and Robertson[23] and Ewing[24] have provided compelling evidence that the conventional V→T mechanism is not sufficiently efficient to yield VP lifetimes of $\tau \sim 10^{-10}$ sec, or shorter, which would provide a significant contribution, $\sim 10^{-2}$ cm^{-1}, to the broadening of the infrared absorption bands of hydrogen-bonded systems. Even though the two intramolecular and intermolecular V→V+T channels are expected to overwhelm the *V*→*T* process, it appears that these additional VP mechanisms are not sufficiently efficient, since a time scale of $\sim 10^{-10}$ sec is required to measurably contribute to the lifetime broadening of the infrared absorption bands of hydrogen-bonded complexes. It appears that Stepanov's original suggestion[22] regarding the spectral broadening of hydrogen-bonded systems should be modified and that the major line-broadening mechanism originates from inhomogeneous broadening effects in the condensed phase.

C. A Comment on VP of Electronically Excited Polyatomic Complexes

The efficient VP of the interesting He–NO_2 molecule in the electronically excited ^2B state of NO_2 [17] is of considerable interest as it provides a novel mechanism for intramolecular dynamics in electronically excited states of medium-sized polyatomic molecules. The conventional V→T process, as well as the intramolecular and intermolecular V→V+T mechanisms, will be rather inefficient for the VP of He–NO_2 as the relevant energy gaps are large. Levy[17] has pointed out that in a medium-sized molecule, such as NO_2, interstate nonadiabatic scrambling between the ^2B electronically excited state and the ^2A ground state is effective, resulting in a manifold of molecular eigenstates (ME). The level density of the MEs is 0.05 cm. these MEs are active in absorption, emission and in intramolecular dynamics (Fig. 19). Conventional optical excitation does not prepare a zero-order state of ^2B but rather, as the MEs are well spaced relative to their widths, optical excitation of NO_2 "prepares" a single ME. The He–NO_2 vdW complex is expected, accordingly, to be characterized by a moderately high density of levels with one state per 10–20 cm^{-1}, all of which are amenable to selective optical excitation. Now, the restrictions imposed by the energy gap law on the efficiency of the VP process can be easily relaxed. Near-resonance VP can occur as, in view of the congested level structure of the MEs, degradation of vibrational energy will occur between close-lying MEs and the kinetic energy of the fragments will be very low, that is, of the order of level spacing between MEs. A quantitative theory of this process involving induced vibrational energy exchange

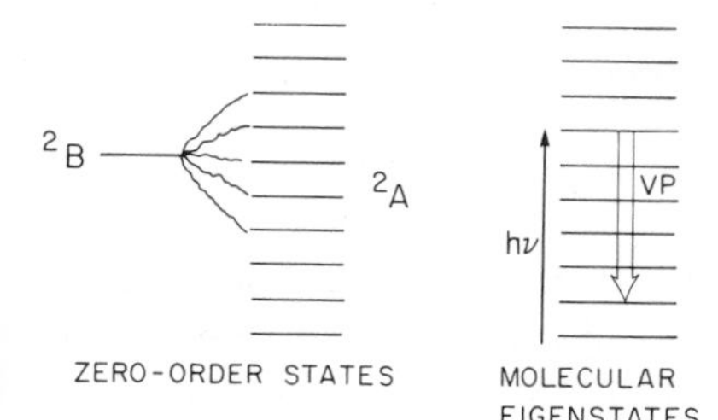

Fig. 19. Energy level scheme corresponding to excited NO_2.

between MEs requires the incorporation of intramolecular anharmonicity effects. This VP process in electronically excited states bears some analogy to collisionally-induced electronic relaxation between electronically excited states of medium-sized molecules. For electronically excited states of polyatomic vdW complexes, where the MEs retain their individuality, the induced VP process can prevail. Systems that come to mind correspond to the intermediate level structure in electronically excited states of medium-sized and some large molecules.

Finally, it should be noted that the VP mechanisms in electronically excited states of vdW complexes, where the excited molecular component corresponds to the statistical limit, are qualitatively and quantitatively different. Now the interstate coupling is weak but the density of background states is huge. Accordingly, VP in the "initially excited" electronic configuration of the complex will involve nuclear dynamics of a single, electronically-excited, potential surface, while the electronic relaxation process provides just a parallel decay channel. This state of affairs, prevailing in the electronically excited statistical limit, is similar to the dynamics on the ground-state potential surface.

V. CONCLUDING REMARKS

We have attempted to advance a conceptual framework for the understanding of VP of vdW molecules, which provide important information regarding the basic problem of intramolecular vibrational energy flow and vibrational energy redistribution in polyatomic molecules. In this context the following basic VP processes were considered with increasing complexity of the molecular constituents of the vdW complex:

1. For a triatomic X—AB vdW molecule, where AB is a conventional diatomic molecule and X represents a rare-gas atom, the VP process involves the conversion of vibrational energy of AB to translational energy of the AB and X fragments, that is, $V \rightarrow T$ process. In the present discussion of VP of linear and of T-shaped vdW molecules, we shall disregard the effects of vibrational-rotational energy conversion, while the VP of the HeI_2 vdW molecule can be semiquantitatively discussed in terms of the V→T process. Such V→R+T processes may be very important in some cases, for example, in the VP of the ArHCl complex.[26]
2. For a vdW dimer AB–CD, where AB and CD are both normal diatomic molecules, the VP channels involve the V→T process as well as the conversion of the vibrational energy of one molecule to vibrational energy of the other, while the energy balance is made up by

translational energy. The V→T process determines the VP mechanism of a homodimer such as Cl_2–Cl_2. Intermolecular vibrational energy exchange, constituting an intermolecular V→V+T process, considerably enhances the VP rate of some vdW heterodimers.

3. For a vdW molecule X–A, where X is a rare-gas atom and A is a polyatomic molecule, two VP channels should be considered. First, the conventional V→T process prevails. Second, one should consider the intramolecular degradation of a vibrational quantum of A to a lower frequency vibration of the same molecule, the energy balance being made up by translation. This process involves an intermolecular V→V +T exchange. This intramolecular vibrational energy process bears a close analogy to collision-induced redistribution in polyatomic molecules.
4. In the case of a vdW complex A–B, consisting of a pair of polyatomic molecules A and B, the VP process can involve conventional V→T as well as additional channels which correspond to vibrational energy exchange. Two types of such VP decay mechanisms accompanied by vibrational energy exchange should be considered. First, in complete analogy to the case of the heterodimer, we can consider the intermolecular conversion of the vibrational energy of A to the vibrational energy of B, that is, an intermolecular V→V+T process. This intermolecular process bears a close analogy to V→V transfer in collisions[58] and was considered also in relation to the broadening of infrared absorption bands in liquids.[59] Second, we can envision the intramolecular energy exchange within the initially excited molecule A, resulting in the degradation of a vibrational quantum of A to a vibration of a lower frequency of the same A molecule, i.e., an intramolecular equivalent to the V→V+T exchange in the VP of X–A considered in point 3.
5. The VP mechanisms involving V→T, V→R, intermolecular V→V+T and intramolecular V→V+T occur on a single nuclear potential surface, which corresponds either to the ground electronic state or to an electronically excited configuration. Novel VP processes can be exhibited in electronically excited states of vdW complexes consisting of medium-sized molecules, or where the electronically excited state corresponds to the intermediate level structure.

We have attempted to provide a general conceptual framework for the understanding of the VP mechanisms and the intramolecular dynamics of small and medium-sized vdW complexes, emphasizing the effects of degradation of vibrational energy into translational energy, the exchange of vibrational energy between subunits of the weakly bound complex, as well as intramolecular vibrational energy exchange induced by VP of the polyatomic complex. Although negative results of theoretical studies are usually of methodological and pedagogical interest, several negative results emerge from our work which are pertinent not only for the elucidation of the interesting problem of VP of vdW molecules but also in the general context of the exploration of several features of intramolecular dynamics of polyatomic molecules. First, in contrast to intuitive preconceived notions, we have shown that the VP of vdW complexes often constitutes a slow process, whereupon intramolecular vibrational flow from a strong molecular bond to a weak vdW bond is not necessarily efficient and ultrafast. Instead, we have demonstrated that the VP rate, which for different systems varies for 20 orders of magnitude, can be adequately rationalized in terms of our energy gap law. Several experimental indications for the stability of some vdW complexes with respect to VP start emerging and, in particular, the striking evidence concerning the slow VP rate of the $(Cl_2)_2$ complex[16] is relevant in this context. It is comforting that to some extent theoretical predictions, which have rationalized this unexpected stability of weak vdW bonds, have preceded experimental work in this interesting area. Second, we have asserted that resonant intramolecular vibrational energy exchange between two subunits of a vdW dimer can be a very slow process. As is evident from our analysis of the intramolecular dynamics of some homodimers, the complex thus does not always like to play a game of musical chairs for exchange of vibrational energy between its subunits, which can be slow on the time scale of VP. This is again a counterintuitive result, as conventional statistical theories[53] will imply that resonant intramolecular vibrational energy exchange is expected to be first. Third, we have advanced compelling evidence for the inapplicability of conventional chemical theories such as the celebrated RRKM theory of unimolecular reactions,[53] to account for the fragmentation of medium-sized polyatomic vdW complexes. This failure of the RRKM theory reflects the breakdown of the basic assumption of effective intramolecular vibrational redistribution within the vdW complex, which is inherent in the statistical theory. The statistical approach was replaced by an exploration of the VP process from the microscopic point of view. From these negative results it is apparent that chemical intuition has to be complemented and supplemented by the results of the microscopic theory of intramolecular dynamics.

VI. RECENT PROGRESS IN VP DYNAMICS

We would like to survey progress in the fast developing area of VP dynamics of vdW molecules, accomplished during the first half of 1980. The most interesting developments in the understanding of “small” triatomic molecules pertain to four aspects: (1) the determination of energetic parameters that provide basic input data for the dynamics, (2) the elucidation of some details of the VP process, such as the role of anharmonicity effects, (3) semiclassical calculations of VP dynamics that offer an alternative to the quantum mechanical approach and that may be extended for larger systems, and (4) the exploration of rotational effects on VP dynamics. Regarding the VP dynamics of "large" vdW complexes, detailed information on the intramolecular and intermolecular vibrational energy exchange during the VP of complexes consisting of a pair of triatomic molecules was gathered. These results provide a first step toward understanding vibrational energy flow in large vdW molecules.

A. Energetics of the Iodine-Rare Gas van der Waals Molecules

Recently[60] the product state distribution of the fragment I_2, produced when an I_2–rare gas van der Waals molecule dissociates, has been used to measure the binding energies of $X \cdots I_2$ complexes. The results are: 13.6–14.8 cm^{-1} for HeI_2, 65–67 cm^{-1} for NeI_2, and 220–226 cm^{-1} for ArI_2. These are now among the very few atom–molecule vdW binding energies that are known. The Chicago group has also measured the energy difference ΔE between two vibrational levels of the van der Waals bond. For HeI_2 this is 5.66 cm^{-1}. If one assumes that this corresponds to excitation of the vdW stretch and that it can be described by a Morse potential, the Morse curve parameters D and α can be determined. The result is $D = 17.5$ cm^{-1} and $\alpha = 0.4$ A^{-1}. This is a surprisingly low value for α. Usually the inverse length parameter α has been estimated in most systems to be in the range 1 and 2 A^{-1}. Also according to the models for vibrational predissociation, the rates for α of the order of 0.4 A^{-1} will be extremely low of the order of 10^6 sec^{-1} in contradiction with the experimental values ($\sim 10^{10}$ sec^{-1}). It will be possible that the excited level they are observing corresponds to bending excitation rather than to a stretching mode. Recently[61] the bending levels for HeI_2 have been calculated using a pairwise potential and the infinite order sudden approximation. An energy levels diagram for HeI_2 including both bending and stretching modes[61] is presented in Fig. 20. This energy scheme provides a self-consistent picture both for the spectroscopy and for the dynamics. This problem deserves further study.

B. Scaling Theoretical Analysis of VP in the H I

Recently Ramaswamy and De Pristo[62] presented a scaling theoretical analysis of VP. The treatment is based on the energy corrected sudden (ECS) scaling theory for inelastic collision processes.[63]

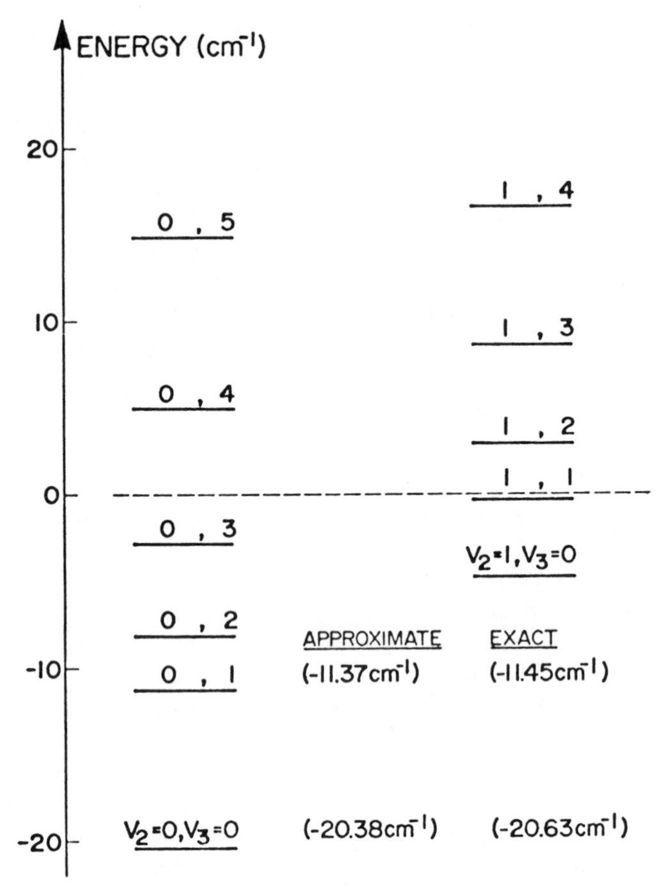

Fig. 20. Energy level diagram for some bound and quasibound states of the van der Waals molecule $HeI_2(B)$. The energy represented is $\varepsilon_{v_2v_3}$ [see Eq. (D41) in the text] and is relative to the vibrational energy of the $I_2(B)$ bond.

The scaling relationship appropriate for a vibration translation (V–T) deexcitation rate was written as

$$\frac{\Gamma(v\to v-1)}{\Gamma(1\to 0)}=\left|I_1^{v,v-1}\right|^2\left[\frac{6+(\omega_{1,0}\tau_C/2)^2}{6+(\omega_{v,v-1}\tau_C/2)^2}\right]^2 \tag{B1}$$

where $\omega_{v,v-1}$ is the frequency difference between the vibrational states v and $v-1$ of the diatomic fragment, τ_C is the average "half-collision" time,

and

$$I_1^{v,v-1} = \int \chi_v^*(r)\chi_{v-1}(r)\chi_1(r)/\chi_0(r)\,dr \tag{B2}$$

where $\chi_v(r)$ is the vibrational part of the wave function for $I_2(B, v)$. The values of $\omega_{v,v-1}$ and $I_1^{v,v-1}$ can be determined from the vibrational constants of the I_2^* molecule. The average of the "half-collision" time τ_C is determined by a fit of the experimental results, and in the case of the HeI_2^* the "best fit" is obtained for $\tau_C = 2\times10^{-13}$ sec. The conclusion of this work is that the details of the van der Waals bond potential are unimportant for the scaling of rates.[62] It should be noted, however, that τ_C is here an adjustable parameter that in any detailed theory should be estimated from the details of the intermolecular potential. The conclusion of this work should be then rephrased as follows: only one parameter characterizing the intermolecular interaction is necessary to describe the vibrational dependence of the rates. It should be noted that this conclusion concurs with the energy gap law derived in Section II.B. Using the definition of the exponential factor α_{XB} of the Morse potential, namely $\alpha_{XB} = \omega_{XB}(\mu_{XB}/2D_{XB})^{1/2}$ in Eq. (2.B.59), we obtain the super linear dependence of the VP rate on the vibrational quantum number

$$\Gamma_{diss} \alpha v \exp(\beta v) \tag{B3}$$

with

$$\beta = \frac{\pi(\hbar\omega_{BC}\mu_{X,BC}/2)^{1/2}}{2K_{BC}\alpha_{XB}} \tag{B4}$$

Therefore only the parameter α_{XB} of the intermolecular potential determines the dependence of the rates on the vibrational quantum number v. It is interesting to note that α_{XB} determines the range of the potential and therefore is related to τ_C by the approximate relationship

$$\tau_C = \alpha_{XB}^{-1}(\epsilon/2\mu_{X,BC})^{-1/2} \tag{B5}$$

From Eqs. (B4) and (B5) we get

$$(\tau_C/\beta) = 4K_{BC}\hbar^{1/2}/(\epsilon\omega_{BC})^{1/2} \tag{B6}$$

and as $\epsilon \sim \omega_{BC}$ we obtain the neat expression

$$\Gamma_{diss} \alpha v \exp[(\omega_{BC}\tau_C/4K_{BC})v] \tag{B7}$$

It will be interesting to confront this result with the "half collision" formalism.

C. Quasiclassical Trajectory Studies of VP

Recently[64] Woodruff and Thompson have studied vibrational predissociation of the $He \cdots I_2(B)$ using quasiclassical trajectories. The study was restricted to collinear motion and comparisons have been made with the quantum mechanical calculations.[27–31] The computed trajectory results show that the unimolecular decay as a function of time obeys the exponential decay law quite well. The computed rates are in accord with the quantum mechanical values.[27–31] This approach is of intrinsic interest as it can be readily extended to handle the VP of large vdW complexes, consisting of a rare gas atom bound to a polyatomic molecule.

D. The Effects of Rotation on the Vibrational Predissociation Rates

1. *The Weak-Coupling Case*

In the weak-coupling case the van der Waals molecules are characterized by very small anisotropic terms in the intermolecular potential, as compared with rotational spacings. The bound ro-vibrational states can be then described fairly accurately by the rotational and vibrational wavefunctions of the "free" constituents. For a triatomic van der Waals molecule the process of vibrational predissociation can then be described by

$$X \cdots BC(v_1, j) \xrightarrow{\Gamma} X + BC(v_1' < v_1, j') \tag{D1}$$

where v_1 is the vibrational quantum number of the BC molecule, while j is the rotational quantum number. So far, all the treatments reviewed here have been restricted to a fixed-angular configuration (colinear or perpendicular)[21–23, 27–31] or to a three-dimensional calculation neglecting the coupling between rotation and vibration[25] (i.e., assuming the potential energy surface to be isotropic). As we have shown in Section II.B these treatments predict that vibrational predissociation rates for molecules like $X \cdots H_2, X \cdots HCl$ etc., where the vibrational frequency of the BC molecule greatly exceeds the frequency of the van der Waals bond vibration, will be exceedingly small ($<1\ sec^{-1}$ and in some cases $<10^5\ sec^{-1}$). These extremely low VP rates originate from the small intermolecular potential couplings and can be considered to reflect the general validity of the adiabatic principle, which precludes efficient energy transfer between degrees of freedom with very different velocities.

A central question arises, however, whether these VP rates would be significantly enhanced if the coupling between rotational and vibrational degrees of freedom is taken into account. In other words, can rotations assist vibrational predissociation in van der Waals molecules? A partial answer to this question has been provided by Ashton and Child[26] by a numerical study of the Ar$\cdots$HCl molecule. They have shown that for this highly anisotropic system the most efficient channels for vibrational predissociation are indeed those involving large changes in rotational quantum number. However, the rates were still very low ($\sim 1\ \mathrm{sec}^{-1}$), being far from the experimentally accessible domain.

An approximate analytic method to calculate VP rates for weak-coupling complexes has been recently provided.[65] This approach rests on the result that the VP rates are essentially determined by the shape of the potential in the region of the attractive well where it can be accurately fitted by Morse functions, providing fully analytic expressions for energies and widths for various spatial configurations. The starting point is the Hamiltonian for the three-dimensional nuclear motion of a triatomic system which, after separation of the center-of-mass motion of the whole system, may be written as

$$H=\frac{\hbar^2}{2\mu_{\mathrm{X,BC}}}\left[-\frac{d^2}{\partial R_{\mathrm{X,BC}}^2}+\frac{\mathbf{l}^2}{R_{\mathrm{X,BC}}^2}\right]+\frac{\hbar^2}{2\mu_{\mathrm{BC}}}\left[-\frac{d^2}{dR_{\mathrm{BC}}^2}+\frac{\mathbf{j}^2}{R_{\mathrm{BC}}^2}\right]$$

$$+U(R_{\mathrm{X,BC}},R_{\mathrm{BC}},\gamma)+V_{\mathrm{BC}}(R_{\mathrm{BC}}) \tag{D2}$$

where as before $R_{\mathrm{X,BC}}$ is the distance between X and the center of the mass of the diatomic molecule BC, R_{BC} is the internuclear distance of the diatomic, while $\mu_{\mathrm{X,BC}}$ and μ_{BC} are the reduced masses, Eq. (2.B 3). In addition there are now two centrifugal terms corresponding to orbital angular momentum $\mathbf{l}$ of the atom X with respect to the center of mass of the molecule BC and the rotational angular momentum $\mathbf{j}$ of the BC molecule. The intermolecular potential is now also dependent on the angle γ between the two vectors $\mathbf{R}_{\mathrm{X,BC}}$ and $\mathbf{R}_{\mathrm{BC}}$. A convenient representation for the intermolecular potential is obtained by an expansion in terms of Legendre polynomials $P_\lambda(\cos\gamma)$ and a power series in the diatomic displacement from the equilibrium distance $\overline{R}_{\mathrm{BC}}$:

$$U(R_{\mathrm{X,BC}},R_{\mathrm{BC}},\gamma)=\sum_{m,\lambda}U_{\lambda m}(R_{\mathrm{X,BC}})\left(R_{\mathrm{BC}}-\overline{R}_{\mathrm{BC}}\right)^m P_\lambda(\cos\gamma) \tag{D3}$$

A zero-order vibrational basis set is then defined by taking the total wavefunction $\psi(\mathbf{R}_{X,BC},\mathbf{R}_{BC})$ in the form:

$$\Psi_{v_1\xi}(\mathbf{R}_{X,BC},\mathbf{R}_{BC})=\chi_{v_1}(R_{BC})\varphi_\xi(\mathbf{R}_{X,BC},\hat{R}_{BC}) \tag{D4}$$

where $\chi_{v_1}(R_{BC})$ is the eigenfunction of the Schrödinger equation corresponding to the vibration of the "free" BC molecule

$$\left[-\frac{\hbar^2}{2\mu_{BC}}\frac{d^2}{dR_{BC}^2}+V_{BC}(R_{BC})\right]\chi_{v_1}(R_{BC})=W_{v_1}\chi_{v_1}(R_{BC}) \tag{D5}$$

with v_1 being the vibrational quantum number, while $\varphi_\xi(\mathbf{R}_{X,BC},\hat{\mathbf{R}}_{BC})$, where $\hat{R}_{BC}$ is the unit vector associated to $\mathbf{R}_{BC}$, is the solution of

$$\left[\frac{\hbar^2}{2\mu_{X,BC}}\left(-\frac{d^2}{dR_{X,BC}^2}+\frac{\mathbf{l}^2}{R_{X,BC}^2}\right)+B\mathbf{j}^2+U\left(R_{X,BC},\bar{R}_{BC},\gamma\right)\right]\varphi_\xi=E_\xi\varphi_\xi \tag{D6}$$

with $B=\hbar^2/(2\mu_{BC}\bar{R}_{BC}^2)$ and ξ being a collective quantum number to be specified.

In order to further specify the wavefunction φ_ξ it is convenient to take a rotational basis set in the "body-fixed"[66] reference system with the z axis lying along the $\mathbf{R}_{X,BC}$ vector. The basis set is

$$\langle\hat{R}_{X,BC},\hat{R}_{BC}|JMj\Omega\rangle=\left[\frac{2J+1}{4\pi}\right]^{1/2}D_{M\Omega}^{J*}(\rho,\theta,0)Y_{j\Omega}(\gamma,\Psi) \tag{D7}$$

where J is the quantum number associated with the total angular momentum $\mathbf{J}=\mathbf{l}^2+\mathbf{j}$, M is the projection of $\mathbf{J}$ on the laboratory z axis, j is the quantum number associated to $\mathbf{j}$, while Ω is the tumbling[66c] angular momentum quantum number associated to the projection of $\mathbf{J}$ on the body-fixed z axis (i.e., on $\mathbf{R}_{X,BC}$). The function $D_{M\Omega}^{J}(\rho,\theta,0)$ is the Wigner-rotation function[67] with θ and ρ being the polar angles of $\mathbf{R}_{X,BC}$ with respect to the laboratory system of reference, while $Y_{j\Omega}(\gamma,\Psi)$ is a spherical harmonic function depending on the polar angles γ and Ψ of $\mathbf{R}_{BC}$ with respect to the body-fixed reference system. All the calculations are performed for fixed values of the quantum numbers J and M, since they are constants of the motion. On the other hand, j and Ω are not strictly good quantum numbers. For example, the operator $\mathbf{l}^2$ is diagonal with respect to

j but nondiagonal with respect to Ω and its matrix elements are given by[66]

$$\langle JMj\Omega|\mathbf{l}^2|JMj\Omega'\rangle = \left[J(J+1)+j(j+1)-2\Omega^2\right]\delta_{\Omega\Omega'} - \{J(J+1)-\Omega(\Omega\pm 1)\}^{1/2}\{j(j+1)-\Omega(\Omega\pm 1)\}^{1/2}\delta_{\Omega\Omega'\pm 1} \tag{D8}$$

The intermolecular potential $U(R_{\mathrm{X,BC}}, R_{\mathrm{BC}}, \gamma)$ is diagonal with respect to but nondiagonal with respect to j. Using the expansion (D3), its matrix elements are given by those of the Legendre polynomials[66]

$$\langle JMj\Omega|P_\lambda|JMj'\Omega\rangle = \left[\frac{2j'+1}{2j+1}\right]^{1/2} C(j'0\lambda 0|j'\lambda j0)C(j'\Omega\lambda 0|j'\lambda j\Omega) \tag{D9}$$

where $C(j_1m_1j_2m_2|j_1j_2j_3m_3)$ are Clebsh-Gordan coefficients.

Neglecting all the off-diagonal coupling terms, a zero-order ro-vibrational wavefunction may be written as

$$\Psi^{(JM)}_{v_1\xi j\Omega}(\mathbf{R}_{\mathrm{X,BC}}, \mathbf{R}_{\mathrm{BC}}) = \chi_{v_1}(R_{\mathrm{BC}})\phi^{(Jj\Omega)}_{\xi}(R_{\mathrm{X,BC}})\langle \hat{R}_{\mathrm{X,BC}}, \hat{R}_{\mathrm{BC}}|JMj\Omega\rangle \tag{D10}$$

where $\phi^{(JMj\Omega)}_{\xi}$ is the eigenfunction of the effective Hamiltonian

$$H^{(JMj\Omega)} = -\frac{\hbar^2}{2\mu_{\mathrm{X,BC}}}\frac{d^2}{dR^2_{\mathrm{X,BC}}} + V^{(JMj\Omega)}(R_{\mathrm{X,BC}}) \tag{D11}$$

with

$$V^{(JMj\Omega)} = \langle JMj\Omega|V(R_{\mathrm{X,BC}}, \bar{R}_{\mathrm{BC}}, \gamma)|JMj\Omega\rangle + \langle JMj\Omega|\mathbf{l}^2|JMj\Omega\rangle/2\mu_{\mathrm{X,BC}}R^2_{\mathrm{X,BC}} \tag{D12}$$

This Hamiltonian has discrete as well as continuum energy spectra corresponding to bound and dissociative states of the van der Waals molecule, respectively. Denoting by $\xi = v_2$ the vibrational quantum number associated to the discrete spectrum and by $\xi = \epsilon$ the continuum quantum number corresponding to the relative kinetic energy of the fragments, the zero-order basis set, Eq. (D10), will then consist of:

1. Discrete bound wavefunctions

$$|\Psi^{(JM)}_{v_1v_2j\Omega}\rangle = |\chi_{v_1}\rangle|\phi^{(Jj\Omega)}_{v_2}\rangle|JMj\Omega\rangle \tag{D13a}$$

with total energy $E_{v_1 v_2 J j \Omega} = W_{v_1} + Bj(j+1) + \epsilon_{v_2}^{(Jj\Omega)}$ where $\epsilon_{v_2}^{(Jj\Omega)}$ is the eigenvalue of the $H^{(Jj\Omega)}$ Hamiltonian of Eq. (13).

2. Continuum dissociative wavefunctions

$$|\Psi_{v_1', \epsilon' j' \Omega'}^{(J'M')}\rangle = |\chi_{v_1'}\rangle |\phi_{\epsilon'}^{(J'j'\Omega')}\rangle |J'M'j'\Omega'\rangle \tag{D13b}$$

with total energy

$$E_{v_1', \epsilon' j'} = W_{v_1'} + Bj'(j'+1) + \epsilon',$$

where ϵ' is the relative kinetic energy of the fragments.

The linewidths for vibrational predissociation is now given by the familiar Golden-rule formula,

$$\Gamma_{v_1 v_2 j \Omega}^{(JM)} = \pi \sum_{v_1' < v_1} \sum_{j'\Omega'} |\langle \Psi_{v_1 v_2 j \Omega}^{(JM)} | U | \Psi_{v_1' \epsilon' j' \Omega'}^{(JM)} \rangle|^2 \tag{D14}$$

provided that the continuum eigenfunction is energy normalized.

In order to obtain analytic expressions for the matrix elements in Eq. (D14) all the functions $U_{\lambda 0}(R_{\mathrm{X,BC}})$ in Eq. (D3) were fitted in terms of two exponential terms:

$$U_{\lambda 0}(R_{\mathrm{X,BC}}) = A_\lambda \exp(-2\alpha R_{\mathrm{X,BC}}) + B_\lambda \exp(-\alpha R_{\mathrm{X,BC}}) \tag{D15}$$

The distorted potentials $\langle JMj\Omega | U | JMj\Omega \rangle$ are then given by

$$\langle JMj\Omega | U | JMj\Omega \rangle = \sum_\lambda \langle JMj\Omega | P_\lambda | JMj\Omega \rangle \left[A_\lambda \exp(-2\alpha R_{\mathrm{X,BC}}) + B_\lambda \exp(-\alpha R_{\mathrm{X,BC}}) \right] \tag{D16}$$

These potentials have a minimum for

$$\bar{R}_{\mathrm{X,BC}}^{(j\Omega)} = \alpha^{-1} \ln \left[\frac{2 \sum_\lambda \langle JMj\Omega | P_\lambda | JMj\Omega \rangle A_2}{\sum_\lambda \langle JMj\Omega | P_\lambda | JMj\Omega \rangle B_\lambda} \right] \tag{D17}$$

Expanding now the centrifugal term in Eq. (D12) around this minimum with the same exponentials:

$$R_{\mathrm{X,BC}}^{-2} \simeq a_{j\Omega} \exp\left[-2\alpha\left(R_{\mathrm{X,BC}} - \bar{R}_{\mathrm{X,BC}}^{(j\Omega)} \right) \right] + b_j \exp\left[-\alpha\left(R_{\mathrm{X,BC}} - \bar{R}_{\mathrm{X,BC}}^{(j\Omega)} \right) \right] \tag{D18}$$

and imposing the condition that the functions on the two sides of Eq. (D18) as well as their first derivatives will be equal at $R_{X,BC}=\bar{R}^{(j\Omega)}_{X,BC}$, one gets

$$a_{j\Omega}=-\left\{\bar{R}^{(j\Omega)}_{X,BC}\right\}^{-2}\left\{1-2/\alpha\bar{R}^{(j\Omega)}_{X,BC}\right\} \tag{D19a}$$

$$b_{j\Omega}=2\left\{\bar{R}^{(j\Omega)}_{X,BC}\right\}^{-2}\left\{1-1/\alpha\bar{R}^{(j\Omega)}_{X,BC}\right\} \tag{D19b}$$

Introducing now Eqs. (D16) and (D18) into (D12) we obtain

$$V^{(Jj\Omega)}=A'_{Jj\Omega}\exp(-2\alpha R_{X,BC})+B'_{Jj\Omega}\exp(-\alpha R_{X,BC}) \tag{D20}$$

with

$$A'_{Jj\Omega}=\sum_{\lambda}A_{\lambda}\langle JMj\Omega|P_{\lambda}|JMj\Omega\rangle + \frac{\hbar^2}{2\mu_{X,BC}}a_{j\Omega}\exp\left(2\alpha\bar{R}^{(j\Omega)}_{X,BC}\right)\langle JMj\Omega|\mathbf{l}^2|JMj\Omega\rangle \tag{D21a}$$

$$B'_{Jj\Omega}=\sum_{\lambda}B_{\lambda}\langle JMj\Omega|P_{\lambda}|JMj\Omega\rangle+\frac{\hbar^2}{2\mu_{X,BC}}b_{j\Omega} \times\exp\left(\alpha\bar{R}^{(j\Omega)}_{X,BC}\right)\langle JMj\Omega|\mathbf{l}^2|JMj\Omega\rangle \tag{D21b}$$

The potentials $V^{(Jj\Omega)}$ defined in Eq. (D20) correspond now to an effective Morse potential of the form

$$V^{(Jj\Omega)}=D_{Jj\Omega}\left\{\exp\left[-2\alpha\left(R_{X,BC}-\bar{R}'^{(Jj\Omega)}_{X,BC}\right)\right] -2\exp\left[-\alpha\left(R_{X,BC}-\bar{R}'^{(Jj\Omega)}_{X,BC}\right)\right]\right\} \tag{D22}$$

with

$$D'_{Jj\Omega}=B'^2_{Jj\Omega}/4A'_{Jj\Omega} \tag{D23a}$$

and

$$D'_{Jj\Omega}=B'^2_{Jj\Omega}/4A'_{Jj\Omega} \tag{D23a}$$

The parameters of this effective potential are determined by the rotational state.

The discrete and continuum wavefunctions for the Morse oscillator are well known. What remains to be done is the calculation of the discrete-continuum matrix elements $\langle\Psi_{v_1v_2j\Omega}|V|\Psi_{v_2'\epsilon'j'\Omega'}\rangle$ which are required in Eq. (D14). Using Eqs. (D3) and (D10) we have

$$\langle\Psi_{v_1v_2j\Omega}|V|\Psi_{v_1'\epsilon'j'\Omega'}\rangle=\sum_{n,\lambda}\langle\phi_{v_2}^{(Jj\Omega)}|U_{\lambda n}|\phi_{\epsilon'}^{(Jj'\Omega')}\rangle$$
$$\langle JMj\Omega|P_\lambda|JMj'\Omega'\rangle\langle\chi_{v_1}|\left(R_{BC}-\bar{R}_{BC}\right)^n|\chi_{v_1'}\rangle \tag{D24}$$

The matrix elements $\langle\chi_{v_1}|(R_{BC}-\bar{R}_{BC})^n|\chi_{v_1'}\rangle$ are well known. The discrete-continuum van der Waals matrix elements $\langle\phi_{v_2}^{(Jj\Omega)}|U_{\lambda n}|\phi_{\epsilon'}^{(Jj'\Omega')}\rangle$ can be evaluated analytically if the functions $U_{\lambda n}$ are fitted by a sum of exponentials[68]

$$U_{\lambda n}(R_{X,BC})=\sum_k C_{\lambda n}^{(k)}\exp\left(-\alpha_{\lambda n}^{(k)}R_{X,BC}\right) \tag{D25}$$

The final result is[63]

$$\langle\phi_{v_2}^{(Jj\Omega)}|U_{\lambda n}|\phi_{\epsilon'}^{(Jj'\Omega')}\rangle=\sum_k C_{\lambda n}^{(k)}\langle\phi_{v_2}^{(Jj\Omega)}|\exp\left(-\alpha_{\lambda n}^{(k)}R_{X,BC}\right)|\phi_{\epsilon'}^{(Jj'\Omega')}\rangle \tag{D26}$$

with

$$\begin{aligned}\langle\phi_{v_2}|\exp(-\eta R_{X,BC})|\phi_\epsilon'\rangle=&(2K)^{-\eta/\alpha}\exp\left(-\eta\bar{R}_{X,BC}\right)\\&\times(\alpha\pi\hbar)^{-1}\left[\frac{v_2!(2K-2v_2-1)}{\Gamma(2K-v_2)}\right]^{1/2}\\&\times(-1)^{v_2}\{\mu_{X,BC}\sinh(2\pi\theta_\epsilon)\}^{1/2}\\&\times|\Gamma\left(\tfrac{1}{2}-K'-i\theta_\epsilon\right)|\sum_{m=0}^{v_2}Q_m^{(v_2)}F_m^{(\eta/\alpha)}\end{aligned} \tag{D27}$$

where

$$\begin{aligned}F_m^{(\eta/\alpha)}=&\left[\frac{z}{1+\beta}\right]^{K-m+\eta/\alpha-\frac{1}{2}}\frac{\Gamma\left(-i\theta_\epsilon+K-m+\eta/\alpha-\frac{1}{2}\right)}{\Gamma\left(i\theta_\epsilon-K'+\frac{1}{2}\right)}\\&\times\left[\frac{2\beta}{1+\beta}\right]^{i\theta_\epsilon}\sum_{p=0}^{\infty}\frac{\Gamma\left(i\theta_\epsilon+K-m+\eta/\alpha-\frac{1}{2}+p\right)}{\Gamma(K-m+\eta/\alpha-K'+p)}\\&\times\Gamma\left(i\theta_\epsilon-K'+\tfrac{1}{2}+p\right)\frac{1}{p!}\left[\frac{1-\beta}{1+\beta}\right]^p\end{aligned} \tag{D28}$$

and

$$Q_m^{(v_2)} = \frac{(2K - v_2)(-1)^{v_2 - m}}{m!(v_2 - m)!(2K - v_2 - m)} \tag{D29}$$

$$K = (2\mu_{X,BC} D)^{1/2}/\hbar\alpha \tag{D30}$$

$$\theta_\epsilon = (2\mu_{X,BC}\epsilon)^{1/2}/\hbar\alpha \tag{D31}$$

$$\beta = (K'/K)e^{-\alpha(\bar{R} - \bar{R}')} \tag{D32}$$

Application of these analytic expressions has been made to the van der Waals molecules $X \cdots H_2$ (X = Ar, Kr, Xe) for which 3-D potential energy surfaces exist.[69] The $U_{\lambda n}$ potential functions of Eq. (D3) have been provided in terms of Buckingham-Corner functions

$$U_{\lambda n} = G_{\lambda n} e^{-\beta R_{X,BC}} - \{C_8^{(\lambda n)}/R_{X,BC}^8 + C_6^{(2n)}/R_{X,BC}^6\} D(R_{X,BC}) \tag{D33}$$

with

$$\begin{aligned} &\exp\{-4(\mathrm{Re}/R_{X,BC} - 1)\} && \text{for} \quad R_{X,BC} < Re \\ &D(R_{X,BC}) = 1 && \text{for} \quad R_{X,BC} \geqslant Re \end{aligned} \tag{D34}$$

The potential energy surface has been provided in terms of four terms corresponding to $\lambda = 0, 2$ and $n = 0, 1$. The initial angular momentum quantum numbers were specified by $J = 0, j = 0$. The process studied is then

$$X \cdots H_2(v_1 = 1, j = 0) \to X + H_2(v_1' = 0, j') \tag{D35}$$

For $j \neq 0$ we would have, in addition to the vibrational predissociation channels, other channels corresponding to pure rotational predissociation

$$X \cdots BC(v_1, j) \to X + BC(v_1, j' < j) \tag{D36}$$

In Table XIII we present the results of these calculations for the process given by Eq. (D35). For the potential energy surface used here only two

TABLE XIII
VP Lifetimes of $X \cdots H_2$ Complexes From $j = 0$ to Different Final j' Rotational States

Molecule	Γ_0 (cm^{-1})	Γ_2 (cm^{-1})	Γ_{tot} (cm^{-1})	τ (sec)
$Ar \cdots H_2$	1.5 10	1.6 10	1.7 10	16
$Kr \cdots H_2$	4.4 10	6.4 10	6.4 10	4
$Xe \cdots H_2$	1.7 10	2.3 10	2.5 10	107

channels ($j'=0,2$) are effective in inducing predissociation. We denote by the partial width associated to each one of these channels. The total width is

$$\Gamma_{tot} = \sum_{j'} \Gamma_{j'} \tag{D37}$$

and the predissociation lifetime is given by $\tau = \hbar/2\Gamma_{tot}$.

The first conclusion which emerges from an inspection of Table 13 is that the predissociation lifetimes are rather long, ranging from a few to a hundred microseconds. This is in agreement with the estimates provided by the colinear[27–31] and isotropic calculations.[25] Second, the effects of rotation are of interest. For ArH_2 the result obtained in the isotropic approximation, i.e., $\Gamma_0 = 1.5\ 10^{-7}\ cm^{-1}$, is almost identical to the total width since Γ_2 is much smaller than Γ_0 in this case. For $Kr\cdots H_2$ and $Xe\cdots H_2$, on the contrary, the channel involving the excitation of the rotational state $j'=2$ of H_2 is the most efficient in inducing the predissociation. As a result the lifetimes for the molecules are one or two order of magnitude smaller than the corresponding "isotropic" results, which neglects rotational effects. Third, the low VP lifetimes ($\sim 10^3\ sec^{-1}$) raise the question whether alternative dynamic channels may be important. Recently[68] it has been shown that for the molecules considered here rotational predissociation rates may be of the order of $10^9 - 10^8\ sec^{-1}$. The vibrational predissociation rates are much smaller and therefore the major decay channels for the adiabatic predissociation process of $X\cdots H_2$ complexes involves predissociation by rotation rather than VP.

2. *The Strong-Coupling Case*

In this case the anisotropy of the potential in the region of the minimum exceeds the rotational spacings and the approximations used above are not valid. For molecules like $He\cdots I_2$, characterized by very slow rotational motion of the I_2 molecule, an approximation[48] has been implemented.[70] Starting from Eqs. (D2), (D4), (D6), the internuclear distance $R_{X,BC}$ in the centrifugal term in Eq. (D6) is replaced by the equilibrium distance $\bar{R}_{X,BC}$ for the bound states of the complex. The functions φ_{v_2} are then written as

$$\varphi_{v_2}(\mathbf{R}_{X,BC}, \hat{\mathbf{R}}_{BC}) = \phi_{v_2}(R_{X,BC};\gamma) F_{v_2}(\hat{R}_{X,BC}, \hat{R}_{BC}) \tag{D38}$$

where $\phi_{v_2}(R_{X,BC};\gamma)$ is the solution of

$$\left[-\frac{\hbar^2}{2\mu_{X,BC}} \frac{d^2}{dR_{X,BC}^2} + V(R_{X,BC}, \gamma) \right] \phi_{v_2}(R_{X,BC} i\gamma) = W_{v_2}(\gamma) \phi_{v_2}(R_{X,BC} i\gamma) \tag{D39}$$

Notice that Eq. (D38) is a rotational "adiabatic" function depending parametrically on γ. The zero-order adiabatic approximation is now

$$\mathbf{l}^2\varphi=\phi\mathbf{l}^2F \tag{D40a}$$

$$\mathbf{j}^2\varphi=\phi\mathbf{j}^2F \tag{D40b}$$

which is equivalent to neglecting the effect of the angular momentum operators on the vibrational wavefunction $\phi(R_{\mathrm{X,BC}};\gamma)$. Introducing now Eq. (D38) into (D6) and using (D40) we obtain, after multiplication by ϕ^* and integration over $R_{\mathrm{X,BC}}$,

$$\left[b\mathbf{l}^2+B\mathbf{j}^2+W_{v_2}(\gamma)\right]F_{v_2,v_3}=\epsilon_{v_2v_3}F_{v_2,v_3} \tag{D41}$$

with

$$b=\hbar^2/2\mu_{\mathrm{X,BC}}\overline{R}^2_{\mathrm{X,BC}} \tag{D42}$$

The new quantum number v_3 appearing in Eq. (D41) corresponds to the bending (libration or hindered rotation) of the complex. The solution of Eq. (D41) is straightforward by expansion of F in any complete rotational basis set and diagonalization. Using the body-fixed[66] reference system the basis set as specified by Eq. (D7). The continuum wavefunctions are determined by the use of the infinite-order sudden approximation.[48] In this method Eq. (D6) is simplified by making the approximations

$$\mathbf{j}^2\Psi=j_0(j_0+1)\Psi \tag{D43a}$$

$$\mathbf{l}^2\Psi=l_0(l_0+1)\Psi \tag{D43b}$$

where j_0 and l_0 are constants whose choice depends on the initial conditions. In our case we are dealing with low rotational excitation so that j_0 and l_0 may be chosen to be zero. We then write the functions φ as

$$\varphi(\mathbf{R}_{\mathrm{X,BC}},\hat{R}_{\mathrm{BC}})=\phi_\epsilon(R_{\mathrm{X,BC}};\gamma)\langle\hat{R}_{\mathrm{X,BC}},\hat{R}_{\mathrm{BC}}|JMj\Omega\rangle \tag{D44}$$

with ϕ_ϵ obeying the Schrödinger equation:

$$\left[-\frac{\hbar^2}{2\mu_{\mathrm{X,BC}}}\frac{d^2}{dR^2_{\mathrm{X,BC}}}+U(R_{\mathrm{X,BC}},\gamma)\right]\phi_\epsilon(R_{\mathrm{X,BC}};\gamma)=\epsilon\phi_\epsilon(R_{\mathrm{X,BC}};\gamma) \tag{D45}$$

where ϵ is the relative kinetic energy of the fragments. Using these wavefunctions the final expression of the linewidth is

$$\Gamma_{v_1v_2v_3}=\pi\sum_{v_1'j'}\left|\sum_{j\Omega}C_{j\Omega}^{(v_1v_3)}\langle JMj\Omega|V_{v_1v_2,v_1'\epsilon}^{\mathrm{d-c}}(\gamma)|JMj'\Omega\rangle\right|^2 \tag{D46}$$

where

$$V_{v_1v_2,v_1'\epsilon}^{\mathrm{d-c}}=\langle\phi_{v_2}\chi_{v_1}|U(R_{\mathrm{X,BC}},R_{\mathrm{BC}},\gamma)|\phi_\epsilon\chi_{v_1'}\rangle \tag{D47}$$

is a "distorted-wave" discrete-continuum coupling. These coupling terms are similar to those of the fixed-configuration calculations.[27-31] In particular for $\gamma=0$, Eq. (D47) is the expression for the "distorted-wave" discrete-continuum coupling in the colinear model.

Finally, the distribution of vibrational and rotational states of the diatomic fragment BC will be given by

$$P_{v_1v_2v_3\to v_1'j'}=\frac{\pi\left|\sum_{j\Omega}C_{j\Omega}^{(v_2v_3)}\langle JMj\Omega|V_{v_1v_2,v_1'\epsilon}^{\mathrm{d-c}}|JMj'\Omega\rangle\right|^2}{\Gamma_{v_1v_2v_3}} \tag{D48}$$

The method has been applied to the HeI_2 molecule in the electronically excited $B^3\pi$ configuration. The van der Waals interaction was specified in terms of two pairwise interactions between the He atom and each one of the iodine atoms, i.e.,

$$U(R_{\mathrm{X,BC}},R_{\mathrm{BC}},\gamma)=U_{\mathrm{HeI}}(R_{\mathrm{XB}})+U_{\mathrm{HeI}}(R_{\mathrm{XC}}) \tag{D49}$$

where the interparticle distances R_{XB} and R_{XC} are related to the dissociation coordinates $R_{\mathrm{X,BC}}$, R_{BC}, γ by the trigonometric relations

$$R_{\mathrm{XB}}=\left[R_{\mathrm{X,BC}}^2+\xi_{\mathrm{C}}^2R_{\mathrm{BC}}^2-2\xi_{\mathrm{C}}R_{\mathrm{BC}}R_{\mathrm{X,BC}}\cos\gamma\right]^{1/2} \tag{D50a}$$

$$R_{\mathrm{XC}}=\left[R_{\mathrm{X,BC}}^2+\xi_{\mathrm{B}}^2R_{\mathrm{BC}}^2+2\xi_{\mathrm{B}}R_{\mathrm{BC}}R_{\mathrm{X,BC}}\cos\gamma\right]^{1/2} \tag{D50b}$$

with

$$\xi_r=\frac{m_r}{m_{\mathrm{B}}+m_{\mathrm{C}}}\qquad r=B\ \text{ or }\ \mathrm{C} \tag{D51}$$

The individual interaction U_{HeI} has been specified by a Morse potential

$$U_{\mathrm{HeI}}(R)=D\{\exp[-2\alpha(R-R_0)]-2\exp[-\alpha(R-R_0)]\} \tag{D52}$$

The calculations have been performed for two sets of values of potential parameters D, α, and R_0 which were used in previous work.[27–31] In Figs. 21 and 22 the linewidths for the $v_2 = v_3 = 0$ level are presented as a function of the vibrational quantum number v_1 of the I_2 bond. The results denoted 3-Dimensional are those obtained with the present method and they are compared with those for the colinear and perpendicular models.[27–31] The effect of rotations is to narrow the line, the colinear model giving the largest linewidths.

Finally, in Fig. 23 the final rotational distribution of the I_2 fragments for the two sets of parameters are given. The distribution has its maximum for $j' = 2$ and a width of about 2, being broader for the potential with larger D

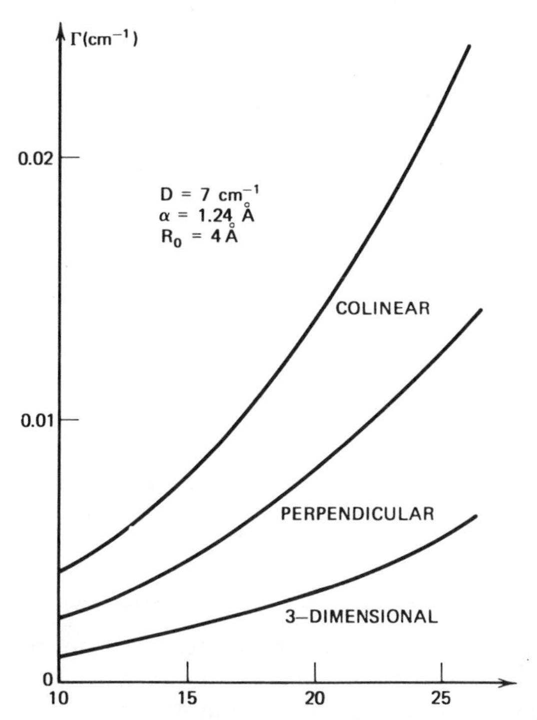

Fig. 21. Vibrational predissociation linewidths for the $HeI_2(B)$ molecule as a function of the stretching quantum number v_1 corresponding to the vibration of the I_2 bond. The other quantum numbers v_2 and v_3 as well as the total angular momentum J are equal to zero. The parameters of the intermolecular potential energy surface are $D = 7$ cm^{-1}, $\alpha = 1.24$ Å^{-1}, $R_0 = 4$ Å and $\bar{R}_{BC} = 3.016$ Å. The frequency and anharmonicity of the I_2 bond are taken to be $\omega_e = 125.273$ cm^{-1}, $\omega_e\chi_e = 0.7016$ cm^{-1}. The curve labeled *3-Dimensional* correspond to the results of the approximate treatment for treating the effect of rotations. The curves labeled *colinear* and *perpendicular* are the results obtained for fixed angular configurations $\gamma = 0$ and $\gamma = \pi/2$.

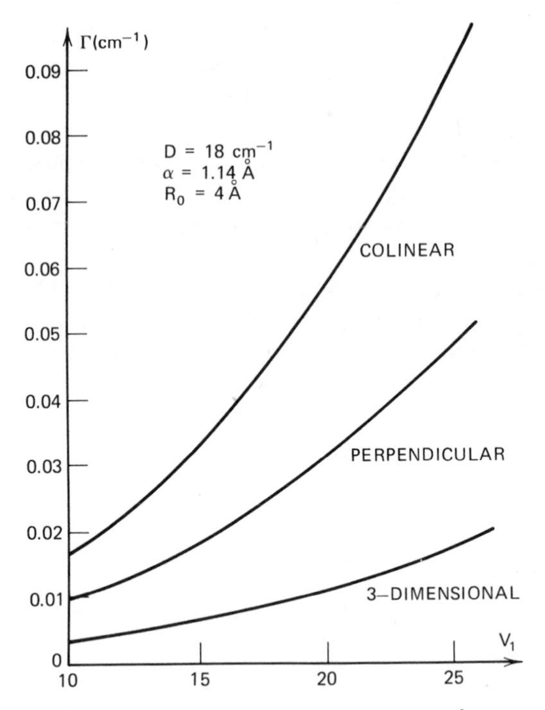

Fig. 22. Same as Fig. 21 for the parameters $D=18\ \text{cm}^{-1}$, $\alpha=1.14\ \text{Å}^{-1}$.

(i.e., with larger anisotropy). The fraction of energy going to rotation is thus very small owing to the very small rotational constant of the I_2 fragment. For the distributions of Fig. 22 the rotational energy is just ~0.3 cm^{-1}. We can conclude that for the case of $\text{He}\cdots I_2$ the adiabatic predissociation involves essentially a $V \to T$ process and rotational effects are minor.

E. VP of VdW Complexes Containing Triatomic Molecules

We have applied[31] the formalism of Section IV to perform detailed numerical calculations of the dynamics of model systems corresponding to the colinear and perpendicular VP of atom–linear triatomic and of linear triatomic–linear triatomic vdW molecules. The relevant schemes are portrayed in Fig. 24, which will be now discussed.

1. *The Colinear* $\text{He}\cdots N_2O$ *Molecule*

The intermolecular interaction is represented in this case by a simple atom-atom interaction between the nearest atoms of the two molecules,

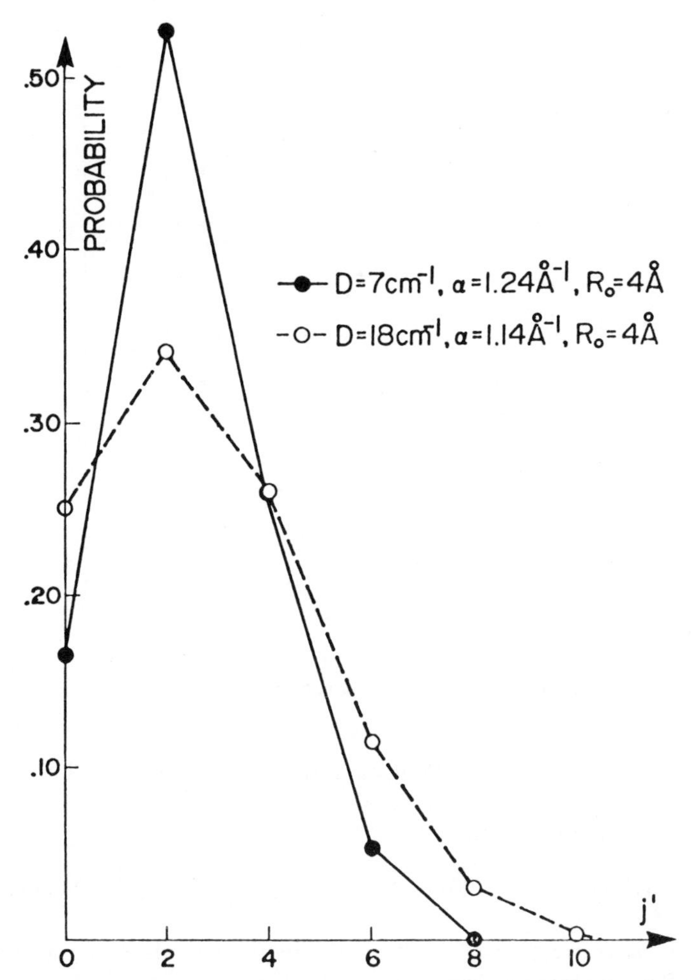

Fig. 23. Final states rotational distribution of the $I_2(B)$ fragments in the vibrational predissociation of the $HeI_2(B)$ molecule. The quantum numbers of the predissociated state are: $V_1=25$, $v_2=0$, $v_3=3$, and $J=0$. The values of the parameters D, α, and R_0 are indicated in the figure. All other constants are as in Fig. 21.

i.e., between atoms 3 and 1. This is sufficient for the understanding of the general qualitative behavior of the VP dynamics, in view of our poor knowledge of the details of the interaction potential. In the normal mode description of N_2O there will then be no coupling between the bending mode and the other degrees of freedom. We consider an initial state after photon absorption corresponding to the excitation of the mode v_3 of N_2O,

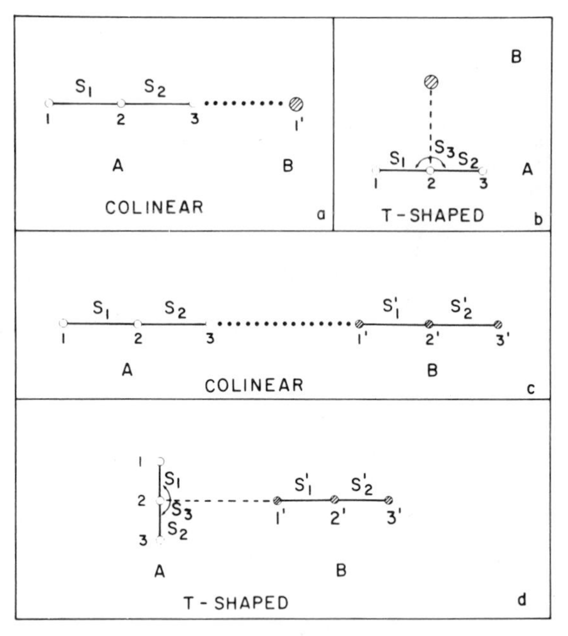

Fig. 24. Model systems of linear and *T*-shaped van der Waals molecule of atom–linear triatomic complex and a dimer of two triatomic molecules.

the other two modes being in their ground states. This state will be denoted by (001); i.e., $v_1 = 0$, $v_2 = 0$ and $v_3 = 1$.

Two VP processes are then possible:

$$V \rightarrow T;\ \mathrm{He} \cdots \mathrm{N_2O}(001) \rightarrow \mathrm{He} + \mathrm{N_2O}(000), \qquad \Delta E = 2224\ \mathrm{cm} \qquad \text{(E1)}$$

involving the conversion of vibrational energy in mode 3 to translational energy of the fragments, and

$$V \rightarrow V + T;\ \mathrm{He} \cdots \mathrm{N_2O}(001) \rightarrow \mathrm{He} + \mathrm{N_2O}(100), \qquad \Delta E = 939\ \mathrm{cm} \quad \text{(E2)}$$

corresponding to the intramolecular degradation of the vibrational quantum in mode 3 to the lower frequency vibration v_1, the excess energy being transformed into translational kinetic energy. Reaction (28) corresponds to a second-order process and, therefore, the intramolecular coefficients $A^{(\rho)}$ Eq. (4.B.19) will be smaller. On the other hand the excess energy ΔE being smaller, in this case the intermolecular terms $B^{(\beta)}$ (see 4.B.20) will be larger according to the energy gap law.[27,37] The competition between these two opposite effects will finally determine which one of the two processes, (E1) or (E2) will provide the most efficient decay channel.

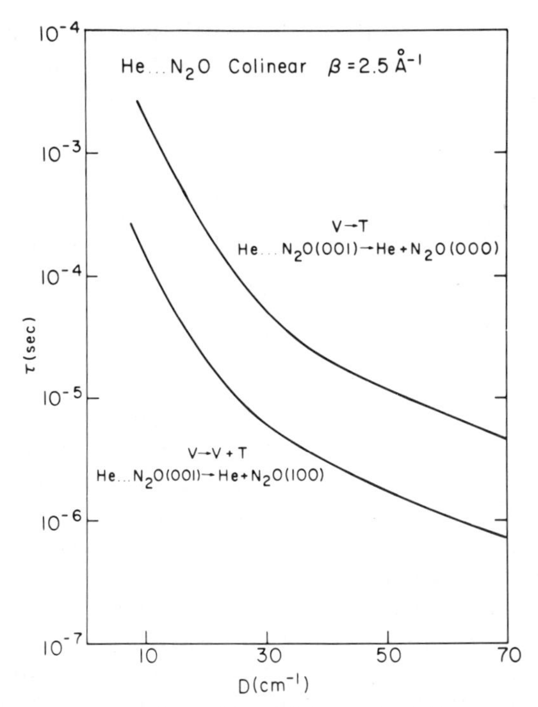

Fig. 25. Lifetimes for VP of collinear $He\cdots N_2O$ complex initially excited at (001) of the N_2O constituent. The potential parameters for the van der Waals bond, are $\beta = 2.5\ A^{-1}$, while the dissociation energy is varied in the range 10–70 cm^{-1}. Calculations were performed for the channels (E1) and (E2).

We have calculated the lifetimes, τ, for processes (E1) and (E2). The results for some typical intermolecular potential parameters are presented in Fig. 25. From the calculations two conclusions emerge. First, we note that the intramolecular $V \to V + T$ process is about an order of magnitude more efficient than the simple first order $V \to T$ process. Second, for reasonable parameters of the vdW bond the calculated VP lifetime for the more efficient $V \to V + T$ process is $\sim 10^{-5} - 10^{-6}$ sec.

2. *The T-shaped* $He\cdots N_2O$ *Molecule*

In this case we cannot get away by considering only the nearest-neighbor interatomic interaction. If we only consider the interaction between nearest atoms, i.e., the two atoms 2 and 1′ in Fig. 24, there will be no coupling between the intermolecular vibration and the Q_3 mode (in the harmonic approximation for N_2O). We have therefore considered both the interaction between the helium atom and the two end atoms of N_2O. We assume the same parameters β and D for all the pairwise Morse potentials. Now, we will have in addition to the $V \rightarrow T$ process (E1) and the intramolecular $V \rightarrow V+T$ process the following additional intramolecular $V \rightarrow V+T$ channels:

$$\mathrm{He}\cdots \mathrm{N_2O}(001) \rightarrow \mathrm{He} + \mathrm{N_2O}(010), \qquad \Delta E = 1635 \text{ cm} \tag{E3}$$

$$\mathrm{He}\cdots \mathrm{N_2O}(001) \rightarrow \mathrm{He} + \mathrm{N_2O}(020), \qquad \Delta E = 1046 \text{ cm} \tag{E4}$$

$$\mathrm{He}\cdots \mathrm{N_2O}(001) \rightarrow \mathrm{He} + \mathrm{N_2O}(030), \qquad \Delta E = 457 \text{ cm} \tag{E5}$$

$$\mathrm{He}\cdots \mathrm{N_2O}(001) \rightarrow \mathrm{He} + \mathrm{N_2O}(110), \qquad \Delta E = 350 \text{ cm} \tag{E6}$$

All of these new intramolecular $V \rightarrow V+T$ channels involve the bending excitation of the N_2O fragment. Process (E3) is a second-order process similar to the one considered before in Eq. (E2), the only difference being that the final state of N_2O involves bending rather than symmetric stretch vibration. Processes (E4) and (E6) are third-order processes in which two quanta are created, while process (E5) is a fourth-order process in which three quanta of the bending vibrational mode are excited. The lifetimes for all these processes as a function of the dissociation energy parameter D are plotted in Fig. 26. The following features of the numerical results should be noted. Firstly, for the T-shaped vdW molecule the first-order $V \rightarrow T$ process, Eq. (E1), is more efficient than any of the second-order $V \rightarrow V+T$ processes, Eqs. (E2)–(E6). The efficiency of the $V \rightarrow T$ channel for the T-shaped configuration is in contrast with the characteristics of the linear molecule where the $V \rightarrow V+T$ process, Eq. (E2), dominates the VP dynamics. Secondly, the VP lifetimes for the $V \rightarrow T$ channels of the T-shaped and of the linear configurations of the $\mathrm{He}\cdots \mathrm{N_2O}$ are very similar. Accordingly, the VP process of the T-shaped vdW molecule is less efficient by about one order of magnitude than that of the linear molecule.

3. *The Collinear* $(N_2O)_2$ *Dimer*

In complete analogy with the $\mathrm{He}\cdots \mathrm{N_2O}$ case, we have the $V \rightarrow T$ process, Eq. (E1), and the $V \rightarrow V+T$ process, Eq; (E2), with the He atom being replaced by another N_2O molecule in its ground state. However, the intermolecular dissociation energy parameter D is much larger in this case. Also, the reduced mass for intermolecular relative motion is larger. These two changes affect the VP lifetimes in opposite ways. While the increase of D usually decreases the VP lifetimes,[27] the increase of the reduced mass results in an increase of the VP lifetimes.[27,37] In Fig. 5 we have presented the two processes, the $V \rightarrow T$ process $(000)\cdots(001) \rightarrow (000)+(000)$ and the

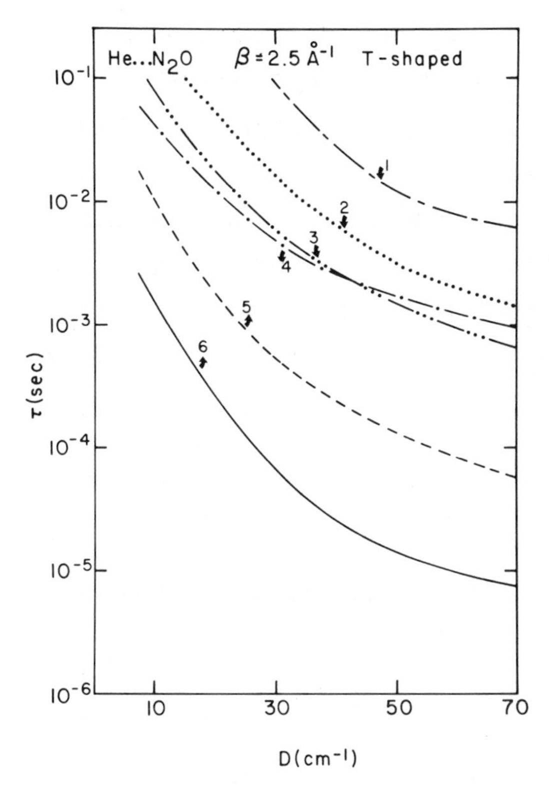

Fig. 26. Lifetimes for VP of a *T*-shaped He$\cdots N_2O$ complex initially excited at (001) of the N_2O constituent. Calculations were performed for the channels (E1)–(E6). The curves correspond to the following processes:

1. $V \rightarrow V+T$: He$\cdots N_2O(001) \rightarrow He+N_2O(030)$
2. $V \rightarrow V+T$: He$\cdots N_2O(001) \rightarrow He+N_2O(020)$
3. $V \rightarrow V+T$: He$\cdots N_2O(001) \rightarrow He+N_2O(100)$
4. $V \rightarrow V+T$: He$\cdots N_2O(001) \rightarrow He+N_2O(110)$
5. $V \rightarrow V+T$: He$\cdots N_2O(001) \rightarrow He+N_2O(010)$
6. $V \rightarrow T$: He$\cdots N_2O(001) \rightarrow He+N_2O(000)$

$V \rightarrow V+T$ process $(000)\cdots(001) \rightarrow (000)+(100)$, as a function of the dissociation energy parameter D. We note that the $V \rightarrow V+T$ process is faster than $V \rightarrow T$ predissociation in complete analogy with the situation for the collinear He$\cdots$N O molecule. On the other hand, the absolute values of the lifetimes for $(N_2O)_2$ collinear dimers are much larger than the corresponding values for collinear He$\cdots N_2O$, demonstrating the strong effect of the reduced mass on the VP dynamics.

4. *The T-shaped* $(N_2O)_2$ *Dimer*

In this case we have the following VP processes

$$(000)\cdots(001)\rightarrow(000)+(000), \qquad \Delta E=2224 \text{ cm} \tag{E7}$$

$$(000)\cdots(001)\rightarrow(010)+(000), \qquad \Delta E=1635 \text{ cm} \tag{E8}$$

$$(000)\cdots(001)\rightarrow(020)+(000), \qquad \Delta E=1046 \text{ cm} \tag{E9}$$

$$(000)\cdots(001)\rightarrow(030)+(000), \qquad \Delta E=457 \text{ cm} \tag{E10}$$

$$(000)\cdots(001)\rightarrow(000)+(100), \qquad \Delta E=939 \text{ cm} \tag{E11}$$

$$(000)\cdots(001)\rightarrow(010)+(100), \qquad \Delta E=350 \text{ cm} \tag{E12}$$

Equation (E1) corresponds to the usual $V \rightarrow T$ process, while Eqs. (E8)–(E10), correspond to intermolecular $V \rightarrow V+T$ processes in which vibrational energy in one molecule is converted in vibrational energy of the

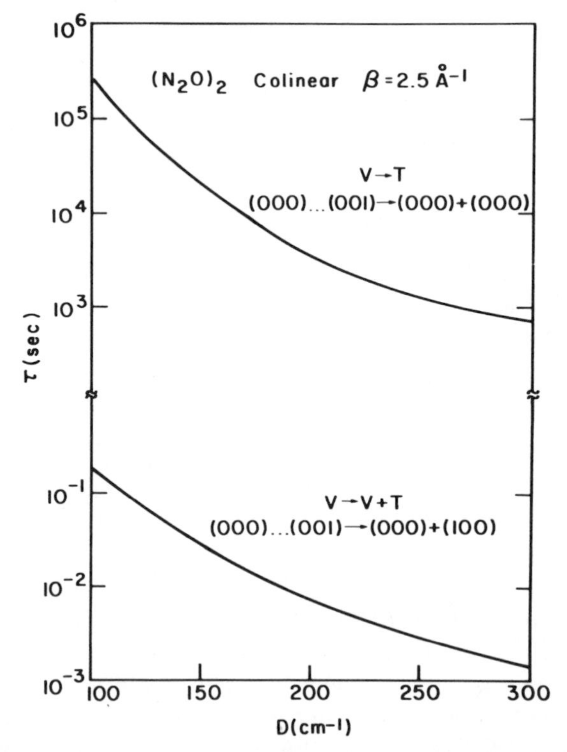

Fig. 27. Lifetimes for VP of the collinear $(N_2O)_2$ dimer initially excited at the (001) state of N_2O. The atom–atom potential parameters, are $\beta=2.5\ A^{-1}$, while the dissociation energy of the van der Waals bond was varied in the range $D=100$–$300\ \text{cm}^{-1}$. Calculations were performed for the $V \rightarrow T$ channel and for the intramolecular $V \rightarrow V+T$ channel.

other molecule forming the dimer. Equation (E11) corresponds to an intramolecular $V \to V+T$ process. Finally, Eq. (E12) describes a mixed $V \to V+T$ process in which the initial vibrational quantum in the mode v_3 is converted into one vibrational quantum in the mode v_1 of the same molecule and in an additional vibrational quantum in mode v_2 of the second N_2O molecule.

In Figs. 28 and 29 are presented the results of model calculations of the VP lifetimes for processes (E7)–(E12) for different intermolecular potential

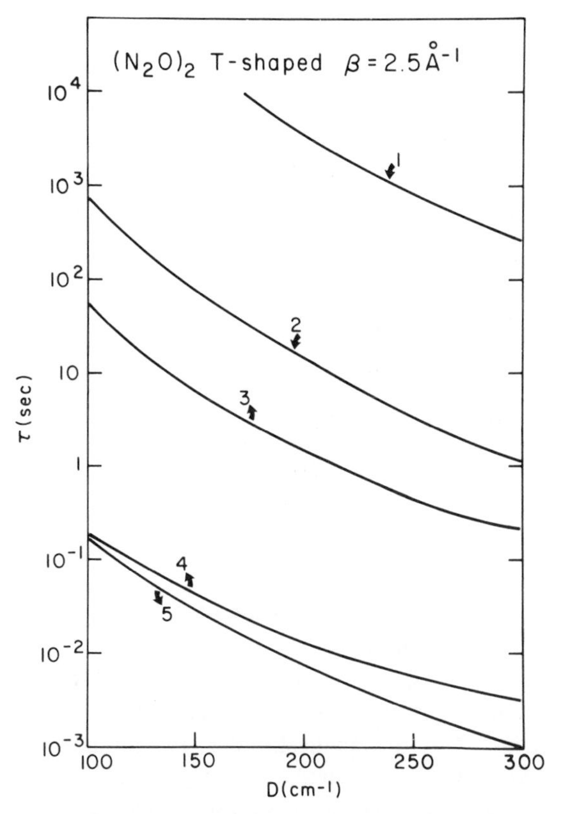

Fig. 28. Lifetimes for VP of the T-shaped $(N_2O)_2$ dimer initially excited at the (001) state of N_2O. The intermolecular potential was constructed from three atom–atom Morse functions, with $\beta = 2.5\ A^{-1}$ and $D = 100$–$300\ \text{cm}^{-1}$. Calculations were performed for the channels (E6)–(E10). The curves correspond to the following processes:

1. $V \to T$: (000)$\cdots$(001)$\to$(000)+(000)
2. $V \to V+T$: (000)$\cdots$(001)$\to$(010)+(000)
3. $V \to V+T$: (000)$\cdots$(001)$\to$(020)+(000)
4. $V \to V+T$: (000)$\cdots$(001)$\to$(030)+(000)
5. $V \to V+T$: (000)$\cdots$(001)$\to$(000)+(100)

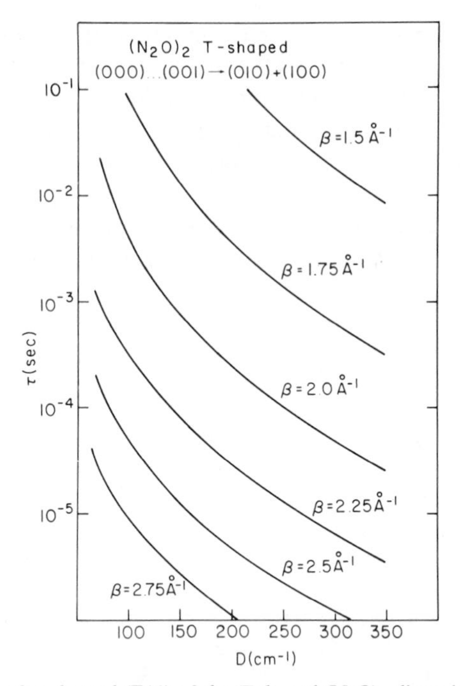

Fig. 29. VP lifetimes for channel (E11) of the *T*-shaped $(N_2O)_2$ dimer initially excited to the (001) state of N_2O. The intermolecular van der Waals potential was constructed as a superposition of three atom–atom Morse potentials. The reciprocal length was varied in the range $\beta = 1.5\ A^{-1}$, while the dissociative energy was varied in the range $D = 100$–$300\ \mathrm{cm}^{-1}$.

parameters. We see that, in general, the processes with smaller excess translational energy ΔE are characterized by shorter lifetimes. One exception, however, is the second-order process (E11) which is faster than the fourth-order process (E10), although it has a value of ΔE which is twice as large. The most efficient VP channel involves the mixed process (E12) which is characterized by the smallest of ΔE. From a cursory examination of the results of the model calculations presented in Figs. 27–29, we conclude that for a wide range of reasonable potential parameters, which specify the vdW bond, the predicted features of the VP dynamics of the $(N_2O)_2$ dimer are:

1. The major decay channel involves the third-order process (E12), which involves simultaneous intermolecular and intramolecular vibrational energy exchange and is characterized by the lowest translational energy, in accordance with the energy gap law.

2. All other decay channels (E7)–(E11) are considerably less efficient than the dominating channel (E11). The branching ratio between the two most efficient channels (E11) and (E12) is 10^{-3}.
3. The VP lifetime of the T-shaped $(N_2O)_2$ vdW complex is $10^{-5}-10^{-6}$ sec.

We now turn to the confrontation of these theoretical results with the experimental evidence. The $(N_2O)_2$ dimer excited to the $N_2O(001)\cdots N_2O(000)$ state exhibits VP on the ground state potential surface.[18] Gough, Miller, and Scoles have measured the infrared spectrum of N_2O dimers in a beam at low temperatures.[18] The VP lifetime of the complex was estimated to be in the range 10^{-12} sec $\leqslant \tau \leqslant 10^{-4}$ sec, and the lower limit was obtained from the assignment of the entire width of the infrared absorption band to lifetime broadening, while the upper limit was estimated from a time of flight experiment. Recently, Bernstein and Kolb[71] performed a detailed analysis of the inhomogeneously broadened absorption lineshapes, showing that there are 10^4-10^5 lines/cm^{-1} at the beam temperature. Bernstein and Kolb[71] attributed the continuous absorption spectrum[18] to arise from thermal broadening rather than from lifetime broadening, and have estimated the VP lifetimes to be in the range $10^{-4}-10^{-6}$ sec. It is apparent that at the present stage of development of this interesting area theory will be extremely useful in providing general outlines for the understanding of the VP dynamics of such vdW complexes. The structure of the $(N_2O)_2$ dimer has not yet been determined, however; Bernstein and Kolb[71] favor the T-shaped configuration on the basis of spectroscopic evidence. The numerical model calculations for the VP of the T-shaped vdW complex provide a theoretical estimate of $\tau=10^{-5}-10^{-6}$ sec for the VP lifetime. This theoretical result is close to the lower limit of τ set by Gough et al.,[18] being in accord with the recent analysis of the available experimental data.[71] We would like to conclude this discussion of the VP dynamics of the $(N_2O)_2$ dimer by pointing out that our model calculations provide evidence for the efficiency of the simultaneous intramolecular and intermolecular vibrational energy redistribution accompanying the VP process and result in explicit predictions regarding the vibrational energy of the products, which can be subjected to an experimental test. Finally, we would like to emphasize that these results for the VP dynamics of triatomic complexes are of intrisic interest as they provide an avenue for the understanding of energy flow, VP dynamics and vibrational relaxation in large vdW complexes.

D. Morales and G. Ewing[72] have also studied the VP process of the $(N_2O)_2$ dimer. Their approach differs from ours both regarding the structure of the dimer and concerning the specification of the nature of the residual interaction which induces the reactive process. They have considered a structure of the vdW dimer consisting of two parallel triatomic molecules, while we prefer the *T*-shaped structure. The residual interaction was taken by Morales and Ewing to consist of dipole–dipole coupling (ddc) involving the infrared transition moments while we use the short range repulsive interactions. The use of ddc to describe nearest interactions between vibrationally excited states is fraught with difficulties. In this context we note that studies of the band structure of vibrational excitons[73] in organic crystals clearly demonstrate that the Davydov splittings and the band width are determined by short range repulsive interactions rather than by dipole–dipole couplings. The Morales and Ewing[72] approach and our approach[31] start from different structures and predict different final vibrational distributions of the triatomic molecules resulting from the VP process. The basic structural assumption as well as the prediction should be subjected to an experimental test.

References

1. L. Pauling, *Nature of the Chemical Bond*, Cornell University Press, Ithaca, New York, 1940.
2. J. H. Jeans, *The Dynamical Theory of Gases*, University Press, Cambridge, 1904.
3. J. O. Hirschfelder, C. F. Curtiss, and R. B. Bird, *Molecular Theory of Gases and Liquids*, Wiley, New York, 1954.
4. A. D. Buckingham and B. D. Utling, *Ann. Rev. Phys. Chem.*, **21**, 287 (1970).
5. P. R. Certain and L. W. Bruch, *MTP Int. Rev. Sci., Phys. Chem. Ser. 1*, **1**, 113 (1972).
6. For a review see G. E. Ewing, *Ann. Rev. Phys. Chem.*, **27**, 553 (1976).
7. (a) E. E. Nikitin, *Opt. Spect.*, **9**, 8 (1960). (b) J. Bellingsley and A. B. Callear, *Trans. Faraday Soc.*, **67**, 257 (1971). (c) J. R. Airey and I. W. M. Smith, *J. Chem. Phys.*, **57**, 1669 (1972). (d) P. Zittel and C. B. Moore, *J. Chem. Phys.*, **59**, 6636 (1973). (e) H. M. Audibert, C. Joffrin, and J. Ducuing, *Chem. Phys. Lett.*, **25**, 158 (1974); (f) R. A. Lucht and T. A. Cool, *J. Chem. Phys.*, **63**, 3962 (1975).
8. (a) C. C. Bouchiat, M. A. Bouchiat, and L. C. L. Pottier, *Phys. Rev.*, **181**, 144 (1969). (b) M. A. Bouchiat, J. Brossel, and L. C. Pottier, *J. Chem. Phys.*, **56**, 3703 (1972).
9. (a) Vu-Hai and B. Vodar, *Z. Electrochem.*, **64**, 756 (1960). (b) D. H. Rank, B. S. Rao, and T. A. Wiggins, *J. Chem. Phys.*, **37**, 2511 (1962). (c) D. H. Rank, P. Sitaram, A. Glickman, and T. A. Wiggins, *J. Chem. Phys.*, **39**, 2673 (1963). (d) D. H. Rank, W. A. Glickman, and T. A. Wiggins, *J. Chem. Phys.*, **43**, 1304 (1965).
10. (a) A. Watanabe and H. L. Welsh, *Phys. Rev. Lett.*, **13**, 810 (1964). (b) A. K. Kudian, H. L. Welsh, and A. Watanabe, *J. Chem. Phys.*, **43**, 3397 (1965). (c) *ibid.*, **47**, 1553 (1967). (d) A. K. Kudian and H. L. Welsh, *Can. J. Phys.*, **49**, 230 (1971) (e) A. R. McKellar and H. L. Welsh, *J. Chem. Phys.*, **55**, 595 (1971). (f) L. Mannick, J. C. Stryland, and H. L. Welsh, *J. Chem. Phys.*, **49**, 3056 (1971). (g) A. R. McKellar and H. L. Welsh, *Can. J. Phys.*, **50**, 1458 (1972). (h) A. R. McKellar, *J. Chem. Phys.*, **61**, 4636 (1974). (i) A. R. McKellar and H. L. Welsh, *Can. J. Phys.*, **52**, 1082 (1974).
11. (a) C. E. Dinerman and G. E. Ewing, *J. Chem. Phys.*, **53**, 626 (1970). (b) C. A. Long, G. Henderson, and G. E. Ewing, *Chem. Phys.*, **2**, 485 (1973). (c) C. A. Long and G. E. Ewing, *J. Chem. Phys.*, **58**, 4824 (1973). (d) G. Henderson and G. E. Ewing, *J. Chem. Phys.*, **59**, 2280 (1973). (e) G. Henderson and G. E. Ewing, *Mol. Phys.*, **27**, 903 (1974).

12. (a) F. T. Green and T. A. Milne, *J. Chem. Phys.*, **39**, 3150 (1963). (b) R. E. Leckenly, E. J. Robbins, and P. A. Trevalion, *Proc. Roy. Soc. (London)*, **A280**, 409 (1964). (c) R. E. Leckenly and E. J. Robbins, *Proc. Roy. Soc. (London)*, **A291**, 389 (1966).
13. For reviews see (a) W. Klemperer, *Ber. Bunsen Gesel. Phys. Chem.*, **78**, 128 (1974); B. J. Howard, *MTP Int. Rev. Phys. Chem. Ser. 2*, **2**, 93 (1975). See also the bibliography at the end of this review.
14. See R. E. Smalley, L. Wharton, and D. H. Levy, *Acct. Chem. Res.*, **10**, 139 (1977) for a review.
15. (a) R. E. Smalley, D. H. Levy, and L. Wharton, *J. Chem. Phys.*, **64**, 3266 (1976). (b) M. S. Kim, R. E. Smalley, L. Wharton, and D. H. Levy, *J. Chem. Phys.*, **65**, 1216 (1976). (c) K. E. Johnson, L. Wharton, and D. H. Levy, *J. Chem. Phys.*, **69**, 2719 (1978). (d) G. Kubiak, P. S. H. Fitch, L. Wharton, and D. H. Levy, *J. Chem. Phys.*, **68**, 4477 (1978). (e) W. Sharfin, K. E. Johnson, L. Wharton, and D. H. Levy, *J. Chem. Phys.*, **71**, 1292 (1979). (f) D. H. Levy, *Advances in Chemical Physics* (this volume).
16. D. A. Dixon and D. R. Herschbach, *Ber. Bunsen Ges. Phys. Chem.*, **81**, 145 (1977).
17. R. E. Smalley, L. Wharton, and D. H. Levy, *J. Chem. Phys.*, **66**, 2750 (1977).
18. T. E. Gough, R. E. Miller, and G. Scoles, *J. Chem. Phys.*, **69**, 1588 (1978).
19. R. Schutz, As. S. Sudb, Y. T. Lee, and Y. R. Yen, *Int. Quantum Electronics Conf.*, Optical Society of America, Atlanta, Ga., 1978.
20. M. A. Hoffbauer, W. R. Gentry, and C. F. Giese, in Laser induced processes in molecules, K. Kompa and S. D. Smith, eds., *Springer Series in Chemical Physics*, Vol. 6, Springer, Berlin, Heidelberg, New York, 1978.
21. N. Rosen, *J. Chem. Phys.*, **1**, 319 (1933).
22. B. I. Stepanov, *Nature*, **157**, 808 (1946).
23. (a) C. A. Coulson and G. N. Robertson, *Proc. Roy. Soc.*, **A337**, 167 (1974). (b) *ibid.*, **A342**, 289 (1975). (c) G. Robertson, *J. Chem. Soc. Faraday Trans. II*, **72**, 1153 (1976).
24. G. E. Ewing, *J. Chem. Phys.*, **72**, 2096 (1980).
25. G. E. Ewing, *Chem. Phys.*, **29**, 253 (1978).
26. (a) G. J. Ashton and M. Child, *Faraday Disc. Chem. Soc.*, **62**, 307 (1977). (b) C. J. Ashton, Thesis, Oxford, 1977.
27. (a) J. A. Beswick and J. Jortner, *Chem. Phys. Lett.*, **49**, 13 (1977). (b) *ibid.*, *J. Chem. Phys.*, **68**, 2277 (1978).
28. J. A. Beswick and J. Jortner, *J. Chem. Phys.*, **69**, 512 (1978).
29. J. A. Beswick, B. Delgado-Barrio, and J. Jortner, *J. Chem. Phys.*, **70**, 3895 (1979).
30. J. A. Beswick and J. Jortner, *Chem. Phys. Lett.*, **65**, 240 (1979); *ibid.*, *J. Chem. Phys.*, **71**, 4737 (1979).
31. J. A. Beswick and J. Jortner, *J. Chem. Phys.* (in press).
32. J. Tusa, Sulkes, and S. A. Rice, *J. Chem. Phys.*, **70**, 3136 (1979).
33. J. Goodman and L. E. Brus, *J. Chem. Phys.*, **67**, 4858 (1977).
34. D. H. Levy, private communication.
35. U. Fano, *Phys. Rev.*, **124**, 1866 (1961).
36. M. Abramowitz and I. A. Stegun, *Handbook of Mathematic Functions*, National Bureau of Standards, Washington, D.C., 1965.
37. G. E. Ewing, *J. Chem. Phys.*, **71**, 3143 (1979).
38. D. Rapp and T. E. Sharp, *J. Chem. Phys.*, **38**, 2641 (1963).
39. F. Mies, *J. Chem. Phys.*, **40**, 523 (1964).
40. D. Secrest and W. Eastes, *J. Chem. Phys.*, **56**, 2502 (1972).
41. S. E. Novick, K. C. Janda, S. L. Holmgren, M. Waldman, and W. Klemperer, *J. Chem. Phys.*, **65**, 1114 (1976).
42. O. Atabek, J. A. Beswick, R. Lefebvre, S. Mukamel, and J. Jortner, *J. Chem. Phys.*, **65**, 4035 (1976).

43. (a) K. E. Holdy, L. C. Klotz, and K. R. Wilson, *J. Chem. Phys.*, **52**, 4588 (1970). (b) M. Shapiro and R. D. Levine, *Chem. Phys. Lett.*, **5**, 499 (1970).
44. A. F. Devonshire, *Proc. Roy. Soc. (London)*, **A158**, 269 (1937).
45. L. Fox, *The Numerical Solution of Two-Point Boundary Value Problems in Ordinary Differential Equations*, Oxford University Press, London, 1957.
46. J. R. Taylor, *Scattering Theory*, Wiley, New York, 1972.
47. G. Herzberg, *Spectra of Diatomic Molecules*, Van Nostrand, New York, 1966.
48. (a) C. F. Curtiss, *J. Chem. Phys.*, **49**, 1952 (1968). (b) T. P. Tsien, G. A. Parker, and R. T. Pack, *J. Chem. Phys.*, **59**, 5373 (1973). (c) D. Secrest, *J. Chem. Phys.*, **62**, 770 (1975). (d) L. W. Hunter, *J. Chem. Phys.*, **62**, 2855 (1975).
49. D. A. Dixon, D. R. Herschbach, and W. Klemperer, *Faraday Disc. Chem. Soc.*, **62** (1977).
50. See for example, R. S. Knox, *Theory of Excitons*, Academic Press, New York, 1963.
51. R. S. Mulliken and W. B. Person, *Molecular Complexes*, Wiley, New York, 1963.
52. M. Bixon, J. Jortner, and Y. Dothan, *Mol. Phys.*, **17**, 109 (1969).
53. P. J. Robinson and K. A. Hollbrook, *Unimolecular Reactions*, Wiley, London, New York, 1972.
54. Miklavc and S. Fisher, *Chem. Phys. Lett.*, **44**, 209 (1976).
55. E. B. Wilson, J. C. Decius, and P. C. Cross, *Molecular Vibrations*, McGraw-Hill, New York, 1955.
56. J. C. Amiot, Thése Université Pierre et Marie Curie, 1976.
57. J. Plivac, *Critical Evaluation of Chemical and Physical Structural Information*, National Academy of Science, Washington, D.C., 1974.
58. S. Lermont and G. W. Flynn, *Ann. Rev. Phys. Chem.*, **28**, 261 (1977).
59. W. G. Rothschild, *J. Chem. Phys.*, **65**, 455 (1976).
60. J. A. Blazy, B. M. Dekoven, T. D. Russell, and D. H. Levy, *J. Chem. Phys.*, **72**, 2439 (1980).
61. J. A. Beswick and G. Delgado-Barrio, *J. Chem. Phys.*, **73**, 3653 (1980).
62. R. Ramaswamy and A. E. De Pristo, *J. Chem. Phys.*, **72**, 770 (1980).
63. A. E. De Pristo, S. D. Augustin, R. Ramaswamy and H. Rabitz, *J. Chem. Phys.*, **71**, 850 (1976).
64. S. B. Woodruff and D. L. Thompson, *J. Chem. Phys.*, **71**, 376 (1979).
65. J. A. Beswick and A. Requena, *J. Chem. Phys.*, **73**, 4347 (1980).
66. (a) C. F. Curtis, J. O. Hirschfelder and F. T. Adler, *J. Chem. Phys.*, **18**, 1683 (1950); (b) R. T. Pack, *J. Chem. Phys.*, **60**, 633 (1974); (c) G. C. Schatz and A. Kupperman, *J. Chem. Phys.*, **65**, 4642 (1976).
67. M. E. Rose, *Elementary Theory of Angular Momentum*, (Wiley, New York, 1957).
68. J. A. Beswick and A. Requena, *J. Chem. Phys.*, **72**, 3018 (1980).
69. (a) R. J. LeRoy, J. S. Carley and J. E. Grabenstetter, *Faraday Disc. Chem. Soc.*, **62**, 169 (1977); (b) J. S. Carley, *Faraday Disc. Chem. Soc.*, **62**, 303 (1977).
70. J. A. Beswick and G. Delgado-Barrio, *J. Chem. Phys.*, **73**, 3653 (1980).
71. L. Bernstein and C. Kolb, *J. Chem. Phys.*, **71**, 2818 (1979).
72. D. Morales and G. Ewing, to be published.
73. D. M. Hanson, R. Kopelman, and G. W. Robinson, *J. Chem. Phys.*, **51**, 212 (1969).

BIBLIOGRAPHY

In what follows we have attempted to provide a comprehensive bibliography of the literature pertaining to the structure, properties, energetics, and dynamics of van der Waals molecules.

General

J. H. Jeans, *The Dynamical Theory of Gases*, Cambridge University Press, London, 1904.

J. O. Hirschfelder, C. F. Curtiss, and R. B. Bird, *Molecular Theory of Gases and Liquids*, Wiley, New York, 1954.

Intermolecular Forces

A. D. Buckingham and B. D. Utling, Intermolecular forces, *Ann. Rev. Phys. Chem.*, **21**, 287 (1970).

P. R. Certain and L. W. Bruch, Intermolecular forces, *MTP Int. Rev. Sci., Phys. Chem. Ser. One*, **1**, 113 (1972).

Reviews (General)

G. E. Ewing, Infrared spectroscopy, *Ann. Rev. Phys. Chem.*, **23**, 141 (1972).

G. E. Ewing, Intermolecular interactions: van der Waals molecules, *Angew. Chem. Int. Ed. Engl.*, **11**, 486 (1972).

G. E. Ewing, Structure and properties of van der Waals molecules, *Acc. Chem. Res.*, 8, 185 (1975).

B. L. Blaney and G. E. Ewing, van der Waals molecules, *Ann. Rev. Phys. Chem.*, **27**, 553 (1976).

G. E. Ewing, The spectroscopy of van der Waals molecules, *Can. J. Phys.*, **54**, 487 (1976).

Reviews on Electric Resonance Spectroscopy in Supersonic Beams

W. Klemperer, Molecular spectroscopy of loosely bound complexes, *Ber. Bunsen. Ges. Phys. Chem.*, **78**, 128 (1974).

B. J. Howard, *MTP Int. Rev. Phys. Chem. Ser.* **2**(2), 93 (1975).

Reviews on Optical Spectroscopy in Supersonic Beams

R. E. Smalley, L. Wharton, and D. H. Levy, Molecular optical spectroscopy with supersonic beams and jets, *Acct. Chem. Res.*, **10**, 139 (1977).

D. H. Levy, Laser spectroscopy of cold gas-phase molecules, *Ann. Rev. Phys. Chem.*, **31**, (1980).

First Experimental Evidence of Clustering in Bulk

Vu-Haī and B. Vodar, *Z. Electrochem.*, **64**, 756 (1960).

Experiments in Bulk

Earlier Spectroscopic Work

D. H. Rank, B. S. Rao, and T. A. Wiggins, Absorption spectra of hydrogen-halide-rare-gas mixtures, *J. Chem. Phys.*, **37**, 2511 (1963).

D. H. Rank, P. Sitaram, A. Glickman, and T. A. Wiggins, Gas-phase complexes in hydrogen chloride, *J. Chem. Phys.*, **39**, 2673 (1963).

D. H. Rank, W. A. Glickman, and T. A. Wiggins, Infrared absorption spectra of isotopic dimeric hydrogen chloride molecules, *J. Chem. Phys.*, **43**, 1304 (1965).

Infrared and Visible Spectra (Long Path Length Spectroscopy)

A. Watanabe and H. L. Welsh, Direct spectroscopic evidence of bound states of $(H_2)_2$ complexes at low temperatures, *Phys. Rev. Lett.*, **13**, 810 (1964).

A. K. Kudian, H. L. Welsh, and A. Watanabe, Direct spectroscopic evidence of bound states of H_2-Ar complexes at 1000 K, *J. Chem. Phys.*, **43**, 3397 (1965).

A. K. Kudian, H. L. Welsh, and A. Watanabe, Spectra of H_2-Ar, H_2-N_2, and H_2-CO van der Waals complexes, *J. Chem. Phys.*, **47**, 1553 (1967).

A. K. Kudian and H. L. Welsh, Spectra of H_2-Ar, H_2-Kr and H_2-Xe van der Waals complexes in pressure-induced infrared absorption, *Can. J. Phys.*, **49**, 230 (1971).

A. R. McKellar and H. L. Welsh, Anisotropic intermolecular force effects in spectra of H_2 and D_2-rare-gas complexes, *J. Chem. Phys.*, **55**, 595 (1971).

L. Mannick, J. C. Stryland, and H. L. Welsh, An infrared spectrum of CO_2 dimers in the "locked" configuration, *Can. J. Phys.*, **49**, 3056 (1971).

A. R. McKellar and H. L. Welsh, Spectra of H_2-Ne and D_2-Ne van der Waals complexes in the collision-induced fundamentals bands of hydrogen and deuterium, *Can. J. Phys.*, **50**, 1458 (1972).

A. R. McKellar, Infrared-spectrum of the HD–Ar van der Waals complex, *J. Chem. Phys.*, **61**, 4636 (1974).

A. R. McKellar and H. L. Welsh, Spectra of $(H_2)_2$, $(D_2)_2$ and H_2-D_2 van der Waals complexes, *Can. J. Phys.*, **52**, 1082 (1974).

C. E. Dinerman and G. E. Ewing, Infrared spectrum, structure and heat of formation of gaseous $(NO)_2$, *J. Chem. Phys.*, **53**, 626 (1970).

C. A. Long, G. Henderson, and G. E. Ewing, the infrared spectrum of $(N_2)_2$ van der Waals molecule, *Chem. Phys.*, **2**, 485 (1973).

C. A. Long and G. E. Ewing, Spectroscopic investigation of van der Waals molecules. I. The infrared and visible spectra of $(O_2)_2$, *J. Chem. Phys.*, **58**, 4824 (1973).

G. Henderson and G. E. Ewing, Infrared spectrum, structure and properties of O_2–Ar van der Waals molecule, *J. Chem. Phys.*, **59**, 2280 (1973).

G. Henderson and G. E. Ewing, Infra-red spectrum, structure and properties of the N_2–Ar van der Waals molecule, *Mol. Phys.*, **27**, 903 (1974).

R. K. Thomas, Hydrogen bonding in the gas phase: The infrared spectra of complexes of hydrogen fluoride with hydrogen cyanide and methyl cyanide, *Proc. Roy. Soc. London Ser.*, **A325**, 133 (1971).

E. W. Boom, D. Frenkel, and J. van der Elsken, A far infrared study of the Ar HCl van der Waals molecule, *J. Chem. Phys.*, **66**, 1826 (1977).

D. Frenkel and J. van der Elsken, Density dependence of the pressure induced shift of HCl rotational lines perturbed by argon, *Chem. Phys. Lett.*, **50**, 116 (1977).

C. E. Morgan and L. Frommhold, *Phys. Rev. Lett.*, **29**, 1053 (1972).

H. B. Levine, Spectroscopy of dimers, *J. Chem. Phys.*, **56**, 2455 (1972).

Vacuum—UV

Y. Tanaka and K. Yoshino, *J. Chem. Phys.*, **50**, 3087 (1969).

Y. Tanaka and K. Yoshino, Absorption spectrum of the argon molecule in the vacuum-uv region, *J. Chem. Phys.*, **59**, 5160 (1973).

Y. Tanaka and K. Yoshino, Absorption spectra of Ne_2 and He–Ne molecules in the vacuum-uv region, *J. Chem. Phys.*, **57**, 29764 (1972).

Y. Tanaka, K. Yoshino, and D. E. Freeman, *J. Chem. Phys.*, **59**, 5160 (1973).

D. E. Freeman, K. Yoshino, and Y. Tanaka, *J. Chem. Phys.*, **61**, 4880 (1974).

J. Billingsley and A. B. Callear, Investigation of the 2050 Å System of the nitric oxide dimer, *Trans. Faraday Soc.*, **67**, 589 (1971).

D. E. Freeman, K. Yoshino, and Y. Tanaka, Emission specturm of rare gas dimers in the vacuum-UV region II. Rotational analysis of band system I for Ar_2, *J. Chem. Phys.*, **71**, 1780 (1979).

Supersonic Beams Experiments

Experimental Evidence of Clustering in Nozzle Beams

F. T. Green and T. A. Milne, Mass spectrometric detection of polymers in supersonic molecular beams, *J. Chem. Phys.*, **39**, 3150 (1963).

R. E. Leckenby, E. J. Robbins, and P. A. Trevalion, Condensation embryos in an expanding gas beam, *Proc. Roy. Soc. (London)*, **A280**, 409 (1964).

R. E. Leckenby and E. J. Robbins, The observation of double molecules in gases, *Proc. Roy. Soc. London Ser.*, **A291**, 389 (1966).

Electric Resonance Experiments

T. R. Dycke, G. R. Tomasevich, W. Klemperer, and W. E. Falconer, Electric resonance spectroscopy of hypersonic molecular beams, *J. Chem. Phys.*, **57**, 2277 (1972).

T. R. Dycke, B. J. Howard, and W. Klemperer, Radiofrequency and microwave spectrum of the hydrogen fluoride dimer; A nonrigid molecule, *J. Chem. Phys.*, **56**, 2442 (1972).

S. E. Novick, P. B. Davies, S. J. Harris, and W. Klemperer, Determination of the structure of Ar–HCl, *J. Chem. Phys.*, **59**, 2273 (1973).

S. J. Harris, S. E. Novick, and W. Klemperer, Determination of the structure of Ar–HF, *J. Chem. Phys.*, **60**, 3208 (1974).

S. J. Harris, S. E. Novick, W. Klemperer, and W. E. Falconer, Intermolecular potential between an atom and a diatomic molecule: The structure of Ar–ClF, *J. Chem. Phys.*, **61**, 193 (1974).

T. R. Dyke and J. S. Muenter, Microwave spectrum and structure of hydrogen bounded water dimer, *J. Chem. Phys.*, **60**, 2929 (1974).

S. J. Harris, K. C. Janda, S. E. Novick, and W. Klemperer, Intermolecular potential between an atom and a linear molecule: The structure of Ar–OCS, *J. Chem. Phys.*, **63**, 881 (1975).

S. E. Novick, S. J. Harris, K. C. Janda, and W. Klemperer, Structure and bounding of Kr–ClF: Intermolecular force fields in van der Waals molecules, *Can. J. Phys.*, **53**, 2007 (1975).

K. C. Janda, W. Klemperer, and S. E. Novick, Measurement of the sign of the dipole moment of ClF, *J. Chem. Phys.*, **64**, 2698 (1976).

S. E. Novick, K. C. Janda, S. L. Holmgrem, M. Waldman, and W. Klemperer, Centrifugal distortion in Ar–HCl, *J. Chem. Phys.*, **65**, 1114 (1976).

W. Klemperer, *Faraday Disc. Chem. Soc.*, **62**, 179 (1977).

K. V. Chance, K. H. Bowen, J. S. Winn, and W. Klemperer, Microwave and radio frequency spectrum of XeHCl, *J. Chem. Phys.*, **70**, 5157 (1979).

T. J. Balle, E. J. Campbell, M. R. Keenan, and W. H. Flygare, "A new method for observing the rotational spectra of weak molecular complexes: KrHCl", *J. Chem. Phys.*, **71**, 2723 (1979).

J. M. Steed, L. S. Bernstein, T. A. Dixon, K. C. Janda and W. Klemperer, "The microwave spectrum of argon methyl chloride", *J. Chem. Phys.*, **71**, 4189 (1979).

R. L. DeLeon, K. M. Mack and J. S. Muenter, "Structure and Properties of the argon-ozone van der Waals molecule", *J. Chem. Phys.*, **71**, 4487 (1979).

J. M. Steed, T. A. Dixon, and W. Klemperer, "Determination of the structure of $ArCO_2$ by radiofrequency and microwave spectroscopy", *J. Chem. Phys.*, **70**, 4095 (1979).

K. C. Janda, J. M. Steed, S. E. Novick and W. Klemperer, "Hydrogen bonding: The structure of HF—HCl", *J. Chem. Phys.*, **67**, 5162 (1977).

T. R. Dyke, K. R. Mack, and J. S. Muenter, "The structure of water dimers from molecular beam electric resonance spectroscopy", *J. Chem. Phys.*, **66**, 498 (1977).

A. E. Barton, T. J. Henderson, P. R. R. Langridge-Smith and B. J. Howard, "The rotational spectrum and structure of van der Waals complexes. II. Krypton-Hydrogen Chloride", *J. Chem. Phys.*, **45**, 429 (1980).

Polarity Determination

S. Novick, J. M. Lehn, and W. Klemperer, On the polarity of 1,3-butadiene, 2,3-dichloro, 1,3-butadiene and their van der Waals adducts with ethylene, *J. Am. Chem. Soc.*, **95**, 8189 (1973).

S. E. Novick, P. B. Davies, T. R. Dycke, and W. Klemperer, Polarity of van der Waals molecules, *J. Am. Chem. Soc.*, **95**, 8547 (1973).

S. J. Harris, S. E. Novick, J. S. Winn, and W. Klemperer, $(Cl_2)_2$: A polar molecule, *J. Chem. Phys.*, **61**, 3866 (1974).

A. E. Barton, A. Chablo, and B. J. Howard, On the structure of the carbon dioxide dimer, *Chem. Phys. Lett.*, **60**, 414 (1979).

Nuclear Magnetic Resonance

E. M. Mattison, D. E. Pritchard, and D. Kleppner, Spin-rotation coupling in the alkali-rare-gas van der Waals molecule KAr, *Phys. Rev. Lett.*, **32**, 507 (1974).

Alignment and Laser-Induced Fluorescence

M. P. Sinha, A. Schultz, and R. N. Zare, Internal state distribution of alkali-dimers in supersonic nozzle beams, *J. Chem. Phys.*, **58**, 549 (1973).

M. P. Sinha, C. D. Caldwell, and R. N. Zare, Alignment of molecules in gaseous transport alkali dimers in supersonic nozzle beams, *J. Chem. Phys.*, **61**, 491 (1974).

Chemical Reactions

D. L. King, D. A. Dixon, and D. R. Herschbach, Molecular beam chemistry. Facile six-center reactions of dimeric chlorine with bromine and with hydrogen iodide, *J. Am. Chem. Soc.*, **96**, 3328 (1974).

D. L. King, D. A. Dixon, and D. R. Herschbach, Molecular beam chemistry. Reactions exchanging van der Waals bonds among three or more halogen molecules, *J. Am. Chem. Soc.*, **97**, 6268 (1975).

D. A. Dixon and D. R. Herschbach, Energy transfer processes involving van der Waals bonds, *Ber. Bunsen. Ges. Phys. Chem.*, **81**, 145 (1977).

A. G. Urena, R. B. Bernstein, and G. R. Phillips, Molecular beam reaction of the van der Waals clusters $(CH_3I)_n$ with alkalis, *J. Chem. Phys.*, **62**, 1818 (1975).

R. Behrems Jr., A. Freedman, R. R. Herm, and T. P. Parr, The $A+B_x$ condensation reaction: Crossed beams of Br_2 and $(Cl_2)_x$ or $(NH_3)_x$ clusters, *J. Chem. Phys.*, **63**, 4622 (1975).

Optical Spectroscopy

R. E. Smalley, D. H. Levy, and L. Wharton, The fluorescence excitation spectrum of the HeI_2 van der Waals complex, *J. Chem. Phys.*, **64**, 3266 (1976).

M. S. Kim, R. E. Smalley, L. Wharton, and D. H. Levy, Energy distribution in the photodissociation products of the van der Waals complex I_2He, *J. Chem. Phys.*, **65**, 1216 (1976).

K. E. Johnson, L. Wharton, and D. H. Levy, The predissociation lifetime of the van der Waals molecule I_2He, *J. Chem. Phys.*, **69**, 2719 (1978).

J. A. Blazy and D. H. Levy, The use of the van der Waals molecules to identify spectroscopic origins, *Chem. Phys. Lett.*, **51**, 395 (1977).

R. E. Smalley, L. Wharton, and D. H. Levy, the fluorescence excitation spectrum of the $He-NO_2$ van der Waals complex, *J. Chem. Phys.*, **66**, 2750 (1977).

R. E. Smalley, D. A. Auerbach, P. S. H. Fitch, D. H. Levy, and L. Wharton, Laser spectroscopic measurement of weakly attractive interatomic potentials: The Na+Ar interaction, *J. Chem. Phys.*, **66**, 3778 (1977).

R. E. Smalley, L. Wharton, D. H. Levy, and D. W. Chandler, The fluorescence excitation spectrum of *s*-tetrazine cooled in a supersonic free jet, *J. Mol. Spect.*, **66**, 375 (1977).

R. E. Smalley, L. Wharton, D. H. Levy, and D. H. Chandler, Fluorescence excitation spectrum of *s*-tetrazine cooled in a supersonic free jet: van der Waals complexes and isotopic species, *J. Chem. Phys.*, **68**, 2487 (1978).

G. Kubiak, P. S. H. Fitch, L . Wharton, and D. H. Levy, The fluorescence excitation spectrum of the ArI_2 van der Waals complex, *J. Chem. Phys.*, **68**, 4477 (1978).

S. M. Beck, M. G. Liverman, D. L. Monts, and R. E. Smalley, Rotational analysis of the $B_{2\mu}(\pi\pi^*)\,^1A_{1g}(6_0^1)$ band of benzene and helium–benzene van der Waals complexes in a supersonic jet, *J. Chem. Phys.*, **70**, 232 (1979).

W. Sharfin, K. E. Johnson, L. Wharton, and D. H. Levy, Energy distribution on the photodissociation products of van der Waals molecules: Iodine–helium complexes, *J. Chem. Phys.*, **71**, 1292 (1979).

A. Amirav, U. Even and J. Jortner, "External heavy atom effect on intramolecular intersystem crossing in a supersonic beam", *J. Chem. Phys. Lett.*, **67**, 9 (1979).

J. E. Kenny, K. E. Johnson, W. Sharfin, and D. H. Levy, "The photodissociation of van der Waals molecules: complexes of iodine, neon, and helium", *J. Chem. Phys.*, **72**, 1109 (1980).

J. A. Blazy, B. M. DeKoven, T. D. Russell, and D. H. Levy, "The binding energy of iodine-rare gas van der Waals molecules", *J. Chem. Phys.*, **72**, 2439 (1980).

Infrared Spectroscopy

T. E. Gough, R. E. Miller, and G. Scoles, Photo-induced vibrational predissociation of the van der Waals molecule $(N_2O)_2$, *J. Chem. Phys.*, **69**, 1588 (1978).

R. Schutz, As. S. Sudb, Y. T. Lee, and Y. R. Yen, International Quantum Electronics Conference, Optical Society of America, Atlanta, Ga. 1978.

M. A. Hoffbauer, W. R. Gentry, and C. F. Giese, Pulsed molecular beam study of ethylene dimer photodissociation with a CO_2 laser, in *Laser Induced Processes in Molecules*, K. Kompa and S. D. Smith, eds., *Springer Series in Chemical Physics*, **6**, Springer Berlin, Heidelberg, 1978.

M. P. Casassa, D. S. Bonise, J. L. Beauchamp and K. C. Janda, Infrared Photochemistry of ethylene clusters, *J. Chem. Phys.*, **72**, 6805 (1980).

Photoionization of van der Waals Dimers

C. Y. Ng, D. J. Trevor, P. W. Tiedemann, S. T. Ceyer, P. L. Kronebusch, B. H. Mahan, and Y. T. Lee, *J. Chem. Phys.*, **67**, 4235 (1977).

S. T. Ceyer, P. W. Tiedemann, B. H. Manan, and Y. T. Lee, *J. Chem. Phys.*, **70**, 14 (1979).

P. W. Tiedemann, S. L. Anderson, S. T. Ceyer, T. Hirooka, C. Y. Ng, B. H. Mahan, and Y. T. Lee, Proton affinities of hydrogen halides determined by the molecular beam photoionization method, *J. Chem. Phys.*, **71**, 605 (1979).

Influence of van der Waals Molecules on Relaxation Processes

Vibrational Relaxation

J. Bellingsley and A. B. Callear, Kinetics and mechanism of the vibrational relaxation of NO($x^2\pi(v=1)$ in the temperature range 100–433 K, *Trans. Faraday Soc.*, **67**, 257 (1971).

J. R. Airey and I. W. M. Smith, Quenching of infrared chemiluminescence: Rates of energy transfer from HF ($v \leqslant 5$) to CO_2 and HF, from DF ($v \leqslant 3$) to CO_2 and HF, *J. Chem. Phys.*, **57**, 1669 (1972).

P. Zittel and C. B. Moore, Vibrational relaxation in HBr and HCl from 144°K to 584°K, *J. Chem. Phys.*, **59**, 6636 (1973).

H. M. Audibert, C. Joffrin, and J. Ducuing, Vibrational relaxation of H_2 in the range 500–40°K, *Chem. Phys. Lett.*, **25**, 158 (1974).

R. A. Lucht and T. A. Cool, Temperature dependence of vibrational relaxation in the HF–DF, HF–CO_2 and DF–CO_2 systems. II, *J. Chem. Phys.*, **63**, 3962 (1975). Paper I. *JCP*, **60**, 1026 (1974). T 295–670°K.

Spin Relaxation

C. C. Bouchiat, M. A. Bouchiat, and L. C. L. Pottier, Evidence for Rb-rare-gas molecules from the relaxation of polarized Rb atoms in a rare-gas theory, *Phys. Rev.*, **181**, 144 (1969).

M. A. Bouchiat, J. Brossel, and L. C. Pottier, Evidence for Rb-rare-gas molecules from the relaxation of polarized Rb atoms in a rare gas. Experimental results, *J. Chem. Phys.*, **56**, 3703 (1972).

Theoretical

Dimerization

A. Kantrowitz and J. Grey, A high intensity source for the molecular beam. I. Theoretical, *Rev. Sci. Inst.*, **22**, 328 (1951).

R. J. Gordon, Y. T. Lee, and D. R. Herschbach, Supersonic molecular beams of alkali-dimers, *J. Chem. Phys.*, **54**, 2393 (1971).

W. C. Schieve and H. W. Harrison, Molecular dynamics study of dimer formation in three dimensions, *J. Chem. Phys.*, **61**, 700 (1974).

Spectroscopy and Determination of Intermolecular Potentials

S. Bratoz and M. L. Martin, Infrared spectra of highly compressed gas mixtures of the type HCl+X. A theoretical study, *J. Chem. Phys.*, **42**, 1051 (1965).

J. K. Cashion, Determination of intermolecular-potential parameters from induced infrared spectra: The complex H_2–Ar, *J. Chem. Phys.*, **45**, 1656 (1966).

R. G. Gordon and J. K. Cashion, Intermolecular potentials and the infrared spectrum of the molecular complex $(H_2)_2$, *J. Chem. Phys.*, **44**, 1190 (1966).

J. K. Cashion, Simple formulas for the vibrational and rotational Eigenvalues of the Lennard–Jones 12–6 potential, *J. Chem. Phys.*, **48**, 94 (1968).

H. B. Levine, Spectroscopy of dimers, *J. Chem. Phys.*, **56**, 2455 (1972).

G. Girardet and D. Robert, Interpretation of the far infrared spectra of the dimers of HCl and DCl trapped in monoatomic solids, *J. Chem. Phys.*, **58**, 4110 (1973).

L. Frommhold, Interpretation of Raman-spectra of van der Waals dimers in argon, *J. Chem. Phys.*, **61**, 2996 (1974).

R. J. Le Roy and H. Kreek, Anisotropic intermolecular potentials from an analysis of spectra of H_2- and D_2- inert gas complexes, *J. Chem. Phys.*, **61**, 4750·(1974).

R. J. Le Roy and J. Van Kranendonk, Anisotropic intermolecular potentials from an analysis of spectra of H_2- and D_2- inert gas complexes, *J. Chem. Phys.*, **61**, 4750 (1974).

A. M. Dunker and R. G. Gordon, Calculations on the HCl–Ar van der Waals complex, *J. Chem. Phys.*, **64**, 354 (1976).

R. L. Ellis, A study of molecular pair potentials: A semiempirical approach, *J. Chem. Phys.*, **64**, 342 (1976).

R. J. Le. Roy, J. S. Carley, and J. E. Grabenstetter, Determining anisotropic intermolecular potential for van der Waals molecules, *Faraday Disc. Chem. Soc.*, **62**, 169 (1977).

A. M. Dunker and R. G. Gordon, Bound atom-diatomic molecule complexes. Anisotropic intermolecular potentials for the hydrogen-rare gas systems, *J. Chem. Phys.*, **68**, 700 (1978).

G. A. Gallup, The intermolecular potential and its angular dependence for two H_2 molecules, *Mol. Phys.*, **33**, 943 (1977).

S. L. Holmgren, M. Waldman, and W. Klemperer, a) Internal dynamics of van der Waals complexes I. Born–Oppenheimer separation of radial and angular motion, *J. Chem. Phys.*, **67**, 4414 (1977); b) II. Determination of a potential surface for ArHCl, *J. Chem. Phys.*, **69**, 1661 (1978).

S. Tarr, H. Rabitz, D. Fitz, and R. A. Marcus, Classical and quantum centrifugal decoupling approximation for HCl–Ar, *J. Chem. Phys.*, **66**, 2854 (1977).

D. E. Stogryn and T. Hirschfelder, Initial pressure dependence of thermal conductivity and viscosity, *J. Chem. Phys.*, **31**, 1545 (1959).

S. K. Kim and J. Ross, Viscosity of moderately dense gases, *J. Chem. Phys.*, **42**, 263 (1965).

E. A. Guggenheim, Dimerization of gaseous nitric oxide, *Mol. Phys.*, **10**, 401 (1966).

G. D. Mahan, Number of bound states in Lennard–Jones potentials, *J. Chem. Phys.*, **52**, 258 (1970).

R. D. Olmsted and C. F. Curtiss, A quantum kinetic theory of moderately dense gases. III. The effect of bound states on the transport coefficients, *J. Chem. Phys.*, **63**, 1966 (1975).

R. D. Olmsted and C. F. Curtiss, *J. Chem. Phys.*, **62**, 903 (1975).

R. D. Olmsted and C. F. Curtiss, *J. Chem. Phys.*, **62**, 3979 (1975).

C. L. Briant and J. J. Burton, Molecular dynamics study of the structure and thermodynamic properties of argon microclusters, *J. Chem. Phys.*, **63**, 2045 (1975).

H. K. Shin, Vibrational and thermodynamic properties of ArHX van der Waals molecules, *Chem. Phys. Lett.*, **49**, 193 (1977).

J. M. Deutch and W. Klemperer, The relation between structure of van der Waals molecular dimers and dielectric second virial coefficients, *J. Chem. Phys.*, **66**, 2753 (1977).

W. Kutzelnigg, Q. M. Calc. of intermolecular potentials, mainly of van der Waals type, *Faraday Disc. Chem. Soc.*, **62**, 185 (1977).

K. T. Tang and J. P. Toennies, A simple theoretical model for the van der Waals potential at intermediate distances. I. Specially symmetric potentials, *J. Chem. Phys.*, **66**, 1496 (1977).

R. J. LeRoy, "Comment regarding potential functions, level spacings and thermodynamical properties of van der Waals molecules", *Chem. Phys. Lett.*, **67**, 207 (1979).

J. H. Goble and J. S. Winn, "Analytic potential functions for weakly bound molecules: The X and A states of NaAr and the A state of NaNe", *J. Chem. Phys.*, **70**, 2051 (1979).

J. H. Goble and J. S. Winn, "Estimation of the Dissociation Energy of Weakly-bound Molecules from Spectroscopic Data", *J. Chem. Phys.*, (to be published).

Vibrational and Rotational Predissociation

N. Rosen, *J. Chem. Phys.*, **1**, 319 (1933).

B. I. Stepanov, *Nature*, **157**, 808 (1946).

M. Shapiro, Dynamics of dissociation. I. Computational investigation of unimolecular breakdown processes, *J. Chem. Phys.*, **56**, 2582 (1972).

C. A. Coulson and G. N. Robertson, A theory of the broadening of the infrared absorption spectra of hydrogen-bonded species. I, *Proc. Roy. Soc.*, **A337**, 167 (1974).

C. A. Coulson and G. N. Robertson, A theory of the broadening of the infrared absorption spectra of hydrogen-bonded species. II, *Proc. Roy. Soc.*, **A342**, 289 (1975).

G. Robertson, Simplified treatment of Stepanov's vibrational predissociation effect in hydrogen-bonded species, *J. Chem. Soc. Faraday Trans. II*, **72**, 1153 (1976).

C. J. Ashton and M. Child, *Faraday Disc. Chem. Soc.*, **62**, 307 (1977).

C. J. Ashton, Thesis, Oxford, 1977.

D. A. Dixon, D. R. Herschbach, and W. Klemperer, *Faraday Disc. Chem. Soc.*, **62**, 341 (1977).

J. A. Beswick and J. Jortner, A model for vibrational predissociation of van der Waals molecules, *Chem. Phys. Lett.*, **49**, 13 (1977).

J. A. Beswick and J. Jortner, Vibrational predissociation of triatomic van der Waals molecules, *J. Chem. Phys.*, **68**, 2277 (1978).

J. A. Beswick and J. Jortner, Comment on vibrational predissociation of polyatomic van der Waals molecules, *J. Chem. Phys.*, **68**, 4455 (1978).

J. A. Beswick and J. Jortner, Perpendicular vibrational predissociation of T-shaped van der Waals molecules, *J. Chem. Phys.*, **69**, 512 (1978).

J. A. Beswick and J. Jortner, Intramolecular dynamics of some van der Waals dimers, *J. Chem. Phys.* **71**, 4337 (1979).

J. A. Beswick, G. Dalgado Barrio, and J. Jortner, Vibrational predissociation lifetimes of the van der Waals molecule He–I_2, *J. Chem. Phys.*, **70**, 3895 (1979).

G. Ewing, Vibrational predissociation in hydrogen bonded complexes, *J. Chem. Phys.*, **71**, 2096 (1980).

G. E. Ewing, A guide to the lifetimes of vibrationally excited van der Waals molecules: The momentum gap, *J. Chem. Phys.*, **71**, 3143 (1979).

S. B. Woodruff and D. L. Thompson, A quasiclassical trajectory study of vibrational predissociation of van der Waals molecules: Collinear He$\cdots I_2(B^3\pi)$, *J. Chem. Phys.*, **71**, 376 (1979).

J. A. Beswick and J. Jortner, Intermolecular V–V transfer in the vibrational predissociation of some van der Waals dimers, *Chem. Phys. Lett.*, **65**, 240 (1979).

J. A. Beswick and J. Jortner, Intermolecular and intramolecular V–V transfer in the vibrational predissociation of some polyatomic van der

Waals complexes, *J. Chem. Phys.*, (in press).

J. A. Beswick and J. Jortner, Effect of the anharmonicity on vibrational predissociation, *Mol. Physics*, *Mol. Phys.*, **39**, 1137 (1979).

J. A. Beswick and A. Requena, "Rotational Predissociation of Triatomic van der Waals molecules", *J. Chem. Phys.*, **72**, 3018 (1979).

J. E. Grabenstetter and R. J. LeRoy, "Widths (lifetimes) and energies for metastable levels of atom-diatom complexes", *Chem. Phys.*, **42**, 41 (1979).

R. Ramaswamy and A. E. DePristo, "Dynamics of van der Waals molecules: A scaling theoretical analysis of I_2He", *J. Chem. Phys.*, **72**, 770 (1980).

J. A. Beswick and G. Delgado-Barrio, "Effects of rotation on the vibrational predissociation of the HeI_2 van der Waals molecule", *J. Chem. Phys.*, **73**, 3653 (1980).

J. A. Beswick and A. Requena, "Rotational effects in the vibrational predissociation of $X..H_2$ van der Waals molecules", *J. Chem. Phys.*, **73**, 4347 (1980).

Dimer Concentration and Contribution to Bulk Properties

G. Glocker, C. P. Roe, and D. L. Fuller, *J. Chem. Phys.*, **1**, 703 (1933).

D. Stogryn and J. Hirschfelder, Contribution of bound, metastable and free molecules to the second virial coefficient and some properties of double molecules, *J. Chem. Phys.*, **31**, 1531 (1959).

Dimers Contribution to Vibrational Relaxation

E. E. Nikitin, *Opt. Spectr.*, **9**, 8 (1960).

G. Ewing, The role of van der Waals molecules in vibrational relaxation processes, *Chem. Phys.*, **29**, 253 (1978).

Section 2

MULTIPHOTON-INDUCED CHEMISTRY

REDUCED EQUATIONS OF MOTION FOR COLLISIONLESS MOLECULAR MULTIPHOTON PROCESSES

SHAUL MUKAMEL

Department of Chemistry, William March Rice University, Houston, Texas 77001

CONTENTS

I. Introduction . . . 509
II. The Systematic Reduction Scheme . . . 513
III. The Model Hamiltonian for Molecular Multiphoton Processes—REM for the Populations . . . 516
IV. Expansion of the REM to Second-Order in the Field . . . 522
V. Evaluation of the Higher Order Correlation Functions . . . 527
VI. REM for Populations and Coherences . . . 535
VII. The Role of Coherences and Intramolecular Dephasing . . . 540
  A. The Early Stages of Region I-Coherent Driving . . . 540
  B. The Perturbative Line-Broadening Limit: Weak Coupling of a "System" and a "Bath" . . . 541
    1. Ordinary Line Shapes . . . 543
    2. Absence of Energy Redistribution . . . 543
    3. Limitations of the Perturbative Approach . . . 543
  C. The Quasicontinuum . . . 544
  D. The Complete Incoherent Driving—Rate Equations Revisited . . . 544
VIII. A Model for the Intramolecular Dipole Correlation Functions . . . 546
Acknowledgments . . . 551
References . . . 551

I. INTRODUCTION

The discovery[1-3] that polyatomic molecules under collision-free conditions may absorb many infrared quanta from a powerful laser and acquire energies of chemical interest (few eV), had triggered considerable experimental and theoretical efforts in recent years.[1-19] The currently available experimental information regarding molecular multiphoton processes (MMP) includes absolute cross-sections for energy absorption,[6] translational,[7] and vibrational[8] distribution of the products of unimolecular decomposition following infrared pumping, total reaction yields and branching ratios of various channels as a function of laser frequency, intensity and fluence,[9] infrared,[10] and visible[11] emission characteristics of the excited molecules, double resonance experiments[12] (both frequency and time resolved), etc.

The following qualitative picture[13–19] has emerged out of the numerous experimental and theoretical studies: The molecular energy levels are separated into three regions. In the lowest energy range (region I) the density of molecular states is very low and the laser field is interacting with isolated molecular states (coherent driving). In this region the laser power is required to overcome the molecular anharmonicities and phenomena such as threshold power, saturation behavior, isotopic selectivity, and multiphoton resonances are accounted for in terms of the molecular level-scheme of region I. After the molecule has absorbed few quanta, the density of molecular states becomes very large, and we can no longer describe the time evolution in terms of few isolated molecular states. This region is denoted region II or the quasicontinuum and a proper description of the molecular time evolution in this range requires a quantitative understanding of the mechanisms of intramolecular energy transfer and line broadening (dephasing)[20] of highly vibrationally excited polyatomic molecules, of which very little is known at present. (We should note, however, that recent developments in overtone spectroscopy,[21, 22] coherent transients,[23] and high-resolution optical spectroscopy[24] are currently yielding novel information regarding intramolecular line broadening.) Finally, when the molecule acquires enough energy for dissociation, it enters region III, where, in addition to all the complications of region II, we have to incorporate also the dynamics of unimolecular decomposition.

Some of the theoretical problems which are underlying the current studies of molecular multiphoton processes are as follows: (1) How much energy is absorbed by a polyatomic molecule interacting with a strong infrared laser, and what is the *intermolecular* distribution of energy as a function of the molecular and laser parameters (molecular size, frequencies, anharmonicities, laser frequency, intensity, and duration)? (2) What is the *intramolecular* energy redistribution rate? How much time does it take for the absorbed energy to flow among the various degrees of freedom? Does the energy randomize, and at what time scales? (i.e., does the molecule exhibit an ergodic behavior?)[25] These questions are essential for the observation of laser-specific nonthermal effects in chemical reactions. (3) Can we use the information extracted from studies of molecular multiphoton processes to study the dynamics of unimolecular reactions and, in particular, to test the validity of the various statistical approaches that are extensively used to predict reaction rates and branching ratios?[26]

It is clear that a complete quantum-mechanical treatment of these problems is extremely complicated and is impractical. The density of molecular states is very rapidly increasing with energy. (For SF_6, e.g., it is $10^3/\mathrm{cm}^{-1}$, $2\times10^8/\mathrm{cm}^{-1}$, and $3\times10^{11}/\mathrm{cm}^{-1}$ at energies of 5000 cm^{-1}, 10,000 cm^{-1}, and 19,000 cm^{-1}, respectively.) Due to the lack of structural information (i.e., potential surfaces) on highly excited polyatomic molecules, we do not know the exact nature and coupling strengths of these states even for a single polyatomic molecule. Moreover, even if we had this structural (static) information, it would have been impossible to solve for the dynamics of about 10^{10} states interacting with a strong laser field. On the other hand, we should bear in mind that the information that is of real interest for us is much less detailed than the knowledge of the complete molecular density matrix, including the amplitudes and phases of all molecular states. In practice we are interested only in a few molecular observables and their time evolution on a coarse-grained time scale ($\sim10^{-9}$ sec) which is much longer than the molecular frequencies (i.e., 10^{-13}–10^{-14} sec). *A complete dynamical treatment of the problem is thus neither feasible nor desirable.*

The current theories of unimolecular reactions[26] avoid these complications by assuming a complete microcanonical redistribution of energy prior to the reaction. This assumption makes it feasible to evaluate reaction rates using equilibrium statistical mechanics. The relative success of these theories in predicting reaction rates does not, however, prove the validity of their basic assumptions, which were very seriously challenged[25] by a variety of recent experimental[27] and theoretical[28] studies.

An inevitable conclusion from the foregoing discussion is that we should adopt a *mesoscopic* level of theoretical description, which is intermediate between the fully dynamical and completely statistical approaches.[29] The basic idea of the mesoscopic level of description is to find some few key molecular variables and to adopt a reduction scheme that will enable us to derive closed reduced equations of motion (REM) describing the approximate time evolution of these variables in the presence of the rest. Formally this is one of the most important problems in irreversible statistical mechanics,[30] and there are many methods for constructing REM, for example, the Langevin approach,[31–33] the stochastic Liouville approach,[34] and projection operator techniques.[35, 36]

It has been suggested phenomenologically[16, 6] and demonstrated in several cases[4, 5] that a simple description of the multiphoton excitation process in terms of ordinary rate equations may be quite adequate (at least for highly excited molecules in the quasicontinuum region). A basic prediction of this postulate,[6] which was experimentally verified, in particular by the recent low-power ICR experiments,[5b] is the dependence of the dissociation yield on the total fluence (and not on the laser power once it is above threshold). The major problem with the phenomenological approach is that apart from a convenient way of fitting experimental data,[7] it does not give us any microscopic interpretation of the observed rate constants.

Early attempts[17, 18] to derive reduced equations of motion (REM) for MMP (which may lead to rate equations in some limits) relied on separating the molecular degrees of freedom into a "system" and a "bath" with weak interaction between them. The few molecular normal modes that interact with the radiation field are taken to be the system, whereas the rest are the bath. This approach is in the spirit of conventional theories of line broadening.[34, 35, 37–39] It is, however, fraught with some difficulties, since the dynamics of highly vibrationally excited polyatomic molecules is not expected to be properly described, using a perturbative approach in anharmonicities.[25] Moreover, perturbative treatments in anharmonicities necessarily become less adequate with the degree of molecular excitation.

In this review we develop a new approach toward the derivation of REM for MMP,[19] which is based on the projection operator formalism of Zwanzig[36] and Mori[40] combined with the representation of the true molecular states. The latter enables us to formulate the problem in a form free of perturbative arguments in any intramolecular interactions.[19, 41, 42] The main steps in this "hydrodynamic-like" approach[19] are: (1) the choice of a few molecular operators whose expectation values are the important variables for the dynamics of MMP; (2) the definition of an appropriate Mori projection operator onto the space spanned by these operators; and (3) the derivation of reduced equations of motion (REM) for the time evolution of these variables.

This procedure is completely general, and formally the choice of the number and type of variables is arbitrary. However, the complexity and usefulness of the resulting REM depend crucially upon a successful choice of variables.

In Section II we present the general systematic reduction scheme, which allows us to construct a closed set of equations of motion for *any* chosen set of molecular variables, starting from the complete Liouville equation. We present two formally different schemes based on different choices of time ordering.[34, 43, 44] In Section III we construct the molecular Hamiltonian for MMP using molecular states "dressed" by the radiation field[15, 19, 45] and define a "minimal set" of reduced variables corresponding to the populations of the various levels. We then derive our REM for these variables. We are able to give closed formal expressions for the complete information that is contained in any multiphoton experiment involving populations only. This information is a hierarchy of k-time intramolecular dipole correlation functions where $k=2,4,6,\ldots$ (29). In Section IV we expand the REM to second-order in the field and show how under quite general conditions (the Markovian limit,[34, 36, 40] where the integrated Rabi frequency is small compared to the energetic spread of the states within the levels) they reduce to simple rate equations. In Section V we consider higher order terms in the expansion of the REM and define an expansion parameter which shows that in the Markovian limit the higher order correlation functions are not important. The conclusion from Sections IV and V is that in the Markovian limit simple rate equations apply, and most of the molecular information contained in the complete set of correlation functions (29) is redundant and is reduced to two numbers per transition. These are the integrated Rabi frequency and a dephasing rate given in terms of the time integral over a two-time dipole correlation function. (In addition, the REM depend on *ratios* of the statistical weights of the various levels.) In Section VI we derive a different set of REM for the populations and the coherence variables. The explicit inclusion of molecular variables corresponding to coherences in our REM is important for the sake of getting a unified description all the way from region I (where they are important) to the quasicontinuum, for establishing the connection between MMP and other spectroscopic techniques, and for getting a better insight on the dynamics of MMP. Intramolecular relaxation of populations (T_1 processes) need not be considered at all in these REM (unlike phenomenological Bloch equations) since they are "buried" in our choice of basis set. The intramolecular dephasing processes (T_2), however, are playing a major role in the continuous transition of the driving from coherent to totally incoherent. It should be pointed out that no perturbative arguments in intramolecular couplings need to be made in order to give a precise definition to the dephasing rates, since they are associated solely with our reduction scheme (i.e., choice of variables).[19, 20] We finally show how in the limit of fast dephasing these REM reduce to the same rate equations obtained in Section IV. This alternative derivation of the simple rate equations gives us a better understanding regarding the "loss of relevant information," which occurs when the REM reduce to simple rate equations. Finally in Section VIII we present a microscopic model for the intramolecular two-time dipole correlation functions which enables us to calculate the dephasing rates appearing in our REM and to compare with existing experimental data.

II. THE SYSTEMATIC REDUCTION SCHEME

We consider a large system with many degrees of freedom characterized by a Hamiltonian H and a density matrix ρ. The time evolution of ρ is given by the Liouville equation:

$$\frac{d\rho}{dt} = -i[H, \rho] \equiv -iL\rho \tag{1}$$

where L is the Liouville (tetradic) operator corresponding to H, i.e.,

$$\frac{d\rho_{ab}}{dt} = -i\sum_{cd} L_{ab,cd}\rho_{cd} \tag{2}$$

and

$$L_{ab,cd} = H_{ac}\delta_{b,d} - H^*_{bd}\delta_{ac} \tag{3}$$

The density matrix $\rho(t)$ contains the complete information regarding the state of the system at time t. In practice, however, much of this information is redundant and we shall be interested only in some projections of $\rho(t)$ on few "relevant" operators A_μ whose nature is determined by the initial conditions and the type of experiments considered. We are thus interested in deriving reduced equations of motion (REM) which will provide us directly with the time evolution of the expectation values of A_μ.

The appropriate formalism to achieve that goal is the projection operator technique of Zwanzig[36] and Mori.[40] The procedure goes as follows: We first have to define the set of molecular operators of interest, A_μ. We then define a scalar product of two operators as

$$(A_\mu, A_\nu) \equiv \mathrm{Tr}(A^\dagger_\mu A_\nu) \tag{4}$$

Throughout the present work we shall assume that the chosen operators A_μ are orthonormal with respect to the scalar product (4), that is,

$$S_{\mu\nu} \equiv (A_\mu, A_\nu) = \delta_{\mu,\nu} \tag{5}$$

although the general reduction procedure does not rely on this property and it could be easily generalized to include a general overlap matrix $S_{\mu\nu}$, as well as a more general definition of the scalar product.[40]

We now define a Mori projection P which projects onto the subspace spanned by our relevant operators A_μ:

$$PB = \sum_\mu (B, A_\mu)A_\mu \tag{6}$$

and the complementary projection

$$Q \equiv 1 - P \tag{6b}$$

Our quantities of interest are the expectation values of A_μ at time t, that is,

$$\sigma_\mu(t) \equiv (A_\mu, \rho(t)) \tag{7}$$

and using the definition (6) it is clear that all our relevant information is contained in the projection $P\rho(t)$, that is,

$$\sigma_\mu(t) = (A_\mu, P\rho(t)) \tag{8}$$

Making use of the Zwanzig[36]–Mori[40] technique we can now derive REM for $P\rho(t)$ and project it on the variables A_μ resulting in the following REM for σ_ν:

$$\frac{d\sigma_\nu}{dt} = -i\sum_\mu \langle L\rangle_{\nu\mu}\sigma_\mu(t) - \int_0^t d\tau \sum_\mu \langle R(t-\tau)\rangle_{\nu\mu}\sigma_\mu(\tau) - i\langle F(t)\rangle_\nu \tag{9}$$

where $R(t-\tau)$ is the tetradic operator

$$R(t-\tau) \equiv L\exp[-iQL(t-\tau)]QL \tag{10}$$

and the tetradic matrix elements $\langle L\rangle_{\nu\mu}$ and $\langle R\rangle_{\nu\mu}$ are defined as

$$\langle Y\rangle_{\nu\mu} \equiv \mathrm{Tr}(A_\nu^\dagger Y A_\mu) \qquad Y = R, L \tag{11}$$

$\langle F\rangle$ is the vector:

$$\langle F(t)\rangle_\nu \equiv \mathrm{Tr}[A_\nu^\dagger L\exp(-iQLt)Q\rho(0)] \tag{12}$$

$\rho(0)$ being the density matrix at time $t=0$. Equations 9 are exact and are valid for an arbitrary choice of dynamical operators A_μ. In practice, however, the memory kernel $\langle R\rangle$ and the vector $\langle F\rangle$ are usually evaluated in some approximate manner using an expansion in a properly chosen parameter. To that end it is sometimes advantageous to use a different form of the REM,[19] that is,

$$\frac{d\sigma_\nu}{dt} = -i\sum_\nu \langle L\rangle_{\nu\mu}\sigma_\nu(t) - \sum_\nu \langle K(t)\rangle_{\nu\mu}\sigma_\mu(t) - i\langle G(t)\rangle_\nu \tag{13}$$

where

$$\langle K(t)\rangle_{\nu\mu} = \sum_{\nu'} W(t)_{\nu\nu'} V_{\nu'\mu}^{-1}(t) \tag{14}$$

$$W_{\nu\nu'}(t) = \mathrm{Tr}[A_\nu^\dagger LQ\exp(-iLt)A_{\nu'}] \tag{15}$$

$$V_{\nu'\mu} = \mathrm{Tr}[A_{\nu'}^\dagger \exp(-iLt)A_\mu] \tag{16}$$

and

$$\langle G(t)\rangle_\mu = \mathrm{Tr}\left[A_\nu^\dagger L \exp(-iLt) Q\rho(0) \right] \tag{17}$$

The form (9) arises naturally when keeping the complete time ordering of the various operators and will be referred to as the COP (chronological ordering prescription). The form (13) uses only partial time ordering and will be referred to as the POP (partial ordering prescription). Equation 13, like (9), is also exact. However, once an expansion is made both equations may have very different predictions. A comparison of the two forms for general-relaxation and line-shape problems was made recently.[43, 44, 46] In the next section we shall perform a formal expansion of both REM for a particular choice of variables for the multiphoton problem, and this will enable us to point out more precisely the differences between the two.

III. THE MODEL HAMILTONIAN FOR MOLECULAR MULTIPHOTON PROCESSES—REM FOR THE POPULATIONS

We consider a polyatomic molecule interacting with a monochromatic infrared laser beam whose frequency is ω_L, under collision-free conditions. We assume that the Schrödinger equation for the isolated molecule (in the absence of the field) has been solved and that we have the complete set of molecular eigenvalues as well as the dipole matrix elements between them. Assuming that the molecule is initially cold ($kT \ll \hbar\omega_L$), then only states with energies around $n\omega_L$, $n=0,1,2,\ldots$ are important for the multiphoton excitation process and need to be considered. We shall therefore group these relevant molecular states into levels and denote them as $\{|n\alpha\rangle\}$ with eigenvalues $E^\circ_{n\alpha}$, where n stands for the level and α runs over the states within the nth level. We further invoke the rotating wave approximation (RWA),[13-15, 45] which is very reasonable for MMP with infrared photons and which amounts to neglecting high-frequency terms in the Hamiltonian, which are not expected to contribute significantly to the molecular time evolution. We can thus write the combined Hamiltonian for the molecule and the field in the time-independent form:[19]

$$H = H_0 + H' \tag{18}$$

where

$$H_0 = \sum_{n,\alpha} |n\alpha\rangle E_{n\alpha} \langle n\alpha| \tag{18a}$$

and

$$H' = \epsilon \sum_{\substack{n\alpha \\ m\beta \\ m=n\pm 1}} |n\alpha\rangle \mu_{nm}^{\alpha\beta} \langle m\beta| \tag{18b}$$

Here the molecular states within the nth level have absorbed n infrared quanta from the field, and $E_{n\alpha} = E_{n\alpha}^{\circ} - n\omega_L$ is the energy of the $|n\alpha\rangle$ state dressed (to zero order) by the field. $\mu_{nm}^{\alpha\beta} = \langle m\beta|\mu|n\alpha\rangle$ is the transition dipole between the $|n\alpha\rangle$ and $|m\beta\rangle$ states and ϵ is the laser field amplitude. The molecular level and coupling scheme is presented in Fig. 1.

We shall now turn to the construction of the set of relevant operators. In a molecular multiphoton excitation experiment the quantities that are of primary interest to us are the populations of the various levels (the probability $P_n(t)$ for the molecule to absorb n photons at time t). It is thus clear that a minimal set of relevant variables should include these populations. In order to derive REM for $P_n(t)$ we shall now introduce the following set of molecular operators:[19]

$$A_{nn} = \frac{1}{\sqrt{d_n}} \sum_{\alpha} |n\alpha\rangle\langle n\alpha| \qquad n = 0, 1, \ldots, N-1 \tag{19}$$

d_n being the number of states within the nth level (the statistical weight of that level) and N is the total number of levels considered. The choice of the states included in A_{nn} (and consequently of d_n) is important since the ratios d_n/d_m enter explicitly in the resulting REM (48). It should thus be made with physical insight and only states that are expected to participate in the dynamics of the MMP in the experimentally relevant time scale should be included in the summation (19). The A_{nn} operators (19) are

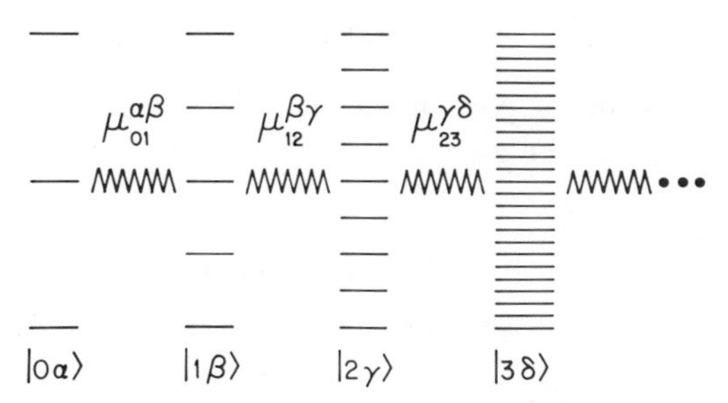

Fig. 1. The coupling scheme for molecular multiphoton processes. The true molecular states are grouped into levels where the nth level includes states $|n\phi\rangle$ which absorbed n photons from the field. μ is the radiative dipole coupling.

orthonormal, that is,

$$(A_{nn}, A_{kk}) = \delta_{n,k} \tag{20}$$

and the populations of the various levels are

$$P_n(t) = \sqrt{d_n}\,(A_{nn}, \rho(t)) = \sqrt{d_n}\,\sigma_n(t) \tag{21}$$

From the definitions (6), (19), and (18) it is clear that for our particular choice of P [(6) with (19)] we have

$$PLP = 0 \tag{22a}$$

and

$$PL_0 = L_0 P = 0 \tag{22b}$$

Here L, L_0 and L' denote the Liouville operators corresponding to H, H_0 and H', respectively. We shall further assume that initially at $t=0$ all the population is in the zeroth level so that

$$\rho(0) = \frac{1}{\sqrt{d_0}} A_{00} \tag{23a}$$

which implies that

$$Q\rho(0) = 0 \tag{23b}$$

Substitution of (23b) in (12) and (17) shows that for this type of initial condition we have

$$\langle F \rangle = \langle G \rangle = 0 \tag{23c}$$

that is, the inhomogeneous part of the REM vanishes.

We shall turn now to the expansion of the REM in a power series of L' using both time-ordering prescriptions.

A. The COP Reduced Equations of Motion

Using (6), (19), and (18) we note that

$$QL = (1-P)(L_0 + L') = L_0 + QL' \tag{24}$$

which enables us to expand $\exp(-iQLt)$ and the memory kernel R eq (10) in a power series in QL', that is,

$$\exp(-iQLt) = \exp(-iL_0 t) - i\int_0^t d\tau \exp[-iL_0(t-\tau)]QL'\exp(-iQL\tau) \tag{25}$$

We note in addition that due to the nature of our P and L', only even powers in QL' will contribute to PRP. We thus have

$$\frac{d\sigma_\nu}{dt} = -\int_0^t d\tau \sum_\mu \langle R(t-\tau)\rangle_{\nu\mu}\sigma_\mu(\tau) \tag{26}$$

where

$$\langle R(t-\tau)\rangle = \langle R^{(2)}(t-\tau)\rangle + \langle R^{(4)}(t-\tau)\rangle + \langle R^{(6)}(t-\tau)\rangle + \cdots \tag{26a}$$

and where

$$\langle R^{(2)}(t-\tau)\rangle = \theta_2(t-\tau,0) \tag{26b}$$

$$\langle R^{(4)}(t-\tau)\rangle = \int_0^{t-\tau} d\tau_1 \int_0^{\tau_1} d\tau_2\, \theta_4(t-\tau,\tau_1,\tau_2,0) \tag{26c}$$

$$\langle R^{(6)}(t-\tau)\rangle = \int_0^{t-\tau} d\tau_1 \int_0^{\tau_1} d\tau_2 \int_0^{\tau_2} d\tau_3 \int_0^{\tau_3} d\tau_4\, \theta_6(t-\tau,\tau_1,\tau_2,\tau_3,\tau_4,0) \tag{26d}$$

and

$$\langle R^{(2k)}(t-\tau)\rangle = \int_0^{t-\tau} d\tau_1 \int_0^{\tau_1} d\tau_2 \cdots \times \int_0^{\tau_{2k-3}} d\tau_{2k-2}\, \theta_{2k}(t-\tau,\tau_1,\tau_2,\ldots,\tau_{2k-2},0) \tag{26e}$$

Here

$$\theta_{2k}(\tau_1,\tau_2,\ldots,\tau_{2k}) = (-1)^{k+1}\langle L'(\tau_1)L'(\tau_2)(1-P)L'(\tau_3)L'(\tau_4) (1-P)\ldots(1-P)L'(\tau_{2k-1})L'(\tau_{2k})\rangle \tag{27}$$

where

$$L'(\tau) = \exp(iL_0\tau)L'\exp(-iL_0\tau) \tag{28}$$

Let us introduce further the kth moment of L' as the k-time correlation function

$$M_k(\tau_1, \tau_2 \ldots \tau_k) = \langle L'(\tau_1) L'(\tau_2) \ldots L'(\tau_k) \rangle \tag{29}$$

in terms of which we may rewrite (27) in the form:

$$\theta_2(\tau_1, \tau_2) = M_2(\tau_1, \tau_2) \tag{30a}$$

$$\theta_4(\tau_1, \tau_2, \tau_3, \tau_4) = -[M_4(\tau_1, \tau_2, \tau_3, \tau_4) - M_2(\tau_1, \tau_2) M_2(\tau_3, \tau_4)] \tag{30b}$$

$$\vdots$$

etc.

The $\langle \ldots \rangle$ in (26)–(30) denotes a tetradic matrix element as defined by eq (11).

Equations 26 together with (27)–(30) constitute our COP reduction scheme. They enable us to derive closed REM for the N populations of the various levels ($P_n = \sqrt{d_n}\ \sigma_n$, $n = 0, 1, \ldots, N-1$) in terms of the tetradic $N \times N$ R matrix. Evaluation of the latter requires the calculation of the intramolecular dipole correlation functions M_k (29) [or θ_k, (30)], $k = 2, 4, \ldots$ which provide us with the complete molecular information needed for the description of all MMP whenever the experimental observables are connected with populations only.

B. The POP Reduced Equations of Motion

Turning now to the POP reduction scheme for our particular choice of P [(6) with (19)], we may use the relation

$$\exp(-iLt) = \exp(-iL_0 t) - i\int_0^t d\tau \exp(-iL(t-\tau)) L' \exp(-iL_0 \tau) \tag{31}$$

to expand $\langle K(t) \rangle$ (14) in a power series in L', that is,

$$\langle K(t) \rangle = \langle L'Q(1 - i\int_0^t d\tau \exp(-iL(t-\tau)) L' \rangle$$

$$\times \left[1 - i\int_0^t d\tau \langle \exp(-iL(t-\tau)) L' \rangle \right]^{-1} \tag{32}$$

resulting in[19]

$$\frac{d\sigma_\nu}{dt} = -\sum_\mu \langle K(t) \rangle_{\nu\mu} \sigma_\nu(t) \tag{33}$$

where

$$\langle K(t)\rangle = \langle K^{(2)}(t)\rangle + \langle K^{(4)}(t)\rangle + \cdots \tag{33a}$$

and where

$$\langle K^{(2)}(t)\rangle = \int_0^t d\tau\, M_2(t-\tau,0) \tag{33b}$$

$$\begin{aligned}\langle K^{(4)}(t)\rangle &= -\int_0^t d\tau_1 \int_0^{\tau_1} d\tau_2 \int_0^{\tau_2} d\tau_3\, M_4(t,\tau_1,\tau_2,\tau_3) + \int_0^t d\tau_1 \int_0^t d\tau_2 \int_0^{\tau_2} d\tau_3 \\ &\quad \times M_2(t,\tau_1) M_2(\tau_2,\tau_3) \\ &= -\int_0^t d\tau_1 \int_0^{\tau_1} d\tau_2 \int_0^{\tau_2} d\tau_3 \big[M_4(t,\tau_1,\tau_2,\tau_3) - M_2(t,\tau_1)M_2(\tau_2,\tau_3) \\ &\quad - M_2(t,\tau_2)M_2(\tau_1,\tau_3) - M_2(t,\tau_3)M_2(\tau,\tau_2)\big] \end{aligned} \tag{33c}$$

$$\vdots$$

The complete molecular information that enters into the POP equations (33) is identical to that used for the COP (26); that is, it consists of the entire set of intramolecular correlation functions M_k (29). However, these correlation functions enter in a different way in each reduction scheme. The expansions (26) and (33) enable us to point out more precisely the differences between the two prescriptions. In principle they are both exact.[43, 44] However, if we truncate the REM at second order, this amounts to setting $\langle R^{(4)}\rangle = \langle R^{(6)}\rangle = \cdots = 0$ in the COP (26) and $\langle K^{(4)}\rangle = \langle K^{(6)}\rangle = \cdots = 0$ in the POP (33). $\langle R^{(4)}\rangle = 0$ implies

$$M_4(\tau_1,\tau_2,\tau_3,\tau_4) = M_2(\tau_1,\tau_2)M_2(\tau_3,\tau_4) \tag{34}$$

whereas $\langle K^{(4)}\rangle = 0$ implies

$$\begin{aligned} M_4(\tau_1,\tau_2,\tau_3,\tau_4) &= M_2(\tau_1,\tau_2)M_2(\tau_3,\tau_4) + M_2(\tau_1,\tau_3)M_2(\tau_2,\tau_4) \\ &\quad + M_2(\tau_1,\tau_4)M_2(\tau_2,\tau_3) \end{aligned} \tag{35}$$

Thus, the two expansions correspond to different statistical properties of the dipole operator. These points and their implications were discussed recently in detail for other line-shape problems.[43, 44]

Finally, we should point out that there is one general type of condition in which both equations reduce to the same form. This is the *Markovian limit* where a separation of time scales exists between the relevant operators A_μ and the dipole correlation functions, whereupon the latter decay on a time scale τ_c which is much more rapid than the evolution of the former. In this case the higher order terms M_4, M_6 have a contribution which is higher order in τ_c compared to that of M_2 and may be neglected, and both REM [(26) and (33)] reduce, on a coarse-grained time scale $t > \tau_c$, to the form:

$$\frac{d\sigma_\nu}{dt} = -\sum_\mu W_{\nu\mu}\sigma_\nu(t) \tag{36}$$

where

$$W = \langle K^{(2)}(\infty)\rangle = \int_0^\infty d\tau \langle R^{(2)}(\tau)\rangle = \int_0^\infty d\tau \langle L'(\tau)L'(0)\rangle \tag{36a}$$

In this case the resulting REM (36) attain a very simple time-independent form and the amount of molecular information necessary for the description of the MMP is considerably reduced [we need consider only W (36a) instead of the complete set of correlation functions M_k (29)]. In the coming sections we shall make use extensively of the Markovian limit condition after justifying it from microscopic considerations for MMP.

IV. EXPANSION OF THE REM TO SECOND-ORDER IN THE FIELD

We shall now apply the formal results of Section III to derive explicit expressions for the REM, to second order in the applied field (H'). We shall adopt here the COP formulation. We should bear in mind, however, that the final rate equations derived here (52) are in the Markovian limit where the POP equations coincide with the COP as implied by (36).

To second-order in the applied field (H'), we set

$$\langle R(t-\tau)\rangle \cong \langle R^{(2)}(t-\tau)\rangle \tag{37}$$

in (26). Substitution of (18) in (26b) results in the following nonzero matrix elements of R:

$$\langle R_{nn,mm}(t-\tau)\rangle = \frac{1}{(d_n d_m)^{1/2}} \sum_{\alpha\beta} \left(L'^{\alpha\alpha,\alpha\beta}_{nn,nm}(t-\tau) L'^{\alpha\beta,\beta\beta}_{nm,mm}(0) + \text{c.c.} \right)$$

$$m = n \pm 1 \tag{38a}$$

and

$$\langle R_{nn,nn}(t-\tau)\rangle = \frac{1}{d_n} \sum_{\substack{\alpha,\beta \\ m=n\pm 1}} \left(L'^{\alpha\alpha,\alpha\beta}_{nn,nm}(t-\tau) L'^{\alpha\beta,\alpha\alpha}_{nm,nn}(0) + \text{c.c.} \right) \tag{38b}$$

Here

$$L'^{\alpha\beta,\gamma\delta}_{ij,kl} - \langle\langle i\alpha, j\beta | L' | k\gamma, l\delta \rangle\rangle \tag{39}$$

comes for the tetradic matrix element of L' (we are using here the double bracket notation[39] whereby the tetradic state corresponding to $|a\rangle\langle b|$ is denoted $|ab\rangle\rangle$). In (38) we have made use of the Liouville conjugation symmetry

$$L'_{ab,cd} = -L'^{*}_{ba,dc} \tag{40}$$

which is valid for any tetradic operator[39] and may be easily verified using (3).

We shall now define the two-time intramolecular dipole correlation function for the nm transition:

$$I_{nm}(t) = \mathrm{Re}\,\tilde{I}_{nm}(t) \tag{41a}$$

where

$$\tilde{I}_{nm}(t) = \frac{\langle \mu_{nm}(0)\mu_{mn}(t)\rangle}{\langle \mu_{nm}(0)\mu_{mn}(0)\rangle} = \frac{1}{\gamma_{nm}^2} \sum_{\alpha,\beta} |\mu_{nm}^{\alpha\beta}|^2 \exp(-i\omega_{n\alpha,m\beta} t) \tag{41b}$$

Here

$$\mu_{nm}(t) = \exp(iH_0 t)\mu_{nm}\exp(-iH_0 t) \tag{42}$$

$$\gamma_{nm}^2 = \langle \mu_{nm}(0)\mu_{mn}(0)\rangle = \sum_{\alpha,\beta} |\mu_{nm}^{\alpha\beta}|^2 \tag{43}$$

and

$$\hbar\omega_{n\alpha,m\beta} = E_{n\alpha} - E_{m\beta} \tag{44}$$

We further define the integrated Rabi frequency for the nm transition

$$\bar{\Omega}_{nm}^2 \equiv 2\epsilon^2 \sum_{\alpha,\beta} \frac{|\mu_{nm}^{\alpha\beta}|^2}{(d_n d_m)^{1/2}} \tag{45}$$

Making use of the quantities (41)–(45) we can express the matrix elements of $\langle R\rangle$ (38) as follows:

$$\langle R^{(2)}_{nn,mm}(t-\tau)\rangle = -\bar{\Omega}_{nm}^2 I_{nm}(t-\tau) \tag{46a}$$

and

$$\langle R^{(2)}_{nn,nn}(t-\tau)\rangle = \sum_{m=n\pm 1} \bar{\Omega}^2_{nm}\left(\frac{d_n}{d_m}\right)^{1/2} I_{nm}(t-\tau) \tag{46b}$$

Upon substitution of (46) in (9) and putting $P_n = (d_n)^{1/2}\sigma_n$, we finally get:

$$\frac{dP_n}{dt} = \sum_{m=n\pm 1} \bar{\Omega}^2_{nm}(d_n d_m)^{1/2}\int_0^t d\tau\, I_{nm}(t-\tau)\left[\frac{P_m(\tau)}{d_m} - \frac{P_n(\tau)}{d_n}\right] \tag{47}$$

or, alternatively,

$$\frac{dP_n}{dt} = \sum_{m=n\pm 1} \bar{\Omega}^2_{nm}\int_0^t d\tau\, I_{nm}(t-\tau)\left[P_m(\tau)\left(\frac{d_n}{d_m}\right)^{1/2} - P_n(\tau)\left(\frac{d_m}{d_n}\right)^{1/2}\right] \tag{48}$$

Equations (47) or (48) reduce to simple rate equations in the following limit: If $I_{nm}(t-\tau)$ has a characteristic time scale $\tau_c = \Gamma^{-1}_{nm}$ such that

$$\bar{\Omega}_{nm} \gg \Gamma_{nm} \tag{49}$$

then we expect $P_n(t)$ to vary on a time scale considerably longer than Γ^{-1}_{nm} (this expectation will be verified later). In this case $I_{nm}(t-\tau)$ acts like a δ function inside the integrals of (47) or (48). We thus have:

$$I_{nm}(t-\tau) = \Gamma^{-1}_{nm}\delta(t-\tau) \tag{50}$$

where

$$\tau_c \equiv \Gamma^{-1}_{nm} = \int_0^\infty d\tau\, I_{nm}(\tau) \tag{51}$$

Substitution of (50) in (48) results in the simple rate equations:

$$\frac{dP_n}{dt} = \sum_{m=n\pm 1} W^{(2)}_{nm}\left[P_m\left(\frac{d_n}{d_m}\right)^{1/2} - P_n\left(\frac{d_m}{d_n}\right)^{1/2}\right] \tag{52}$$

where

$$W^{(2)}_{nm} = \frac{\bar{\Omega}^2_{nm}}{\Gamma_{nm}} \tag{52a}$$

Equation 52 may be recast in the form:

$$\frac{dP_n}{dt} = \sum_{m=n\pm 1} \mathrm{K}^{(2)}_{nm}P_m(t) - \mathrm{K}^{(2)}_{nn}P_n(t) \tag{53}$$

where

$$K_{nm}^{(2)} = \left(\frac{d_n}{d_m} \right)^{1/2} \frac{\bar{\Omega}_{nm}^2}{\Gamma_{nm}} \tag{53a}$$

and

$$K_{nn}^{(2)} = \sum_{m=n\pm1} K_{mn}^{(2)} \tag{53b}$$

We note that K_{nm} satisfies the detailed balance relation

$$\frac{K_{nm}}{K_{mn}} = \frac{d_n}{d_m} \tag{53c}$$

which implies that the radiation field is tending to establish a distribution of molecular states where all radiatively coupled states are equally populated. Thus the ratio (53c) is equal to the ratio of the number of effectively coupled states in each level. The superscript (2) in (52) and (53) signifies that $W^{(2)}$ and $K^{(2)}$ are evaluated to second order in $H'(\mu)$. Utilizing (53a) and (52a) we have

$$W_{nm}^{(2)} = \left(K_{nm}^{(2)} K_{mn}^{(2)} \right)^{1/2} \tag{54}$$

Condition (49) is the *Markovian limit*[34–36, 40, 19] and implies that a separation of time scales exists in our problem such that our chosen set of variables (i.e., the P space) is slowly varying compared to the other variables (the Q space). Equations 52 could have been obtained directly from (36) where we have derived the Markovian form in a formal way. A posteriori we can now justify the substitution (50). Using (52) we notice that the characteristic rate of change of P_n is $W_{nm} = \bar{\Omega}_{nm}^2 / \Gamma_{nm}$. Condition (49) thus implies also that $W_{nm} \ll \Gamma_{nm}$, which provides a consistency check to our assumption that P_n are varying on a much slower time scale than Γ_{nm}^{-1} [which led to (50)].

We note that the only molecular information that enters the rate equations (53) is

1. $\bar{\Omega}_{nm}$, the integrated dipole for the *nm* transition times the field amplitude;
2. Γ_{nm}, inverse correlation time (i.e. a dephasing rate); and
3. *Ratios* of statistical weights of the levels d_n / d_m.

The relevant molecular information thus reduces essentially to two numbers ($\bar{\Omega}^2_{nm}/\Gamma_{nm}$ and d_n/d_m) per transition in the Markovian limit where rate equations apply.

The significance of $\bar{\Omega}_{nm}$ as defined in (45) lies in the reasonable assumption[19] that for large molecules where only few degrees of freedom are coupled with the radiation field, $\bar{\Omega}_{nm}$ will be approximately independent of the number of states involved (i.e., d_n and d_m). This is expected, since any quantity of the form

$$\sum_{\alpha\beta} \frac{1}{d_n} |\langle n\alpha|\mu|m\beta\rangle|^2 = |\mu_{nm}|^2 \tag{55a}$$

or

$$\sum_{\alpha\beta} \frac{1}{d_m} |\langle n\alpha|\mu|m\beta\rangle|^2 = |\mu_{nm}|^2 \tag{55b}$$

is independent on the addition of degrees of freedom that do not couple with μ. The dipole sum rule (55a) implies $\Omega^2_{nm} \equiv \Sigma_{\alpha\beta}|\langle n\alpha|\mu|m\beta\rangle|^2 \propto d_n$ and (55b) implies $\Omega^2_{nm} \propto d_m$. It is thus fair to assume that Ω^2_{nm} is proportional to $(d_n d_m)^{1/2}$, which implies that $\bar{\Omega}_{nm}$ is independent of d_n and d_m. Regarding the dephasing rate Γ_{nm} it is clear from its definition (51) that it is associated with the energetic spread of the states within the n and m molecular levels. For the sake of illustration, we shall consider now a simple model whereby $|\mu^{\alpha\beta}_{nm}|^2$ is constant, independent on α and β, that is,

$$|\mu^{\alpha\beta}_{nm}|^2 = \begin{cases} |V|^2 & -\Delta < E_{n\alpha}, E_{m\beta} < \Delta \\ 0 & \text{else} \end{cases} \tag{56}$$

Assuming that the number of states d_n, d_m is sufficiently large and that they are uniformly distributed throughout the n, m manifolds, we can replace the summation (41) by an integration, resulting in

$$I_{nm}(t) = \frac{\sin^2(\Delta t)}{(\Delta t)^2} \tag{57a}$$

and

$$\bar{\Omega}^2_{nm} = 2\epsilon^2 |V|^2 (d_n d_m)^{1/2} \tag{57b}$$

The characteristic time scale of $I_{nm}(t)$ is thus Δ^{-1}. The condition for the

Markovian limit (49) is then equivalent to

$$\bar{\Omega}_{nm} \ll \Delta \tag{58}$$

and we have

$$\Gamma_{nm}^{-1} = \int_0^{\infty} d\tau\, I_{nm}(\tau) = \frac{\pi}{2\Delta} \tag{59}$$

Finally, it should be pointed out that the molecular driving rates $\mathrm{K}_{nm}^{(2)}$ may be recast in the form

$$\mathrm{K}_{nm}^{(2)} = \frac{\bar{\Omega}_{nm}^2}{\Gamma_{nm}} \left(\frac{d_n}{d_m} \right)^{1/2} = 2\pi\epsilon^2 |V|^2 \rho_n \tag{60}$$

where

$$\rho_n = \frac{d_n}{2\Delta} \tag{61}$$

is the density of states in the $|n\alpha\rangle$ manifold. Equation 60 is the familiar golden-rule-type expression. We found it convenient to express the REM in terms of the quantities $\bar{\Omega}_{nm}$ and Γ_{nm} rather than $|V_{nm}|^2$ and ρ_n, since $\bar{\Omega}_{nm}$ and Γ_{nm} are the two actual time scales of the problem and their relative magnitude determines the validity of the Markovian assumption. Also, when adding more modes to the molecule, which do not couple with the radiation field, then ρ_n is changing and $|V_{nm}|^2$ will depend strongly on ρ_n and ρ_m. Thus $|V_{nm}|^2$ and ρ_n are not the natural independent parameters for the problem for MMP. Ω_{nm} and Γ_{nm}, however, may be considered independent.

V. EVALUATION OF THE HIGHER ORDER CORRELATION FUNCTIONS

In order to get an estimate of the approximations involved in the expansion of the REM to second-order in H' (as was done in Section IV), we shall now proceed to the evaluation of the fourth-order contribution to $\langle R \rangle$. This will enable us to define a dimensionless expansion parameter for the series (26a) and as a result to define precisely the general conditions for its truncation.

We shall consider first $M_4(t-\tau, \tau_1, \tau_2, 0)$ which is required for the evaluation of $\langle R^{(4)}(t-\tau)\rangle_{mm,nn}$, where $m=n$, $n\pm 2$. To that end we take

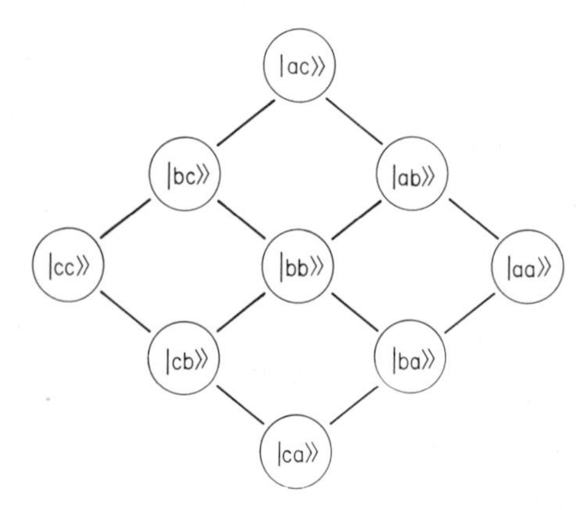

Fig. 2. Diagrammatic representation of the Liouville space terms contributing to M_4. Note that there are six pathways to go from $|aa\rangle\rangle$ to $|cc\rangle\rangle$ in fourther order.

three consecutive levels n, l, and m. For abbreviating the notation we shall throughout the present section substitute a, b, b', and c for $n\alpha$, $l\beta$, $l\beta'$, and $m\gamma$, respectively. Similarly Σ_a will substitute Σ_α, etc. Figure 2 presents diagrammatically the coupling scheme required for the evaluation of $(M_4)_{cc,aa}$. Using Fig. 2. we notice that there are six pathways which lead from $|aa\rangle\rangle$ to $|cc\rangle\rangle$ in fourth-order. However, by virtue of the Liouville conjugation symmetry (40) we need consider only three independent paths and the other three have a contribution which is simply their complex conjugate. We thus have:

$$M_4(t-\tau,\tau_1,\tau_2,0)_{cc,aa} = \text{①} + \text{②} + \text{③} + \text{c.c.} \tag{62}$$

where

$$\text{①} = \frac{1}{(d_a d_c)^{1/2}} \sum_{\substack{ab\\b'c}} L'_{cc,b'c}(t-\tau) L'_{b'c,ac}(\tau_1) L'_{ac,ab}(\tau_2) L'_{ab,aa}(0)$$

$$= \frac{1}{(d_a d_c)^{1/2}} \sum_{\substack{ab\\b'c}} \mu_{b'a}\mu_{ab}\mu_{bc}\mu_{cb'} \exp[i\omega_{cb'}(t-\tau) + i\omega_{b'a}\tau_1 + i\omega_{bc}\tau_2] \tag{62a}$$

$$\text{②} = \frac{1}{(d_a d_c)^{1/2}} \sum_{\substack{ab\\b'c}} L'_{cc,b'c}(t-\tau) L'_{b'c,b'b}(\tau_1) L'_{b'b,ab}(\tau_2) L'_{ab,aa}(0)$$

$$= \frac{1}{(d_a d_c)^{1/2}} \sum_{\substack{ab\\b'c}} \mu_{b'a}\mu_{ab}\mu_{bc}\mu_{cb'} \exp[i\omega_{cb'}(t-\tau) + i\omega_{bc}\tau_1 + i\omega_{b'a}\tau_2] \tag{62b}$$

and

$$\text{(III)} = \frac{1}{(d_a d_c)^{1/2}} \sum_{\substack{ab \\ b'c}} L'_{cc,cb}(t-\tau) L'_{cb,b'b}(\tau_1) L'_{b'b,ab}(\tau_2) L'_{ab,aa}(0)$$

$$= \frac{1}{(d_a d_c)^{1/2}} \sum_{\substack{ab \\ b'c}} \mu_{b'a}\mu_{ab}\mu_{bc}\mu_{cb'} \exp[i\omega_{bc}(t-\tau) + i\omega_{cb'}\tau_1 + i\omega_{b'a}\tau_2] \tag{62c}$$

At this stage we introduce a simplifying assumption which makes use of the complexity of our system. The various μ's are expected to vary randomly and have an arbitrary phase. As a result, we anticipate that

$$\sum_a \mu_{b'a}\mu_{ab} \cong \sum_a |\mu_{ab}|^2 \delta_{b,b'} \tag{63a}$$

and

$$\sum_c \mu_{bc}\mu_{cb'} \cong \sum_c |\mu_{bc}|^2 \delta_{bb'} \tag{63b}$$

This is a form of the statistical random phase approximation (RPA). Making use of this assumption we can omit the b' summation in (62) and set $b=b'$. It is now clear that when (62) holds, then diagrams (II) and (III) do not contribute at all to θ_4 (27) (and to $R^{(4)}$). This arises since they both pass through $A_{bb} = (1/\sqrt{d_l})\Sigma_\beta |l\beta\rangle\langle l\beta|$, and by construction of the P projection (6) we have

$$(1-P)A_{bb} = 0 \tag{64}$$

An alternative way to see this is by looking at (30b). We then note that the contribution to $M_4(t-\tau, \tau_1, \tau_2, 0)$ from these diagrams exactly equals that of $M_2(t-\tau, \tau_1)M_2(\tau_2, 0)$ and as a result their net contribution to θ_4 vanishes. Using (30), (62), and (63) we thus have:

$$\theta_4(t-\tau, \tau_1, \tau_2, 0) = -[\text{(I)} + \text{c.c.}]$$

$$= \frac{-1}{(d_a d_c)^{1/2}} \sum_{\substack{a,b \\ c}} |\mu_{ab}|^2 |\mu_{bc}|^2 \exp[i\omega_{cb}(t-\tau-\tau_2) + i\omega_{ba}\tau_1] + \text{c.c.} \tag{65}$$

Within the random phase approximation we do not expect $|\mu_{ab}|^2$ and $|\mu_{bc}|^2$ to be correlated. It is thus reasonable to assume that (65) may be further factorized in the form

$$\theta_4=\frac{-1}{\left(d_a d_c d_b^2\right)^{1/2}}\left[\sum_{ab}|\mu_{ab}|^2\exp(i\omega_{ba}\tau_1)\sum_{bc}|\mu_{bc}|^2\exp(i\omega_{cb}(t-\tau-\tau_2))+\text{c.c.}\right] \tag{66}$$

i.e.

$$\theta_4(t-\tau,\tau_1,\tau_2,0)=-\bar{\Omega}_{ab}^2\bar{\Omega}_{bc}^2\cdot\text{Re}\left[\tilde{I}_{ab}(\tau_1)\tilde{I}_{bc}(t-\tau-\tau_2)\right] \tag{67}$$

We are now in a position to evaluate $\langle R^{(4)}\rangle$,

$$\langle R^{(4)}_{cc,aa}(t-\tau)\rangle=\int_0^{t-\tau}d\tau_1\int_0^{\tau_1}d\tau_2\,\theta_4(t-\tau,\tau_1,\tau_2,0) \tag{68}$$

From the definition (41) we expect $\tilde{I}(\tau)$ to be finite only over a limited time scale $\tau\lesssim\tau_c=\Gamma^{-1}$, where Γ Is a measure of the energy spread of the levels involved. Using (67) it is clear that θ_4 (and consequently also $\langle R^{(4)}_{cc,aa}(t-\tau)\rangle$ (26c)) vanishes when $t-\tau\gtrsim\tau_c$. Thus the characteristic time scale of R is equal to that of $\tilde{I}$. Usually we are interested in the time evolution of P_n which occurs on a much longer time scale than Γ^{-1}. This is again a manifestation of the Markovian assumption (49). We can thus substitute in (26)

$$R^{(4)}(t-\tau)=-W^{(4)}\delta(t-\tau) \tag{69}$$

where

$$\begin{aligned}W^{(4)}_{cc,aa}&=-\int_0^{\infty}dt\int_0^{t}d\tau_1\int_0^{\tau_1}d\tau_2\,\theta_4(t,\tau_1,\tau_2,0)\\&=\bar{\Omega}_{ab}^2\bar{\Omega}_{bc}^2\int_0^{\infty}dt\int_0^{t}d\tau_1\int_0^{\tau_1}d\tau_2\text{Re}\left[\tilde{I}_{ab}(\tau_1)\tilde{I}_{bc}(t-\tau-\tau_2)\right]\end{aligned} \tag{70}$$

Equation (69) together with (50) and (26) result in the following rate equations:

$$\begin{aligned}\frac{dP_n}{dt}=&\sum_{m=n\pm1}W^{(2)}_{nm}\left[P_m\left(\frac{d_n}{d_m}\right)^{1/2}-P_n\left(\frac{d_m}{d_n}\right)^{1/2}\right]\\&+\sum_{\substack{m=n\pm1\\n\pm2}}W^{(4)}_{nm}\left[P_m\left(\frac{d_n}{d_m}\right)^{1/2}-P_n\left(\frac{d_m}{d_n}\right)^{1/2}\right]\end{aligned} \tag{71}$$

The evaluation of $W_{nm}^{(4)}$ where $m=n\pm 1$ proceeds along the same lines and the results are similar to (70). For brevity we shall not consider these terms here.

We shall now turn to an order of magnitude estimate of $W^{(4)}$. From the previous arguments, it is clear that the only contribution to the integral (68) comes from the region:

$$0<\tau_1, \tau_2, t<\tau_c=\Gamma^{-1} \tag{72}$$

since otherwise the integrand vanishes. We thus expect that

$$W^{(4)} \cong \frac{\bar{\Omega}_{ab}^2 \bar{\Omega}_{bc}^2}{\Gamma^3} \tag{73}$$

For the sake of illustration, let us take

$$\tilde{I}(\tau)=\exp(-\Gamma\tau) \tag{74}$$

which yields

$$\langle R^{(4)}(t-\tau)\rangle = \bar{\Omega}_{ab}^2 \bar{\Omega}_{bc}^2 \int_0^{t-\tau} d\tau_1 \int_0^{\tau_1} d\tau_2 \exp(-\Gamma\tau_1) \exp[-\Gamma(t-\tau-\tau_2)] \tag{75}$$

resulting in

$$\langle R^{(4)}(t-\tau)\rangle_{cc,aa} = \frac{\bar{\Omega}_{ab}^2 \bar{\Omega}_{bc}^2}{\Gamma^2} \exp[-\Gamma(t-\tau)] \{\exp[-\Gamma(t-\tau)]+\Gamma(t-\tau)-1\} \tag{76a}$$

and

$$W_{cc,aa}^{(4)} = -\int_0^{\infty} dt\, R^{(4)}(t)_{cc,aa} = \frac{1}{2} \frac{\bar{\Omega}_{ab}^2 \bar{\Omega}_{bc}^2}{\Gamma^3} \tag{76b}$$

Equations (76) demonstrate the validity of our general argument leading to (73), which is not restricted to the particular form (74). Assuming that $\bar{\Omega}_{ab} \cong \bar{\Omega}_{bc} = \bar{\Omega}$ we recall from (52a) that

$$W^{(2)} = \frac{\bar{\Omega}^2}{\Gamma} \tag{77}$$

We thus get, using (73) and (77)

$$\frac{W^{(4)}}{W^{(2)}}=\frac{\mathrm{K}^{(4)}}{\mathrm{K}^{(2)}}=(\bar{\Omega}/\Gamma)^2=\frac{W^{(2)}}{\Gamma} \tag{78}$$

which shows that the expansion parameter for the series

$$W=W^{(2)}+W^{(4)}+\dots \tag{79}$$

is

$$\eta=\left(\frac{\bar{\Omega}}{\Gamma}\right)^2=\frac{W^{(2)}}{\Gamma} \tag{80}$$

Thus

$$W^{(2k+2)}\cong W^{(2)}\eta^k \tag{81}$$

We further note that the condition for the validity of the Markovian limit [i.e., (49)] is $\eta \ll 1$, which implies that *in this limit the higher order terms in the expansion* (79) *become unimportant*!

In concluding this section we note that using the present formalism we were able to define precisely the complete molecular information that is relevant for MMP and to see how this information is reduced considerably in the *Markovian limit* leading to simple rate equations. We shall now summarize the basic steps leading to the rate equations and discuss the validity of the Markovian limit for typical molecular multiphoton experiments. Our general derivation of the rate equations goes as follows:

1. We have grouped the molecular states relevant for the dynamics of MMP into levels, and making use of the Mori–Zwanzig[36,40] projection operator formalism we have derived our most general set of REM [(26) or (33)] for the populations of these levels. The complete molecular information that is required for our REM is a hierarchy of k-time intramolecular dipole correlation functions M_k where $k=2,4,6,\dots$.
2. The two-time correlation functions $M_2(t-\tau,0)$ are factorized in the form

$$M_2(t-\tau,0)_{nn,mm}=-\bar{\Omega}^2_{nm}I_{nm}(t-\tau) \tag{82}$$

where $\bar{\Omega}^2_{nm}=M_2(0,0)$ is the square of the integrated Rabi frequency for the nm transition, is proportional to the incoming laser intensity, and is roughly independent of the addition of molecular degrees of freedom that do not couple directly with the radiation field. $I_{nm}(t-\tau)$ is a dimensionless correlation function that decays from one to zero on a characteristic correlation time $\tau_c = \Gamma_{nm}^{-1}$, which has to do with the energy spread of the states within the n and m levels.

3. The Markovian limit is defined whenever

$$\sqrt{\eta} = \frac{\bar{\Omega}}{\Gamma_{nm}} \ll 1 \tag{83}$$

For a typical infrared transition dipole (0.1 Debye) and taking an intense laser field of 10 MW cm^{-2} we have $\bar{\Omega} \sim 1$ cm^{-1}. Γ_{nm} may be estimated from the observed multiphoton cross-sections for energy absorption[4, 5] and it is typically around 10–100 cm^{-1} for highly vibrationally excited polyatomic molecules (see Section VIII). Thus condition (83) is expected usually to hold for real life MMP even for very intense laser fields. In this limit we have the following:

i. On the relevant time scale for the evolution of populations we may write

$$I_{nm}(t-\tau) = \Gamma_{nm}\delta(t-\tau) \tag{84}$$

where

$$\Gamma_{nm}^{-1} = \int_0^{\infty} d\tau \, I_{nm}(\tau) \tag{85}$$

and the REM reduce to simple rate equations of the form (55) with the rate matrix

$$W = W^{(2)} + W^{(4)} + W^{(6)} + \ldots \tag{86}$$

ii. The higher order terms in the expansion (86) are much smaller than $W^{(2)}$. In fact

$$W^{(2k+2)} \cong W^{(2)}\eta^k \ll W^{(2)} \tag{87}$$

iii. The molecular information that is required for a proper description of the multiphoton excitation process is greatly reduced. Not only that the high-order correlation functions M_{2k}, $k=2,3,\ldots$ are not important, but even M_2 enters only via $\bar{\Omega}_{nm}$ and $\Gamma_{nm}^{-1} = \int_0^{\infty} d\tau \, I_{nm}(\tau)$. Thus even the details of the two-time correlation functions $I_{nm}(\tau)$ (41) are also irrelevant for the dynamics of the MMP. The rate equations thus depend only on $\bar{\Omega}_{nm}$, Γ_{nm} and *ratios* of the statistical weights of the levels d_n/d_m.

iv. The COP and POP equations are the same and the present analysis justifies the formal derivation of (36), which is equivalent to (52).

4. Finally, we would like to make a few comments regarding a constant coupling model, since it was extensively studied in the past in connection with other molecular relaxation problems,[47, 48] although this model is physically unrealistic for MMP. In the constant coupling model we take $\mu_{nm}^{\alpha\beta}$ to be independent of α and β (no randomness in phase). We thus assume:

$$\mu_{nm}^{\alpha\beta}=\begin{cases} V & -\Gamma<E_{n\alpha},\ E_{m\beta}<\Gamma \\ 0 & \text{else} \end{cases} \tag{88}$$

We further assume that Γ^{-1} is much faster than our relevant time scale for P_n (the Markovian assumption) and that $d\gg 1$. For this model the REM will assume the same form [i.e., (71)] as for the previous coupling model (56) with the random phase approximations, and $W_{ab}^{(2)}$ attains the same value, that is,

$$\tilde{W}_{ab}^{(2)}=W_{ab}^{(2)}=\frac{\bar{\Omega}_{ab}^2}{\Gamma_{ab}}=2\pi\epsilon^2|V|^2\sqrt{\rho_a\rho_b} \tag{89a}$$

where

$$\rho_i=\frac{d_i}{2\Gamma} \qquad i=a,b \tag{89b}$$

(We add ˜ to quantities corresponding to the present, constant coupling, model to distinguish them from the previous quantities correponding to (56) with the random phase assumption.) $\tilde{W}^{(4)}$, however, will be much larger than $W^{(4)}$ (73), and simple insertion of (88) in (62) yields

$$\tilde{W}_{cc,aa}^{(4)}=\frac{\bar{\Omega}_{ab}^2\bar{\Omega}_{bc}^2}{\Gamma^3}d_b=W_{cc,aa}^{(4)}d_b \tag{89c}$$

Thus, for the constant coupling model we have $\tilde{\eta}\equiv\tilde{W}^{(4)}/\tilde{W}^{(2)}=(\bar{\Omega}/\Gamma)^2 d$ whereas for our random phase model we had $\eta=W^{(4)}/W^{(2)}=(\bar{\Omega}/\Gamma)^2$ (78). This arises simply, since the effective number of pathways to go from

$|aa\rangle\rangle$ to $|cc\rangle\rangle$ is d times larger for $\tilde{W}^{(4)}$.[19] For real life MMP, $(\bar{\Omega}/\Gamma)^2$ is a small parameter $\sim 10^{-2}-10^{-4}$ (see #3 above), whereas $(\bar{\Omega}/\Gamma)^2 d$ need not be small. When $(\bar{\Omega}/\Gamma)^2 d \gg 1$, then our expansion (86) does not converge and the resulting time evolution is then very different from that predicted by the simple rate equations (52).[47, 49] We should bear in mind, however, that the random phase assumptions (63), (66), and (67) are much more physically realistic for MMP than the constant coupling (88). This is indeed verified by the applicability of the simple rate equations (52) to actual MMP experiments.[4, 5]

VI. REM FOR POPULATIONS AND COHERENCES

In Sections III–V we have derived REM for the populations in MMP and showed how in the Markovian limit they reduce to simple rate equations. The basic reduction procedure of Section II is, however, more general and enables us to derive a closed set of REM for *any* arbitrary set of chosen variables. As we have already pointed out, the choice of the right number and kind of variables is a crucial step in the derivation of the REM since their simplicity and applicability depend on a successful choice. In particular it is desirable (if possible) to choose a complete set of slow variables whose evolution occurs on a much longer time scale compared to the rest, since then the REM attain a simple time-independent form [the Markovian limit (36)].[19] Thus a non-Markovian equation may become Markovian with the addition of a few more variables.[31] On the other hand, if we consider also fast variables, this will complicate the REM, may force us to make unnecessary simplifying assumptions; and we may end up with less accurate and oversimplified equations. Thus the flexibility of the general formulation of Section II should be utilized to match the number of variables to the problem.

In this section we shall construct and analyze a different set of REM for MMP by the addition of more variables corresponding to coherences. This is done due to several reasons:

1. It is clear that at the early stages of the molecular driving (region I), the expansion of $\langle R\rangle$ to second-order in the field (37) and the Markovian limit of (49) do not hold and we should in fact solve the exact Schrödinger equation with few states and coherent driving. Although we can in principle retain the populations as our only variables and expand the evolution operator $\langle R\rangle$ in higher powers in the field, it is much better to add few variables corresponding to coherences and get a simple Markovian equation, which will be in the form of a general (multilevel) Bloch equation.[50] Thus the explicit

inclusion of coherences as variables in region I results in a considerable simplification of the description in this region.

2. Studies of multilevel systems in contact with a bath and subject to coherent driving are usually carried out by introducing a Zwanzig projection operator which projects out the bath degrees of freedom and results in a set of variables consisting of a complete set of system operators.[32, 33] It is thus of interest to see the connection between the present and the more common formulations. We shall be able to show how in the case of a weak coupling of a system and a bath, our REM reduce to the familiar line broadening formulations.
3. In spectroscopic studies other than multiphoton excitations (ordinary line shapes,[34, 44] double resonance,[12] coherent transients,[23] resonance fluorescence[51]) the experimentalist usually probes directly the time evolution of coherences and their damping (dephasing) rates (e.g., an ordinary line shape is the Fourier transform of the correlation function of the molecular coherence).[34] In the present REM for the populations, the dephasing rates are "buried" inside the kernel $\langle R \rangle$ (or $\langle K \rangle$). By using a less-reduced description including coherences, we are able to see clearly the role of coherences in the dynamics of MMP and we can use the results of other spectroscopic experiments to evaluate the parameters appearing in our REM.
4. Conceptually, the addition of coherences enables us to look at MMP from a different viewpoint and to gain a better insight into the meaning of the Markovian limit and the "reduction of information" that occurs there. We shall be able to provide an alternative derivation to the rate equations (52), which will demonstrate how the explicit inclusion of coherence variables becomes redundant in this case.

We shall now turn to the construction of the relevant set of variables for our new REM. The first group of variables consists of the population variables A_{nn}, which were introduced in (19). These variables should, of course, be included in any REM for MMP, since they contain the significant information which is of primary interest to us, that is, the intermolecular distribution of energy as a function of time. We next define a set of operators which correspond to the time derivatives of A_{nn}, i.e., $[H', A_{nn}]$ [H' was defined in (18b)]. We thus introduce the operators corresponding to single quantum coherences:

$$A_{nm} = \frac{1}{\gamma_{nm}} \sum_{\alpha\beta} |n\alpha\rangle \mu_{\alpha\beta}^{nm} \langle m\beta| \tag{90}$$

where γ_{nm} was introduced in (43). We can now continue the process of constructing new variables by adding operators of the form $[H',[H', A_{nn}]]$, $[H',[H',[H', A_{nn}]]]$, etc. which correspond to the second, third, etc. time derivatives of A_{nn}. We thus get the following set of N^2 operators corresponding to populations (A_{nn}) and multiquantum coherences defined as follows:

$$A_{nm} = \frac{1}{\gamma_{nm}} \sum_{\alpha\beta} \nu_{\alpha\beta}^{nm} |n\alpha\rangle\langle m\beta| \qquad n, m = 0, 1, \ldots, N-1 \tag{91}$$

where (taking $m > n$)

$$\nu_{\alpha\beta}^{nm} = \sum_{\substack{\gamma, \delta \ldots \\ \beta'}} \mu_{\alpha\gamma}^{n, n+1} \mu_{\gamma, \delta \ldots}^{n+1, n+2} \mu_{\beta'\beta}^{m-1, m} \tag{91a}$$

and

$$\gamma_{nm}^2 = \sum_{\alpha\beta} |\mu_{\alpha\beta}^{nm}|^2 \tag{91b}$$

(For $m < n$ we have $A_{nm} = A_{mn}^{\dagger}$.) When $|n-m| = 1$, the definition (91) coincides with (90). We should note at this point that the set (91) contains only certain projections of the high order derivatives of A_{nm}. Also the truncation at $N-1$ is arbitrary and the level of theoretical description may be easily varied within the present formulation by changing the number of relevant operators. The attempts[14, 17, 18] to provide a phenomenological description for MMP in terms of $N^2 \times N^2$ generalized Bloch equations (for an N-level system) are intimately related to the picture of a system and a bath that are *weakly* interacting (since N^2 is the size of a complete set of "system" operators) (see Section VII). From the present approach, however, it is clear that the number N^2 does not play any special role. A complete description of the molecular density matrix requires much more than N^2 operators, but in practice we may construct a convenient set of REM whereby the number of variables is significantly smaller than N^2; for example, it may turn out that only single-quantum coherences are sufficient for a complete simple description of the molecular evolution all the way up to dissociation, and that we could ignore the effects of multiquantum coherences. Using the set of relevant operators (19) and (91) we may now evaluate the REM making use of the formulation of Section II. This was done in detail in Ref. 19b, making use of the POP formulation. (We note that for this choice of operators the Mori projection P does not commute with L_0 and $PL_0 \neq 0$. As a result the expansion of $\exp(-iLt)$ in powers of L' becomes more convenient than the expansion of $\exp(-iQLt)$, and this is the reason for the adoption of the POP formulation in this case.) The resulting equations, to first-order in H' and after invoking some simplifying assumptions, are[19b]

$$\frac{dP_n}{dt} = \frac{-i\bar{\Omega}_{n,n+1}}{\sqrt{2}}(\sigma'_{n+1,n} - \sigma'_{n,n+1}) - \frac{i\bar{\Omega}_{n,n-1}}{\sqrt{2}}(\sigma'_{n-1,n} - \sigma'_{n,n-1}) \tag{92a}$$

$$\begin{aligned}\frac{d\sigma'_{n,n+1}}{dt} &= \left[-i\bar{\omega}_{n,n+1} - \bar{\Gamma}_{n,n+1}\right]\sigma'_{n,n+1} - \frac{i\bar{\Omega}_{n,n+1}}{\sqrt{2}} \\ &\quad\times\left[P_{n+1}\left(\frac{d_n}{d_{n+1}}\right)^{1/2} - P_n\left(\frac{d_{n+1}}{d_n}\right)^{1/2}\right] \\ &\quad+ \frac{i\bar{\Omega}_{n+1,n+2}}{\sqrt{2}}\sigma'_{n,n+2} - \frac{i\bar{\Omega}_{n,n-1}}{\sqrt{2}}\sigma'_{n-1,n+1}\end{aligned} \tag{92b}$$

$$\begin{aligned}\frac{d\sigma'_{nm}}{dt} &= \left[-i\bar{\omega}_{nm} - \bar{\Gamma}_{nm}\right]\sigma'_{nm} + \frac{i\bar{\Omega}_{m,m+1}}{\sqrt{2}}\sigma'_{n,m+1} \\ &\quad+ \frac{i\bar{\Omega}_{m,m-1}}{\sqrt{2}}\theta^n_{m,m-1}\left(\frac{d_m}{d_{m-1}}\right)^{1/2}\sigma'_{n,m-1} \\ &\quad- \frac{i\bar{\Omega}_{n,n+1}}{\sqrt{2}}\theta^m_{n,n+1}\left(\frac{d_n}{d_{n+1}}\right)^{1/2}\sigma'_{n+1,m} \\ &\quad- \frac{i\bar{\Omega}_{n,n-1}}{\sqrt{2}}\sigma'_{n-1,m} \qquad (m>n+1)\end{aligned} \tag{92c}$$

[In (92c) we have taken $m>n+1$. The REM for $\sigma'_{nm}(m<n-1)$ and for $\sigma'_{n+1,n}$ are simply the complex conjugates of (92c) and (92b).] Here

$$\sigma'_{nm} = \left(d_n d^2_{n+1} d^2_{n+2} \cdots d^2_{m-1} d_m\right)^{1/4}\sigma_{nm} \tag{92d}$$

σ_{nm} being the expectation value of the coherence operator A_{nm}, that is,

$$\sigma_{nm}(t) = (A_{nm}, \rho(t)) \tag{93}$$

The quantity inside the parentheses in (92d), which scales the coherences σ'_{nm}, is the *statistical weight* of the *nm* coherence (the number of possible pathways to go from n to m). The θ_{ab}^{c} factors are defined as

$$\theta_{ab}^{c} = \frac{\overline{\Gamma}_{ac} - \overline{\Gamma}_{bc} + i\overline{\omega}_{ab}}{\overline{\Gamma}_{ab} + i\overline{\omega}_{ab}} \tag{94}$$

$\overline{\omega}_{nm}$ and $\overline{\Gamma}_{nm}$ are the off-resonant frequency and the dephasing rate of the nm coherence and are defined in terms of the asymptotic behavior of the two-time dipole correlation function $\tilde{I}_{nm}(t)$ (41), that is,

$$-i\overline{\omega}_{nm} - \overline{\Gamma}_{nm} = \lim_{t\to\infty} \frac{1}{\tilde{I}_{nm}} \frac{d\tilde{I}_{nm}}{dt} = \lim_{t\to\infty} \frac{d}{dt} \ln \tilde{I}_{nm}(t) \tag{95}$$

$\overline{\Omega}_{nm}$ and d_n were defined in Section III. The more general equations derived in Ref. 19b have the same form as (92); however, $\overline{\Gamma}_{nm}$ and $\overline{\omega}_{nm}$ are defined by (95) without the limit $t\to\infty$ and are thus time-dependent. Also the $\overline{\Omega}_{nm}$ factors in (92b) and (92c) are replaced by more complicated time-dependent factors, which include also higher order (three-time) dipole correlation functions, which are direct generalizations of $\tilde{I}_{nm}$, that is,

$$J_{abc}(\tau_1, \tau_2) = \langle v_{ab}(\tau_1) v_{ba}(0) v_{ac}(0) v_{ca}(\tau_2) \rangle \tag{96}$$

In order to get the form (92) we have assumed[19b] that $\tilde{I}_{nm}(t)$ has a time scale τ_c much shorter than that of our variables P_n. This is the Markovian limit of the present REM. In Section VIII we shall demonstrate that τ_c^{-1} is of the order of the molecular frequencies that are much larger than the multiphoton rates and that provide a justification to the Markov assumption. (We recall that the Markovian limit actually means separation of time scales between the P and Q variables, and it has a different meaning when we change the definition of P. Thus invoking the Markov assumption in the present equations is not equivalent to the Markov assumption in the previous REM.) Furthermore, we have assumed that asymptotically for long times $\tilde{I}_{nm}(t)$ exhibits an exponential behavior,

$$\tilde{I}_{nm}(t) \sim \exp\left(-i\overline{\omega}_{nm}t - \overline{\Gamma}_{nm}t\right) \tag{97}$$

This type of behavior is reasonable for dipole correlation functions as verified by some solvable models corresponding, for example, to impurities in solids[52] and pressure broadening[53] (see also Section VIII). In addition we have factorized the three-time correlation functions in the form

$$J_{abc}=\langle v_{ab}(\tau_1)v_{ba}(0)v_{ac}(0)v_{ca}(\tau_2)\rangle \cong \frac{\langle v_{ab}(\tau_1)v_{ba}(0)\rangle\langle v_{ac}(0)v_{ca}(\tau_2)\rangle}{d_a} \tag{98}$$

This factorization is consistent with a random-phase approximation which is expected to hold in the quasicontinuum. In the next section we shall discuss (92) and their connection with the previous REM (52).

VII. THE ROLE OF COHERENCES AND INTRAMOLECULAR DEPHASING

Our general REM (92) provide a unified description for MMP, which is valid for weak and strong driving fields and interpolate continuously all the way from the coherent to the totally incoherent limits of the molecular driving. They may be thus used to describe the evolution of a polyatomic molecule starting in region I up to the dissociation. We note that due to our adoption of a basis set of true molecular states, all the anharmonicities are properly (nonperturbatively) incorporated in our REM (92). As a result, no relaxation of population (T_1 type) terms need to be considered. The T_1 terms which appear in the perturbative approaches[50, 17, 18] couple different zero-order states and allow for energy exchange between the "system" and the "bath." In the present formulation they are included in the dephasing operators. The equivalence of T_1 and T_2, depending on the choice of a basis set, was discussed recently for intra- and intermolecular interactions.[20, 54]

In this section we shall analyze the behavior of the REM (92) throughout the multiphoton pumping process.

A. The Early Stages of Region I-Coherent Driving

At the early stages of the molecular excitation there is no reduction and each level contains only one state. We thus have $\tilde{I}_{nm}(t)=\exp(-i\bar{\omega}_{nm}t)$ so that $\bar{\Gamma}_{nm}=0$, there is no dephasing and (92) become equivalent to the complete Schrödinger equation for the driven molecule (coherent driving). This behavior demonstrates how, within the present formulation, the dephasing is a direct consequence of our *reduced description* of the molecular dynamics in terms of few variables.[20] The dephasing rates in our REM $\bar{\Gamma}_{nm}$ (95) are expressed in terms of microscopic intramolecular dipole correlation functions. They are independent of the dipole strength but rather depend merely on the *functional form* of the dipole operator. *No perturbative arguments regarding intramolecular interactions are required in order to give a precise definition of these terms*. Thus intramolecular interactions are rigorously treated by using the true molecular basis set. This state of affairs is in contrast to ordinary line-broadening formulations[35, 38, 39] where the dephasing is treated perturbatively in the system-bath interactions (see Section VII.B).

B. The Perturbative Line-Broadening Limit: Weak Coupling of a "System" and a "Bath"

As the molecule absorbs more photons, the density of molecular states rapidly increases and the reduction starts to play a role. But as long as the total molecular energy is not too high, the normal mode picture for the molecule is quite adequate and the anharmonicities may be treated as weak perturbations. In this case the dephasing operator assumes the well-known form from perturbative line-broadening theories.[35, 38, 39] We shall now analyze the behavior of the REM in the weak intramolecular interaction limit. To that purpose we assume separation of our degrees of freedom into "system" and "bath." We further assume that only one "system" degree of freedom interacts directly with the radiation field. However, it is coupled to the bath by a weak perturbation V. The molecular Hamiltonian (18) thus assumes the form

$$H=\sum_{m}|m\rangle E_m\langle m|+\sum_{\alpha}|\alpha\rangle E_\alpha\langle\alpha|\sum_{\substack{m,\alpha,\beta\\ \beta\neq\alpha}}|m\alpha\rangle V_{m\alpha,m\beta}\langle m\beta| \tag{99}$$

Here m is the system quantum number, whereas α comes for the collection of all bath quantum numbers. As in (18), $|m\alpha\rangle$ are molecular states "dressed" by the radiation field (m photons were absorbed at $|m\alpha\rangle$). Therefore the intramolecular coupling cannot connect molecular states with different $|m\rangle$

$$V_{n\alpha,m\beta}=V_{n\alpha,n\beta}\delta_{m,n} \tag{100}$$

The radiation field is assumed to interact only with the system thus

$$\langle n\alpha|\mu|m\beta\rangle=\mu_{nm}\delta_{\alpha,\beta} \tag{101}$$

In order to evaluate the dephasing terms in our REM perturbatively in V, let us first diagonalize the molecular states to first-order in V. We thus get

$$|m\beta^+\rangle=|m\beta\rangle+\sum_{\alpha\neq\beta}\frac{V_{m\beta,m\alpha}}{\omega_{\beta\alpha}+i\eta}|m\alpha\rangle \tag{102a}$$

and

$$\langle n\alpha^{+}| = \langle n\alpha| + \sum_{\alpha\neq\beta} \langle n\beta| \frac{V_{n\beta,n\alpha}}{\omega_{\alpha\beta} - i\eta} \tag{102b}$$

Substitution of (102) in (101) results in

$$\langle n\alpha^{+}|\mu|m\beta^{+}\rangle = \mu_{nm}\left[\delta_{\alpha,\beta} + \frac{\Delta V_{\beta\alpha}^{mn}}{\omega_{\beta\alpha} + i\eta}(1-\delta_{\alpha\beta})\right] \tag{103}$$

where we have defined

$$\Delta V_{\beta\alpha}^{mn} \equiv V_{m\beta,m\alpha} - V_{n\beta,n\alpha} \tag{104}$$

Substitution of (103) into (41) results in (where for the sake of simplicity we take $\Delta V_{\alpha\alpha}^{mn} = 0$)

$$\begin{aligned}\tilde{I}_{nm}(t) &= \frac{1}{\gamma_{nm}^{2}} \sum_{\alpha\beta} |\langle n\alpha^{+}|\mu|m\beta^{+}\rangle|^{2} \exp(-i\omega_{\alpha\beta}t) \\ &= |\mu_{nm}|^{2}\left[\sum_{\alpha} 1 + \sum_{\alpha\neq\beta} \frac{|\Delta V_{\beta\alpha}^{mn}|^{2}}{\omega_{\alpha\beta}^{2}} \exp(-i\omega_{\alpha\beta}t)\right]\end{aligned} \tag{105a}$$

From (95) and (105) we thus get, to second order in V:

$$G_{nm}(t) \equiv -\frac{1}{\tilde{I}_{nm}} \frac{d\tilde{I}_{nm}}{dt} = \frac{i}{d} \sum_{\alpha\neq\beta} \frac{|\Delta V_{\beta\alpha}^{mn}|^{2}}{\omega_{\alpha\beta}} \exp(-i\omega_{\alpha\beta}t) \tag{106}$$

where d is the number of relevant bath states. Since $G_{nm}(0) = 0$, (106) can be recast in the form

$$G_{nm}(t) = \int_{0}^{t} d\tau\, \chi_{nm}(\tau) \tag{107}$$

Where $\chi_{nm}(\tau)$ is a dynamical line-width function

$$\chi_{nm}(\tau) = \frac{1}{d} \sum_{\alpha,\beta} |\Delta V_{\beta\alpha}^{nm}|^{2} \cos \omega_{\alpha\beta}\tau \tag{108}$$

and we have

$$\bar{\Gamma}_{nm} = G_{nm}(\infty) \tag{109}$$

These are the familiar expressions from the theories of line broadening.[35, 38, 39]

In conclusion we note the following:

1. Ordinary Line Shapes

In ordinary line-shape studies, the driving field is weak and is switched adiabatically. If the molecule is initially at the nth level, then the line shape predicted by our REM is the Fourier transform of the molecular coherence correlation function, which is the solution of our REM for $\sigma'_{nm}(t)$ with $\bar{\Omega}=0$ and $\sigma'_{nm}(0)=1$.[34] We thus get for the absorption line shape

$$\Phi_{nm}(\omega_0)=\int_0^\infty d\tau\cos(\omega_{nm}\tau)\exp\left[-\int_0^t d\tau(t-\tau)\chi_{nm}(\tau)\right] \qquad (110)$$

[In the derivation of (110) we have used the non-Markovian version of (92), which in the perturbative limit simply amounts to replacing $i\bar{\omega}_{nm}+\bar{\Gamma}_{nm}$ by $G_{nm}(t)$.[19] Only in the Markovian limit we have $G_{nm}(t)\cong G_{nm}(\infty)$, $\int_0^t d\tau(t-\tau)\chi_{nm}(\tau)\to t\int_0^\infty d\tau\,\chi_{nm}(\tau)$, and the line shape (110) assumes a simple Lorentzian form with a width of $\int_0^\infty d\tau\,\chi_{nm}(\tau)$.][19, 34]

2. Absence of Energy Redistribution

In the weak perturbation limit, the system mode is being pumped by the radiation field, and the bath merely causes a dephasing but does not induce any relaxation of population (T_1) in the system (all energy absorbed from the field remains in the pumped mode). This arises since in our "dressed" picture the bath cannot couple states which belong to different levels (100) and in the perturbative limit each level is associated with a definite state of the system. We could take account for T_1 within a perturbative approach by adding more variables (each level could be split into several groups with the same total energy but with different energy in the pumped mode). This will result in a large increase in the number of variables.[17, 18]

3. Limitations of the Perturbative Approach

The usage of a zero-order (harmonic or local-mode)[22, 55, 56] basis set with intramolecular couplings may be advantageous provided we can get along with few states (say, when only one state in each level is carrying oscillator strength to the previous level). In such a case we can, in the Markovian limit, attribute a width of $2\pi|V|^2\rho_f$ to the various levels (where ρ_f is the density of final molecular states), and this provides a very convenient framework for the description of molecular radiative phenomena.[57] This is the case in ordinary optical line-shape[24] and transient experiments[23] in electronically excited states of polyatomic molecules where it is possible to find a well-defined "doorway state."[57] Another type of related experiments where such a zero-order basis set was proved useful is the novel gas-phase CH stretch overtone spectroscopy in benzene done by Bray and Berry.[21, 22] For these experiments, by adopting a local mode picture we may again consider a single (local-mode) doorway state and perform a dynamical line-shape analysis by considering its coupling to the rest of the modes.[22] This is not the case, however, for MMP in the quasicontinuum where we do not expect a perturbative treatment in intramolecular interactions to hold. This is why in the present work we have chosen a basis set ($|n\alpha\rangle$) of the true molecular states for the description of the highly excited molecules.

C. The Quasicontinuum

At high degrees of excitation we expect the Markovian limit [which led to (92)] to hold very early (see Section VIII) so that we are left with the general REM (92) where each transition is characterized by a frequency $\bar{\omega}_{nm}$, integrated Rabi frequency $\bar{\Omega}_{nm}$ (which is roughly independent on the molecular size) and a dephasing rate $\bar{\Gamma}_{nm}$ given by the asymptotic behavior of the logarithmic derivative of $\tilde{I}_{nm}(t)$. [The perturbative expression (109) no longer holds.] In addition the REM (92) include *ratios* of statistical weights (d_n/d_m) of the various levels.

D. The Complete Incoherent Driving—Rate Equations Revisited

If the dephasing rates $\bar{\Gamma}_{nm}$ are fast compared to the driving $\bar{\Omega}_{nm}$, we may invoke a steady state assumption for the coherences, (i.e. set $d\sigma'_{nm}/dt=0$ in (92)), solve for σ'_{nm} and substitute back into the equations for the populations. As a result our REM assume the form of simple rate equations corresponding to incoherent driving[58]

$$\frac{dP_n}{dt} = \sum_{m\neq n} \mathrm{K}_{nm}P_m - \mathrm{K}_{nn}P_n \tag{111}$$

If we consider only single quantum coherences (i.e., set $\sigma'_{nm}=0$ for $|n-m|>1$) we have

$$\sigma'_{n,n+1} = \frac{-i\bar{\Omega}_{n,n+1}}{\sqrt{2}\left(i\bar{\omega}_{n,n+1}+\bar{\Gamma}_{n,n+1}\right)} \times\left[P_{n+1}\left(\frac{d_n}{d_{n+1}}\right)^{1/2} - P_n\left(\frac{d_{n+1}}{d_n}\right)^{1/2}\right] \tag{112}$$

which when substituted back into (92a) results in

$$\frac{dP_n}{dt} = \sum_{m=n\pm 1} \frac{\overline{\Gamma}_{nm}\overline{\Omega}_{nm}^2}{\overline{\Gamma}_{nm}^2 + \overline{\omega}_{nm}^2} \left[P_m \left(\frac{d_n}{d_m} \right)^{1/2} - P_n \left(\frac{d_m}{d_n} \right)^{1/2} \right] \tag{113}$$

We recall that in the derivation of (92) we have assumed

$$\tilde{I}_{nm}(t) = \exp(-i\overline{\omega}_{nm}t - \overline{\Gamma}_{nm}t) \tag{114}$$

Substitution of (114) in (51) results in

$$\Gamma_{nm}^{-1} = \mathrm{Re} \int_0^\infty d\tau \, \tilde{I}_{nm}(\tau) = \frac{\overline{\Gamma}_{nm}}{\overline{\omega}_{nm}^2 + \overline{\Gamma}_{nm}^2} \tag{115}$$

which shows that *the simple rate equations* (113) *derived from our new REM* (92) *are identical with* (52) *derived from the previous REM* (48).

The present alternative derivation of (52) provides us with a new insight regarding the significance of the simple rate equations. As is clearly seen from (92), the dynamics of molecular multiphoton processes is governed by the competition between the driving terms $\overline{\Omega}_{nm}$, which tend to build higher coherences and the dephasing rates $\overline{\Gamma}_{nm}$, which tend to destroy them. If

$$\beta \equiv \frac{\overline{\Omega}^2}{\overline{\omega}^2 + \overline{\Gamma}^2} \ll 1 \tag{116}$$

we can solve iteratively for the steady state of (92), that is, we can substitute $\sigma'_{n,n+1}$(112) in (92c) to generate $\sigma'_{n,n+2}$, etc. This simple solution of (92) (perturbative in β) reveals that we have a hierarchy of multiquantum coherences where $\sigma'_{n,n+k} = 0(\beta^{k/2})$, $k > 0$. The contribution of $\sigma'_{n,n+k}$ to the rate equation (111) for the populations will be $0(\beta^k)$. Equations (112) and (113) are the simplest demonstration of this where for $k=1$, $\sigma'_{n,n+1} = 0(\beta^{1/2})$ and $\mathrm{K}_{n,n+1} = 0(\beta)$. When $\beta \ll 1$ this means that the steady-state values of the high-order coherences will be very small, and when we ignore all $\sigma'_{n,n+k}$ except for $k=1$, we get our simple rate equations (113). These arguments show how most of the information regarding the molecular dynamics (i.e., the dynamics of the higher order coherences $k > 1$) becomes irrelevant for the dynamics of MMP in the Markovian limit $\beta \ll 1$ where simple rate equations apply. In the previous derivation of (52) from (48) all the extra dynamical information was hidden in the higher order correlation functions contributing to $\langle R \rangle$.

It is obvious from the comparison of the two derivations of the REM (52) and (92) that the k quantum coherences in the latter $\sigma'_{n,n+k}$ play the role of $W^{(2k)}$ in the former. If we take, for example, $W^{(4)}_{cc,aa}$, we can see from Fig. 2 that it corresponds to the path $|cc\rangle\rangle-|bc\rangle\rangle-|ac\rangle\rangle-|ab\rangle\rangle-|aa\rangle\rangle$, which leads from $|aa\rangle\rangle$ to $|cc\rangle\rangle$ via the two quantum coherences $|ac\rangle\rangle$. The neglect of $W^{(4)}$ in (79) is thus equivalent to ignoring $\sigma'_{n,n+2}$ in (92). The same rate equations thus result from the neglect of all higher order coherences in (92) or of all higher order correlation functions in (48). Furthermore, the expansion parameter β (116) corresponds to η (80) (from (115) we see that if $\bar{\omega}=0$ then $\eta=\beta$).

VIII. A MODEL FOR THE INTRAMOLECULAR DIPOLE CORRELATION FUNCTIONS

In the preceding sections we have developed a complete theory for MMP where in the Markovian limit the resulting REM (52) or (92) are expressed in terms of the two-time intramolecular dipole correlation functions $\tilde{I}_{nm}(t)$ (41).

In this section we shall develop a simple microscopic model for $\tilde{I}_{nm}(t)$, which will enable us to relate the dephasing times (and the whole dynamics of MMP) to real molecular parameters (size, frequencies, anharmonicities, masses, etc.). We start with the molecular Hamiltonian

$$H_M = \sum_{\nu} H_{0\nu}(q_\nu) + H' \tag{117}$$

where $H_{0\nu}(q_\nu)$ is a harmonic Hamiltonian for the νth normal mode and q_ν is its dimensionless coordinate. H' is the anharmonic part of the Hamiltonian and includes terms cubic and higher in $\mathbf{q}$. Our expressions for the dephasing rates are given in terms of the true molecular states. We thus need a way for obtaining a reasonable approximation for these states. In fact, since the dephasing is essentially a spreading process of a wavepacket of molecular states on the energy shell, we need to have a "mean field" Hamiltonian that will describe correctly the motion only on the energy shell. The simplest way to achieve that is to expand H' to linear terms in q_ν, that is,

$$H' = \sum_{\nu} F_\nu(\mathbf{q}) q_\nu \tag{118}$$

and to replace each $F_\nu(\mathbf{q})$ by its microcanonial average at energy E, that is,

$$\Delta_\nu(E) \equiv \langle F_\nu \rangle_E = \frac{\mathrm{Tr}[F_\nu \delta(E-H)]}{\mathrm{Tr}\,\delta(E-H)} \tag{119}$$

We thus get

$$H(E)=\sum_{\nu} H_{0\nu}(q_\nu)+\sum_{\nu}\Delta_\nu(E)q_\nu \tag{120}$$

(Note that Δ_ν and q_ν are dimensionless.) We have thus established a simple picture of a collection of harmonic oscillators whose equilibrium position is being shifted as a function of the molecular energy. At $E=0$, $\langle q_\nu\rangle=0$ and we recover the normal-mode Hamiltonian. $\Delta_\nu(E)$ are related to the anharmonicities to lowest order as

$$\Delta_\nu(E)=\sum_{\nu'}\alpha_{\nu\nu'\nu'}\langle q_{\nu'}^2\rangle_E \tag{121}$$

where $\alpha_{\nu\nu'\nu'}$ are cubic anharmonicities and $\langle q_{\nu'}^2\rangle_E$ is the microcanonial mean square displacement of the νth mode.

Let us consider now the following *microcanonial* correlation function

$$\tilde{I}_{nm}(E,t)=\frac{1}{\gamma_{nm}^2 W(E)}\sum_{\alpha,\beta}|\mu_{nm}^{\alpha\beta}|^2\exp(-i\omega_{n\alpha,m\beta}t)\delta(E_{n\alpha}-E) \tag{122}$$

where $W(E)$ is the density of molecular states at energy E. In terms of this correlation function we have

$$\tilde{I}_{nm}(t)=\tilde{I}_{nm}(\bar{E}_n,t) \tag{122a}$$

where $\bar{E}_n$ is the mean energy of the nth level. We further assume that the dipole operator is coupled only with one normal mode (ν_s)

$$\mu=\bar{\mu}q_s \tag{123}$$

The quantity that may be easily evaluated is, however, the *canonial* correlation function[59–62]

$$\hat{I}_{nm}(\beta,t)=\frac{1}{Q(\beta)}\int_0^\infty dE\,\tilde{I}_{nm}(E,t)W(E)\exp(-\beta E) \tag{124}$$

where $Q(\beta)$ is the partition function

$$Q(\beta)=\int_0^\infty dE\,W(E)\exp(-\beta E) \tag{125}$$

$I_{nm}(E,t)$ may then be evaluated by the inverse Laplace transform[49, 50]

$$\tilde{I}_{nm}(E,t)=\frac{\int_{\lambda-i\infty}^{\lambda+i\infty}d\beta\exp(\beta E)Q(\beta)\hat{I}_{nm}(\beta,t)}{\int_{\lambda-i\infty}^{\lambda+i\infty}d\beta\exp(\beta E)Q(\beta)} \tag{126}$$

The anharmonicity of q_s, $\Delta_s(E)$ has a special role in determining $\bar{\omega}_{nm}$ (the mean frequencies of the transitions); however, for the dephasing it contributes just as any other mode. For the sake of simplicity we shall assume $\Delta_s=0$. (Incorporating Δ_s will not affect substantially our final expressions.) We then get[59–62]

$$\hat{I}_{nm}(\beta,t)=\exp[-S(\beta)]\exp[S_+(\beta,t)+S_-(\beta,t)] \tag{127}$$

where

$$S_+(\beta,t)=\sum_\nu \tfrac{1}{2}|\Delta_\nu^{(nm)}|^2(\bar{n}_\nu+1)\exp(i\omega_\nu t) \tag{127a}$$

$$S_-(\beta,t)=\sum_\nu \tfrac{1}{2}|\Delta_\nu^{nm}|^2\bar{n}_\nu\exp(-i\omega_\nu t) \tag{127b}$$

and

$$S(\beta)=S_+(\beta,0)+S_-(\beta,0)=\sum_\nu \tfrac{1}{2}|\Delta_\nu^{nm}|^2(2\bar{n}_\nu+1) \tag{127c}$$

Here

$$\Delta_\nu^{(nm)}=\Delta_\nu(E_n)-\Delta_\nu(E_m) \tag{128}$$

and $\bar{n}_\nu$ is the mean occupation number of the νth oscillator at temperature $\beta^{-1}=kT$, that is,

$$\bar{n}_\nu=[\exp(\beta\hbar\omega_\nu)-1]^{-1} \tag{129}$$

Evaluation of the inverse Laplace transform (126) should now be made in order to evaluate $\tilde{I}_{nm}(E,t)$. We recall that under quite general conditions[62, 63] the inversion may be achieved by using the saddle-point method. This results in the extremely simple relation

$$\tilde{I}(E,t)\simeq\hat{I}(\beta^*,t) \tag{130}$$

where β^* is the saddle point, obtained from the solution of

$$\sum_{\nu} \bar{n}_\nu(\beta^*)\hbar\omega_\nu = E \tag{131}$$

Using this result we get

$$\tilde{I}(E,t) = \exp(-S(\beta^*))\exp[S_+(\beta^*,t)+S_-(\beta^*,t)] \tag{132}$$

The exact equation (122) satisfies $\tilde{I}_{nm}(t) = \tilde{I}^*_{mn}$. This is no longer the case due to our use of the saddle point; we thus take

$$\tilde{I}_{nm}(t) = \tilde{I}_{nm}\left(\frac{\bar{E}_n + \bar{E}_m}{2}, t\right) \tag{132a}$$

Substituting (132) in (106) we get

$$G_{nm}(t) = -\frac{1}{\tilde{I}_{nm}}\frac{d\tilde{I}_{nm}}{dt} = -\frac{d}{dt}[S_+(\beta^*,t)+S_-(\beta^*,t)]$$

$$= -\frac{d}{dt}S(\beta^*,t)\Big|_{t=0} + \int_0^t d\tau\, \chi_{nm}(\tau) \tag{133}$$

where

$$\chi_{nm}(\tau) = \chi'_{nm}(\tau) + i\chi''_{nm}(\tau) \tag{134}$$

$$\chi'_{nm}(\tau) = \sum_{\nu} \frac{|\Delta_\nu^{mn}|^2}{2}\omega_\nu^2(2\bar{n}_\nu + 1)\cos\omega_\nu t \tag{135a}$$

$$\chi''_{nm}(\tau) = \sum_{\nu} \frac{|\Delta_\nu^{nm}|^2}{2}\omega_\nu^2 \sin\omega_\nu t \tag{135b}$$

$$\frac{d}{dt}S(\beta^*,t)\Big|_{t=0} = i\sum_{\nu} \frac{|\Delta_\nu^{nm}|^2}{2}(2\bar{n}_\nu + 1)\omega_\nu \tag{135c}$$

And the dephasing rate Γ_{nm} in the Markovian limit (95) assumes the form

$$\bar{\Gamma}_{nm} = \tfrac{1}{2}\int_0^\infty d\tau \sum_{\nu} |\Delta_\nu^{nm}|^2(2\bar{n}_\nu + 1)\omega_\nu^2 \cos\omega_\nu\tau \tag{136}$$

which corresponds to a Lorentzian line profile.

Let us now have a rough estimate of Γ_{nm}. The integral (136) in the high-temperature limit is over a wavepacket whose frequency width is the spread in molecular frequencies that is of the same magnitude as the molecular frequencies. We thus have

$$\Delta_\nu^{nm}(E) \sim \frac{\alpha}{2}\frac{\omega_0}{\langle\omega\rangle} \tag{137a}$$

and

$$\bar{\Gamma}_{nm} \sim \frac{1}{2}|\Delta|^2(2\bar{n}+1)\frac{\langle\omega\rangle^2}{\langle\omega\rangle} = \frac{1}{2}\left(\frac{\alpha}{2}\right)^2(2\bar{n}+1)\omega_0 \tag{137b}$$

Here $\langle\omega\rangle$ is a typical molecular frequency, ω_0 is the laser frequency, and $\bar{n}$ is the total number of absorbed quanta. α is a dimensionless cubic anharmonicity and is typically $\alpha \sim 10^{-1} - 10^{-2}$.

Taking the typical values $\bar{n}=40$ and $\omega_0 = 1000$ cm^{-1}, we get $\bar{\Gamma}_{nm}=1$ cm^{-1} assuming $\alpha = 10^{-2}$ and $\bar{\Gamma}_{nm} = 100$ cm^{-1} assuming $\alpha = 10^{-1}$. These are very reasonable values for MMP. If we consider the experimental data of SF_6, the absorption cross-sections were fitted to experiment assuming a rate equation (113),[6, 4] resulting in a cross-section of $\sigma \sim 2 \times 10^{-20}$ cm^2 at $\bar{n}=40$. Using (113) the cross-section is given by

$$\frac{\sigma\Phi}{\hbar\omega_0} = \frac{\bar{\Omega}_{nm}^2}{\bar{\Gamma}_{nm}} \tag{138}$$

where Φ is the incoming laser flux and we have taken $\bar{\omega}_{nm}=0$. Assuming a diluted oscillator strength of $\mu \sim 0.03-0.1$ Debye we get $\Gamma_{nm} \sim 15-150$ cm^{-1}, which agrees very nicely with the above estimates.

Furthermore, the multiphoton absorption data[7] indicate that the effective multiphoton absorption cross section defined as $d\bar{n}/dI$ is decreasing with Φ (and hence with the degree of excitation). This type of behavior is predicted by our REM due to the roughly linear increase of $\bar{\Gamma}_{nm}$ with $\langle n\rangle$ (137) and the dilution of $\bar{\Omega}_{nm}$, which result in a gradual decrease of the effective absorption cross-section with $\bar{n}$. From these estimates we can also verify the validity of the Markovian assumption leading to (136) as the relation $\Gamma \ll \tau_c^{-1} \sim \langle\omega\rangle$ is equivalent to (49). Rate equations of the form (111) and (113) were used by several authors to fit experimental data.[5c, 6, 7, 16] We should bear in mind, however, that these equations are only the final stage in the reduction hierarchy described in this work and only at the higher energy part of the quasicontinuum are we allowed to use the simple rate equation (113).

Furthermore, in the actual calculations[7] it was assumed that

$$\frac{K_{nm}}{K_{mn}} = \frac{\rho_n}{\rho_m} \tag{139}$$

where ρ is the density of molecular states. From the present derivation it is clear that ρ_n/ρ_m should be actually d_n/d_m, that is, the ratio of effectively radiatively coupled states, which may be very different. This may crucially affect the intermolecular energy distributions and the conclusions drawn by Grant et al.[7] should be thus treated with caution. Fitting of the present REM with experimental data may thus provide a clue for understanding the dynamics of highly excited polyatomic molecules, by providing us with $\bar{\Omega}_{nm}$ and $\tilde{I}_{nm}(t)$ as a function of the molecular degree of excitation. The experimental data available at present are not sufficiently detailed to allow for an unambiguous quantitative study, and this is the reason that different authors are able to fit their data using completely different assumptions.[46] For that reason it is necessary to use data from other types of experiments, especially regarding the intramolecular dephasing times, which will eliminate the number of unknown parameters in the REM. Great progress has been recently achieved in that direction by various techniques.[21–24]

Finally we should note that although M_k were defined in Section III using the true molecular eigenstates, their microscopic evaluation does not necessarily require the complete knowledge of the molecular eigenstates. It is possible to calculate M_k semiclassically directly from the Hamiltonian and to avoid the reference to the molecular eigenstates altogether. Such semiclassical methods were recently developed for absorption and fluorescence spectra[64] (i.e. one and two photon processed, M_2 and M_4) and they may be easily extended towards the evaluation of any intramolecular dipole correlation function M_k.

Acknowledgments

This work was supported by The Robert A. Welch Foundation (Grant C-727), the National Science Foundation (Grant CHE 7822104) and the Westinghouse Educational Foundation Grant of the Research Corporation. This support is gratefully acknowledged.

I also wish to thank Professor J. Jortner for very useful discussions.

References

1. (a) N. R. Isenor and M. C. Richardson, *Appl. Phys. Lett.*, **18**, 224 (1971), and *Optics Commun.*, **3**, 360 (1971). (b) N. R. Isenor, V. Merchant, R. S. Hallsworth, and M. C. Richardson, *Can. J. Phys.*, **51**, 1281 (1973).
2. (a) R. V. Ambartsumyan, V. S. Letokhov, E. A. Ryabov, and N. V. Chekalin, *JETP Lett.*, **20**, 273 (1974). (b) R. V. Ambartsumyan, Yu. A. Gorkhov, V. S. Letokhov, and G. N. Makarov, *JETP Lett.*, **21**, 171 (1975).

3. J. L. Lyman, R. J. Jensen, J. Rink, C. P. Robinson, and S. D. Rockwood, *Apply. Phys. Lett.*, **27**, 87 (1975).
4. See papers in: (a) *Tunable Lasers and Applications*, edited by A. Mooradian, T. Jaeger, and P. Stokseth, Springer, New York, 1976. (b) *Multiphoton Processes*, edited by J. H. Eberly and P. Lambropolous, Wiley, New York, 1978.
5. For recent reviews see: (a) C. D. Cantrell, S. M. Freund, and J. L. Lyman, "Laser Induced Chemical Reactions and Isotope Separation," in *Laser Handbook* Vol. IIIb, edited by M. L. Stitch, North-Holland, Amsterdam, 1980. (b) R. L. Woodin, D. S. Bomse, and J. L. Beauchamp, in *Chemical and Biochemical Applications of Lasers*, V. IV, edited by C. B. Moore, Academic, New York, 1979. (c) N. Bloembergen and E. Yablonovitch, *Phys. Today*, May, p. 23 (1978).
6. (a) J. G. Black, E. Yablonovitch, N. Bloembergen, and S. Mukamel, *Phys. Rev. Lett.*, **38**, 1131 (1977). (b) K. L. Kompa in this volume.
7. (a) M. J. Coggiola, P. A. Schultz, Y. T. Lee, and Y. R. Shen, *Phys. Rev. Lett.*, **38**, 17 (1977). (b) E. R. Grant, P. A. Schultz, As. S. Sudbo, Y. R. Shen, and Y. T. Lee, *Phys. Rev. Lett.*, **40**, 115 (1978).
8. (a) D. J. King and J. C. Stephenson, *Chem. Phys. Lett.*, **51**, 48 (1977) (b) J. C. Stephenson and D. J. King, *J. Chem. Phys.*, **69**, 1485 (1978).
9. (a) R. V. Ambartsumyan in 4(a). (b) R. B. Hall and A. Kaldor, *J. Chem. Phys.*, **70**, 4027 (1979).
10. J. D. McDonald, private communication.
11. (a) Z. Karny, A. Gupta, R. N. Zare, S. T. Lin, J. Nieman, and A. M. Ronn (to be published). (b) I. Burak, T. J. Quelly, and J. I. Steinfeld, *J. Chem. Phys.*, **70**, 334 (1979).
12. (a) D. S. Frankel, *J. Chem. Phys.*, **65**, 1696 (1976). (b) See papers by E. Yablonovitch and by K. L. Kompa in this volume.
13. (a) N. Bloembergen, *Opt. Commun.*, **15**, 416 (1975). (b) N. Bloembergen, C. D. Cantrell, and D. M. Larsen in 4(a).
14. M. F. Goodman, J. Stone, and D. A. Dows, *J. Chem. Phys.*, **65**, 5052 (1976).
15. S. Mukamel and J. Jortner, *J. Chem. Phys.*, **65**, 5204 (1976).
16. J. L. Lyman, *J. Chem. Phys.*, **67**, 1868 (1977).
17. D. P. Hodgkinson and J. S. Briggs, *Chem. Phys. Lett.*, **43**, 451 (1976); *J. Phys. B.*, **10**, 2583 (1977).
18. (a) C. D. Cantrell, in *Laser Spectroscopy II*, edited by J. L. Hall and J. L. Carlsten, Springer-Verlag, New York, 1977, p. 109. (b) C. D. Cantrell, H. W. Galbraith, and J. R. Ackerhalt in 4(b).
19. (a) S. Mukamel, *Phys. Rev. Lett.*, **42**, 168 (1979). (b) S. Mukamel, *J. Chem. Phys.*, **70**, 5834 (1979); *ibid.*, **71**, 2012 (1979).
20. S. Mukamel, *Chem. Phys.*, **31**, 327 (1978).
21. R. Bray and M. J. Berry, *J. Chem. Phys.* **71**, 4909 1979.
22. D. F. Heller and S. Mukamel, *J. Chem. Phys.*, **70**, 463 (1979).
23. (a) R. G. Brewer and R. L. Shoemaker, *Phys. Rev. Lett.*, **27**, 631 (1971). (b) T. E. Orlowski and A. Zewail, *J. Chem. Phys.*, **70**, 1390 (1979). (c) T. J. Aartsma, J. Morsink, and D. A. Wiersma, *Chem. Phys. Lett.*, **42**, 520 (1976); **47**, 425 (1977); **49**, 34 (1977).
24. (a) S. M. Beck, D. L. Monts, M. G. Liverman, and R. E. Smalley, *J. Chem. Phys.*, **70**, 1062 (1979). (b) P. S. H. Fitch, L. Wharton, and D. H. Levy, *J. Chem. Phys.*, **70**, 2018 (1979).
25. S. A. Rice in *Excited States*, Vol. 2, edited by E. C. Lim Academic Press, New York, 1975.
26. (a) D. L. Bunker, *Theory of Elementary Gas Reaction Rates*, Pergamon, New York, 1966. (b) P. J. Robinson and K. A. Holbrook, *Unimolecular Reactions*, Wiley-Interscience, New York, 1972.
27. (a) J. F. Meagher, K. J. Chao, J. R. Barker, and B. S. Rabinovitch, *J. Phys. Chem.*, **78**,

2535 (1974). (b) J. M. Parson, K. Shobatake, Y. T. Lee, and S. A. Rice, *J. Chem. Phys.*, **59**, 1402 (1973). (c) J. G. Moehlman, J. T. Gleaves, J. W. Hudgens, and J. D. McDonald, *J. Chem. Phys.*, **60**, 4790 (1974).
28. J. M. McDonald and R. A. Marcus, *J. Chem. Phys.*, **65**, 2180 (1976).
29. N. Van Kampen, *Adv. Chem. Phys.*, **34**, 245 (1976).
30. I. Oppenheim, K. E. Shuler, and G. H. Weiss, *Stochastic Processes in Chemical Physics: The master Equation*, MIT Press, Cambridge, Mass., 1977.
31. M. W. Wang and G. E. Uhlenbeck, *Rev. Mod. Phys.*, **17**, 323 (1945).
32. M. Lax, *Phys. Rev.*, **109**, 1921 (1958); *Phys. Rev.*, **129**, 2342 (1963); *J. Phys. Chem. Solids*, **25**, 487 (1964); *Phys. Rev.*, **145**, 110 (1966).
33. W. H. Louisell, *Quantum Statistical Properties of Radiation*, Wiley, New York, 1973.
34. (a) R. Kubo in *Fluctuation, Relaxation and Resonance in Magnetic Systems*, edited by D. Ter Haar (Oliver & Boyd, Edinburgh, 1962). (b) R. Kubo, *Adv. Chem. Phys.*, **XV** (1969). (c) R. Kubo, *J. Math. Phys.*, **4**, 174 (1963).
35. U. Fano, *Phys. Rev.*, **131**, 259 (1963).
36. R. Zwanzig, *Physica*, **30**, 1109 (1964).
37. R. Karplus and J. Schwinger, *Phys. Rev.*, **73**, 1020 (1948).
38. P. W. Anderson, *Phys. Rev.*, **80**, 511 (1950).
39. A. Ben-Reuven, *Adv. Chem. Phys.*, **33**, 235 (1975).
40. (a) H. Mori, *Prog. Theoret. Phys.*, **33**, 423 (1965). (b) M. Tokuyama and H. Mori, *ibid.*, **55**, 411 (1976).
41. J. Jortner, *Adv. Laser Spectr.*, **113**, 88 (1977).
42. M. Quack, *J. Chem. Phys.*, **69**, 1282 (1978).
43. S. Mukamel, I. Oppenheim, and J. Ross, *Phys. Rev.*, **A17**, 1988 (1978).
44. S. Mukamel, *Chem. Phys.*, **37**, 33 (1979).
45. (a) C. Cohen Tannoudji in *Cargese Lectures in Physics*, Vol. 2, edited by M. Levy, Gordon & Breach, New York, 1967, p. 347. (b) C. Cohen Tannoudji and S. Reynaud, in *Multiphoton Processes*, edited by J. H. Eberly and P. Lambropolous, Wiley, New York, 1978.
46. B. Yoon, J. M. Deutch, and J. H. Freed, *J. Chem. Phys.*, **62**, 4687 (1975).
47. A. Nitzan, J. Jortner, and B. Berne, *Mol. Phys.*, **26**, 281 (1973).
48. E. J. Heller and S. A. Rice, *J. Chem. Phys.*, **61**, 936 (1974).
49. B. Carmeli and A. Nitzan, *Chem. Phys. Lett.*, **58**, 310 (1978).
50. A. Abragam, *The Principles of Nuclear Magnetism*, Oxford University Press, London, 1961.
51. J. Fiutak and J. Van Kranendonk, *Can. J. Phys.*, **40**, 1085 (1962).
52. V. Hizhnyakov and I. Tehver, *Phys. Stat. Sol.*, **21**, 755 (1967).
53. S. Mukamel, *J. Chem. Phys.*, **71**, 2884 (1979).
54. S. Mukamel, *Chem. Phys. Lett.*, **60**, 310 (1979).
55. B. R. Henry, *Acct. Chem. Res.*, **10**, 297 (1977).
56. M. L. Elert, P. R. Stannard, and W. M. Gelbert, *J. Chem. Phys.*, **67**, 5395 (1977).
57. See, e.g., J. Jortner and S. Mukamel, in *The World of Quantum Chemistry*, edited by R. Daudel and R. Pullman, D. Reidel, 1973.
58. (a) J. R. Ackerhalt and J. H. Eberly, *Phys. Rev.*, **A14**, 1705 (1976). (b) J. R. Ackerhalt and B. W. Shore, *Phys. Rev.*, **A16**, 277 (1977).
59. R. Kubo and Y. Toyozawa, *Prog. Theor. Phys.*, **13**, 160 (1955).
60. R. Englman and J. Jortner, *Mol. Phys.*, **18**, 145 (1970).
61. K. F. Freed, in *Topics in Applied Physics*, Vol. 15, edited by F. K. Fong, Springer, Berlin, 1976.
62. S. H. Lin, *J. Chem. Phys.*, **58**, 5760 (1973).
63. M. R. Hoare and Th. W. Riuuirok, *J. Chem. Phys.*, **52**, 113 (1970).
64. S. Mukamel, *J. Chem. Phys.* (to be published).

N-LEVEL MULTIPLE RESONANCE

ABRAHAM BEN-REUVEN AND YITZHAK RABIN

Institute of Chemistry, Tel Aviv University, Tel Aviv, Israel

Abstract

An algebraic solution is presented for the steady-state populations and attenuation rates of coherently-driven thermally-relaxing N-level multiple resonances in which each resonance transition is coupled only to one radiation mode of arbitrary strength. Included in it are all cross relaxation rates and coherence damping rates. The theory is extended to subsets of larger sets of energy levels with back relaxation (repopulation), or open subsets. The algebraic solutions are expressed in terms of a matrix of single-passage rates which play the role of radiation-induced rate coefficients in the reduction to master equations. Properties of multiple resonances are illustrated by the three distinct configurations of the three-level double resonance. Extensions of the theory are briefly outlined.

CONTENTS

I. Introduction 555
II. Multiple Resonances 557
III. Steady-State Solutions 559
IV. Open Subsets 564
V. Master Equations 567
VI. Three-Level Double Resonance 569
  A. The Ladder Double Resonance 570
  B. The Raman Double Resonance 573
  C. The Inverted-Raman Double Resonance 574
VII. Discussion 575
  A. Corrections to the RWA 575
  B. Multiply-Connected Resonances 576
  C. Disconnected Multiple Resonances 576
  D. Pulsed Radiation 576
  E. Field-Modified Relaxation 577
  F. Inhomogeneous Broadening 577
  G. Memory Effects 578
  H. Resonance Scattering 578
References 579

I. INTRODUCTION

The use of present-day tunable laser sources, with their characteristic monochromaticity, coherence, and high intensity, has brought about a surge of developments in various fields, such as high-resolution laser spectroscopy,[1-4] mode-selective multiphoton chemistry (dissociation, ionization, and isomerization),[5-8] and resonance conversion of laser energy from one spectral range to another.[9-12] In all these processes multiple resonance phenomena play an important role.

These new developments pose new challenges to theory. Under conditions where the electromagnetic radiation can no longer be treated as a weak probe, nonperturbative solutions of the coupled Maxwell–Schrödinger equations are required.

The behavior of simple molecular systems coupled to the radiation is complicated by the presence of homogeneous broadening effects (such as radiative damping, collision broadening, or intramolecular relaxation),[13-26] which are usually described by coupling to a large thermal reservoir, or by the presence of inhomogeneous broadening effects (such as motionally dependent effects).[1-3, 9, 27-31] Thermal bath effects are usually incorporated through the Bloch–Redfield equations, and their generalizations. Motional effects can be incorporated with the help of generalized kinetic equations.

The rather complicated nature of coherent radiative coupling is reflected in the Bloch–Redfield equations by the excitation of off-diagonal elements of the molecular density matrix ("coherences"). Under certain conditions this complication can be removed, and the Bloch–Redfield equations can be replaced by master (or rate) equations involving level populations only.[26, 32-34] The nature of these conditions is closely associated with the relaxation mechanism, such as the intramolecular dynamics of polyatomic molecules in the so-called quasicontinuum region of energy levels.[34-36]

Most theoretical studies of multiple resonance have dealt with the coupling of radiation modes of constant intensity to a discrete set of molecular states. These theories must be modified[37-40] to take into account the pulsed nature of the laser signal prevailing in most experiments in which high-power lasers are used.

The aim of this work is to present a nonperturbative method for calculating steady-state populations and radiation attenuation rates (absorption spectra) in N-level multiple resonances, to study the resulting algebraic solution, with illustrations using the three-level double resonance, and to discuss the applicability of various versions of the equations of motion and the reduction to master equations.

The model of multiple resonance studied here, and the conditions of its validity, are presented in Section II. The method of solution is outlined in Section III, following Ref. 25, for a set of discrete levels closed under relaxation (i.e., with conservation of population). In Section IV, the problem of open subsets is discussed. In particular, the case is studied where the energy levels resonantly coupled to the radiation form an open subset of a larger closed set of discrete levels, with possible back relaxation (repopulation). It is shown that such subsets can nevertheless be treated separately by introducing modified relaxation rates.

Section V discusses the derivation of the master equations, and its relation to the single-passage rates introduced in Section III. Some results concerning the three-level double resonance (in its three possible configurations) are presented in Section VI, as an illustrative example.

Possible extensions of the present theory are briefly discussed in the concluding Section, with regard to such problems as resonance transitions coupled to more than one mode, breakdown of the resonance condition (rotating wave approximation), pulsed radiation, field-modified relaxation (e.g., optical collisions), inhomogeneous motional effects, and resonance scattering.

II. MULTIPLE RESONANCES

Consider the multiple resonance excitation, by multimode monochromatic coherent radiation, of a molecular system coupled to a thermal bath (by radiative decay, collisions, etc.). By "multiple resonance" we mean that the energy taken from, or given to, the field modes at every step in the radiative process is approximately equal to a change in the molecular energy. The application of coherent monochromatic external fields excites "coherences," or off-diagonal elements of the density matrix, in the molecular system. The effect of the thermal bath is to introduce damping to these coherences, expressed in the form of T_2 rates. In addition, "populations," or diagonal elements of the density matrix are also affected by the bath, in the form of T_1 relaxation rates.

Assume that the molecules undergo only *homogeneous relaxation*, independent of their positional, or translational states. The behavior of such systems, including coherent radiative couplings, T_2 rates, and T_1 rates, should be generally described by the *Bloch–Redfield equations*,[41, 42]

$$i\frac{\partial}{\partial t}\rho(t)=\hbar^{-1}[(H_0+V),\rho(t)]+\Sigma\rho(t) \tag{1}$$

where ρ is the molecular density matrix, H_0 is the molecular Hamiltonian, and V is the coherent radiative coupling (assumed here to have only nondiagonal elements). In the dipole approximation V is given by

$$V_{ji}(t)=-\mu_{ji}E(t) \tag{2}$$

where (for several-mode fields)

$$E(t)=\sum_k\left(\mathcal{E}_k e^{-i\omega_k t}+\mathcal{E}_k^* e^{i\omega_k t}\right) \tag{3}$$

is the electric field, and μ_{ji} is the dipole transition element. Neglecting line-shifting effects, $i\Sigma$ is a real supermatrix (or tetradic matrix) of relaxation rates connecting the various elements of ρ. Equation 1 is generally valid for field intensities sufficiently strong to cause multiphoton transitions and saturation effects but insufficient to affect the relaxation rates (which are determined by the equilibrium properties of the system). Under such conditions, Σ is reducible into a T_1 part, connecting only populations, and a T_2 part, affecting only coherences. The T_2 part is further reducible, as we shall see below.

Solution of the Bloch–Redfield equations with harmonic radiative couplings usually involves an expansion of ρ in harmonics of the applied field frequencies[15, 43]

$$\rho(t)=\sum_{n}\rho^{(n)}(t)e^{in\cdot\omega t}\left(n\cdot\omega=\sum_{k}n_k\omega_k \qquad n_k=0,\pm1,\ldots\right) \tag{4}$$

One obtains in this method, known as the Floquet method,[44] a set of equations with time-independent coefficients, at the price of enlarging the number of independent variables. This complication is avoided, however, if the following conditions hold:

1. Only frequency combinations $n\cdot\omega$ close to a molecular resonance frequency are retained (the so-called rotating-wave approximation, or RWA).
2. Only *one* such combination fits this resonance condition, for a given transition.

To this category belong the *simply-connected multiple resonances* illustrated by Fig. 1a, b, provided the mode frequencies are sufficiently close to resonance, and nearly-degenerate resonance transitions are excited by the same field mode. In such a simply-connected chain there is only one resonance, or succession of resonances, connecting one level to the other.

Under the two conditions stated above only one term of the Floquet expansion should be retained for each element ρ_{ji} of the density matrix—the one with the frequency combination $n\cdot\omega$ near resonance with the molecular resonance frequency $\nu_{ji}=\hbar^{-1}(E_j-E_i)$. We shall therefore omit in such cases the superscript n. The remaining term still retains the slow time variation produced by the (now time-independent) coefficients of the Bloch–Redfield equations.

The retaining of a single Floquet term has another important consequence. Owing to the assumption that the relaxation rates are independent of the fields, the supermatrix Σ cannot couple elements of ρ with different

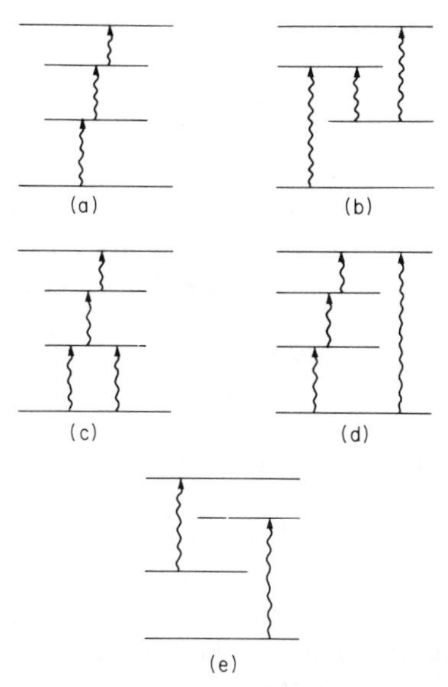

Fig. 1. Examples of (a, b) simply-connected, (c, d) multiply-connected, and (e) disconnected multiple resonances.

sets of Floquet numbers n. Thus, for example, a coherence cannot be coupled to a population, a one-photon coherence to a two-photon coherence, etc., or a coherence excited by one mode to a coherence excited by another.

The Bloch–Redfield equations are generally valid under conditions of homogeneous broadening. Inhomogeneous broadening effects (such as Doppler broadening in gases, combined with velocity-changing collisions) require a modification of the equations. Equations 1 are applicable to complete sets of molecular states, closed under cross-relaxation, or to open subsets with irreversible leaking (as in predissociative systems coupled to a continuum). They can be reduced to subsets of states (such as the subset of levels in resonance with the applied fields) by modifying the Σ elements to take into account repopulation of the reduced subset by back relaxation. These modifications are discussed in Section IV below.

III. STEADY-STATE SOLUTIONS

A steady-state solution to (1) exists for closed sets of levels in which the total population is conserved (and consequently for subsets of such systems). Denoting as Γ the part of $i\Sigma$ connecting populations, such systems obey the *conservation law*

$$\sum_j \Gamma_{ji} \equiv \sum_j (i\Sigma_{jj,ii}) = 0 \tag{5}$$

The requirement that Γ be determined by equilibrium properties entails that in the absence of external fields ρ must reduce to the *equilibrium distribution* $\rho^{\rm eq}$ which, owing to *detailed balance*, obeys

$$\sum_i \Gamma_{ji} \rho_{ii}^{\rm eq} = 0 \tag{6}$$

that is, $\rho^{\rm eq}$ forms an eigenvector of Γ with zero eigenvalue in a closed system.

Algebraic expressions can be obtained for the steady-state solutions in a closed set of N levels by using the method of retarded Green's functions (in a manner analogous to the one used in scattering theory, adapted to the density-matrix formalism). This method is based on an infinite-series expansion and provides means ("renormalization schemes") for the summation of these series. The method is fully described in Ref. 25 and is only outlined here. The uninterested reader can skip the mathematical detail and consider only the definitions and the final results.

The solution as presented in Ref. 25, explicitly refers to the N-level "ladder," in which the radiation successively couples levels of increasing energy. However, it can be easily extended to any "folded" resonance chain, as will be seen below.

The infinite series of terms involved in the calculation of the transition rates is represented by the diagram in Fig. 2. In this diagram, the vertex

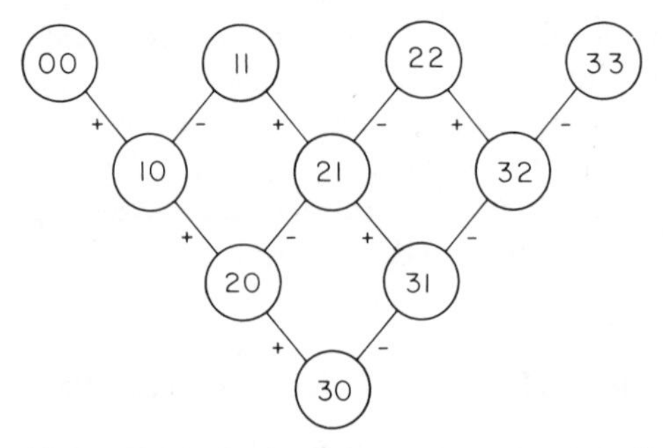

Fig. 2. Diagrammatic representation of infinite series and renormalization scheme for a N-level multiple resonance. Circles denote the various populations (ii) and coherences (ji). Segments denote radiative couplings with their appropriate signs.

(shown as a circle) labeled *ij* corresponds to the element ρ_{ij}, with $i, j = 0, 1, \ldots, N-1$, denoting the position of each level along the simply-connected chain of resonances *irrespective* of their ordering on the energy scale. The line segments represent couplings by the relevant field modes. Cross-relaxation, which can also connect different vertices, is not shown explicitly.

The calculation of the transition rate W_{ji} from i to j involves a summation over all possible trajectories connecting vertex *ii* to *jj*. To each intermediate vertex we assign a resolvent (or Green's function) and to each segment—a coupling coefficient $\pm\beta_i$, where

$$\beta_i \equiv |V_{i,i-1}|\hbar^{-1} \tag{7}$$

the sign depending on whether V operates on the row or the column index of ρ. All vertices, including those pertaining to populations, may appear as intermediate vertices any number of times. The analysis is simplified by introducing an auxiliary matrix of "*single passage*" rates, R_{ji}, which are defined as the sum over all trajectories leading from *ii* to *jj* *without* going through any intermediate population vertex. Using matrix notation we can then write

$$\begin{aligned} W_{ji} &= R_{ji} + (RGR)_{ji} + \cdots \\ &= (R + RGW)_{ji} \end{aligned} \tag{8}$$

where G_{kl} is an element of a matrix of Green's functions (resolvents) confined to the populations alone,

$$G = (\Gamma + \epsilon I)^{-1} \qquad (\epsilon \to +0) \tag{9}$$

with I being the unit matrix. The diagonal elements of Γ,

$$\Gamma_{ii} \equiv \gamma_i \tag{10}$$

are the *inverse lifetimes* of the levels. Its off-diagonal elements are

$$\Gamma_{ji} \equiv -\eta_{ji} \tag{11}$$

where η_{ji} is the *cross-relaxation* rate from i to j. The increment ϵ must be retained finite to the end of the calculation, since Γ is a singular matrix by (5).

In the case of the simply-connected N-level resonance chain, a closed algebraic expression can be obtained for the rank-N R matrix in the RWA.

Figure 2 illustrates the presence of a hierarchy of coherences excited by the applied fields. The first row of vertices below the row of populations represents one-photon coherences $\rho_{i,i-1}$, the second row, two-photon coherences $\rho_{i,i-2}$, and so on. (The coherences ρ_{ij} $(i<j)$ are not shown explicitly; their contribution to R being simply the complex conjugate of that of the ρ_{ji}'s.)

The infinite series representing all possible repetitions of intermediate coherences can be summed up by a simple renormalization scheme, unfolding the diagram from the single N-photon vertex $\rho_{N-1,0}$ upward. This scheme results in an expression of the R elements as N-step continued fractions. To successive denominators in the continued fraction (represented by corresponding rows in Fig. 2) we assign matrices of Green's functions of successively decreasing rank. Thus, to the mth row (consisting of the m-photon coherences $\rho_{j,j-m}$) we assign the rank-$(N-m)$ matrix $g^{(m)}$, the inverse of which is

$$\left[g^{(m)}\right]_{jk}^{-1}=(-i\Delta_{j,j-m}+\gamma_{j,j-m})\delta_{jk}+\sigma_{jk}^{(m)}$$
$$(m=1,\ldots,N-1 \qquad j,k=m,m+1,\ldots,N-1) \tag{12}$$

Here δ_{jk} is the Kronecker symbol,

$$\Delta_{ji}\equiv n\cdot\omega-\nu_{ji} \tag{13}$$

is the off-resonance *frequency detuning* corresponding to the ρ_{ji} coherence, and

$$\gamma_{ji}\equiv i\Sigma_{ji;\,ji}=\tfrac{1}{2}(\gamma_i+\gamma_j)+\xi_{ji} \tag{14}$$

is the corresponding T_2 relaxation rate, composed of a T_1 contribution and the proper-T_2 contribution ξ_{ji} (frequently referred to as the *dephasing rate*). The renormalization by coupling to lower rows is represented by

$$\sigma_{jk}^{(m)}=\sum_{il}\chi_{jk;\,il}^{(m)}g_{il}^{(m+1)} \qquad (m=1,\ldots,N-2)$$
$$=0 \qquad (m=N-1) \tag{15}$$

where

$$\chi_{jk;\,il}^{(m)}=\delta_{j+1,i}\delta_{k+1,l}\beta_{j+1}\beta_{k+1}+\delta_{ji}\delta_{kl}\beta_{j-m}\beta_{k-m}$$
$$-\delta_{j+1,i}\delta_{kl}\beta_{j+1}\beta_{k-m}-\delta_{ji}\delta_{k+1,l}\beta_{j-m}\beta_{k+1}$$
$$\times(j,k=m,\ldots,N-1 \qquad i,l=m+1,\ldots,N-1) \tag{16}$$

The single-passage rates are then given by

$$R_{jk} = -2\,\mathrm{Re}\sum_{il} \chi^{(0)}_{jk;\,il} g^{(1)}_{il} \tag{17}$$

As stated above, to each ν_{ji} corresponds just one frequency combination $n\cdot\omega$ in the simply-connected chain. There is, however, no limitation on having several ν_{ji}'s corresponding to the same $n\cdot\omega$ as, for example, in the case of the nearly-harmonic ladder of energy levels excited by a single radiation mode. In such cases the element $\gamma_{j,\,j-m}\delta_{jk}$ in (12) should be replaced by the supermatrix element $i\Sigma_{j,\,j-m;\,k,\,k-m}$. The elements $j\neq k$ representing *cross-coherence relaxation* generally depend on the relative positions of the levels on the energy scale. If, however, cross-coherence relaxation can be neglected, R attains a very useful property in the RWA: Aside from a possible change in the sign of some Δ_{ji}, which reflects the sign of ν_{ji}, the calculation of R is independent of the positioning of the levels on the energy scale (given a set of values of the γ_{ji}'s). Thus, for example, the two chains (a) and (b) in Fig. 1 have the same R matrix (up to a change of sign of Δ_{21}), provided the levels are labeled by their position along the chain, and not by their energy. The major difference between various foldings of the chain is in the matrix G, which, owing to detailed balance, depends on the relative energies of the levels. However, G is simple to calculate, whereas R is much more complicated.

The one-photon coherence $\rho_{i,\,i-1}$ couples either to the population $\rho_{i-1,\,i-1}$ with the coupling coefficient β_i, or to ρ_{ii} with $-\beta_i$. From this symmetry it follows that R is a singular matrix, since

$$\sum_i R_{ij} = 0 \tag{18}$$

Likewise, Γ is singular because of (5) and therefore G diverges as $\epsilon\to 0$. However, the coupling symmetry allows for a reduction of the rank of the matrices G, R, and W, replacing them by the nonsingular nondivergent rank-$(N-1)$ matrices $\hat{G}$, $\hat{R}$ and $\hat{W}$, defined on the space of population differences $(\rho_{ii}-\rho_{i-1,\,i-1})$ by

$$\hat{G}_{kl} = \sum_{ij} G_{ij}\theta_{ij;\,kl} \tag{19}$$

$$R_{ij} = \sum_{kl} \hat{R}_{kl}\theta_{ij;\,kl} \qquad W_{ij} = \sum_{kl} \hat{W}_{kl}\theta_{ij;\,kl}$$

where

$$\theta_{ij;\,kl} = (\delta_{ik}-\delta_{i,\,k-1})(\delta_{jl}-\delta_{j,\,l-1})$$
$$(i,j=0,1,\ldots,N-1 \qquad k,l=1,\ldots,N-1) \tag{20}$$

To the matrix element $\hat{W}_{ji}$ in the simply-connected chain corresponds a transition starting with the resonance excitation of the $\rho_{i,i-1}$ coherence and ending with the $\rho_{j,j-1}$ coherence. To each such $j, j-1$ coherence corresponds a single field mode k_j. The steady-state *attenuation rate* of the k_j mode (in photons per molecule per unit time) is given by

$$A_{k_j} = \sum_i s_{j,j-1} \hat{W}_{ji} (\rho_{ii}^{eq} - \rho_{i-1,i-1}^{eq}) \tag{21}$$

where $s_{j,j-1}$ is the *sign* of the energy difference $\hbar \nu_{j,j-1}$. In case the same field mode k_i couples to several $j, j-1$ pairs, Eq. (21) should be summed over all such pairs. The steady-state populations of a closed set of levels are given by

$$\rho_{jj}^{st} - \rho_{j-1,j-1}^{st} = \sum_i (\hat{I} + \hat{G}\hat{W})_{ji} (\rho_{ii}^{eq} - \rho_{i-1,i-1}^{eq}) \tag{22}$$

where $\hat{I}$ is the rank-$(N-1)$ unit matrix. $\hat{W}$ can be calculated by the matrix relation

$$\hat{W} = (\hat{R}^{-1} - \hat{G})^{-1} \tag{23}$$

Equations 21 and 22 depend on ρ^{eq} because it is presumably the only eigenvector of Γ with a zero eigenvalue. They are valid irrespective of the initial conditions imposed on ρ in a closed system.

IV. OPEN SUBSETS

In actual applications, the multiple resonance process involves only a small subset of internal states. Even if the complete set of molecular states is closed under relaxation (no coupling to a continuum), the *resonance subset* is open to relaxation to, or through, other levels. It is nevertheless possible to show that this subset can be treated as a separate set, provided the matrix Γ (defined on the complete set) is replaced by a Γ^{eff} (defined on the resonance subset) in which the relaxation rates are modified for relaxation through nonresonant states. This correction can be obtained by the use of projection operator techniques, in a manner analogous to the one used in derivations of optical potentials.

Let

$$i, j \in P \qquad \alpha, \beta \in Q \tag{24}$$

where P and Q denote, respectively, the resonance subset and the complementary subset of nonresonant states. Then[26]

$$\Gamma_{ji}^{\text{eff}}=\Gamma_{ji}-\sum_{\alpha\beta}\Gamma_{j\alpha}(\Gamma_Q+\epsilon I_Q)_{\alpha\beta}^{-1}\Gamma_{\beta i}\qquad(\epsilon\to+0)\tag{25}$$

where Γ_Q is the part of the Γ matrix confined to the Q subset, and I_Q is the corresponding unit matrix.

It is important to keep track of the increment ϵ in (25). As a result, the sum rule (5) is not obeyed by Γ^{eff}, to order of ϵ, on the P subset. Since Γ^{eff} becomes a singular matrix in the limit $\epsilon\to 0$, the order-ϵ terms yield nonvanishing contributions to $\hat{G}$. The total steady-state population of the P subset is therefore not equal to the total equilibrium population, and the rank-N matrices R and G must be used in order to calculate at least one population, through

$$\rho_{jj}^{\text{st}}=\sum_i (I+GW)_{ji}\rho_{ii}^{\text{eq}}\tag{26}$$

One may thus write

$$\begin{aligned}\gamma_i^{\text{eff}}&=\gamma_i-\phi_i+\epsilon r_{ii}\\ \eta_{ji}^{\text{eff}}&=\eta_{ji}+\eta_{ji}^{Q}-\epsilon r_{ji}\end{aligned}\tag{27}$$

Here ϕ_i is the repopulation rate of level i, owing to backward relaxation from levels in the Q subset to which it relaxes; η_{ji}^{Q} is the indirect decay rate from i to j through intermediate Q states; r_{ji} consist of ratios of cross-relaxation rates.

Consider, for example the two-level set 0, 1, in resonance with a single radiation mode with the coupling coefficient β. Assuming that relaxation may occur only down the energy scale ($\gamma_0=0$, $\gamma_1=\eta_{01}$), the rank-1 matrices $\hat{R}$ and $\hat{G}$ are given by

$$\hat{R}_{11}=-\frac{2\beta^2\gamma_{10}}{(\omega-\nu_{10})^2+\gamma_{10}^2}\qquad \hat{G}_{11}=\frac{2}{\gamma_1}\tag{28}$$

If, however, level 1 can also relax to 0 through an intermediate state 2 ($\gamma_2=\eta_{02}$, $\gamma_1=\eta_{01}+\eta_{21}$), we get $\eta_{01}^{Q}=\eta_{21}$ and $r_{01}=\eta_{21}/\gamma_2$, and hence $\hat{G}_{11}$ should be multiplied by $(1+\frac{1}{2}\eta_{21}\gamma_2^{-1})$. Indirect relaxation thus effectively lengthens the apparent lifetime of the excited level 1. Repopulation of level 1 by back relaxation from neighboring levels results in a similar effect.

The use of the effective relaxation rates introduces non-Markovian retardation effects into the Bloch–Redfield equations defined on the reduced P set. The Green's functions used in the steady-state solution are generally defined as functions of a complex variable z. Steady-state solutions involve their value at $z=i\epsilon$ ($\epsilon\to+0$), that is, their limit as z approaches the origin on the complex plane from above. Time-evolution operators, as functions of the time ($t>0$) are obtained by Fourier-transforming the Green's functions on the real axis ($z=x+i\epsilon$). Therefore, in order to describe the time-resolved behavior of the reduced set, one must replace ϵ by $\epsilon-ix$ before proceeding with the calculation of the Fourier transforms. The resulting equations for the populations of the reduced set, $\rho_{ii}(t)$, are

$$i\frac{\partial}{\partial t}\rho_{jj}=\hbar^{-1}[(H_0+V),\rho(t)]_{jj}$$
$$-i\sum_i\int_{-\infty}^{\infty}\Phi_{ji}(t-t')[\rho_{ii}(t')-\rho_{ii}^{\text{eq}}]\,dt' \tag{29}$$

where

$$\Phi_{ji}(t)=\Gamma_{ji}\delta(t)-\sum_{\alpha\beta}\Gamma_{j\alpha}(e^{-\Gamma_Q^0 t})_{\alpha\beta}\Gamma_{\beta i}\theta(t) \tag{30}$$

is the Fourier transform of Γ^{eff}. Here $\theta(t)$ is the Heaviside step function, and $\delta(t)$ is the Dirac delta function. The term involving the equilibrium populations can be subtracted, since Γ^{eff} and its Fourier transform obey the detailed balance law on the reduced set ρ_{jj}^{eq} ($j\in P$), if Γ obeys it on the complete set.

If the resonance subset P is imbedded in a much larger set of closely spaced energy levels (as in large or medium size molecules), repopulation and indirect relaxation may be a very slow process, and therefore, *on the time scale of the level decay times* the non-Markovian addition to Γ on (29) may be neglected. If no direct cross-relaxation occurs between the levels, the *Bloch equations* may be used,

$$i\frac{\partial}{\partial t}\rho_P(t)=\hbar^{-1}[(H_0+V),\rho_P(t)]+\Sigma_P[\rho_P(t)-\rho_P^{\text{eq}}] \tag{31}$$

where ρ_P and Σ_P are, respectively, the parts of ρ and Σ confined to the P subset, with $(\Gamma_P)_{ij}=\gamma_i\delta_{ij}$. Quasi-steady-state solutions of (31) can be calculated by using (26) with G_P instead of G, where G_P is obtained from (9), upon replacing Γ with Γ_P. These quasi-steady-state solutions (not to be confused with the true steady-state solutions) are valid only on a time scale long compared to the level lifetimes but short compared to the repopulation time. The total population of the P subset is generally not conserved (unless Γ_P is a constant matrix, $\Gamma_P = \gamma I_P$).

The situation is qualitatively different if the resonance subset forms a truly open set by direct (or indirect) irreversible decay to a continuum (as in photodissociation or photoionization). The total population will generally decay under a continuous-wave irradiation.

All the foregoing arguments regarding the level population dynamics do not affect the single-passage matrix. As will be seen in the next section, it plays an important role in the time-resolved behavior of level populations in open as well as in closed sets.

V. MASTER EQUATIONS

Master equations can be derived in certain limiting situations,[26, 32, 34, 45] by a reduction of the equations of motion to populations, or diagonal elements of $\rho(t)$. As shown below, such a reduction can be performed on the Bloch–Redfield equations, introducing the single-passage rates R_{ji} in the role of field-induced rate coefficients. This reduction, too, can be accomplished by Green's function techniques. Define $R(z)$, replacing ϵ by $-iz$ in (9), and adding $-iz\delta_{jk}$ to (12) in the calculation of R, so that $R = R$ $(z = i\epsilon)$ in the steady-state solutions. A time-resolved matrix $\tilde{R}(t)$ is defined by the Fourier transform

$$\tilde{R}(t) = \left(\frac{1}{2\pi}\right)\int_{-\infty}^{\infty} e^{-ixt} R(x + i\epsilon)\,dx \qquad (\epsilon \to +0,\, t > 0) \tag{32}$$

The reduction of the equations of motion by projecting onto the diagonal part of ρ leads to

$$\frac{\partial}{\partial t}\rho_{ii}(t) = -\sum_j \left[\Gamma_{ij}\rho_{jj}(t) - \int \tilde{R}_{ij}(t-t')\rho_{jj}(t')\,dt'\right] \tag{33}$$

The steady-state populations are readily obtained from this equation by noting that ρ^{eq} is the zero-eigenvalue eigenvector of Γ, and that

$$R_{ij} = \int_{-\infty}^{t} \tilde{R}_{ij}(t-t')\,dt' \tag{34}$$

Equations 33 are non-Markovian rate equations with a memory kernel $\tilde{R}(t-t')$. Markovian equations are obtained as an approximation in case the time variation of $\tilde{R}(t)$ is much more rapid than that of $\rho(t)$. Replacing

$\rho(t')$ by $\rho(t)$ under the integral sign we then get, using (34),

$$\frac{\partial}{\partial t}\rho_{ii}(t) = -\sum_j (\Gamma_{ij} - R_{ij})\rho_{jj}(t) \tag{35}$$

It can be shown that off-diagonal elements of R are definite-positive. As R obeys (18), (35) forms a set of *master equations*, with conservation of populations in closed systems, under the combined action of Γ and R. These equations also apply to open systems, in which Γ does not obey the conservation law (5), or to resonance subsets, replacing Γ by the integral operator of (29), R being confined to the resonance subset.

In the limit of weak coupling to the radiation,

$$\beta_i \ll \gamma_{i,i-1} \qquad (i=1,\ldots,N-1) \tag{36}$$

R_{ij} can be calculated by perturbation theory. To lowest order, $R_{i\pm m,i}$ (with $m>0$) are proportional to the product of intensities of the m modes directly connecting i to j in the simply-connected multiple resonance. Thus, for example, in the nearly-harmonic ladder excited by a single radiation mode (k) with the *intensity* I_k,

$$R_{i\pm m,i} \propto I_k^m \qquad (m>0) \tag{37}$$

The master equations are dominated in the limit of (36) by the elements R_{ji} with $j-i=0, \pm 1$, which are proportional to the intensity I_k, and the solutions of the equations than depend on the *fluence*[46] $I_k t$. At higher field intensities the power law (37) becomes moderated by saturation, as described by the renormalization scheme discussed in Section III, ultimately leading to an upper bound for R_{ji} determined by the T_2 rates.

As implied by (36), the weak coupling limit is valid when the radiative couplings β_i are much weaker than the T_2 rates. However, in this limit the time variation of $\tilde{R}(t)$ (with the frequency detunings sufficiently close to resonance) is governed by the T_2 rates. The change in $\rho(t)$ produced by R is comparably slower. Let, in addition, the T_1 rates be much weaker than the T_2 rates,

$$\eta_{ij} \ll \gamma_{kl} \qquad (\text{for all } i, j, k, l) \tag{38}$$

The overall variation of $\rho(t)$ will then be much slower than that of $\tilde{R}(t)$, and the Markovian approximation (35) will hold. Condition (38) is observed, for example, in some models of mode-selective vibrational excitation and photodissociation of polyatomic molecules. In these models,[34, 35, 47] the T_1 rates are actually assumed to vanish, and intermode coupling is represented as a T_2 process. In view of the large T_2 rates associated with such models, the weak coupling condition (36) is generally observed with experimentally applied field intensities.

VI. THREE-LEVEL DOUBLE RESONANCE

The formal results obtained in Section III regarding the steady-state behavior of N-level multiple resonances are studied here by application to the simplest multiple resonance—the three-level system.[2, 9, 18, 21, 23, 25, 48] Many of the general features of N-level multiple resonances can be induced from this example.

There are three possible configurations to the three-level double resonance, illustrated in Fig. 3: the "ladder double resonance" (LDR), "Raman double resonance" (RDR), and "inverted-Raman double resonance" (IDR). With the levels in all three configurations labeled according to their position in the chain of resonances (as in Fig. 3), the single-passage rates for the three configurations are related to each other by a change of sign of one of the frequency detunings. Let the 1–0 resonance be excited by the radiation mode $k=1$, with its frequency detuned off resonance by the amount Δ_1, and the 2–1 resonance by mode $k=2$, with its frequency detuned by Δ_2. Then, given a set of T_2 rates,

$$\begin{aligned} R_{\mathrm{LDR}} &= R_{\mathrm{LDR}}(\Delta_1, \Delta_2) \\ R_{\mathrm{RDR}} &= R_{\mathrm{LDR}}(\Delta_1, -\Delta_2) \\ R_{\mathrm{IDR}} &= R_{\mathrm{LDR}}(-\Delta_1, \Delta_2) \end{aligned} \tag{39}$$

In all three configurations, R is invariant under change of sign of *both* detunings.

The major difference between the three configurations lies in the matrix G of T_1 rates. Let us consider here only spectra in the *optical range*, where energy spacings are much higher than the mean thermal energy.

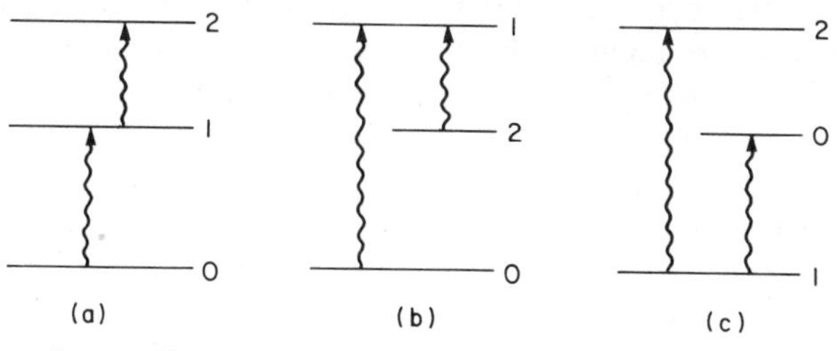

Fig. 3. The various configurations of the three-level double resonance: (a) the "ladder" (LDR), (b) the "Raman" (RDR), and (c) the "inverted-Raman" (IDR).

Equilibrium population is then confined to the lowest resonance level, and cross relaxation occurs only to lower levels.

A. The Ladder Double Resonance

Extensive work has been done on the LDR configuration[2, 3, 49] (Fig. 3a). Here,

$$\Gamma_{\mathrm{LDR}}=\begin{bmatrix} \gamma_2 & 0 & 0 \\ -\eta_{12} & \gamma_1 & 0 \\ -\eta_{02} & -\eta_{01} & 0 \end{bmatrix} \tag{40}$$

(arranging the row and column indices in descending order). Assume that the two resonances are nondegenerate; that is, $|\nu_{10}-\nu_{21}|$ is much larger than all T_1 and T_2 rates, radiative couplings, and frequency detunings. In such a situation, the RWA applies, and the formal theory of Section III can be used to obtain algebraic expressions for steady-state populations and attenuation rates, in terms of the 10 independent parameters η_{01}, η_{12}, η_{02}, ξ_{10}, ξ_{21}, ξ_{20}, β_1, β_2, Δ_1, and Δ_2. Let us further assume here that cross-relaxation occurs only by cascading down to the nearest level below; that is,

$$\eta_{j-1,j}>0 \qquad \eta_{ij}=0 \qquad (i\neq j-1) \tag{41}$$

Then[21, 25]

$$A_j=\gamma_j\rho_j^{\mathrm{st}} \qquad (j=1,\ldots,N-1) \tag{42}$$

for the N-level ladder.

Figure 4 illustrates the absorption spectrum of each mode, A_1 vs. Δ_1, and A_2 vs. Δ_2, at several fixed detunings of the other mode, in a strongly saturated double resonance, in the absence of dephasing ($\xi_{ij}=0$). The coupling coefficients β_1 and β_2 are here in the saturation range ($\beta_i \gg \gamma_{i,i-1}$), where A_1 and A_2 are bounded by the magnitude of the T_1 rates. We have therefore used a system of inverse time units where the smaller of $\frac{1}{2}\gamma_i$ ($i=1$ or 2) serves as unity. Thus, for example, $\gamma_1=\gamma_2=2$ in Fig. 4. Quite obviously, the absorption of radiation occurs in two possible modes: *one photon resonance absorption* (OPRA), centered around $\Delta_1=0$, and *two-photon resonance absorption* (TPRA), around $\Delta_1+\Delta_2=0$. The absorption A_2 occurs only through the latter mode, when Δ_1 is detuned far off resonance, whereas A_1 occurs through both modes. The width of the OPRA is dominated by saturation broadening, while the TPRA peak is much

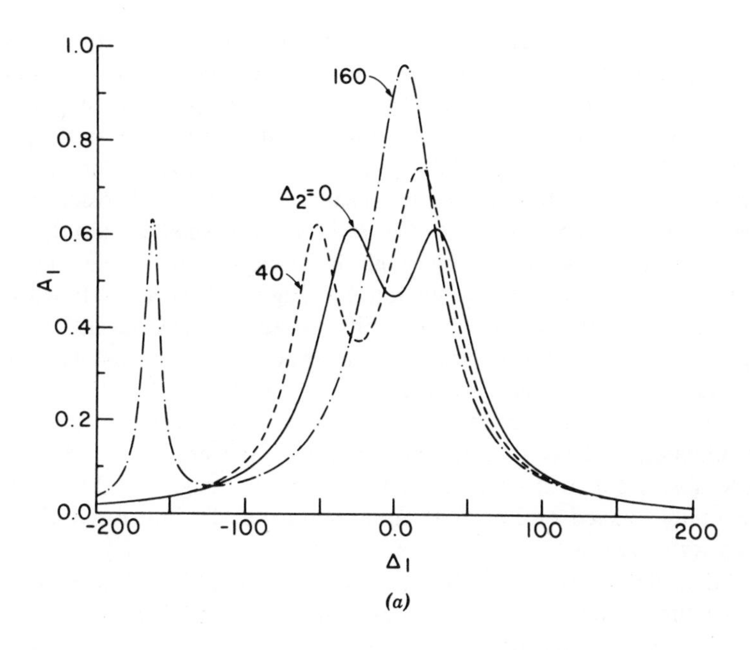

(a)

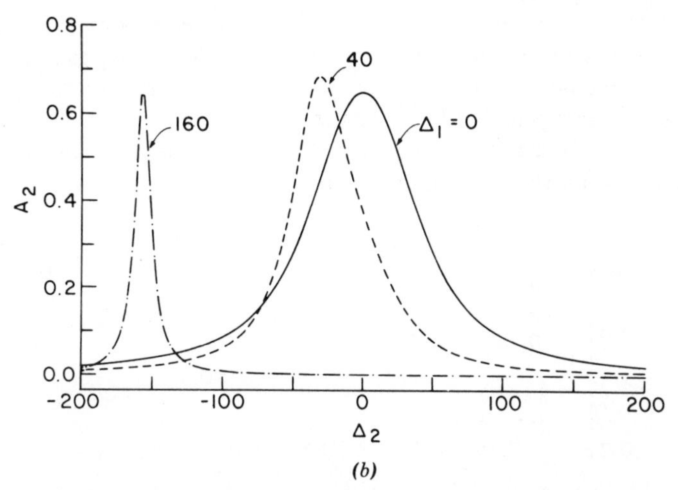

(b)

Fig. 4. LDR attenuation rates without dephasing: (a) A_1 vs. Δ_1, with $\Delta_2=0$, 40, 160. (b) A_2 vs. Δ_2, with $\Delta_1=0$, 40, 160. $\beta_1=20$, $\beta_2=30$, $\gamma_0=0$, $\gamma_1=\gamma_2=2$, $\eta_{02}=0$ (all in units of $\frac{1}{2}\gamma_2$).

narrower, its width decreasing as the detuning increases. With $\beta_2 > \beta_1$, the A_1 spectrum is split by dynamic Stark doubling[50] as Δ_2 is tuned to resonance. This doubling generally occurs in the spectrum of the weaker beam when the stronger beam is in the saturation range. Also, the position of the TPRA peak is slightly shifted to a nonzero value of $\Delta_1 + \Delta_2$ (dynamic Stark shift) when the two beams are detuned off resonance. The sign of the shift depends on whether β_2 is larger or smaller than β_1.

The effect of adding dephasing, or proper-T_2 rates (e.g., by elastic collisions in gas spectra), is illustrated in Fig. 5. Here A_2 is shown vs. Δ_2, with Δ_1 fixed at 240 (in same units as above), for several values of $\xi_{10} = \xi_{21} = \xi_{20}$. The major feature is the appearance of a new OPRA peak around $\Delta_2 = 0$. This peak is similar in origin to the "redistribution" line in resonance fluorescence,[51–53] resulting from the population of level 1 by the introduction of dephasing. As the dephasing rates are increased, more and more absorption will occur in this fashion. Although the TPRA peak is eventually washed out, even at large detunings, the integrated absorption intensity increases.

Redistribution peaks, in the general N-level ladder, will appear around zero detuning, irrespective of the detunings of other beam modes. This is an expression of the *incoherent absorption* associated with the Markovian chain of radiative jumps produced by the master equations (35).

The intensity of the redistribution peaks nevertheless depends on preceding steps in the ladder. In particular, it depends on the T_2 damping rates of resonances down the ladder. Thus, for example, Fig. 6 illustrates the sensitivity of the redistribution intensity to the magnitude of ξ_{10}, and its relative insensitivity to the other dephasing rates.

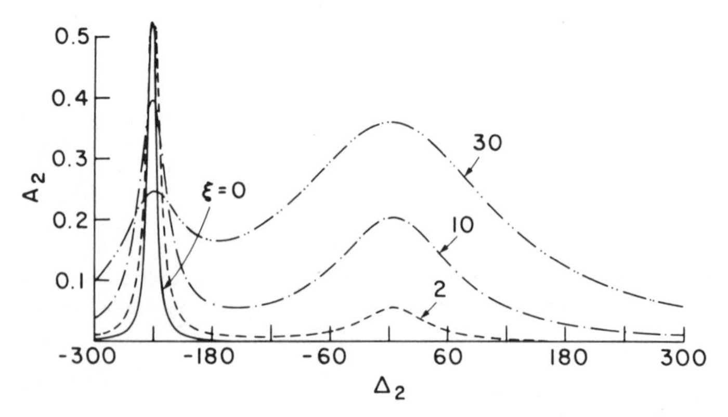

Fig. 5. LDR attenuation rates with dephasing: A_2 vs. Δ_2, with $\Delta_1 = 240$, $\beta_1 = 30$, $\beta_2 = 20$, $\gamma_0 = 0$, $\gamma_1 = \gamma_2 = 2$, $\eta_{02} = 0$, $\xi_{10} = \xi_{21} = \xi_{20} \equiv \xi = 0$, 2, 10, and 30 (all in units of $\frac{1}{2}\gamma_2$).

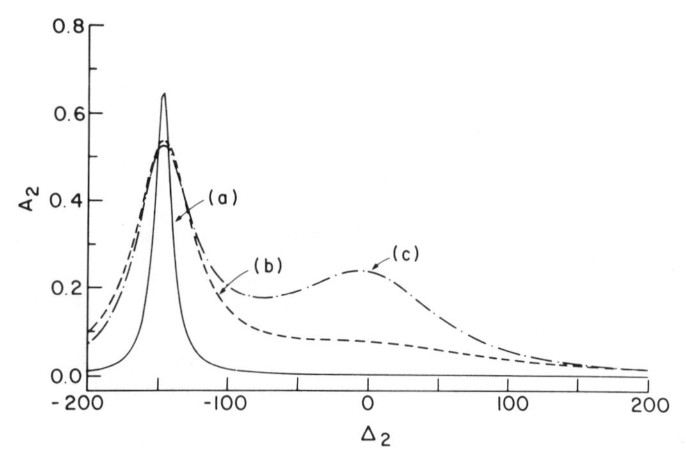

Fig. 6. LDR attenuation rates with unequal dephasing: A_2 vs. Δ_2, with $\Delta_1=150$, $\beta_1=20$, $\beta_2=30$, $\gamma_0=0$, $\gamma_1=\gamma_2=2$, $\eta_{02}=0$. (a) $\xi_{10}=\xi_{21}=\xi_{20}=0$; (b) $\xi_{10}=2$, $\xi_{21}=\xi_{20}=10$; (c) $\xi_{10}=\xi_{20}=10$, $\xi_{21}=2$ (all in units of $\frac{1}{2}\gamma_2$).

B. The Raman Double Resonance

In the RDR configuration (Fig. 3b),

$$\Gamma_{RDR}=\begin{bmatrix} \gamma_2 & -\eta_{21} & 0 \\ 0 & \gamma_1 & 0 \\ -\eta_{02} & -\eta_{01} & 0 \end{bmatrix} \tag{43}$$

We distinguish between the inducing (or "pump") mode $k=1$, which is always absorbed, and the Stokes mode $k=2$, which under almost all conditions is *amplified* ($A_2<0$). Figure 7 shows A_2 vs. Δ_2, at several fixed values of Δ_1, with both β_i's in the saturation range, $\gamma_1=\gamma_2=2$, and $\eta_{01}=2$, with no dephasing. Here, too, A_1 occurs through both OPRA and TPRA, while A_2 occurs through TPRA only when Δ_1 is detuned off-resonance. The conversion from beam 1 to 2 is especially efficient at large detunings, where (in the absence of dephasing) $A_1 \approx -A_2$ (i.e., one-to-one photon conversion). This process forms the basis of the double-resonance Raman amplifier, or photon convertor.[9, 10, 12, 48, 54–62] The stronger field mode imposes a dynamic Stark shift on the weaker field mode, to which the TPRA is very sensitive (although the macroscopic conversion process obtained by collinear propagation along the absorption cell is nevertheless most efficient at equal detunings[12, 63] $\Delta_1=\Delta_2$).

The TPRA at large detunings leads to equal sharing of the steady-state populations between levels 0 and 2, leaving the higher level 1 empty. This

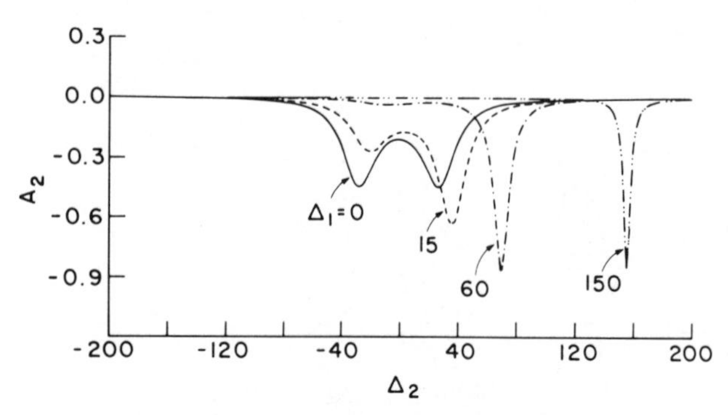

Fig. 7. RDR amplification (negative attenuation) without dephasing: A_2 vs. Δ_2, with $\Delta_1=0$, 15, 60, 150. $\beta_1=30$, $\beta_2=10$, $\gamma_0=0$, $\gamma_1=\gamma_2=2$, $\eta_{01}=2$ (all in units of $\frac{1}{2}\gamma_2$).

situation is totally opposed to stimulated-emission processes induced by population inversion, in which both pumping and emission are typically one-photon processes. The TPRA is therefore actually independent of γ_1. At zero detunings, the amplification becomes sensitive to the dynamics of level 1, and increases with increasing branching ratio η_{01}/η_{02}.

The amplification process by TPRA is badly damaged both by proper-T_2 processes, and by inhomogeneous-broadening effects (such as Doppler broadening). In particular, it is affected by the broadening of the two-photon coherence (ξ_{20}), and proportionately much less by that of the other coherences.

C. The Inverted-Raman Double-Resonance

In the IDR configuration (Fig. 3c),

$$\Gamma_{\mathrm{IDR}}=\begin{pmatrix} \gamma_2 & 0 & 0 \\ -\eta_{12} & 0 & -\eta_{10} \\ -\eta_{02} & 0 & \gamma_0 \end{pmatrix} \tag{44}$$

The higher-frequency mode $k=2$ is always absorbed. The other mode, however, may be amplified under certain conditions.[2, 64, 65] The occurrence of amplification is particularly sensitive to the branching ratio η_{02}/η_{12} of the decay of level 2, as illustrated by Fig. 8. In this example, again, the two beams are in the saturation range, and no dephasing occurs. The $k=1$ mode is absorbed when level 2 decays directly to the lower level 1, and is amplified when it decays through the intermediately lying level 0. In contrast to the RDR, this amplification process requires the repopulation of level 0 by cross-relaxation from 2, and is therefore confined to small

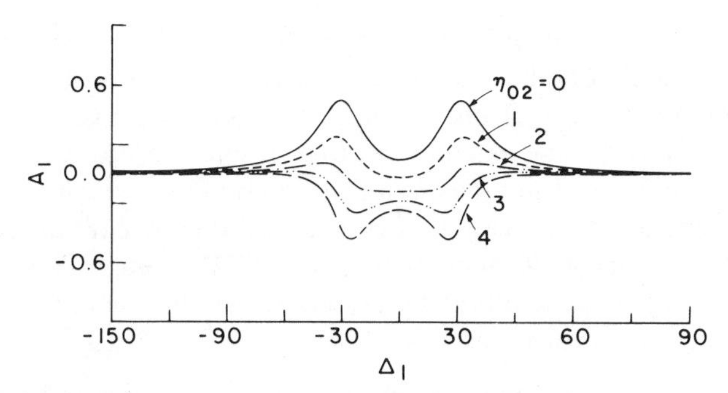

Fig. 8. IDR attenuation without dephasing: A_1 vs. Δ_1, with $\Delta_2=0$, $\beta_1=10$, $\beta_2=30$, $\gamma_0=2$, $\gamma_1=0$, $\gamma_2=4$, $\eta_{12}=0$, 1, 2, 3 and 4 (all in units of $\frac{1}{2}\gamma_0$).

detunings Δ_2. (The RDR amplification by TPRA occurs best, as one may recall, at large detunings.)

The kind of behavior illustrated by Fig. 8 clearly demonstrates the advantage of the Bloch–Redfield equations (1) over the widely used form of the Bloch equations (31) when individual cross-relaxation rates play an important role.

VII. DISCUSSION

The analysis in previous sections was limited by the following requirements: (1) Validity of the resonance condition (and hence applicability of the rotating-wave approximation). Application to (2) simply (3) connected multiple resonances. (4) Constant field intensities. (5) Assumption that relaxation rates are not affected by the applied fields. (6) Homogeneous broadening with (7) short-memory relaxation. (8) Restriction to applied fields, as opposed to scattered fields. We shall briefly discuss here possible extensions beyond these limitations.

A. Corrections to the RWA

Numerous attempts have been made to incorporate corrections to the RWA, mostly concerning the two-level single-resonance.[66]

The approximation methods used are most valuable when a single Floquet term $\rho_{ji}^{(n)}$ still dominates but is renormalized by coupling to other terms with n not obeying the resonance condition. Such renormalizations are usually expressed as infinitely continued fractions, and lead to further splitting and shifting (Bloch–Siegert shifts) of the resonance peaks.[67] These renormalization schemes should be applied directly to the density-matrix formalism[43] (and not to the Schrödinger equation), if one wants to incorporate cross-relaxation and proper-T_2 rates in a straightforward manner. All applications of the Green's function method to the density-matrix formalism are based on the idea that elements of ρ form a Hilbert space (dyadic, or double space) on which perturbation expansions can be defined. The RWA can be extended to treat resonance transitions involving simultaneously several photons (with no intermediate resonance levels) as single "elementary" processes. Such transitions occur in two-photon absorption, ordinary Raman scattering, CARS, etc. The interaction Hamiltonian (2) is generalized by adding terms representing two-photon, three-photon, etc., transitions.

B. Multiply-Connected Resonances

In multiply-connected resonances of the kind shown in Fig. 1, the resonance condition

$$n\cdot\omega\approx\nu_{ji} \qquad [n=(n_1,n_2,\ldots)] \tag{45}$$

has more than one solution. The simplest case is the two-level double resonance, with $\omega_1\approx\omega_2\approx\nu_{10}$. In all such cases there exists a nontrivial combination $m=(m_1, m_2,\ldots)$, with no common divisor, such that $m\cdot\omega\approx 0$, and therefore, if n obeys (45), then all $n+km$, with $k=0, \pm 1,\ldots,$ obey it too. For example, in the two-level double resonance, $m=(1, -1)$. Elements of R and G then form operators defined on the infinite manifold of $\rho_{ji}^{(n+km)}$ labeled by k, with G diagonal in k but R connecting terms with different k. The various resolvents such as (12) should be renormalized in order to include all the frequency combinations allowed by the RWA, $\Delta_{ji}+km\cdot\omega$. This renormalization can lead, again, to infinitely-continued fractions.

C. Disconnected Multiple Resonances

The best-known examples of disconnected multiple resonances are four-level double resonances, such as Fig. 1e. This is not really an extension of the case studied here but rather a special case, since they can be obtained from a connected resonance chain in the limit when some of the radiative couplings vanish. Some elements of R then vanish, but the corresponding elements of G may not.

D. Pulsed Radiation

Pulse shapes can be formally incorporated in (1) by imposing on $E(t)$ a time-dependent envelope. There is no general method of attacking this problem even for the two-level single resonance (although the differential equations can be transformed into integral equations that are more amenable to iterative solutions). For slowly varying envelopes, an adiabatic-following approximation[37] is valid at finite detunings. In the two-level case, an exact solution[68] (the so-called Rabi solution) exists at $\Delta = 0$. Given special types of envelopes (e.g., exponential functions[38]), solutions can be found in the form of hypergeometric functions, for arbitrary detunings, in absence of relaxation.

E. Field-Modified Relaxation

Under sufficiently strong radiation, the supermatrix Σ of relaxation rates can not be considered as independent of the applied field. Well-known examples of field-dependent rate processes are the so-called radiative collisions.[69–77] One can extend the definition of field-dependent relaxation rates to the general theory of spectral lineshapes. The basic idea is to calculate the thermal bath effects (e.g., collision cross-sections) as if they occur not with the free molecule, but with the dressed molecule, the eigenstates of which are modified by the interaction with the applied field. As a result, the reduction symmetry with respect to the Floquet components described in Section II generally breaks down, and the number of allowed elements of Σ proliferates.

The modified rates

$$\Sigma(\beta) \equiv \Sigma(\beta_1, \beta_2, \ldots) \tag{46}$$

do not maintain various properties of the equilibrium rates, such as detailed balance. However, their action is confined[26] to deviations of ρ from equilibrium. Therefore (1) should be replaced by

$$i\frac{\partial}{\partial t}\rho(t) = [(H_0 + V), \rho(t)] + \Sigma(\beta)[\rho(t) - \rho^{eq}] \tag{47}$$

A useful representation for expressing solutions in this case is the so-called *dressed-atom representation*,[22, 78] in which eigenstates of $H_0 + V$ are used. It is particularly useful when the splitting between such eigenvalues (Rabi shifts) appreciably exceeds the magnitude of corresponding elements of Σ. Certain elements (such as cross-coherence relaxations between well-isolated resonance peaks) can then be neglected.

F. Inhomogeneous Broadening

Molecular states require, in order to be specified completely, a set of quantum numbers pertaining to both internal and positional (translational) degrees of freedom. While the earlier form a discrete set (below the dissociation, or ionization, threshold), the latter usually form a continuous set (e.g., momentum states in gases) or a quasicontinuous set (e.g., in lattices). In general, relaxation processes mix the positional states (velocity-changing collisions in gases, or exchange phenomena in solids), and the Bloch–Redfield equations must be replaced by more general kinetic equations.[79] In the approximation that such exchange phenomena can be ignored, the various positional states will form a continuum (or quasicontinuum) of invariant manifolds. On each such manifold, a Bloch–Redfield equation applies, with possibly varying values of the resonance frequencies and the relaxation rates. The spectral distribution of the attenuation rates will then be obtained by a convolution of single-manifold solutions with the distribution of such manifolds. This idea is behind most works on saturation (or Lamb-dip) spectroscopy in gases,[27–31, 80] and is valid provided the homogeneously broadened single-manifold spectra are considerably narrower than the distribution of manifolds (such as the Gaussian distribution of Doppler shifts in gases).

G. Memory Effects

The derivation of the Bloch–Redfield equations, as ordinary differential equations, is based on the approximation that the correlation (or memory) times associated with the coupling to the thermal bath are extremely short. Taking into account their finite duration, the supermatrix Σ becomes dependent on the frequency detunings in the steady-state solution. This problem, known as the "line-wing" problem, has been extensively studied in single-photon atomic spectra[81] and had some recent implications on resonance fluorescence.[82] Linear-response theories of spectral line shapes can be readily extended to multiple resonances.

H. Resonance Scattering

The methods described above for obtaining steady-state solutions can be easily extended to the calculation of scattering rates or differential cross-sections. In an ideally resolved experiment we can consider a single scattered field mode as part of the "relevant" system, and add its states (in second-quantized representation) to the definition of the basis on which ρ is specified. In scattering induced by strong incident fields many scattered photons can be coincidentally emitted in various modes ("cascading"). However (unless we do coincidence measurements), emission in all "irrelevant" modes may be incorporated into the relaxation rates by a proper definition of the radiative "thermal bath."[26]

The emission mode (as opposed to the applied fields) can be considered as a "weak probe,"[14, 16] and treated by perturbation methods, whereas the applied fields may still require non-perturbative treatments. The dressed-atom representation[22] is particularly useful in deriving the spectral distribution of the scattered radiation in this situation.

Acknowledgments

The authors are grateful to Mr. Michael Berman for assistance in the computations.

The authors are grateful to Dr. Myron F. Goodman and Dr. James Stone for drawing their attention to their works on the N-level problem. The algorithm for calculating the single-passage matrix for the N-level ladder appeared in J. Stone, E. Thiele, and M. F. Goodman, *J. Chem. Phys.* **59**, 2909 (1973). Proper-T_2 (dephasing) effects, as well as a detailed study of the master equations, were included in a more recent work [J. Stone and M. F. Goodman, *Phys. Rev.* **A18**, 2618 and 2642 (1978)]. Stone and Goodman did not state, however, the topological arguments generalizing the solution to "folded" single-chain multiple resonances. It should be noted, at this point, that closed algebraic solutions exist also for simply-connected "branched-chain" multiple resonances (such as the 4-level system in which one level is resonantly coupled to all three others), although the resulting algorithm is quite different [A. Ben-Reuven and Y. Rabin (unpublished work)].

References

1. V. S. Letokhov, in *High Resolution Laser Spectroscopy*, K. Shimoda, ed., Springer, Berlin, 1976, p. 95.
2. V. P. Chebotaev, *ibid.*, p. 201.
3. N. Bloembergen and M. D. Levenson, *ibid.*, p. 315.
4. V. S. Letokhov and V. P. Chebotaev, *Nonlinear Laser Spectroscopy*, D. L. MacAdam, ed., Springer, Berlin, 1977.
5. V. S. Letokhov and C. B. Moore, in *Chemical and Biochemical Applications of Lasers*, Vol. 3, C. B. Moore, ed., Academic Press, New York, 1978.
6. R. V. Ambartzumian and V. S. Letokhov, *ibid.*
7. C. D. Cantrell, S. M. Freund, and J. L. Lyman, in *Laser Handbook IIIb*, North-Holland, Amsterdam, 1978.
8. J. Jortner, *Adv. Lasers Spect.*, **113**, 88 (1977).
9. Th. Hänch and P. Toschek, *Z. Phys.*, **236**, 213 (1970).
10. Z. Drozdowicz, P. Woskoboinikov, K. Isobe, D. R. Cohn, R. J. Temkin, K. J. Button, and J. Waldman, *IEEE J. Quant. Electron.* **QE-13**, 413 (1977).
11. R. J. Temkin, *IEEE J. Quant. Electron.* **QE-13**, 450 (1977).
12. Y. Rabin, M. Berman, and A. Ben-Reuven, *J. Phys.*, **B 13**, 2127 (1980).
13. H. C. Torrey, *Phys. Rev.*, **76**, 1059 (1949).
14. B. R. Mollow, *Phys. Rev.*, **188**, 1969 (1969); *Phys. Rev.*, **A2**, 76 (1970).
15. A. Ben-Reuven and L. Klein, *Phys. Rev.*, **A4**, 753 (1971).
16. B. R. Mollow, *Phys. Rev.*, **A5**, 2217 (1972).
17. G. S. Agarwal, *Quantum Optics, Springer Tracts in Modern Physics*, Vol. 70, Springer, Heidelberg, 1974.
18. R. G. Brewer and E. L. Hahn, *Phys. Rev.*, **A11**, 1641 (1975).
19. B. R. Mollow, *Phys. Rev.*, **A12**, 1919 (1975).
20. H. J. Kimble and L. Mandel, *Phys. Rev.*, **A13**, 2123 (1976).
21. R. M. Whitley and C. R. Stroud Jr., *Phys. Rev.*, **A14**, 1498 (1976).
22. C. Cohen-Tannoudji and S. Reynaud, *J. Phys.*, **B10**, 345 (1977); **10**, 2311 (1977).
23. Z. Bialynicka-Birula and I. Bialynicki-Birula, *Phys. Rev.*, **A16**, 1318 (1977).
24. B. R. Mollow, *Phys. Rev.*, **A15**, 1023 (1977).
25. Y. Rabin and A. Ben-Reuven, *Phys. Rev.*, **A19**, 1697 (1979).
26. A. Ben-Reuven and Y. Rabin, *Phys. Rev.*, **A19**, 2056 (1979).
27. B. J. Feldman and M. S. Feld, *Phys. Rev.*, **A1**, 1375 (1970).
28. S. Haroche and F. Hartmann, *Phys. Rev.*, **A6**, 1280 (1972).

29. E. V. Baklanov and V. P. Chebotaev, *Sov. Phys. JETP*, **34**, 490 (1972).
30. L. Klein, M. Giraud, and A. Ben-Reuven, *Phys. Rev.*, **A16**, 289 (1977).
31. C. Delsart and J. C. Keller, *J. Phys.*, **39**, 350 (1978).
32. J. R. Ackerhalt and J. H. Eberly, *Phys. Rev.*, **A14**, 1705 (1976).
33. J. R. Ackerhalt and B. W. Shore, *Phys. Rev.*, **A16**, 277 (1977).
34. S. Mukamel, *J. Chem. Phys.*, **71**, 2012 (1979).
35. J. G. Black, E. Yablonovich, N. Bloembergen, and S. Mukamel, *Phys. Rev. Lett.*, **38**, 1131 (1977).
36. B. Carmeli and A. Nitzan, *J. Chem. Phys.*, **72**, 2054 (1980); **72**, 2070 (1980).
37. D. Grischkowsky, M. M. T. Loy, and P. F. Liao, *Phys. Rev.*, **A12**, 2514 (1975).
38. A. D. Wilson and H. Friedmann, *Chem. Phys.*, **23**, 105 (1977).
39. T. K. Yee and T. K. Gustafson, *Phys. Rev.*, **A18**, 1597 (1978).
40. J. R. Ackerhalt, *Phys. Rev.*, **A17**, 293 (1978).
41. F. Bloch, *Phys. Rev.*, **105**, 1206 (1957).
42. A. G. Redfield, *Adv. Mag. Resonance*, **1**, 1 (1965).
43. L. Klein, M. Giraud, and A. Ben-Reuven, *Phys. Rev.*, **A10**, 682 (1974).
44. J. H. Shirley, *Phys. Rev.*, **B138**, 979 (1965).
45. G. S. Agarwal, *Phys. Rev.*, **A18**, 1618 (1978).
46. P. Kolodner, C. Winterfeld, and E. Yablonovitch, *Opt. Commun.*, **20**, 119 (1977).
47. N. Bloembergen and E. Yablonovitch, *Phys. Today*, **May 1978**, p. 23.
48. M. S. Feld and A. Javan, *Phys. Rev.*, **177**, 540 (1969).
49. I. M. Beterov and V. P. Chebotaev, in *Progress in Quantum Electronics*, Vol. 3, J. H. Sanders and S. Stenholm, eds., Pergamon Press, New York, 1974.
50. H. R. Gray and C. R. Stroud Jr., *Opt. Commun.*, **25**, 359 (1978).
51. D. L. Huber, *Phys. Rev.*, **158**, 843 (1967); **170**, 418 (1968); **178**, 93 (1969).
52. A. Omont, E. W. Smith, and J. Cooper, *Astrophys. J.*, **175**, 185 (1972); **182**, 283 (1973).
53. Y. R. Shen, *Phys. Rev.*, **B9**, 622 (1974).
54. J. L. Carlsten and P. C. Dunn, *Opt. Commun.*, **14**, 8 (1975).
55. D. Cotter, D. C. Hanna, P. A. Kärkäinen, and R. Wyatt, *Opt. Commun.*, **15**, 143 (1975).
56. D. Cotter and D. C. Hanna, *Opt. Quant. Electron.*, **9**, 509 (1977).
57. S. J. Petuchowski, A. T. Rosenberger, and T. A. De-Temple, *IEEE J. Quant. Electron.*, **QE-13**, 476 (1977).
58. D. Seligson, M. Ducloy, J. R. R. Leite, A. Sanchez, and M. S. Feld, *IEEE J. Quant. Electron.*, **QE-13**, 468 (1977).
59. J. Heppner and C. O. Weiss, *Opt. Commun.*, **21**, 381 (1977).
60. J. Heppner, C. O. Weiss, and P. Plainchamp, *Opt. Commun.*, **23**, 381 (1977).
61. D. Cotter and D. C. Hanna, *IEEE J. Quant. Electron.*, **QE-14**, 184 (1978).
62. D. Cotter and W. Zapka, *Opt. Commun.*, **26**, 251 (1978).
63. R. Frey, F. Pradere, and J. Ducuing, *Opt. Commun.*, **23**, 65 (1977).
64. I. M. Beterov and V. P. Chebotaev, *Sov. Phys. JETP Lett.*, **9**, 127 (1969).
65. S. M. Freund and T. Oka, *Phys. Rev.*, **A13**, 2178 (1976).
66. D. R. Dion and J. O. Hirschfelder, *Adv. Chem. Phys.*, **35**, 265 (1976).
67. P. R. Berman and J. Ziegler, *Phys. Rev.*, **A15**, 2042 (1977).
68. J. Allen and J. H. Eberly, *Optical Resonance and Two Level Atoms*, Wiley, New York, 1975.
69. L. I. Gudzenko and S. I. Yakovlenko, *Sov. Phys. JETP*, **35**, 877 (1972).
70. V. S. Lisitsa and S. I. Yakovlenko, *Sov. Phys. JETP*, **39**, 759 (1974); **41**, 233 (1975).
71. S. E. Harris and D. B. Lidow, *Phys. Rev. Lett.*, **33**, 674 (1974).
72. A. M. F. Lau, *Phys. Rev.*, **A13**, 139 (1976).
73. N. M. Kroll and K. M. Watson, *Phys. Rev.*, **A13**, 1018 (1976).

74. D. B. Lidow, R. W. Falcone, J. F. Young, and S. E. Harris, *Phys. Rev. Lett.*, **36**, 472 (1976).
75. J. I. Gersten and M. H. Mittleman, *J. Phys.*, **B9**, 383 (1976).
76. T. F. George, J. M. Yuan, I. H. Zimmerman, and J. R. Laing, *Discuss. Faraday Soc.*, **62**, 246 (1977).
77. J. Light and A. Szöke, *Phys. Rev.*, **A18**, 1363 (1978).
78. C. Cohen-Tannoudji, in *Frontiers in Laser Spectroscopy*, Vol. 1, R. Balian, S. Haroche, and S. Liberman, eds., North-Holland, Amsterdam, 1977.
79. E. G. Pestov and S. G. Rautian, *Sov. Phys. JETP*, **29**, 488 (1969).
80. P. R. Berman, *Phys. Rep.*, **43**, 103 (1978); *Adv. At. Mol. Phys.*, **13**, 57 (1978).
81. *Bibliography on Atomic Line Shapes and Shifts*, J. R. Fuhr et al., eds., U.S. National Bureau of Standards Special Publ. No. 366, 1972; and Supplements 1, 1974; 2, 1975; and 3, 1978.
82. J. Carlsten, A. Szöke, and M. G. Raymer, *Phys. Rev.*, **15**, 1029 (1977).

LASER EXCITATION OF SF_6: SPECTROSCOPY AND COHERENT PULSE PROPAGATION EFFECTS*

C. D. CANTRELL

Center for Quantum Electronics and Applications, The University of Texas at Dallas, Richardson, Texas 75080

A. A. MAKAROV

Institute of Spectroscopy, USSR Academy of Sciences, Troitzk, Podol'sky Raion, Moscow Region 142092

and

W. H. LOUISELL

Department of Physics, University of Southern California, Los Angeles, California 90007

CONTENTS

I. Introduction 584
II. High-Resolution Spectroscopy of the ν_3 Band of SF_6 586
  A. Vibration-Rotation Basis 592
  B. Vibration-Rotation Hamiltonian 594
  C. Fitting of Spectroscopic Parameters 596
  D. Transition Moments 598
  E. A Model for the ν_3 Mode of SF_6 601
III. Coherent Propagation Effects in SF_6 603
  A. Electromagnetic Field Equations 603
  B. Calculation of the Polarization Density 604
  C. Thin-Sample Approximation 606
  D. Spectral Content of the Radiated Field 609
  E. Generation of Sidebands at Resonant Molecular Transition Frequencies 611
  F. Effects of Spatial (M) Degeneracy 613
IV. Summary and Discussion 617
Acknowledgments 621
References 622

*Portions of this research were supported by the United States Department of Energy.

Abstract

In this paper we summarize recent theoretical studies of coherent propagation effects in SF_6 and other polyatomic molecules, beginning with an account of relevant aspects of the high-resolution spectroscopy of the ν_3 band of SF_6. We show that a laser pulse propagating in a molecular gas can acquire new frequencies which were not initially present in the pulse, and that, in fact, a wave is coherently generated at the frequency of every molecular transition accessible from the initial molecular energy levels. We discuss the possible consequences of coherent generation of sidebands for the multiple-photon excitation of SF_6 and other polyatomic molecules.

I. INTRODUCTION

This chapter is a brief account of recent theoretical developments in the theory of propagation of laser pulses through a molecular vapor, and of closely related topics in high-resolution molecular spectroscopy, which may help provide some insight into the role of coherence in the laser-driven multiple-photon excitation of SF_6 and other polyatomic molecules. High-resolution infrared spectroscopy and coherent propagation effects are closely linked in SF_6, both conceptually and historically. Early experimental[1-5] and theoretical[6-9] studies of coherent propagation effects in SF_6 either suggested or depended upon specific models of the participating SF_6 energy levels and transition moments. Recent high-resolution spectroscopic studies[10-13] have provided assignments of thousands of transitions in the ν_3 fundamental of SF_6, and have helped provide a framework for the still speculative discussion[14] of excited-state transitions in SF_6. The experimental linewidth observed in saturation spectroscopy of SF_6[13] is less than 10 kHz, which we may remark appears to be inconsistent with a postulated[21] intramolecular thermalization time of 30 ps. In Section II we shall summarize the current state of knowledge of the ν_3 fundamental ($v_3=0 \to v_3=1$) of SF_6. Although we cannot yet fully characterize the energy levels and transition moments of the states of the ν_3 mode of SF_6 with more than one vibrational quantum, we shall describe a model which we believe to possess many of the qualitative features of the energy levels and transition moments of the real SF_6 molecule.

In Section III of this chapter, we shall direct most of our attention to certain coherent propagation effects, which may have a major influence on the development of the spectrum of an initially monochromatic, nearly resonant laser pulse as it propagates through a vapor of polyatomic molecules. Physically, the process of optical propagation consists of the creation of a coherent, macroscopic electromagnetic polarization by the incident optical electric field, and the interference of the optical electric field radiated by this macroscopic polarization with the incident field. The total field produced by this interference acts on the molecular system, and the macroscopic polarization produced thereby must be self-consistent with the total field. Such a self-consistent, nonlinear coupling is well known from Lamb's theory of the laser[22] and theories of laser pulse propagation in simplified two-level systems developed by Hopf and Scully[23] and Icsevgi and Lamb.[24] We shall summarize a general derivation[25] of the equations governing the propagation of a laser pulse interacting with an ensemble of multilevel molecular systems, within the framework of the slowly varying amplitude and phase approximation (SVAPA) and the rotating-wave approximation (RWA). We shall apply this formalism to the specific case of propagation of laser pulses in SF_6 vapor, in the limit of an optically thin sample. The coherent effects which arise in optically thin samples are optical nutation,[6] optical free induction decay,[26] and photon echoes.[2, 27, 28] Optical nutation[6] arises from the fact that the macroscopic polarization produced by the incident optical electric field contains, in addition to the frequency ω of the incident laser field, new frequencies higher or lower than ω by an amount equal to the Rabi frequency of the transitions excited by the incident field. The macroscopic polarization then radiates a field which contains Rabi sidebands; the interference of the radiated field with the incident field makes the sidebands evident as a temporal oscillation of intensity of the total field. If a system is pumped nonresonantly by a laser pulse, the Rabi frequency is very nearly equal to the detuning, so that the frequency of one of the sidebands very nearly coincides with the resonant transition frequency of the system. The molecular excitation produced by this nearly resonant, coherently generated field may, of course, greatly exceed the excitation which would be produced by the nonresonant incident field acting alone, depending on the magnitude of the new field coherently generated in the medium.[25, 29] As is to be expected for coherent effects, the intensity of the field radiated by the macroscopic molecular polarization is proportional to the square of both the molecular number density N and the distance z traveled in the sample. Order-of-magnitude estimates presented below suggest that Rabi sidebands may well be of significant intensity for conditions often encountered in experiments on multiple-photon excitation of polyatomic molecules (pressure$\sim$0.1 torr, $z\sim 10$ cm).

The major conclusion of the work summarized in this paper following our initial suggestion[29] is that the optical field coherently radiated by a molecular vapor subjected to an incident optical field contains a rich spectrum of sidebands, covering essentially the full vibration–rotation band with which the incident field interacts. The new, coherently generated field causes coherent cycling of population between the states radiatively coupled by the incident optical field. In a two-level system not subject to relaxation processes (e.g., collisions), the occurrence of coherent cycling of population would mean that, on a time scale long compared to the resonant Rabi period in the coherently generated field, approximately half the population would appear in the upper state and half in the lower state. This is a qualitatively important effect, which can greatly increase the theoretically predicted effectiveness of multiple-photon excitation of polyatomic molecules. The excitation of many rotational levels in SF_6 at surprisingly low laser intensities has been observed experimentally.[30]

It is, of course, possible to give a phenomenological interpretation of the strong excitation of many molecular energy levels not resonant with the incident laser field as being the result of rapid collisionless intramolecular energy transfer.[21] Such an intramolecular phenomenon would be completely independent of N and z, so that it should in principle be possible to distinguish unimolecular from collective coherent phenomena experimentally by a properly conducted study of the dependence of laser energy deposition in the sample (for example) on N and z. Effects due to coherent generation of Rabi sidebands should be a function of the product Nz, for optically thin samples, and for a given (fixed) incident laser intensity. It is, of course, certain that some effects of sideband generation have already been observed experimentally, but it is very easy to ascribe these effects to other causes. For example, effects due to the increase of sideband electric field as N (for fixed z) could be identified as effects of collisions among the molecules pumped by the laser. However, the effects of sideband generation will also depend on z (for fixed N), and are thus distinguishable in principle from collisional effects.

Although we are aware that the ideas about coherent propagation effects described in this paper are at odds with some published concepts of multiple-photon excitation, we hope that our work will at least stimulate new experiments, and new interpretations of already published data.

II. HIGH-RESOLUTION SPECTROSCOPY OF THE ν_3 BAND OF SF_6

In this section we shall summarize the current state of high-resolution spectroscopy of the ν_3 fundamental band ($v_3 = 0 \rightarrow v_3 = 1$) of SF_6, and shall indicate some current ideas on the structure of SF_6 states with two or more ν_3 quanta. The spectra of the infrared-active modes of tetrahedral and octahedral spherical-top molecules are highly complex, owing in large part to the fact that these vibrational modes are triply degenerate. In octahedral spherical-top molecules, the two triply-degenerate infrared-active modes, ν_3 and ν_4, both belong to the (three-dimensional) F_{1u} representation of the octahedral point group O_h (Fig. 1). The ν_3 mode in SF_6 involves primarily stretching motions, while the ν_4 mode involves both stretching and bending

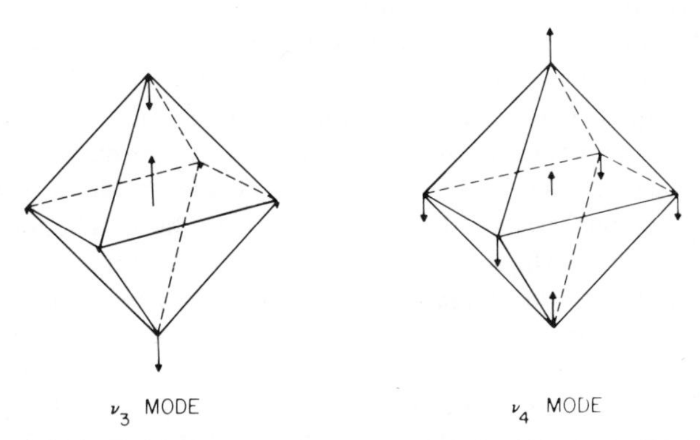

Fig. 1. Atomic displacements in one of the ν_3 and one of the ν_4 modes of SF_6, as determined from the force-field analysis of McDowell et al.[31]. (Courtesy of R. S. McDowell.)

motions.[31] The complexity of the vibration-rotation spectra of these triply-degenerate modes is the result of many physical effects: (1) splitting of levels with two or more vibrational quanta by vibrational anharmonic effects; (2) Teller-Coriolis splitting (and Coriolis interaction between different vibrational states) due to interactions between vibrational and rotational angular momenta; (3) splitting of each rotational level (which is $(2J+1)$-fold degenerate in the molecule-fixed frame) into as many states as are allowed by the molecular point-group symmetry, due to tensor vibration–rotation interactions; (4) nuclear hyperfine splitting. All of these effects are significant in understanding and assigning the experimentally observed high-resolution spectra of the SF_6 ν_3 band. Even the nuclear hyperfine splitting has been resolved, and some additional vibration–rotation spectroscopic constants determined, in recent sub-Doppler studies of SF_6.[13]

However, essentially nothing is firmly established with regard to the spectroscopic constants of states in SF_6 with two or more vibrational quanta. In view of this fact, all we can attempt in this brief review with regard to vibrational overtone states is to outline a model which possesses some important qualitative features of the overtone states and excited-state transition moments of the real SF_6 molecule. The model we shall outline, which has the virtues of computational convenience and physical reasonableness, has been used in the numerical studies of coherent propagation effects in SF_6 which we shall describe in Section III of this chapter.

The derivation of the quantum-mechanical Hamiltonian for a vibrating, rotating polyatomic molecule has been the subject of discussion and study for many years; we refer the reader to a small subset of the literature for a detailed summary.[32–35] A full derivation of the effective vibration–rotation Hamiltonian for a single (degenerate) mode of a spherical-top molecule, starting with a power-series expansion of the vibration potential energy, involves a sequence of contact transformations to bring operators of successively higher order to approximately diagonal form. Such a transformation has been carried out for a triply-degenerate mode of a tetrahedral molecule,[36] but has not yet been attempted in the octahedral case. An alternative approach which is very useful for the assignment of spectra and the determination of spectroscopic constants is to expand the vibration–rotation Hamiltonian in terms of all operators (up to a given order in the vibrational normal coordinates, vibrational and rotational angular momenta, etc.) which are allowed by the molecular symmetry group.[37] The phenomenological constants that multiply the different operators appearing in such an expansion can, of course, be expressed in terms of the parameters characterizing the molecular force field, the equilibrium moment of inertia, etc., and such expressions are known for the tetrahedral case.[38] Because of the far greater complexity of the vibrational Hamiltonian for octahedral molecules,[39] only the phenomenological approach of regarding the constants which appear in the vibration–rotation Hamiltonian as independent parameters, and adjusting these parameters to give a least-squares fit to experimentally measured spectral line positions, has been followed for SF_6 and other octahedral molecules.[10–12]

The amount of spectroscopic detail which can usefully be studied theoretically depends to a great extent on the degree of resolution that can be achieved experimentally. Grating spectra[40] of SF_6 at room temperature show a smooth band contour uninterrupted by rotational structure. A grating spectrum of SF_6 at a temperature of 153 K[41] (Fig. 2), at which 80% of the SF_6 molecules are in the vibrational ground state, shows an irregular contour which is not noise, but is also not recognizable as rotational structure. A real understanding of the SF_6 spectrum depended on obtaining experimental spectra with a resolution limited only by the SF_6 Doppler width (30 MHz at room temperature). The application of semiconductor diode lasers to vibration–rotation spectroscopy[42–44] resulted in Doppler-limited spectra of SF_6 covering a frequency range of approximately ± 1 GHz near the CO_2 laser lines overlapping the SF_6 ν_3 band.[45] Although these spectra went unassigned at that time, they proved to be very useful later[10–12] owing to the fact that they had been calibrated by heterodyning the tunable semiconductor diode laser with a fixed-frequency CO_2 laser stabilized at the center of the CO_2 laser line.[45] This technique gives a direct

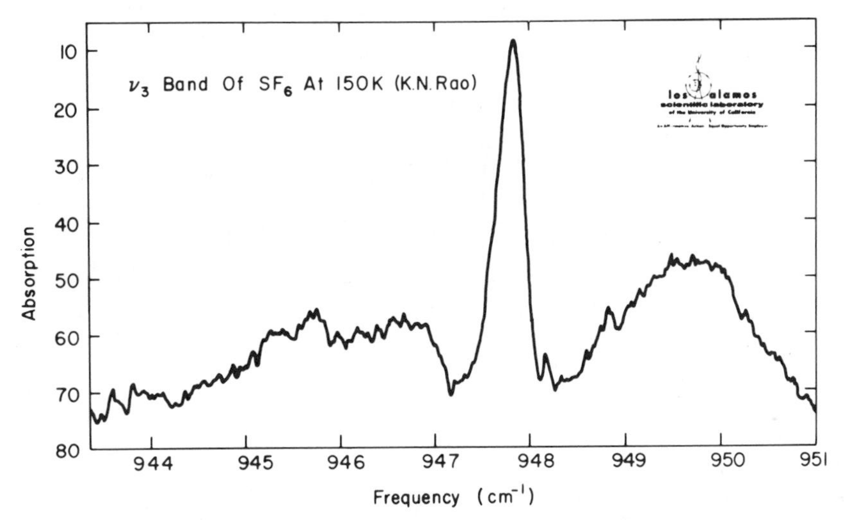

Fig. 2. The absorption spectrum of the ν_3 band of SF_6 at a temperature $T = 153$ K. (Courtesy of K. N. Rao, S. Hurlock, P. L. Houston, and J. I. Steinfeld.[41]) The P, Q, and R branches are clearly evident, but no rotational structure is visible.

measurement of the frequency difference between the center of the CO_2 laser line and the spectroscopic feature to which the semiconductor diode laser is tuned, and is thus considerably superior in accuracy to the technique of calibration by etalon fringes which is more commonly used in laser diode spectroscopy.[46] Much greater accuracy can now be achieved by sub-Doppler saturation spectroscopy[13] than by heterodyne calibration of Doppler-limited spectra.

Assignment of the SF_6 ν_3 band[10–13] was made possible by Doppler-limited spectra covering a large range near the ν_3 band center[47] (Fig. 3) in SF_6 at temperatures where more than 90% of the SF_6 molecules are in the ground vibrational state, and by knowledge of the nuclear spin-statistical weights,[48, 49] which made it possible to recognize the pattern or "fingerprint" of fine-structure lines associated with each value of the total angular momentum J. At first only the $\nu_3 P$ and R branches could be assigned;[10] later the Q branch was also assigned[11] (Fig. 4). More recently, saturation spectroscopy of the ν_3 band of SF_6 has made it possible to determine a more accurate set of vibration-rotation spectroscopic parameters, and to uncover effects of nuclear hyperfine splitting.[13]

Prior to the publication of the assignment of the ν_3 band of SF_6, a number of studies of coherent propagation effects suggested, or depended upon, certain qualitative aspects of the spectroscopy of the ν_3 band. Observations of optical nutation[3, 5] were interpreted to give rather different

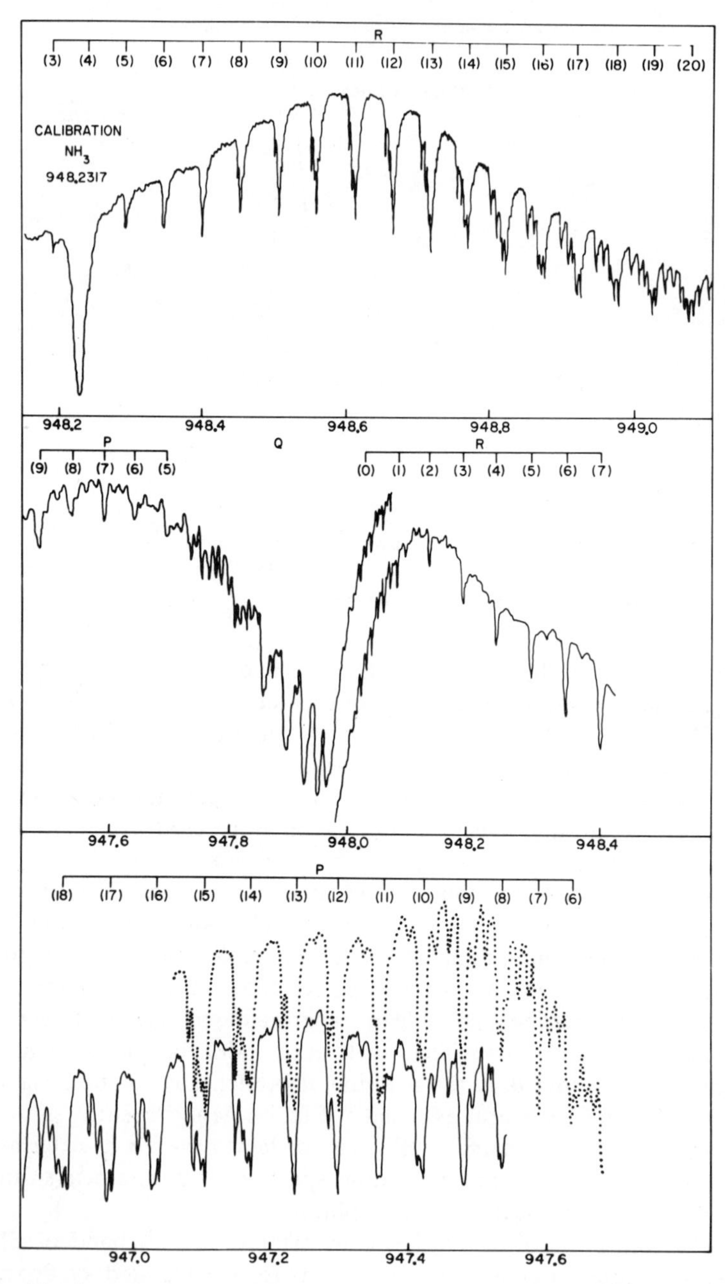

Fig. 3. The absorption spectrum of the central portion of the ν_3 band, of SF_6 near 948 cm^{-1} at a temperature of 135 K, obtained with a semiconductor diode laser.[47] The center trace shows the Q branch.

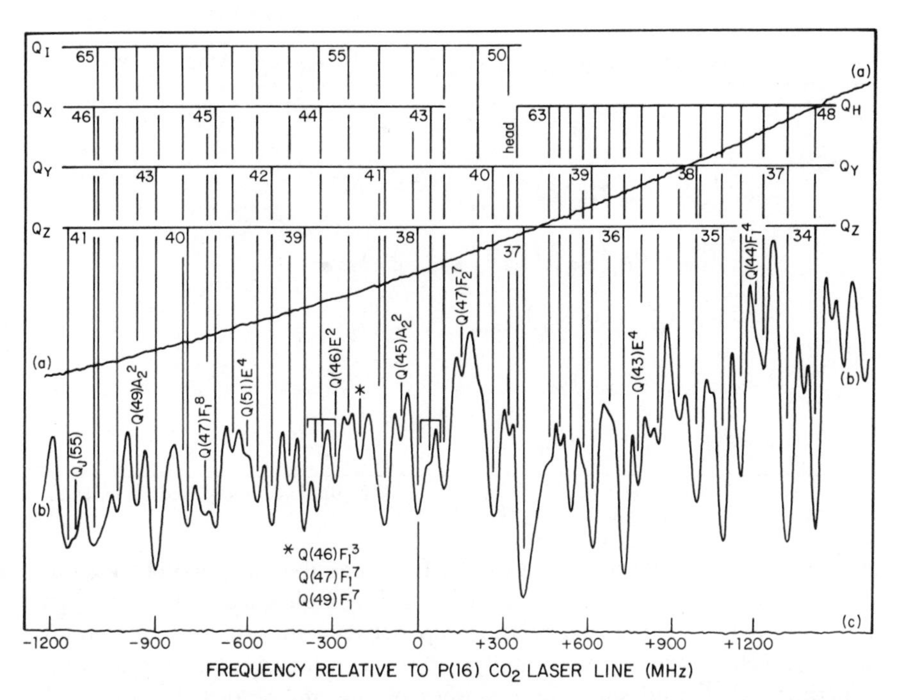

Fig. 4. The absorption spectrum of room-temperature SF_6 near the $P(16)$ CO_2 laser line at 947.7417 cm^{-1}, obtained with a semiconductor diode laser and calibrated by heterodyning with a stable CO_2 laser operating on the $P(16)$ line.[11] SF_6 spectroscopic assignments are indicated. Total tuning range approximately ± 1 GHz.

values of the ν_3 transition dipole moment, one of which[5] compared favorably with later determinations based on the ν_3 band strength[50] and the intensity of single lines in the high-resolution spectrum.[51] Self-induced transparency[52] was inferred to occur in SF_6 irradiated by a pulsed CO_2 laser;[1] although some details of the process of inference were not in agreement with theoretical predictions for two-level systems,[7] self-induced transparency was later unambiguously observed[3] in SF_6 irradiated by some (but not all) of the CO_2 laser lines overlapping the SF_6 ν_3 band (Table I). At first it was argued that self-induced transparency (and coherent propagation effects in general) could occur in SF_6 only if the value of the total angular momentum of the initial state were very low. It was later shown theoretically[8, 9] that coherent propagation effects can occur even for very high values of J, because of a "clustering" of the transition moments for different initial values of the spatial quantum number M. We shall discuss this point in greater detail below, after deriving the form of the transition

TABLE I
Vibration–Rotation Hamiltonian for a Triply Degenerate Mode

$$H = \omega_s T^{10}_{000} + X_{ss} T^{20}_{000} + G_{ss} \vec{\ell}^2_s + T_{ss} T_{404} + B_0 T^{01}_{000} + 2\sqrt{3} B\zeta_s T_{110}$$
$$- \alpha T^{11}_{000} + \alpha_{220} T_{220} + \alpha_{224} T_{224}$$
$$- D_0 T^{02}_{000} - D_{t0} T_{044} + F_{110} T^{01}_{110} + F_{134} T_{134}$$
$$- (D\text{-}D_0) T^{12}_{000} - (D_t - D_{t0}) T^{10}_{044} + G_{220} T^{01}_{224}$$
$$+ G_{244} T_{244} + G_{246} T_{246}$$
$$+ HT^{03}_{000} - H_{4t} T^{01}_{044} + H_{6t} T_{066}$$

K. T. Hecht, *J. Mol. Spectr.*, **5**, 355 (1960); A. G. Robiette, D. L. Gray, and F. W. Birss, *Mol. Phys.*, **38h2**, 1591 (1976)

moments for a spherical-top molecule. Experimental and theoretical studies[2, 28] of photon echoes in SF_6 have led to incomplete agreement between the observed and predicted relationship between the pump polarizations and the echo polarization. Finally, infrared–infrared double resonance experiments[53–55] and shock-tube studies of the SF_6 absorption contour[56] gave important semiquantitative information on the magnitude of SF_6 J values for states interacting with the various CO_2 laser lines. More recent double-resonance studies of SF_6 have provided data which challenge theoretical efforts at assignment.[20, 57]

A. Vibration–Rotation Basis

The vibration–rotation basis states for a spherical-top molecule may conveniently be taken as linear combinations of products of vibrational wavefunctions and rotational wavefunctions, within the framework of the Born–Oppenheimer approximation. Two vibration–rotation bases that are widely used in the theory of spherical-top molecules employ vibrational wavefunctions ϕ^{vl}_m, which are adapted to spherical symmetry, and rigid-rotor wavefunctions D^J_{KM}. In the coupled spherical basis,[38] a linear combination of the products $D^J_{KM}\phi^{vl}_m$ is taken, which results in coupled angular-momentum basis functions ϕ^{vlJR}_{KM}. In the symmetry-adapted basis introduced by Moret–Bailly,[58] a different linear combination of the products $D^J_{KM}\phi^{vl}_m$ is taken, such that the sum belongs to one row of an irreducible representation of the molecular point group. Both of these bases are important for a discussion of spherical-top energy levels and transition moments.

The vibrational wavefunctions ϕ_m^{vl} in the spherical basis are eigenfunctions of the total number of vibrational quanta (v), the vibrational angular momentum ($l=v, v-2,\ldots,1$ or 0) and the projection of the vibrational angular momentum along the molecule-fixed z axis ($m=-l,-l+1,\ldots,l$). Explicitly, as functions of the normal coordinates q_1, q_2, q_3 of a triply-degenerate vibrational mode, they are[59]

$$\phi_m^{vl} = N_{vl} Y_{lm}\left(\frac{\mathbf{q}}{|\mathbf{q}|}\right) e^{-q^2/2}(q^2)^{l/2} L_{(n-l)/2}^{(l+1/2)}(q^2) \tag{1}$$

where

$$q^2 = \sum_{j=1}^{3} q_j^2 \equiv |\mathbf{q}|^2 \tag{2}$$

$$\mathbf{q} = (q_1, q_2, q_3) \tag{3}$$

$$N_{vl} = \left\{2\left[\tfrac{1}{2}(v-l)\right]!\right\}^{1/2} \left\{\left[\tfrac{1}{2}(v+l+3)\right]!\right\}^{-1/2} \tag{4}$$

and where Y_{lm} is a spherical harmonic; $L_p^{(\alpha)}$ is an associated Laguerre polynomial.[60]

The rigid-rotor wavefunctions for a given total angular momentum J, with the projection of J along the molecule-fixed axis and space-fixed axis being, respectively, K and M, are proportional to the elements of the representation D^J of the rotation group,

$$\psi_{KM}^J = \left[\frac{2J+1}{8\pi^2}\right]^{1/2} D_{KM}^J(\Omega^{-1}) \tag{5}$$

where Ω is the 3×3 matrix which rotates the components of a vector in the laboratory-fixed frame into the components of the same vector in the molecule-fixed frame:

$$\mathbf{x}_{\text{mol}} = \Omega \mathbf{x}_{\text{lab}} \tag{6}$$

Equation 5 is a direct consequence of the symmetry of a spherical-top molecule, and may be established by purely group-theoretical arguments.[61, 62]

The coupled spherical vibration-rotation basis vectors are

$$\phi_{K_R M}^{vlJR} = \sum_m \langle lJ; mK \mid RK_R \rangle \left[\phi_m^{vl}\right]^* \psi_{KM}^J \tag{7}$$

where $K_R = K - m$. To obtain (7) one must *subtract* the vibrational angular momentum $\mathbf{l}$ from the total angular momentum $\mathbf{J}$ to obtain the purely rotational angular momentum of the molecular framework:

$$\mathbf{R} = \mathbf{J} - \mathbf{l} \tag{8}$$

The symmetry-adapted basis functions are eigenfunctions of the rotational angular momentum R and the laboratory-fixed z component of total angular momentum M, and also belong to a specific row of a specific irreducible point-group representation. The symmetry-adapted basis may be obtained[58, 64] from the coupled spherical basis by a unitary transformation matrix ${}^{(R)}G$:

$$\psi_{pM}^{vlJR} = \sum_{K_R} {}^{(R)}G_p^{K_R} \phi_{K_R M}^{vlJR} \tag{9}$$

In (9) the indices

$$C = (\Gamma\gamma) \quad \text{and} \quad p = C^{(n)} \tag{10}$$

are composite labels consisting of the irreducible representation Γ, the row γ of that representation, and an index n which distinguishes the different state vectors with the same point-group symmetry, which arise in the reduction of D^R into irreducible representations of the molecular point group. The elements of ${}^{(R)}G$ have been calculated for R up to 130,[64, 65] and are tabulated[58] for $R \leqslant 20$.

B. Vibration–Rotation Hamiltonian

It was shown in the pioneering work of Hecht,[38] Moret–Bailly,[65] and Louck[66] that many advantages accrue from expressing the vibration–rotation Hamiltonian for a spherical-top molecule in terms of spherical tensor operators. The greatest advantage is that the Racah–Wigner angular momentum calculus allows one to take the coupled-basis matrix elements of the qth component $T_q^{(k_1 k_2 k)}$ of a spherical tensor operator $T^{(k_1 k_2 k)}$ of rank k (formed by coupling a purely vibrational tensor operator $T_{\text{vib}}^{(k_1)}$ of rank k_1 to a purely rotational operator $T_{\text{rot}}^{(k_2)}$ of rank k_2) in a completely systematic manner:

$$\begin{aligned}\left(\phi_{K_R' M'}^{v'l'JR'}, T_q^{(k_1 k_2 k)} \phi_{K_R M}^{vlJR}\right) &= \delta_{M'M}(-1)^{R+K_R'} \begin{pmatrix} R & k & R' \\ K_R & q & -K_R' \end{pmatrix} \left[(2k+1)(2R+1)(2R'+1)\right]^{1/2} \\ &\quad \cdot \begin{Bmatrix} l' & l & k_1 \\ J & J & k_2 \\ R' & R & k \end{Bmatrix} \langle v'l' \| T_{\text{vib}}^{(k_1)} \| vl \rangle \langle J \| T_{\text{rot}}^{(k_2)} \| J \rangle \end{aligned} \tag{11}$$

Only the reduced matrix elements need to be evaluated (once!) by a calculation involving the explicit forms of the operators $T_{\text{vib}}^{(k_1)}$ and $T_{\text{rot}}^{(k_2)}$. The reduced matrix elements needed for most operators in the SF_6 vibration–rotation Hamiltonian are tabulated by Hecht[38] and Robiette et al.[67]

To construct a vibration–rotation Hamiltonian which is invariant under the operations of the molecular point group, it is in general necessary to form linear combinations of the spherical tensor components $T_q^{(k_1k_2k)}$. In fact, the octahedrally (or tetrahedrally) invariant linear combinations are

$$T_{A_1}^{(k_1k_2k)} = \sum_q {}^{(k)}G_{A_1}^q T_q^{(k_1k_2k)} \tag{12}$$

where A_1 is the totally symmetric representation of the point group. In terms of the symmetry-adapted basis (9), we can use (12) to reduce (11) to the form

$$\begin{aligned}\left(\psi_{p'M'}^{v'l'JR'}, T_{A_1}^{(k_1k_2k)}\psi_{pM}^{vlJR}\right) &= \delta_{MM'}\begin{Bmatrix} l' & l & k_1 \\ J & J & k_2 \\ R' & R & k \end{Bmatrix}\left[(2k+1)(2R+1)(2R'+1)\right]^{1/2} \\ &\times \langle v'l' \| T_{\text{vib}}^{(k_1)} \| vl\rangle \langle J \| T_{\text{rot}}^{(k_2)} \| J\rangle (-1)^R F_{A_1p'p}^{(kR'R)}\end{aligned} \tag{13}$$

where the Moret–Bailly $F^{(k)}$ coefficient[58] is defined as

$$F_{A_1p'p}^{(kR'R)} = \sum_{m_1m_2m_3} {}^{(k)}G_{A_1}^{m_1}\, {}^{(R')}G_{p'}^{m_2}\, {}^{(R)}G_p^{m_3} \begin{pmatrix} R' & k & R \\ m_2 & m_1 & m_3 \end{pmatrix} \tag{14}$$

The off-diagonal $(R' \neq R)$ $F^{(4)}$ and $F^{(6)}$ coefficients have been calculated numerically by Krohn[64] for $R' \leqslant 98$, $R \leqslant 98$, and the diagonal $(R' = R)$ $F^{(4)}$ and $F^{(6)}$ coefficients have been calculated[64] for $R \leqslant 120$.

In agreement with Robiette et al.[67] we adopt the following notation for the tensor operators appearing in the vibration–rotation Hamiltonian:

$$T_{k_1k_2k;\,q}^{mn} = (\hat{n})^m (\hat{\mathbf{J}}^2)^n T_q^{(k_1k_2k)} \tag{15}$$

where $\hat{n}$ is the operator giving the total number of ν_3 vibrational quanta. In this notation, the vibration–rotation Hamiltonian for the ν_3 mode of SF_6 including all allowed operators up to and including rank $k=6$, is shown in Table I. The matrix elements of this Hamiltonian may be calculated using (11) or (13) (or using other bases), and the Hamiltonian may be diagonalized numerically.

C. Fitting of Spectroscopic Parameters

It is evident from (11) or (13) that the Hamiltonian will, in general, not be diagonal in the rotational angular momentum R. For not too large values of J (in SF_6, for $J \lesssim 25$) the off-diagonal elements in (13) lead to negligible effects on the energy eigenvalues. In that case the diagonal contribution to the energy eigenvalue E_{pM}^{vlJR} of a particular tensor operator may be written down directly from (13). Evaluation of the 9J symbols and reduced matrix elements for the operators appearing in Table I leads to the parametrization of the fundamental transition frequencies

$$\begin{aligned}
\nu_R(R,p) &= E_{pM}^{1,1,R+1,R} - E_{pM}^{0,0,R,R} \qquad (R \text{ branch}, J \to J+1)\\
\nu_Q(R,p) &= E_{pM}^{1,1,R,R} - E_{pM}^{0,0,R,R} \qquad (Q \text{ branch}, J \to J)\\
\nu_P(R,p) &= E_{pM}^{1,1,R-1,R} - E_{pM}^{0,0,R,R} \qquad (P \text{ branch}, J \to J-1) \qquad (16)
\end{aligned}$$

shown in Table II.[68] These expressions have been extensively used in fitting Doppler-limited spectra of SF_6.[10–12] The relationship between the Hamiltonian parameters appearing in Table I and the spectroscopic parameters appearing in Table II is complex and will be discussed in a future publication.[69] Also, in Table II, the quantities $\Delta_P(R,p)$, $\Delta_Q(R,p)$, and $\Delta_R(R,p)$ are, by definition, the difference between the transition frequencies determined through exact diagonalization of the Hamiltonian including matrix elements off-diagonal in R, and the transition frequencies calculated using approximate energies calculated using the purely diagonal (in R) form of (13).

The assignment of an actual spectrum falls into three states: (1) provisional assignment, using approximate parameters obtained by a hand calculation, etc.; (2) determination of preliminary constants given a provisional assignment; (3) assignment of additional lines and final determination of constants. Stages (1) and (2) may be accomplished using the modified Bobin – Fox parametrization[68] shown in Table II, with $\Delta_{P,Q,R}(R,p)=0$. The values of $\Delta_{P,Q,R}(R,p)$ may then be determined by an exact diagonalization (including matrix elements off-diagonal in R) using the parameters determined in stage (2). The resulting numerically determined values of $\Delta_{P,Q,R}(R,p)$ may be inserted in the modified Bobin–Fox expressions (Table II), new constants determined, and so on until the iteration converges on a consistent set of parameters.[10–12] The values and standard deviations of the SF_6 spectroscopic parameters obtained using line positions measured by high-resolution saturation spectroscopy are shown in Table III.[13]

TABLE II

Bobin-Fox Parametrization of Spherical-Top Transition Frequencies

- NOTATION: R = ROTATIONAL ANGULAR MOMENTUM
  p = OCTAHEDRAL OR TETRAHEDRAL SYMMETRY SPECIES AND INDEX
  $F^{(kRR)}_{A_1pp}$ = NORMALIZED EIGENVALUE OF DIAGONAL BLOCK OF HAMILTONIAN (MORET-BAILLY) INCLUDING OPERATORS OF RANK k IN R
  M = −R(P BRANCH) OR R+1 (R BRANCH)
  $\Delta(R,p)$ = OFF-DIAGONAL CORRECTION (IN R)

- P AND R BRANCHES:

$$\nu_{P,R}(R,p) = m + nM + pM^2 + qM^3 + sM^4 + tM^5 + xM^6$$

$$+ (g - hM + kM^2 + \ell M^3 + jM^4)\frac{(-1)^R}{D,F(R)}F^{(4RR)}_{A_1pp}$$

$$+ (z' + z''M + z'''M^2)\frac{(-1)^R}{D',F'(R)}F^{(6RR)}_{A_1pp}$$

$$+ \Delta_{P,R}(R,p)$$

- Q BRANCH

$$\nu_Q(R,p) = m + vR(R+1) + w[R(R+1)]^2$$

$$+ \left\{-2g + uR(R+1) + z[R(R+1)]^2\right\}\frac{(-1)^R}{E(R)}F^{(4RR)}_{A_1pp}$$

$$+ [-2z' + z'''R(R+1)]\frac{(-1)^R}{E'(R)}F^{(6RR)}_{A_1pp}$$

$$+ \Delta_Q(R,p)$$

FUNCTIONS APPEARING IN BOBIN–FOX PARAMETRIZATION

$$D(R) = \frac{2R(2R-1)}{[(2R-3)\cdots(2R+5)]^{1/2}}$$

$$D'(R) = \frac{2R(2R-1)}{[(2R-5)\cdots(2R+7)]^{1/2}}$$

$$E(R) = \frac{(2R+2)(2R+3)}{[(2R-3)\cdots(2R+5)]^{1/2}}$$

$$E'(R) = \frac{(2R+2)(2R+3)}{[(2R-5)\cdots(2R+7)]^{1/2}}$$

$$F(R) = \frac{2R(2R+2)}{[(2R-3)\cdots(2R+5)]^{1/2}}$$

$$F'(R) = \frac{2R(2R+2)}{[(2R-5)\cdots(2R+7)]^{1/2}}$$

TABLE III
Spectroscopic Parameters for the ν_3 Fundamental of SF_6^{13}

Parameter (see table II)	Value (cm^{-1}) and standard deviation
m	947.9763307(62)
n	$5.581731(39)\times 10^{-2}$
p	$-1.615414(89)\times 10^{-4}$
q	$1.236(47)\times 10^{-8}$
s	$-6.77(23)\times 10^{-11}$
t	$-5.8(9.0)\times 10^{-14}$
v	$-6.9876(46)\times 10^{-5}$
w	$-0.9(1.3)\times 10^{-11}$
g	$-2.45621(32)\times 10^{-5}$
h	$-1.910(12)\times 10^{-9}$
k	$-1.54(11)\times 10^{-11}$
u	$0.4(1.0)\times 10^{-11}$

In Fig. 5 we show a saturation spectrum[13] of a portion of the frequency interval [near the $P(16)$ CO_2 laser line[70] at 947.7417363 cm^{-1}] included at lower resolution in Figs. 3 and 4. The positions of the lines marked with a bullet in Fig. 5 were measured by the technique of frequency offset locking[71] to an accuracy of a few kHz (with respect to the reference CO_2 laser line). The total tuning range in Fig. 5 is approximately ± 250 MHz. The experimental linewidth is less than 10 kHz.

D. Transition Moments

To complete our survey of spherical-top spectroscopic theory, we give a brief derivation of the transition moments for dipole-allowed transitions in which one vibrational quantum changes, in the bases (7) and (9).[72–75] The spherical components of the dipole transition operator in the laboratory-fixed frame, μ_σ, are physically the quantities which interact with an externally applied optical electric field E, through the dipole Hamiltonian. Consequently we shall calculate the matrix elements of the laboratory-fixed components μ_σ, which are related to the molecule-fixed components of the vibrational normal coordinates by the equation

$$\mu_\sigma = A\sum_\tau D^1_{\tau\sigma}(\Omega^{-1})q_\tau \tag{17}$$

where A is a constant characteristic of the vibrational mode. For an individual with extensive experience in tensor operators it is easy to see from (17) that the dipole operator is of tensor form, with $k_1=1$, $k_2=1$, and $k=0$, so that the matrix elements follow directly from (11) or (13). A more

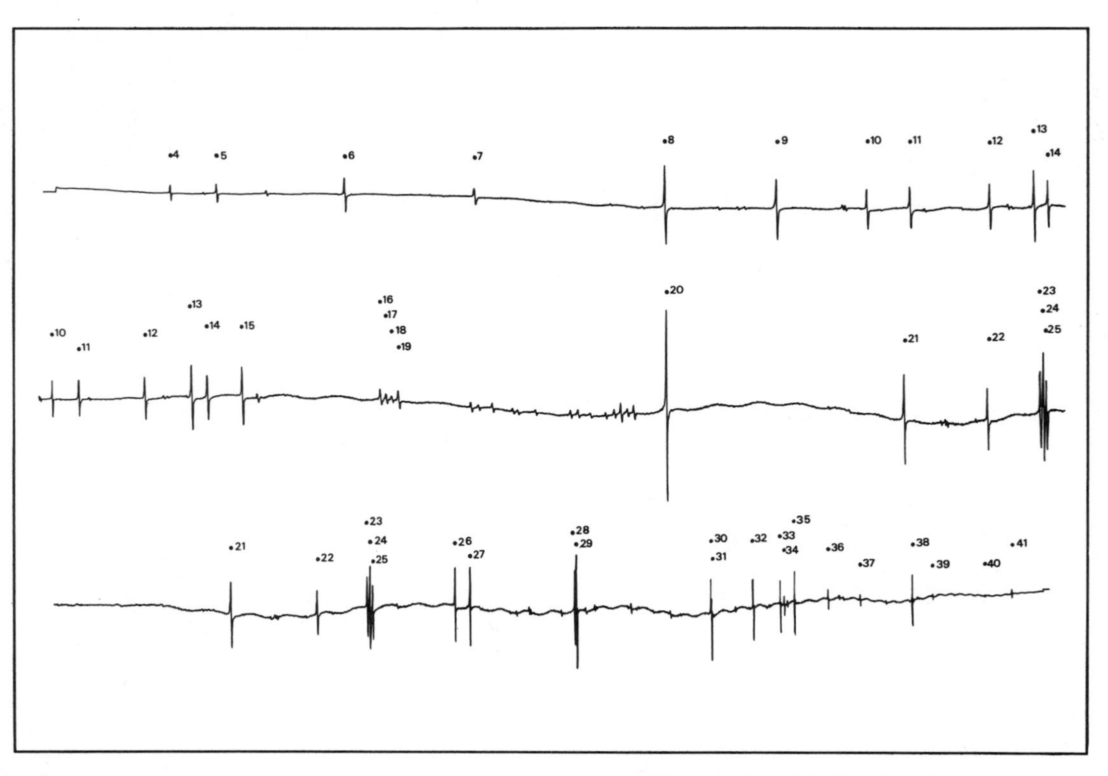

Fig. 5. The saturation spectrum of room-temperature SF_6 near the $P(16)$ CO_2 laser line at 947.7417363 cm^{-1}, calibrated by offset frequency looking to a stable CO_2 laser locked on an SF_6 absorption line.[71] Total tuning range approximately −205 MHz to +212 MHz.

pedestrian approach is the following[74, 75]: express the states ϕ_{KM}^{vlJR} using (7); use the Wigner–Eckart theorem to evaluate $(\phi_{m'}^{v'l'}, q_\tau \phi_m^{vl})$; evaluate the rotational matrix elements using the well-known integral over a product of three D^J's; and resum all the resulting $3J$ symbols to $6J$ or $9J$ symbols. The results are[73–75]

$$\left(\phi_{p'M'}^{v'l'J'R'}, \mu_\sigma \phi_{pM}^{vlJR}\right) = A\delta_{RR'}\delta_{pp'}(-1)^{J+M'+1}\begin{pmatrix} J & 1 & J' \\ M & \sigma & -M' \end{pmatrix} \cdot \left[3(2R+1)(2J+1)(2J'+1)\right]^{1/2}\langle vl \| q \| v'l' \rangle \cdot \begin{Bmatrix} l' & l & 1 \\ J' & J & 1 \\ R' & R & 0 \end{Bmatrix} \tag{18}$$

The reduced matrix element is

$$\langle vl \| q \| v'l' \rangle = \begin{cases} \left[(v+l+3)(l+1)/2\right]^{1/2} & \text{when } v'=v+1,\, l'=l+1 \\ \left[(v-l+2)l/2\right]^{1/2} & \text{when } v'=v+1,\, l'=l-1 \end{cases} \tag{19}$$

The selection rules evident from (19) are:

$$R' = R \tag{20}$$

$$p' = p \tag{21}$$

$$v' = v \pm 1 \tag{22}$$

$$l' = l \pm 1 \tag{23}$$

$$J' = J \text{ or } J \pm 1 \tag{24}$$

$$M' = M + \sigma \tag{25}$$

Selection rules (22) and (23) are expected to hold for all dipole-allowed infrared transitions; physically they imply that one vibrational quantum is changed in the transition. Selection rules (20) and (21) are applicable only when the functions ϕ_{pM}^{vlJR} are nearly eigenfunctions of the exact Hamiltonian. However, several of the tensor operators in Table I, including the possibly important vibrational anharmonic operator T_{404}, mix states with different R. When this happens, (20) no longer applies. Also, the nuclear hyperfine interaction mixes states with different point-group symmetry type p (particularly in the ground vibrational state[13]), so that (21) is also an approximate selection rule. Transitions which do not obey the selection rules (20)–(21) may have a substantial effect on the process of multiple-photon excitation,[76] through nearly resonant enhancement of multiphoton transitions. It is not clear that transitions which violate (20)–(21) are necessarily orders of magnitude weaker than transitions which obey (20)–(21), since the mixture of states with different values of R and p need not be negligible.

It should be noted from (18) that the transition moment is always proportional to the $3J$ symbol

$$\begin{pmatrix} J & 1 & J' \\ M & \sigma & -M' \end{pmatrix} \tag{26}$$

regardless of the mixture of states produced by the vibration–rotation interaction. (If the nuclear hyperfine interaction is strong, J and M must be replaced by F and M_F.) The presence of the $3J$ symbol in (18) is a direct consequence of spherical symmetry.

E. A Model for the ν_3 Mode of SF_6

In this work we have dedicated ourselves to a study of coherent effects in multiple-photon excitation, and to an investigation of whether processes which in the past have been ascribed to unimolecular or collisional relaxation are in fact the result of coherent processes. The details of our model are conditioned by this physical approach. We assume that the molecule remains in the ν_3 vibrational mode, i.e., that there is no intramolecular relaxation of population. In this case the basic physics of the interaction of a single molecule with a laser field is multiphoton excitation.[77–80] The molecule goes from $v_3=0$ to $v_3=N$ by absorption of N photons, without necessarily making resonant single-photon transitions along the way. The resonant laser frequency for an N-photon transition from $v_3=0$ to $v_3=N$ is given by

$$\omega_{pM}^{NlJR}=N^{-1}\left[E_{pM}^{NlJR}-E_{pM}^{0,0,R,R}\right] \tag{27}$$

As the proliferation of indices in (27) indicates, there are many nearby N-photon resonances (for all the allowed values of l, J, R, and p) with a wide range of resonant frequencies. The population which can be excited to $v_3=N$ in an N-photon resonance can, in principle, be large; the highest and lowest levels can engage in coherent Rabi oscillations, much like the upper and lower levels of a two-level system. If levels between $v_3=0$ and $v_3=N$ are also significantly excited (due to a near-coincidence of ω_{pM}^{NlJR} with one or more single-photon transition frequencies, for example) then the analogy with a two-level system is no longer applicable, but the total number of molecules excited from $v_3=0$ may be higher. Certain physical effects can, in fact, lead to a high probability of coincidence of single-photon transition frequencies with multiphoton resonant frequencies, and thereby compensate the expected anharmonic decrease of the transition frequencies from v_3 to v_3+1 with increasing v_3. Both rotational energy[77, 81] and vibrational anharmonic splitting[14] can at least partially compensate the effects of vibrational anharmonicity on single-photon transition frequencies. Both of these effects are present in our model. We have included rotations by including the terms

$$BT^{01}_{000}+2\sqrt{3}\,B\zeta_3 T_{110}=B\mathbf{J}^2-2B\zeta_3\mathbf{J}\cdot\mathbf{l} \tag{28}$$

from the first line of Table II. Anharmonic splitting is provided in our model by the spherically symmetric anharmonic contribution

$$G_{33}\mathbf{l}_3^2 \tag{29}$$

from Table II. We have neglected the octahedrally symmetric anharmonic contribution

$$T_{33}T_{404} \tag{30}$$

in Table II, which also produces anharmonic splitting, because this term couples states ψ^{vlJR}_{pM} with different values of l (and R) and thereby complicates the model. Neglecting (30) entails a neglect of transitions which violate selection rule (20).

In our numerical calculations we at first[82] lumped together in a single "effective state"[83] $|vlJR\rangle$ all the states ψ^{vlJR}_{pM} with given (fixed) values of v, l, J, and R. This reduces the number of states involved in the computations, but forces one to employ instead of (18) a "typical" transition moment, which we take to be the root mean square of (18):

$$\langle v'l'J'R'|\mu|vlJR\rangle_{\text{typ}}=A\langle vl\|q\|v'l'\rangle W(l'lJ'J;1R) \tag{31}$$

where W is a Racah coefficient. In other words, this approach makes a simplification in assuming that both energy levels and transition moments are independent of p and M.

In more recent computations we have employed the effective states $|vlJR;\,M\rangle$, in which states with different octahedral symmetry types (p)

but the same quantum numbers v, l, J, R, and M are lumped together. In this case the state-to-state transition moments (18), which are independent of p, must be used. The approximation made in this case is to neglect the dependence of the energy levels on p; that is, we neglect the vibration–rotation fine structure. In both cases we have, of course, also neglected nuclear hyperfine effects, which would mix states with different symmetry type p; these are already lumped together in our effective-states approaches.

III. COHERENT PROPAGATION EFFECTS IN SF_6

A. Electromagnetic Field Equations

We consider an infinite plane, quasimonochromatic electromagnetic wave traveling in the $+z$ direction, with a real electric field

$$\mathbf{E}=\hat{e}E(z,t) \tag{32}$$

to be incident on a uniform gaseous medium with molecular number density N. In (32), $\hat{e}$ is a unit polarization vector, such that

$$\hat{e}^*\cdot\hat{e}=1 \tag{33}$$

The incident field sets up a polarization density

$$\mathbf{P}=\hat{e}P(z,t) \tag{34}$$

The propagation of **E** in the medium is described by the one-dimensional wave equation (in MKS units)

$$\frac{\partial^2 E}{\partial z^2}-\frac{1}{c^2}\frac{\partial^2 E}{\partial t^2}+\frac{c\kappa}{2}\frac{\partial E}{\partial t}=-\mu_0\frac{\partial^2 P}{\partial t^2} \tag{35}$$

In (35), κ is a linear attenuation coefficient introduced to allow for scattering losses, etc. In order to make further progress, we make the slowly varying amplitude and phase approximation (SVAPA): we assume that the field and polarization are of the form

$$E(z,t)=E'(z,t)\cos\zeta(z,t) \tag{36}$$

$$P(z,t)=C(z,t)\cos\zeta(z,t)+S(z,t)\sin\zeta(z,t) \tag{37}$$

$$\zeta(z,t)=kz-\omega t+\phi(z,t) \tag{38}$$

where E', C, S, and ϕ are slowly varying in the sense that

$$\left|\frac{\partial f}{\partial z}\right| \ll |kf|, \qquad \left|\frac{\partial f}{\partial t}\right| \ll |\omega f| \tag{39}$$

where f is E', C, S, or ϕ. If we insert (36)–(38) in (35), and make the SVAPA (39), then we find that

$$\left[\frac{\partial}{\partial z} + \frac{1}{c}\frac{\partial}{\partial t} + \frac{\kappa}{2}\right]\mathcal{E} = \frac{k}{2\epsilon_0}\mathcal{P} \tag{40}$$

where we have introduced the complex electric field

$$\mathcal{E}(z,t) = E'(z,t)e^{i\phi(z,t)} \tag{41}$$

and complex polarization density

$$\mathcal{P}(z,t) = [S(z,t) + iC(z,t)]e^{i\phi(z,t)} \tag{42}$$

In terms of the retarded-time variables

$$z' = z, \qquad t' = t - z/c \tag{43}$$

(40) becomes

$$\left[\frac{\partial}{\partial z'} + \frac{\kappa}{2}\right]\mathcal{E} = \frac{k}{2\epsilon_0}\mathcal{P} \tag{44}$$

As it stands (44) is simply an approximate form of Maxwell's equations adapted to a particular physical situation and is therefore incomplete. To complete (44) we must give a prescription for calculating $\mathcal{P}$, which we shall do below. In general $\mathcal{P}$ depends on $\mathcal{E}$ in a highly nonlinear manner, so that (44) is one of a set of coupled, nonlinear partial differential equations. A self-consistent solution of these equations can ordinarily be obtained only by numerical calculations.

B. Calculation of the Polarization Density

The polarization in a dilute medium is simply the product of the molecular number density N and the quantum-mechanical expectation value of the dipole operator:

$$P = N\,\mathrm{tr}(\rho\boldsymbol{\mu}\cdot\hat{e}^*) = N\,\mathrm{tr}(\rho\mu_\sigma) \tag{45}$$

where ρ is the reduced density matrix for the ν_3 mode of SF_6.[84] In terms of a basis of states $|vA\rangle$, where v is the vibrational quantum number and A denotes the remaining quantum numbers, (45) becomes

$$P=N\sum_{v,u}\sum_{A,B}\rho_{vA,uB}\mu_{vA,uB} \tag{46}$$

In keeping with the SVAPA, which removes harmonics from Maxwell's equations, we shall employ the rotating-wave approximation (RWA), which removes harmonics from the equation of motion for the density matrix ρ. Let

$$\rho_{vA,uB}=e^{i(\Omega_u-\Omega_v)t'}\tilde{\rho}_{vA,uB} \tag{47}$$

where

$$\Omega_v=v\omega \tag{48}$$

Since $-\omega t'=kz-\omega t$ by (43), (47) removes the rapid spatial and temporal oscillation of ρ created by the incident field. The equation of motion for ρ is[84]

$$\begin{aligned}\frac{\partial\rho_{vA,uB}}{\partial t}=&\left\{-\frac{i}{\hbar}(E^{vA}-E^{uB})-\gamma_{vA,uB}(1-\delta_{vu}\delta_{AB})\right\}\rho_{vA,uB}\\&+\frac{i}{\hbar}E(z,t)\sum_{qC}\left[\mu_{vA,qC}\rho_{qC,uB}-\rho_{vA,qC}\mu_{qC,uB}\right]\\&-\delta_{uv}\delta_{AB}\sum_{qC}\left[\rho_{vA,vA}w_{vA,qC}-\rho_{qC,qC}w_{qC,vA}\right]\end{aligned} \tag{49}$$

where $\gamma_{vA,uB}=\gamma_{uB,vA}$ is the dephasing rate of an off-diagonal density-matrix element and $w_{vA,uB}$ is the rate of population transfer (per molecule) from vA to uB. For a two-level system, the relationship between the γ's and w's and the more familiar relaxation constants T_1 and T_2 is

$$(T_1)^{-1}=\tfrac{1}{2}(w_{12}+w_{21}) \tag{50}$$

$$(T_2)^{-1}=\gamma_{12}=\gamma_{21} \tag{51}$$

We regard the physical origin of the γ's and w's as collisions, not intramolecular transitions.

We implement the RWA by substituting (47) into (49), using (36), (41), and (48), and discarding terms which vary as $\exp(\pm 2i\omega t')$. The result is

$$\begin{aligned}\frac{\partial\tilde{\rho}_{vA,uB}}{\partial t} = &\{i[(\Omega_v - E^{vA}/\hbar)-(\Omega_u - E^{uB}/\hbar)]\\ &-\gamma_{vA,uB}(1-\delta_{vu}\delta_{AB})\}\tilde{\rho}_{vA,uB}\\ &+\frac{i}{2\hbar}\sum_C \{\mu_{vA,(v+1)C}\tilde{\rho}_{(v+1)C,uB}\mathcal{E}^*\\ &+\mu_{vA,(v-1)C}\tilde{\rho}_{(v-1)C,uB}\mathcal{E}\\ &-\tilde{\rho}_{vA,(v-1)C}\mu_{(v-1)C,uB}\mathcal{E}^*\\ &-\tilde{\rho}_{vA,(v+1)C}\mu_{(v+1)C,uB}\mathcal{E}\}\\ &-\delta_{vu}\delta_{AB}\sum_{qC}\{\tilde{\rho}_{vA,vA}w_{vA,qC}-\tilde{\rho}_{qC,qC}w_{qC,vA}\}\end{aligned} \tag{52}$$

Only slowly varying quantities appear in (52).

The transformation (47) also enables us to express the slowly varying complex polarization directly in terms of $\tilde{\rho}$:

$$\mathcal{P}(z',t') = 2iN\sum_{A,B}\tilde{\rho}_{vA,(v-1)B}\mu_{vA,(v-1)B} \tag{53}$$

[Equation (53) follows directly from (46), after some algebra.] Equations (52) and (53), together with initial conditions for $\tilde{\rho}$, define the complex polarization, which appears in the field propagation equation (44).

C. Thin-Sample Approximation

The field equation (44) has the formal solution

$$\begin{aligned}\mathcal{E}(z',t') = &\left[\exp\left(\frac{-\kappa z'}{2}\right)\right]\mathcal{E}(0,t')\\ &+\frac{k}{2\epsilon_0}\int_0^{z'}\left\{\exp\left[\frac{\kappa(z''-z')}{2}\right]\right\}\mathcal{P}(z'',t')\,dz''\end{aligned} \tag{54}$$

$$\equiv \mathcal{E}_{\text{inc}}(z',t') + \mathcal{E}_{\text{rad}}(z',t') \tag{55}$$

Physically, the first term in (54) is the incident field, and the second term is the field radiated by the macroscopic polarization $\mathcal{P}$; their sum is the total field $\mathcal{E}$. The spectral content of $\mathcal{E}$ will differ from that of $\mathcal{E}_{\text{inc}}$ to the extent that $\mathcal{E}_{\text{rad}}$ contains frequencies (or, equivalently, a time variation) not present in $\mathcal{E}_{\text{inc}}$. We shall call the components of $\mathcal{E}_{\text{rad}}$ at these new frequencies *sidebands*.

To facilitate our analytical and numerical studies of the alteration of the spectrum of the field through propagation, we consider the limit in which

$$|\mathcal{E}_{\text{rad}}| \ll |\mathcal{E}_{\text{inc}}| \tag{56}$$

so that we may take[6]

$$\mathcal{E}(z', t') \cong \mathcal{E}_{\text{inc}}(z', t') \tag{57}$$

in calculating the polarization $\mathcal{P}$ by use of (52) and (53). Then (54) implies

$$\mathcal{E}_{\text{rad}}(z', t') \cong \frac{k}{\epsilon_0} \frac{(1 - e^{-\kappa z'/2})}{\kappa} \mathcal{P}(0, t') \tag{58}$$

If the linear attenuation κ is small (i.e., $\kappa z' \ll 1$) then

$$\mathcal{E}_{\text{rad}}(z', t') \cong \frac{kz'}{2\epsilon_0} \mathcal{P}(0, t') \tag{59}$$

Equation (59) is fundamental for making analytical (and numerical) estimates of the magnitude of the sidebands.

To continue our investigation, and to make the presence of sidebands in $\mathcal{E}_{\text{rad}}$ obvious, we make the additional assumption that $\mathcal{E}_{\text{inc}}$ is a step pulse:

$$\mathcal{E}_{\text{inc}}(0, t') = \begin{cases} 0 & \text{for } t' < 0 \\ E_0 & \text{for } t' > 0 \end{cases} \tag{60}$$

In what follows we shall also set the linear attenuation κ equal to zero. In view of (60), the coefficients in the density-matrix equation of motion (52) are independent of time, suggesting an eigenfunction expansion of $\tilde{\rho}$. We shall write (52) in the finite-dimensional matrix form

$$\frac{d\tilde{\rho}}{dt} = A\tilde{\rho} \tag{61}$$

where we assume the basis $|vA\rangle$ has been truncated to M vectors; we regard $\tilde{\rho}$ as a $(M^2 \times 1)$ column vector, and A as an $(M^2 \times M^2)$ matrix whose components may be read off from (52). Since (61) is linear in $\tilde{\rho}$ and first-order in time, it possesses a solution of the form

$$\tilde{\rho}(t') = V(t', t_0')\tilde{\rho}(t_0') \tag{62}$$

where V must satisfy the initial condition

$$V(t_0', t_0')=1 \tag{63}$$

If we put (62) into (61) and note that $\tilde{\rho}(t_0')$ may be completely arbitrary, we find the following equation for V:

$$\frac{\partial V}{\partial t}=AV \tag{64}$$

The form of (64) suggests we expand V as

$$V_{v'A',u'B';vA,uB}(t',t_0')=\sum_{\lambda} e^{-\lambda(t'-t_0')}C_{v'A',u'B'}(\lambda)D_{vA,uB}(\lambda) \tag{65}$$

This expansion is in the same spirit as the eigenfunction expansion of $\tilde{\rho}$ used by Goodman and Thiele.[85] The initial condition (63) implies that

$$\sum_{\lambda} C_{v'A',u'B'}(\lambda)D_{vA,uB}(\lambda)=\delta_{v'v}\delta_{u'u}\delta_{A'A}\delta_{B'B} \tag{66}$$

that is, that the matrix $D_{vA,uB}(\lambda)$ (with rows ordered by λ and columns by (vA, uB)) is a right inverse of the matrix $C_{v'A',u'B'}(\lambda)$ (with columns ordered by λ and rows ordered by $(v'A', u'B')$). Since a right-inverse matrix is also a left inverse,

$$\sum_{v,u}\sum_{A,B} C_{vA,uB}(\lambda)D_{vA,uB}(\lambda')=\delta_{\lambda\lambda'} \tag{67}$$

Relations (66)–(67) are to be expected for an eigenfunction expansion of the evolution operator V.

We now indicate how the matrix C in (65) may be determined. If we put (65) into (62), we find the following eigenvector–eigenvalue equation:

$$-\lambda C(\lambda)=AC(\lambda) \tag{68}$$

where $C(\lambda)$ is a column vector as noted above.

If we assume the eigenvalue problem (68) has been solved, then the time development of the reduced density matrix $\tilde{\rho}(t')$ (considered as a column vector) is given by (62) and (65):

$$\tilde{\rho}_{vA,uB}(t')=\sum_{\lambda} e^{-\lambda(t'-t_0')}\sum_{v'u'}\sum_{A'B'} C_{vA,uB}(\lambda)D_{v'A',u'B'}(\lambda) \cdot\tilde{\rho}_{v'A',u'B'}(t_0') \tag{69}$$

From (53) and (69), the slowly varying complex polarization $\mathcal{P}$ is

$$\mathcal{P}(0,t')=\sum_{\lambda} e^{-\lambda(t'-t_0')}\mathcal{P}_{\lambda} \tag{70}$$

where

$$\mathcal{P}_{\lambda}=2iN\sum_{v,A,B}\mu_{vB,(v-1)A}\sum_{v'u'}\sum_{A'B'}C_{(v-1)A,vB}(\lambda)D_{v'A',u'B'}(\lambda) \cdot\tilde{\rho}_{v'A',u'B'}(t_0') \tag{71}$$

Equations (70)–(71) are a fundamental result of this chapter.

D. Spectral Content of the Radiated Field

It is evident from (70)–(71) that sidebands will be present in $\mathcal{E}$ whenever the eigenvalues λ have a nonzero imaginary part. We shall now investigate under what physical circumstances this will occur. First, we note that the eigenvalue $\lambda=0$ corresponds to a steady-state solution of (52) or (61), that is, a solution in which radiative pumping by the field is exactly balanced by relaxation due to collisions. (In the steady state, $\partial\tilde{\rho}^{ss}/\partial t'=0$.) Consequently we may identify the eigenvector $C(0)$ corresponding to the zero eigenvalue as the steady-state density matrix [this establishes the normalization of $C(0)$]:

$$\tilde{\rho}^{ss}=C(0) \tag{72}$$

It follows from (72), (69), and (67) that if $\tilde{\rho}(t_0')=\tilde{\rho}^{ss}$, then

$$\begin{aligned}\tilde{\rho}_{vA,uB}(t')&=\sum_{\lambda}e^{-\lambda(t'-t_0')}\sum_{v'u'}\sum_{A'B'}C_{vA,uB}(\lambda)D_{v'A',u'B'}(\lambda)\\&\quad\cdot C_{v'A',u'B'}(0)\\&=\sum_{\lambda}\delta_{\lambda 0}e^{-\lambda(t'-t_0')}C_{vA,uB}(\lambda)\\&=\tilde{\rho}^{ss}_{vA,uB}\end{aligned} \tag{73}$$

Equation 73 expresses the physically reasonable statement that if the system is in the steady state, then it stays in the steady state. Equation 73 also implies, however, that there are no sidebands when the system is in the steady state. It follows that the existence of sidebands is a transient phenomenon, and that the sidebands persist only while the system is evolving toward the steady state.

To calculate the spectral content of the real optical electric field (36), which is expressed in terms of $\mathcal{E}$ as

$$E(z',t')=\mathrm{Re}\{e^{-i\omega t'}\mathcal{E}(z',t')\} \tag{74}$$

we calculate the autocorrelation function

$$G(\tau)=\frac{1}{T}\int_0^T\left[E(z',t'+\tau)E(z',t')-E(z',t')^2\right]dt' \tag{75}$$

which, according to the Wiener–Khinchin theorem, is the Fourier transform of the power spectrum of E. Since the random process represented by E is not stationary, T should not be taken to be longer than the length of the laser pulse. From (74), (70), and (58), we find

$$G(\tau)=\frac{1}{4T}\left[\frac{kz'}{2\epsilon_0}\right]^2\left\{e^{-i\omega\tau}\sum_{\lambda}{}'\sum_{\lambda'}{}'\int_0^T e^{-\lambda^*(t'+\tau)-\lambda't'}\mathcal{P}_\lambda^*\mathcal{P}_{\lambda'}\,dt'\right.$$
$$\left.+\text{complex conjugate}\right\} \tag{76}$$

where Σ_λ' means a sum on all the eigenvalues

$$\lambda=\lambda_r+i\lambda_i \tag{77}$$

for which $\lambda_i\neq 0$. In (77), λ_r is a relaxation rate, and λ_i is the frequency of the sideband corresponding to the eigenvalue λ. Equation 76 holds in the limit in which

$$|\lambda_r T|\ll 1 \tag{78}$$

(i.e., the laser pulse length T is short compared to relaxation times) and

$$|\lambda_i T|\gg 1 \tag{79}$$

(i.e., the laser pulse length T is long compared to the period of any of the sideband frequencies). Carrying out the integral in (76), we obtain

$$G(\tau)=\frac{1}{4T}\left[\frac{kz'}{2\epsilon_0}\right]^2\sum_{\lambda}{}'\frac{|\mathcal{P}_\lambda|^2}{\lambda+\lambda^*}\left(1-e^{-(\lambda+\lambda^*)T}\right)\left(e^{(-\lambda^*+i\omega)\tau}+e^{(-\lambda-i\omega)\tau}-2\right) \tag{80}$$

provided we assume that

$$|\lambda_i - \lambda_i'|T \gg 1 \quad \text{for } \lambda \neq \lambda' \tag{81}$$

(i.e., the sideband frequencies are all sufficiently different that none of their beat notes are comparable to T^{-1}).

Physically, (80) says that the intensity of the sideband with frequency $(\omega+\lambda_i)$ is

$$\left[\frac{kz'}{2\epsilon_0}\right]^2 \frac{1}{4\lambda_r T}(1-e^{-2\lambda_r T})|\mathcal{P}_\lambda|^2 e^{-\lambda_r \tau} \tag{82}$$

This result clearly shows the decay of the sideband intensity to zero for laser pulses which are long compared to the relaxation times of the system. This may be rather simply understood: the system gradually evolves toward the steady state, with time constants which are given by λ_r, according to (69); and in the steady state there are no sidebands.

If the laser pulse is switched on adiabatically starting with $E_0=0$, then the system will always be in the steady state corresponding to the present value of E_0. Although the eigenvalues and eigenvectors will also evolve adiabatically, and although there may be some eigenvalues with nonzero imaginary part, there will nevertheless be no observable sidebands because $\mathcal{P}_\lambda$ will vanish for $\lambda \neq 0$.

E. Generation of Sidebands at Resonant Molecular Transition Frequencies

The sideband frequencies are $(\omega+\lambda_i)$ according to (80). For a qualitative study of λ_i, we turn to simple systems with only a few levels, which are supposed to represent the one $v_3=0$ and three $v_3=1$ levels shown in Fig. 6. If we label the $v_3=0$ level as 0 and the three $v_3=1$ levels as 1, 2, and 3 in order of increasing energy, and if we assume the laser frequency ω is such that the $0\to1$ transition [$P(J_0)$ in Fig. (6)] is resonant, then the detunings $2\Delta_2$ and $2\Delta_3$ of levels 2 and 3, respectively, are such that $\Delta_3>\Delta_2>0$ and $\Delta_3 \cong 2\Delta_2$. Further, we shall assume for simplicity that the "typical" transition moments (31) for the $0\to1$, $0\to2$ and $0\to3$ transitions are all equal to μ. We shall also assume that all relaxation rates are negligible compared to the resonant Rabi frequency (in Hz)

$$\Omega = \frac{\mu E_0}{2h} \tag{83}$$

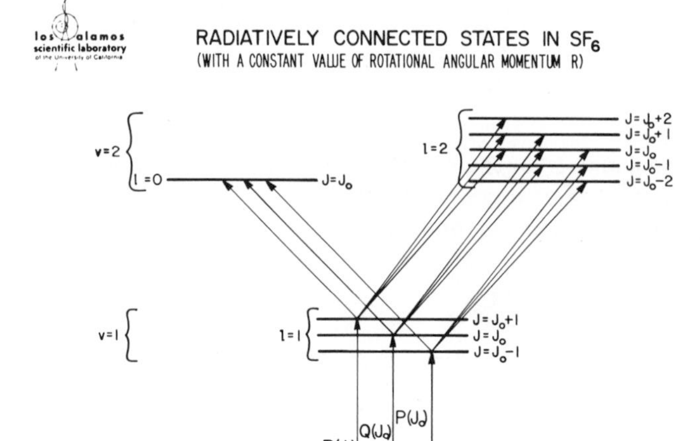

Fig. 6. States in SF_6 radiatively connected to one value of $J(=J_0)$ in the vibrational ground state. For clarity only $v_3=0$, 1, and 2 are shown.

The sideband frequencies in the limit

$$\Omega \ll \Delta_2, \qquad \Omega \ll \Delta_3 \tag{84}$$

are shown in Table IV, along with the associated polarizations $\mathcal{P}_\lambda$, as determined by an approximate eigenvalue analysis. It will be noticed that the strongest sidebands correspond to $\omega \pm 2\Omega$, and that the next strongest sidebands are at frequencies which (up to the AC Stark splitting of $\pm\Omega$) are resonant with the frequencies of the transitions $0\to2$ and $0\to3$. The electric field strengths at these nearly resonant transition frequencies are

$$|\mathcal{E}_{\text{rad}}^{(j)}| \cong \frac{Nkz'\mu\Omega}{4\epsilon_0\Delta_j} \tag{85}$$

where $j=2$ or 3.

TABLE IV
Approximate Eigenvalues and Associated Polarizations for an Undamped Four-Level System

Sideband frequency $\omega+\lambda_i$	$\mathcal{P}_\lambda$
$\omega\pm2\Omega$	$\mp iN\mu/2$
$\omega+2\Delta_2\pm\Omega+0(\Omega^2/\Delta_2)$	$-iN\mu\Omega/2\Delta_2$
$\omega+2\Delta_3\pm\Omega+0(\Omega^2/\Delta_3)$	$-iN\mu\Omega/2\Delta_3$

If we assume, for example, that $E_0 = 3\times 10^7$ V/m (corresponding to an incident intensity of $\sim 1.2\times 10^8$ W/cm^2), then $\Omega/c \sim 1$ cm^{-1}. Let us assume $2\Delta_2/c \sim 4$ cm^{-1}, as would be the case for pumping an SF_6 P-branch transition with the $P(20)$ laser line of CO_2 ($\omega/2\pi c \cong 944$ cm^{-1}). We take the transition moment to be $\mu \cong 0.3$ Debye (10^{-30} MKS). We assume a total molecular number density $N \cong 3\times 10^{15}$ cm^{-3}, corresponding to a pressure of $\cong 0.1$ torr. We also assume that only those transitions well within a frequency interval Ω of the resonantly pumped line, with roughly $g \cong 10^{-1}$ of the total population, contribute to $\mathcal{E}_{\text{rad}}^{(j)}$ at the frequency $\omega + 2\Delta_j$, so that the population to be used in (85) is gN rather than N. Then

$$|\mathcal{E}_{\text{rad}}^{(2)}| \sim 2\times 10^5 \text{ V/m} \tag{86}$$

We shall give a brief qualitative discussion of the consequences of the phenomenon of resonant sideband generation for multiple-photon excitation in Section IV.

F. Effects of Spatial (M) Degeneracy

The dependence of the transition dipole moment between the effective states $|vlJR; M\rangle$ or the molecular eigenstates ψ_{pM}^{vlJR} on M is given by the $3J$ symbol in (26). In view of the M selection rule (25), the levels which are radiatively connected to an initial M_0 are not radiatively connected to a different initial $M_0' \neq M_0$. Let $I \equiv I(v_0 l_0 J_0 R_0; M_0)$ be the set of levels which are radiatively connected to the initial state with quantum numbers v_0, l_0, J_0, R_0, and M_0. Then by definition

$$\mu_{vA,uB} = \delta_{I(vA),I(uB)} \mu_{vA,uB} \tag{87}$$

where δ is the Kronecker delta. The sum in (46) then becomes

$$P = N \sum_I \sum_{vA\in I} \sum_{uB\in I} \rho_{vA,uB} \mu_{vA,uB} \tag{88}$$

Physically, (88) implies that the total polarization is the sum of the polarizations due to each subset I of radiatively connected states. In view of (69) and the normalization condition

$$\sum_I \sum_{vA\in I} \rho_{vA,vA} = \sum_I \sum_{vA\in I} \tilde{\rho}_{vA,vA} = 1 \tag{89}$$

it is convenient to define the population fraction for the Ith subset

$$g_I = \sum_{vA\in I} \rho_{vA,vA} = \sum_{vA\in I} \tilde{\rho}_{vA,vA} \tag{90}$$

and the normalized density matrix for the Ith subset

$$\tilde{\rho}^{(I)}_{vA,uB}=(g_I)^{-1}\tilde{\rho}_{vA,uB} \quad (\text{where } vA, uB \in I) \tag{91}$$

With these definitions (88) becomes

$$P=N\sum_I g_I \sum_{vA\in I}\sum_{uB\in I}\rho^{(I)}_{vA,uB}\mu_{vA;uB} \tag{92}$$

where ρ^I and $\tilde{\rho}^I$ may be calculated as in our previous examples. It may happen, of course, that collisional relaxation couples $\rho^{(I)}$ with $\rho^{(I')}$; for the sake of simplicity we have excluded this possibility in deriving (92), although this is not an essential assumption in our general formalism.

One expects that when all molecules are initially in the vibrational ground state,

$$g_I=\frac{1}{2J_0+1}g(J_0) \tag{93}$$

where J_0 is the initial value of J, and $g(J_0)$ is the fraction of molecules with ground-state angular momentum J_0.

Since the transition dipole moment (18) appears to depend strongly on M, and since coherent effects in a resonantly pumped system depend on Rabi frequencies which are directly proportional to the transition dipole moment, it was initially believed that no coherent effects could appear in a system with a large value of J_0. Subsequently, however, it was pointed out by Hopf, Rhodes, and Szöke[8] and by Gibbs, McCall, and Salamo[9] that the $3J$ symbol (26), to which the transition dipole moment (18) is proportional, has the property that for certain values of J' and σ many of the values of (26) are very close to one another. This "clustering" of transition moments means that the Rabi frequencies for many M values (for a resonantly pumped two-level system, for example) are very nearly equal, thus enabling the different contributions (I) in (92) to oscillate in phase for a large number of cycles.

The values of the $3J$ symbol (26) for $J'=J, J+1$, and for $\sigma=1$ (circularly polarized light) and $\sigma=0$ (linearly polarized light) are shown in Table V. For $J'=J+1$, $\sigma=0$, for example, the transition moment is proportional to the function

$$f(M)=\left[(J+1)^2-M^2\right]^{1/2} \tag{94}$$

It is easy to see that $f(M)$ is a slowly varying function of M near $M=0$, so

TABLE V
Values of the $3J$ Symbol[62]

J'	σ	$\begin{pmatrix} J & 1 & J' \\ M & \sigma & -(M+\sigma) \end{pmatrix}$
J	1	$(-1)^{J-M-1}\left[\dfrac{(J-M)(J+M+1)}{2J(J+1)(2J+1)}\right]^{1/2}$
J	0	$(-1)^{J-M-1}\dfrac{M}{[J(J+1)(2J+1)]^{1/2}}$
$J+1$	1	$(-1)^{J-M-1}\left[\dfrac{(J+M+1)(J+M+2)}{(2J+1)(2J+2)(2J+3)}\right]^{1/2}$
$J+1$	0	$(-1)^{J-M}\left[\dfrac{(J-M+1)(J+M+1)}{(J+1)(2J+1)(2J+3)}\right]^{1/2}$

that many of the values of $f(M)$ for $M=-J,\ldots,J$ are nearly equal to $f(0)$. It should be noted from Table V that the clustering of transition moments depends on the change of angular momentum $|J'-J|$ and the polarization σ of the incident field. For P or R branch transitions ($|J'-J|=1$) the transition moments cluster for linearly polarized light ($\sigma=0$) but not for circularly polarized light ($|\sigma|=1$). For Q branch transitions ($J'=J$) the reverse is true: the transition moments cluster for circularly polarized light but not for linearly polarized light.

The molecular transition frequencies (shown in the last two lines of Table IV for the example of a four-level system) at which resonant sidebands develop are only weakly affected by the M dependence of the transition moments, while the resonant Rabi frequencies (the first line of Table IV) are strongly affected. Consequently the conclusions drawn above in (85)–(86) are essentially unaltered by the M dependence of the transition moments.

The values of $\mathcal{P}_\lambda/i$ summed over initial values of M for a system without damping are shown in Fig. 7. The system employed was our second effective-states model of SF_6 ($|vlJR;M\rangle$), with $\sigma=0$, $J_0=68$, $E_0=3\times10^5$ V/m, $\omega/2\pi c=944.2$ cm^{-2}, $\nu_3=948$ cm^{-1}, $X_{33}=-2.54$ cm^{-1}, $G_{33}=0.303$ cm^{-1}, $B=.0907$ cm^{-1}, $\zeta_3=0.693$.[18] The values of $\mathcal{P}_\lambda/i$ shown in Fig. 7, obtained by numerical diagonalization followed by summation over all $(2J_0+1)=137$ values of M_0, clearly show the clustering of resonant Rabi frequencies Ω (84), which directly reflects the clustering of the dipole transition moments (18). Under these conditions E_0 is sufficiently weak that the model SF_6 molecule behaves nearly as a two-level system, as far as the sidebands in the first line of Table IV are concerned.

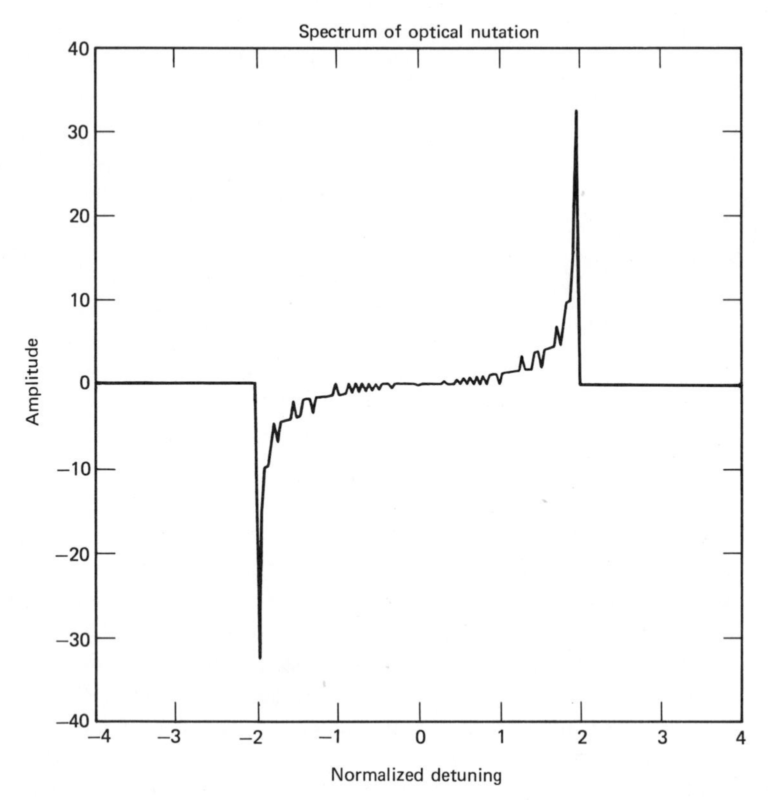

Fig. 7. Sideband amplitudes ($\mathcal{P}_\lambda/i$) summed over all initial M values for a single ground-state angular momentum $J_0 = 68$, plotted as a function of detuning (λ_i). All relaxation rates were zero in this calculation. Other parameters are given in the text.

The values of $\mathcal{P}_\lambda/i$ summed over initial values of J_0 and M_0, under conditions of stronger radiative driving ($E_0 = 3\times 10^7$ V/m), are shown in Fig. 8. The sidebands at small detuning correspond roughly to the first line of Table IV; the sidebands at larger detunings are approximately at resonant molecular transition frequencies. The calculations which resulted in Fig. 8 involved a summation over 49 values of J_0 and 11 values of M_0 for each J_0.

A discussion of sidebands would not be complete without some mention of related, but different calculations already performed by others. Calculations of the spectrum of light scattered spontaneously by a two-level system by Mollow[86] and Freed and Villaeys[87] show peaks in the spectrum of scattered light at the same frequencies at which sidebands are generated in the transmitted field.[29] Another analysis by Mollow shows that a weak

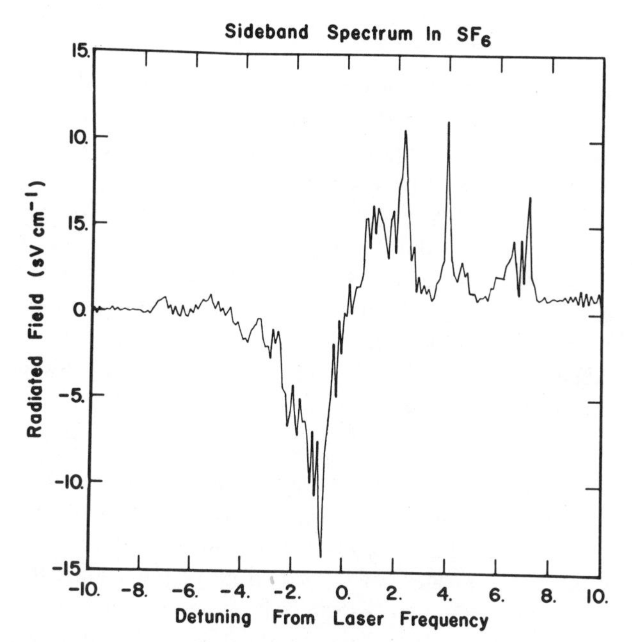

Fig. 8. Sideband amplitudes ($\mathcal{P}_\lambda/i$) averaged over 11 initial M values from $M=-J$ to $+J$ and over 49 initial J values from $J_0=5$ to 101, plotted as a function of detuning (λ_i).[29] All relaxation rates were zero in this calculation. Other parameters are given in the text. Total frequency range; -6.91 to $+6.91$ cm^{-1}. The peak at $+4$ on the horizontal scale is a sideband in resonance with the Q branch. The peak at $+7$ is in resonance with R-branch transitions originating near $J_0=68$.

probe field can be amplified by a two-level system driven by a strong pump field, for certain values of the detunings of the pump and probe.[88] Similar conclusions were reached by McCall, using a semiclassical formalism for homogeneously and inhomogeneously broaded two-level systems.[89] The existence of amplification at certain frequencies of a weak probe beam was pointed out for nondegenerate equidistant multilevel systems by Cohen-Tannoudji and Reynaud,[90] using a dressed-state description. We are actively investigating the possibility of using the phenomenon of amplification at the sideband frequencies as a possible experimental technique for demonstrating the existence of sideband generation in complex systems.

IV. SUMMARY AND DISCUSSION

We have established[29] the following points with regard to the optical field radiated by the macroscopic polarization which is set up in a molecular gas irradiated by a pulsed laser field:

1. The sideband field is proportional to $\mu Nkz'$ and the sideband intensity is proportional to $(\mu Nkz')^2$, in the limit where the radiated field is small compared to the incident field, and where no significant reabsorption of the radiated field occurs.
2. The sideband intensity approaches zero as the molecules approach equilibrium between radiative pumping and collisional (or intramolecular) relaxation. Sidebands will be observable only if the risetime of the laser pulse is short compared to the shortest collisional (or intramolecular) relaxation time.
3. Sidebands are generated in near resonance with every molecular transition accessible from the initial state of the molecules.

The most qualitatively important effect from the point of view of multiple-photon excitation of polyatomic molecules is Point 3, since this result implies that a laser pulse generates essentially every transition frequency of the molecular vibration–rotation band with which it interacts, as a result of the process of propagation. In particular, this collective generation of new frequencies[29] is the only process known to us which can explain the very large number of rotational states in SF_6 pumped by a CO_2 laser at rather modest laser intensities.[30] To illustrate this point, let us consider the Rabi frequency at which molecules will cycle population between levels 0 and 2 as a result of the coherently generated field (86). The Rabi frequency for population cycling is

$$\frac{\mu|\mathcal{E}_{\rm rad}^{(2)}|}{h} \sim 3\times 10^8 \ {\rm sec}^{-1} \tag{95}$$

i.e., the Rabi period is $\sim$3 nsec, under the conditions assumed for (86). Averaged over in times long compared to 3 nsec, then, half the molecular population in state 0 will appear in state 2, under the influence of the field $\mathcal{E}_{\rm rad}^{(2)}$. This time for nearly saturated pumping of level 2 is less than the 50–100 nsec CO_2 laser pulse lengths often employed in experiments on multiple-photon excitation of SF_6. The predicted time-averaged probability of 0.5 for excitation of level 2 by the field $\mathcal{E}_{\rm rad}^{(2)}$ in this case should be contrasted with the time-averaged probability of excitation of level 2 produced by the incident field $\mathcal{E}_{\rm inc}$ acting alone, which is

$$\left[\frac{\Omega}{2\Delta_2}\right]^2 \sim 6\times 10^{-2} \tag{96}$$

For weaker incident fields, $(\Omega/2\Delta_2)^2$ is smaller (in proportion to the incident intensity), and the enhancement of excitation by $\mathcal{E}_{\text{rad}}^{(2)}$ (for sufficiently long laser pulses and low relaxation rates) is even greater than in this example.

Although the estimates just presented depend on the assumption of no collisional or unimolecular relaxation during the time of the laser pulse (according to point 2), this assumption should be fulfilled for collisional effects at the pressure ($\sim$0.1 torr) assumed in deriving (86), for laser pulses of length $T < 100$ nsec. In applying result 2, it should also be noted that any major change in laser pulse amplitude or frequency will have the effect of disequilibrating the molecular populations and enabling a renewed generation of sidebands. Even unstabilized CW lasers may be able to generate sidebands in SF_6 and other molecular gases, due to frequency jitter and a consequent lack of equilibrium between laser pumping and collisional relaxation.

The conditions for efficient experimental observation of sidebands are evidently

$$T \ll \text{shortest relaxation time} \tag{97}$$

$$T \ll |h/\mu \mathcal{E}_{\text{rad}}^{(j)}| \tag{98}$$

$$T \gg (2\Delta_j)^{-1} \tag{99}$$

where T is the laser pulse length. Condition (97) is required by point 2. Condition (98) means physically that the radiated field does not strongly excite any molecules, and thus is able to escape from the sample cell. Condition (99), which will be fulfilled in practice for all but the shortest pulses, simply means that the sidebands can execute many periods of oscillation with respect to the incident frequency, so that there is a well-defined frequency for either spectral or temporal observation.

The functional dependence of sideband-induced effects on the product $\mu Nkz'$ as stated in point 1 provides a way to distinguish the effects of collective, coherent generation of sidebands from unimolecular or collisional relaxation. For unimolecular relaxation there should be no dependence on N or z'. For collisional relaxation, the increase of energy absorbed per molecule due to collisions might be expected to be proportional to N. In certain limits the molecular excitation produced by (for example) $\mathcal{E}_{\text{rad}}^{(2)}$ (85) will be roughly proportional to N, for fixed z'; this dependence could easily be mistaken for a collisional effect.[29] True collisional effects can, in principle, be distinguished from coherent propagation effects under the conditions (97)–(99) by a study of the dependence on z', for a fixed N.

In view of point 2, the experimental observation of sidebands in SF_6 (or other polyatomic molecules) would be evidence that the collisional or intramolecular relaxation time is not much shorter in order to magnitude than the laser pulse length T. Clearly, if (for example) SF_6 relaxes to thermodynamic equilibrium with resonant laser light in a very short time, as some have claimed,[21] then no sidebands should be observed according to (73). Experiments to observe sideband production may, in fact, provide a way to distinguish unambiguously between coherent excitation (which produces sidebands) and unselective laser heating of the molecules (which produces no sidebands). Quantitative measurements of the energy transferred to sidebands, and comparison with theoretical results such as (80), may be a useful technique for measuring (or for putting a lower limit on) intramolecular relaxation times.

Sidebands may, in principle, be observed either spectroscopically or temporally. Temporal observation (i.e., observation of optical nutation) depends on the fact that the intensity of the total field, which is proportional to

$$|\mathcal{E}(z',t')|^2=|\mathcal{E}_{\text{inc}}(t')|^2+2\,\text{Re}(\mathcal{E}_{\text{inc}}(t')\mathcal{E}_{\text{rad}}(z',t')^*)$$
$$+|\mathcal{E}_{\text{rad}}(t')|^2 \tag{100}$$

contains an oscillatory interference term between the incident and radiated fields. The advantage of temporal observation is that the heterodyne term $2\,\text{Re}(\mathcal{E}_{\text{inc}}\mathcal{E}^*_{\text{rad}})$ in (100), which will display temporal oscillations at the dominant sideband frequency (Fig. 9), provides a substantial amplication of the weak field $\mathcal{E}_{\text{rad}}$. The disadvantage of temporal observation is that the presence of many different sideband frequencies leads gradually to destructive interference, causing the oscillations to decay in amplitude. Such a decay is a well-known consequence of the range of Rabi frequencies induced by Doppler broadening.[6] Similar phenomena in optical free induction decay have recently been discussed theoretically in the context of electronic molecular transitions.[91] The advantage of spectroscopic observation is, of course, that the sidebands may be observed even if the range of sideband frequencies is so great that only a very few temporal oscillations appear. The disadvantage is that the sideband intensity may be very weak, except for the relatively strong sidebands at the resonant Rabi frequency (see the first line of Table IV).

A number of points which we have not addressed in this work deserve further study, and will be subjects of our continuing research. First, we have not given any quantitative treatment of the effects of sideband generation on the energy deposited in a molecular gas by an incident laser

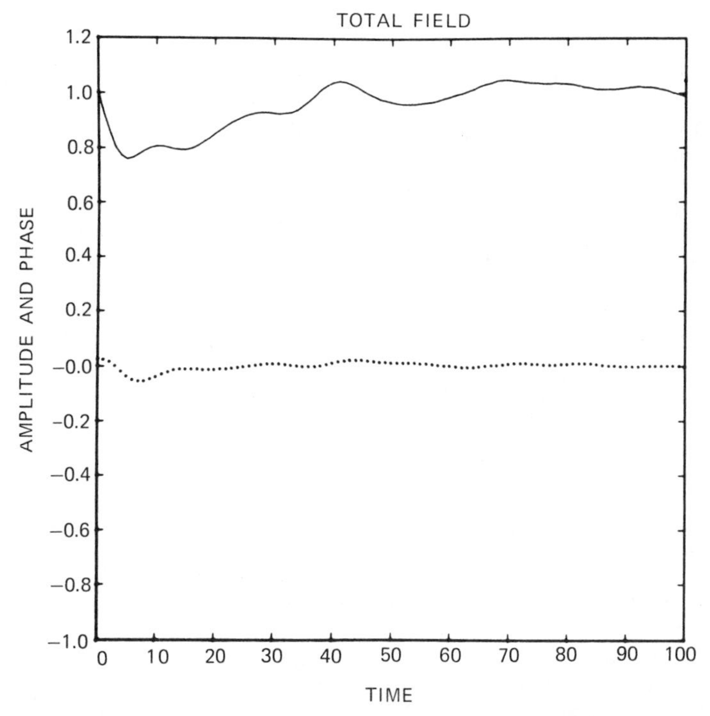

Fig. 9. Amplitude $|\mathcal{E}|$ and phase ϕ of the total field (55) at $z' = 10$ cm as a function of retarded time t'. Conditions same as for Fig. 8. Total time interval: 0 to 48.3 psec.

pulse. This promises to be a challenging problem in radiative transport. Second, we have considered no coherent propagation effects other than sideband generation.[92] In particular, we have not yet addressed the question of explaining the photon-echo data in SF_6.[28] Although we have made numerical studies of optical free induction decay, space has not permitted us to discuss them here. Finally, we have not discussed effects due to the transverse variation of the incident field.

We believe that our calculations reported in this paper have established that the consequences of sideband generation for multiple-photon excitation of polyatomic molecules deserve serious study. We expect to report our continuing studies of this problem in future publications.

ACKNOWLEDGMENTS

We thank R. V. Ambartzumian, K. Boyer, C. B. Collins, B. J. Feldman, R. A. Fisher, D. Ham, J. Jortner, O. P. Judd, and V. S. Letokhov for helpful discussions.

References

1. C. K. N. Patel and R. E. Slusher, *Phys. Rev. Lett.*, **19**, 1019 (1967).
2. C. K. N. Patel and R. E. Slusher, *Phys. Rev. Lett.*, **20**, 1087 (1968).
3. G. B. Hocker and C. L. Tang, *Phys. Rev.*, **184**, 356 (1969).
4. A. Zembrod and Th. Grühl, *Phys. Rev. Lett.*, **27**, 287 (1971).
5. S. S. Alimpiev and N. V. Karlov, *Sov. Phys. JETP*, **39**, 260 (1974).
6. C. L. Tang and B. D. Silverman, in *Physics of Quantum Electronics*, P. L. Kelly, B. Lax, and P. E. Tannenwald, eds., McGraw-Hill, New York, 1966, pp. 280–293.
7. F. A. Hopf and M. O. Scully, *Phys. Rev. B*, **1**, 50 (1970).
8. F. A. Hopf, C. K. Rhodes, and A. Szöke, *Phys. Rev. B*, **1**, 2833 (1970).
9. H. M. Gibbs, S. L. McCall, and G. J. Salamo, *Phys. Rev. A*, **12**, 1032 (1975).
10. R. S. McDowell, H. W. Galbraith, B. J. Krohn, C. D. Cantrell, and E. D. Hinkley, *Opt. Commun.*, **17**, 178 (1976).
11. R. S. McDowell, H. W. Galbraith, C. D. Cantrell, N. G. Nereson, and E. D. Hinkley, *J. Mol. Spectr.*, **68**, 288 (1977).
12. R. S. McDowell, H. W. Galbraith, C. D. Cantrell, N. G. Nereson, P. F. Moulton, and E. D. Hinkley, *Opt. Lett.*, **2**, 97 (1978).
13. Ch. J. Bordé, M. Ouhayonn, A. VanLerberghe, C. Salomon, S. Avrillier, C. D. Cantrell, and J. Bordé, in *Laser Spectroscopy IV*, H. Walther and K. W. Rothe, eds., Springer-Verlag, Berlin, 1979, p. 142.
14. C. D. Cantrell and H. W. Galbraith, *Opt. Commun.*, **18**, 513 (1976); **21**, 374 (1977).
15. V. M. Akulin, S. S. Alimpiev, N. V. Karlov, B. G. Sartakov, and L. A. Shelepin, *Zh. ETF*, **71**, 454 (1976).
16. C. D. Cantrell, H. W. Galbraith and J. R. Ackerhalt, in *Multiphoton Processes*, J. H. Eberly and P. Lambropoulos, eds., Wiley, New York, 1978, pp. 307–330.
17. H. W. Galbraith and C. D. Cantrell, in *The Significance of Nonlinearity in the Natural Sciences*, B. Kursunoglu, A. Perlmutter and L. F. Scott, eds., Plenum, New York, 1977, pp. 227–264.
18. C. D. Cantrell and K. Fox, *Opt. Lett.*, **2**, 151 (1978).
19. J. R. Ackerhalt and H. W. Galbraith, *J. Chem. Phys.*, **69**, 1200 (1978).
20. P. F. Moulton and A. Mooradian, in *Laser-Induced Processes in Molecules*, K. L. Kompa and S. D. Smith, eds., Springer-Verlag, Berlin, 1979, pp. 37–42.
21. H. S. Kwok and E. Yablonovitch, *Phys. Rev. Lett.*, **41**, 745 (1978).
22. W. E. Lamb, Jr., *Phys. Rev.*, **134**, A1429 (1964).
23. F. A. Hopf and M. O. Scully, *Phys. Rev.*, **179**, 399 (1969).
24. A. Icsevgi and W. E. Lamb, Jr., *Phys. Rev.*, **185**, 517 (1969).
25. C. D. Cantrell, A. A. Makarov, and W. H. Louisell (to be published).
26. R. L. Shoemaker, in *Laser Applications to Optics and Spectroscopy*, S. F. Jacobs, M. Sargent III, J. F. Scott and M. O. Scully, eds., Addison-Wesley, Reading, Mass., 1975, pp. 453–504.
27. N. A. Kurnit, I. D. Abella, and S. R. Hartmann, *Phys. Rev. Lett.*, **13**, 567 (1964); *Phys. Rev.*, **141**, 391 (1964).
28. C. V. Heer and R. J. Nordstrom, *Phys. Rev. A*, **11**, 536 (1975); W. M. Gutman, and C. V. Heer, *Phys. Rev. A*, **16**, 659 (1977).
29. A. A. Makarov, C. D. Cantrell, and W. H. Louisell, *Opt. Commun.*, **31**, 31 (1979).
30. S. S. Alimpiev, V. N. Bagratashvili, N. V. Karlov, V. S. Letokhov, V. V. Lobko, A. A. Makarov, B. G. Sartakov, and E. M. Khokhlov, *Zh. ETF Pis. Red.*, **25**, 582 (1977); A. S. Akhmanov, V. N. Bagratashvili, V. Yu. Baranov, Yu. R. Kolomisky, V. S. Letokhov, V. D. Pismenny, and E. A. Ryabov, *Opt. Commun.*, **23**, 357 (1977).

31. R. S. McDowell, J. P. Aldridge, and R. F. Holland, *J. Phys. Chem.*, **80**, 1203 (1976).
32. B. T. Darling and D. M. Dennison, *Phys. Rev.*, **57**, 128 (1940).
33. J. K. G. Watson, *Mol. Phys.*, **15**, 479 (1968).
34. J. D. Louck, *J. Mol. Spectr.*, **61**, 107 (1976).
35. Yu. S. Makushkin and O. N. Ulenikov, *J. Mol. Spectr.*, **68**, 1 (1977).
36. W. H. Shaffer, H. H. Nielsen and L. H. Thomas, *Phys. Rev.*, **56**, 895 and 1051 (1939).
37. J. Moret-Bailly, *J. Mol. Spectr.*, **15**, 344 (1965).
38. K. T. Hecht, *J. Mol. Spectr.*, **5**, 355 (1960).
39. K. Fox, B. J. Krohn, and W. H. Shaffer, *J. Chem. Phys.*, (in press).
40. H. Brunet and M. Perez, *J. Mol. Spectr.*, **29**, 472 (1969); H. Brunet, *IEEE J. Quant. Elect.*, **QE-6**, 678 (1970).
41. P. L. Houston and J. I. Steinfeld, *J. Mol. Spectr.*, **54**, 335 (1975); K. N. Rao and S. Hurlock, private communication.
42. E. D. Hinkley, *Appl. Phys. Lett.*, **13**, 49 (1968).
43. E. D. Hinkley and P. L. Kelley, *Science*, **171**, 635 (1971).
44. K. W. Nill, F. A. Blum, A. R. Calawa, and T. C. Harman, *Appl. Phys. Lett.*, **19**, 79 (1971).
45. E. D. Hinkley, *Appl. Phys. Lett.*, **16**, 351 (1970).
46. R. S. McDowell, in *Advances in Infrared and Raman Spectroscopy*, R. J. H. Clark and R. E. Hester, eds., Heyden & Son, London, 1977.
47. J. P. Aldridge, H. Filip, H. Flicker, R. F. Holland, R. S. McDowell, N. G. Nereson, and K. Fox, *J. Mol. Spectr.*, **58**, 165 (1975).
48. C. D. Cantrell, Los Alamos Scientific Laboratory Informal Report LA-5464-MS (1973).
49. C. D. Cantrell and H. W. Galbraith, *J. Mol. Spectr.*, **58**, 158 (1975).
50. K. Fox and W. B. Person, *J. Chem. Phys.*, **64**, 5218 (1976).
51. K. Fox, *Opt. Commun.*, **19**, 397 (1976).
52. S. L. McCall and E. L. Hahn, *Phys. Rev.*, **183**, 457 (1969).
53. I. Burak, A. V. Nowak, J. I. Steinfeld, and D. G. Sutton, *J. Chem. Phys.*, **51**, 2275 (1969).
54. J. I. Steinfeld, I. Burak, D. G. Sutton, and A. V. Nowak, *J. Chem. Phys.*, **52**, 5421 (1970).
55. D. S. Frankel and J. I. Steinfeld, *J. Chem. Phys.*, **62**, 3358 (1975).
56. A. V. Nowak and J. L. Lyman, *J. Quant. Spectr. Radiat. Transf.*, **15**, 945 (1975).
57. P. F. Moulton, D. M. Larsen, J. N. Walpole, and A. Mooradian, *Opt. Lett.*, **1**, 51 (1977).
58. J. Moret-Bailly, L. Gautier, and J. Montagutelli, *J. Mol. Spectr.*, **15**, 355 (1965).
59. H. H. Nielsen, in *Encyclopedia of Physics*, Vol. 37/1, S. Flügge, ed., Springer-Verlag, Berlin, 1959, pp. 173–313.
60. M. Abramowitz and I. A. Stegun, *Handbook of Mathematical Functions*, U.S. Government Printing Office, Washington, 1964.
61. E. P. Wigner, *Group Theory and Its Application to the Quantum Mechanics of Atomic Structure*, tr. by J. J. Griffin, Academic Press, New York, 1959, Ch. 19.
62. D. M. Brink and G. R. Satchler, *Angular Momentum*, 2nd ed., Clarendon Press, Oxford, 1971.
63. K. Fox and B. J. Krohn, *J. Comp. Phys.*, **25**, 386 (1977).
64. B. J. Krohn, private communication.
65. J. Moret-Bailly, *Can. Phys.*, **15**, 237 (1961).
66. J. D. Louck, Ph.D. thesis (unpublished).
67. A. G. Robiette, D. L. Gray, and F. W. Birss, *Mol. Phys.*, **32**, 1591 (1976).
68. B. Bobin and K. Fox, *J. Phys. (Paris)*, **34**, 571 (1973).
69. C. D. Cantrell, S. Avrillier, Ch. Salomon, and Ch. J. Bordé (to be published).
70. F. R. Petersen, D. G. McDonald, J. D. Cupp, and B. L. Danielson, in *Laser Spectroscopy*, R. G. Brewer and A. Mooradian, eds., Plenum, New York, 1974, p. 555.

71. A. Van Lerberghe, S. Avrillier, and Ch. J. Bordé, *IEEE J. Quant. Elect.*, **QE-14**, 481 (1978).
72. K. Fox, *Opt. Lett.*, **1**, 214 (1977).
73. H. W. Galbraith, *Opt. Lett.*, **3**, 154 (1978).
74. C. D. Cantrell (unpublished).
75. J. S. Briggs (unpublished).
76. I. N. Knyazev, V. S. Letokhov, and V. V. Lobko, *Opt. Commun.*, **25**, 337 (1978).
77. D. M. Larsen and N. Bloembergen, *Opt. Commun.*, **17**, 254 (1976).
78. D. M. Larsen, *Opt. Commun.*, **19**, 404 (1976).
79. C. J. Elliott and B. J. Feldman, *Bull. Am. Phys. Soc.*, **20**, 1282 (1975).
80. S. Mukamel and J. Jortner, *Chem. Phys. Lett.*, **40**, 150 (1976).
81. R. V. Ambartzumian, Yu. A. Gorokhov, V. S. Letokhov, G. N. Makarov, and A. A. Puretzky, *Zh. ETF Pis. Red.*, **23**, 26 (1976). [*JETP Lett.*, **23**, 22 (1976).]
82. C. D. Cantrell, W. H. Louisell, and J. F. Lam, in *Laser-Induced Processes in Molecules*, K. L. Kompa and S. D. Smith, eds., Springer-Verlag, Berlin, 1979, pp. 138–141.
83. J. Stone, M. F. Goodman, and D. A. Dows, *J. Chem. Phys.*, **65**, 5062 (1976).
84. C. D. Cantrell, S. M. Freund, and J. L. Lyman, in *Laser Handbook*, Vol. III, M. L. Stitch, ed., North-Holland, Amsterdam, 1979, p. 485.
85. M. F. Goodman and E. Thiele, *Phys. Rev. A*, **5**, 1355 (1972).
86. B. R. Mollow, *Phys. Rev.*, **188**, 1969 (1969).
87. K. F. Freed and A. A. Villaeys, *J. Chem. Phys.*, **70**, 3071 (1979).
88. B. R. Mollow, *Phys. Rev. A*, **5**, 2217 (1972).
89. S. L. McCall, *Phys. Rev. A*, **9**, 1515 (1974).
90. C. Cohen-Tannoudji and S. Reynaud, *J. Phys. B*, **10**, 345 (1977).
91. J. Jortner and J. Kommandeur, *Chem. Phys.*, **28**, 273 (1978).
92. O. P. Judd (to be published).

INITIATION OF ATOM–MOLECULE REACTIONS BY INFRARED MULTIPHOTON DISSOCIATION

PAUL L. HOUSTON*

Department of Chemistry, Cornell University, Ithaca, New York 14853

Abstract

A new technique for the measurement of rate constants for atom-molecule reactions is evaluated. The technique involves initiation of the reaction by infrared multiphoton dissociation of an atom precursor molecule. In the present version of the method, the reaction rate constant is determined by monitoring the appearance rate of vibrational fluorescence emitted by excited product molecules. The limitations of this technique are discussed. Multiphoton initiation appears to be a promising tool not only for the study of atom-molecule reactions but also for the investigation of reactions of heretofore elusive radicals.

CONTENTS

I. Introduction . . . 625
II. Experimental . . . 628
III. Results . . . 629
IV. Discussion—Limitations of the Method . . . 633
V. Conclusion . . . 637
Acknowledgments . . . 637
References . . . 637

I. INTRODUCTION

Atom–molecule reactions are among the simplest of all chemical processes. These reactions form a testing ground for theories of chemical interaction and reaction dynamics. They are important as well to atmospheric chemistry, chemical lasers, and combustion. In short, atom–molecule reactions form the very basis of our understanding of chemical kinetics.

Because of their wide ranging importance, it is not surprising that the quest for understanding of atom–molecule reactions has led historically to many of the experimental techniques that are now commonplace in the field of chemical kinetics. These include flame techniques,[1] molecular

*Alfred P. Sloan Research Fellow.

beams,[2] discharge flow techniques,[3] chemical lasers,[4] shock tubes,[5] chemiluminescence methods,[6] and flash photolysis.[7] Many of these techniques and the results they provide have been reviewed elsewhere.[3] The purpose of this paper is to evaluate a newly developed technique for the measurement of atom–molecule rate constants.

The technique to be described below is based on initiation of the reaction by infrared multiphoton dissociation of an atom precursor. While it is most similar to the flash photolysis method developed by Norrish and Porter in 1949,[7] it draws on recent developments in which visible or ultraviolet lasers have replaced the flashlamp used in earlier studies. Moore and his co-workers,[8] for example, have initiated halogen atom reactions by pulsed dissociation of Br_2 or Cl_2 with a visible or ultraviolet laser. In the technique to be presented here, the visible/ultraviolet laser is replaced by a pulsed infrared laser which dissociates the atom precursor in a multiphoton process. Although multiphoton initiation of atom–molecule reactions was first used by Quick and Wittig[9] and by Preses, Weston, and Flynn[10] to estimate the translational energy of the fluorine atom dissociated from SF_6, much development of the method as a general tool for measuring rate constants has occurred in our own laboratory.[11–14] Since a variety of atoms or reactive radicals can be created by infrared multiphoton dissociation, this method is expected to find a wide range of applications in chemical kinetics.

The elementary kinetics for the reaction of an atom X with a molecule M to form product(s) P is described by the equation

$$X + M \overset{k_r(i,T)}{\rightarrow} P(i) \tag{1}$$

where the index i represents the set of quantum numbers for the product molecule(s) and $k_r(i,T)$ is the temperature-dependent rate constant for creation of the product in its ith internal state. For all the reactions to be discussed below, the distribution of energy in the reactants will be a Boltzmann distribution at the temperature T. The total rate of disappearance of X is given by the differential equation

$$\frac{-d[X]}{dt} = k_r(T)[X][M] \tag{2}$$

where $k_r(T) = \Sigma k_r(i,T)$ and the summation extends over all possible internal states of the product(s). Typically, the experiment is performed under pseudo-first-order conditions, so that $[M] = [M]_0$ is a constant. The solution to (2) is then

$$[X]=[X]_0\exp\{-k_r(T)[M]_0t\} \tag{3}$$

where $[X]_0$ is the initial concentration of the atom.

In many cases the reaction creates products which are vibrationally excited. Since these can be monitored relatively easily by observation of their vibrational fluorescence, it is of interest to learn how their time dependence is related to the rate constant $k_r(T)$. We assume in what follows that rotational relaxation takes place on a time scale short compared to the reaction. Vibrational deactivation of the products typically proceeds by energy transfer to the molecule M,

$$P(v)+M \overset{k_d(v,T)}{\rightarrow} P(v-1)+M+\Delta E \tag{4}$$

Solution of the kinetic scheme corresponding to (1) and (4) shows that the time dependence of [X] is still given by (3), while that of $[P(v)]$ is given by

$$[P(v)]=\frac{k_r(v,T)[X]_0}{k_r(T)-k_d(v,T)}\{\exp(-k_d(v,T)[M]_0t) - \exp(-k_r(T)[M]_0t)\} \tag{5}$$

In the typical case where $k_r(T)>k_d(v,T)$, the concentration of $P(v)$, and hence its fluorescence, rises exponentially with the rate $k_r(T)[M]_0$ and decays exponentially with the rate $k_d(v,T)[M]_0$. It is important to note that, while the *amplitude* of $[P(v)]$ is proportional to the individual rate constant $k_r(v,T)$, the *appearance rate* of $[P(v)]$ is proportional to the total rate constant $k_r(T)$.

Many reactions populate more than one vibrationally excited state of the product. The total fluorescence intensity for the $\Delta v=-1$ band will then be proportional to

$$I(t)=\sum_{v=1}^{v_{max}} A_{v,v-1}[P(v)] \tag{6}$$

where $A_{v,v-1}$ is the Einstein coefficient for spontaneous emission and v_{max} is the highest vibrational energy level formed by the reaction. Insertion of the value for $[P(v)]$ from (5) into (6) shows that, when $k_r(T)>k_d(v,T)$, the fluorescence intensity still rises exponentially with the single rate $k_r(T)[M]_0$. However, it decays with what may be a multiple exponential composed of the various rates $k_d(v, T)[M]_0$. It should, perhaps, be noted that when $k_d(v, T) > k_r(T)$ the identification of the rise and decay with the reaction deactivation rate constant is reversed, that is, $k_r(T)$ is derived from the fluorescence decay. In either case, it is a simple task to extract the quantity $k_r(T)[M]_0$ from the time dependence of the fluorescence signal. A plot of this quantity vs the partial pressure of M gives a line whose slope is the rate constant $k_r(T)$.

The remainder of this paper will describe the apparatus for obtaining the rate constants $k_r(T)$ and will evaluate the results obtained to date by the method of infrared multiphoton initiation of atom-molecule reactions. Section IV will attempt to delineate the limitations of the technique, while Section V will suggest further directions which might be pursued using this method.

II. EXPERIMENTAL

The experimental apparatus used for initiation of atom–molecule reactions by infrared multiphoton dissociation has been described elsewhere.[11-13] A schematic diagram is provided in Fig. 1. The output of a TEA CO_2 laser is focused by a sodium chloride or germanium lens into a cell containing argon, the atom precursor, and the molecular reactant. Most of the reactions studied thus far have been hydrogen abstraction reactions by fluorine atoms. For these studies, SF_6 is a useful atom precursor. Argon or another inert gas is included to thermalize the fluorine atoms before they react with the hydrogen donor. A discussion of the optimal argon pressure for this purpose is deferred until Section IV. Infrared fluorescence from the vibrationally excited product molecules is monitored through a quartz window in a direction perpendicular to that of the laser beam. Fluorescence is isolated with an infrared interference filter centered on the transition of interest and passed to the element of an infrared detector. The electronic signal from this detector is amplified, digitized by a transient recorder, and averaged in a hardwired signal analyzer. The time constant for the detector and associated electronics is 200 nsec. After the laser has been pulsed a sufficient number of times so that an acceptable signal-to-noise ratio is obtained, the signal is transferred to a computer for analysis, plotting, and storage. It is usually found that a function of the form $A[\exp(-t/\tau_d) - \exp(-t/\tau_r)]$ provides an accurate fit to the fluorescence signal. A nonlinear least squares fitting routine is used to obtain the best values of A, τ_d, and τ_r. Occasionally, it is necessary to fit the decay of the fluorescence signal to a multiple exponential.

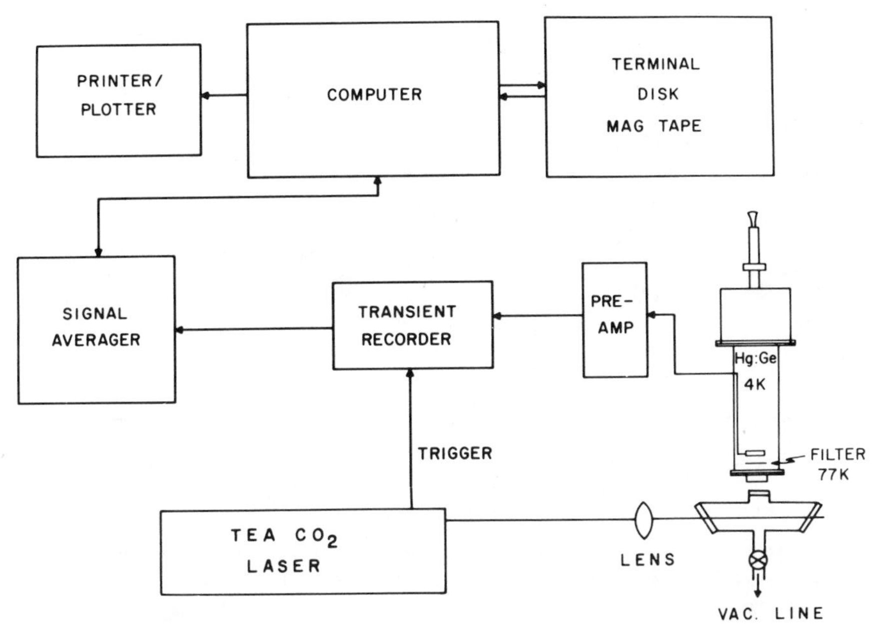

Fig. 1. Schematic diagram of apparatus for initiation of atom–molecule reactions by multiphoton dissociation. Fluorescence from vibrationally excited products is detected as a function of time to determine the rate constant for reaction.

III. RESULTS

Typical results obtained from application of the multiphoton initiation technique to fluorine atom reactions are provided by the example of F + HI→HF + I.[11, 13] Figure 2 displays the HF fluorescence signal observed following the photolysis of 0.050 torr of SF_6 in 0.100 torr of HI and 5.2 torr of argon. The signal is the average of 8 laser pulses. The rise time of the signal was independent of SF_6 pressure in the 0.03–0.10 torr range and independent of laser fluence in the 4–35 J/cm^2 range. Equation 5 shows that the reciprocal of this rise time should be equal to $k_r(T)P_{HI}$, so that a plot of τ_r^{-1} vs P_{HI} should give a straight line with a slope equal to the reaction rate $k_r(T)$. Figure 3 displays the lines obtained by this procedure at a variety of cell temperatures. The least squares value of the slope at 293 K is $(4.16 \pm 0.11) \times 10^{-11}$ cm^3 sec^{-1} where the uncertainty represents a precision of 2 standard deviations. An Arrhenius plot for the temperature dependence of $k_r(T)$ is given in Fig. 4. The variation of the rate constant with temperature is plotted as the lower curve, while the variation of the

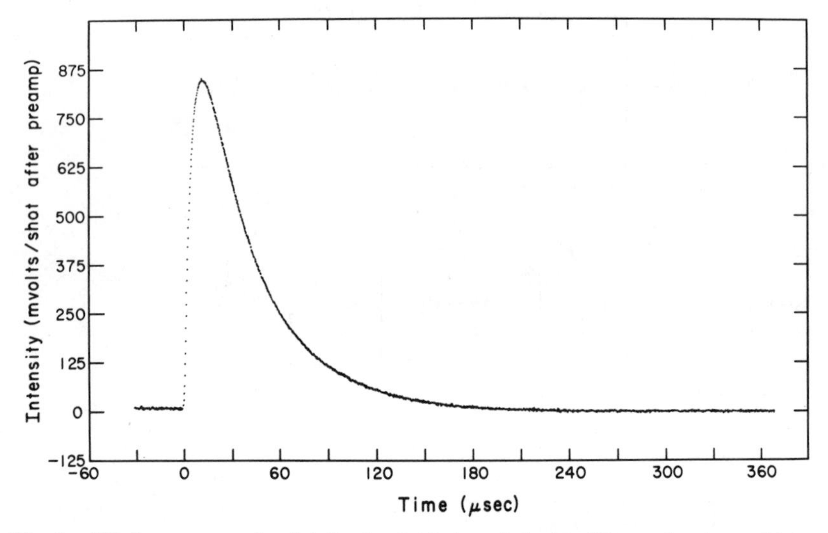

Fig. 2. HF fluorescence signal following initiation of the F + HI reaction by multiphoton dissociation of SF_6. The signal is the average of 8 laser pulses. (From Ref. 11.)

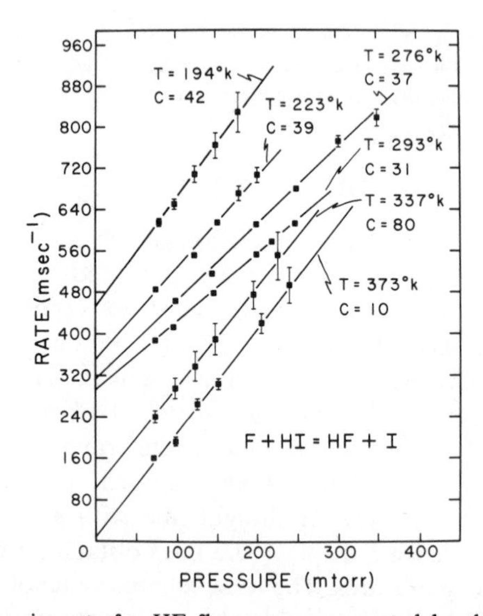

Fig. 3. Plots of the rise rate for HF fluorescence generated by the F + HI reaction as a function of HI pressure and temperature. The lines have been displaced by arbitrary amounts in the vertical direction for clarity. The value of C gives the actual intercept.

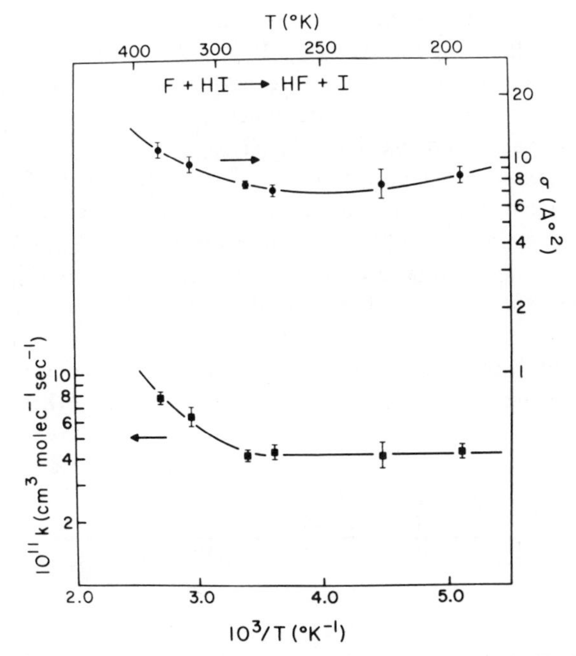

Fig. 4. Arrhenius plot of the rate constant and cross section for the F + HI reaction.

TABLE I
Rate Constants for Hydrogen Abstraction by Fluorine Atoms

Reactant	Rate constant (cm^3 $molec^{-1}$ sec^{-1})	Ref.
HI	$(4.16 \pm 0.11) \times 10^{-11}$	13
HBr	$(4.50 \pm 0.40) \times 10^{-11}$	13
HCl	$(8.07 \pm 0.53) \times 10^{-12}$	13
DI	$(2.10 \pm 0.18) \times 10^{-11}$	11
DBr	$(2.63 \pm 0.20) \times 10^{-11}$	11
DCl	$(6.00 \pm 0.68) \times 10^{-12}$	11
H_2	$(2.27 \pm 0.18) \times 10^{-11}$	13
D_2	$(1.42 \pm 0.03) \times 10^{-11}$	13
H_2O	$(7.11 \pm 0.74) \times 10^{-12}$	14
D_2O	$(4.01 \pm 0.50) \times 10^{-12}$	14
H_2S	$(1.38 \pm 0.13) \times 10^{-10}$	14
H_2Se	$(1.04 \pm 0.06) \times 10^{-10}$	14
NH_3	$(3.94 \pm 0.71) \times 10^{-11}$	14

cross-section is plotted as the upper curve. The cross-section changes by less than 40% in the temperature range 194–337 K.

The rate constant for the F+DI reaction has also been measured by observing the rise time of DF fluorescence as a function of P_{DI}. At room temperature it was found that $k_H/k_D = 1.27$.[11]

Although the example above has discussed the results of the F+HI reaction, similar results have been obtained for a variety of other reactions involving hydrogen abstraction by fluorine atoms. Table I provides a summary of room temperature rate constants. The temperature dependence of the rate constants for several of these reactions has been determined.[13]

One reaction involving hydrogen abstraction by bromine has been investigated using this technique. The rise time of HBr fluorescence

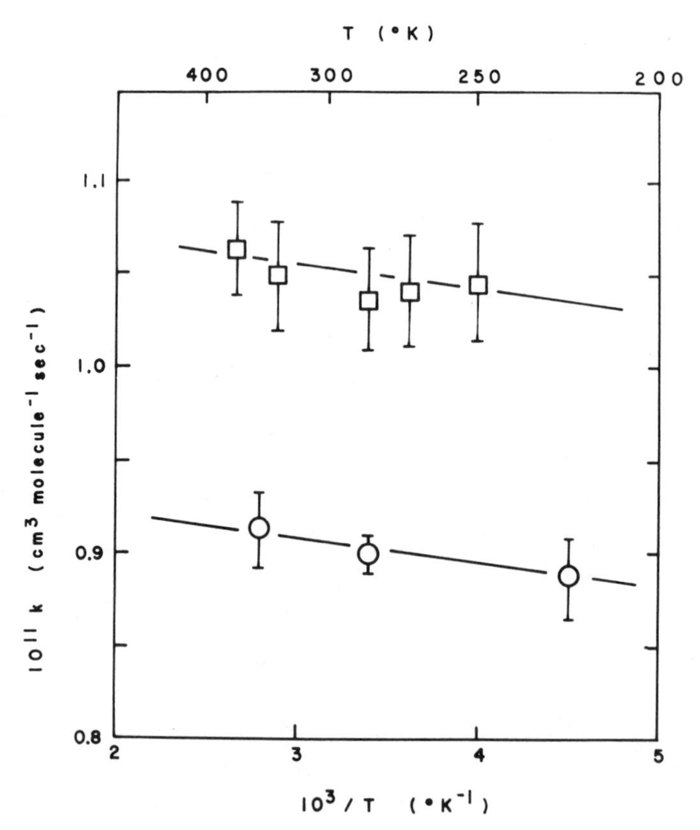

Fig. 5. Plot of the temperature dependence of the rate constant for the Br+HI reaction. The circles are data obtained from the infrared multiphoton initiation technique, while the squares are data obtained by single photon ultraviolet dissociation using the method of Ref. 8d.

produced by the Br + HI reaction has been measured as a function of P_{HI} at three temperatures. CF_3Br was used as a parent compound for the dissociation. The results are shown in Fig. 5 which compares the values obtained by the infrared multiphoton technique (circles) to those obtained by Mei and Moore (squares), who used visible single photon dissociation of Br_2 to initiate the reaction.[8d] The two experimental techniques agree to within 20% over the temperature range 250–350 K.

IV. DISCUSSION—LIMITATIONS OF THE METHOD

Although the multiphoton dissociation method for initiating atom–molecule reactions appears to provide direct and accurate rate constant determinations, it is important at the outset to evaluate its limitations and potential pitfalls. The most serious limitation involves transient heating of the sample by the dissociation laser. We approach this problem first by discussing the time scale for heating and then by evaluating the extent of heating.

Multiphoton excitation of molecules is typically accomplished by use of a TEA CO_2 laser whose pulse consists of a 100 nsec spike containing about half the energy followed by a 1 μsec tail containing the remainder. Excitation of the atomic precursor is, therefore, complete within 1 μsec. In addition to becoming excited by the laser, the atomic precursor is also subject to two other processes. The first involves dissociation from high-lying vibrational levels and occurs with a rate that depends on the internal vibrational energy and on the molecular size. For precursor molecules of less than 7–10 atoms, the dissociation rate increases so rapidly with internal energy that the dissociation lifetime is usually shorter than 10^{-9} sec for levels only a few photons above the dissociation limit. If the molecule is pumped hard enough, most of the dissociation will take place from these levels, while only a small fraction of the dissociation will take place from levels close enough to the dissociation limit to have lifetimes longer than 10^{-6} sec. We conclude that the vast majority of atomic fragments is produced on a time scale shorter than 10^{-6} sec. In a few cases there is direct experimental evidence to support this conclusion.[15, 16] The second process to which the excited molecules are subject is collisional deactivation. We will be most concerned about those atomic precursors which, after not receiving quite enough energy to dissociate, are subsequently relaxed by collisions with the other molecules in the cell. It is this relaxation which gives rise to heating. The time scale for the collisional relaxation depends on the nature of the molecules and on their partial pressures. It may range anywhere from 10 to 1000 μsec for systems of interest to the study of atom–molecule reactions. Ideally, one would like the atom–molecule reaction to occur on a time scale longer than that during which the atoms are produced (1 μsec) but shorter than that during which vibrational relaxation might lead to temperature changes. While it is usually possible to make the reaction time longer than the production time by going to low enough partial pressure of the molecular reactant, changing the pressure cannot change the ratio of reaction to relaxation, since both depend linearly on pressure. Therefore, the multiphoton initiation method will be best for those systems in which the rate constant for reaction of the atom with the molecule is large compared to the rate constant for relaxation of the precursor by the molecule. Since this favorable condition will not always be met for systems of interest, it is necessary to evaluate how much heating might be caused by the vibrational relaxation.

In order to determine the degree of heating we first calculate the amount of energy absorbed from the laser field and then divide by the heat capacity of the system to obtain the maximum temperature change. For most precursors a significant amount of dissociation (>1%) is achieved when the average number of photons absorbed is on the order of half the number of photons which corresponds to the dissociation energy. If D is the dissociation energy, then the average precursor will have absorbed an energy equivalent to $D/2$ in producing this amount of dissociation. Although some of this energy will be used for the dissociation itself, we can assume as a worse case that the total amount of energy available for vibrational relaxation is $E=(D/2)N_pV$, where N_p is the number density of precursor molecules and V is the volume in which the multiphoton dissociation takes place. The total heat capacity of the molecules in the volume V is given by the weighted sum of the heat capacities of the molecules present, $C=\Sigma N_iC_iV$, where N_i is the number density of component i and C_i is its molecular heat capacity. Neglecting the effect of thermal diffusion provides an upper limit on the change in temperature following vibrational relaxation: $\Delta T \leqslant E/C=(D/2)N_p/(\Sigma N_iC_i)$. Since most of the total heat capacity is provided by the partial pressure of the inert gas used to thermalize the atoms, $\Sigma N_iC_i \approx N_{\text{inert}}C_{\text{inert}} \approx N_{\text{inert}}(3R/2)$. Consequently,

$$\Delta T \leqslant \frac{D}{3R}\frac{N_P}{N_{\text{inert}}} \tag{7}$$

Typical operating pressures might be 0.030 torr of the precursor, 1.0 torr of the molecular reactant, and 10 torr of inert gas. For a typical dissociation energy of 50 kcal/mole, (7) predicts that the maximum change of temperature due to vibrational relaxation would be about 25 K. The actual value is likely to be lower since some of the absorbed energy is used in dissociating the precursor and since the actual heat capacity is larger than that for just the inert gas. For determination of most rate constants, an accuracy of $\Delta T = \pm 10$–20 K is sufficient.

The role of the inert gas is crucial to this experimental scheme. First, it may happen that the dissociation produces translationally hot or cold atomic fragments. Although there is good evidence against such translation disequilibrium, [9, 10, 15] collisions with the excess of inert gas would, in any case, ensure that the atoms are thermalized before they react. Second, the inert gas is important in minimizing the change in temperature following vibrational relaxation. Equation 7 shows that this change is proportional to N_p/N_{inert}. If all other things were equal, the method would work best for the highest partial pressure of inert gas and for the lowest partial pressure of precursor which would still give a reasonable signal-to-noise ratio. However, a large amount of inert gas also shortens the time scale for vibrational relaxation, so that there is usually a trade off between the benefit of having the reaction occur before vibrational relaxation and that of having a small change in temperature. It should also be noted that, in many cases, the pressure of inert gas affects the efficiency of the multiphoton dissociation process and the amount of energy absorbed.

Experimental evidence suggests that the vibrational relaxation does not affect the measured rate constant under normal conditions. For the room temperature F+HI reaction cited above, the measured rise time for HF fluorescence was independent of SF_6 pressure in the 0.03–0.10 torr range. According to (7), this pressure range corresponds to a threefold increase in ΔT. By referring to Fig. 4 and by assuming that a change in k of 20% could have been observed, it may be estimated that ΔT at 0.03 torr is less than 15 K. The HF fluorescence rise time was independent also of laser fluence in a range (4–35 J/cm^2) over which the average number of absorbed photons changes by roughly a factor of 3.[17] Again, this invariance implies that ΔT is less than about 15 K during the reaction. Unfortunately, the F+HI reaction is not the best test case because the rate constant is rather insensitive to temperature and because vibrational relaxation is probably somewhat slower than the reaction. Other tests have also been performed, but they are no more stringent. For example, the Br+HI rate constant obtained by the multiphoton initiation method is in good agreement with that obtained by a method in which ΔT is known to be small (Fig. 4).[8d] However, the rate constant for this reaction is also rather insensitive to temperature. Further experimental work will be needed to confirm the conclusion of (7) that ΔT is small for multiphoton initiation of atom–molecule reactions.

In addition to the heating problem discussed above, there are some other less serious limitations to the multiphoton initiation technique. For example, there might occur systems in which both the atom precursor and the molecular reactant would absorb the laser light. If this were the case, the energy distribution of the reactants could no longer be described by a Boltzmann distribution at a single temperature, and marked deviations in the Arrhenius behavior might be observed. Usually this pitfall can be avoided by suitable choice of laser line and atom precursor. A more insidious problem might occur if there were appreciable vibrational-to-vibrational energy transfer from the atom precursor to the molecule on the time scale of the reaction. Low partial pressure of the precursor and low levels of excitation can help to minimize the effects of any such transfer. Finally, it is possible that the radical fragment produced by dissociation of the atom precursor could interfere with the measurement by fluorescing in the spectral region of observation or by reacting itself with the molecular component. Fortunately, the wide choice of compounds that undergo multiphoton dissociation helps to alleviate such problems.

A brief discussion of the limitations of the infrared fluorescence detection scheme is also in order. One problem which we have recently encountered in our laboratory[14] is that there are systems for which the vibrational relaxation of the product is more rapid than the reaction rate. With reference to the discussion of Section I, these are systems for which $k_d(v,T) > k_r(T)$. If it is known that this is the case there is no problem, since $k_r(T)$ can be obtained from the fluorescence decay. However, the relative rates of reaction and vibrational relaxation are not always known. In the case of $F + NH_3$, for example, the rise and decay rates are almost equal. Only by independently measuring the vibrational relaxation of HF by NH_3 could we distinguish which rate corresponded to reaction and which rate corresponded to relaxation.[14] A second and more major problem occurs when the product molecules are not vibrationally excited or when they have an Einstein emission coefficient that is negligible or zero. One of two approaches may be used to overcome this problem. The desired reaction might be measured in concurrent competition with another reaction which does produce vibrational fluorescence. The time dependence of that fluorescence would still give the total disappearance rate of the atoms. When measured vs pressure of the appropriate molecular reactant, the desired rate constant could still be extracted from the data. A second approach might be to find an alternate method for monitoring the time-dependent concentration of reactants or products. We have had some success in monitoring the $F + NH_3$ reaction by using laser-induced fluorescence to detect the NH_2 product. While more difficult in this case than monitoring HF vibrational fluorescence, this method has the advantage that the vibrational distribution of the NH_2 product can also be obtained.

Other likely candidates for study using this technique are the $F + H_2O$ and $F + H_2S$ reactions.

In summary, although there are still questions concerning heating by vibrational relaxation, it appears that initiation of chemical reactions by multiphoton dissociation is a versatile and useful technique. When coupled with the detection of infrared fluorescence from vibrationally excited products, this technique provides accurate values of chemical rate constants.

V. CONCLUSION

Infrared multiphoton initiation of atom–molecule reactions provides an attractive alternative to the more traditional initiation by single photon visible or ultraviolet photolysis. When coupled with time-resolved detection of vibrational emission from excited product molecules, this method has provided rate constants for a variety of reactions involving hydrogen abstraction by fluorine atoms (Table I). The major limitation of the technique stems from the change in temperature following vibrational relaxation of laser excited species. However, it has been shown that this heating should cause less than a 25 K temperature change for typical operating conditions. Although the method has been illustrated for atom–molecule reactions, its greatest promise may lie in the investigation of reactions of more elusive radicals. For example, Filseth et al.[18] have investigated the reaction of $C_2(a\,^3\Pi_u)$ with O_2, while Reisler et al.[19] have examined the similar reaction with NO. The C_2 radical was generated by infrared multiphoton dissociation of C_2H_3CN in both cases. The same precursor has also been used to prepare $CN(X^1\Sigma^+)$,[20] $C_2(X^1\Sigma_g^+)$,[20] and $C_3(X^1\Sigma_g^+)$.[21] Finally, Hartford[22] has monitored the reaction of $ND(a^1\Delta)$ with DN_3. Because of the wide variety of molecules that undergo infrared multiphoton dissociation, it is expected that this technique will make possible the investigation of a large number of reactions between radicals or atoms and stable molecules.

Acknowledgments

This article has been based on the work reported in Ref. 11–14. It is a pleasure in particular to acknowledge the collaboration of Dr. A. J. Grimley and of Dr. E. Würzberg. Support for this work by the Air Force Office of Scientific Research (AFOSR-78-3513) and by the Standard Oil Company of Ohio is also gratefully acknowledged.

References

1. M. Polanyi, *Atomic Reactions*, Williams and Norgate, 1932.
2. E. H. Taylor and S. Datz, *J. Chem. Phys.*, **23**, 1711 (1955);
   D. R. Herschbach, *Disc. Farad. Soc.*, **33**, 149 (1962); J. Ross, ed., *Adv. Chem. Phys.*, **10**, *Molecular Beams*, Wiley-Interscience, New York, 1966.

3. See for example, J. Wolfrum, Atom reactions, in *Physical Chemistry, An Advanced Treatise*, Vol. VIB, W. Jost, ed., Academic Press, New York, 1975.
4. J. V. V. Kasper and G. C. Pimentel, *Phys. Rev. Lett.*, **14**, 352 (1965).
5. R. Becker, *Z. Physik*, **8**, 321 (1922); J. N. Bradley, *Shock Waves in Chemistry and Physics*, Wiley, New York, 1962; E. F. Greene and J. P. Toennies, *Chemical Reactions in Shock Waves*, Arnold, London, 1964; S. H. Bauer, *Science*, **141**, 3584 (1963).
6. K. G. Anlauf, P. J. Kuntz, D. H. Maylotte, P. D. Pacey, and J. C. Polanyi, *Disc. Farad. Soc.*, **44**, 183 (1967).
7. R. G. W. Norrish and G. Porter, *Nature*, **164** 658 (1949).
8. a) F. J. Wodarczyk and C. B. Moore, *J. Chem. Phys.*, **62**, 484 (1974);
   b) K. Bergmann and C. B. Moore, *J. Chem. Phys.*, **63**, 643 (1975);
   c) C.-C. Mei and C. B. Moore, *J. Chem. Phys.*, **67**, 3936 (1977);
   d) C.-C. Mei and C. B. Moore, *J. Chem. Phys.*, **70**, 1759 (1979).
9. C. R. Quick and C. Wittig, *Chem. Phys. Lett.*, **48**, 420 (1977).
10. J. M. Preses, R. E. Weston, Jr., and G. W. Flynn, *Chem. Phys. Lett.*, **48**, 425 (1977).
11. E. Würzberg, A. J. Grimley, and P. L. Houston, *Chem. Phys. Lett.*, **57**, 373 (1978).
12. E. Würzberg, L. J. Kovalenko, and P. L. Houston, *Chem. Phys.*, **35**, 311 (1978).
13. a) E. Würzberg and P. L. Houston, *J. Chem. Phys.*, **72**, 4811 (1980);
    b) E. Würzberg and P. L. Houston, *J. Chem. Phys.*, **72**, 5915 (1980).
14. A. J. Grimley and P. L. Houston, in preparation.
15. E. R. Grant, M. J. Coggiola, Y. T. Lee, P. A. Schulz, Aa. S. Sudbø, and Y. R. Shen, *Chem. Phys. Lett.*, **52**, 595 (1977); Aa. S. Sudbø, P. A. Schulz, Y. R. Shen, and Y. T. Lee, *J. Chem. Phys.*, **69**, 2312 (1978); Aa. S. Sudbø, P. A. Schulz, E. R. Grant, Y. R. Shen, and Y. T. Lee, *J. Chem. Phys.*, **70**, 912 (1979).
16. J. C. Stephenson and D. S. King, *J. Chem. Phys.*, **69**, 1485 (1978); J. C. Stephenson, D. S. King, M. F. Goodman, and J. Stone, *J. Chem. Phys.*, **70**, 4496 (1979).
17. J. G. Black, E. Yablonovitch, and N. Bloembergen, *Phys. Rev. Lett.*, **38**, 1131 (1977).
18. S. V. Filseth, G. Hancock, J. Fournier, and K. Meier, *Chem. Phys. Lett.*, **61**, 288 (1979).
19. H. Reisler, M. Mangir, and C. Wittig, *J. Chem. Phys.*, **71**, 2109 (1979).
20. H. Reisler, M. Mangir, and C. Wittig, *Chem. Phys.*, **47**, 49 (1980).
21. M. L. Lesiecki, K. W. Hicks, A. Orenstein, and W. A. Guillory, *Chem. Phys. Lett.*, **71**, 72 (1980).
22. A. Hartford, *Chem. Phys. Lett.*, **57**, 352 (1978).

INFRARED LASER CHEMISTRY OF COMPLEX MOLECULES

R. B. HALL, A. KALDOR, D. M. COX, J. A. HORSLEY, P. RABINOWITZ, G. M. KRAMER, R. G. BRAY, AND E. T. MAAS, JR.

Exxon Research and Engineering Company, Corporate Research Laboratories, Linden, New Jersey 07036

CONTENTS

I. The Molecules 641
A. Laser Dissociation Experiments in a Molecular Beam 644
B. Average Number of Photons Absorbed 647
II. The Model 648
III. Real-Time Reaction Monitors 652
IV. Discussion 655
References 658

The prospect of controlled laser excitation of specific molecular vibrations to induce or enhance preselected chemical reactions has stimulated the imagination of the chemistry and physics community for many years.[1] In the past decade or so this concept has been explored in both bimolecular and unimolecular reactions. The pioneering work of Letokhov et al.[2] on infrared multiple photon dissociation (MPD) showed at an early stage that such reactions can be species specific, and the obvious application to laser isotope separation stimulated considerable research in that area.[3] Indeed, IR-MPD is now recognized not only as important in laser isotope separation but as an important tool in the study of reaction dynamics in general. Much of the work since 1974 has been done on SF_6, OsO_4, BCl_3, and other closely related compounds.[1, 3] There has been extensive, but less systematic work on small hydrocarbons[4–10] and, during the past three years, work on progressively more complex molecules has become common.[11–18]

As part of our laser isotope separation research, we have started a systematic study of the thermal and infrared laser chemistry of a homologous series of uranyl compounds. The molecules, consisting of from 34 to 62 atoms, are more complex than those typically studied. The results suggest that the multiple photon excitation of such complex molecules differs from the excitation of smaller molecules in some important ways. We believe that many of the laser photochemical characteristics of the uranyl molecules will prove to be typical of complex molecules in general.

To explain the means by which a single-frequency laser can be used to excite an anharmonic molecule to the dissociation limit and to account for the experimental data, various models have been developed.[19–25] These models have some common features. Discrete molecular rovibrational levels in resonance with the incident laser radiation are responsible for the absorption of the first few photons. As the molecules are excited through the discrete levels, resonance is maintained by Rabi-broadening,[21, 26] accidental overlap of spectral fine structure components,[27] multiphoton resonances,[20] rotational state compensation,[28] hot band excitation,[1b] or some combination thereof. As the energy content increases, the density of states increases until eventually a quasicontinuum of rovibrational states is reached. Allowed transitions in the quasicontinuum account for the continued absorption of photons.[22, 28, 29] Eventually the energy content of the molecule exceeds the energy necessary for dissociation with further energy deposition limited by the unimolecular decomposition of the energized molecules. In most instances the energy gap between the initial thermal energy of the molecules and the energy necessary for dissociation is at least 10,000 cm^{-1}, thus more than ten CO_2 laser photons must be absorbed to achieve a significant dissociation rate.

While MPD is species selective,[3] there is considerable controversy about whether infrared laser excitation is significantly different in any other characteristic from a "conventional" heat source.[11, 24, 30–32] With few exceptions the present experimental data base suggests that any differences are not pronounced. The pivotal aspect of this question is the extent to which the excitation deposited by the laser can remain localized. If the time for intramolecular energy randomization (IVR) is short compared to the time for reaction, the potential for laser control of the reaction pathway is severely limited. Only the statistically favored pathways will be taken.

A useful formalism, suggested by Goodman and Stone,[33] is that in the absence of collisions the energy deposited in the pumped vibrational mode relaxes into an isoenergy shell made up of all possible combinations of vibrational quantum numbers at that energy. If the relaxation is rapid ($\sim 10^{-12}$ sec), then all of the modes will be involved and the laser-induced reaction will be statistical. Thiele, Goodman, and Stone have recently extended the model by including a finite IVR rate between elements on the isoenergy shell.[32] The extended model is appealing in that it provides for the expectation that a given vibrational mode will couple more strongly to some modes than to other modes. This treatment adds a mathematical formality to a descriptive picture proposed by Hall and Kaldor to explain what appears to be nonstatistical behavior in the laser-induced reactions of cyclopropane.[11] It provides for the existence of groups of vibrational modes such that for modes within a group the coupling is strong and the energy redistribution time is short, but between different groups the coupling may be weak. This weak coupling would be the rate-determining factor for IVR. Most interestingly, Thiele et al. have shown that several experiments generally assumed to demonstrate that IVR occurs on a time scale of 10^{-12} sec are not sufficiently sensitive to support such a conclusion.[32] For example, the measurement of the product translational energies in molecular beam experiments[4] does not, without further information, determine the distribution of internal energy, and the measurement of the RRK low pressure falloff[34] sets only a lower bound on the time for IVR of roughly 10^{-9} sec. It is likely that laser-induced reactions can compete with energy randomization on this latter time scale.

In our work on the uranyl compounds we have attempted to examine the question of energy localization in a number of experiments. Most of our results can be accounted for by a simple rate equation model with dissociation rates given by RRK theory. This suggests that the laser-induced reaction of these complex molecules is statistical. It is in the long time regime (greater than 10^{-6} μsec), however, where the model works best. There is some evidence that, on the nanosecond time scale, energy randomization may not be complete.

I. THE MOLECULES

A series of molecules of the generic type $UO_2(hfacac)_2 \cdot B$, uranylbishexafluoroacetylacetonate·B, where B is a neutral Lewis base, have been synthesized. Fig. 1 shows the structure of one of these molecules, the tetrahydrofuran (THF) compound. The preparation and characterization of this compound is reported elsewhere.[35] We have studied other uranyl compounds with bases such as acetone (ACE); dimethysulfoxide (DMSO); dimethyltetrahydrofuran (DMTHF); water (H_2O); trimethyl phosphate (TMP); tributyl phosphate (TBP), and pyridine (PYR). All of these molecules are volatile (the THF compound has about 1 torr vapor pressure at 100°C) and are reasonably stable at moderate temperatures. When the vapor is heated the molecules undergo a reversible loss of the Lewis base, that is,

$$UO_2(hfacac)_2B \rightleftharpoons UO_2(hfacac)_2 + B \qquad (1)$$

For the THF compound, the ΔH for this equilibrium is about 30 kcal

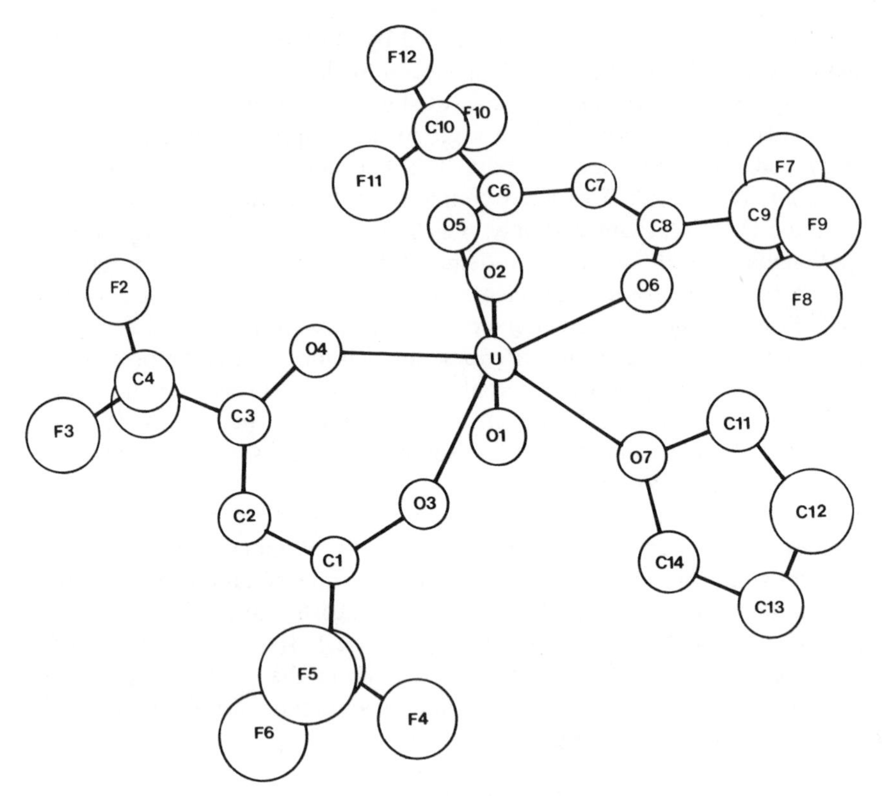

Fig. 1. The structure of $UO_2(hfacac)_2 \cdot THF$. The labels refer to the atomic constituents of the molecule.

mol^{-1}, for the TMP compound it is about 32 kcal mol^{-1}.[36] This dissociation is the same reaction that occurs when the UO_2 asymmetric stretch is pumped by the CO_2 laser. Since the recombination of the neutral base and the coordinatively unsaturated uranyl is not likely to have a significant energy barrier, we take the equilibrium ΔH value as the energy barrier for the laser induced dissociation. The energy barrier for dissociation varies only slightly for many of the Lewis bases. Since in changing the complexity of the leaving group the number of vibrational modes is changed, this series affords the opportunity to investigate the role of the thermal vibrational energy content and heat capacity of the leaving group on the laser-induced dissociation.

The asymmetric stretch, ν_3, of the uranyl moiety occurs at around 960 cm^{-1}. The absorption profile is typically 8 cm^{-1} FWHM and exhibits no resolvable fine structure. With the chelating groups attached, the rotational contribution to the linewidth is small. The molecule however has a significant thermal energy content, about 10,000 cm^{-1} at 100°C, and this can influence the linewidth in several ways. Since many other vibrational modes are populated, the hot band contribution to the width of the absorption profile may be significant. In addition, the density of states is so high at $v=1$ of ν_3 (about 10^{23} states per cm^{-1}) that the molecule is

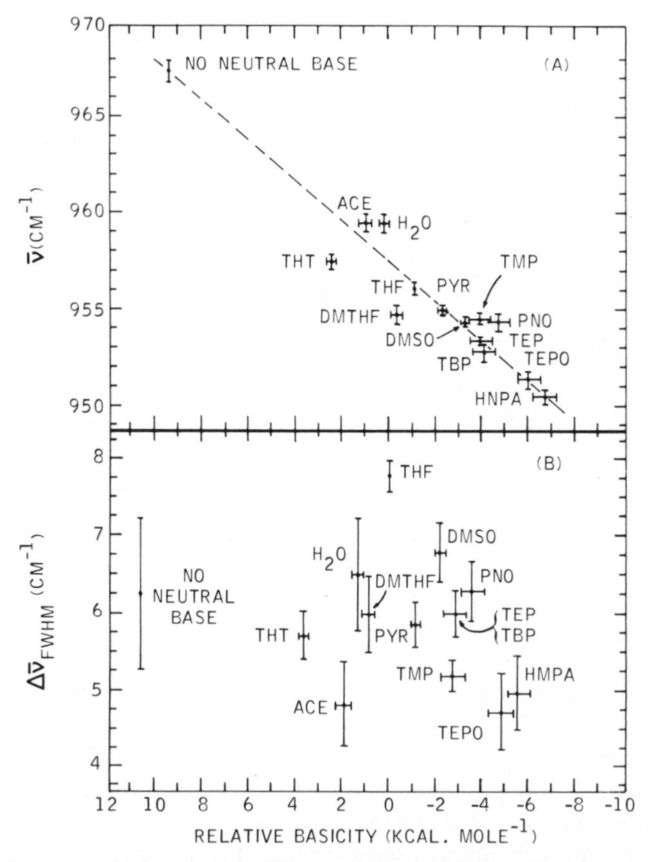

Fig. 2. Dependence of gas phase UO_2 asymetric stretch transition frequencies (a) and linewidths (FWHM) (b) on relative base strength for $UO_2(hfacac)_2 \cdot B$ [B=neutral base] complexes. The base strengths have been measured in solution (relative to B=THF) and the gas phase values are expected to be comparible. The approximate linear dependence in the upper panel has been extrapolated to define the relative base strength for $UO_2(hfacac)_2$. Abbreviations for neutral bases: ACE=acetone; DMSO=dimethylsulphoxide; DMTHF= dimethyltetrahydrofuran; H_2O = water; HMPA = hexamethylphosphoramide; PNO = pyridine-N-oxide; PYR=pyridine; TBP=tributylphosphate; TEP=triethylphosphate; TEPO = triethylphosphineoxide; THF = tetrahydrofuran; THT = tetrahydrothiophene; TMP = trimethylphosphate.

already in the quasicontinuum. If the coupling between ν_3 and the quasi-continuum is so strong that the lifetime of $\nu_3(v=1)$ is on the order of 10^{-12} sec, the observed linewidth could be attributable to lifetime broadening. It appears that the latter effect is important and in fact determines the mechanism of the multiple photon excitation.

The variation of the ν_3 linewidth and position of the line center as a function of the basicity of the leaving group is shown in Fig. 2. The variation of the line center reflects changes in the force constant of the uranyl stretch due to perturbations of the electronic force field by the Lewis base.[37] The linewidth variation does not show a clear systematic trend. If lifetime broadening is assumed to be a major contribution to the linewidth, then variation in linewidth is due primarily to changes in the relaxation rate as a function of base size and complexity. We have attempted to analyze the absorption profiles in terms of a Lorentzian lineshape (generally indicative of a homogeneous transition) but find that only the wings of the line are Lorentzian; the peak of the line is more nearly Gaussian. The absorption profile is thus mixed and the variation in linewidth cannot be attributed simply to coupling variations. As indicated below, the IR-MPD measurements provide additional information on this question. An understanding of the lineshape, and the deconvolution of the various contributions to it are objectives of our present research.

A. Laser Dissociation Experiments in a Molecular Beam

The CO_2 laser-induced dissociation was examined in an effusive molecular beam. The experimental apparatus is shown in Fig. 3; details are described elsewhere.[38] The results we discuss in this paper are obtained by monitoring the depletion of the parent molecular ion due to unimolecular dissociation induced either by a pulsed or by a cw CO_2 laser. With the pulsed laser, the irradiation time is determined by the pulse width, roughly 2 μsec including the N_2 tail. With the cw laser, the irradiation time is determined by the flight time of the molecules through the laser beam, roughly 5 μsec for the geometry used. From the point of interaction with the laser beam, the molecular beam traverses either 50 cm or 10 cm before entering the ionizer of the quadrupole mass spectrometer system depending on which of two configurations is used. The mean times of flight are 5 and 1 msec, respectively. All dissociation that occurs during the time of flight is detected. The values for the yield are integrated over this time interval. We are not able to measure the actual reaction rate with this technique.

As shown in Fig. 4, the dissociation yield for $UO_2(hfacac)_2 \cdot THF$ varies linearly with laser fluence to a point where significant depletion of the

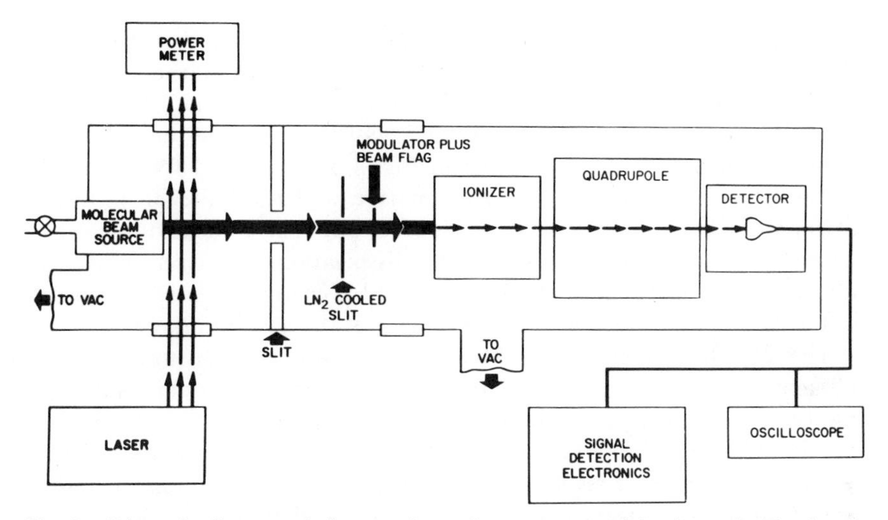

Fig. 3. Schematic diagram of the experimental apparatus used in the molecular beam dissociation experiments.

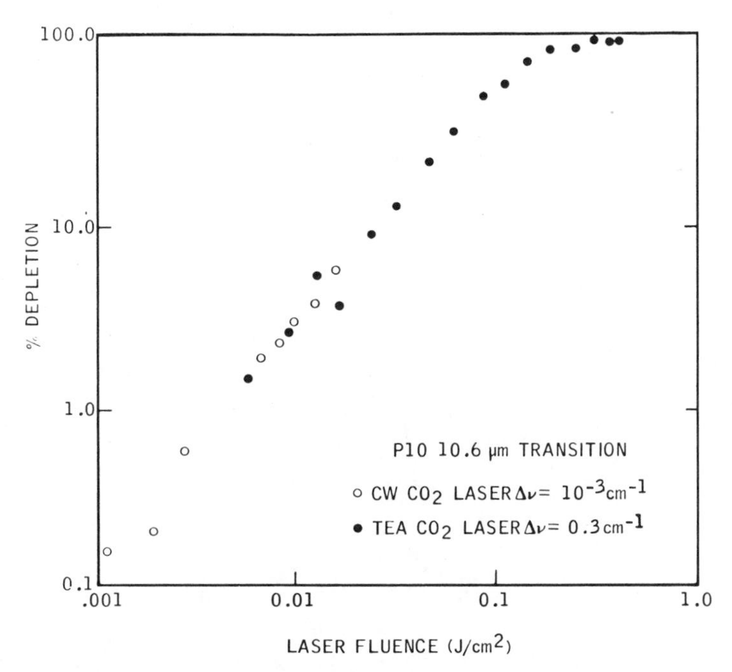

Fig. 4. Fluence dependence of the CO_2 laser-induced unimolecular dissociation of $UO_2(hfacac)_2 \cdot THF$.

molecules in the interaction volume occurs, roughly 60% dissociation. In the pulsed laser experiments the actual depletion is reduced by partial refilling of the 1.2 cm segment initially burned out of the molecular beam because of the thermal spread of velocities of the molecules on either side of this segment. The values shown in Fig. 4 are corrected for this filling factor. It is assumed that dissociation is complete at the high fluence asymptote where no detectable ion signal is left. There does not appear to be any significant hole burning in the dissociation. The laser apparently interacts with all the molecules in the ensemble even though the bandwidth of the laser is very much less than the width of the absorption profile.

The depletion data obtained with the cw laser are identical to those obtained with the pulsed laser at the same fluence. This is significant, since the bandwidth and peak power of the two lasers are quite different (see Fig. 4). This result implies that the contribution to the dissociation due to spectral fine structure resolved by the cw laser is negligible. Similar results are obtained at any wavelength within the absorption profile. The lack of a dependence on the laser bandwidth and the absence of hole burning in the

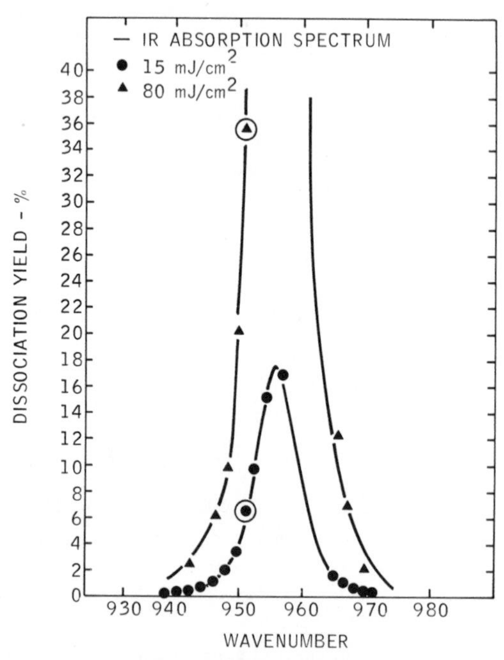

Fig. 5. Frequence dependence of the pulsed CO_2 TEA laser-induced unimolecular dissociation of $UO_2(hfacac)_2 \cdot THF$ in a molecular beam. The dissociation data (symbols) is normalized to the FTIR absorption spectrum (solid line) at the points circled.

dissociation suggest that the absorption profile is homogeneous with respect to dissociation. A model that accounts for these observations is given below.

The wavelength dependence for the dissociation of the THF complex is shown in Fig. 5. Both at low laser fluence (low yield) and at high laser fluence (high yield) the dissociation spectrum matches the absorption spectrum. This is unusual; most small molecules exhibit relatively higher dissociation at laser frequencies to the red of the peak in the absorption spectrum.[1b] For the uranyl molecules, however, a red shift is not observed. Similar results have been obtained with other complex molecules.[12, 16]

B. Average Number of Photons Absorbed

The average number of photons absorbed by $UO_2(hfacac)_2 \cdot TMP$ as a function of laser fluence is shown in Fig. 6. The values are obtained by measuring the fraction of the TEA laser pulse (450 nsec FWHM) transmitted through a 30-cm gas cell. The laser beam is collimated so that the field intensity along the absorption path is nearly constant. The analysis of the data is similar to that used earlier[38] except that the correction for the radial distribution of intensity is not necessary because the transmitted beam is measured through an aperture that samples only the uniform field around the center of the TEM_{00} mode radial distribution.

The deviation from linearity (constant absorption cross-section) is due to the shift of the uranyl absorption band out of resonance with the laser as the molecule reacts. There is no absorption at the laser frequency by either the coordinatively unsaturated $UO_2(hfacac)_2$ or the free TMP. At the

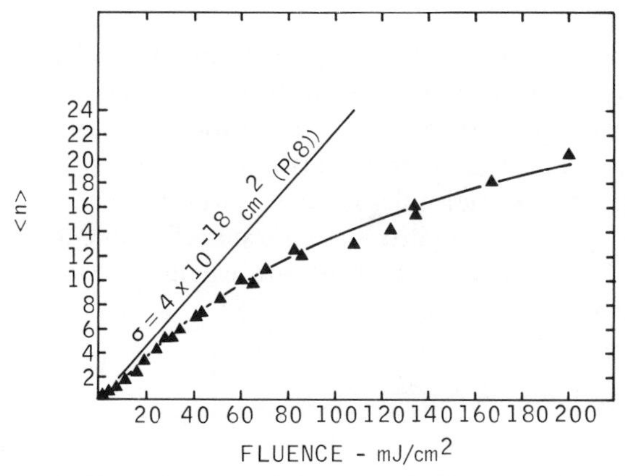

Fig. 6. Average number of photons absorbed, $\langle n \rangle$, per $UO_2(hfacac)_2 \cdot TMP$ molecule. The straight line indicates the absorption expected if the molecules absorbed photons indefinately at the small signal absorption cross-section.

pressure used, 100 mtorr, there is a single gas kinetic collision during the laser pulse; experiments are now underway to repeat these measurements at shorter laser pulse widths (50 nsec, and ~1 nsec) to determine the effects of excitation rate on the absorption process.

II. THE MODEL

The experimental evidence suggests that for molecules as complex as those studied here, the quasicontinuum extends down to the first vibrational level of the pumped mode. In our model, laser excitation to $v=1$ is followed by rapid intramolecular relaxation into a group of modes strongly coupled to the pumped mode. This group of modes is referred to as the heat bath. Because the heat bath occupies a considerably larger region of phase space than the pumped mode, this relaxation effectively returns the pumped mode to the ground state. This process can then be repeated until adequate energy is deposited in the molecule for it to dissociate. The cross-section for the repeated absorption of photons is assumed to remain constant, since, as shown in Fig. 5, the red shift of the absorption profile associated with the MPD of smaller molecules does not occur.

For this type of excitation, the MPD process may be described by a set of coupled rate equations. The formalism is described in detail elsewhere[39]; for clarity a brief description is included here. The change in population of vibrational level with time is given by:

$$\frac{dN}{dt}=\sigma_{j-1}F\left(N_{j-1}-\frac{g_{j-1}}{g_j}N_j\right)+\sigma_jF\left(\frac{g_j}{g_{j+1}}N_{j+1}-N_j\right)-k_jN_j \tag{2}$$

where N_j=population of level j; $\sigma_{j-1}=\sigma_j$=experimental absorption cross section (assumed constant); F=photon flux; g_j=density of states at level j; and k_j=dissociation rate at level j. Excitation is thus described by the usual stimulated absorption and emission terms. As the heat bath is excited, eventually the molecule acquires sufficient energy to dissociate. This is treated by calculating dissociation rates corresponding to each level of excitation. Dissociative decay rates are calculated using the quantum RRK expression

$$k=A\frac{n!(n-m+s-1)!}{(n-m)!(n+s-1)!} \tag{3}$$

where k is the dissociation rate constant for a molecule containing n quanta in s modes, m quanta being necessary for dissociation. The constant A may be identified with the preexponential A-factor in the Arrhenius expression for the rate constant of the thermal dissociation at high pressure.[41] Modes with a frequency greater than $\sim 3kT$ (800 cm^{-1} at 100°C) were assumed to make a negligible contribution in the experimental temperature range and so were omitted from the calculation. The average vibrational frequency of those modes with a frequency below 800 cm^{-1} was estimated to be 250 cm^{-1}, and this was taken as the size of the quantum. The value of m was determined by taking the barrier to dissociation to be 25 kcal/mol and a 250 cm^{-1} quantum. Values for A and s were then adjusted so that calculated yield data reproduced the molecular beam results in Fig. 4.

The initial thermal energy content of the molecules plays an important role in the laser-induced dissociation. The distribution of internal energy in the ensemble of molecules follows a Boltzmann distribution. This distribution is of course, continuous, but we make the approximation of grouping together all molecules within one photon of the dissociative levels in one "level," all those within two photons in the next "level" down, etc., so that a histogram of the initial populations in the various levels mimics the shape of the Boltzmann distribution. A schematic of the distribution before and after laser excitation is indicated in Fig. 7. The thermal energy content of the molecules contributes to the energy required for dissociation so that the laser need only make up the difference. Taking the initial distribution into account, the values required to fit the molecular beam data are

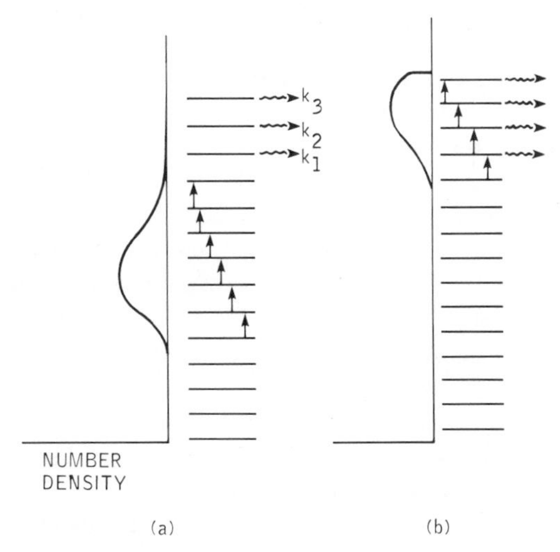

Fig. 7. The thermal distribution of internal energy for the ensemble of molecules; (a) before laser excitation, (b) after laser excitation.

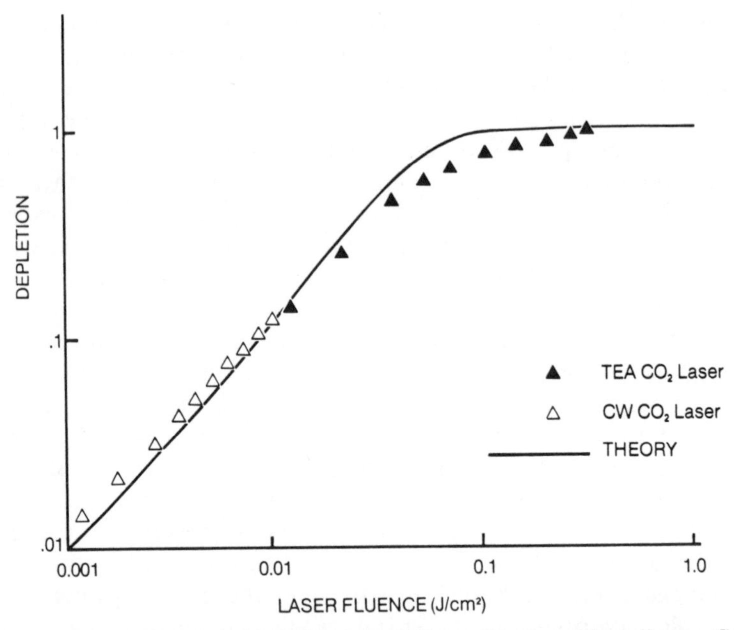

Fig. 8. Fractional depletion of $UO_2(hfacac)_2 \cdot THF$ as a function of laser fluence. Comparison of data (symbols) and calculation (solid line) with $A = 10^{18}$, $s = 60$ is indicated.

$A = 10^{18}$ sec^{-1} and $s = 60$ modes. The high value for A may result from the entropy released in the dissociation.[40] The value for s is the same value that is predicted from quantum statistics. As shown in Fig. 8, the agreement with the experimental data is reasonably good.

Using the same values for A and s, it is also possible to reproduce the temperature dependence of the laser-induced dissociation. As indicated in Fig. 9, the laser-induced decomposition yield varies exponentially with the initial temperature of the molecules at low yields where the laser is dissociating the exponential tail of the Boltzmann distribution. The variation is linear at higher yields where the laser dissociates molecules in the nearly linear portion of the distribution (Fig. 10). The fit of the model to these experimental results is encouraging.

While this model does not fit the experimental data exactly, it does point out the critical role played by the thermal energy in the kinetics of the laser-induced decomposition process. It points out that the temperature dependence of the laser-induced reaction in fact provides a novel way to map out the thermal distribution of the molecular ensemble. As mentioned earlier, these complex molecules already possess some 10,000 cm^{-1} of

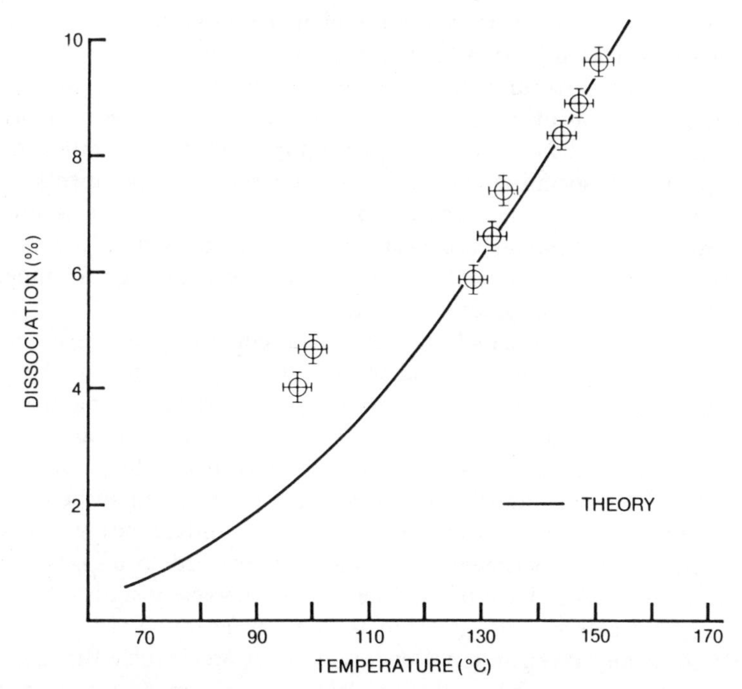

Fig. 9. Temperature dependence of the laser-induced dissociation yield of $UO_2(hfacac)_2 \cdot THF$ at low yields (using a cw CO_2 laser).

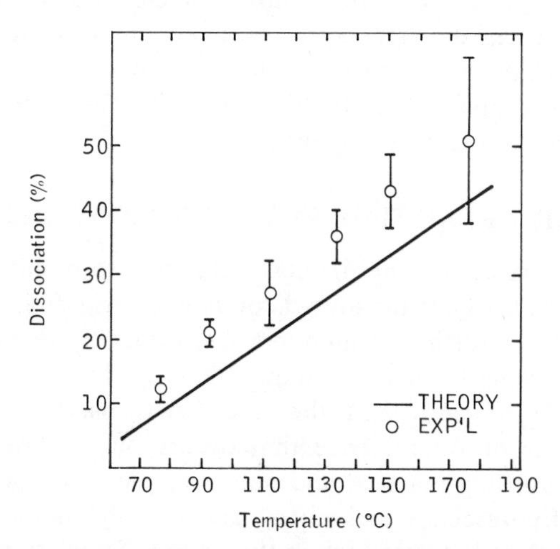

Fig. 10. Temperature dependence of the laser-induced dissociation yield of $UO_2(hfacac)_2 \cdot THF$ at high yields (using a pulsed CO_2 laser).

internal energy on the average, but the high energy tail of the distribution extends to 15,000 cm^{-1}. It is for this reason that, provided the A-factor is 10^{18} sec^{-1}, the absorption of a single photon is sufficient to cause a measurable fraction of molecules to dissociate. The absence of a fluence threshold for the dissociation implied in Fig. 4 is directly attributed to single photon dissociation. We are now undertaking experiments to examine the behavior of the uranyl system as the temperature is lowered substantially. Experiments are also underway to determine the thermal unimolecular reaction rates. These experiments will provide an independent check on the real values for A and s.

We have also investigated the effects of depositing the laser energy initially in a vibrational mode other than the UO_2 stretch. In the TMP compound the CO stretches of the trimethyl phosphate leaving group are accessible to the CO_2 laser. These modes are at the extremities of the leaving group and are thus spatially removed from the UO_2 stretch. Though the couplings and bath modes might be expected to be different for the CO stretch relative to the UO_2 stretch, no differences in molecular beam dissociation experiments that could be ascribed to a nonstatistical distribution of energy have been observed. A completely statistical model, as above, that assumes all modes of the UO_2 $(hfacac)_2 \cdot TMP$ that can be populated at any given internal energy content are involved in the laser induced dissociation predicts this result. It is important to recognize that, as mentioned, the molecular beam experiments integrate over all processes that occur with lifetimes of less than roughly 1 msec. Thus nonstatistical processes that might occur in the nanosecond time regime may be masked by statistical processes that occur at later times. In order to detect nonstatistical behavior, it is necessary to follow the laser-induced processes on a time scale comparable to the time for IVR. Real-time measurements are the only way this can be achieved.

III. REAL-TIME REACTION MONITORS

We have explored two experimental techniques to monitor the real-time laser-induced chemistry of the uranyl compounds. The first is laser-induced fluorescence. The starting compound $UO_2(hfacac)_2 \cdot B$ fluoresces when excited at wavelengths shorter than 5000 Å, but the coordinatively unsaturated $UO_2(hfacac)_2$ and the free Lewis bases do not.[18, 42] We attribute the loss of fluorescence that occurs on IR laser excitation to reaction of the starting material and not to simple vibrational heating.[18, 42] Although the fluorescence technique may actually monitor some as yet unidentified intermediate product on the way to dissociation, it is nevertheless possible to trace out the real time rate of loss of starting material with this technique.

The experiments are carried out in a static gas cell at pressures between 10 and 500 mtorr. The data for roughly 15 mtorr of $UO_2(hfacac)_2 \cdot TMP$ at several different laser fluences is shown in Fig. 11. The asymptotic values for the fractional depletion of starting material are similar to the yields observed in the molecular beam experiments. However, the time dependence is also measured. The rate of disappearance is given by the slope of the curves in Fig. 11. The time resolution is limited in this experiment by the slow rise time (~50 nsec) of the CO_2 laser pulse and by the 10 nsec uncertainty in synchronization of the CO_2 laser pulse and the interrogating dye laser pulse. A precise value for the rate of disappearance cannot be deconvoluted from the temporal shape of the CO_2 laser pulse, but if the rate was slower than 10^7-10^8 sec^{-1}, it would be detected. We take this value as a lower bound for the rate of disappearance at these fluence levels.

The fluence required to produce a fixed yield as the peak power is increased can be obtained from the same data. The fraction of the laser pulse used at the points of intersection of a horizontal line with the integral yield curves indicates the fluence required to produce that yield. If yields below roughly 30% are considered the correction for the reduced reactant concentration is small. The results are summarized in Fig. 12. As the peak

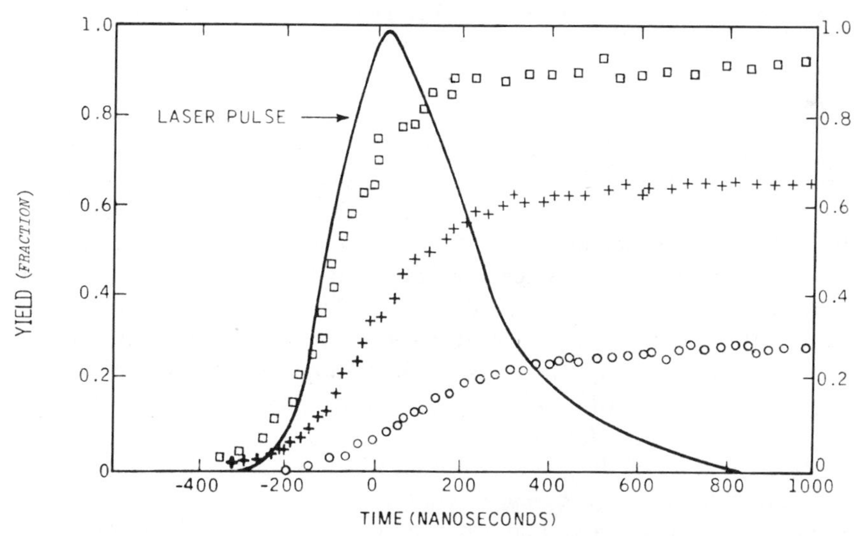

Fig. 11. Time dependence of the CO_2 laser induced fluorescence depletion of $UO_2(hfacac)_2 \cdot TMP$ at several laser fluences. The fractional yield is defined as $\{1-I(t)/I_0\}$ where, I_0 = fluorescence intensity in the absence of CO_2 laser irradiation; $I(t)$ = fluorescence intensity at a delay time t from the peak of the CO_2 laser pulse. (0) 29mJ/cm^2; (+) 69 mJ/cm^2; (□) 140 mJ/cm^2.

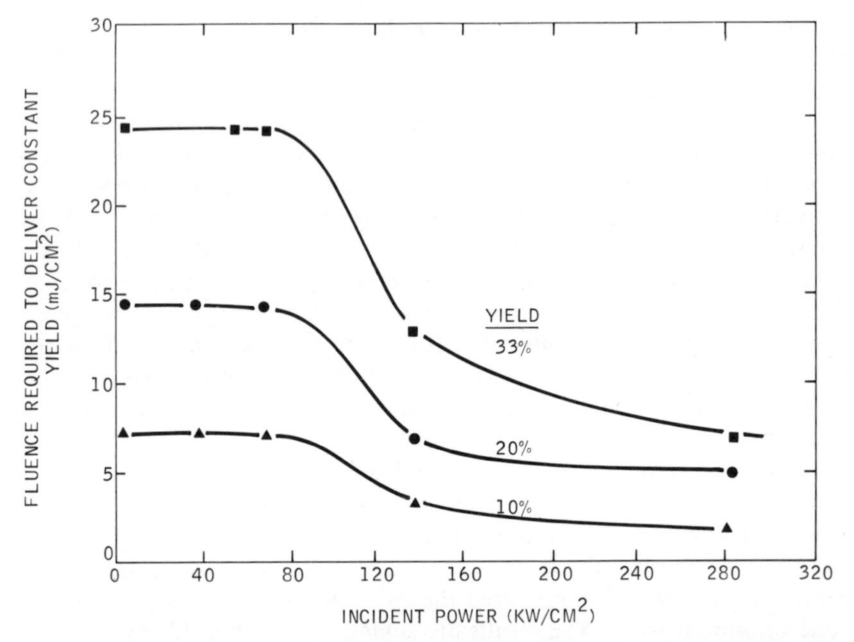

Fig. 12. CO_2 laser fluence required to produce a fixed yield as a function of incident laser power. The points at 1 kW/cm^2 are taken from the asymptotic yield resulting from cw CO_2 laser dissociation in a molecular beam. The remaining points are taken from data as in Fig. 11. (See text.)

power of the laser is increased, the excitation rate becomes faster. As the excitation rate exceeds a certain point, the fluence required to produce a given yield decreases. This is unusual intensity dependence. As discussed below this intensity dependence is not attributable to an anharmonicity bottleneck in the excitation process. It is also unlikely to be the result of some collisional process, since even at the low fluence limits where the approach to the asymptotic yield requires the longest time, no pressure dependence of the asymptotic yields has been detected between 10 and 500 mtorr. One possible explanation is that excitation and reaction are competing with IVR at the $10^{-8}-10^{-9}$ sec timescale. A phenomenological model for this is suggested in the next section.

One limitation of the fluorescence technique is that only parent depletion is monitored. It is not clear if there is an intermediate involved or even if parent depletion can be mapped one to one into product formation. To address this issue we are now adopting a second technique, infrared–infrared double resonance. Unlike SF_6, or other related molecules for which infrared double resonance failed to make much of an impact, in the uranyl system the prospects are more attractive because in these complex molecules the spectral features are relatively simple. Experiments are now underway to explore the details of the laser-induced chemistry.[45]

IV. DISCUSSION

Multiple photon excitation of molecules as complex as the uranyl compounds differ from excitation of simpler molecules in some important ways. In simpler molecules, such as SF_6 or small halocarbons, the factor that determines many of the properties of the dissociation is the mechanism by which the anharmonicity of the discrete vibrational levels is overcome.[1b, 19] When, for example, a specific change of rotational state is necessary to compensate for the anharmonicity, the so called PQR mechanism,[28] only those molecules in the appropriate rotational levels will dissociate. The absorption profile for an ensemble of molecules will thus be inhomogeneous with respect to laser-induced dissociation. When laser power broadening is the critical factor in overcoming anharmonicity, a power threshold exists below which no dissociation will occur. Both of these properties are typical of low-level multiple-photon excitation of small molecules.[1b, 19] Once the molecule has been excited beyond the discrete levels, estimated to be $v=3$ to $v=6$, the density of states is sufficiently high that the molecule becomes its own heat bath. In this quasicontinuum region, energy in the pumped vibrational mode is rapidly transferred to the other modes of the molecule, permitting further excitation. The dissociation yield should therefore depend on the laser fluence and not on the peak power once the quasicontinuum can be reached. This has been demonstrated for SF_6.[43]

It is unlikely that any of the conventional mechanisms for overcoming anharmonicity are applicable to the uranyl compounds. The anharmonicity of the pumped mode is roughly 5 cm^{-1} per vibrational level. The width of the rotational distribution is considerably smaller than this, power broadening is only about 0.1 cm^{-1}, and there is no spectroscopic fine structure (degenerate vibrational level splittings) to shift states into resonance. On the other hand, the molecules have a density of states sufficiently high that even at the first vibrational level they are already in the quasicontinuum. The mechanism for multiple photon excitation is therefore the repeated excitation of the $\nu_3(v=0 \text{ to } v=1)$ transition. Energy from $\nu_3(v=1)$ is rapidly transferred to a set of a states in a time T_1, effectively returning the molecule to the $\nu_3(v=0)$ state. Thus the saturation and induced transparency expected in a two level system are avoided as long as the excitation rate remains slower than T_1^{-1}.

Because the quasicontinuum extends down to the first vibrational level, the laser excitation of these molecules differs from that of simpler molecules. Most importantly, there is no anharmonicity bottleneck in the excitation process. Thus, no power threshold for reaction exists. This conclusion is supported by the yield curve of Fig. 4 and by the observation, in a molecular beam, that the dissociation can be driven with a cw laser and that the yield is proportional to the irradiation time.[38] Similar results have been observed for other large molecules,[17] and it is likely this will prove to be a general characteristic of IR laser chemistry of complex molecules.

This mechanism further predicts that the absorption profile is determined by the lifetime, T_1. If so, the absorption profile for an ensemble of molecules should be homogeneous with respect to dissociation. Thus, even though the absorption profile is approximately 6 cm^{-1} FWHM and the laser linewidth is 0.1 cm^{-1} (including power broadening), it should be possible to dissociate 100% of the molecules irradiating anywhere within the absorptive profile. This, in fact, appears to be the case. With the laser tuned to the edge of the absorption profile, a yield curve similar to that in Fig. 4, shifted to an appropriately higher fluence, is obtained.[38]

As there is no anharmonicity barrier in the excitation process, one would expect that the reaction yield should depend only on the laser fluence, not on the laser power. As indicated in Fig. 12, this is indeed true over a considerable range. From 1 kW/cm^2 (cw laser) up to roughly 100 kW/cm^2 the yield shows no power dependence. This differs from the results of Ashfold et al. who used a similar laser-induced fluorescence technique to study the laser-induced dissociation of the "small" molecule CH_3NH_2.[44] They find that for CH_3NH_2 there is also power dependence. They attribute the dependence to the power broadening required to overcome the anharmonicity barrier. They are most likely correct. For the uranyls, however, power broadening is not necessary, and in the threshold region where such effects should be most evident, only a fluence dependence is observed.

In the region beyond 100 kW/cm^2, however, the yield is no longer dependent simply on fluence. As indicated in Fig. 12, as the power is increased above 100 kW/cm^2 the fluence required to produce a given yield decreases. This is an unusual result, not at all typical of multiple photon dissociation of small molecules for which a power dependence is normally observed only near threshold. Furthermore, the only power dependence predicted by the model is an increase in the fluence required to produce a given yield due to a bottleneck in the energy transfer as the excitation rate (power) begins to approach T_1^{-1}. The opposite trend, however, is observed.

A highly speculative, but possible explanation for the increased photon efficiency at high power is that excitation and reaction within a subset of the vibrational modes competes with randomization (IVR) to include the remaining modes. The phenomenological model suggested here is similar to one proposed earlier.[11, 18, 32] In this simplified picture, the pumped vibrational mode is coupled only to a subset of the total number of modes in the molecule. This subset consists of those (unidentified) modes strongly coupled to the pumped mode, labeled S_{strong}. Modes that are only weakly or indirectly coupled make up another subset, S_{weak}. The transfer time T_1 from the pumped mode to S_{strong} is very rapid (picoseconds) because of the strong coupling. The randomization of energy within this subset is also likely to be rapid. Thus the molecule is still in a quasicontinuum at $\nu_3(v=1)$, but not all of the modes are immediately involved. Energy is transferred to S_{weak} on a relatively slow time scale. Thus the rate of transfer between subsets is the rate determining step in the intramolecular vibrational energy randomization. It is assumed that this randomization can be characterized by a single rate, k_{IVR}. (Modes involved only on a time scale long relative to the reaction time can be considered to be uncoupled.)

With the reaction coordinate contained within S_{strong}, the photon efficiency will be highest when only this subset of modes is excited. In the absence of a T_1 bottleneck, the rate of excitation of the molecule, k_{excite}, is proportional to the laser power. In the limit of very slow excitation (e.g., low power cw laser radiation), $k_{excite} \ll k_{IVR}$ and the effective heat bath is the combined set, $S_{strong}+S_{weak}$. In this limit both subsets must be excited to reach the reaction threshold. In the limit of fast excitation such that $k_{excite} > k_{IVR}$, the subset S_{strong} is preferentially excited. If S_{strong} can be sufficiently heated that the reaction rate, k_{RXN}, is faster than k_{IVR}, then less energy is required because less energy is diverted into S_{weak}.

If k_{excit} and k_{RXN} were known as a function of laser power, it would be possible to estimate k_{IVR}. To date, we have only lower bounds on the sum, $(k_{excite}+k_{RXN})$, at several power levels. With $(k_{excite}+k_{RXN})>10^7$ sec^{-1} at power levels below that at which the increased photon efficiency is observed and perhaps an order of magnitude or two faster at power levels where the increased efficiency is observed, the time regime for IVR would seem to fall in the nanosecond time scale.

It is expected that selective laser chemistry should be observable if the critical energy randomization time is on the order of nanoseconds. We are currently continuing our investigation into the laser chemistry of uranyl compounds and other complex molecules to resolve this issue. We hope eventually to develop general propensity rules for IVR rates across various functional groups. It should then be possible, for a given molecule, to predict the potential for selective laser chemistry on the basis of the molecular structure.

References

1. For recent reviews, see: (a) S. Kimel and S. Speiser, *Chem. Rev.*, **77**, 437, 1977. (b) R. Ambartzumian and V. Letokhov, *Accts. Chem. Res.*, **10**, 61 (1977), and *Chemical and Biochemical Applications of Lasers*, Vol. III, edited by C. B. Moore, Academic Press, New York, 1977. (c) N. Bloembergen and E. Yablonovitch, *Phys. Today*, **21**, 23 (1978). (d) V. N. Panfilov and Yu. N. Molin, *Russian Chem. Rev.*, **47**, 503 (1978).
2. R. V. Ambartzumian, V. S. Letokhov, Z. A. Ryabov, and N. V. Chekalin, *JETP Lett.*, **20**, 279 (1974).
3. V. S. Letokhov and C. B. Moore, *Chemical and Biochemical Applications of Lasers*, Vol. III, edited by C. B. Moore, Academic Press, New York, 1977.
4. A. S. Sudbo, P. A. Schultz, E. R. Grant, Y. R. Shen, and Y. T. Lee, *J. Chem. Phys.*, **70**, 912 (1979).
5. S. Bralkowski and W. Guillory, *J. Chem. Phys.*, **66**, 2061 (1977).
6. J. H. Hall, M. L. Lesiecki, and W. Guillory, *J. Chem. Phys.*, **68**, 2247 (1978).
7. C. Reiser, F. Lussier, C. Jensen, and J. Steinfeld, *JACS*, **101**, 350 (1979).
8. J. Campbell, M. H. Yu, and C. Wittig, *Appl. Phys. Lett.*, **32**, 413 (1978).
9. L. Selwyn, R. Back, and C. Willis, *Chem. Phys.*, **32**, 323 (1978).
10. J. C. Stevenson, D. S. King, M. Goodman, and J. Stone, *J. Chem. Phys.*, **70**, 4496 (1979).
11. R. Hall and A. Kaldor, *J. Chem. Phys.*, **70**, 4027 (1979).
12. W. Danen, W. Munslow, and D. Setser, *JACS*, **99**, 6961 (1977).
13. D. M. Brenner, *Chem. Phys. Lett.*, **57**, 357 (1978).
14. A. Yogev and R. M. J. Benmer, *Chem. Phys. Lett.*, **46**, 290 (1977).
15. D. Gutman, W. Braun, and W. Tsang, *J. Chem. Phys.*, **67**, 4291 (1977).
16. J. Lyman, W. Danen, A. Nilsson, and A. Nowak, *J. Chem. Phys.*, **71**, 1206 (1979).
17. R. Woodin, D. Bomse, and J. Beauchamp, *JACS*, **100**, 3248 (1978).
18. A. Kaldor, R. Hall, D. Cox, J. Horsley, P. Rabinowitz, and G. Kramer, *JACS*, **101**, 4465 (1979).
19. For an excellent summary of the current state of the theory, see: J. Jortner, *Proc. Soc. Photo-Opt. Instrum. Eng.*, **113**, 88 (1978).
20. (a) M. Goodman, J. Stone, and E. Thiele, Multiple-photon excitation and dissociation of polyatomic molecules, *Topics in Current Physics*, edited by C. D. Cantrell, Springer-Verlag, New York, 1979. (b) Hai Lung Dai, A. H. King and C. B. Moore, *Phys. Rev. Lett.*, **43**, 761 (1979).
21. N. Bloembergen, *Opt. Comm.*, **15**, 416 (1975).
22. S. Mukamel and J. Jortner, *Chem. Phys. Lett.*, **40**, 150 (1976) and *J. Chem. Phys.*, **65**, 5204 (1976).
23. D. M. Larsen and N. Bloembergen, *Opt. Comm.*, **17**, 254 (1976).
24. E. Grant, P. Schultz, A. Sudbo, Y. Shen, and Y. Lee, *Phys. Rev. Lett.*, **40**, 115 (1978).
25. J. Horsley, J. Stone, M. Goodman, and D. Dows, *Chem. Phys. Lett.*, **66**, 461 (1979).
26. S. Haroche, Ann de Physique, **6**, 189 (1971).
27. C. Cantrell and H. Galbraith, *Opt. Comm.*, **18**, 513 (1976).
28. R. Ambartzumian, Y. Gorokhov, V. Letokhov, G. Makarov, and A. Puretzkii, *Pis'ma Zh. ETF*, **23**, 26 1976; *Opt. Comm.*, **17**, 250 1976.
29. J. G. Black, E. Yablonovitch, N. Bloembergen, and S. Mukamel, *Phys. Rev. Lett.*, **38**, 1131 (1977).

30. I. Oref and B. Rabinovitch, *Acct. Chem. Res.*, **12**, 166 1979.
31. M. J. Schultz, E. Yablonovitch, *J. Chem. Phys.*, **68**, 3007 (1977).
32. E. Thiele, M. Goodman, and J. Stone, Laser applications to chemistry, *Opt. Eng.*, **Jan/Feb** (1980).
33. M. Goodman and J. Stone, *J. Chem. Phys.*, **71**, 408 (1979).
34. H. O. Pritchard, R. G. Snowden, and A. F. Trotman-Dickenson, *Proc. Roy. Soc.* (*A*), **217**, 563 (1953); E. W. Schlag and B. S. Rabinovitch, *JACS*, **82**, 5990 (1960). (See also Ref. 41.)
35. G. Kramer, M. Dines, R. Hall, A. Kaldor, A. Jacobsen, and J. Scanlon, *Inorg. Chem.*, **19**, 1340 (1979); G. Kramer and E. Maas, National ACS Meeting, Washington, D.C. Sept. 1979.
36. R. Woodin, D. Cox, R. Hall, and A. Kaldor, to be published.
37. J. P. Day and L. M. Venanzi, *J. Chem. Soc.* (*A*), **1966**, 1363.
38. D. Cox, R. Hall, J. Horsley, G. Kramer, P. Rabinowitz, and A. Kaldor, *Science*, **205**, 390 (1979).
39. J. G. Black, P. Kolodner, M. J. Schulz, E. Yablonovitch, and N. Bloembergen, *Phys. Rev. A*, **19**, 704 (1979).
40. D. M. Cox, J. A. Horsley, *J. Chem. Phys.*, **72**, 864 (1980).
41. P. J. Robinson and K. A. Holbrook, *Unimolecular Reactions*, Wiley, New York, 1972.
42. R. B. Hall, to be published.
43. P. Kolodner, C. Winterfield, and E. Yablonovitch, *Opt. Comm.*, **20**, 119 (1977).
44. M. N. R. Ashfod, G. Hancock, and G. Ketley, Kinetics of state selected species, Faraday Disc., **67**, (April, 1979).
45. E. B. Priestley, to be published.

LUMINESCENCE OF PARENT MOLECULE INDUCED BY MULTIPHOTON INFRARED EXCITATION

AVIGDOR M. RONN

Chemistry Department, City University of New York at Brooklyn College, Brooklyn, New York, 11210

CONTENTS

I. Introduction 661
II. Experimental 666
III. Results 666
  A. The 6300 Å Region 668
  B. The 5300 Å Region 671
  C. The 4000 Å Region 672
IV. Discussion 673
Acknowledgments 676
References 676

I. INTRODUCTION

Visible or ultraviolet emissions that accompany infrared multiphoton absorption (MPA) and dissociation (MPD) of molecules have been documented in a large number of cases.[1-10] Since, for the most part, the interest in MPD centered on the subsequent chemical changes, only speculative comments regarding the origin of such luminescences surfaced in the literature. Recently, however, interest in the luminescence per se was rekindled by reports that such emissions occur with very short laser pulses and low enough pressure so as to earn the title "collisionless fluorescence."[11, 12] The basic phenomenon of visible emission induced via TEA IR laser excitation of a molecule has been explained by a number of mechanisms. First, if the fluence level of the laser was great enough to initiate optical breakdown such electronic luminescence is expected. Second, such emission can be caused by radical recombination or by subsequent chemical reaction of the fragments. Third, such emission can be due to the production of electronically excited products. Fourth, it can arise from charged species recombination processes. Fifth, and the subject of this chapter, such fluorescence can arise from the production of electronically excited parent species in the absence of collisions.

Before proceeding to describe in detail the experimental observations and their interpretation it is perhaps instructive to try and answer the rather fundamental question regarding the fifth case, namely, what observable behavior differentiates electronically excited parent molecules fluorescence from the other mechanisms of emission.

Avoiding the rather obvious regime of high-fluence and high-pressure conditions insures the absence of optical breakdown with its associated plasma-like emissions. Careful measurements of the emissions wavelengths and their dependence on laser fluence, excitation wavelength, molecular pressure, rare gas diluent pressures, and excitation pulse duration do allow for some obvious and some not so obvious distinctions between the remaining four mechanisms suggested earlier.

For example, the emission may well be documented to follow the exciting laser pulse precisely and yet no clear-cut decision can be made regarding its origin. Exactly the same problem remains when the emission is found to appear under "collisionless conditions." That is so, since the MPA is normally brought about via laser pulses of some 100–2000 nsec duration, a time sufficiently long to mask primary events from careful investigations. On the surface at least, it appears that, while a large number of measurements both physical and chemical must be carried out in order to fully document the existence of parent fluorescence, there exists one experimental observation unique to that mechanism. That observation is the comparison of the observed MPA-induced electronic spectrum with the spectrum observed via conventional excitation schemes. Once such a documentation is available no question regarding the origin of the emission remains. Thus, so as to cast the tone of this article at the outset, Fig. 1 shows the molecular luminescence spectrum of CrO_2Cl_2 subsequent to irradiation by a CO_2 TEA laser. This spectrum corresponds exactly to the emission of CrO_2Cl_2 subsequent to excitation via argon or dye lasers both in the matrix and the gas phase.[13–16] With this positive identification of the emitting species it is instructive to review previous investigations of the system so as to gain some insight into the complex nature of this seemingly simple molecule.

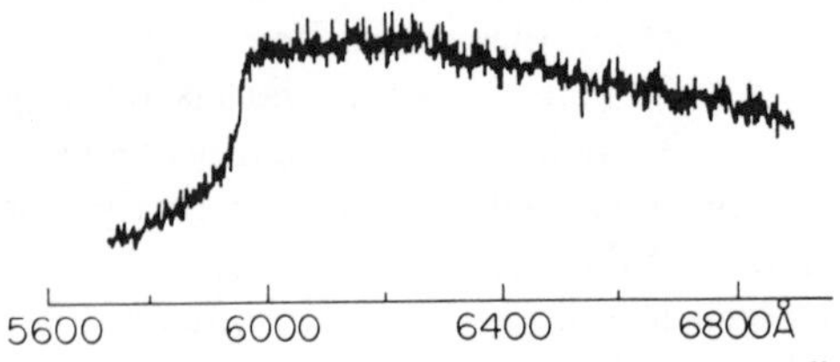

Fig. 1. Molecular luminescence spectrum of CrO_2Cl_2 following irradiation by a TEA CO_2 laser.

TABLE I
Published Vibrational Frequencies of CrO_2Cl_2 (Ground State)[a]

Symmetry (C_{2v})	Mode	v (cm^{-1})	Description
$A_1(z)$	v_1	995[b]	Cr—O stretch
	v_2	475[b]	Cr—Cl stretch
	v_3	356[b]	CrO_2 bend
	v_4	140[c]	$CrCl_2$ bend
A_2	v_5	224[c]	O_2—Cl_2 torsion
$B_1(x)$	v_6	1000[b]	Cr—O stretch
	v_7	215[c]	CrO_2 rock
$B_2(y)$	v_8	500[b]	Cr—Cl stretch
	v_9	257[c]	$CrCl_2$ rock

[a]As taken from Ref. 16.
[b]Vapor values 18.
[c]Liquid values 19, 20.

Chromyl chloride was investigated spectroscopically as early as 1933;[17] vibrational gas-phase assignments were relatively easy, since the molecule is nearly tetrahedral, yet retains the C_{2V} symmetry.[18] The vibrational bands are listed in Table I[19, 20] and shown in Fig. 2 under low resolution. The strong and only absorption near 10 μm is due to both the ν_1, (A_1) symmetric Cr—O stretch at 995 cm^{-1} and the $\nu_6(B_1)$ asymmetric Cr—O stretch at 1002 cm^{-1}. The existence of this strong absorption, the high vapor pressure of the system, and the early spectroscopic observation of resolvable vibrational (and rotational) structure in the electronic spectrum combined to make CrO_2Cl_2 a unique candidate for MPA experiments.

In recent years both the dynamics and electronic spectroscopic investigations of CrO_2Cl_2 in gas and solid phases have been extensive. Dunn and co-workers[14] have measured both the absorption and emission spectra of the solid as single crystal at 1.7 K and in argon matrices at 4 K. Their observations culminated in three observed band systems in the region 3800–6000 Å a weak system whose origin was at 5891 Å, a stronger one at 5796 Å, and a structureless system beginning at 4400 Å. The 5891 and 5796 systems were assigned as transitions to the lowest triplet and singlet, respectively. The 5796 Å singlet was further identified as that giving rise to the early gas-phase absorption study by Kronig.[17]

More recent studies by McDonald[15] concentrated on both the dynamics and spectroscopy in the gas phase. His work shows that previous claims that the molecule photodissociates at excitation wavelengths shorter than 5650 Å are not necessarily correct, since radiationless decay easily accounts for the breakoff in fluorescent intensity. McDonald shows that, even though the lowest vibrational state of the first excited singlet is free of

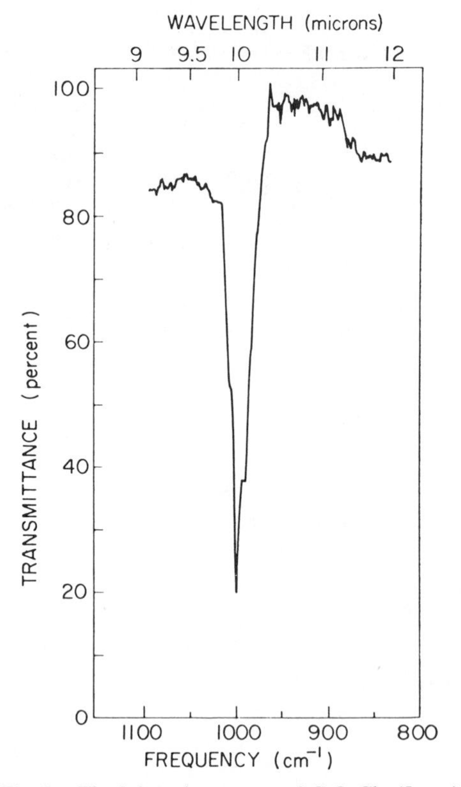

Fig. 2. **The infrared spectrum of CrO_2Cl_2 (5 torr).**

internal conversion, no clear-cut study of the radiationless decay can be performed, since sequence congestion as well as rotational and isotopic structure obscure the accessible bands.

Molecular orbital calculations[21, 22] assign the highest occupied molecular orbitals as B_1, B_2, and A_2 in symmetry (C_{2V}) all lying very close in energy. The first virtual orbital is thus A_1^*, which is strongly Cr—Cl antibonding and weakly Cr—O bonding. Thus the three expected excited electronic states are A_2, B_1, and B_2. The origin of the first excited systems is near 17,200 cm^{-1}, and the energy discrepancy between these near-lying states is thus around 200–300 cm^{-1}. An additionally interesting semiclassical calculation shows that the density of ground-state vibrational levels near the 17,200 cm^{-1} origin is approximately 3×10^6 cm^{-1}.

The energy requirements for Cr—Cl bond rupture is estimated to be 83 kcal/mole from mass spectrometric studies[23] and 68 kcal/mole from the appearance of photodecomposition studies.[24] Thus for either case the dissociation limit clearly exceeds the origin of the first excited electronic state by 19–34 kcal/mole.

It appears that with all the available spectroscopic, thermodynamic, and dynamic data presently available for CrO_2Cl_2 that a clear cut case can be visualized here for electronic excitation via MPA from a TEA CO_2 laser. Clearly, although excitation is delivered into the Cr—O stretching motions, one expects the Cr—Cl to dissociate due to its inherent low value as well as due to the nature of the $d\pi^*$ orbital of the chromium (the A_1^* at 17,200) which is strongly Cr—Cl antibonding. The large discrepancy between the energy required for bond dissociation (83 or 68) and the electronic origin (47 kcal) does, however, point strongly toward population of that state as the favored channel over the dissociative one. In terms of a MPD or MPA experiment our present concept is shown in Fig. 3 below as the potential surface along the Cr—Cl reaction coordinate.

McDonald's work showed a lifetime in the ($v=0$) of the first singlet to be 1.34 μsec (gas) with excitation at 5802 Å. If MPA "prepared" the system in the same "state" as the single photon excitation then lifetimes in the microsecond regime are to be expected. This work then was undertaken with the built-in concept of achieving electronically excited species via MPA as a probe to the chemistry and dynamics of such an event. The

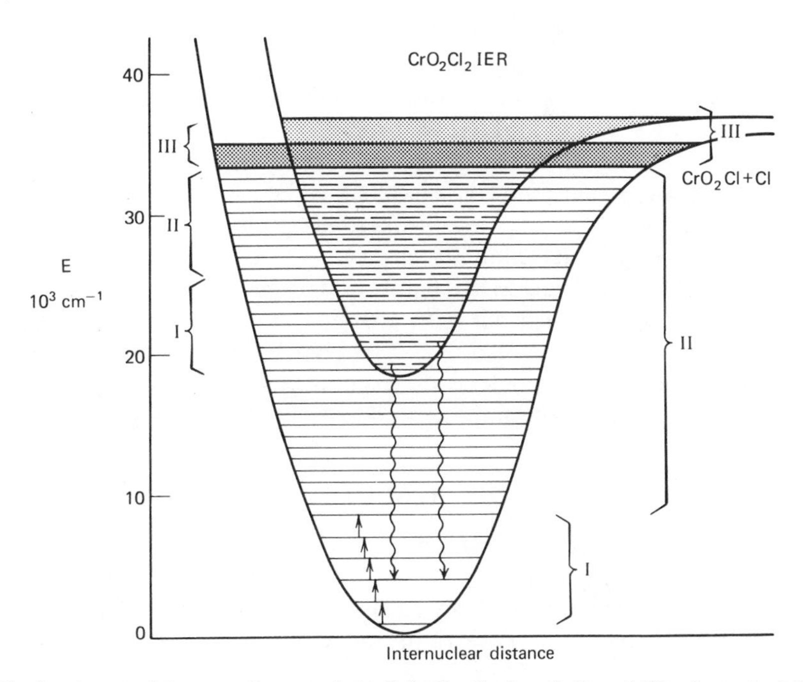

Fig. 3. A potential energy diagram of the CrO_2Cl_2. Regions I, II, and III refer to the MPD nomenclature of the discrete, quasicontinuum, and dissociative continuum regimes.

rarity of electronic excitation in infrared multiphoton absorption reflects a "general rule" that the ground state of molecules correlates adiabatically with ground-state products. However, CrO_2Cl_2 was to be the first example of a class of exceptions in which bound states are located below the ground-state dissociation limit thus allowing that excited state to be populated substantially and resulting in electronic luminescence.

II. EXPERIMENTAL

The experimental system was a conventional MPA one. The CO_2 TEA laser system was grating-tuned and provided line-selected energies of 0.1–2.5 J. Pulse shapes were selected based on N_2-rich and -lean mixtures. In N_2-rich mixtures the sharp rise of 200 nsec ended with a 2 μsec long tail, while with no N_2 a 200 nsec FWHH pulse was available. The CrO_2Cl_2 was obtained from Research Organic/Inorganic Chemical Corporation, 99.5% purity, and underwent freeze–thaw cleanout with each experiment. The cells utilized varied in size but were all Pyrex body and NaCl windows. Both static and flow cells were used and the diameter varied from 1 to 2.5 in. and lengths from 10 to 100 cm. Laser focusing was achieved by a number of lenses both AR-coated germanium of 10-in. and 20-in. focal length and ZnSe 5-in. focal length were used. The molecular emission was observed perpendicularly to the laser axis and viewed through either narrow band filters (Oriel laser filters) or a $\frac{1}{4}$ m Jarrel-Ash monochromator. Photodetection was achieved with three phototubes, an RCA C31034, a Hamamatsu R955, and an EMI 9816KB.

The phototube signal was amplified and displayed on a Tektronix 7704 oscilloscope. Laser pulses were measured with a HgTe:CdTe detector having a rise time of 5 nsec and a photondrag detector. The electronic response time of the detection system was ~5 nsec. Pressure measurements were made with an MKS Baratron and laser energy and fluence measurements were carried out with a Scientech 36-001 and a Coherent 201 power meter in conjunction with laser pulse duration measurements.

In both static and flow systems the experiments consisted of measuring the molecular emission, its time dependence, as well as its pressure and fluence dependence.

III. RESULTS

The CrO_2Cl_2 absorption spectrum in the IR shown in Fig. 2 shows a fairly wide potentially absorbing region for a tunable CO_2 laser. A mapping of such absorptions was carried out and is shown in Fig. 4. The absorption coefficient was measured for the $R(30)$ line of the 10.6 μm band

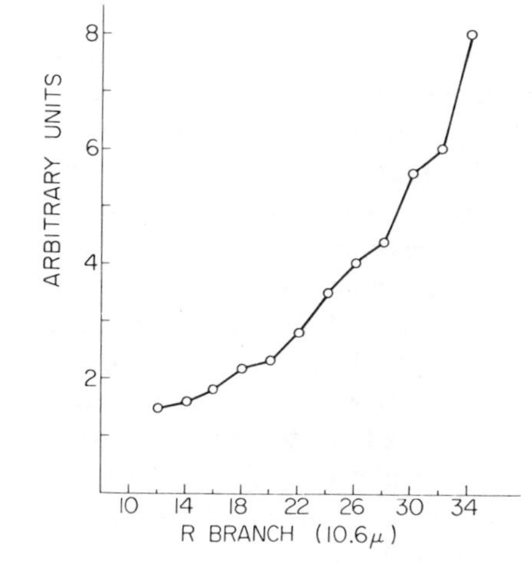

Fig. 4. The optoacoustic signal from CrO_2Cl_2 (unfocused geometry).

at three different laser fluences and yielded $\alpha = 0.0074 \text{ cm}^{-1}$. No measurable absorption was detected when the laser was tuned to the *P*-branch transitions of the 10.6 branch. The luminescence intensity as observed at different laser lines is not strongly dependent on the selected exciting line. Figure 5 presents those results. It is readily apparent that utilization of the *R*(22) or *R*(32) line results in virtually identical luminescence intensity while the absorption coefficient varies by a factor of 2. The different dependence simply reflects the fact that the threshold for luminescence is not strongly dependent on the absorption coefficient.

Our first report[25] on the luminescence from CrO_2Cl_2 stressed the fact that the emission was linearly dependent on laser fluence in the range of

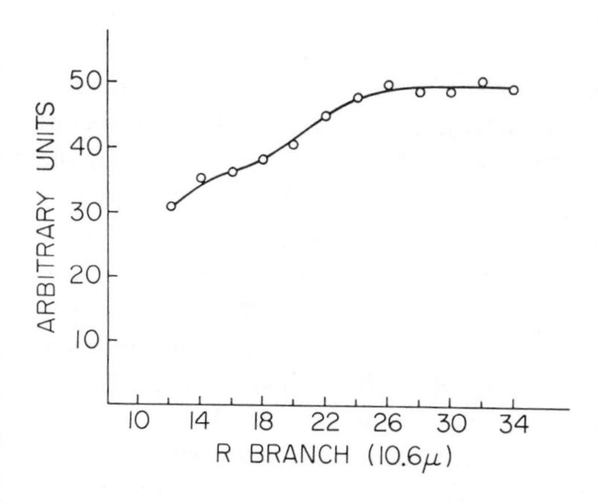

Fig. 5. Molecular luminescence spectrum vs. CO_2 *R*-branch laser lines (10.6 μ).

0.2–1.1 J/p. Beyond this range of fluence significant dissociation occurs as well as spectral broadening of the emission. It was speculated at that time that emission from $CrO_2Cl\cdot$ and CrO_2 fragments may well contribute to the parent fluorescence and broaden the spectral range to 3400 Å in the blue and 6900 Å in the red. With these early results a systematic study was instituted in which three particular wavelength regions were isolated for fluence, pressure, and time-dependent investigations. These wavelengths were 6300 Å, 5300 Å, and 4000 Å and corresponded to CrO_2Cl_2, CrO_2, and $CrO_2Cl,\cdot$ respectively. It was postulated that if different dependencies were observed for these wavelengths positive proof for at least two if not three distinct species would be available. Not only were the studies divided by wavelength regions but each study was carried out to completion with both a long and a short pulse so as to completely and unequivocally demonstrate the distinct behavior of collisional and collisionless components of the signal. Studies in terms of the wavelength regions investigated will be discussed in the following sections.

A. The 6300 Å Region

The molecular luminescence of CrO_2Cl_2 as shown in Fig. 1, as well as other work, shows a broad maximum between 6000–6300 Å. At 6000 Å significant overlap with potential emission from CrO_2 can occur, and thus the 6300 region was selected to observe only CrO_2Cl_2 fluorescence. The fluorescence was isolated either with a monochromator or with a laser line filter centered at 6328 having a spectral width of ±10 Å.

Figure 6 shows the dependence of the parent molecule fluorescence on laser fluence and Fig. 7 shows the dependence on pressure. The fluence dependence measurements were carried out with both short and long laser pulses.

Luminescence was observed, as described earlier, only in focused geometry. No parent emission was detectable with unfocused laser irradiation up to 2.5 J/p in energy. In our focused geometry the peak fluorescence intensity at 6300 Å was linearly proportional to laser energy within the fluence regime indicated above (0.1–1.5 J/p). As can be seen from Fig. 8 the quenching rate associated with the CrO_2Cl_2 fluorescence is $\sim 10^7$ sec^{-1} $torr^{-1}$. This rate is in excellent agreement with McDonald's data (1.8×10^7 sec^{-1} $torr^{-1}$). The behavior of the luminescence at higher pressure is in total accord with collisional quenching accompanied by collision-induced dissociation. In all 6300 Å data that are amplitude dependent it is clear that a maximum is reached at 2.5–3 torr and a gradual decline ensues. Thus, since fluorescence amplitude basically measures a number density of emitting species, a plateau or decline is directly interpretable in terms of

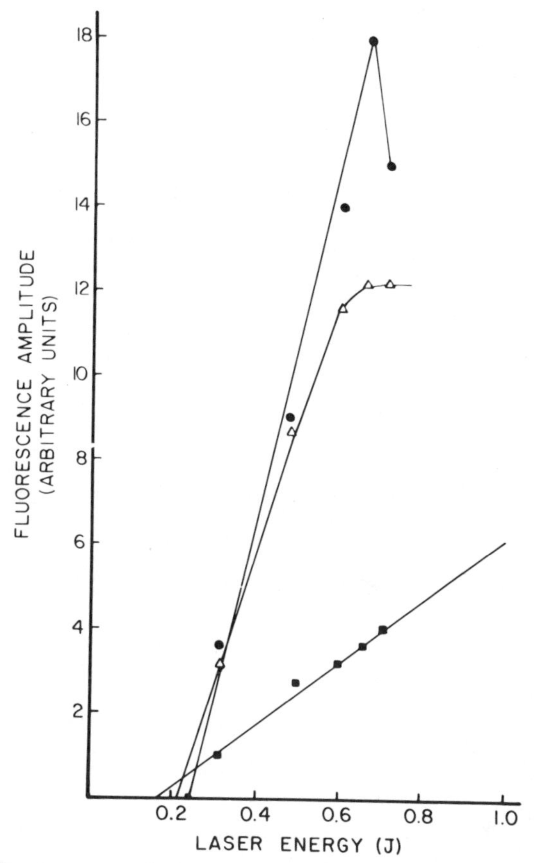

Fig. 6. Plot of fluorescence amplitude vs laser energy for three wavelength regions. (■) 6300 Å; (△) 5300 Å; (●) 4000 Å.

such species' disappearance. The rise time of the fluorescence signal does not, at any given pressure regime, deviate from the rise time of the laser pulse. No strong dependence on laser fluence is noted either, although it is conceivable that slight changes in the rise time do occur at higher fluences. If such changes occur they would be reflected within the 50-nsec uncertainty in our detection systems. The dependence on fluence is linear as can be seen from Fig. 6. This measurement has been verified throughout the pressure regime 10^{-4}–1 torr. The intercept, clearly observable for a number of such fluence measurements yields a minimum threshold for fluorescence observation at 6300 Å. This threshold with the given absorption coefficient yields 19–22 photons per molecule absorbed on the average

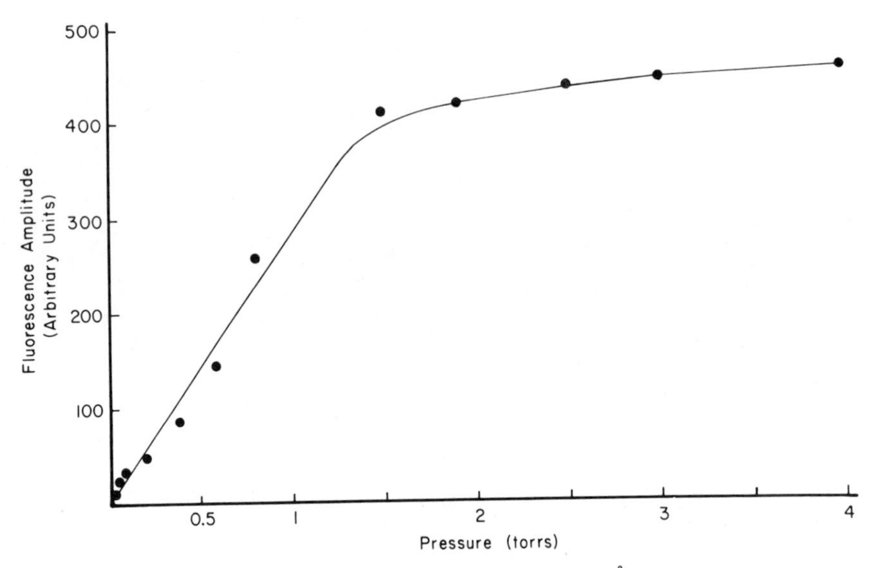

Fig. 7. Fluorescence amplitude of parent molecule at 6300 Å plotted vs. pressure.

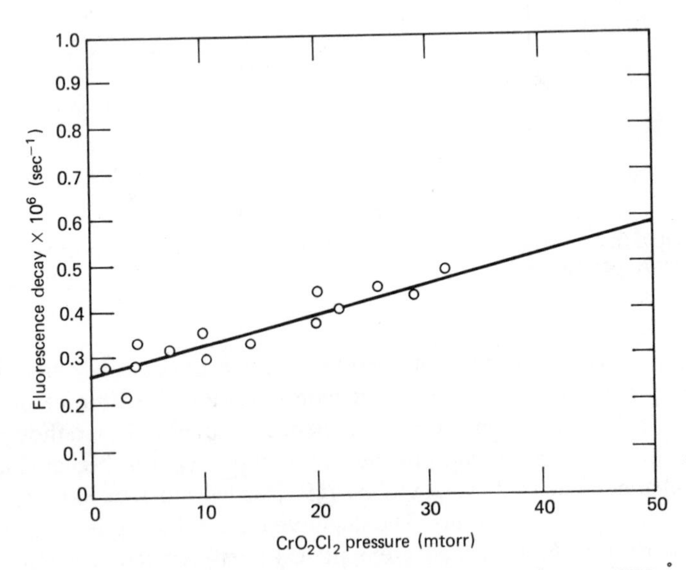

Fig. 8. Stern–Volmer plot for the CrO_2Cl_2 fluorescence decay at 6300 Å.

($\langle n \rangle$) in remarkable agreement with the known origin of the first excited electronic state.

The low-pressure decay time extrapolated to zero pressure yields a collisionless decay time of 3.85 ± 0.5 μsec. This value is significantly longer than that reported by McDonald (1.34 μsec). While this is of some concern when trying to assign the emission to the parent species, it is rather difficult to postulate a collisional mechanism or an electronically excited product as being responsible for the longer lifetime observed. It is worthy to note now, and I shall return to this point later, that the quenching rate of the 6300 Å emission with helium, argon, and chlorine is basically the same. If the emission was due to a product, the most obvious being $CrO_2Cl\cdot$, one would expect very efficient quenching by chlorine. This was indeed observed for the 4000 Å band, as will be shown later, but not for the 6300 Å emission.

B. The 5300 Å Region

As was discussed earlier in the introduction, prior gas-phase work showed negligible fluorescence blue of 5650 Å. More recent work by Levy et al.[26] extended this region to 5550 Å and assign such emission to more highly excited members of a vibrational progression. Their success apparently depended on the method of excitation namely that in their free jet expansion a relatively collisionless jet, vibrationally and rotationally cold CrO_2Cl_2 molecules were available for dye laser excitation. In our excitation scheme, a room temperature Boltzmann distribution prevails, and yet significant emission is observed to the blue of 5650 Å at higher excitation fluences. The spectrum shown in Fig. 1 indeed tails off at around 5600 Å. However, at laser energies exceeding 0.5 J, and with tight focusing a significant spectral broadening occurs through 3400 Å.

The fluence dependence of the emission at 5300 Å is shown in Fig. 6 below. The minimum threshold for observation of emission at 5300 Å is approximately 0.3 J/p. At 1–1.5 J/p strong emission is observed, which is characterized by a totally different slope than the emission at 6300 Å. In addition, the emission reaches a peak amplitude at 1.0 J/p and begins a sharp decline at higher energies. The slope is approximately 4 times steeper than the 6300 Å emission and the 1.5 power dependence on laser energy is clear.

At low pressures 10^{-4}–10^{-2} torr this emission is significantly lower in amplitude than the parent emission at 6300 Å. It is barely discernible at 10^{-4} torr and 0.1 J/p short pulse (200 FWHH). The rise time of the signal is basically identical to that observed at 6300, that is, follows the laser's rise time (Fig. 9). The fluorescence quenching of this signal was measured as $9 \pm 1 \times 10^6$ sec^{-1} torr^{-1} and its collision-free lifetime as obtained from these studies was 4.5 ± 0.5 μsec.

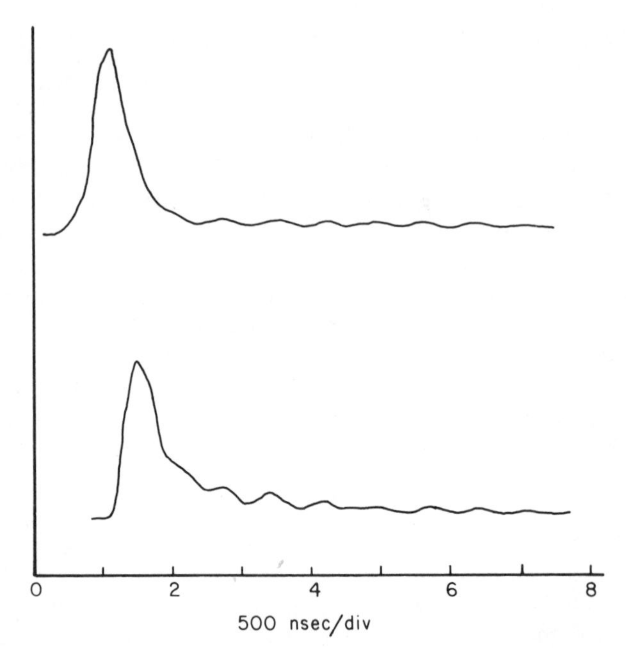

Fig. 9. Scope photographs of the laser pulse (top) and the fluorescence emission at 6300 Å (bottom). CrO_2Cl_2 pressure is 0.400 torr.

To check the validity of the a priori assumption, that is, that the 5300 Å is due to emission from CrO_2, studies of the quenching rate as a function of chlorine were performed. Both amplitude and lifetimes were severely affected by chlorine, by comparison with the same study at 6300 Å. The quenching rate was estimated to be in excess of an order of magnitude higher than for CrO_2Cl_2 namely $\sim 10^8$ sec^{-1} torr^{-1}. The intensity of the emission was also severely reduced, much more so than that of the 6300 Å emission.

C. The 4000 Å Region

No previous investigation has shown significant parent fluorescence at wavelength shorter than 5550 Å.[26] It is thus a priori attractive to postulate that whatever emission is observed at 4000 Å is undoubtedly due to some product. In the early work of Halonbrenner the spectra of $CrO_2Cl\cdot$ was tentatively assigned to this wavelength regime. Based both on such an assignment and on the rather clear-cut MPD work of others as well as our own it is attractive to postulate $CrO_2Cl\cdot$ as the primary dissociation product of CrO_2Cl_2 and the species of interest in this emission. In this context it was anticipated that chlorine quenching would again be most effective.

Basically parallel observations to those at 5300 Å were noted. No observation of the emission was noted at laser energies lower than 0.1 J/p with 10^{-4}–10^{-2} pressure. Very strong fluence dependence, rising sharply and turning over at fluences exceeding 1 J/p were recorded and shown in Fig. 6.

No particularly significant differences in rise time were noted in line with nanosecond or faster dissociation rates. The decay rate of the fluorescence at both 5300 Å and 4000 Å was studied as a function of pressure. The slope of rates vs. pressure yielded a collisional quenching rate of $1.0 \pm 0.2 \times 10^{-7}$ sec^{-1} $torr^{-1}$ ($\sigma = 220$ (Å)^2) and an intercept of 3.5 ± 0.5 μsec as the collisionless lifetime. While these numbers are close to the values observed for the 6300 Å, chlorine deactivation data and the fluence behaviors clearly indicate that both the 5300 Å and 4000 Å signals do not originate from parent molecules.

These results summarize research that was completed before the Ein-Bokek conference (December 1978). At that time, fluorescence lifetime work concentrated on the pressure regime of 5–100 mtorr. Consequently the collision-free lifetimes were obtained from the intercepts of Stern–Volmer plots based on higher pressures than those pertaining to truly collisionless conditions. The very large cross-section exhibited by CrO_2Cl_2 coupled with our laser pulse duration demands a pressure regime of less than 1 mtorr for truly collisionless conditions. This research is still being conducted, and all the data are not yet reduced. However, there is a clear indication that at pressures of 5 mtorr and below an intercept in excess of 10 μsec and a quenching rate of at least 6×10^7 sec^{-1} $torr^{-1}$ prevail for the 6300 Å emission. The lifetime data for the 5300 Å and 4000 Å also show significant lengthening vis-à-vis our earlier results. A complete account of this more recent work will be published elsewhere shortly.

IV. DISCUSSION

Within the context of the introduction it is implied that one can divide the study of MPD and MPA in CrO_2Cl_2 into two regimes. The first, dealing strictly with the collisionless production of electronically excited parent species in the first excited state, is the lower fluence regime demanding only an 18 photon absorption per molecule. The second and higher fluence regime is one in which a 25 and higher photon absorption leads to MPD and the creation of $CrO_2Cl\cdot$ and ultimately CrO_2 and Cl_2.

It is tempting to deal with each regime separately, yet it is quite difficult to ascertain that crossover from region to region does not occur even at intermediate conditions. An added complexity to this problem is the fact that the origin of the second excited state has been assigned to lie some 270 cm^{-1} above that of the first excited state. The density of ground-state vibrational levels at the origin of the 1st excited state is $3\times10^6/cm^{-1}$ and at the 2nd state's origin $6.5\times10^6/cm^{-1}$.[27] At an energy of 1000 cm^{-1} (one CO_2 photon) above the origin of the 2nd excited state the density of the 1st excited state vibrational levels is on the order of $1/cm^{-1}$. While by comparison this density is quite low, reasonably efficient coupling can occur even in this region. It is thus clear that MPA to levels somewhat higher than the 1st state's origin can not only result in nonclassical coupling between excited state levels and high-lying ground-state levels but also to coupling between the 1st and 2nd states' vibronic levels.

If the mechanism by which MPA collisionlessly populates the 1st excited electronic state is identical to that encountered by a single photon fluorescence excitation spectrum, then the dynamics of both should parallel. In our work, however, the measured collision-free lifetime is 3.85 ± 0.5 μsec as compared to McDonald's number of 1.34 μsec. The bimolecular quenching rates are, however, remarkedly similar both being nearly $\sim10^7$ $torr^{-1}$ sec^{-1}.

From Fig. 3 one can easily visualize that the intermediate level structure in the ground-state quasicontinuum is communicating with the rather sparse level structure of the 1st excited state (at low fluences). Initially our interpretation of the electronic emission invoked active infrared transitions between high-lying vibrational states of the ground-state and low-lying vibronic levels of the 1st state. This notion is incomplete, since it only probes interstate coupling. Intrastate radiative coupling within the high vibrational levels of the ground state should also be considered for a complete description of the parent molecule's fluorescence emission. The concept and consequence of intrastate radiative coupling bear close analogy to intramolecular *V–E* processes. This concept has been developed fully elsewhere and was shown to embody a complete treatment of the phenomenon renamed inverse electronic relaxation (IER).[28]

Basically it is shown that it is the ratio-of-state densities that closely controls the IER rate. The process involves the one-photon spontaneous radiative decay of the electronically excited molecular eigenstates reached by nonadiabatic coupling with the ground-state quasicontinuum. The longer collision-free lifetime observed in this work vis-à-vis the one-photon excitation may just be a demonstration of the "time sharing" between states encountered when a sparse excited state manifold interacts with the high density of ground vibrational levels. If one also considers excitation into the 2nd excited state as a possible dilution factor then long lifetimes, as have been measured in our case, are even more compatible with ratio-of-state densities.

The time evolution of the luminescence signal, its pressure dependence and fluence dependence and particularly its spectral characteristics clearly demonstrate its origin as being parent fluorescence. The dynamics of each of the three separate spectral regions strongly support a MPA and MPD scheme as is shown here:

$$\begin{array}{ccc} CrO_2Cl_2 & \xrightarrow{nh\nu} & CrO_2Cl_2^* \\ & \searrow & \downarrow mh\nu \\ & & CrO_2Cl\cdot + Cl \\ & & \downarrow kh\nu \\ & & CrO_2 + Cl \end{array}$$

In sequential order then it is the formation of CrO_2Cl_2 in the 1st excited state that is obtained collisionlessly on absorption of some 18 IR photons. Photofragmentation of CrO_2Cl_2 can and does occur simultaneously on absorption of some 24 IR photons, resulting in highly excited $CrO_2Cl\cdot$. In direct agreement with the now celebrated SF_6 case, the photofragment $CrO_2Cl\cdot$ absorbs an additional number of photons, thus acquiring sufficient total energy to either emit or photofragment further to CrO_2. The latter represents a stable thermodynamic entity and signals the end of the process.

No observable delay in the onset of the 4000 Å emission due to $CrO_2Cl\cdot$ is detectable with respect to the onset of the parent emission at 6300 Å. Again by analogy to a dozen other MPD experiments it is expected that photofragmentation will occur within the pico- or nanosecond range. The dependence on fluence and quenching partner (CrO_2Cl_2, Cl_2, He) of the 5300 and 4000 spectral regimes is distinctly different than that of the 6300 Å region, thus characterizing parent and products as different chemical species.

In particular it is worth renoting that no earlier investigation of the parent's emission has ever recorded a spectral response shorter than 5550 Å.[26] Thus such emission must originate in CrO_2Cl_2 photofragments. The clear observation of Halonbrenner that CrO_2Cl_2 decomposed on excitation below 4200 Å via initial Cl atom formation and consequent chain reaction culminating in Cl_2 and CrO_2 as final products further support this development.

Since this electronic luminescence represents the first case of direct multiphoton infrared pumping into a species' electronic manifold, the burden of proof is heavier than in other more acceptable MPD experiments. In the next paragraph or so I should like to summarize the experimental observations that appear to us to be firmly convincing of this phenomenon.

At pressures of 10^{-3} torr and lower the fluorescence spectra is a replica of the well-assigned spectrum of CrO_2Cl_2. At these same pressures little or no fluorescence is observed to the blue of 5550 Å. The peak fluorescence intensity at 6300 Å is linearly dependent on pressure at low to intermediate pressures. The lifetime dependence on pressure is linear and shows a distinct increase at lower pressure. The quenching cross-section is very high in both the neat gas and with helium or chlorine as collision partners.

The behavior of the 5300 and 4000 Å fluorescent signals with both fluence and pressure changes is quite distinct from the 6300 Å behavior. The fluence dependence in particular is convincingly steeper and saturates more quickly.[29] No 5300 or 4000 Å signal can be observed at very low laser fluences for which the 6300 Å fluorescence is clearly measured. It is on these grounds as well as the known spectra that the shorter wavelength fluorescence emission are ascribed to $CrO_2Cl\cdot$ and CrO_2.

On theoretical grounds it appears that the concept of inverse electronic relaxation is directly applicable to the observations contained here. CrO_2Cl_2 is not a large polyatomic but does have sufficient level density in the first excited (and perhaps the second) state to allow for microsecond IER rates.[28] The observed collision-free decay times certainly reflect the fact that a MPA mechanism does not parallel that observed in conventional gas-phase single photon excitation.

On theoretical grounds there is no uniqueness to the CrO_2Cl_2 species. Any polyatomic species possessing a low-lying electronic state, a reasonable absorption cross-section to a CO_2 TEA laser and a ground-state level density of 10^{4-8} cm^{-1} can be similarly excited and show IER decay. Work in progress at this laboratory is directed toward this goal.

Acknowledgments

I would like to express my thanks to Mr. J. Nieman, Dr. Y. Langsam and Mr. A. Schwebel for the experimental work and the inumerable group discussions. I wish to thank the Research Foundation of the City University of New York and the FRAP for partial support of this work.

Special thanks also go to Prof. Joshua Jortner for the invitation to participate in a most exciting conference on Laser Chemistry at Ein-Bokek, Israel.

References

1. C. Borde, A. Henry, and L. Henry, *Compt. Rend. Acad. Sci. (Paris)*, **B 262**, 1389 (1966).
2. V. V. Losev, V. P. Papulovski, V. P. Tischinski, and G. A. Fedina, *High Energy Chem.*, **8**, 331 (1969).

3. N. V. Karlov, Yu. N. Petrov, A. M. Prokhorov, and O. M. Stel'makh, *JETP Lett.*, **11**, 135 (1970).
4. N. R. Isenor and M. C. Richardson, *Appl. Phys. Lett.*, **18**, 224 (1971).
5. V. S. Letokhov, E. A. Ryabov, and O. A. Tumanov, *Opt. Commun.*, **5**, 168 (1972).
6. N. R. Isenor, V. Merchant, R. S. Hallsworth, and M. C. Richardson, *Can. J. Phys.*, **51**, 1281 (1973).
7. R. V. Ambartzumian, N. V. Chekalin, V. S. Doljikov, V. S. Letokhov, and E. A. Ryabov, *Chem. Phys. Lett.*, **25**, 515 (1974).
8. R. V. Ambartzumian, N. V. Chekalin, V. S. Doljikov, V. S. Letokhov, and V. N. Lokhman, *J. Photochem.*, **6**, 55 (1976).
9. R. V. Ambartzumian, Yu. A. Gorokhov, G. N. Makarov, A. A. Puretski, and N. P. Furzikov, *Chem. Phys. Lett.*, **45**, 231 (1977).
10. M. L. Lesiecki and W. A. Guillory, *J. Chem. Phys.*, **66**, 4317 (1977).
11. Y. Haas and G. Yahav, *Chem. Phys. Lett.*, **48**, 63 (1977).
12. Y. Haas and G. Yahav, *Chem. Phys.*, **35**, 41 (1978).
13. M. Spoliti, J. H. Thirtle, and T. M. Dunn, *J. Mol. Spectrosc.*, **52**, 146 (1974).
14. T. M. Dunn and A. H. Francis, *J. Mol. Spectrosc.*, **25**, 86 (1968).
15. J. R. McDonald, *Chem. Phys.*, **9**, 423 (1975).
16. R. N. Dixon and C. R. Webster, *J. Mol. Spectrosc.*, **62**, 271 (1976).
17. R. DeKronig, A. Schaafsma, and P. K. Peerlkamp, *Z. Physik. Chem.*, **22**, 323 (1933).
18. W. E. Hobbs, *J. Chem. Phys.*, **28**, 1220 (1958).
19. H. Stammreich, K. Kawai, and Y. Tavares, *Spectrochim. Acta*, **9**, 738 (1959).
20. F. A. Miller, G. L. Carlson, and W. B. White, *Spectrochim. Acta*, **9**, 709 (1959).
21. J. P. Jasinski, S. L. Holt, J. H. Wood, and L. B. Asprey, *J. Chem. Phys.*, **63**, 757 (1975).
22. T. H. Lee and J. W. Rabalais, *Chem. Phys. Lett.*, **34**, 135 (1975).
23. G. D. Flesch, R. M. White and H. J. Svec, *Int. J. Mass Spectrom. Ion Phys.*, **3**, 339 (1969).
24. R. Halonbrenner, J. R. Huber, U. Wild, and H. H. Gunthard, *J. Phys. Chem.*, **72**, 3929 (1968).
25. Z. Karny, A. Gupta, R. N. Zare, S. T. Lin, J. Nieman, and A. M. Ronn, *Chem. Phys.*, **37**, 15 (1979).
26. Joseph A. Blazy and Donald H. Levy *J. Chem. Phys.*, **69**, 2901 (1978).
27. P. C. Haarhoff, *Mol. Phys.*, **7**, 101 (1963).
28. Abraham Nitzan and Joshua Jortner a) *Chem. Phys. Lett.*, **60**, 1 (1979). b) *J. Chem. Phys.*, **71**, 3524 (1979).
29. K. M. Leary, J. L. Lyman, L. B. Asprey, and S. M. Freund, *J. Chem. Phys.*, **68**, 1671 (1978).

ELECTRONIC LUMINESCENCE RESULTING FROM INFRARED MULTIPLE PHOTON EXCITATION*

HANNA REISLER AND CURT WITTIG

Departments of Electrical Engineering, Physics, and Chemistry, University of Southern California, University Park, Los Angeles, California 90007

CONTENTS

I. Introduction 679
II. Collisionless and Collisional Production of Electronic Luminescence via Multiple Photon Excitation 681
  A. Survey of Experimental Results 681
  B. Kinetics 687
    1. Collisionless Mechanisms 687
    2. Collisional Mechanisms 689
  C. Mechanisms 692
III. Chemiluminescent Reactions Initiated via IR Multiple Photon Dissociation 694
  A. $C_2(a^3\Pi_u)+NO$ 695
  B. $C_2(X^1\Sigma_g^+, a^3\Pi_u)+O_2$ 701
  C. Other Reactions 705
IV. Concluding Remarks 707
References 710

I. INTRODUCTION

The electronic luminescence that often accompanies IR multiple photon excitation (MPE) of species in the gas phase was one of the first aspects of IR MPE to be reported.[1-3] The appearance of "prompt" luminescence from electronically excited fragments such as C_2^*, CH*, etc., concomitant with the IR multiple photon dissociation (MPD) of hydrocarbons, led initially to the conclusion that these electronically excited species were formed in the absence of collisions. Had this been the case, it would have

*Research supported by the USC Joint Services Electronics Program and the Air Force Office of Scientific Research.

suggested that reactions may occur that are different from those corresponding to the lowest thermally accessible channels. These early results led to speculation concerning a molecule's ability to accumulate a considerable amount of energy above the dissociation threshold prior to dissociation, and nonrandomization of the vibrational energy of the molecule. Subsequently it was established, using molecular beam techniques, that the products of IR MPD can be accounted for by rather conventional theories. Thus, the interpretation of the observed emissions in terms compatible with our present understanding of MPE is of major importance, and the question of whether or not these observed emissions are indeed produced under collisionless conditions has become the subject of considerable debate. It is evident that in order to offer a satisfactory explanation for the experimentally observed electronically excited species produced via MPE, it is imperative first to distinguish unambiguously between collisionless and collisional production of the luminescence.

In this chapter we describe the different types of electronic emissions observed in IR MPE and the diagnostic techniques that can be employed to sort out the mechanisms responsible for the production of the emitting species. Roughly, we can divide the observed emissions into three categories:

1. Collisionless production of luminescence from parent molecules or from dissociation products.
2. Production of luminescence via collisions between the offsprings of the parent molecules (e.g., vibrationally excited parent molecules, free radical photofragments, etc.).
3. Production of chemiluminescence via reactions of photofragments, produced via MPD, with unexcited parent molecules or with selected added molecules.

The first two processes are discussed in Section II, with special emphasis on the distinction between collisional and collisionless processes, and on the mechanisms of the production of luminescence. In Section III, we describe several chemiluminescent reactions of radicals which are produced by IR MPD. In view of the lack of any "state selective chemistry" occurring in the photodissociation process, this aspect of IR MPD may prove to be one of the most useful. It provides a convenient means for studying reactions of radicals which heretofore have been very difficult to generate under clean and well characterized conditions. Finally, in Section IV, we summarize the paper and we indicate what we feel are fruitful areas for future endeavor.

II. COLLISIONLESS AND COLLISIONAL PRODUCTION OF ELECTRONIC LUMINESCENCE VIA MULTIPLE PHOTON EXCITATION

A. Survey of Experimental Results

The first reports of visible luminescence, concomitant with pulsed CO_2 laser MPE, at power levels insufficient to produce optical breakdown, were published as early as 1971–72.[4, 5] The characteristic features of the MPE induced luminescence of molecular gases have, since then, been investigated in a variety of systems.

Visible luminescence has been observed concomitant with the MPD of BCl_3 with the output from a pulsed CO_2 laser.[6] The authors report "instantaneous" luminescence at pressures as low as 50 mtorr and laser intensities $\geqslant 2\times 10^8$ W cm^{-2}, with a rise time, which, at low pressures, is independent of BCl_3 pressure.[7] A delayed component of the luminescence is evident at powers $\geqslant 10^7$ W cm^{-2}. Both the delay and the rise time of the delayed feature, which appears only when the tail in the CO_2 laser pulse is present, are pressure dependent. In a later study,[8] it was found that the dissociation is accompanied by an unidentified "instantaneous" continuum emission in the region 440–600 nm, and by delayed intense emission bands of the $A^1\Pi \rightarrow X^1\Sigma^+$ system of BCl. The peak intensity of the BCl emission showed a quadratic dependence on BCl_3 pressure and a linear dependence on laser power. It is concluded that excited BCl is formed via collisions. The pressure dependences of the rise and decay portions of the luminescence signals suggest that consecutive kinetic events are responsible for the emitting species. The mechanisms whereby the "prompt" luminescence is produced were not investigated in detail.

Luminescence has also been reported in the region 350–400 nm concomitant with the IR MPD of OsO_4 at pressures $\geqslant 0.1$ torr and laser intensities $\geqslant 5\times 10^6$ W cm^{-2}.[9] The authors note that the OsO_4 luminescence, I_{OsO_4}, obeys the relation

$$I_{OsO_4} \propto I_{laser}^7 \qquad \text{for } I_{laser} = (5-20)10^6 \text{ W cm}^{-2}$$

$$I_{OsO_4} \propto I_{laser}^4 \qquad \text{for } I_{laser} = (20-40)10^6 \text{ W cm}^{-2}$$

where I_{laser} is the intensity of the CO_2 laser. At an OsO_4 pressure of 150 mtorr, the rise time of the fluorescence is longer than the CO_2 laser pulse, and the luminescence in this pressure region is attributed to collisional processes.

Luminescence from diatomic species such as C_2^*, CH^*, OH^*, CCl^*, and CN^* was first observed upon the IR MPD of molecules such as C_2H_4,

C_2F_3Cl, CH_3OH, CH_3NO_2, CH_3CN, and CF_2Cl_2 at pressures $\geqslant 1$ torr. The detection system was set to accept that portion of the luminescence which is produced $\leqslant 150$ nsec after the onset of the CO_2 laser.[10] As no details of the time or pressure dependence of the luminescence were reported in this early work, no definite conclusions regarding the origin of these emissions could be reached. Subsequently, the luminescence from electronically excited C_2^* and CH* radicals produced in the MPD of $C_2H_2Cl_2$ was studied in detail.[11] Both the rise and decay of the time-resolved luminescence are pressure dependent and its peak intensity depends quadratically on pressure in the range 0.2–2 torr (Fig. 1). Thus, collisional processes are responsible for these emissions. In the MPD of CH_3CN[12] and CH_3OH[13], emissions from CN* and C_2^* in the case of the former precursor, and from CH*, OH*, and C_2^* in the case of the latter

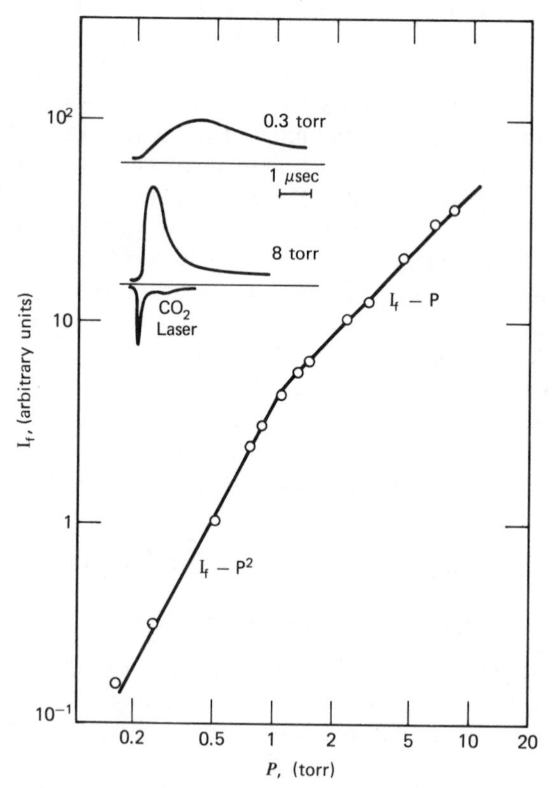

Fig. 1. Dependence of luminescence intensity on $C_2H_2Cl_2$ pressure ($\nu_{laser} = 935$ cm^{-1}, energy density at focus $= 800$ J cm^{-2}) and shape of luminescence pulses at different pressures of $C_2H_2Cl_2$.

precursor have been observed in the pressure range 0.1–5 torr. Based on the pressure-dependent rise and decay rates and on the quadratic pressure dependence of the peak luminescence intensity (Fig. 2), it was concluded that collisional processes are important in the production of electronically excited diatomics in both systems.

Luminescence concomitant with the MPD of SiF_4 in the presence of the focused output from a CO_2 TEA laser is also well documented.[4, 14, 15] The time evolution of the visible luminescence has been studied in the range 0.12–13 torr,[15] and it has been suggested that the luminescence in this pressure range is a consequence of collisional processes.

Recently, several publications have appeared in which the collisionless emission that accompanies MPE was studied in some detail. MPD of tetramethyldioxetane (TMD) produces intense blue emission.[16, 17] The prompt appearance and short lifetime of the luminescence, which prevail at pressures as low as 70 mtorr, suggest collisionless production. A delayed luminescence feature, persisting several microseconds after the termination of the laser pulse, has been explained in terms of a collisional process. The prompt luminescence follows closely the laser profile, but its onset is delayed with respect to the beginning of the laser pulse. The delay is a function of laser energy. This very detailed study is reported elsewhere in this volume and will not be discussed further here.

Electronic emission was observed in the MPE of CrO_2Cl_2 in the pressure range 5×10^{-4}–15 torr.[18] The authors ascribe the emission to electronically excited parent molecules and note that at high pressures both collisionless and collisional components appear in the time-resolved fluorescence. This

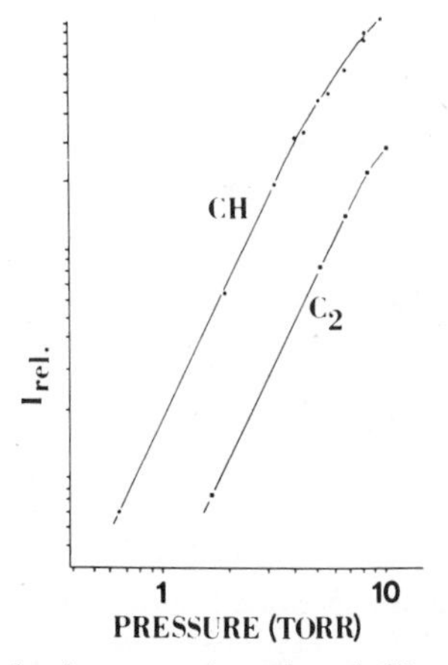

Fig. 2. Plot of the relative luminescence intensity of C_2^* and CH* vs. CH_3OH pressure (log–log scale). The luminescence was observed 0.3 μsec after the laser pulse.

system, too, is described elsewhere in this book and will not be treated in detail here.

Luminescence has been observed upon irradiation of vinyl cyanide (C_2H_3CN) with the focused output from a CO_2 TEA laser in the pressure range 10^{-5}–10^{-1} torr.[19] This luminescence shows two broad structureless features, at 390 nm and in the IR (Fig. 3). The pressure dependence of the amplitude of the luminescence for these two spectral features is linear in the pressure range 0.5–10 mtorr (Fig. 4). At higher pressures, the pressure dependence is higher than linear, indicating a collisional contribution. This work clearly demonstrates that in order to identify a collisionless process it is imperative to carry the experiments to very low pressures and to determine the pressure dependence of the luminescence peak intensity, since collisional channels are detected in this system at pressures as low as 15 mtorr. Inspection of the time resolved luminescence at low pressures (< 10 mtorr) reveals a fast rise, which is never longer than the CO_2 laser pulse duration and a much slower decay (Fig. 5). The emitting species has a long collisionless lifetime (18–28 μsec) but its luminescence is quenched very efficiently both by vinyl cyanide ($k=(4.5\pm0.5)10^6$ sec^{-1} torr^{-1}) and by Ar ($k=(1.8\pm0.2)10^6$ sec^{-1} torr^{-1}). The identity of the emitting species is not yet clear. Simple diatomics can be ruled out on the basis of their known emission features. The lack of structure in the emission spectra suggests that a molecular fragment or fragments larger than diatomic are involved. C_2H and C_2CN are but two of the possibilities, and further work

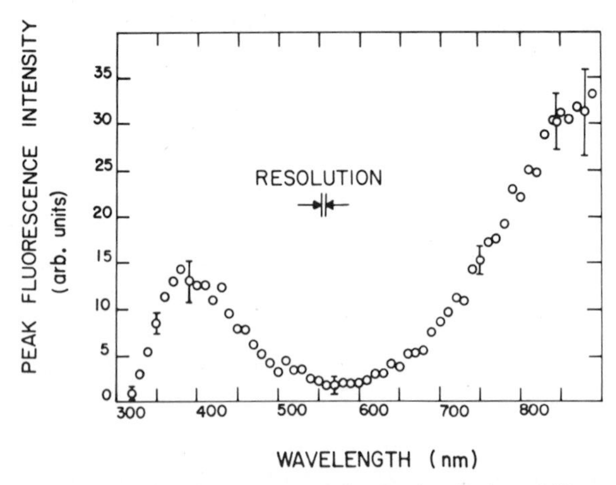

Fig. 3. Peak luminescence intensity spectrum following irradiation of 19 mtorr of C_2H_3CN with the focused output from a CO_2 TEA laser. The resolution is 4 nm, and each point represents the sum of 8 shots. The spectrum is corrected for the wavelength response of the detection system. Error estimates are given for the different parts of the spectrum.

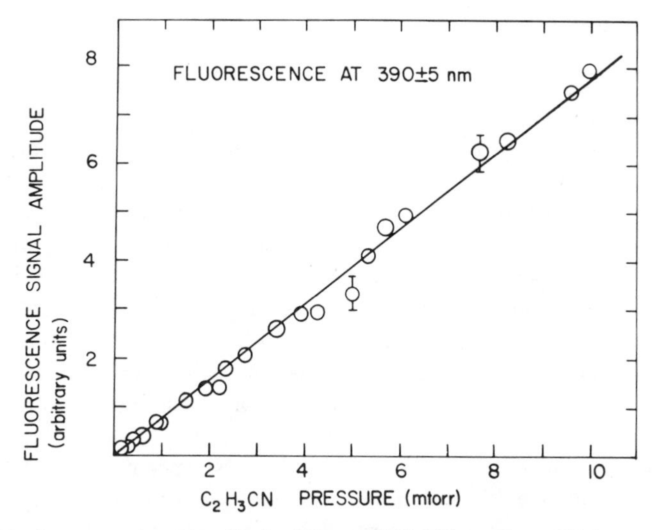

Fig. 4. Luminescence signal amplitude, $I(0)$, vs. C_2H_3CN pressure. $I(0)$ is proportional to the concentration of the emitting species produced by the CO_2 laser pulse. Each point represents an average of 8 to 64 shots. Typical uncertainties are indicated by the error bars.

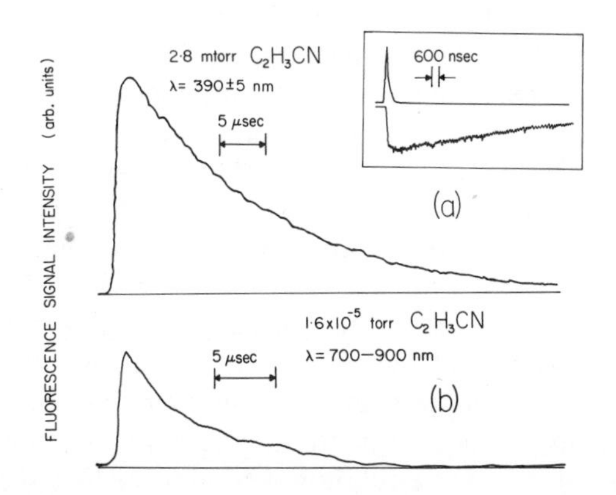

Fig. 5. (a) Typical signal averaged near uv luminescence showing the decay of the emission when 2.8 mtorr of C_2H_3CN is irradiated with the focused output from a CO_2 TEA laser. The inset shows the rise of the fluorescence (lower trace) and the CO_2 laser pulse shape (upper trace). (b) Typical signal averaged ir (700–900 nm) luminescence showing the decay of the emission when 1.6×10^{-5} torr C_2H_3CN is irradiated with the focused output from the TEA laser. Here, the pressure is measured with an ion gauge that has been calibrated against a capacitance manometer at 10^{-4} torr. The short decay time in (b) is due to the small area of the GaAs PMT which was used. When an RCA C31000 PMT was used, comparable decay times were observed for the near UV and IR spectral regions.

is currently being carried out to elucidate the nature of the emitting species.

It now appears that collisionless emission accompanies the MPD of many other hydrocarbons as well. Although no emission from other olefins was detected in the work described above,[19] this was simply a consequence of smaller absorption cross-sections at the CO_2 laser frequency. When improving the sensitivity of the detection system, collisionless emission, concomitant with the MPE of such molecules as ethylene, propylene, vinyl chloride, and propenal, was detected at pressures as low as 0.5 mtorr.[20] With all of these molecules, as well as vinyl cyanide, the ratio of the number of C_2 radicals produced (measured by laser-induced fluorescence) to the number of species producing collisionless luminescence is comparable. This suggests that the small absorption cross-section of the other olefins, as compared to vinyl cyanide, is the reason for the less efficient production of collisionless luminescence in these species. In each case, the pressure dependence of the luminescence has been measured and is linear up to 10 mtorr. There are several common features of the luminescent species produced in these photolysis experiments. They all have rather long collisionless lifetimes (5–20 μsec), they are quenched efficiently even by inert gases such as Ar, and their spectra are always broad and structureless and cannot be assigned to any known diatomic. The spectra and lifetime, however, differ from one case to the next.

Electronic luminescence concomitant with the MPD of propenal has also been observed recently in the pressure range 0.5–5 torr.[21] A set of interference filters was used to establish that the emission bands extend throughout the UV, visible, and near-IR as in Ref. 20. The luminescence is thought to originate from a dissociation product. The luminescence decays with a rate that is a linear function of propenal pressure. The slope of the line is independent of the CO_2 laser intensity. The intercept, however, is proportional to the square of the laser fluence over the range $(1.6–3.4)10^9$ W cm^{-2}. The authors suggest that the pressure-independent intercept reflects the intramolecular energy transfer rate in the isolated molecule prior to dissociation. We point out, however, that the lowest pressures used by the authors were 500 mtorr, whereas we have observed collisional effects at pressures >10 mtorr. Thus, the extrapolation from the torr region to the collisionless region may not be warranted.

Only very recently have preliminary results on the luminescence observed in the IR MPE of F_2CO been reported.[22] The fluorescence has been assigned to the $\tilde{A}^1A_2 \rightarrow \tilde{X}^1A_1$ system of F_2CO. The pressure range employed in these experiments was 10–750 mtorr, and a collisionless lifetime of 12 ± 3 μsec was obtained. The luminescence is quenched efficiently by F_2CO with a rate coefficient of $(5.6 \pm 0.8)10^6$ sec^{-1} $torr^{-1}$.

We believe that as the number of molecules dissociated by an intense IR laser field increases, so will the experimental observations of collisionless emission. In fact, it will not be surprising at all if luminescence is found to be a general phenomenon accompanying IR MPE.

Much of the discussion and speculation during the past several years regarding the luminescence which accompanies IR MPE has concentrated on two issues: the collisionless or collisional nature of the luminescence, and the mechanism(s) by which such emission can be brought about. It is therefore pertinent, in the context of the present review, to elaborate on these questions. In the next section, kinetic analyses of the time-resolved emissions are discussed for several representative cases, and experiments which distinguish between collisionless and collisional processes are outlined. In Section II.B, several possible routes, by which such luminescence can be brought about, will be discussed.

B. Kinetics

Much of the controversy regarding the collisionless or collisional nature of the luminescence has focused on the combination of pressures and times that can reasonably be called collisionless. Opinions range from advocacy that observation of luminescence 100 nsec after the onset of the laser pulse at pressures <1 torr is collisionless, to the gospel that only in molecular beams can collisions be safely ignored.

Time resolved observation of the emission usually shows rise and fall portions of the signal. Much attention has been paid to the rise time and its pressure dependence, and a collisionless process was often inferred solely on the basis of the "prompt" appearance of luminescence, even at rather high pressures. To illustrate the pitfalls inherent in such reasoning, four representative cases describing collisional and collisionless production of luminescence are discussed here in some detail.

1. Collisionless Mechanisms

Case (a) The emitting species, M*, is produced directly via MPE of a molecule, R, in the absence of collisions, and is subsequently removed by spontaneous emission or collisions with other species.

$$\mathrm{R} \xrightarrow{k_1} \mathrm{M}^* \tag{1}$$

$$\mathrm{M}^* \xrightarrow{\tau_s^{-1}} \mathrm{M} + h\nu \tag{2}$$

$$\mathrm{M}^* + Q_i \xrightarrow{k_{Qi}} \mathrm{M} + Q_i \tag{3}$$

where τ_s^{-1} is the rate of spontaneous emission, and k_{Q_i} is the rate coefficient for the quenching of M* by Q_i. Assuming that the rate of pumping of species R can be represented by the function $k_1 \exp(-t/\tau_{L1})$ where k_1 and τ_{L1} depend on the CO_2 laser,[23] we have

$$\frac{d}{dt}[M^*]=k_1[R]\exp(-t/\tau_{L1})-\tau_{M^*}^{-1}[M^*] \tag{4}$$

where $\tau_{M^*}^{-1}=(\tau_s^{-1}+\Sigma_i k_{Q_i}[Q_i]+\tau_d^{-1})$, and τ_d^{-1} is the rate of diffusion. The solution of (4) gives

$$[M^*](t)=\frac{k_1[R]}{\tau_{L1}^{-1}-\tau_{M^*}^{-1}}[\exp(-t/\tau_{M^*})-\exp(-t/\tau_{L1})] \tag{5}$$

The rise of the luminescence signal will reflect the faster process and the fall will be dominated by the slower process. At low pressure, where $k_{Qi}[Q_i]\ll\tau_s^{-1}$, both the rise and fall will be independent of pressure.

Case (b) A species is produced which cannot be represented as either a "pure" vibrational or electronic state. Here, there is a strong coupling between vibrational and electronic degrees of freedom so that the true states can be represented as linear combinations of the form

$$|\psi\rangle=a|M^\dagger\rangle+b|M^*\rangle \tag{6}$$

where a and b are normalized state densities of $|M^\dagger\rangle$ and $|M^*\rangle$, respectively. Since $a\gg b$ (due to the state densities), $|\psi\rangle$ resembles $|M^\dagger\rangle$ with the important addition that electronic luminescence, normally associated with $|M^*\rangle$, is allowed. The oscillator strength of the $|M^*\rangle$ luminescence is diluted considerably by the mixing represented by (6), and the spontaneous decay rate will be considerably smaller than if a "pure" state M* were excited. This is discussed further in Section II.C. For the sake of simplicity, we will represent these mixed species as $M^\dagger$. Subsequent to its production, $M^\dagger$ either dissociates, is deactivated collisionally by parent molecules, diffuses out of the field of view of the detector, or radiates with an emission rate, τ_s^{-1}, which reflects the mixing represented by (6).

$$R\xrightarrow{k_7}M \tag{7}$$

$$M^\dagger\xrightarrow{k_8}\text{products} \tag{8}$$

$$M^\dagger+R\xrightarrow{k_9}\text{products} \tag{9}$$

$$M^\dagger\xrightarrow{\tau_d^{-1}}\text{physical removal} \tag{10}$$

$$M^\dagger\xrightarrow{\tau_s^{-1}}M \tag{11}$$

The rate equation describing the fate of $M^{\dagger}$ is

$$\frac{d}{dt}[M^{\dagger}] = k_7[R]\exp - t/\tau_{L7} - \tau_{M^{\dagger}}^{-1}[M^{\dagger}] \tag{12}$$

where $\tau_{M^{\dagger}}^{-1}$ is given by

$$\tau_{M^{\dagger}}^{-1} = (k_8 + k_9[R] + \tau_s^{-1} + \tau_d^{-1}) \tag{13}$$

Solving (12) yields

$$[M^{\dagger}](t) = \frac{k_7[R]}{\tau_{L7}^{-1} - \tau_{M^{\dagger}}^{-1}}\{\exp(-t/\tau_{M^{\dagger}}) - \exp(-t/\tau_{L7})\} \tag{14}$$

Thus, the decay of the luminescence reflects $\tau_{M^{\dagger}}$ and is pressure dependent at high pressures.

There are many possibilities available under case (b), and we have only described one limiting situation in order to illustrate the dependence of [M*] on the various parameters that are critical to the experiments.

2. *Collisional Mechanisms*

Case (c) A MPE product, P, (dissociation fragment or vibrationally excited molecule) is produced in the absence of collisions and collides with the parent molecule, thereby producing M*, which decays as per (2) and (3).

$$R \xrightarrow{k_{15}} P \tag{15}$$

$$P + R \xrightarrow{k_{16}} M^* \tag{16}$$

$$P + R \xrightarrow{k_{17}} \text{products} \tag{17}$$

The rate equations describing these processes are

$$\frac{d}{dt}[P] = k_{15}[R]\exp(-t/\tau_{L15}) - k_{17}[R][P] - \tau_d^{-1}[P] \tag{18}$$

$$\frac{d}{dt}[M^*] = k_{16}[R][P] - \tau_{M^*}^{-1}[M^*] \tag{19}$$

These equations are similar to (12), and the solution for $[M^*](t)$ is

$$[M^*](t) = \frac{k_{15}k_{16}[R]^2}{(\tau_{L15}^{-1} - \tau_P^{-1})(\tau_{M^*}^{-1} - \tau_P^{-1})(\tau_{M^*}^{-1} - \tau_{L15}^{-1})}\{(\tau_{M^*}^{-1} - \tau_{L15}^{-1})\exp(-t/\tau_P) - (\tau_{M^*}^{-1} - \tau_P^{-1})\exp(-t/\tau_{L15}) + (\tau_{L15}^{-1} - \tau_P^{-1})\exp(-t/\tau_{M^*})\} \tag{20}$$

where $\tau_P^{-1} = k_{17}[\mathrm{R}] + \tau_d^{-1}$. At low pressures, τ_P^{-1} is the slowest rate and therefore is associated with the fall of the luminescence signal. The rise will reflect the production of P and the radiative decay of M*, and will be very rapid and nearly independent of pressure, as in case (b).

Case (d) Luminescence is produced via reaction or energy transfer involving two species, each of which are offspring of the parent.

$$\begin{gathered} \mathrm{R} \xrightarrow{k_{21a}} \mathrm{P}_1 \\ \mathrm{R} \xrightarrow{k_{21b}} \mathrm{P}_2 \end{gathered} \tag{21}$$

$$\mathrm{P}_1 + \mathrm{R} \xrightarrow{k_{22}} \text{products} \tag{22}$$

$$\mathrm{P}_2 + \mathrm{R} \xrightarrow{k_{23}} \text{products} \tag{23}$$

$$\mathrm{P}_1 + \mathrm{P}_2 \xrightarrow{k_{24}} \mathrm{M}^* \tag{24}$$

The general solution of the rate equations that describe (21)–(24) is tedious, since the production of M* cannot be described by a first-order process. When $\tau_{\mathrm{L}}^{-1} \gg \tau_P^{-1}$, $[\mathrm{P}_1](t)$ and $[\mathrm{P}_2](t)$ are given by

$$[\mathrm{P}_{1,2}](t) = \frac{k_{21a,b}[\mathrm{R}]}{\tau_{\mathrm{L}21a,b}^{-1}} \exp(-t/\tau_{\mathrm{P}_{1,2}}) \tag{25}$$

Thus, under these conditions, the rate equation describing (24) becomes

$$\frac{d}{dt}[\mathrm{M}^*] = \frac{k_{24}k_{21a}k_{21b}[\mathrm{R}]^2}{\tau_{\mathrm{L}21a}^{-1}\tau_{\mathrm{L}21b}^{-1}} \left\{\exp\left[-\left(\tau_{\mathrm{P}1}^{-1} + \tau_{\mathrm{P}2}^{-1}\right)t\right]\right\} - \tau_{\mathrm{M}^*}^{-1}[\mathrm{M}^*] \tag{26}$$

and the solution is

$$[\mathrm{M}^*](t) = \frac{k_{24}k_{21a}k_{21b}[\mathrm{R}]^2}{\tau_{\mathrm{L}21a}^{-1}\tau_{\mathrm{L}21b}^{-1}\left\{\tau_{\mathrm{M}^*}^{-1} - \left(\tau_{\mathrm{P}_1}^{-1} + \tau_{\mathrm{P}_2}^{-1}\right)\right\}} \times \left\{\exp\left[-\left(\tau_{\mathrm{P}_1}^{-1} + \tau_{\mathrm{P}_2}^{-1}\right)t\right] - \exp\left[-t/\tau_{\mathrm{M}^*}\right]\right\} \tag{27}$$

Again, the decay portion reflects $(\tau_{\mathrm{P}_1}^{-1} + \tau_{\mathrm{P}_2}^{-1})$ and is pressure dependent, while the rise time will be fast and nearly independent of pressure.

Equations (5), (14), (20), and (27) indicate clearly that the observation of a prompt and pressure independent risetime is not proof of the collisionless production of the luminescence. Inspection of the decay rates reveals that, for collisional processes [cases (c) and (d)], a plot of the decay rate vs. concentration will have an intercept which is equal to τ_d^{-1}, the "fly out" rate of the reactants from the field of view of the detector. In case (a), at low pressures τ_{M*}^{-1} should be nearly independent of pressure and equal to the sum of the radiative rate of the emitting species and τ_d^{-1}. In case (b), the luminescence decay rate is determined either by the laser pulse duration or $\tau_{M\dagger}^{-1}$. When τ_d^{-1} is small compared to the other decay rates, it is possible to distinguish between collisional and collisionless processes by measuring decay rates at low pressures. In practice, however, it is usually difficult to determine τ_d^{-1} accurately, and the interpretation of the low-pressure decay rate becomes ambiguous. We conclude, therefore, that observation of the luminescence in real time is not always sufficient to distinguish between collisionless and collisional production mechanisms.

Inspection of the different expressions for [M*](t) indicates that by extrapolating the decay portion of the luminescence curve to $t=0$, we obtain a preexponential factor, [M*](0) that depends linearly on [R] for collisionless processes [(5) and (14)]. Collisional production [(20) and (27)] gives rise to a higher than linear dependence of [M*](0) on [R]. The pressure dependence will be quadratic in those cases in which τ_L^{-1} and τ_{M*}^{-1} are much larger than the collisional production rate of the emitting species.

We believe that only by determining the pressure dependence of [M*](0) as described above at sufficiently low pressures (usually $<10^{-2}$ torr), and by observing both the rise and decay of the time resolved luminescence, can the collisionless nature of the luminescence be unequivocally demonstrated.

Considering the above, it is surprising that pressure dependence measurements of the luminescence intensity have not been carried out in most of the studies reported to date. Of the work published thus far, only in the study of the luminescence observed via MPD of vinyl cyanide were the experiments done at low enough pressure (10^{-5} torr) while both the pressure dependence and the time evolution of the luminescence were measured (Figs. 4 and 5). The dependence of [M*](0) on pressure is linear at low pressures. However, collisional processes manifest themselves at pressures >10 mtorr thus underscoring the importance of working at sufficiently low pressures. Similar linear dependences of the luminescence signals on pressure were observed in the MPE of C_2H_4, C_2H_3CHO, and C_3H_6 [20] in the pressure range 0.5–10 mtorr. Of the other luminescent systems reported to date, we feel that the short lived prompt luminescence observed via the IR MPD of TMD is collisionless, as suggested by the large decay rates observed at ~70 mtorr. The CrO_2Cl_2 luminescence is also collisionless in nature at the lowest pressures employed (5×10^{-4} torr), since the decay rate here is pressure independent (giving a collisionless lifetime of 5–15 μsec), and the pressure dependence of the peak luminescence signal is linear.[24]

In the collisional pressure regime (>0.1 torr), a quadratic pressure dependence was measured for the production of luminescence from C_2^* and CH* produced via the MPD of CH_2Cl_2[8] and CH_3OH,[13] and from CN* produced via the MPD of CH_3CN.[12] We believe that the appearance of excited diatomics such as CH*, C_2^*, and CN* in the IR MPD of other hydrocarbons is also due to collisions despite the rapid rise of their emissions. We have looked for emissions from excited diatomics in our studies of the IR MPD of C_2H_4, C_2H_3CN, C_2H_3Cl, and CH_3CN, and found none at pressures <100 mtorr.

Of the other luminescent systems reported in the literature, we feel that the broad-band emissions reported in the MPD of BCl_3 and OsO_4 may have a collisionless component at low pressures. Although the lowest pressures reported in the literature are 50 mtorr (OsO_4 luminesces at lower pressures as well[25]), both have broad, structureless emission features, which resemble those observed in the MPE of vinyl cyanide and CrO_2Cl_2. These molecules are also good candidates for the observation of V–E energy transfer according to the criteria given in Ref. 26. Thus, it would be quite valuable to perform experiments with these species at low pressures and to determine the pressure dependence of the luminescence intensity as well as details of the kinetics.

C. Mechanisms

By far, the most exciting feature of the luminescence produced via IR MPE is its appearance in the absence of collisions. Possible mechanisms for the production of electronically excited species via MPE have been discussed in several recent publications.[18, 19, 22, 26] Molecules which have been excited to high vibrational levels may undergo transitions to a nearby excited electronic state, which lies below the dissociation limit. The density of states, however, is higher for the ground electronic state manifold than for the excited electronic state manifold at any excitation energy where such a transition between the two manifolds may occur. Thus, the species may only spend a small fraction of their time in the electronically excited manifold. Since, in the absence of collisions, the vibrationally excited species in the ground electronic manifold are metastable, "sampling" of the electronic excited state may occur for a long time reflecting the "dilution factor" of the two states. Case (b) in Section II.C.1 describes the kinetics of this process.

The conditions under which it may be favorable to observe experimentally this "inverse electronic relaxation" have been recently discussed.[26] The authors distinguish between two internal energy regions: (1) a region of sparse density of the excited electronic state, and (2) a higher energy region where the ground electronic state quasicontinuum is coupled to an electronically excited state quasicontinuum. In (1), only isolated groups of vibrational states in the ground electronic manifold are coupled to vibrational states in the excited electronic manifold. The authors conclude that in (1), inverse electronic relaxation is amenable to experimental observation in medium size molecules in which the density of effectively coupled background states is not too large. They point out that CrO_2Cl_2 falls into this category. In large molecules, where the ground and excited state quasicontinuum are coupled effectively, the inverse electronic relaxation rate is proportional to the ratio of the densities of states and the process becomes more efficient at high vibrational excitations.

The long lifetimes that have been observed for several of the emitting species at low pressures[18–22, 24] are compatible with such an intramolecular V–E energy transfer process. The fast quenching of the fluorescence, even by inert gases, is interpreted as reflecting the quenching of the vibrationally excited ground electronic state molecule by collisions (vibration→ translation energy transfer).

Inverse electronic relaxation can occur via MPE either in the parent molecule or in a dissociation fragment. It is believed that the MPD of a large molecule to produce small (e.g., diatomic) fragments proceeds via the sequential dissociation of species that are produced during the laser pulse. Primary dissociation products are further excited by the IR laser and, depending on the number of photons absorbed, either dissociate further or may undergo inverse electronic relaxation. The fluorescence observed in the dissociation of vinyl cyanide, ethylene, etc.[19, 20] may fall into this category.

In the case of collisionless emission from a dissociation fragment that is not further excited via MPE, direct dissociation along a potential surface which correlates with electronically excited products may be the source of the emission. Such a mechanism may be responsible for the luminescence observed in the MPD of TMD[16, 17] where the emitting species has a very short lifetime, which is similar to the lifetime of the excited state of acetone.

We believe that as more molecules are dissociated, and the emitting species become identified, more information concerning the relative importance of these processes will be forthcoming. We should emphasize, however, that the processes discussed above, which may lead to an efficient production of luminescence, do not imply that the molecules are excited in a way that is non-statistical or bypasses the lowest energy channels.

The appearance of luminescence which is a consequence of collisions is apparently due to a variety of routes specific to the particular system under consideration. It is very difficult to sort out the exact sequences of reactions responsible for the observed emissions, as the experiments are usually not carried out under well controlled conditions. The multitude of possible vibrationally excited species and reactive free radicals which may participate in reactions, and the possibility of more than single collisions, further complicate matters. Various mechanisms which have been advanced in the literature include (1) collision-assisted V–E energy transfer,[19] (2) collisions between two vibrationally excited species to produce an electronically excited species or fragment[12, 15–17], (3) reaction of a fragment produced in the primary dissociation stage with the parent molecule,[27] (4) exothermic reactions between two radicals that produce electronically excited species,[12, 13] and (5) complex schemes of consecutive and parallel reactions that lead to electronically excited products.[8]

III. CHEMILUMINESCENT REACTIONS INITIATED VIA IR MULTIPLE PHOTON DISSOCIATION

It is now well documented that MPD is accompanied by complete randomization of energy prior to dissociation, and that the distribution of dissociation products can, to a great extent, be predicted using RRKM theory. However, due to the sequential nature of the dissociation processes in which primary fragments are further excited by the same CO_2 laser pulse and dissociate, MPD offers a unique opportunity for producing free radicals, which in the past could not be prepared in a well-controlled experimental environment. Such radicals include, among others, C_2, C_3, and CH, radicals whose detailed kinetic behavior has eluded scientists for many years. These radicals are in general very reactive and we expect many more studies involving their reactions to be carried out using IR photolysis. Here, we shall concentrate only on those exothermic reactions which are known to give rise to chemiluminescent products.

The experimental technique typically uses a CO_2 TEA laser to photodissociate a polyatomic molecule, thereby producing the radical of interest. A tunable dye laser monitors the free radicals via laser-induced fluorescence (LIF), thereby identifying the specific electronic and vibrational states of interest. The dynamics and mechanisms of the reactions are deduced either from observations of LIF from specific product states or by monitoring time resolved luminescence from an electronically excited product, thereby obtaining information about energy partitioning in the products. A typical experimental arrangement is shown in Fig. 6.

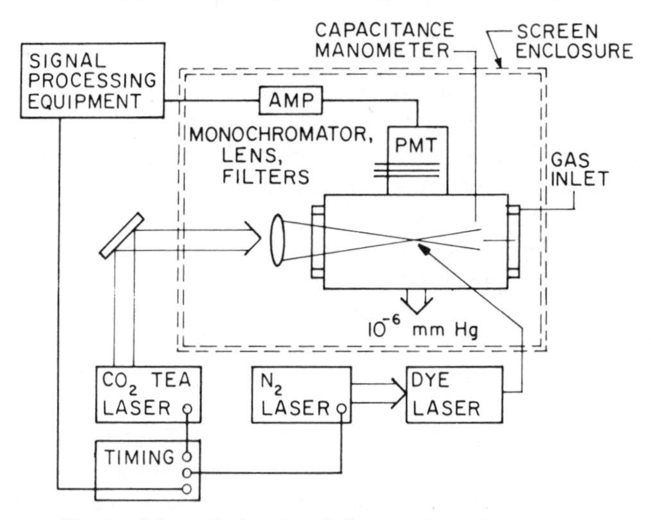

Fig. 6. Schematic drawing of the experimental arrangement.

A. $C_2(a^3\Pi_u)+NO$

The most detailed study to date concerning a chemiluminescent reaction using IR laser photolysis to produce the free radical reactant, involves the reaction of C_2 in the $a^3\Pi_u$ state (which lies 610 cm^{-1} above the ground state) with NO.[28] To the best of our knowledge, this is the first study of a reaction of C_2 molecules, in which both the kinetics and the energy partitioning in some of the products have been investigated. The reaction can proceed via a number of routes, among which are the following channels leading to electronically excited CN:

$$C_2(a^3\Pi_u)+NO(X^2\Pi)\rightarrow CN(A^2\Pi)+CO(X^1\Sigma^+)+4.73\text{ eV} \qquad (28)$$

$$\rightarrow CN(B^2\Sigma^+)+CO(X^1\Sigma^+)+2.86\text{ eV} \qquad (29)$$

$C_2(a^3\Pi_u)$ can be very conveniently prepared using a number of source molecules such as C_2H_3CN, C_2H_4, etc. It can be monitored directly via LIF using the $C_2(d^3\Pi_g\rightarrow a^3\Pi_u)$ system. The nascent distributions of vibrational, rotational, and translational energies in $C_2(a^3\Pi_u)$ have been determined in separate experiments.[29] Upon photolysis of mixtures of C_2H_4 (or C_2H_3CN), NO, and Ar (Ar is used to slow diffusion) intense emissions identified as the $CN(B^2\Sigma^+\rightarrow X^2\Sigma^+)$ and $CN(A^2\Pi\rightarrow X^2\Sigma^+)$ emission systems are observed.

By measuring C_2 peak LIF intensities as the delay between the dye and CO_2 lasers is varied, the time evolution of the $C_2(a^3\Pi_u)$ concentration can be obtained. The pseudo-first-order rate coefficient for the quenching of

$C_2(a^3\Pi_u)$ by NO is determined from the slope of a plot of the quenching rate vs. [NO]. The quenching rate coefficients for $C_2(a^3\Pi_u)$ by NO have been determined, using C_2H_3CN and C_2H_4 as $C_2(a^3\Pi_u)$ precursors, by monitoring either $C_2(a^3\Pi_u)$ LIF or CN(B→X) and CN(A→X) time resolved emissions. All the rate coefficients thus obtained are the same [$k=(7.5\pm0.3)10^{-11}$ cm^3 $molecule^{-1}$ sec^{-1}] establishing the reaction between $C_2(a^3\Pi_u)$ and NO as the source of the CN emissions. Representative plots are shown in Fig. 7. Preliminary results from the authors' laboratory indicate that C_2 molecules in the ground $X^1\Sigma_g^+$ state react much more rapidly with NO than do C_2 molecules in the $a^3\Pi_u$ state. This rate is quite different from the rate obtained from the single exponential chemiluminescent decay curves.[30]

Energy disposal within the CN $B^2\Sigma^+$ and $A^2\Pi$ states has been determined from the spectrally resolved emissions. Typical traces of the $\Delta v=0$ and $\Delta v=-1$ sequences of the $CN(B_2\Sigma^+\rightarrow X^2\Sigma^+)$ emission system, and of the $CN(A^2\Pi\rightarrow X^2\Sigma^+)$ system are shown in Figs. 8 and 9. The relative populations of the vibrational and rotational levels in the B state were estimated using a computer routine to simulate the $\Delta v=0$ and $\Delta v=-1$ emission bands. Boltzmann distributions of rotational levels were

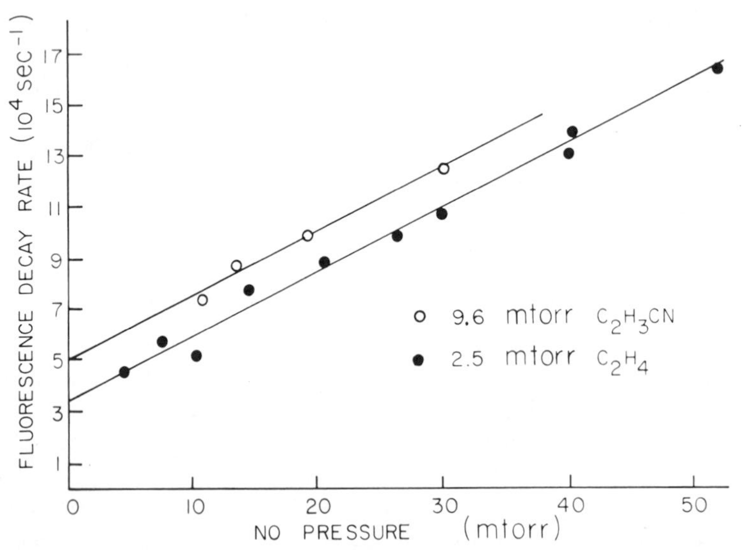

Fig. 7. Plots of $\tau_{C_2}^{-1}$ vs. NO pressure. The slope gives the quenching rate coefficient of $C_2(a^2\Pi_u)$ by NO, $k_1=(2.45\pm0.1)10^6$ sec^{-1} $torr^{-1}$. The intercept gives the sum of the quenching rate of $C_2(a^3\Pi_u)$ by the precursor molecules and τ_d^{-1}. Each point represents the average of 8–16 shots. The sources of $C_2(a^3\Pi_u)$ in these plots are: ○ 9.6 mtorr vinyl cyanide, and ● 2.5 mtorr ethylene.

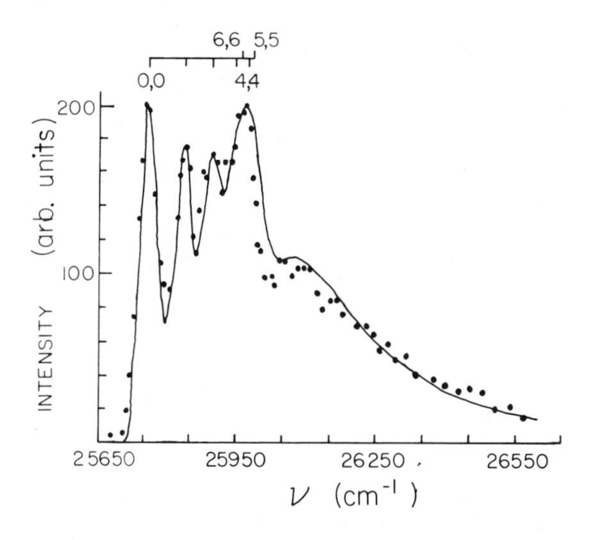

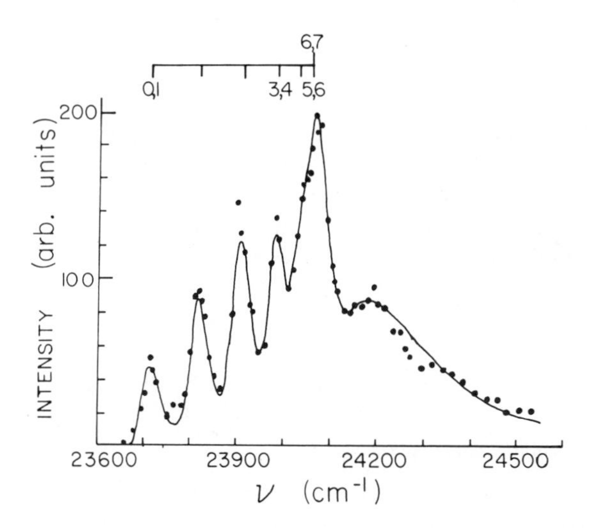

Fig. 8. The $\Delta v=0$ and $\Delta v=-1$ sequences of the $CN(B^2\Sigma^+ \rightarrow X^2\Sigma^+)$ violet band system produced in a mixture of vinyl cyanide (9 mtorr), NO (112 mtorr) and Ar (200 mtorr). The spectra were taken in second order, point by point, at intervals of 0.1–0.2 nm, with a resolution of 0.3 nm. Each point is the average of 8–32 shots. Spectra are not corrected for detector response. The dots are the experimental values. The solid line is the simulated spectrum obtained using the vibrational populations and rotational temperatures given in Table I.

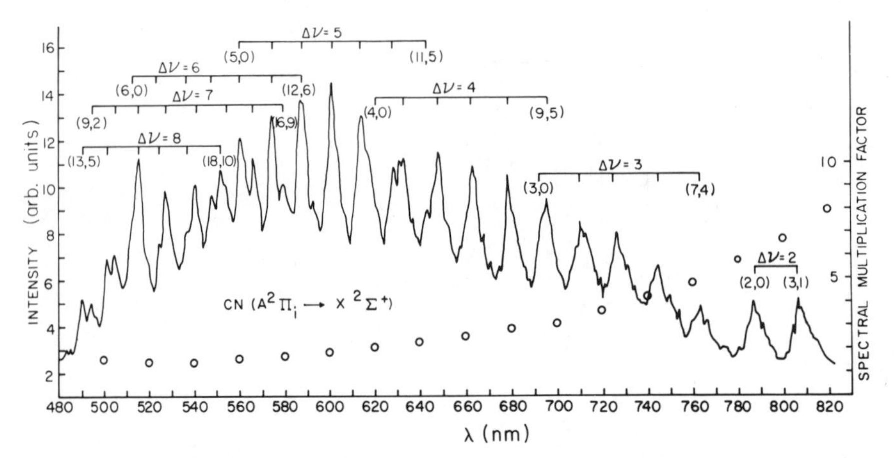

Fig. 9. $CN(A^2\Pi \rightarrow X^2\Sigma)$ red band emissions. The spectrum was taken point by point at intervals of 1 nm with a resolution of 1 nm. Each point is the average of 8–16 shots. The spectrum is uncorrected for monochromator/PMT response. The experimental correction factors by which the intensity at each wavelength should be multiplied are given as open circles by the right-hand scale. The spectrum was taken with 18 mtorr C_2H_3CN, 200 mtorr NO, and 300 mtorr Ar.

assumed in the simulation. Published spectroscopic constants and Franck–Condon factors were used, as well as the exponential monochromator resolution and emission intensities. The best fit was obtained for monotonically decreasing vibrational populations and rotational temperatures, with increasing vibrational excitation. The calculated "best-fit" vibrational populations and rotational temperatures obtained from the $\Delta v=0$ and $\Delta v=-1$ sequences are listed in Table I. It should be noted that, although

TABLE I
Rotational Temperatures and Relative Vibrational Populations in $CN(B^2\Sigma^+)$

Vibrational	*Relative vibrational populations*			Rotational temperature	
level (v')	From $\Delta v=0$ sequence	From $\Delta v=-1$ sequence	Average	$T_R(v')$ (K)	$\langle f_R \rangle = \frac{kT_R(v')}{E_t - E_{v'}}$
0	1.0	1.0	1.0	7000 ± 1000	0.21
1	0.64	0.79	0.70 ± 0.09	6000 ± 1000	0.21
2	0.48	0.63	0.55 ± 0.08	5000 ± 1000	0.19
3	0.32	0.42	0.37 ± 0.06	4200 ± 800	0.19
4	0.23	0.37	0.30 ± 0.07	3600 ± 700	0.18
5	0.18	0.26	0.22 ± 0.04	3100 ± 700	0.19
6	0.13	0.21	0.17 ± 0.04	2700 ± 500	0.20

the simulation was not sensitive to changes in rotational temperatures within 1000 K, a decrease in rotational temperature with increasing vibrational excitation was always required in order to reproduce the experimental data. The simulation program was much more sensitive to changes in the vibrational populations, especially in the $\Delta v = -1$ sequence where the bands are well resolved. Since the (5, 6) and (6, 7) bands are nearly overlapped, it is very difficult to distinguish between them in the simulation and an equally good fit can be obtained by using vibrational levels up to $v' = 5$, and approximately doubling the $v' = 5$ population. Therefore, the relative populations of $v' = 5$ and $v' = 6$ should be viewed with caution. The discrepancies between the relative vibrational populations obtained from the $\Delta v = 0$ and $\Delta v = -1$ sequences reflect experimental uncertainties in the measured intensities. The average vibrational populations listed in Table I decrease exponentially with vibrational energy and can be describe by an "effective" temperature of $(10.5 \pm 1.5)10^3$ K.

Intense emission bands of the $A^2\Pi \rightarrow X^2\Sigma^+$ red system of CN were observed in the region 460–810 nm, as shown in Fig. 9. Vibrational levels up to $v' = 16$ are easily identified. Higher vibrational levels may be populated, but they are overlapped by bands originating from lower v' levels. Approximate relative vibrational populations have been determined from integrated band intensities assuming the electronic transition moment, $[R_e(\bar{r})]^2$, is constant, and using calculated Franck–Condon factors. They decrease monotonically, and, given the accuracy of the measurements, show no evidence of a population inversion for $v' > 2$. The observed distribution shows a nearly exponential decrease with increasing v', corresponding to an "effective" temperature of $(13 \pm 3)10^3$ K (Fig. 10). Since the populations of $v' = 0$, and $v' = 1$ could not be measured, the possibility of a population inverstion for these levels cannot be discounted.

By looking at the time resolved fluorescence of the entire $A^2\Pi \rightarrow X^2\Sigma^+$ and $B^2\Sigma^+ \rightarrow X^2\Sigma^+$ band systems, the branching ratio between the A and B states can be estimated. To do this, the observed emission is related to the number of excited species by:

$$I_0(\text{obs}) = \left[R_e(\bar{r}) \right]^2 hcN_0' \sum_{v'v''} P_{v'} \nu^4_{v',v''} \cdot q_{v',v''} \cdot T_F(\nu) \cdot S(\nu) \tag{30}$$

where $[R_e(\bar{r})]^2$ is the electronic transition moment, N_0' is the number of molecules formed in the excited state, $P_{v'}$ is the probability of excitation of the v' vibrational level, $\nu_{v',v''}$ is the frequency of the $v' \rightarrow v''$ transition, $q_{v',v''}$ is the corresponding Franck–Condon factor, and $T_F(\nu)$ and $S(\nu)$ are the filter transmission and the photomultiplier sensitivity, respectively, at wave number ν. $T_F(\nu)$ and $S(\nu)$ have been determined experimentally. By

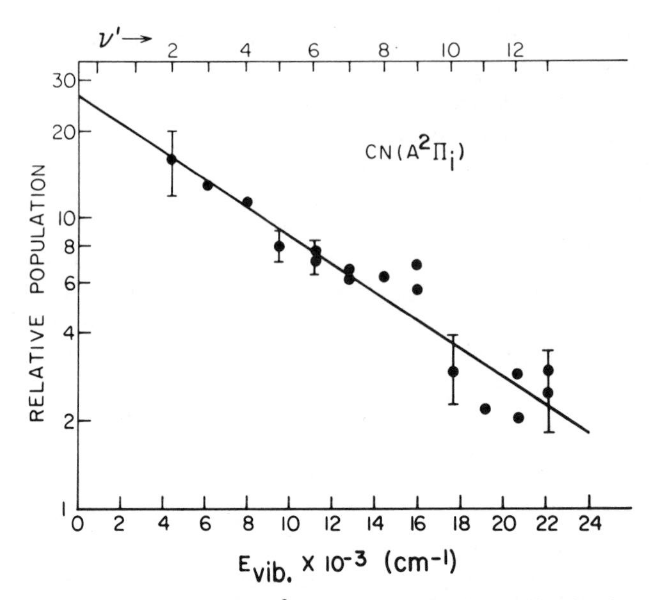

Fig. 10. Relative populations of $CN(A^2\Pi)$ vibrational levels vs. vibrational energy. Vibrational levels are indicated at the top. The straight line represents $T = 13{,}000°K$. Typical uncertainties are indicated by error bars.

using the measured $P_{v'}$ values and published electronic transition moments, it was found that the A state is more populated than the B state by a factor of 7. Systematic, rather than random, errors are the main source of uncertainty in this estimate. These involve mainly the uncertainties in the published experimental values of $[R_e(\bar{r})]^2$ and in the measured $P_{v'}$ values. Thus, the value quoted here for the branching ratio may be uncertain by as much as a factor of 2. It is of interest, however, to note that this value corresponds to an effective electronic temperature of $\cong 10^4$ K, similar to the vibrational temperatures obtained for the A and B states.

The measured vibrational and rotational distributions for the B and A states of CN are similar to the nascent distributions produced in the reaction. For CN in the B state (lifetime ~60 nsec), the likelihood of relaxation other than spontaneous emission is very small. In addition, the good agreement between the vibrational and rotational distributions obtained in Ref. 28, and a recent crossed molecular beam study of the $C_2 + NO$ reaction[31] attest to the unrelaxed nature of the B state distributions obtained in Ref. 28. The lifetime of the A state is 6–7 μsec, and some collisional relaxation, especially of high vibrational levels, may occur. However, it has been observed experimentally that vibrational levels in the

A state are not significantly relaxed even at pressures ~1 torr.[32] Perturbation between the B and A states were shown to be insignificant in populating the B state. Likewise, cascading from the B to the A state is unimportant in populating the A state. It is concluded, therefore, that the A and B states of CN are populated directly in the reaction. The fractions of the available energy deposited in the B and A states is found to be 0.21 and 0.22, respectively, and the fraction of the available energy deposited in rotation for each vibrational band of the B state is 0.20 ± 0.02 for $v' = 0$–6 (see Table I). As noted before, the vibrational and rotational energy distributions deduced for the B state are in satisfactory agreement with those obtained by Krause,[31] who also found rather good agreement between his experimental values and the "prior" distribution which he calculated.

The data presented in Ref. 28 are compatible with the existence of a long-lived complex with a lifetime of at least several vibrations. Intuitively, one might anticipate such a complex in a reaction where only one bond is formed initially, and subsequently the C_2NO transition complex rotates and rearranges itself so that a second bond is formed and the products separate. Such a mechanism is rather speculative at this stage, and data concerning the branching ratio for the different chemical species which may be formed in the reaction of C_2 with NO will be very illuminating. If, for example, the rate coefficient for formation of CN is similar to the total quenching rate coefficient of $C_2(a^3\Pi_u)$, which is nearly gas kinetic, then no serious geometric or steric barriers are expected in the reaction, favoring the mechanism described above. However, if the probability of formation of CN in the reaction is small, it implies that the relative orientation of the colliding molecules may be of importance and a four-center concerted mechanism would have to be considered as well.

The possibility that CN radicals are formed, not in a single step reaction of C_2 with NO, but rather via consecutive reactions (i.e., $C_2 + NO \rightarrow C_2O^* + N$; $C_2O^* + NO \rightarrow CN(B, A) + CO_2$) was also considered. A solution of the rate equations for such consecutive reactions, however, shows that the chemiluminescence signal will display a rise whose rate depends on [NO]. Inspection of the chemiluminescence signals of ref. 28 reveals a very fast rise whose rate is independent of [NO]. It is concluded, therefore, that CN molecules in the B and A states are produced by the single-step reactions, (28) and (29), apparently via a long-lived transition complex.

B. $C_2(X^1\Sigma_g^+, a^3\Pi_u) + O_2$

Another reaction which has been the subject of recent study is the reaction between C_2 radicals and O_2. C_2 was again produced via the MPD of C_2H_3CN in samples containing 0.2–10 mtorr C_2H_3CN, 5–300 mtorr O_2, and 2–5 torr Ar. Chemiluminescence was observed in the vacuum UV (VUV),[33] and in the entire UV, visible, and near-IR spectral regions.[34] Using different filter gases interposed between the photomultiplier tube and the fluorescence cell, the VUV emission was assigned to the $CO(A^1\Pi \rightarrow X^1\Sigma^+)$ 4th positive system.[33] The emission in the spectral region from the UV to the near-IR was resolved using a monochromator with 3 nm resolution (Fig. 11). Three band systems of CO were identified: the Asundi ($a'^3\Sigma^+ \rightarrow a^3\Pi$; $v' = 5$–11), the Triplet ($d^3\Delta i \rightarrow a^3\Pi$; $v' = 1$–9), and the Herman ($e^3\Sigma^- \rightarrow a^3\Pi$; $v' = 3$–7). These channels are all energetically possible, as the energies available for the reactions leading to ground-state products

$$C_2(a^3\Pi_u) + O_2(X^3\Sigma^-) \rightarrow 2CO(X^1\Sigma^+) \tag{31a}$$

$$C_2(X^1\Sigma_g^+) + O_2(X^3\Sigma^-) \rightarrow 2CO(X^1\Sigma^+) \tag{31b}$$

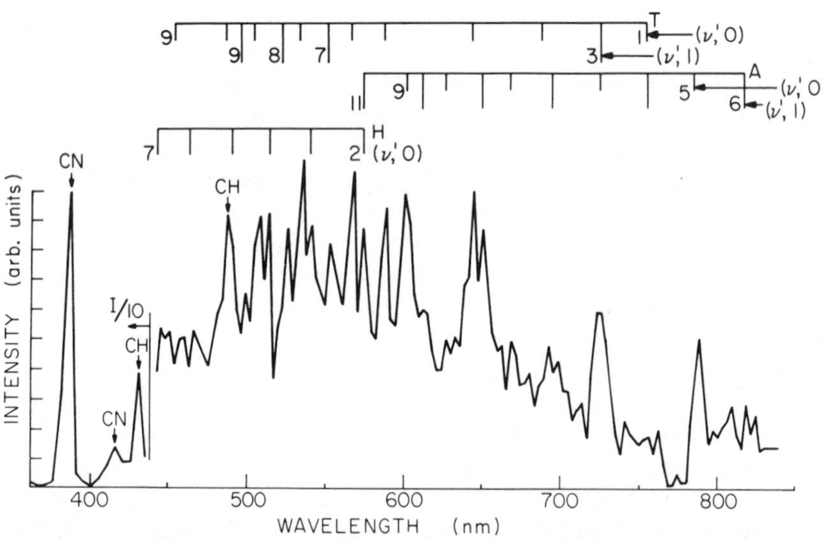

Fig. 11. Chemiluminescence spectrum that results from the reaction of O_2 with the ir mpd photofragments of C_2H_3CN. In repeated experiments, a monochromator (2 nm resolution) was scanned in 1 nm steps. This spectrum was taken with 28 mtorr C_2H_3CN, 540 mtorr O_2, and 710 mtorr Ar. A, T, and H denote the Asundi, Triplet, and Herman ($a'^3\Sigma$, $d^3\Delta$, $e^3\Pi$, $\rightarrow a^3\Pi$) emission systems of the CO molecule. Vibrational bands were identified using published band heads. $CN(B^2\Sigma^+ \rightarrow X^2\Sigma^+)$ and $CH(A^2\Delta \rightarrow X^2\Pi)$ transitions are also indicated. Their intensity is divided by a factor of 10. The spectrum is not corrected for PMT/monochromator response. However, the underlying broad band luminescence from the photolysis of vinyl cyanide is subtracted.

are 252 and 250 kcal/mole, respectively. The decay of the time resolved chemiluminescence involving these excited CO states follows closely the removal of $C_2(a^3\Pi_u)$[33-35] and $C_2(X^1\Sigma_g^+)$[34]. Plots of the observed decay rate as a function of $[O_2]$ give a rate coefficient of $(3.0 \pm 0.2)10^{-12}$ cm^3 molecule^{-1} sec^{-1} for the quenching of $C_2(a^3\Pi_u)$ and $C_2(X^1\Sigma_g^+)$ by O_2 (Fig. 12). Presently, it is impossible to distinguish between the contributions of $C_2(a^3\Pi_u)$ and $C_2(X^1\Sigma_g^+)$ reactions to the observed emissions, as suitable scavengers have not yet been identified.

Other reactions that may be of importance in the quenching of C_2 by O_2 are

$$C_2 + O_2 \rightarrow CO_2 + C \qquad \Delta H = -124 \text{ kcal/mole} \tag{32}$$

$$\rightarrow C_2O + O \qquad \Delta H = -74 \text{ kcal/mole} \tag{33}$$

C_2O has very recently been observed as a primary reaction product.[35] No branching ratios for the production of the various products have been reported to date. The subsequent reactions of $C(^1S_0)$ [an energetically permitted product in reaction (32)] and/or C_2O [which may be produced

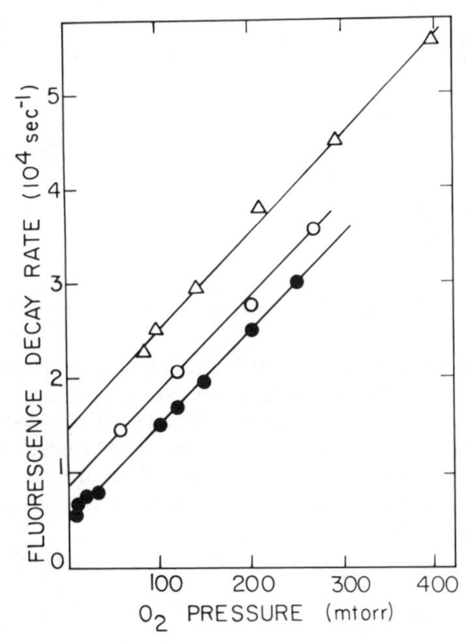

Fig. 12. The decay rate of CO fluorescence at 560 nm ●, and the removal rate, measured by LIF, of $C_2(a^3\Pi_u)$ ○, and $C_2(X^1\Sigma_g^+)$ △ vs. O_2 pressure. The slope gives the rate coefficient for the $C_2 + O_2$ reaction: $k = 3 \times 10^{-12}$ cm^3 molecule^{-1} sec^{-1}.

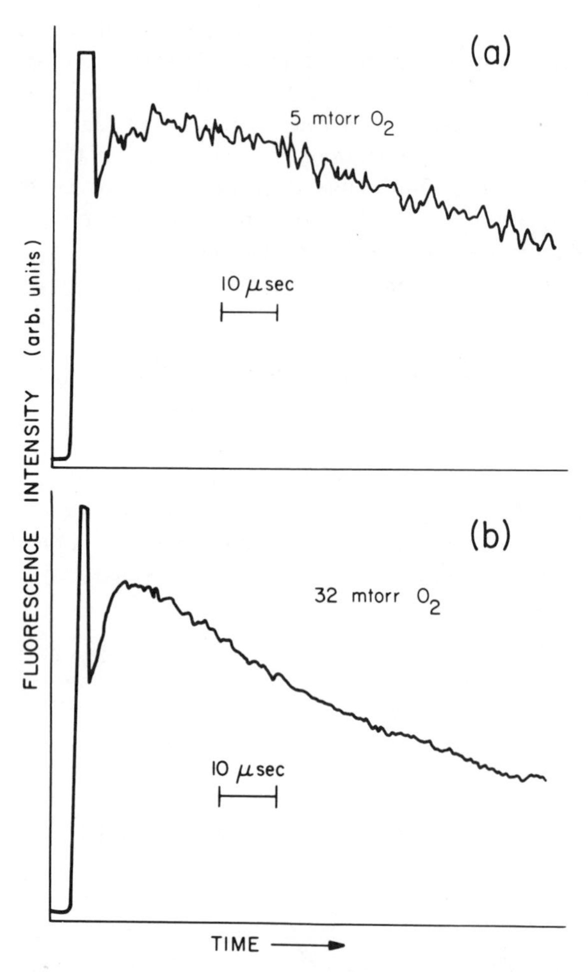

Fig. 13. Time resolved fluorescence signal following CO_2 laser irradiation (FWHM 0.2 μsec) of 0.5 mtorr C_2H_3CN and 2000 mtorr Ar, in the presence of (a) 5 mtorr O_2, (b) 32 mtorr O_2. Fluorescence was observed with an interference filter centered at 560 nm, with a 10 nm spectral width. Each curve was obtained by signal averaging 64 pulses. The initial spike is due to luminescence from an unknown fragment of C_2H_3CN, being quenched at Ar. Note that the risetimes in both cases are shorter than 10 μsec, and the decay times are dependent on O_2 pressure.

in an excited state via reaction (33)] with O_2 are sufficiently exothermic to account for the observed chemiluminescence. However, as discussed in Section III.A, a kinetic scheme for consecutive reactions will lead to luminescence signals that are characterized by rise and fall times that are both dependent on $[O_2]$. Consequently, a careful look at the rise portion of the chemiluminescence signal obtained in the photolysis of 0.5 mtorr of vinyl cyanide in the presence of 5–35 mtorr of O_2 was undertaken.[34] A rise time of several microseconds was observed, which was only weakly dependent on $[O_2]$ (Fig. 13). This rise is interpreted as reflecting the average decay rate of the excited triplet states, and is given by

$$\tau_{\text{obs}}^{-1} = \sum_{T} \left(\tau_{sT}^{-1} + k_{qT}[O_2] \right) \tag{34}$$

where T denotes an excited triplet state with radiative decay rate $= \tau_{sT}^{-1}$, k_{qT} is a quenching rate coefficient, and the summation is over participating triplet states. If consecutive reactions are important in the production of excited CO molecules, the rate of the fast, second reaction should be faster than the observed rise. The reaction $C(^1S_0) + O_2$ is too slow to account for the experimental observation.[36] The rate coefficient for the $C_2O + O_2$ reaction is unknown. However it is hard to imagine that it will be faster than 5×10^{-10} cm^3 molecule^{-1} sec^{-1}, which is the rate coefficient required to account for the experimental results. It is therefore suggested that the excited CO molecules that are observed are formed directly in a single step reaction of C_2 with O_2.

C. Other Reactions

In addition to CO emissions, the IR photolysis of mixtures of vinyl cyanide and O_2 also produces emissions from the CN($B^2\Sigma^+ \rightarrow X^2\Sigma^+$) and CH($A^2\Delta \rightarrow X^2\Pi$) band systems.[34] The time behavior of the former is complex, consisting of several exponentials, and has not been studied in detail. The latter exhibits a single exponential decay with a rate which is proportional to $[O_2]$. The rate coefficient deduced from the slope of the observed rate vs. $[O_2]$ is $(2.4 \pm 0.3)10^{-11}$ cm^3 molecule^{-1} sec^{-1}. The reaction responsible for the emission is not known with certainty at this time. The emission appears with olefins as the precursors of the reactive species and only in the presence of oxygen. The rate coefficient derived from CH emissions is the same irrespective of the olefin precursor; (C_2H_4, C_2H_3CN, C_2H_3Cl, and C_2H_3CHO all give the same rate coefficient). In none of these molecules is CH a primary photolytic product,[34] ruling out the possibility that energy transfer from excited CO molecules (produced in the C_2+O_2 reaction) to CH is responsible for the CH emissions. Dissociation of CH_3CN gives CH as a primary photolytic product. However, no CH emission was observed upon addition of O_2.

One possible route, which is energetically possible, is the reaction of C_2H with O_2:

$$C_2H(\tilde{X}^2\Sigma^+)+O_2(X^3\Sigma_g^-)\rightarrow CH(A^2\Delta)+CO_2(\tilde{X}^1\Sigma^+) \qquad \Delta H=-7\ \text{kcal/mole} \tag{35}$$

The reaction is symmetry allowed for a planar transition state of C_s symmetry. As yet, however, there is no direct proof that C_2H is produced in the MPD of olefins, nor is any information available concerning its reactions. Another possible mechanism for the formation of CH* is via a complex kinetic scheme involving radical–radical reactions [see case (d) in Section II.B]. However, the low pressures of vinyl cyanide and oxygen employed in these experiments (10–200 mtorr) would mitigate against radical–radical reactions. The lack of any pressure dependence of the rise argues against consecutive reactions.

In another study, BO emissions have been observed upon the mpd of BCl_3 in the presence of oxygen.[14] The rate coefficient deduced for the reaction is $(2.4\pm1.2)10^{-11}$ cm^3 molecule^{-1} sec^{-1}. The BCl_3 fragment involved in the chemiluminescent reaction has not been identified. However, the time-resolved luminescence has a rise which depends on $[O_2]$ in a complex way, and suggests a kinetic scheme involving consecutive reactions or radical–radical reactions. The emissions have been observed at high pressures (several torr) thus increasing the probability of radical–radical reactions.

Emissions were also observed in the mpd of HN_3 and DN_3, and were identified as the $(\tilde{A}^2A_1\rightarrow\tilde{X}^2B_1)$ emission systems of NH_2 and ND_2.[27] The emission is thought to be the result of the reactions:

$$NH(D)(a^1\Delta)+H(D)N_3\rightarrow NH_2(D_2)(^2A_1)+N_3 \tag{36}$$

The rate coefficient for the quenching of $ND(^1\Delta)$ by DN_3 was obtained from the time-resolved emission of ND_2 ($k=6.4\times10^{-11}$ cm^3 molecule^{-1} sec^{-1}).

Very recently, emissions were observed in the reaction of CH radicals with oxygen.[37] The emissions were identified as the (0,0) and (1,1) bands of the OH $(A^2\Sigma^+\rightarrow X^2\Pi)$ system. CH radicals were produced by MPD of CH_3NH_2 and CH_3CN. The rate of removal of CH radicals is equal to the decay rate of the chemiluminescence signal and is proportional to O_2 pressure. The rate coefficient for the reaction of CH ($X^2\Pi$) with O_2, obtained both from CH LIF signals and from the time resolved OH emissions is $(3.2 \pm 0.4)10^{-11}$ cm^3 molecule^{-1} sec^{-1}.

The number of chemiluminescent reactions which have been generated via IR MPD is not large at this time. We expect that this will change in the future, as the number of radicals that can be efficiently produced via MPD is growing rapidly [e.g., CH (from CH_3OH,[13] and CH_3CN[12]), C_2 (from C_2H_4,[38] and C_2H_3CN[28]), C_3 (from allene[39]), NH (from HN_3[27]), NH_2 (from N_2H_4, NH_3[40]), CH_2 (from $(CH_3CO)_2O$[41]), CF_2 (from CF_2Br_2, CF_2Cl_2[42]), etc.]. Many of these radicals are very reactive and lead to various exothermic reactions, from which emission may be observed. The energy partitioning among vibrational and rotational levels of the product's excited states may be obtained from the spectrally resolved emissions, and integrated emission intensity measurements can give information about the branching ratios in the reaction, thus shedding light on the mechanism and dynamics of the reaction. Quenching of the reactive radical may be followed directly using LIF, which has proven to be a very sensitive diagnostic for probing the kinetics and reactivity of selected rovibronic states of the reactant. For example, the $a^3\Pi_u$ and the $X^1\Sigma_g^+$ states of C_2 are separated by only 610 cm^{-1}. However, they may react quite differently, since one is a triplet and the other is a singlet. Reactions of $C_2(a^3\Pi_u)$ are now being observed directly using LIF on the Swan bands. Similarly, preliminary results have been obtained regarding reactions of $C_2(X^1\Sigma_g^+)$ molecules by monitoring $C_2(X^1\Sigma_g^+)$ directly via LIF using the Phillips system.[34, 35b] Thus, the detailed, state specific, kinetics of C_2 in the singlet and triplet states may be elucidated for the first time.

IV. CONCLUDING REMARKS

Research concerning electronic luminescence produced via IR MPE is presently only in its initial stages. The early experiments which, as it turned out, were done under collisional conditions, demonstrated that collision induced luminescence is a general phenomenon associated with MPE, not because of any specific interaction of the IR field with the molecules, but simply because production of small radicals is a general characteristic of MPD and exothermic reactions of these radicals often lead to luminescence. The importance of the initial work was to spur interest in the possible appearance of luminescence under collisionless conditions. Only during the past two years has careful work, specifically concerned with collisionless luminescence, been carried out. Much of this work has not yet appeared in print at the time of this writing, and from the point of view of a reviewer, it would have been more advantageous to write a critical survey of the literature two years from now. Thus, in this chapter we have emphasized the type of experiments which should be done in order to establish definitively the occurrence of luminescence in the absence of collisions and the mechanisms which lead to its formation. We point out that collisional contributions to the luminescence signals have been observed at pressure as low as 15 mtorr, underscoring the need to perform experiments at low pressures, to observe the luminescence in real time, and to establish the dependence of its peak intensity on pressure.

Another experimental difficulty is the identification of the emitting species. In all of the cases reported to date, the observed luminescence is broad band and almost structureless, originating from species larger than diatomic. The assignments, in those cases in which they have been attempted, have been based on the envelope of the electronic luminescence bands and not on any fine spectroscopic features. MPE is generally accompanied by the dissociation of the parent molecule to numerous fragments, which are formed within the duration of the laser pulse, and which may fluoresce in a spectral region similar to that of the parent molecule. The published assignments, although sometimes rather convincing, are not yet definitive in our opinion. We are still awaiting the report of experiments carried out with a molecule whose emission spectrum has been well established by other excitation techniques, whose dissociation fragments are known to have different emission spectra, and which luminesces via IR MPE. Such a molecule will enable us to investigate in detail the coupling of the ground electronic state vibrational manifold to the excited electronic state, a phenomenon which is the source of the observed emissions, and will provide us with a better tool for studying energy transfer processes in vibrationally excited molecules.

We also expect that theoretical work will appear which will be based on more definitive experiments, and which will provide insights and predictions as to which species are amenable to experimental observations of collisionless luminescence. Such theoretical work has only recently begun, and we expect more of it in the near future. The appearance of luminescence is not just an exotic feature of IR MPE, but is related to the general area of relaxation and internal conversion in large molecules, and as such is of lasting value.

On a more practical level, the study of chemiluminescence which is generated by the reactions of specific radicals which are produced via IR MPD has already provided valuable new information concerning the chemistry of C_2 molecules. These studies are state specific and provide very detailed information concerning elementary processes involving radical species such as C_2. Rate coefficients obtained using IR MPD as a source of radicals are in excellent agreement with very recent results using

VUV photolysis,[35] and establish the generality and accuracy of both techniques. These studies are continuing and will shed much light on the chemistry of small radicals which are important in combustion, air pollution, chemistry of the upper atmosphere, etc. For example, the ability to state selectively monitor reactants enables us to measure differences in reactivity between singlet and triplet C_2 molecules, and to compare these results with those obtained for singlet and triplet C atoms and CH_2 radicals. By observing chemiluminescence from products, energy disposal into the product degrees of freedom is monitored, and information concerning details of the reaction dynamics may be obtained.

Note added in proof: Since submission of this chapter, several additional relevant publications have appeared which are briefly summarized below. Collisionless luminescence in the 300–750 nm region was observed in the MPD of $C_2F_4S_2$ at pressures 1 to 1000 mtorr.[43] The decay rate of the luminescence was 4 ± 1 microsec^{-1}. The emitting species is most probably a dissociation product (possibly CF_2S). Collisionless emission, which probably originates from a dissociation product, has also been observed in benzaldehyde.[44] Further efforts have been directed towards the identification of the emitting species in MPE. A method for the determination of the translational energy of long-lived luminescing species, by observation of their quenching by the walls as a function of time, has been described.[45] The method was applied to the MPD of O_SO_4 and it was concluded, based on decay times and spectroscopic observations, that the emitting species is the parent molecule. A more direct and unambiguous identification can be achieved by using molecular beam techniques and determining both the velocity and the angular distribution of the emitting species. Using this technique, it was shown that the emitting species in the MPD of C_2H_3CN and C_2HCl_3 is a dissociation product.[46] In the case of CrO_2Cl_2, some doubt has recently been cast on the assignment of the emission to the parent molecule.[47]

Electronic luminescence resulting from MPE has very recently been used to probe the quasicontinuum region of large molecules, to determine the probability distribution of the absorbed photons, and to gain insight into such phenomena as intramolecular energy transfer and MPE within electronically excited states. In these experiments, both a dye laser and a CO_2 laser were used to prepare molecules in high vibrational levels of the ground or electronically excited states.[48–51]

Regarding the chemiluminescent reactions of C_2, it is now known that $C_2(X^1\Sigma_g^+)$ reacts more rapidly with NO than does $C_2(a^3\Pi_u)$.[30] Excited CN molecules in the A and B states are formed predominantly in reactions of $C_2(a^3\Pi_u)$. $C_2(X^1\Sigma_g^+)$ results (mainly) in ground state CN, as expected from adiabatic state correlations. It has also been found that the probability of forming CN(X, A, B) is apparently large, lending support to a *sequential* four-center reaction mechanism.[30] Regarding the reaction of C_2 with O_2, it has been shown that O_2 collisionally induces fast intersystem crossing between $C_2(a^3\Pi_u)$ and $C_2(X^1\Sigma_g^+)$.[52] At 300 K this crossing is much faster than reaction. The previously measured reaction rate coefficient for both singlet and triplet C_2 removal by O_2 is reinterpreted as the rate coefficient for removal of *equilibrated* C_2 molecules. It has also been shown that the source of the observed emissions from excited triplet states of CO is mainly reactions of $C_2(X^1\Sigma_g^+)$ with O_2.[52]

Acknowledgments

The authors wish to thank M. Mangir for many stimulating discussions and J. W. Hudgens, S. V. Filseth, and A. M. Ronn for providing results prior to publication.

References

1. In the text that follows, it is understood that the word "luminescence" means electronic luminescence. We will not deal with the vibrational luminescence that accompanies MPE.
2. We use the words "multiple photon" in order to distinguish MPE from multiphoton excitation involving nonstationary intermediate states. MPE may involve both stationary and nonstationary intermediate states.
3. See, for example, R. V. Ambartzumian and V. S. Letokhov, in *Chemical and Biochemical Applications of Lasers*, Vol. III, C. B. Moore, ed., Academic Press, New York, 1977, and references cited therein.
4. R. N. Isenor and M. C. Richardson, *Appl. Phys. Lett.*, **18**, 224 (1971).
5. V. S. Letokhov, E. A. Ryabov, and O. A. Tumanov, *Opt. Comm.* **5**, 168 (1972).
6. R. V. Ambartzumian, N. V. Chekalin, V. S. Doljikhov, and E. A. Ryabov, *Chem. Phys. Lett.*, **25**, 515 (1974).
7. In much of the early work in this field, the word "intensity" is used incorrectly. What actually was measured was fluence (J cm^{-2}) and this number was divided by the pulse duration in order to obtain an approximate intensity.
8. V. N. Bourimov, V. S. Letokhov, and E. A. Ryabov, *J. Photochem.*, **5**, 49 (1976).
9. R. V. Ambartzumian, Yu A. Gorokhov, G. N. Makarov, A. A. Puretzky, and N. P. Furzikov, *Chem. Phys. Lett.*, **45**, 231 (1977).
10. R. V. Ambartzumian, N. V. Chekalin, V. S. Letokhov, and E. A. Ryabov, *Chem. Phys. Lett.*, **36**, 301 (1975).
11. R. V. Ambartzumian, N. V. Chekalin, V. S. Dolzikhov, V. S. Letokhov, and V. N. Lokhman, *J. Photochem.*, **6**, 55 (1976/77).
12. M. L. Lesiecki and W. A. Guillory, (a) *Chem. Phys. Lett.*, **49**, 92 (1977); (b) *J. Chem. Phys.*, **69**, 4572 (1978).
13. S. E. Bialkowski and W. A. Guillory, *J. Chem. Phys.*, **67**, 2061 (1977).
14. N. R. Isenor, V. E. Merchant, R. S. Hallsworth, and M. C. Richardson, *Can. J. Phys.*, **51**, 1281 (1973).
15. V. E. Merchant, *Opt. Comm.*, **25**, 259 (1978).
16. Y. Haas and G. Yahav, *Chem. Phys. Lett.*, **48**, 63 (1977).
17. G. Yahav and Y. Haas, *Chem. Phys.*, **35**, 41 (1978).

18. Z. Karny, A. Gupta, R. N. Zare, S. T. Lin, J. Nieman, and A. M. Ronn, *Chem. Phys.*, **37**, 15 (1979).
19. M. H. Yu, H. Reisler, M. Mangir, and C. Wittig, *Chem. Phys. Lett.*, **62**, 439 (1979).
20. M. Mangir, H. Reisler, and C. Wittig, unpublished.
21. S. I. Blinov, G. A. Zaleskaya, and A. A. Kotov, *Bull. Acad. Sci. USSR, Phys. Ser.*, **42**, 130 (1978).
22. J. W. Hudgens, J. L. Durant, Jr., D. J. Bogan, and R. A. Coveleskie, (a) *Bull. Am. Phys. Soc.*, **24**, 638 (1979); (b) *J. Chem. Phys.*, **24**, 638 (1979).
23. This particular functional form is used to represent laser pumping simply for mathematical simplicity, since we can straightforwardly obtain closed form solutions for $[M^{\dagger}](t)$ and $[M^{*}](t)$.
24. J. Nieman and A. M. Ronn, *Opt. Eng.*, **19**, 39 (1980).
25. R. V. Ambartzumian, private communication.
26. A. Nitzan and J. Jortner, (a) *Chem. Phys. Lett.*, **60**, 1 (1978); (b) *J. Chem. Phys.*, **71**, 3524 (1979).
27. A. Hartford, *Chem. Phys. Lett.*, **57**, 352 (1978).
28. H. Reisler, M. Mangir, and C. Wittig, *J. Chem. Phys.*, **71**, 2109 (1979).
29. (a) J. D. Campbell, M. H. Yu, and C. Wittig, *J. Chem. Phys.*, **69**, 3854 (1978); (b) M. R. Levy, H. Reisler, M. Mangir, M. H. Yu, and C. Wittig, *Faraday Disc. Chem. Soc.*, **67**, 243 (1979).
30. H. Reisler, M. Mangir, and C. Wittig, *J. Chem. Phys.*, **73**, 2280 (1980).
31. H. Krause, *J. Chem. Phys.*, **70**, 3871 (1979).
32. J. A. Coxon, D. W. Setser, and W. H. Duewer, *J. Chem. Phys.*, **58**, 2244 (1973).
33. S. V. Filseth, G. Hancock, J. Fournier, and K. Meier, *Chem. Phys. Lett.*, **61**, 288 (1979).
34. H. Reisler, M. Mangir, and C. Wittig, *Chem. Phys.*, **47**, 49 (1980).
35. (a) V. M. Donnelly and L. Pasternack, *Chem. Phys.*, **39**, 427 (1979). (b) J. R. McDonald, private communication.
36. D. Husain and P. E. Norris, *Faraday Disc. Chem. Soc.*, **67**, 273 (1979).
37. I. Messing, S. V. Filseth, and C. M. Sadowski, *Chem. Phys. Lett.*, **66**, 95 (1979).
38. N. V. Chekalin, V. S. Letokhov, V. N. Lokhman, and A. N. Shibanov, *Chem. Phys.*, **36**, 415 (1979).
39. M. L. Lesiecki, K. W. Hicks, A. Orenstein, and W. A. Guillory, *Chem. Phys. Lett.*, **71**, 72 (1980).
40. J. D. Campbell, G. Hancock, J. B. Halpern, and K. H. Welge, *Chem. Phys. Lett.*, **44**, 404 (1976).
41. D. Feldman, K. Meier, R. Schmiedl, and K. H. Welge, *Chem. Phys. Lett.*, **60**, 30 (1978).
42. J. C. Stephenson and D. S. King, *J. Chem. Phys.*, **69**, 1485 (1978).
43. C. N. Plum and P. C. Houston, *Chem. Phys.*, **45**, 159 (1980).
44. J. T. Yardley and D. F. Heller (to be published).
45. R. V. Ambartzumian, G. N. Makarov, and A. A. Puretzky, *Appl. Phys.*, **22**, 77 (1980).
46. T. Watson, M. S. Mangir, M. R. Levy, and C. Wittig, (to be published).
47. (a) S. Kimel, private communication, (1980); (b) D. F. Heller, private communication, (1980).
48. I. Burak, T. J. Quelly and J. I. Steinfeld, *J. Chem. Phys.*, **70**, 334 (1979).
49. I. Burak, J. Tsao, Y. Prior, and E. Yablonovitch, *Chem. Phys. Lett.*, **68**, 31 (1979).
50. I. Burak, J. Tsao, and E. Yablonovitch, (to be published).
51. D. F. Heller, G. A. West, *Chem. Phys. Lett.*, **69**, 419 (1980).
52. M. S. Mangir, H. Reisler, and C. Wittig, *J. Chem. Phys.* **73**, 829 (1980).

ELECTRONICALLY EXCITED FRAGMENTS FORMED BY UNIMOLECULAR MULTIPLE PHOTON DISSOCIATION

YEHUDA HAAS

Department of Physical Chemistry, The Hebrew University, Jerusalem

CONTENTS

I. Introduction 713
A. Experimental Methods for Studying Unimolecular LIMD Reactions 714
B. Mechanism of LIMD 715
II. Collision-free LIMD Chemiluminescent Reactions 716
III. Multiphoton Dissociation of Tetramethyldioxetane 719
A. Introductory Notes 719
B. Time Behavior of TMD Chemiluminescence 722
C. Nature of the Emitting Species, Comparison with Thermal Reactions 726
IV. Further Experiments of SiF_4 729
Acknowledgments 732
References 732

I. INTRODUCTION

Electronically excited species (atoms, radicals, molecules) are produced in a gas sample whenever a powerful enough laser beam is focused into it. They are easily detected by their luminescence, in the visible or in the ultraviolet. The phenomenon is particularly striking when an infrared laser is used, as the net effect is conversion of low-energy photons into high-energy ones. In this chapter we shall focus on a special case: Production of electronically excited fragments by unimolecular laser induced multiphoton dissociation (LIMD). Thus, we exclude from our discussion dielectric breakdown and chain reactions involving free radicals formed by LIMD. To place chemiluminescent (CL) reactions in a proper perspective with other LIMD experimental methods, we shall briefly review available experimental evidence for the unimolecular nature of the process. The currently accepted theoretical approach is rather extensively dealt with elsewhere in this volume. Its limited discussion in this chapter is brought up only in order to point out past, and possible future, contributions of chemiluminescent reactions to the elucidation of LIMD mechanism. Unfortunately, the usefulness of CL results is often restricted by ambiguities concerning the nature of the emitting species. In one case, that of tetramethyldioxetane, the reaction has been extensively studied by usual thermal methods. This molecule was thus chosen for more detailed study, and is used as an example of using LIMD methods for time resolved studies of a thermal reaction.

A. Experimental Methods for Studying Unimolecular LIMD Reactions

The proof that LIMD is sometimes not a collision-assisted process derives basically from low pressure, highly time-resolved studies. In this context, three experimental techniques played a decisive role.

1. Molecular beam studies, monitoring directly the primary products.[1, 2]
2. Laser-induced fluorescence (LIF) studies in very low-pressure bulk material. The method was most extensively applied to electronic excitation of fragments by tunable dye laser.[3–6]
3. Chemiluminescent reactions. Possibly due to experimental simplicity, these provided the first generally accepted evidence for the unimolecular character of LIMD.[7–9] This chapter deals only with electronic excitation, but vibrational chemiluminescence was also studied.[10]

A common feature of these methods is the direct monitoring of nascent products. Other methods were also frequently cited as indicating a unimolecular character. These include isotope separation experiments, and particularly the pressure dependence thereof,[11, 12] branching ratios between different possible products as a function of laser fluence[13] and ultrashort laser pulses techniques[14] (see also Prof. Yablonovitch's chapter in this book). These latter-mentioned methods are indirect in the sense that final product distribution, rather than nascent, is monitored. Quite commonly, LIMD leads initially to formation of unstable species. Mechanistic interpretations based on analysis of final products, consequently necessarily involve a larger degree of uncertainty than in the more direct methods.

The direct methods are in a way complementary. Molecular beam studies, by virtue of using mass spectrometric detection, are of general applicability. They provide information on the branching ratio of different reaction channels, and products' angular and translational energy distribution. The method has not yet been used for determining the initial internal energy distribution in the products. LIF and chemiluminescence (CL), on the other hand, are limited to the detection of fluorescent species only. A further restriction on CL is, that they must be formed directly by the reaction in an excited state. Both methods are characterized by excellent time resolution, often limited only by the laser pulsewidth. They can thus be used to determine internal energy distributions, as well as finer details of the excitation process during the laser pulse. Furthermore, real time comparison between unimolecular and collision induced processes is readily available, providing a direct way to assess their relative contributions to overall yields. Before proceeding, a short summary of LIMD reaction mechanism is required, primarily in order to introduce concepts to be used later on.

B. Mechanism of LIMD

In a typical experiment a light source emitting 1000 cm^{-1} photons is used to break a chemical bond with a bond energy of $20–40\times10^3$ cm^{-1}. The many photon process leading to dissociation is often described by reference to a model proposed by Bloembergen et al.[15] Three regions are distinguished in the energy level diagram of the substrate. Region I, extending up to about $3–4\times10^3$ cm^{-1} is called the discrete region. Level density is rather low, and interaction with electromagnetic field is described by usual spectroscopic methods. One of the vibrational modes (LIMD has been so far demonstrated only for polyatomic molecules) is in near resonance with the laser frequency, permitting excitation of $v=1$ in that mode (the "pumped" mode). Further excitation is restricted by anharmonicity, but can be achieved up to $v=2$ or 3 by rotational compensation, aided by power broadening. The next region, lying between ~3000 cm^{-1} and the dissociation limit is termed region II. Energy-level density is rather high, level spacing being of the same order of magnitude as level width. This region is accordingly called the quasicontinuum region. The quasicontinuous background absorption often observed in polyatomic molecules is due to states in region II. Coupling between the pumped mode and other modes leads to population of quasicontinuum states. These can absorb more photons, as the resonance condition is satisfied at practically any frequency. Photons are thus consecutively absorbed in the molecule, building up the internal energy, up to the dissociation limit. Note that in this model excitation is restricted to a single vibrational mode only initially: Photons absorbed by region II states lead to unlocalized excitation. An alternative picture envisions "dumping" of the excitation from the pumped mode to the quasicontinuum, and further absorption in the pumped mode. The net result is the same—energy is distributed in the entire molecular vibrational manifold. The third region is above the dissociation limit. In a polyatomic molecule, many dissociation channels are possible. The lowest one is taken to define the beginning of region III.

Molecular dynamics beyond this limit are treated by standard statistical methods (such as RRKM theory).

The laser bandwidth provides a yardstick for distinguishing between regions I and II. The onset of region II may be taken at the energy E for which the density of states $\rho(E)$ is larger than the effective laser bandwidth, Δ. Increasing laser power leads to an increase in Δ, due to power broadening. Therefore, the onset of region II is determined by both molecular and laser field parameters. In large polyatomic molecules, region II may begin at energies as low as 1000 cm^{-1}, that is, it can be reached by absorption of a single photon.

The model evidently predicts that LIMD reactions will lead to the same product distribution as thermal reactions. In particular, "mode-selective" chemistry is impossible. Furthermore, coherent effects need not be important, so that the energy fluence (J/cm^2) is the decisive factor rather than power density (W/cm^2). So far, these predictions have been found to be consistent with experimental results. Quantitative treatments are thus generally applied using a rate equation approach,[1a, 16, 17] whose justification and limitations were discussed by many authors (see, for example, Jortner, Nitzan and Mukamel's paper in this volume]. As pointed out before,[1] the rate equation approach predicts that translational and internal energies of the products may depend on laser fluence and pulse duration. This arises from competition between a truly unimolecular process (decomposition) and a photon-induced process, (further absorption by levels beyond the dissociation limit). Another prediction is that energy buildup in the molecules is not instantaneous and may in principle be experimentally observed. CL methods are, apparently, uniquely suited to test these predictions. Indeed, results obtained to date, qualitatively support the current model. It is hoped that in the near future more quantitative treatments will become available.

II. COLLISION-FREE LIMD CHEMILUMINESCENT REACTIONS

As mentioned above, almost any molecule will yield electronically excited fragments when subjected to large enough IR laser intensities. Notwithstanding, the number of cases for which a truly unimolecular character was confirmed is rather limited. They include SiF_4,[8] BCl_3,[9] C_2H_3CN,[18] and tetramethyldioxetane (TMD).[19] A further example of somewhat different character, CrO_2Cl_2, is discussed by A. M. Ronn elsewhere in this volume. The unimolecular character of these reactions is usually inferred from the following observations:

1. Luminescence is obtained instantaneously with the IR laser pulse, the rise time being independent of substrate pressure.
2. Luminescence spectrum at very low pressures is usually different from that observed at higher pressures. In all the cases cited above, the unimolecular reaction results in a broad, structureless emission spectrum, while in the "high" pressure spectrum characteristic frequencies of diatomic species are often dominant (BCl, SiF, C_2, etc.). Both spectra are easily distinguished from that obtained by causing a dielectric breakdown—the latter is much more intense, and contains a large contribution of UV radiation.
3. Luminescence intensity increases linearly with substrate pressure.

These reactions played an important role in the early days of LIMD, providing the first evidence for its unimolecular character. Eventually, it became apparent that they involved some peculiarities that could not be easily resolved. A foremost one is the often ill-defined nature of the emitting species. The model discussed above predicts LIMD to yield the energetically most probable products, that is, those formed by the least exothermic route. None of the molecules mentioned above (with the exception of TMD) is known to yield electronically excited fragments on thermal decomposition. As an example, Fig. 1 shows the lower dissociation channels of BCl_3. It is seen that the preferred route would be $BCl_3 \rightarrow BCl_2 + Cl$, a nonluminescent reaction. Unless a very large barrier favors formation of electronically excited states, it is hard to reconcile the observation of strong luminescence with the model of Section 1. It is conceivable that excited species account for an extremely small fraction of the products, and are revealed only due to the excellent sensitivity of the method. This contention is further supported by the very large threshold fluence required to produce the prompt luminescence[9]: 10^9 W/cm^2, or about 300 J/cm^2.

Another puzzling feature is the shape of the emission spectra. Figure 2 shows the chemiluminescence spectrum of SiF_4 irradiated at 10.2 μm. Similarly broad and structureless spectra were obtained with C_2H_3CN[18] and BCl_3. This structure suggests a polyatomic emitting molecule. Even the smallest polyatomic, that is, a triatomic molecule might exhibit such a spectrum if the emission is due to a broad distribution of vibrational states. No attempt was yet made to assign the emission spectra of SiF_4, C_2H_3CN, or BCl_3. In the absence of such assignments, mechanistic interpretations cannot be given. These difficulties considerably reduce the usefulness of the CL method, unless a better defined luminescent system is found.

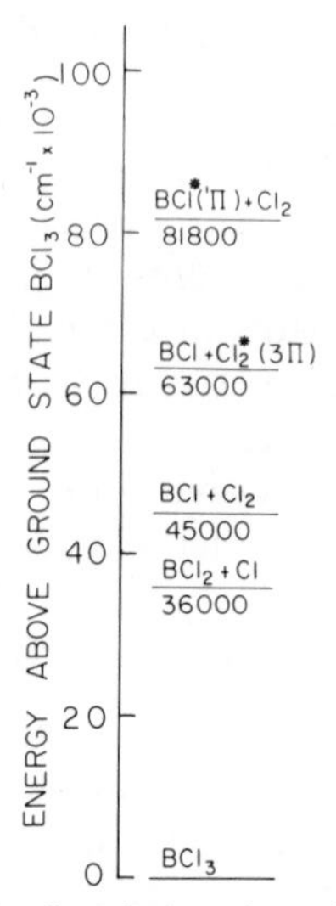

Fig. 1. Energetics of possible BCl_3 dissociation channels. For ground-state products, enthalpy of formation at 298 K is shown. Excited states energies are from G. Herzberg, *Molecular Spectra and Molecular Structure* I. *Spectra of Diatomic Molecules*, Van Nostrand, Princeton, N.J., 1950. Figures are rounded to 1000 cm^{-1}. The position of electronically excited BCl_2 is not known.

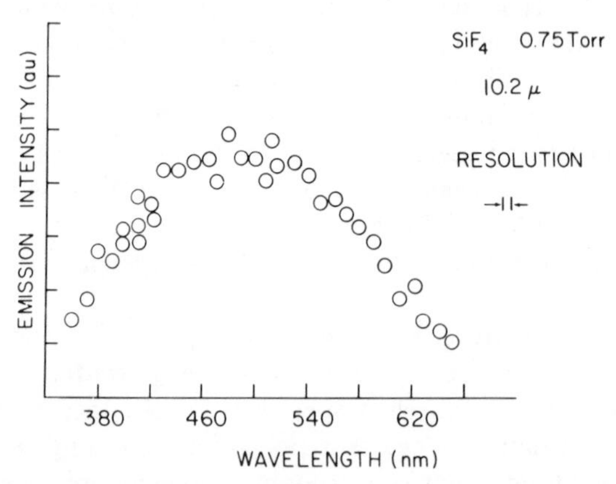

Fig. 2. Chemiluminescence spectrum of SiF_4 dissociated by 10.2 μm CO_2 laser radiation.

In searching for such systems, it is reasonable to try known thermal chemiluminescent reactions. Such reactions are quite common, particularly those involving oxidation (e.g., in flames). A further requirement is that the reaction be unimolecular, and preferably with a stable parent compound. These latter restrictions, narrow the selection down to a rather small group of compounds. An obvious choice is TMD, the best studied dioxetane. Dioxetanes[20] are four-membered ring peroxides, that decompose thermally according to

```
    O — O             O      O
    |   |       →     ||  +  ||     +hν (λ≃420 nm)
  — C — C —           C      C
    |   |            / \    / \
```

The reaction is usually carried out under mild conditions (60–70°C) and is associated with visible light emission, due to the formation of an electronically excited state. Most studies involved liquid solutions, but some of the more volatile dioxetanes (including TMD) have been shown to luminesce also in the vapor phase. Emission[21] and chemical sensitization[22] studies show that the product carbonyl compounds are preferably formed in the triplet state, not involving a singlet precursor. This result should be compared with optical excitation of acetone: Light absorption directly populates the excited singlet, which transforms by intersystem crossing to the triplet. Direct formation of the triplet is spin forbidden. For LIMD purposes, the important observation is that formation of electronically excited products is a *major* pathway, and that the chemiluminescence spectrum is well known. Also of interest is the fact that thermal activation energy is about 25 kcal/mole. This gives a rough estimate for the barrier in unimolecular decomposition, and corresponds to about 9000 cm^{-1}, or ~9 CO_2 laser photons. These data, and a known infrared absorption band at 10.2 μm, make TMD an attractive LIMD-CL candidate.

III. MULTIPHOTON DISSOCIATION OF TETRAMETHYLDIOXETANE

A. Introductory Notes

Irradiation of TMD vapor with the appropriate frequencies of a pulsed, high-power CO_2 laser does indeed yield intense chemiluminescence. Experimental details can be found elsewhere.[19, 23, 24] Here we shall concentrate on the following topics:

1. Comparison between LIMD and thermal reactions.
2. Role of collision induced processes.

3. Effect of laser fluence on the unimolecular kinetics and product distribution.

It turns out that separation of the discussion along these lines is fairly arbitrary. Thus, in order to characterize the nature of the emitting species one has first to analyze the kinetics of the process, requiring, *inter alia*, clear distinction between collisional and unimolecular processes. That, in turn, calls for some knowledge of the nature of the emitter (e.g., its radiative lifetime). We find it convenient to start by describing possible routes for LIMD reactions and then comparing the experimental results with expectations.

Figure 3 shows a schematic energy level diagram for the TMD–acetone system. Some of the parameters are not accurately known, for example, the barrier for TMD dissociation. Thus, Fig. 3 is basically a qualitative guide in discussing the chemiluminescent reaction. In thermal reactions, activation occurs by collisions. RRKM theory postulates rapid randomization of energy in the molecule, at energies close to and beyond the dissociation limit. The molecule is presumed to dissociate as soon as energy accumulates in the critical coordinate, so that the rate can be calculated by statistical methods. Calculations on TMD are not available, but it is likely to behave similarly to molecules of similar size. At 0.1 torr, the thermal reaction is in the "low-pressure" limit,[25, 26] and the dissociation rate is determined by collisional activation. Under these conditions, the concentration of molecules excited beyond the dissociation threshold is very small. They will either dissociate, or be deenergized by collision. It is assumed that the time between collisions ($\sim 10^{-6}$ sec) is long enough for the energy to be randomized in the molecule and cause dissociation. At thermal equilibrium, the observed rate constant is calculated by averaging $k(E)$ over the Maxwell–Boltzmann distribution, $k(E)$ being the unimolecular dissociation rate constant of molecules whose internal energy is E. For our purposes suffice it to note that at relatively low temperature, the overall rate and product energy distribution are dominated by molecules with little excess energy. This is to be compared with a laser-induced reaction, where a different situation may prevail, depending on laser fluence. Suppose a molecule with internal energy E is produced by many photon absorption. Being in region III, it decays (by reaction), with a rate constant $k(E)$. According to the LIMD model, this molecule may be further excited during its lifetime by absorption of laser light. The rate constant for this process, k(photon), is linearly dependent on laser intensity, I, and the absorption cross-section, σ: $k(\text{photon}) = I\sigma$. For instance, with $\sigma = 2 \times 10^{-19}$ cm^2 and $I = 10^8$ W/cm^2 (5×10^{26} photons cm^{-2} sec^{-1}), $k(\text{photon}) = 10^8$ sec^{-1}. If $k(E)$ is smaller than this value, the molecule will

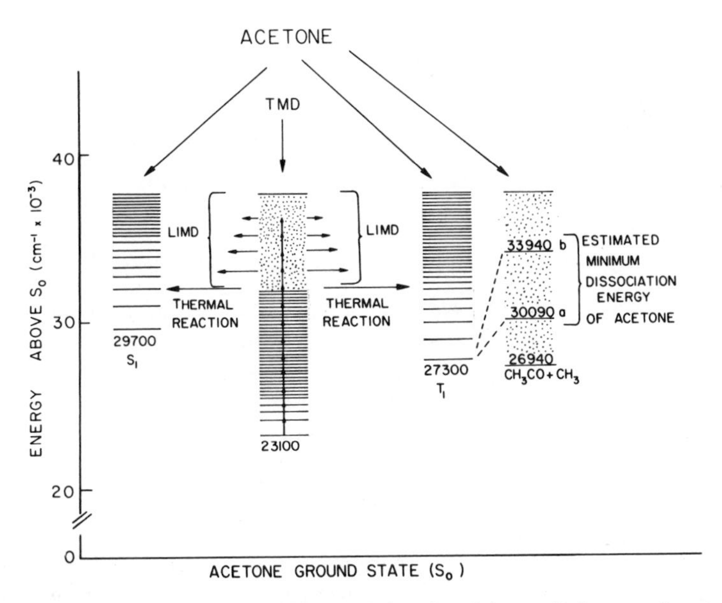

Fig. 3. Energy level diagram for TMD dissociation. Low-lying excited states of acetone are shown, as well as the lowest dissociation channel (to CH_3CO+CH_3). In distinction with photoexcitation, both singlet and triplet acetone may be populated simultaneously by TMD dissociation. This is illustrated in the figure symbolically by placing TMD energy levels in a symmetric position between singlet and triplet acetone levels. Thermal dissociation occurs as soon as the barrier for dissociation is reached, not necessarily via the triplet state. Laser excitation is shown by the vertical arrows. At high fluence levels, excitation to energy levels beyond the dissociation limit is possible during the lifetime of the excited states. This may lead to higher vibrational excitation in the product, compared to thermal reaction. Dissociation of acetone to CH_3CO+CH_3 is energetically feasible at 26,940 cm^{-1}, but has not been experimentally observed below 30,000 cm^{-1}. Estimates for the minimum required energy for the process are also shown. Data for TMD energetics are from Ref. 20. Other references are (a) R. B. Cundall and A. S. Davies, *Prog. React. Kin.*, **4**, 149 (1967); (b) S. W. Larson and H. E. O'Neal, *J. Phys. Chem.*, **70**, 2475 (1966).

not dissociate, but rather be excited to higher states in region III. Increasing I to 10^{10} W/cm^2 (a typical value for many LIMD reactions) will cause appreciable population of even more energetic states. Thus, the dissociation rate and the energy excess that must be carried by the fragments, will in general depend on laser fluence and power. In the case of TMD, different vibrational states of electronically excited acetone will be formed (see Fig. 3). This may lead to different emission spectra, provided vibrational relaxation is slower than fluorescence lifetime. Thermal reactions are known to produce primarily triplet acetone.[20–22] According to Fig. 3, both singlet and triplet acetone may be formed, with a possible preference of the triplet due to higher density of states. This topic, and its implications on the emission spectra observed in both LIMD and thermal reactions is further discussed in Section III.C.

Time-resolved studies of TMD dissociation in the submicrosecond range were first realized using LIMD techniques. They are discussed in the next section.

B. Time Behavior of TMD Chemiluminescence

CL reactions producing *electronically* excited states may be used to monitor dissociation kinetics with 10^{-9} sec time resolution, or better. In this respect they are unique. *Vibrationally* excited HF has been monitored,[10] but time response of IR detectors limits resolution to about 10^{-7} sec. TMD has also been studied by indirect CO_2 laser excitation, using CH_3F as donor.[27] In that study excitation involved collisions and unimolecular processes were not probed.

The luminescence observed by LIMD of TMD is shown in Fig. 4 on three different time scales. The shape of the laser pulse is also given as a reference. It is clear from the figure that part of the luminescence follows closely the laser pulse. This component will henceforth be termed the prompt luminescence. Another part is seen to last much longer than the laser pulse (note the $\times 10$ magnified laser signal decays to practically zero after 2.5 μsec). This emission will be referred to as the delayed luminescence. It is seen to consist of at least two different components: The more

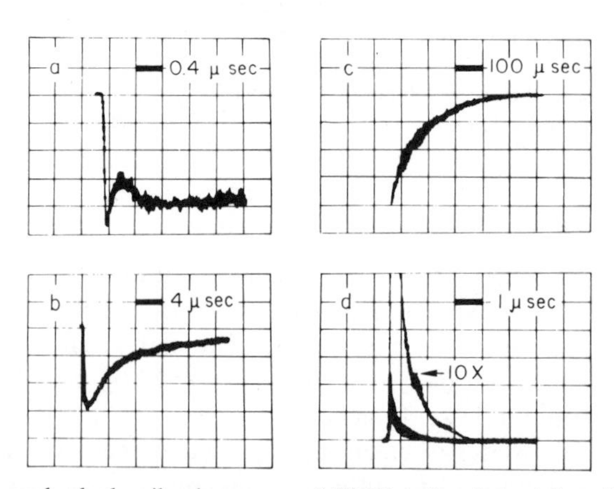

Fig. 4. Time-resolved chemiluminescence of TMD induced by infrared laser radiation. Figures (a–c) display the luminescence on different time scales; (d) shows the laser pulse.

intense one, with a decay time of about 10 μsec, is found to be strongly dependent on inert gas pressure. It is ascribed to collision induced excitation.[19, 28] The weaker one (amplification is 25 times larger in Fig. 4c than in Fig. 4b) is less sensitive to inert gas addition but is completely quenched by oxygen addition. These characteristics, as well as the observed decay time, indicate pure triplet acetone emission. The effect of N_2 and O_2 addition on delayed TMD chemiluminescence is demonstrated in Figs. 5 and 6.

The prompt (and possibly also the collision-induced delayed) emission is associated with a short-lived emitting state. In this context short-lived means shorter than the laser pulsewidth (100 nsec). This was checked by varying the laser pulse profile and by addition of inert gases, changing the decay kinetics. All results were consistent with the assumption that the time dependence of the reaction is reflected by the temporal behavior of the luminescence (as would be expected for a short-lived emitting species). We defer discussion of the nature of the emitting species to the next subsection. For now, suffice it to note that a short-lived species is known to exist in the system: S_1 lifetime in low-lying vibrational states is 2.7 nsec.[29]

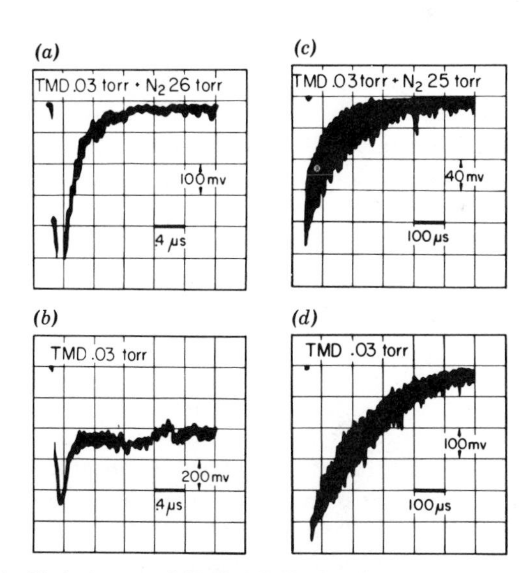

Fig. 5. Effect of added nitrogen of TMD LIMD chemiluminescence. (a) and (b) show that the prompt luminescence is little affected by nitrogen, while the collision-induced process is essentially eliminated. The effect on the long-lived delay emission, due to triplet acetone is shown in (c) and (d). The signal amplification in this case was ×25 higher than in (a) and (b). Reduction in triplet amplitude on addition of nitrogen is due to elimination of triplet formation by collision-induced processes.

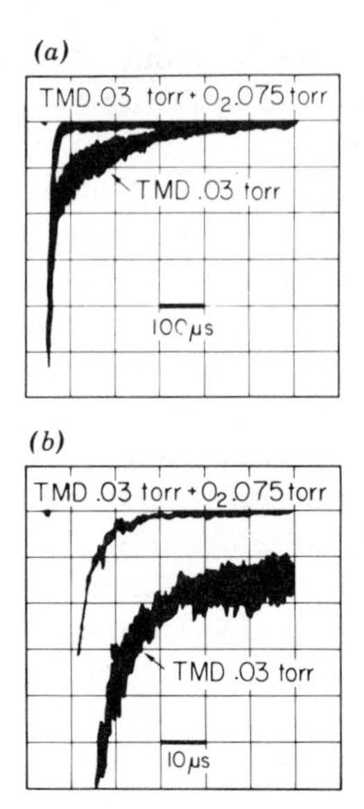

Fig. 6. Effect of oxygen on TMD LIMD chemiluminescence. Triplet emission is seen to be completely quenched.

As Figs. 4 and 5 show, collision-induced processes are clearly recognized in chemiluminescent LIMD reactions. They have been observed in many cases and discussed extensively (see, e.g., Refs. 28 and 30). A simple mechanism resulting in collisional activation is the following:

$$D + nh\nu \rightarrow D^* \tag{1}$$

$$D + n'h\nu \rightarrow D^{**} \tag{2}$$

$$D^* + D^* \rightarrow D^{**} \tag{3}$$

$$D^* + M \rightarrow D \tag{4}$$

$$D^{**} \rightarrow \text{products} \tag{5}$$

Here D is the unexcited substrate, D* and D** are vibrationally excited species below and above the dissociation limit, respectively, and M is an inert gas molecule. Process 3 is producing D** (and therefore chemiluminescence) by collisional activation. M can either be cold TMD or an added gas molecule (see Fig. 5). Collisions affect the decay rate of hot molecules either by inducing reactions, or by vibrational cooling (*V–T* transfer). Using low laser fluence, the latter is expected to dominate. The measured rate constant under these conditions was found to be 2.8×10^6 torr^{-1} sec^{-1}, a reasonable value for a *V–T* process.[20]

Since reaction proceeds by both a unimolecular and a collision-induced process, it is interesting to assess their relative contribution to overall product formation. Note that with a 10^{-7} sec laser pulsewidth and 10^{-6} sec average interval between collisions, one might expect collisionless processes to dominate. It turns out that with TMD, this is definitely not the case. Assuming that light emission intensity is proportional to the concentration of product molecules, one can estimate the total yield by measuring time integrated light intensity. This was done separately for the prompt emission, and for the collision-induced emission. It is found that reaction yield due to the collision-induced process is much larger than the unimolecular yield. The latter's contribution increases with laser fluence, but even with rather high energy densities (50 J/cm^2), it accounts for less than 10% of the yield. This seemingly surprising result may be unique to reactions with low activation energy: Here a rather large fraction of the vibrationally equilibrated population participates in the reaction even at low temperatures. In any case, it should be borne in mind that using a short laser pulse does not guarantee collision-free formation of products.

Turning now to the prompt chemiluminescence, we show a synchronous recording of the laser pulse and the luminescence at 5 nsec resolution in Fig. 7. It is evident that the onset of light emission (i.e., dissociation) is delayed with respect to the onset of the laser pulse train. According to the discussion in Section III.A this delay reflects the finite time interval required to accumulate sufficient energy in a single molecule by successive absorption of photons and to the actual "lifetime" of the superexcited molecule. In any case, the delay duration is expected to *increase* on *decreasing* laser fluence. This expectation is borne out by experiment[19]: The delay period increases, for a particular experimental set-up from 35 nsec at 142 mJ/pulse to 70 nsec at 75 mJ/pulse (cf. Fig. 7). Moreover, the energy content of the laser pulse during the delay period is found to be essentially constant (about 30 mJ) in all experiment, regardless of the total pulse energy. No prompt luminescence is observed when pulse energy is below 30 mJ. The observed incubation period may be looked upon as direct experimental evidence for the successive nature of the photon absorption process. Model calculations, in which this delay is used to estimate absorption coefficients in the quasi-continuum region (region II) are underway. A further result of the experiment is that peak power is not an important factor in LIMD of TMD: With total pulse energy of 70 mJ, the onset of luminescence occurs only *after* peak power was attained (see Fig. 7).

It will be noted that with high laser fluences, the delay is shortened to within experimental error limits (5–10 nsec) and may not be observed. The effect is most clearly displayed at low laser power, which unfortunately

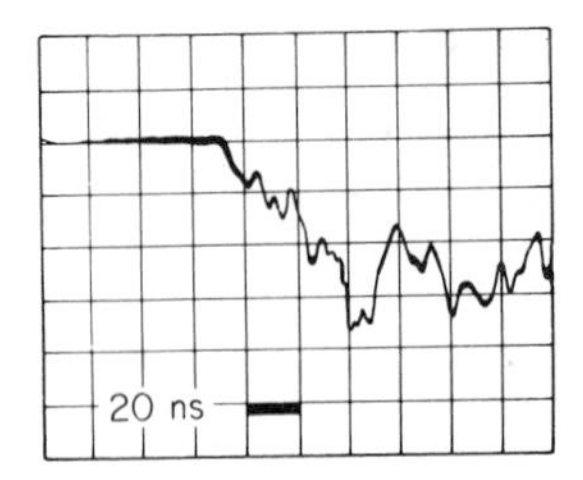

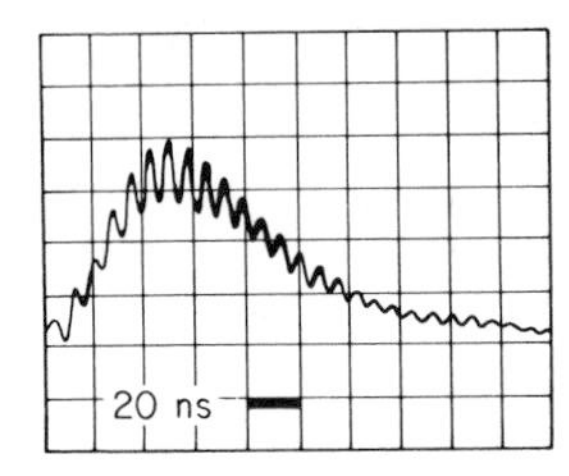

Fig. 7. A synchronous recording of the laser pulse (lower figure) and the chemiluminescence (upper figure). The delay period as defined in the text is 70 nsec. In this case (pulse energy is 75 mJ) the onset of the luminescence is also delayed with respect to the laser peak power.

involves lower dissociation yields. The speed and sensitivity of the CL method are thus fully exploited in this application. In Section IV we show that the appearance of a delay is not specific to TMD decomposition.

C. Nature of the Emitting Species. Comparison with Thermal Reactions

In the previous section kinetic evidence was brought forth to show that the long-lived delayed luminescence is due to triplet acetone. This assignment is further supported by spectral evidence: Fig. 8 shows the spectral distribution of TMD laser induced emission, along with acetone fluorescence and phosphorescence. The long-lived component is seen to overlap reasonably well with the phosphorescence spectrum. Assignment of the prompt component proves to be less straightforward. By the results of Fig. 8 it can evidently not be due to either pure singlet or pure triplet acetone. Its emission is seen to be fairly similar to the chemiluminescence spectrum obtained by thermal dissociation,[31] although a shift to the blue may be noted in the LIMD spectrum. In view of the discussion in Section II.A (cf. Fig. 3), one expects the luminescence to be composed of a mixture of singlet and triplet acetone emissions. The simplest assignment would be to consider the emission as a superposition of the two. This approach is best suited for the interpretation of usual thermal reactions: Luminescence is observed under steady state conditions, so that singlet and triplet state contributions are proportional to their populations multiplied by quantum

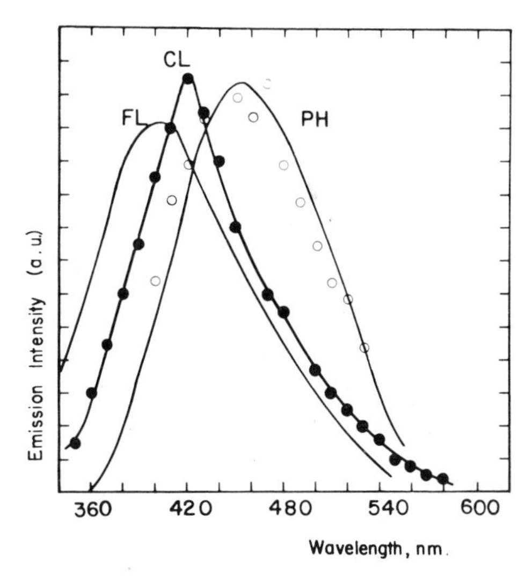

Fig. 8. TMD dissociation chemiluminescence spectra: (●) the prompt luminescence; (○) the spectrum 140 μsec after the laser pulse. The curved marked FL and PH are acetone fluorescence and phosphorescence, respectively.

efficiences, regardless of the actual lifetimes involved. In aerated solutions, dissolved oxygen completely quenches triplet emission, the resulting spectrum being essentially due to fluorescence.[21] In thoroughly deaerated solutions, one expects to obtain essentially pure phosphorescence, as the triplet is formed in large excess over the singlet. Usually, it is very difficult to eliminate traces of oxygen and other impurities, so that some triplet quenching is present even in deaerated solutions. Indeed, the spectrum of such solutions is shifted to the red and coincides with that of photoexcited acetone in similarly deaerated solutions.[21]

In LIMD experiments, a different situation holds. The prompt emission is recorded within about 10^{-7} sec. This means that (1) vibrational relaxation is not complete within the observation time (at pressure below 1 torr), (2) only a small fraction of triplet acetone emission (with a decay lifetime of $\sim 10^{-4}$ sec) can be detected. On the other hand, all of the singlet emission is completed within the said period, defined by the CO_2 laser pulsewidth. If the observed spectrum is still thought to consist of a mixture of singlet and triplet emissions, their relative contribution would be proportional to the emission quantum yield ϕ, which may be expressed as τ/τ_r where τ is the actual fluorescence lifetime and τ_r the radiative lifetime. In the case where τ is longer than the observation period, τ_{obs}, ϕ is replaced by $\phi_{corr} = \phi \cdot \tau_{obs}/\tau = \tau_{obs}/\tau_r$. For singlet acetone we have $\tau = 2.7 \times 10^{-9}$ sec

$\tau_r = 1.3 \times 10^{-6}$ sec,[29] so that $\tau < \tau_{obs}$, $\phi^S \simeq 2 \times 10^{-3}$. For triplet acetone we have $\tau = 2.8 \times 10^{-4}$ sec,[32] $\tau_r = 7 \times 10^{-2}$ sec,[34, 35] thus $\phi^T_{corr} = 1.4 \times 10^{-6}$. For triplet acetone to appreciably affect the prompt spectrum it must thus be formed in about 1400 $(= \phi^S/\phi^T_{corr})$ fold excess over the singlet. The data of Fig. 5 and similar experiments may be used to obtain the triplet contribution to the emission intensity during the laser pulse. This is simply done by extrapolating the triplet emission intensity to $t = 0$, that is, the onset of the reaction, and comparing the obtained value with the actual signal observed. As triplet decay time is much longer than the laser pulsewidth, the latter can be considered as a delta function for this purpose. This procedure leads to an estimated ratio of about 1 : 1 (singlet : triplet). If that indeed were the case, one would expect to observe essentially pure singlet acetone spectrum.

Our task now is to assign the prompt LIMD emission spectrum, and to account for the apparent discrepancy between the relatively low triplet yield obtained by our extrapolation procedure, and the large (50-fold) excess triplet formed in thermal excitation.[22]

We shall develop the arguments by reference to Fig. 3. It should be borne in mind that data on the excited states of acetone were so far obtained by optical excitation only, at relatively high pressures. Recent studies on formaldehyde,[35] biacetyl,[36] and other carbonyl compounds show that many important effects are masked when operating at high pressures. Thus, biacetyl fluorescence was found to be multiexponential at very low pressures, the well-known nanosecond emission being just the fastest decaying component.[36] In TMD LIMD studies, relatively low pressures ($\leqslant 0.1$ torr) were used. Furthermore, the initially formed state is not a pure singlet—it could be thought of as a superposition of zero-order states belonging to the triplet and singlet manifolds. As discussed in Section III.A increased laser fluence should favor formation of highly vibrationally excited states. These may be seen from Fig. 3 to lie above the barrier for acetone decomposition. Thus, in the presence of a strong field the lifetime of the excited acetone is expected to be reduced due to molecular dissociation. The observed spectrum would be that of vibrationally excited acetone. It may again be thought of as a mixture of singlet and triplet acetone, with the important difference, that they are not vibrationally relaxed. Kommandeur et al.[36] showed that in the case of biacetyl, this situation leads to a complicated emission spectrum, consisting of fast and slow fluorescent and of "hot" and thermalized phosphorescence. The slow fluorescence and the "hot" phosphorescence have the same decay rate constant, and can be distinguished only by their spectrum. The situation in the infrared laser induced chemical preparation of excited acetone may be similar, except that the overall decay rate is governed by either decomposition or vibrational relaxation. The latter eventually leads to the formation of the thermalized triplet. This description of TMD LIMD reactions appears to be consistent with results obtained so far. The spectrum is expected to be similar to that obtained thermally, but not identical to it. The lifetime of the emitting species is very short ($\sim 10^{-9}$ sec). Prompt formation of thermalized triplet is negligible, explaining its small contribution to the emission during the laser pulse. In order to further check the proposed description, the following predictions should be tested:

1. Reducing laser fluence, one should obtain products with lower excess energy. The spectrum should correspondingly shift to the red, and extrapolate to pure phosphorescence at zero laser power. Preliminary experiments show that a red shift is indeed obtained on using lower fluences.
2. Vibrational relaxation could compete with dissociation, provided the pressure were high enough. As Fig. 5 shows, addition of 26 torr of nitrogen eliminates collision induced TMD dissociation, but apparently does not affect either the prompt or the thermalized triplet emissions. A reasonable value for vibrational relaxation by nitrogen would be 10^7 s^{-1} $torr^{-1}$. Taken together, these values indicate that the unimolecular rate constant controlling the decay of the emitting species is about 10^9 sec^{-1}, in agreement with optical excitation studies. Increasing nitrogen pressure to a few hundred torr should suffice to cause an increase in the population of the thermalized triplet.

IV. FURTHER EXPERIMENTS OF SiF_4

Properties of LIMD chemiluminescent reactions as revealed in our TMD studies, may possibly be of general applicability. In particular, the delay in the appearance of the prompt emission could be a common occurrence. The nature of the emitting species need not be known to investigate it: all that is required is direct formation of an electronically excited product. We chose to restudy the dissociation of SiF_4, the first molecule claimed to dissociate unimolecular under intense laser radiation.[8] One reason for our choice was a recent challenge to the previous interpretation. Initiating the reaction by a nanosecond laser,[37] the luminescence was shown to be due to collisional activation [cf. (1)–(5)]. In both studies, the laser was operated at 10.6 μm, that is, about 100 cm^{-1} off resonance from the strong 9.6 μm SiF_4 ν_3 fundamental. This may explain in part the relatively high pressures used in both studies.[8, 37] We employed the 10.2 and 9.6 μm bands, and were able to extend the pressure range to about 10^{-3} torr.

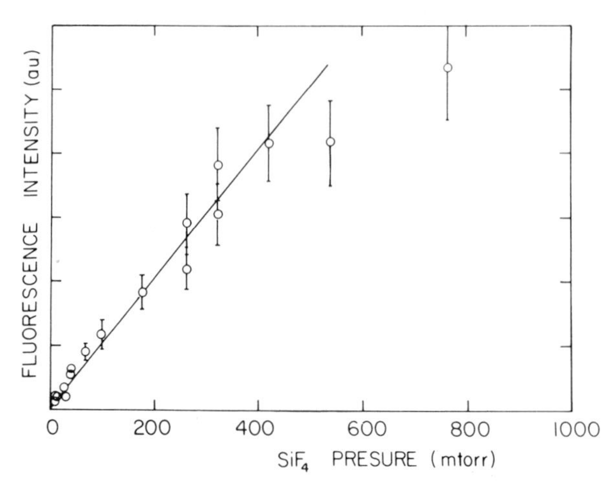

Fig. 9. Intensity of prompt SiF_4 chemiluminescence vs gas pressure. Laser energy was 120 mJ at 10.2 μm.

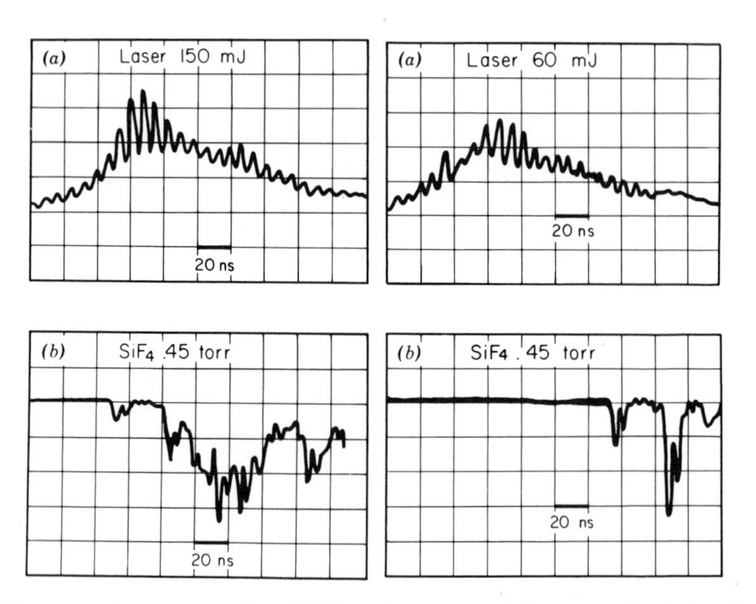

Fig. 10. A synchronous recording of SiF_4 luminescence and the laser pulse. At low fluence level the delay is seen to be much longer than at high levels. 150 mJ and 60 mJ correspond to about 75 J/cm^2 and 30 J/cm^2, respectively.

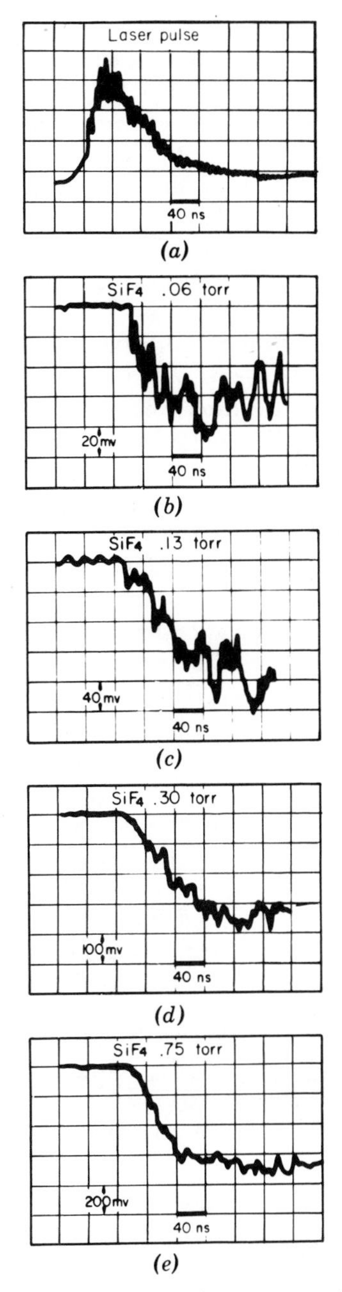

Fig. 11. Time resolved SiF_4 luminescence at various pressures, irradiation with a constant fluence. Note the change in vertical scale sensitivity. The delay is independent of the pressure.

Our results confirm the early assignment of SiF_4 CL as unimolecular; the pressure dependence of the fluorescence signal is shown in Fig. 9. It is seen that up to about 0.5 torr, the signal is linear with pressure. At higher pressures the slope is seen to decrease, perhaps indicating pressure quenching of the fluorescence. Figure 10 shows that, as in TMD, a delay is observed between the onset of the laser pulse and the onset of the luminescence signal. Once again, its duration is independent of substrate pressure, as shown in Fig. 11. Signal intensity is seen to increase considerably (note the changes in vertical scale), but the delay is practically constant.

A detailed discussion of the SiF_4 LIMD CL reactions is not feasible yet, in view of the inherent problems outlined in Section II. The identity and relative abundance of the emitting species are not known. In spite of these difficulties, the unimolecular nature of the process appears to be well established in view of the data of Figs. 9 and 11. The similarity to TMD delay characteristics suggests that an energy level diagram similar to Fig. 3 may be relevant to SiF_4 as well. On the other hand, a different mechanism cannot be ruled out: dissociation of SiF_4 to yield SiF_3, that is further dissociated by the same laser pulse to yield SiF_2—the actual emitting species. Such sequential dissociation has been suggested before.[1] If it holds for SiF_4, the incubation period as a function of laser fluence is readily accounted for. Preliminary experiments on inert gas (N_2) effect on the luminescence intensity are consistent with such a mechanism.

Situations where no delay is to be expected can also arise in LIMD reactions. Thus, it is interesting to consider a case where the rate determining step in the energy buildup process occurs in region I. If such a "bottleneck" exists, any molecule that reaches region II is immediately dissociated. In this case, an incubation period is not expected and a rather large intensity threshold is expected. This situation is extremely unlikely in large molecules, but may be found in small ones, such as OCS, O_3,[38] and perhaps BCl_3.

Acknowledgments

I wish to thank my students, Giora Yahav and Sandy Ruhman who carried out much of the experimental work. I am grateful to Professors C. B. Moore and C. Wittig for sending me preprints of their papers. This work was supported by the Israel Commission for Basic Research.

References

1. (a) M. J. Coggiola, P. A. Schulz, Y. T. Lee, and Y. R. Shen, *Phys. Rev. Lett.*, **38**, 17 (1977). (b) Aa. S. Sudbø, P. A. Schultz, Y. R. Shen, and Y. T. Lee, *J. Chem. Phys.*, **69**, 2312 (1978). (c) Aa. S. Sudbø, P. A. Schultz, E. R. Grant, Y. R. Shen, and Y. T. Lee, *J. Chem. Phys.*, **70**, 912 (1979).

2. F. Brummer, T. P. Cotter, K. L. Kompa, and D. Proch, *J. Chem. Phys.*, **67**, 1547 (1977).
3. J. D. Campbell, G. Hancock, J. B. Halpern, and K. H. Welge, *Opt. Comm.*, **17**, 38 (1976).
4. S. E. Bialkoswki and W. A. Guillory, *J. Chem. Phys.*, **68**, 3339 (1978).
5. J. D. Campbell, M. H. Yu, M. Mangir, and C. Wittig, *J. Chem. Phys.*, **69**, 3854 (1978).
6. J. C. Stephenson and D. S. King, *J. Chem. Phys.*, **69**, 1485 (1978).
7. R. V. Ambartzumian and V. S. Letokhov, *Acc. Chem. Res.*, **10**, 61 (1977).
8. N. R. Isenor, V. Merchant, R. S. Hallsworth, and M. C. Richardson, *Can. J. Phys.*, **51**, 128 (1973).
9. R. V. Ambartzumian, N. V. Chekalin, V. S. Doljikov, V. S. Letokhov, and E. A. Ryabov, *Chem. Phys. Lett.*, **25**, 515 (1974).
10. See, e.g., C. R. Quick, Jr. and C. Wittig, *Chem. Phys.*, **32**, 75 (1978).
11. V. S. Letokhov and C. B. Moore, *Sov. J. Quant. Electron.*, **6**, 259 (1976).
12. M. C. Gower and K. W. Billman, *Opt. Comm.*, **20**, 123 (1977).
13. D. M. Brenner, *Chem. Phys. Lett.*, **57**, 357 (1978).
14. P. Kolodner, C. Winterfeld, and E. Yablonovitch, *Opt. Comm.*, **20**, 119 (1977).
15. See, e.g., N. Bloembergen and E. Yablonovitch, *Phys. Today* (May 1978), p. 28.
16. J. L. Lyman, *J. Chem. Phys.*, **67**, 1868 (1977).
17. W. Fuss, *Chem. Phys.*, **36**, 135 (1979).
18. M. H. Yu, H. Reisler, M. Mangir, and C. Wittig, *Chem. Phys. Lett.*, **62**, 439 (1979). See also H. Reisler and C. Wittig's paper in this volume.
19. G. Yahav and Y. Haas, *Chem. Phys.*, **35**, 41 (1978).
20. For a review, see, e.g., T. Wilson in *Int. Rev. Sci.*, Ser. 2, Vol. 9, *Chemical Kinetics*, ed. D. R. Herschbach, Butterworth, London, 1976, p. 265.
21. N. J. Turro, H. C. Steinmetzer, and A. Yekta, *J. Am. Chem. Soc.*, **95**, 6468 (1973).
22. N. J. Turro and P. Lechtken, *J. Am. Chem. Soc.*, **94**, 2886 (1970).
23. Y. Haas and G. Yahav, *J. Am. Chem. soc.*, **100**, 4885 (1978).
24. Y. Haas and G. Yahav, *Chem. Phys. Lett.*, **48**, 63 (1977).
25. W. Forst, *Theory of Unimolecular Reactions*, Academic Press, New York, 1973.
26. P. T. Robinson and K. A. Holbrook, *Unimolecular Reactions*, Wiley, New York, 1972.
27. W. E. Farneth, G. Flynn, R. Slater, and N. J. Turro, *J. Am. Chem. Soc.*, **98**, 7877 (1976).
28. M. L. Lesiecky and W. A. Guillory, *Chem. Phys. Lett.*, **49**, 92 (1977).
29. G. M. Breuer and E. K. C. Lee, *J. Phys. Chem.*, **75**, 989 (1971).
30. W. Fuss and T. P. Cotter, *Appl. Phys.*, **12**, 265 (1977).
31. S. P. Schmidt and G. B. Schuster, *J. Am. Chem. Soc.*, **100**, 5559 (1978).
32. A. Gandini and P. A. Hackett, *J. Am. Chem. Soc.*, **9**, 6197 (1977).
33. E. H. Gilmore, G. E. Gibson, and D. S. McClure, *J. Chem. Phys.*, **20**, 829 (1952), **23**, 399 (1955).
34. R. F. Borkman and D. R. Kearns, *J. Chem. Phys.*, **44**, 945 (1966).
35. J. C. Weisshar and C. B. Moore: "Collisionless Non-radiative Decay Rates of Single Rotational Levels of S_1 Formaldehyde", LBL 8766 preprint.
36. R. Van Der Werf and J. Kommandeur, *Chem. Phys.*, **16**, 125 (1976).
37. V. E. Merchant, *Opt. Comm.*, **25**, 259 (1978).
38. D. Proch and H. Schroder, *Chem. Phys. Lett.*, **61**, 426 (1979).

AUTHOR INDEX

Numbers in parentheses are reference numbers and indicate that the author's work is referred to although his name is not mentioned in the text. Numbers in italics show the pages on which the complete references are listed.

Aartsma, T. J., 15 (123, 124), 66 (123, 124), 69 (123, 124), 82 (123, 124), 84 (124), *108*, 510 (23c), 536 (23c), 551 (23c), *552*
Abella, I. D., 15 (115), 66 (115), *108*, 585 (27), *622*
Abragam, A., 535 (50), 540 (50), 548 (50), *553*
Abraham, R., 20 (16), 202 (16), 209 (16), 210 (16), 219 (16), *235*
Abramowitz, M., 384 (36), 386 (36), *493*, 593 (60), *623*
Ackerhalt, J. R., 53 (210), 55 (227), 70 (304, 305, 308), 72 (304, 305, 308), 73 (315), 76 (308), 81 (315), *110*, *112*, *113*, 509 (18b), 512 (18b), 537 (18b), 540 (18b), 543 (18b), 544 (58a, 58b), *552*, *553*, 556 (32, 33, 40), 567 (16, 19, 32), *622*
Aczel, J., *292*
Adler, F. T., 472 (66a), 473 (66a), 479 (66a), *494*
Afraimovich, V. S., 202 (8), *235*
Agarwal, G. S., 556 (17), 567 (45), *579*, *580*
Agmon, N., 252 (59), 286 (140), *290*, *292*
Ahmad-Bitar, R., 331 (12), *361*
Airey, J. R., 364 (7c), *492*
Akhmanov, A. S., 586 (30), 618 (30), *622*
Akulin, V. M., *622*
Albrecht, A. C., 55 (224), *110*, 293 (2, 3, 4, 5), 294 (2, 3, 4, 5, 26a, 26b), 295 (27), 298 (2, 3, 4, 5, 26a, 26b), 302 (26b), 303 (43), 305 (26a, 26b), 306 (26a, 26b), 307 (26a, 26b), 309 (26a, 26b), 310 (26a, 26b), 313 (26a, 26b), *321*, *322*
Aldridge, J. P., 587 (31), 589 (47), 590 (47), *623*
Alfano, R. R., 9 (91), *107*
Alhassid, Y., 54 (221), *110*, 252 (59), 264 (82, 83, 89), 265 (89), 267 (89, 99), 268 (89), 270 (89), 275 (89, 122), 281 (89), 285 (122), 286 (89), *290*, *291*, *292*
Alimpiev, S. S., 584 (5), 586 (30), 591 (5), 618 (30), *622*
Allen, J., 577 (68), *580*
Allen, L., 4 (55), 66 (55), 82 (55), *106*
Alterman, E. B., 241 (18), *289*
Ambartzumian, R. V., 3 (43), 5 (43), 12 (95), 18 (43), 51 (43), 70 (43, 95), 72 (43), *106*, *107*, 293 (21, 22a), *321*,

602 (81), *624*, 639 (1b, 2), 640 (28), 647 (1b), 655 (1b), *658*, 661 (7, 8, 9), *677*, 679 (3), 681 (6, 9), 682 (10, 11), 692 (25), 709 (45), *710*, *711*, 714 (7, 9), 716 (9), 717 (9), *733*
Amiot, J. C., 460 (56), *494*
Amirav, A., 6 (69, 75, 76, 77), 7 (77), 11 (69, 76), 15 (133), 44 (69, 76, 77), 46 (69), 75 (69, 75, 76, 77, 133, 319), 90 (360), *106*, *107*, *108*, *113*, *114*
Anderson, A. B., 296 (37), *322*
Anderson, P. W., 512 (38), 541 (38), 543 (38), *553*
Anderson, V. M., 4 (58), *106*
Anlauf, K. G., 626 (6), *638*
Antonov, V. S., 244 (69), 257 (69), *290*
Applebury, M. L., 9 (86), 98 (86), 99 (86), *107*
Arecchi, F. T., 3 (34), *105*
Arez, A., 202 (16), 206 (16), 209 (16),

210 (16), 219 (16), *235*
Armstrong, J. A., 3 (46), 18 (46), *106*
Arnold, V. I., 55 (226), *110,* 124 (8), 161 (8), *198,* 202 (16), 206 (16), 209 (16), 210 (16), 219 (16), *235*
Aronson, E. B., 277 (128), *292*
Ash, R., 274 (119), 275 (119), *292*
Ashford, M. N. R., 656 (44), *659*
Ashmore, P. G., 58 (254), *111*
Ashton, G. J., 367 (26a, 26b), 393 (26b), 464 (26a, 26b), 471 (26a, 26b), *493*
Asprey, L. B., 664 (21), 676 (29), *677*
Atabek, O., 17 (136), 58 (247), *108, 111,* 388 (42), 396 (42), 403 (42), *493*
Audibert, H. M., 364 (7e), *492*
Auerbach, D. A., 328 (7), 331 (11), 334 (11), *361*
Augustin, S., 54 (222), *110*
Augustin, S. D., 467 (63), 476 (63), *494*
Austin, R. H., 91 (377), 98 (377), *114*
Avez, A., 16 (8), 55 (226), *110,* 124 (8), 161 (8), *198*
Avoires, P. H., 10 (373), 98 (373), 99 (373), 100 (373), 101 (373), *114*
Avrillier, S., 584 (13, 69), 587 (13), 589 (13), 598 (13, 71), 599 (71), *622, 623, 624*
Azuel, A., *114*
Azumi, T., 90 (359), *114*

Back, R., 639 (9), 658
Bagratashvili, V. N., 586 (30), 618 (30), *622*
Baklanov, E. V., 556 (29), 578 (29), *580*
Baliam, R., 4 (56), 15 (56), 66 (56), 82 (56), *106,* 577 (78), *581*
Band, Y. B., 58 (244), *111*
Baranov, V. Yu., 586 (30), 618 (30), *622*
Barbanis, B., 150 (34), *199,* 218 (42), *236*
Barker, J. R., 57 (233), 81 (323), *111, 113,* 511 (27a), *552*
Baronavski, A., 244 (58), 245 (58), *290*
Bartlett, J. H., 202 (3), 211 (30), *235, 236*
Bartram, R. H., 97 (383), *114*
Bar Ziv, E., 63 (265), *111*
Bauer, S. H., 58 (251), *111,* 242 (34), *289,* 626 (5), *638*
Beauchamp, J., 639 (17), 656 (17), *658*
Beauchamp, J. L., 3 (53), 73 (53), *106,* 509 (5b), 511 (5b), 512 (5b), 533 (5b), 535 (5b), *552*
Beck, S. M., 6 (71, 72), 44 (71), *106, 107,* 331 (13), 341 (29), *361,* 510 (24a), 543 (24a), 551 (24a), *552*
Becker, R., 626 (5), *638*
Beeson, K. W., 91 (377), 98 (377), *114*
Behlen, F. M., 6 (73), *107*
Bellingsley, J., 364 (7b), *492*
Benettin, G., 118 (4), 124 (4), 141 (4), *198,* 213 (34), 214 (34), 225 (53), 233 (78), *236, 237, 238,* 241 (17), 242 (17), *289*
Benmair, R. M. J., 3 (47), 70 (47), *106*
Benmer, R. M. J., 639 (14), *658*
Ben Reuven, A., 18 (143), 95 (143), *108,* 512 (39), 523 (39), 541 (39), 543 (39), *553,* 556 (12, 15, 25, 26, 30), 558 (15, 43), 569 (25), 570 (25), 573 (12), 575 (43), *579, 580*
Benshaul, A., 2 (22), *105,* 244 (44, 46), 246 (44, 46), 251 (44, 46), 253 (44, 46), 255 (46), 264 (44, 46), 288 (44, 46), *290*
Berg, J. O., 51 (199), *110*
Bergman, K., 65 (273), *112,* 626 (8b), *638*
Berkowitz, M., 85 (341), *113*
Berman, M., 556 (12), 573 (12), *579, 580*
Berman, P. R., 575 (67), 578 (80), *580, 581*
Bernard, L., 15 (110), *107*
Berne, B., 534 (47), 535 (47), *553*
Bernstein, H. J., 15 (109), *107*
Bernstein, L., 491 (71), *494*
Bernstein, R. B., 2 (2), 24 (165, 166), 51 (198), 52 (198), 58 (252), 85 (2), *104, 109, 111,* 231 (59), *237,* 241 (20), 242 (22, 23, 33), 243 (22, 40), 244 (33, 43, 45, 68, 70), 246 (33), 257 (22), 264 (33), 277 (22), *289, 290*
Berry, M. V., 53 (211, 212), 54 (212), 58 (245), *109, 110, 111,* 118 (2), 134 (15), 142 (2), 169 (46, 51), 176 (46), 196 (92), *198, 199,* 204 (18, 22), 205 (17e), 207 (17e), 210 (17e), 215 (17e), *235, 236,* 242 (24, 26), 244 (24, 26,

49, 52), 246 (26), 247 (24, 26, 49), 248 (26, 52), 253 (24), 254 (61), *289*, *290*, 293 (6, 7), 294 (6, 7), 298 (6, 7), 300 (7), 315 (6, 7), 316 (6, 7, 56), 319 (56), *322*, 510, (21), 544 (21), 551 (21), *552*
Bersohn, R., 15 (113), 58 (240, 242), 59 (242), *108, 111*
Beswick, J. A., 17 (136), 18 (144), 25 (170), 26 (170), 39 (170), 86 (170), *108, 109*, 346 (37), 349 (37), 353 (48), 357 (37), *362*, 367 (27a), 368 (28, 29, 30, 31), 373 (27a), 387 (27a), 388 (42), 396 (27a, 42, 70), 403 (42), 413 (28), 418 (28), 419 (28), 421 (29), 422 (28), 427 (27a), 428 (27a), 429 (30), 430 (29), 433 (27a, 30), 441 (30), 442 (30), 451(27a), 459 (31), 462 (31), 467 (61), 470 (27a, 28, 29, 30, 31), 471 (65), 476 (68), 478 (27a, 28, 29, 30, 31), 480 (27a, 28, 29, 30, 31), 481 (27a, 28, 29, 30, 31), 482 (31), 484 (27a), 486 (27a), *493, 494*
Beterov, I. M., 570 (49), 574 (64), *580*
Bialkowski, S. E., 682 (13), 692 (13), 694 (13), 707 (13), *710*, 714 (4), *733*
Bialynicka-Birula, Z., 556 (23), 569 (23), *579*
Bialynicki-Birula, I., 556 (23), 569 (23), *579*
Billman, K. W., 714 (12), *733*
Bird, R. B., 324 (1), *360*, 364 (3), 429 (3), *492*
Birkhoff, G. D., 149 (16), 150 (16), *198*
Birss, F. W., 595 (67), *623*
Bixon, M., 14 (102), 29 (174), 30 (176), 31 (174), 32 (176), 35 (174), *107, 109*, 315 (54), *322*, 446 (52), *494*
Black, J. G., 52 (206), 58 (206), 70 (310), 72 (206, 310), 74 (310), 76 (310), 81 (310), *110, 113*, 241 (9), 244 (9), *289*, 509 (6a), 511 (6a), 550 (6a), 552, 556 (35), 568 (35), *580*, 635 (17), *638*, 640 (29), 648 (39), *658, 659*
Blackwell, B. A., 59 (257), *111*, 286 (139), *292*
Blaney, B. L., 324 (2), *360*
Blazy, J. A., 349 (41, 42), 350 (43), 354 (43), 467 (60), *494*, 671 (26), 672 (26), 675 (26), *677*
Blinov, S. I., 686 (21), 693 (21), *711*
Bloch, F., 557 (41), *580*
Bloembergen, N., 3 (40), 5 (40), 11 (40), 18 (40), 21 (40), 30 (40), 52 (206), 58 (206), 70 (40, 283, 287, 292, 310), 72 (40, 206, 283, 287, 310), 73 (40), 74 (40, 310), 76 (40, 283, 310), 80 (40), 81 (40, 310), *105, 110, 112, 113*, 241 (9), 244 (9), *289*, 293 (22c), *321*, 509 (5c, 6a, 13a, 13b), 511 (5c, 6a), 533 (5c), 535 (5c), 550 (5c, 6a), *552*, 556 (3, 35), 568 (35, 47), 570 (3), *579*, 601 (77), 602 (77), *624*, 635 (17), *638*, 639 (1c), 640 (21, 23, 29), 648 (39), *659*, 715 (15), *733*
Blum, F. A., 588 (44), *623*
Bobin, B., *623*
Boese, R. W., 37 (177), *109*
Bogan, B. J., 47 (192), *109*
Bogan, D., 244 (51), 248 (51), *290*
Bogan, D. J., 244 (48, 50), 246 (48), 248 (50), 249 (50), 253 (50), *290*, 686 (22), 692 (22), 693 (22), *711*
Bogoliubov, N. N., 149 (26), *199*
Boltzmann, L., 54 (214), *110, 292*
Bomse, D., 639 (17), 656 (17), *658*
Bomse, D. S., 509 (5b), 512 (5b), 533 (5b), 535 (5b), *552*
Bondybey, V. E., 84 (336), 85 (339), 89 (352), *113, 114*
Boquillon, J. P., 12 (99), *107*
Borde, C., 661 (1), *676*
Borde, ch, J., 584 (13), 587 (13), 589 (13), 598 (13), 599 (71), *622, 623, 624*
Borkman, R. F., 728 (34), *733*
Born, M., *199*
Bornse, D. S., 3 (53), 73 (53), *106*
Bott, J. F., 3 (35), *105*
Bouchiat, C. C., 364 (8a), *492*
Bouchiat, M. A., 364 (8a, 8b), *492*
Bountis, T., 152 (35), 179 (35), *199*
Bourger, H., 18 (138), *108*
Bourimov, V. N., 681 (8), 692 (8), 694 (8), *710*
Bradley, J. N., 626 (5), *638*
Brady, J. W., 104 (393), *114*
Bralkowski, S., 639 (5), *658*
Brauman, J. I., 57 (233), *111*

Braun, W., 639 (15), *658*
Bray, R., 510 (21), 544 (21), 551 (21), *552*
Bray, R. G., 53 (211), *109, 110*, 293 (6, 7), 294 (6, 7), 298 (6, 7), 300 (7), 315 (6, 7), 316 (6, 7), *321*
Brejot, M., 15 (110), *107*
Brenner, D. M., 639 (13), *658,* 714 (13), *733*
Breuer, G. M., 723 (29), 728 (29), *733*
Brewer, R. G., 4 (56), 15 (56, 116, 117, 118, 119, 120, 132), 66 (56, 116, 117, 118, 119, 120), 82 (56, 326), *106, 108, 113,* 510 (23a), 536 (23a), 551 (23a), *552,* 556 (18), 569 (18), *579*, 598 (70), *623*
Briggs, J. S., 70 (300, 307), 72 (307), 76 (307), *112,* 509 (17), 510 (17), 512 (17), 537 (17), 540 (17), 543 (17), *552*, 598 (75), 600 (75), *624*
Brink, D. M., 593 (62), 615 (62), *623*
Brnstein, R. B., 246 (43, 45), 253 (43, 45), 256 (70), 257 (68, 70), 264 (43, 45), 269 (45), 273 (114), 282 (134), 283 (134), *292*
Brooks, P. R., 63 (267), 64 (267, 269), *111*
Brossel, J., 364 (8b), *492*
Broucke, R., 202 (10), *235*
Brown, R. F., 104 (392), *114*
Bruch, L. W., 364 (5), 371 (5), *492*
Brueck, R. J., 12 (98), *107*
Brumer, P., 118 (3), 124 (10), 177 (55), *198, 200,* 212 (32), 214 (36), 215 (38, 40a, 40b), 216 (32, 41), 217 (40a, 40b), 219 (3b), 221 (40a, 40b, 51), 223 (40a, 40b), 224 (40a, 40b), 225 (53), 231 (40b, 58b), 232 (40b), 233 (40a, 40b, 58b, 69, 79), 234 (32, 40a, 40b, 41, 81), *237, 238,* 241 (19), 242 (19), *289*
Brummer, F., 714 (2), *733*
Brunet, H., 588 (40), *623*
Brunner, F., 81 (321), *113*
Brus, L. E., 9 (82), 84 (336), 85 (338, 339, 343), 88 (349), *107, 113,* 369 (33), *493*
Buckingham, A. D., 2 (16), 3 (16), 5 (16), 14 (16), 17 (16), 32 (16), 39 (16), 40 (16), 41 (16), 42 (16), 43 (16), 89 (16), 90 (16), *105*, 364 (4), *492*
Buhks, E., 94 (380), 101 (387), 102 (387), *114*
Bunker, D. L., 241 (11, 14), *289,* 510 (26a), 511 (26a), *552*
Burak, I., 509 (11b), 510 (11b), *552,* 592 (53, 54), *623,* 709 (48, 49, 50), *711*
Burberry, M. S., 293 (4, 5), 294 (4, 5, 26a, 26b), 295 (26b, 27), 298 (4, 5, 26a, 26b), 302 (26b), 303 (43), 305 (26a, 26b), 306 (26a, 26b), 307 (26a, 26b), 309 (26a, 26b), 310 (26a, 26b), 313 (26a, 26b), *321, 322*
Busch, G. E., *111*
Buss, R. J., 244 (56), 248 (56), *290*
Butler, J. N., 241 (4), 244 (4), *288*
Button, K. J., 556 (10), 573 (10), *579*
Bykov, V. V., 202 (8), *235*

Calawa, A. R., 3 (36), *105,* 588 (44), *623*
Callamon, J. H., 4 (58), *106*
Callear, A. B., 364 (7b), *492*
Callender, R. H., 9 (91), *107*
Campbell, J., 639 (8), *658*
Campbell, J. D., 695 (29a), 707 (40), *711,* 714 (3, 5), *733*
Cantrell, C. D., 3 (41), 5 (41), 11 (41), 15 (134), 18 (41), 21 (41), 70 (41, 292, 297, 298, 303, 308), 72 (41, 134, 303, 308), 74 (41), 75 (134), 76 (308), *105, 108, 112,* 293 (22b), 320 (22b), *321,* 509 (5a, 13b, 18a, 18b), 537 (18a, 18b), 540 (18a, 18b), 543 (18a, 18b), *552,* 556 (7), *579,* 584 (10, 11, 12, 13, 16, 17), 585 (25, 29), 587 (13), 588 (10, 11, 12), 589 (10, 11, 12, 13, 48, 49), 591 (11), 598 (13, 74), 600 (13, 74), 602 (14, 69), 615 (18), *622, 623, 624,* 640 (20, 27), *658*
Carley, J. S., 334 (15, 16), *361,* 477 (69a, 69b), *494*
Carlson, G. L., 663 (20), *677*
Carlsten, J. L., 509 (18a), 510 (18a), 512 (18a), 537 (18a), 540 (18a), 543 (18a), *552,* 573 (54), 578 (82), *580, 581*
Carmeli, B., 40 (181, 187, 188), 41 (187, 188), 70 (187, 188), 72 (187, 188), 75 (188), 76 (187, 188), 77 (187, 188), 78 (187, 188), 80 (188), 81 (187, 188), *109,* 320 (62), *322,* 535 (49), 548 (49), *553,* 556 (36), *580*
Carney, G. D., 295 (33), *322*
Carter, D., 234 (81), *238*

Casartelli, M., 214 (35a), 216 (35a), 219 (47), *236*
Casati, G., 218 (45), 219 (47), 233 (77), *236, 238*
Case, D. A., 277 (125), 286 (125), *292*
Certain, P. R., 364 (5), 371 (5), *492*
Chaiken, J., 14 (103), *107,* 233 (66), *237*
Chalek, C. L., 257 (75), 258 (75), *291*
Chance, B., 9 (85, 92), 91 (85), 98 (374), 99 (374), 101 (374), *107, 114*
Chandler, D. W., 341 (30), *362*
Chank, C. V., 2 (19), 3 (19), 4 (19), *105*
Chao, K. J., 241 (5), 244 (5), *288,* 511 (27a), *552*
Chapman, S., 169 (47), 171 (47), *199*
Chebotaev, V. P., 556 (2, 4, 29), 569 (2), 570 (49), 574 (2, 64), *579, 580*
Chekalin, H. V., 12 (95), 70 (95), *107,* 509 (2a), *551,* 639 (2), *658,* 661 (7, 8), *677,* 681 (6), 682 (10, 11), 707 (38), *710, 711,* 714 (9), 716 (9), 717 (9), *733*
Chen, Y. R., 244 (64), 254 (64), *290*
Chernoff, D. A., 74 (318), *113*
Cheshnovsky, O., 88 (350), *113*
Child, M., 367 (26a), 393 (26a), 464 (26a), 471 (26a), *493*
Child, M. S., 2 (3), *104,* 233 (76), *238,* 301 (39), 307 (39), *322*
Chirikov, B. V., 132 (13), *198,* 209 (28), 226 (28, 31), 229 (28), 230 (31), 233 (77), *236, 238*
Christie, J. R., 4 (58), *106*
Churchill, R. C., 205 (24), 219 (24), *236*
Clark, R. J. H., 589 (46), *623*
Clary, D. C., 261 (80), 262 (80), 277 (124), 284 (80), *291*
Clayton, R. K., 9 (89), 98 (89), 99 (89), 100 (89), *107*
Cogdell, R. J., 98 (372), 99 (372), 100 (372), 101 (372), *114*
Coggiola, M. J., 5 (62a), 12 (62), 52 (205), 58 (205), *106, 110,* 241 (8), 244 (8, 56), 248 (56), *288, 290,* 509 (7a), 512 (7a), 550 (7a), 551 (7a), *552,* 633 (15), 635 (15), *638,* 714 (1a), 716 (1a), 728 (1a), *732*
Cohen, C., 512 (45a, 45b), 516 (45b), *553*
Cohen, E. G. D., 118 (1), *198,* 215 (17a, 17g), *235*
Cohen-Tannoudji, C., 556 (22), 577 (22, 78), *579, 581,* 617 (90), *624*
Cohn, D. R., 556 (10), 573 (10), *579*
Condon, E. U., 95 (159), *109*
Connor, J. N. L., 277 (124), *292*
Contoupolos, G., 120 (7), 150 (33), 161 (37), *198, 199,* 214 (36), 215 (17b), *235*
Cool, T. A., 63 (265), *111,* 364 (7f), *492*
Cooper, D. E., 82 (281), *112*
Cotter, D., 573 (55, 56, 61, 62), *580*
Cotter, T. P., 70 (291), *112,* 714 (2), 724 (30), *733*
Coulson, C. A., 2 (16), 3 (16), 5 (16), 14 (16), 17 (16), 32 (16), 34 (16), 39 (16), 40 (16), 41 (16), 42 (16), 43 (16), 89 (16), 90 (16), *105,* 366 (23a), 369 (23a), 462 (23a), *493*
Covaleskie, R. A., 47 (192), *109,* 233 (65), *237,* 686 (22), 692 (22), 693 (22), *711*
Cox, D., 639 (18), 642 (36), 644 (38), 647 (38), 652 (18), 656 (38), 657 (18), *658, 659*
Cox, D. M., 3 (52a, 52b), 73 (52a, 52b), *106,* 650 (40), *659*
Coxon, J. A., 331 (13), *361,* 701 (32), *711*
Cronin, J., 202 (5), *235*
Cross, P. C., 455 (55), 460 (55), *494*
Cuellar, E., 59 (259), *111*
Cupp, J. D., 598 (70), *623*
Curtiss, C. F., 324 (1), *360,* 364 (3), 426 (48a), 429 (3), 472 (66a), 473 (66a), 478 (48a), 479 (48a, 66a), *492, 494*

Dagdigian, P. J., 12 (97), *107*
Dai, Hai Lung., 640 (20b), *658*
Danby, J. M. A., 150 (32), *199*
Danen, W., 639 (12, 16), 647 (12, 16), *658*
Danielson, B. L., 598 (70), *623*
da Paixano, F. J., 295 (34), *322*
Darling, B. T., 588 (32), *623*
Datz, S., 626 (2), *637*
Daudel, R., 2 (13), 3 (13), 5 (13), 17 (13), 31 (13), 32 (13), 34 (13), 39 (13), 40 (13), *105,* 543 (57), 544 (57), *553*
Davies, P., 335 (22), *361*
Davis, E. A., 19 (151), 91 (151), *108*
Davis, M. J., 233 (70, 75), *237, 238*

Dawson, D. R., 15 (125), 66 (125), 69 (125), 82 (125), *108*
Day, J. P., 644 (37), *659*
de Bries, H., 83 (329), *113*
Decius, J. C., 455 (55), 460 (55), *494*
de Koven, B. M., 350 (43), 354 (43), *361,* 467 (60), *494*
Dekronig, R., 663 (17), *677*
Delahay, P., 19 (148), 91 (148), *108*
Delgado-Barrio, B., 368 (29, 70), 421 (29), 430 (29), 467 (61), 470 (29), 478 (29), 480 (29), 481 (29), *493, 494*
Delgado-Barrio, G., 346 (37), 349 (37), 357 (37), *362*
Delory, J. M., 40 (182), *109*
Delos, J. B., 169 (49), 173 (49), *199*
del Rio, J. L., 54 (220), *110*
Delsart, C., 556 (31), 578 (31), *580*
Dennison, D. M., 588 (32), *623*
De pristo, A. E., 467 (62, 63), 476 (63), *494*
Deprit, A., 149 (30), *199*
De-Temple, T. A., 573 (57), *580*
Devonshire, A. F., 411 (44), 412 (44), *494*
Deutch, J. M., 516 (46), 551 (46), *553*
Deutch, T. F., 12 (98), *107*
De Vault, D., 9 (92), 98 (374), 99 (374), 101 (374), *107, 114*
de Vries, H., 15 (128), 66 (128), 69 (128), 82 (128), 84 (128), *108*
De Vries, P. L., 65 (279), *112*
Dexter, D., 97 (382), *114*
Dexter, D. L., 19 (154), 91 (154), *108*
Diana, E., 214 (35a), 219 (47), *236*
Di Bartolo, B., 2 (29), 84 (29), *105, 114*
Dieke, G. H., 311 (49), *322*
Dinerman, L. E., 364 (11a), *492*
Dines, M., 641 (35), *659*
Ding, A. M. G., 59 (258), 61 (262), *111*
Dion, D. R., 575 (66), *580*
Dirac, P. A. M., 267 (95), *291*
Dispert, H. H., 63 (267), 64 (267), *111*
Dixon, D. A., 25 (169), 26 (169), *109*
Dixon, R. N., 662 (16), 663 (16), *677*
Dixon, T. A., 336 (27), *361,* 366 (16), 368 (16), 427 (16, 49), 448 (16), 451 (16), 466 (16), *493, 494*
Doljikov, V. S., 661 (7, 8), *677,* 681 (16), *710*
Doll, J. D., 104 (393), *114*
Dolson, D. A., 233 (65), *237*
Dolzikhov, V. S., 682 (11), *710,* 714 (9), 716 (9), 717 (9), *733*
Donnelly, V. M., 703 (35a), 709 (35a), *711*
Donovan, R. J., 58 (254), *111*
Dorer, F. H., 241 (2), 244 (2), *288*
Dothan, Y., 14 (102), *107,* 277 (128), *292,* 446 (52), *494*
Douglas, A. E., 6 (64), 36 (64a), 37 (64a), 44 (64a), 75 (64a), 104 (394), *106, 114*
Douglas, D. J., 64 (268), *111*
Dows, D. A., 53 (210), 70 (295, 296), 72 (309), 76 (309), *110, 112,* 509 (14), 510 (14), 516 (14), 537 (14), *552,* 602 (83), *624,* 640 (25), *658*
Drozdowicz, Z., 556 (10), 573 (10), *579*
Ducloy, M., 573 (58), *580*
Ducuing, J., 364 (7e), *492,* 573 (63), *580*
Duewer, W. H., 701 (32), *711*
Duff, J., 215 (38), 216 (38), *236*
Duff, J. W., 177 (55), *200,* 214 (36), 215 (40a, 40b), 216 (41), 217 (40a, 40b), 219 (37), 221 (40a, 40b), 223 (40a, 40b), 224 (40a, 40b, 41), 225 (53), 231 (40b, 41, 58b), 232 (40b), 233 (40a, 40b, 41, 58b), 234 (40a, 40b, 41), *236, 237,* 241 (19), 242 (19), *289*
Dunn, P. C., 573 (54), *580*
Dunn, T. M., 662 (13, 14), 663 (14), *677*
Durana, J. F., 242 (32), 244 (32), *289*
Durant, J. L., Jr., 47 (192), *109,* 686 (22), 692 (22), 693 (22), *711*
Dutton, P. J., 98 (371), 99 (371), 100 (371), 101 (371), *114*
Dutton, P. L., 9 (87), 98 (87), 99 (87), 100 (87), *107*
Dworetsky, S. H., 14 (108), *107*
Dyke, T. R., 337 (28), *361*
Dynkin, E. B., 272 (112), *292*

Eastes, W., 169 (48), *199, 493*
Eberly, J. H., 3 (45, 46), 4 (55), 18 (46, 141), 66 (55), 70 (308), 72 (308), 73 (315), 76 (308), 81 (315), 82 (55),

106, 108, 112, 113, 509 (4), 512 (45b), 516 (45b), 533 (4), 535 (4), 544 (58a), 550 (4), *552, 553,* 556 (32), 567 (32), 577 (68), *580, 622*
Einstein, A., 54 (217), *110*
Eisentein, G., 91 (377), 98 (377), *114*
Elbert, M. L., 40 (185), *109*
Elert, M. L., 294 (25), 296 (25), *321,* 543 (56), *553*
Elliott, C. J., 601 (79), *624*
El-sayed, M. A., 51 (199), *110*
Engelman, R., 41 (189), 91 (189), 93 (189), 94 (189), *109,* 547 (60), 548 (60), *553*
Esherick, P., 3 (46), 18 (46), *106*
Estler, R. C., 64 (270, 272), *112,* 286 (139), *292*
Even, U., 6 (69, 75, 76, 77), 7 (77), 11 (69, 76), 15 (133), 44 (69, 76, 77), 46 (69, 75), 75 (69, 75, 76, 77, 133, 319), 90 (360), *108, 113, 114*
Ewing, G., 18 (145), 87 (344), *108, 113*
Ewing, G. E., 324 (2), 334 (17, 18, 19, 20), 353 (48), *360, 361, 362,* 364 (6, 11a, 11b, 11c, 11d, 11e), 367 (24, 25), 370 (24), 387 (37), 433 (25), 435 (37), 438 (37), 462 (24), 470 (25), 478 (25), 484 (37), 486 (37), 491 (72), *492, 494*

Faisal, F. H. M., 70 (288), *112*
Faist, M. B., 257 (76), *291*
Falcone, R. W., 577 (74), *581*
Falconer, W. E., 335 (24), *361*
Fano, U., 14 (104, 105), *107,* 265 (92, 93), 267 (92), 281 (92), *291,* 317 (57), *322, 493,* 511 (35), 512 (35), 525 (35), 541 (35), 543 (35), *553*
Farneth, W. E., 722 (27), *733*
Farrar, J. M., 244 (55), 248 (56), *290*
Fayer, M. D., 82 (281), *112*
Fedina, G. A., 661 (2), *676*
Feher, G., 9 (90), *107*
Feld, M. S., 556 (27), 569 (48), 573 (48, 58), *579, 580*
Feldman, B. J., 556 (27), 578 (27), *579,* 601 (79), *624*
Feldman, D., 707 (41), *711*
Feldman, D. L., *106*
Fermi, E., 19 (158), *109*
Filip, H., 589 (47), 590 (47), *623*
Filseth, S. V., 637 (18), *638,* 702 (33), 703 (33), 706 (37), *711*
Fischer, S. F., 2 (9), 5 (9), 17 (9), 39 (9), *104*
Fisher, S., *494*
Fitch, P. S. H., 6 (68, 74), 44 (68), 75 (74), *106, 107,* 328 (7), 329 (9), 348 (39), *361, 362,* 365 (15d), 367 (15d), 370 (15d), 375 (15d), 422 (15d), 427 (15d), *493,* 510 (24b), 543 (24b), 551 (24b), *552*
Fiutak, J., 536 (51), *553*
Flesch, G. D., 664 (23), *677*
Flicker, H., 589 (47), 590 (47), *623*
Flynn, G., 722 (27), *733*
Flynn, G. W., 49 (196), 53 (208), 65 (277), 74 (317), *109, 110, 112, 113,* 242 (27), 244 (27), 247 (27), *289,* 465 (58), *494,* 626 (10), 635 (10), *638*
Fong, F. K., 2 (17), 3 (17), 5 (17), 17 (17), 34 (17), 39 (17), 40 (17), 41 (17), 42 (17), 43 (17), 89 (17), *105,* 547 (61), 548 (61), *553*
Ford, J., 118 (1), 126 (11), 132 (12), 146 (18), *198,* 210 (29), 218 (17a, 44, 45), 226 (46), 227 (46), 230 (57), 233 (77), 234 (77), *235, 236, 238*
Forst, W., 57 (234), *111,* 720 (25), *733*
Forster, T., 19 (153), 91 (153), *108*
Forte, B., *292*
Fournier, J., 637 (18), *638,* 702 (33), 703 (33), *711*
Fox, K., 70 (303), 72 (303), *112,* 588 (39), 589 (47, 63, 68), 590 (47), 591 (50, 51), 598 (72), 615 (18), *622, 623, 624*
Fox, L., *494*
Francis, A. H., 662 (14), 663 (14), *677*
Francisco, J., 57 (237), *111*
Frankel, D. S., 509 (12a), 510 (12a), 536 (12a), *552,* 592 (55), *623*
Frauenfelder, H., 91 (377), 98 (377), *114*
Fraunfelder, H., 9 (85), 91 (85), *107*
Freed, J. H., 414 (46), 416 (46), 516 (46), 551 (46), *553*

Freed, K. F., 2 (10, 17), 3 (17), 5 (10, 17), 17 (10, 17), 34 (17), 39 (10, 17), 40 (17), 41 (17), 42 (17), 43 (17), 58 (244), 85 (342), 89 (17), *104, 105, 113,* 193 (66), 194 (68), 195 (69), *200,* 547 (61), 548 (61), *553,* 616 (87), *624*
Freedman, A., 58 (242), 59 (242), *111*
Freund, S. M., 3 (41), 5 (41), 11 (41), 18 (41), 21 (41), 63 (266), 70 (41), 72 (41), 74 (41), *105, 111,* 293 (22b), 320 (22b), *321,* 509 (5a), 511 (5a), 533 (5a), 535 (5a), *552,* 556 (7), 574 (65), *579, 580,* 605 (84), 624, 676 (29), *677*
Frey, R., 573 (63), *580*
Friedmann, H., 556 (38), 577 (38), *580*
Frieman, E. A., 148 (22), *199*
Froeschle, C., 118 (4), 124 (4), 141 (4), *198*
Frosch, R. P., 90 (358), *114*
Fuhr, J. R., 578 (81), *581*
Furzikov, N. P., 661 (9), *677,* 681 (9), *710*
Fuss, W., 11 (93), 30 (93, 175), 70 (291), *107, 109, 112,* 716 (17), 724 (30), *733*

Galbraith, H. W., 53 (210), 55 (227), 70 (297, 298, 304, 305, 308), 72 (304, 305, 308), 76 (308), *110, 112,* 509 (18b), 510 (18b), 512 (18b), 537 (18b), 540 (18b), 543 (18b), *552,* 584 (10, 11, 12), 588 (10, 11, 12, 16, 17, 19), 589 (10, 11, 12, 49), 591 (11), 598 (73), 600 (73), 602 (14), *622, 623, 624,* 640 (27), *658*
Galgani, L., 161 (37), *199,* 213 (34), 214 (34, 35a, 36), 219 (47), 225 (53), 233 (78), *236, 237, 238,* 241 (17), 242 (17), *289*
Gamow, G., 20 (160), 95 (160), *109*
Gandini, A., 728 (32), *733*
Garcia-Colin, L. S., 54 (220), *110*
Garcialus, I. C., 91 (377), 98 (377), *114*
Garrett, B. C., 169 (47), 171 (47), *199*
Gautier, L., 592 (58), 594 (58), 595 (58), *623*
Gebelein, H., 27 (173), 28 (173), *109*
Gentry, W. R., 366 (20), 453 (20), *493*
Geis, M. W., 63 (267), 64 (267), *111*
Gelbart, W. M., 40 (185), *109,* 193 (66), 194 (66), *200,* 294 (25), 296 (25), *321,* 543 (56), *553*
Gemack, A. Z., 15 (119, 120), 66 (119, 120), *108*
George, T. F., 65 (279), *112,* 577 (76), *581*
Gerber, R. B., 85 (341), *113*
Gersten, J. I., 577 (75), *581*
Giardini-Guidon, A., 5 (62b), 12 (62b), *106*
Gibbs, H. M., 584 (9), 591 (9), 614 (9), *622*
Gibbs, W. J., 267 (98), 270 (98), *291*
Gibson, G. E., *733*
Giese, C. F., 366 (20), 453 (20), *493*
Gilmore, E. H., *733*
Giorgilli, A., 161 (37), *199,* 214 (36), 219 (36), 233 (78), *236, 238*
Giraud, M., 556 (30), 558 (43), 575 (43), *580*
Giver, L. P., 37 (177), *109*
Glass, L., 202 (15), *235*
Glatt, I., 3 (48), *106*
Gleaves, J. T., 242 (32), 244 (32), *289,* 293 (19), *321,* 511 (27c), *553*
Glickman, W. A., 364 (9c, 9d), *492*
Godzik, K., 66 (282), *112*
Goldberger, M. L., 2 (24), 40 (24), *105*
Golden, D. M., 57 (233), 81 (323), *111, 113*
Gole, J. L., 257 (75), 258 (75), *291*
Gollub, J. P., 202 (14), *235*
Golomb, D., 104 (392), *114*
Good, I. J., 274 (117), *292*
Good, R. E., 104 (392), *114*
Goodman, J., 9 (82), 85 (343), 88 (349), *107, 113,* 369 (33), *493*
Goodman, M., 639 (10), 640 (20a, 25, 32, 33), *658*
Goodman, M. F., 53 (210), 55 (228), 70 (284, 285, 286, 295, 296, 309), 72 (309), 76 (309), *110, 112,* 509 (14), 510 (14), 516 (14), 537 (14), *552,* 608 (85), *624,* 633 (16), *638*
Gorchakov, V. I., 70 (299), *112*
Gordon, J. P., 75 (316), *113*
Gordon, R. J., 63 (265), *111*
Gorokhov, Yu, A., 293 (21), *321,* 509

(2b), *551*, 602 (81), *624*, 640 (28), 655 (28), *658*, 661 (9), *677*, 681 (9), *710*
Gorokhovskii, A. A., 83 (328), *113*
Gough, T. E., 330 (10), *361*, 366 (18), 453 (18), 462 (18), 491 (18), *493*
Gower, M. C., 714 (12), *733*
Grabenstetter, J. E., 334 (15), *361*, 477 (69a), *494*
Grant, E., 640 (24), *658*
Grant, E. R., 241 (8, 14), 244 (8), *288*, *289*, 509 (7b), 512 (7b), 550 (7b), 551 (7b), *552*, 633 (15), *638*, 639 (4), 641 (4), *658*, 714 (1c), 716 (1c), 728 (1c), *732*
Gray, D. L., 595 (67), *623*
Gray, H. R., 572 (50), *580*
Green, F. T., 365 (12a), *493*
Green, W. R., 65 (278), *112*
Greene, E. F., 626 (5), *638*
Greene, F. T., 104 (390), *114*
Greene, J. M., 230 (57), *237*
Greene, R. N., 18 (137), *108*
Greenlay, W. R. A., 293 (15), 294 (15), 295 (15), 298 (15), 307 (15), *321*
Gresillon, D., 233 (78), *238*
Grey, T., 5 (63), *106*
Griffin, J. J., 593 (61), *623*
Grimley, A. J., 626 (11, 14), 628 (11), 636 (14), 637 (11, 14), *638*
Grischkowsky, D., 556 (37), 576 (37), *580*
Gross, R. W. F., 3 (35), *105*
Groves, S. H., 3 (36), *105*
Gruhl, Th., 584 (4), *622*
Gudzenko, L. I., 577 (69), *580*
Guillory, W., 639 (5), *658*
Guillory, W. A., 637 (20), *638*, 682 (12, 13), 692 (12, 13), 694 (12, 13), 707 (12, 13, 39), *710*, *711*, 714 (4), 723 (28), *733*
Gunthard, H. H., 664 (24), *677*
Gupta, A., 47 (191), 50 (191), 70 (191), *109*, 509 (11a), 510 (11a), *552*, 667 (25), *677*, 683 (18), 692 (18), 693 (18), *711*
Gurney, R. W., 95 (159), *109*
Gurnick, M., 233 (66), *237*
Gustafson, T. K., 556 (39), *580*
Gustavson, F. G., 157 (36), 173 (36), *199*
Gutman, D., 639 (15), *658*
Gutman, W. M., 585 (28), 621 (28), *622*
Gutzwiller, M. C., 169 (50), *199*

Haarhoff, P. C., 674 (27), *677*
Haas, Y., 2 (22), 12 (96), 70 (96), *105*, *107*, 244 (46), 246 (46), 254 (61), 255 (46), 264 (46), *290*, 661 (11, 12), *677*, 683 (16, 17), 693 (16, 17), 694 (16, 17), *710*, 716 (19), 719 (19, 23, 24), 723 (19), 725 (19), *733*
Hackett, P. A., 728 (32), *733*
Haensel, R., 19 (156), 88 (156), *108*
Haeusler, C., 325 (3), *360*
Hahn, E. L., 15 (132), 72 (132), 75 (132), *108*, 556 (18), 569 (18), *579*, 591 (52), *623*
Haken, H., 202 (6), *235*
Halavee, U., 58 (249), *111*
Hall, J. H., 639 (6), *658*
Hall, J. L., 509 (18a), 510 (18a), 512 (18a), 537 (18a), 540 (18a), 543 (18a), *552*
Hall, R., 639 (11, 18), 641 (11, 35), 642 (36), 644 (38), 647 (38), 652 (18), 657 (18, 38), *658*, *659*
Hall, R. B., 3 (52a, 52b), 53 (213), 73 (52a, 52b), *106*, *110*, 509 (9b), 510 (9b), *552*, 652 (42), *659*
Hallsworth, R. S., 509 (1b), *551*, 661 (6), *677*, 683 (14), *710*, 714 (8), 716 (8), 729 (8), *733*
Halonbrenner, R., 664 (24), *677*
Halpern, J. B., 707 (40), *711*, 714 (3), *733*
Hamilton, I., 212 (32), 215 (38,79), 216 (32, 38), 224 (32), 231 (32), *236*, *238*
Hanch, Th., 556 (9), 569 (9), 573 (9), *579*
Hancock, G., 637 (18), *638*, 656 (44), *659*, 702 (33), 703 (33), 707 (40), *711*, 714 (3), *733*
Handy, N. C., 162 (39), 165 (39), 176 (39), 182 (39), *199*, 233 (68), *237*
Hanna, D. C., 573 (55, 56, 61), *58*
Hansel, K. D., 70 (302), *112*, 214 (35b), 216 (35b), *236*, 241 (16), *289*
Hanson, D. M., *494*
Hardwidge, E. A., 241 (2), 244 (2), *288*

Harman, T. C., 3 (36), *105,* 588 (44), *623*
Haroche, S., 4 (56), 15 (56), 66 (56), 82 (56), *106,* 556 (28), 577 (78), *579, 581*, 640 (26), *658*
Harris, S. E., 18 (142), 65 (278), 95 (142), *108, 112,* 577 (71, 74), *580, 581*
Harris, S. J., 335 (22, 23, 24, 25), 336 (26), *361*
Harrman, A., 3 (49), *106*
Harter, R. J., 241 (18), *289*
Harter, W. G., 295 (34), *322*
Hartford, A., 637 (21), *638,* 694 (27), 706 (27), 707 (27), *711*
Hartman, J., 11 (93), 30 (93, 175), *107, 109*
Hartman, R., 15 (115), 66 (115), *108*
Hartmann, F., 556 (28), 578 (28), *579*
Hartmann, S. R., 585 (27), *622*
Hase, E., 202 (2), *235*
Hase, W. L., 241 (13, 15, 18), *289*
Hayes, A. K., 3 (31), *105*
Hayli, A., 215 (17b), *235*
Hays, A. K., 331 (13, 14), *361*
Hayward, R. J., 293 (10, 11, 12, 13), 294 (10, 11, 12, 13), 298 (10, 11, 12, 13), *321*
Hecht, K. T., 588 (38), 592 (38), 594 (38), 595 (38), *623*
Heer, C. V., 585 (28), 621 (28), *622*
Heiles, C., 124 (9), 141 (9), *198,* 218 (60), *237*
Heitler, W., 2 (15), 3 (15), 5 (15), 17 (15), 32 (15), 34 (15), 39 (15), 40 (15), *105*
Heller, D. B., 709 (47b), *711*
Heller, D. F., 313 (52, 53), 319 (52, 53), *322,* 510 (22), 543 (22), 544 (22), 551 (22), *552,* 709 (44, 51), *711*
Heller, E., 204 (23a), 205 (23a, 23b), *236*
Heller, E. J., 40 (180, 185), *109,* 184 (56), 192 (64), 196 (71), 197 (71), *200,* 233 (70, 75), *237, 238,* 534 (48), *553*
Hellman, R., 152 (35), 179 (35), *199*
Hendershort, M. C., 202 (7), *235*
Henderson, D., 58 (250), *111*
Henderson, G., 334 (17, 18, 20), *361,* 364 (11b, 11d)
Henon, M., 124 (9), 141 (9), *198*, 218 (60), *237*
Henry, A., 661 (1), *676*
Henry, B. R., 2 (7, 11), 5 (11), 17 (7, 11), 39 (11), 55 (223), *104, 105, 110*, 293 (9, 10, 11, 12, 13, 14, 15, 16, 17), 294 (9, 10, 11, 12, 13, 14, 15, 16, 17, 24), 295 (15), 298 (9, 10, 11, 12, 13, 14, 15, 16, 17, 24), 307 (14, 15, 17), *321*, 543 (55), *553*
Henry, C. H., 19 (150), 91 (150), *108*
Henry, L., 661 (1), *676*
Heppner, J., 573 (59, 60), *580*
Herman, I. P., 254 (61), *290*
Herschbach, D. R., 7 (79), 22 (79), 25 (169), 26 (169), 58 (238), *107, 109, 111,* 243 (40), 277 (125, 126), *290, 292,* 366 (16), 368 (16), 427 (19), 448 (16), 451 (16), 466 (16), *493, 494,* 626 (2), *637,* 719 (20), 722 (20), 724 (20), *733*
Herzberg, G., 2 (25, 26, 27), 16 (25, 27, 135), 19 (25, 27), 21 (27, 161), 25 (25), *105, 108, 109*, 343 (33), 344 (34), *362*, 421 (47), 448 (47), *494*
Hesselink, W. H., 66 (280), 83 (331), 84 (331), 87 (280, 331), *112, 113*
Hester, R. E., 589 (46), *623*
Hicks, K. W., 637 (20), *638*, 707 (39), *711*
Himes, J. L., 325 (3), *360*
Hinkley, E. D., 584 (10, 11, 12), 588 (42, 43, 45), 589 (10, 11, 12), 591 (11), *622, 623*
Hinze, J., *322*
Hirschfelder, J. O., 324 (1), *360*, 364 (3), 429 (3), 472 (66a), 473 (66a), 479 (66a), *492, 494*, 575 (66), *580*
Hizhnyakov, V., 83 (332), *113*, 539 (52), *553*
Hoare, M. R., 548 (63), *553*
Hobbs, W. E., 663 (18), *677*
Hochstrasser, R. M., 2 (8), 8 (81), 9 (81), 17 (8), 18 (137), 29 (8), 31 (8), 34 (8), 36 (8), 39 (8), 88 (347, 348), 89 (8, 354, 355), 90 (348), 94 (354), *104, 108, 113, 114*
Hocker, G. B., 584 (3), 589 (3), 591 (3), *622*
Hodgkinson, D. P., 70 (300, 307), 72 (307), 76 (307), *112*, 509 (17), 510 (17), 512 (17), 537 (17), 540 (17), 543 (17), *552*

Hodgson, P. E., 263 (81), *291*
Hofacker, G. L., 277 (124), *292*
Hoffbauer, M. A., 366 (20), 453 (20), *493*
Hoffman, J. M., 3 (31), *105*, *361*
Hohla, K., 244 (67), 257 (67), *290*
Holbrook, K. A., 56 (232), 58 (232), *110*, 452 (53), 466 (53), *494*, 510 (26b), 511 (26b), *552*, 649 (41), *659*, 720 (26), *733*
Holdy, K. E., 58 (248), *111*, 396 (43a), 427 (43a), *494*
Holland, R. F., 587 (31), 589 (47), 590 (47), *623*
Holloway, G., 202 (7), *235*
Holmes, B. E., 58 (253), 59 (253), *111*, 241 (3), 242 (31), 244 (3, 31), *288*, *289*
Holmes, P. J., 202 (11), *235*
Holmgren, S. L., 335 (22), *361*, *493*
Holstein, T., 90 (369), *114*
Holt, S. L., 664 (21), *677*
Holten, D., 9 (88), 98 (88), 99 (88), 100 (88), *107*
Hopf, F. A., 15 (121, 122), 66 (121, 122), *108*, 584 (7, 8), 585 (23), 591 (7, 8), 614 (8), *622*
Hopfield, J. J., 98 (375), 99 (375), 102 (388), *114*
Hori, G. I., 149 (29), *199*
Horning, D. F., 311 (50), *322*
Horsley, J. A., 3 (52a, 52b), 53 (210), 73 (52a, 52b), *106*, *110*, 639 (18), 640 (25), 644 (38), 647 (38), 650 (47), 652 (18), 656 (18, 38), 657 (18), *658*, *659*
Houdeau, J. P., 325 (3), *360*
Houle, F. A., 241 (13), *289*
Houston, P. C., 709 (43), *711*
Houston, P. L., 588 (41), 589 (41), *623*, 626 (11, 12, 13a, 13b, 14), 628 (11, 12, 13a, 13b), 629 (11, 12, 13a, 13b), 632 (13a, 13b), 636 (14), 637 (11, 12, 13a, 13b, 14), *638*
Howard, B. J., 365 (13a)
Hoy, A. R., 295 (30), *322*
Huang, K., 90 (366), *114*
Huber, D. L., 572 (51), *580*
Huber, J. R., 664 (24), *677*
Huber, K. P., 6 (64a), 36 (64a), 37 (64a), 44 (64a), 75 (64a), *106*
Hudgens, J. W., 47 (192), *109*, 293 (19), *321*, 511 (27c), *553*, 686 (22), 692 (22), 693 (22), *711*
Hui, A. K., 331 (14), *361*
Hui, K. K., 63 (265), *111*
Hung, I. F., 293 (14, 17), 294 (14, 17), 298 (14, 17), 307 (14, 17), *321*
Hunter, L. W., 426 (48d), 478 (48d), 479 (48d)
Huppert, D., 3 (37), 4 (37), 90 (362), *105*, *114*
Hurlock, S., 588 (41), 589 (41), *623*
Husain, D., 705 (36), *711*
Hutchinson, J. S., 176 (54), *199*, 231 (58a), 233 (73), 234 (58a), *237*, 277 (124), *292*
Hynes, J. T., 220 (49), *236*

Icsevgi, A., 585 (24), *622*
Ippen, E. I., 2 (19), 3 (19), 4 (19), *105*
Ippen, E. P., 90 (363), *114*
Ireton, R. C., 241 (2), 244 (2), *288*
Isenor, N. R., 3 (39, 45), 18 (45), 19 (39), 21 (39), 70 (39), *105*, *106*, 509 (1a, 16), *551*, 661 (4, 6), *677*, 683 (14), *710*, 714 (8), 716 (8), 729 (8), *733*
Isenor, R. N., 681 (4), 683 (4), *710*
Isobe, K., 556 (10), 573 (10), *579*

Jacobsen, A., 641 (35), *659*
Jacon, M., 15 (110), *107*
Jaeger, T., 509 (4a), 511 (4a), 533 (4a), 535 (4a), 550 (4a), *552*
Jaffe, C., 162 (41), 169 (41), 171 (41), *199*, 233 (74), *237*
Jaffe, S., 233 (74), *237*
Jakubetz, W., 277 (124), *292*
Janda, K. C., 335 (22, 25), 336 (26, 41), *361*, *493*
Jasinski, J. P., 664 (21), *677*
Javan, A., 569 (48), 573 (48), *580*
Jaynes, E. T., 264 (85, 115), 265 (85, 91), 274 (85), *291*, *292*
Jeans, J. H., 364 (2), *492*
Jeffreys, H., 272 (112), *292*
Jensen, C., 639 (7), *658*
Jensen, C. C., 241 (6), 257 (6), *288*
Jensen, R. J., 509 (3), *552*

Joffrin, C., 364 (7e), *492*
Johnson, B. R., 24 (165), 51 (197), *109*
Johnson, K. E., 328 (8), 346 (36), 348 (38), 352 (46), 353 (49), 355 (38), 358 (49), *361*, *362*, 365 (15c, 15e), 367 (15c, 15e), 370 (15c, 15e), 371 (15c), 375 (15c, 15e), 382 (15c), 383 (15c), 387 (15c), 422 (15c, 15e), *493*
Johnson, S. A., *112*
Jonah, C., 15 (114), *108*
Jones, C. A., 282 (133), *292*
Jones, R. P., 9 (83), 88 (83), *107*
Jorna, S., 118 (2), 142 (2), 196 (2), *198*, 205 (17e), 207 (17e), 210 (17e), 215 (17e), *235*
Jortner, J., 2 (1, 8, 13, 16), 3 (13, 16, 37), 4 (37, 57), 6 (69, 75, 76, 77), 7 (77), 9 (83), 11 (69, 76), 14 (16, 100, 101, 102, 107), 15 (111, 131, 133), 17 (8, 13, 16, 136), 18 (143, 144), 19 (149, 152, 156), 24 (144), 25 (170), 26 (170), 27 (173), 29 (8, 174), 30 (176), 31 (8, 174), 32 (13, 16, 176), 34 (8, 13, 16), 35 (174), 36 (8), 39 (8, 13, 16, 170), 40 (13, 16, 186, 188), 41 (16, 189), 42 (16, 190), 43 (16), 44 (69, 76, 77), 47 (193, 194), 48 (195), 49 (194), 50 (194), 58 (246), 66 (131, 282), 68 (131), 69 (131), 70 (289, 290, 306), 72 (195, 289, 290, 306), 73 (195, 306), 74 (195), 75 (69, 75, 76, 77, 133, 319), 80 (320), 85 (337), 88 (350), 89 (16), 90 (16, 360, 362), 91 (149, 152, 376), 92 (364), 94 (152, 379, 380, 381), 97 (364), 98 (376), 101 (387), 102 (364, 386, 387, 389), *104*, *105*, *106*, *107*, *108*, *109*, *111*, *112*, *113*, *114*, 295 (28, 29), 296 (38), 303 (44), 304 (44), 311 (28, 51), 313 (28, 38), 314 (38), 315 (54, 55), 317 (38), 320 (58, 59, 60), 346 (37), 353 (48), 357 (37), *361*, *362*, 367 (27a), 368 (28, 29, 30, 31), 373 (27a), 387 (27a), 388 (42), 396 (27a, 42), 403 (42), 413 (28), 418 (28), 419 (28), 421 (29), 422 (28), 427 (27a), 428 (27a), 429 (30), 430 (29), 433 (27a, 30), 441 (30), 442 (30), 446 (52), 451 (27a), 459 (31), 462 (31), 470 (27a, 28, 29, 30, 31), 478 (27a, 28, 29, 30, 31), 480 (27a, 28, 29, 30, 31), 481 (27a, 28, 29, 30, 31), 482 (31), 484 (27a), 486 (27a), *493*, *494*, 509 (15), 512 (41), 516 (15), 534 (47), 535 (47), 543 (57), 544 (27), 548 (60), 556 (8), *552*, *579*, 601 (80), 620 (91), *624* 640 (19, 22), 655 (19), *658*, 674 (28), 676 (28), *677*, 692 (26), 693 (26), *711*
Jost, W., 58 (250), *111*
Judd, O. P., 621 (92), *624*

Kaarli, R. K., 83 (328), *113*
Kafri, A., *292*
Kaiser, W., 86 (327), 87 (327, 345), *113*, 293 (20), *321*
Kaldor, A., 3 (52a, 52b), 53 (213), 73 (52a, 52b), *106*, *110*, 509 (96), 510 (9b), *552*, 639 (11, 18), 641 (35), 642 (36), 644 (38), 647 (38), 652 (18), 656 (38), 657 (18), *658*, *659*
Kamel, A. A., 149 (31), *199*
Kantrowitz, A., 5 (63), *106*
Kaplan, H., 59 (255), *111*, 253 (60), *290*
Kapral, R., 220 (49), *236*
Karjnovich, D. J., 58 (231), *110*, 244 (63), 254 (63), *290*
Karlov, N. V., 584 (5, 15), 586 (30), 591 (5), 618 (30), *622*, 661 (3), *677*
Karny, Z., 47 (191), 50 (191), 64 (270), 70 (191), *109*, *111*, *112*, 286 (139), *292*, 509 (11a), 510 (11a), *552*, 667 (25), *677*, 683 (18), 692 (18), 693 (18), *711*
Karp., J. S., 264 (82, 83), 283 (83), *291*
Karplus, R., 512 (37), *553*
Kasha, M., 2 (7), 17 (7), 31 (7), 89 (356), 90 (356), *104*, *114*
Kasper, J. V. V., 626 (4), *638*
Kassel, L. S., 52 (201), 56 (201), *110*, 243 (37), *289*
Katz, A., 264 (88), 265 (88), 271 (88), *291*
Kaufman, A. N., 230 (56), *237*
Kaufman, F., 244 (42), *290*
Kaufman, J. J., 282 (133), *292*
Kaufman, K. J., 98 (371), 99 (371),

100 (371), 101 (371), *114*
Kawai, K., 663 (19), *677*
Kawasaki, M., 58 (242), 59 (242), *111*
Kay, K. G., 40 (184), 41 (184), *109*, 192 (63), 193 (67), 197 (72), *200*, 233 (72), *237*
Kearns, D. R., 728 (34), *733*
Keifer, W., 15 (109), *107*
Keller, J. C., 556 (31), 578 (31), *580*
Kelly, P. L., 584 (6), 588 (43), 607 (6), *622*, *623*
Kemble, E. C., 295 (32), *322*
Kenney, J. E., 328 (8), 348 (38), 355 (38), *361*, *362*
Kent, J. E., 89 (353), 94 (353), *114*
Kern, C. W., 295 (33), *322*
Kernkre, V. N., 97 (385), *114*
Kestner, N. R., 19 (149), 91 (149), *108*
Ketley, G., 656 (44), *659*
Khokhlov, E. M., 586 (30), 618 (30), *622*
Kim, M. S., 24 (164), *109*, 334 (11), *361*, 365 (15b), 367 (15b), 370 (15b), 375 (15b), 422 (15b), *493*
Kimble, H. J., 556 (20)
Kimel, S., 639 (1a), *658*, 709 (47a), *711*
Kimoshita, M., 90 (359), *114*
King, A. H., 640 (20b), *658*
King, D. J., 509 (8a, 8b), *552*
King, D. S., 81 (322), *113*, 242 (29), 244 (29), *289*, 633 (16), *638*, 639 (10), *658*, 707 (42), *711*, 714 (6), *733*
Kinsey, J. I., 244 (45), 246 (45), 251 (45), 253 (45), 264 (45), 269 (45), 288 (45), *290*
Kinsey, J. L., 58 (241), *111*, 276 (123), *292*
Kirsch, L. J., 61 (262), *111*
Kistiakowski, G. B., 241 (4), 244 (4), *288*
Klein, L., 556 (15, 30), 558 (43), 575 (43), *579*, *580*
Klein, M. J., 54 (216), *110*
Kleinman, L., 15 (133), 75 (133), *108*
Klemperer, W., 324 (2), 335 (21, 22, 23, 24, 25), 336 (26, 27), *360*, *361*, 365 (13a, 41), 427 (49), *493*, *494*
Klick, C. C., 97 (382), *114*
Klotz, L. C., 58 (248), *111*, 396 (43a), 427 (43a), *494*
Knight, A. E. W., 65 (276), *112*
Knott, R. B., 57 (235), *111*
Knox, R. S., 443 (50), *494*
Knyazev, I. N., 244 (69), 257 (69), *290*, 601 (76), *624*
Ko, A. N., 241 (5), 244 (5), *288*
Kobrinsky, P. C., 331 (13), *361*
Koch, E. E., 19 (156), 88 (156), *108*
Kolander, P., 70 (310), 72 (310), 74 (310), 76 (310), 81 (310), *113*
Kolb, C., 491 (71), *494*
Kolodner, P., 568 (46), *580*, 648 (39), 655 (43), *659*, 714 (14), *733*
Kolomisky, Yu. R., 586 (30), 618 (30), *622*
Kommandeur, J., 15 (131), 38 (178), 66 (131), 68 (131), 69 (131), *108*, *109*, 620 (91), *624*, 728 (36), *733*
Kompa, K., 366 (20), 453 (20), *493*
Kompa, K. L., 2 (20, 22), 3 (33, 38, 51), 5 (62b), 11 (93), 12 (99), 15 (134), 30 (93), 65 (274), 70 (291), 72 (134), 73 (51), 75 (134), *105*, *106*, *108*, *112*, 244 (46, 67), 246 (46), 253 (46), 255 (46), 264 (46), *290*, 509 (6b, 12), 511 (6b), 536 (12b), 550 (6b), *552*, 592 (20), 602 (82), *622*, *624*, 714 (2), *733*
Koopman, B. O., 272 (111), *292*
Kopelman, R., *494*
Koski, W. S., 282 (133), *292*
Kosloff, R., 184 (57), 186 (59), 197 (74), *200*, 233 (71), *237*, 283 (135), *292*
Koszykowski, M. L., 168 (44), 196 (70), 197 (70), *199*, *200*, 208 (25, 80), 216 (25), *236*, *238*
Koszykowski, M. St., 70 (302), *112*
Kotov, A. A., 686 (21), 693 (21), *711*
Kovalenko, L. J., 626 (12), 628 (12), 629 (12), 637 (12), *638*
Kovarskii, V. A., 18 (140), *108*
Krajnovich, D. J., 5 (62b), 12 (62b), *106*, 233 (63), *237*
Kramer, G., 639 (18), 641 (35), 644 (38), 647 (38), 652 (18), 656 (38), 657 (18), *658*, *659*

Kramer, G. M., 3 (52b), 73 (52b), *104*
Krause, H., 700 (31), 701 (31), *711*
Krohn, B. J., 584 (10), 588 (39, 63), 594 (64), 595 (64), *622*, *623*
Kroll, N. M., 577 (73), *580*
Kruskal, M., 149 (27), *199*
Krylov, H., 149 (26), *199*
Kubiak, G., 348 (39), *362*, 365 (15G), 367 (15G), 370 (15G), 375 (15G), 422 (15G), *493*
Kubo, R., 90 (367, 368), 91 (368), 93 (368), 95 (367), *114*, 272 (108), *291*, 511 (34a, 34b, 34c) 34c), 513 (34a, 34b, 34c), 34b, 34c), 536 (34a, 34b, 34c), 543 (34a, 34b, 34c), 547 (59), 548 (59), *553*
Kudian, A., 326 (4), *360*
Kudian, A. K., 326 (4), *360*, 364 (10b, 10d), *492*
Kuhn, T. S., 54 (215), *110*
Kuntz, P. J., 626 (6), *638*
Kunz, C., 19 (156), 88 (156), *108*
Kupperman, A., 189 (61), *200*, 241 (18), 261 (79), *289*, *291*, 472 (66c), 473 (66c), 479 (66c), *494*
Kurmit, N. A., 15 (115), 66 (115), *108*
Kurnit, N. A., 585 (27), *622*
Kursunoglu, B., *622*
Kusunoki, I., 282 (132), *292*
Kuttner, H. G., *108*
Kwei, G. H., 241 (21), 242 (21), *289*
Kwok, H. S., 4 (54), 74 (54), (63), *237*, 244 (63), 254 (63), *290*, 584 (21), 586 (21), 620 (21), *622*

Lacey, A. R., 4 (58), *106*
Laing, J. R., 65 (279), *112*, 577 (76), *581*
Lam, J. F., 15 (134), 72 (134), 75 (134), *108*, 602 (82), *624*
Lamb, W. E., Jr., 2 (23), *105*, 585 (22, 24), *622*
Lambert, A., 84 (334), *113*
Lambert, J. D., *109*
Lambropoulos, P., 3 (45), 18 (45), 70 (308), 72 (308), 76 (308), *106*, *112*, 509 (4), 512 (45b), 516 (45b), 533 (4), 535 (4), 550 (4), *552*, *553*, *622*
Landau, L., *109*
Landau, L. D., 97 (384), *114*
Landman, U., 202 (9), *235*
Lang, D. V., 19 (150), 91 (150), *108*
Lapatovich, W. P., *361*
Larsen, D. M., 70 (287, 292, 293), 72 (287), *112*, 509 (13b), 510 (13b), 516 (13b), *552*, 592 (57), 601 (77, 78), 602 (77), *623*, *624*, 640 (23), *658*
Larvor, M., 325 (3), *360*
Lau, A. M. F., 577 (72), *580*
Laubereau, A., 86 (327), 87 (327, 345), *113*, 293 (20), *321*
Laval, G., 233 (78), *238*
Lawton, R. T., 233 (76), *237*, 301 (39), 307 (39), *322*
Lax, B., 584 (6), 585 (6), 607 (6), 620 (6), *622*
Lax, M., 511 (32), 536 (32), *553*
Leach, S., 4 (57), 15 (57), *106*
Leary, K. M., 676 (29), *677*
Lebowitz, J. L., 209 (27), *236*
Lechtken, P., 719 (22), 722 (22), 728 (22), *733*
Leckenby, R. E., 104 (391), *114*
Leckenly, R. E., 365 (12b, 12c) *493*
Lee, E. K. C., 723 (29), 728 (29), *733*
Lee, T. H., 664 (22), *677*
Lee, Y. T., 5 (62a, 62b), 12 (62a, 62b), 52 (205), 53 (207), 58 (205, 231), *06*, *110*, 233 (63), *237*, 241 (8), 244 (8, 55, 56, 62, 63, 64), 248 (55, 56), 254 (62, 63, 64), *288*, *290*, 366 (19), 453 (19), *493*, 509 (7a, 7b), 511 (27b), *553*, 633 (15), 635 (15), *638*, 639 (4), 640 (24), 641 (4), *658*, 714 (1a, 1b, 1c), 716 (1a, 1b, 1c), 728 (1a, 1b, 1c), *732*
Lees, A. B., 241 (21), 242 (21), *289*
Lefebvre, R., 15 (130), 17 (136), 58 (247), 66 (130), 67 (130), *108*, *111*, 388 (42), 396 (42), 403 (42), *493*
Legay, F., 84 (335), 85 (335), 87 (335), *113*
Lehman, J. C., 18 (138), *108*

Leign, J. S., 91 (370), 98 (370, 371), 99 (370, 371), 100 (370, 371), 101 (370, 371), *114*
Leite, J. R. R., 573 (58), *580*
Lengel, R. K., *106*
Leone, S. R., 65 (273), *112*
Lermont, S., 465 (58), *494*
Le Roy, R. J., 334 (15, 16), *361*, 477 (69a), *494*
Lesiecki, M. L., 637 (20), *638*, 639 (6), *658*, 661 (10), *677*, 682 (12), 692 (12), 694 (12), 707 (12,39), *710*, *711*, 723 (28), 724 (28), *733*
Lethokov, V. S., 3 (42, 43), 5 (42, 43), 11 (42), 12 (42, 95), 18 (42, 43), 21 (42, 43), 70 (42, 43, 95), 72 (42, 43), *105*, *106*, *107*
Letokhov, V. S., 70 (294), *112*, 244 (69), 257 (69), *290*, 293 (21, 22a), *321*, 509 (2a, 2b), *551*, 556 (1, 4, 5, 6), *579*, 586 (30), 591 (5), 601 (76), 602 (81), 618 (30), *622*, *624*, 639 (1b, 2, 3), 640 (1b, 3), 647 (1b), 655 (1b), 661 (5, 7, 8), *677*, 679 (3), 681 (5, 8), 682 (10, 11), 692 (8), 694 (8), 707 (38), *710*, *711*, 714 (7, 9, 11), 716 (9), 717 (9), *733*
Leutwyler, S., 3 (49), *106*
Levenson, M. D., 556 (3), 570 (3), *579*
Levich, V. G., 19 (148), 91 (148), *108*
Levine, R. D., 2 (1, 2, 6), 3 (37), 4 (37), 22 (162), 24 (165), 40 (6), 53 (209), 56 (230), 58 (221, 248, 249, 252), 59 (255, 256), 61 (263, 264), 62 (263), 63 (264), 70 (311), 72 (311), 76 (311), *104*, *105*, *109*, *110*, *111*, *113*, 231 (59), *237*, 241 (7, 20), 242 (22, 25, 26, 33, 35), 243 (22), 244 (7, 25, 26, 33, 35, 41, 42, 43, 44, 45, 46), 246 (26, 33, 43, 44, 45, 46), 247 (25, 26), 248 (26), 253 (43, 44, 45, 46, 60), 254 (60, 61), 255 (25), 257 (22, 71, 76), 260 (77, 78), 261 (77), 264 (33, 43, 44, 45, 83, 84, 89, 90), 265 (89, 90), 267 (99), 268 (89, 100, 102, 103), 270 (89, 104), 271 (106), 272 (109), 273 (114), 275 (89, 99, 103, 122), 277 (22, 122, 124), 279 (99, 122), 281 (99, 122, 130, 131), 282 (134), 283 (134, 135, 136), 285 (122, 137), 286 (89, 136, 138, 140), *291*, *292*, 315 (55), *322*, 396 (43b), 427 (43b), *494*
Levy, D. H., 6 (65a, 65b, 66, 67, 68, 74), 11 (76), 23 (66, 67), 24 (66, 67, 164a, 164b, 164c), 25 (168), 26 (171), 36 (65a, 65b), 37 (65a, 65b), 44 (65a, 65b, 66, 67, 68, 76), 75 (65a, 65b, 66, 67, 74, 76), *109*, 326 (5), 327 (6), 328 (7, 8), 329 (9), 331 (7), 334 (11), 341 (30, 31), 345 (35), 346 (36), 348 (38, 39), 349 (40, 41, 42), 350 (43), 352 (46), 353 (49), 354 (43), 355 (38), 358 (49), *361*, *362* 365 (14, 15a, 15b, 15c, 15d, 15e, 15f), 366 (17), 367 (15a, 15b, 15c, 15d, 15e, 15f), 370 (15a, 15b, 15c, 15d, 15e, 15f), 371 (15a, 15c, 34), 375 (15a, 15b, 15c, 15d, 15e, 15f), 382 (15c, 34), 383 (15c), 387 (15c), 413 (34), 422 (15a, 15b, 15c, 15d, 15e, 15f), 424 (15a, 34), 427 (15a, 15d), 453 (17), 463 (17), 467 (60), *493*, *494*, 510 (24b), 543 (24b), 551 (24b), *552*, 671 (26), 672 (26), 675 (26), *677*
Levy, M., 512 (45a), 516 (45a), *553*
Levy, M. R., 695 (29b), 709 (46), *711*
Li, T. Y., 88 (348), 90 (348), *113*
Liao, P. F., 556 (37), 576 (37), *580*
Liberman, S., 4 (56), 15 (56), 66 (56), 82 (56), *106*, 577 (78), *581*
Lichtin, D. A., 244 (68), 257 (68), *290*
Lide, D. R., 296 (35), *322*
Lidow, D. B., 18 (142), 95 (142), *108*, 577 (71, 74), *580*, *581*
Lie, G. C., *322*
Lifschitz, E. M., 97 (384), *114*
Light, J., 577 (77), *581*
Light, J. C., 231 (59), *237*, 243 (40), 255 (66), *290*
Lim, E. C., 2 (5, 12, 14), 3 (14), 5 (12, 14), 8 (81), 9 (81), 17 (12, 14), 34 (14), 39 (12, 14), 40 (14), *104*, *105*, *107*, 191 (62), *200*, 202 (1), *235*, 293 (1), *321*, 510 (25), 511 (25), 512 (25), *552*
Lim, S. T., 47 (191), 50 (191), 70 (191), *109*
Lin, H. M., 65 (276), *112*
Lin, M. C., 244 (47, 57, 58), 245 (47, 57, 58), 246 (47), *290*
Lin, S. H., 15 (113), 84 (336), *108*, *113*, 547 (62), 548 (62), *553*

Lin, S. T., 509 (11a), 510 (11a), *552*, 667 (25), *677*, 683 (18), 692 (18), 693 (18), *711*
Lin, Y. K., 202 (11), *235*
Lindbland, G., 270 (105), *291*
Lindstedt, A., *199*
Liouville, L., 267 (97), *291*
Lisitsa, V. S., 577 (70), *580*
Liu, B., *322*
Liu, K., 257 (74), 258 (74), 262 (74), *291*
Liverman, M. G., 6 (71, 72), 44 (71), *106*, *107*, 331 (13), 341 (29), *361*, *552*
Lobko, V. V., 586 (30), 601 (76), 618 (30), *622*, *624*
Logan, J., 19 (149), 91 (149), *108*
Lokhman, V. N., 661 (8), *677*, 682 (11), 707 (38), *710*, *711*
Long, C. A., 334 (19, 20), *361*, 364 (11b, 11c)
Long, M. E., 293 (2, 3, 4), 294 (2, 3, 4), 298 (2, 3, 4), *321*
Lorenz, E. N., 202 (12), *235*
Losev, V. V., 661 (2), *676*
Louck, J. D., 588 (34), 594 (66), *623*
Louisell, W. H., 15(134), 75 *291*, 511 (33), 536 (33), *553*, 585 (25, 29), 602 (82), 616 (29), 617 (29), 618 (29), 619 (29), *622*, *624*
Loy, M. M. T., 556 (37), 576 (37), *580*
Luban, D. M., 5 (61), *106*
Lubman, D. M., 51 (198), 52 (198), *110*
Lucht, R. A., 364 (7f), *492*
Lukasik, J., 65 (278), *112*
Lunsford, G. H., 132 (12), *198*, 230 (57), *237*
Lussier, F., 639 (7), *658*
Lutz, H., 89 (355), *114*
Lyman, J., 639 (16), *658*
Lyman, J. L., 3 (41, 51), 5 (41), 11 (41), 18 (41), 21 (41), 70 (41), 72 (41), 73 (51), 74 (41), *105*, *106*, 293 (226), 320 (226), *321*, 509 (3, 4, 16), 511 (16), 533 (4), 535 (4), 550 (4, 16), *552*, 556 (7), *579*, 592 (56), 605 (84), *623*, *624*, 676 (29), *677*, 716 (16), *733*

MacAdam, D. L., 556 (4), *579*
McCall, S. L., 584 (9), 591 (52), 614 (9), 617 (89), *622*, *623*, *624*
McClelland, G. M., 7 (79), 22 (79), *107*, 277 (12b), *292*
McClure, D. S., *733*
McDonald, D. G., 598 (70), *623*
McDonald, J. D., 14 (103), *107*, 233 (66), *237*, 241 (13), 242 (32), 244 (32, 54), 248 (54), *289*, *290*, 293 (19), *321*, 509 (10), 511 (27c), *552*, *553*
McDonald, J. M., 511 (28), *553*
McDonald, J. P., 70 (302), *112*
McDonald, J. R.. 662 (15), 663 (15), *677*, 703 (35b), 707 (35b), 709 (35b), *711*
MacDonald, R. G., 65 (273), *112*
McDonald, R. G., 233 (63), *237*
McDowell, R. S., 584 (10, 11, 12), 587 (31), 588 (10, 11, 12), 589 (10, 11, 12, 46, 47), 590 (47), 591 (11), *622*,*623*
MacFarlane, M. R., 83 (330), 87 (330), *113*
MacFarlane, R. M., 15 (120), 66 (120), *108*
McGlynn, S. P., 90 (359), *114*
Mack, K. M., 337 (28), *361*
Mackay, M. C., 202 (15), *235*
McKee, T. M., 3 (32), *105*
McKellar, A. R., 364 (10e, 10g, 10h, 10i), *492*
McKellar, A. R. W., 326 (4), *360*
MacPhail, R. A., 293 (17), 294 (17), 298 (17), 307 (17), *321*
McWeeny, R., 295 (31), *322*
Maier, J. P., 87 (345), *113*, 293 (20), *321*
Makarov, A. A., 585 (25, 29), 586 (30), 616 (29), 618 (29, 30), 619 (29), *622*
Makarov, G., 640 (28), 655 (28), *658*
Makarov, G. N., 293 (21), *321*, 509 (2b), *551*, 602 (81), *624*, 661 (9), *677*, 681 (9), 709 (45), *710*, *711*
Makarow, A. A., 70 (294), *112*
Makushkin, Yu. S., 588 (35), *638*
Malkin, I. A., 277 (128), *292*
Mandel, L., 556 (20), *579*
Manger, T. G., 9 (91), *107*
Mangir, M., 637 (19), *638*, 684 (19), 686 (19, 20), 692 (19), 693 (19, 20), 694 (19), 695 (28, 29b), 696 (30), 702 (34), 703 (34), 705 (34), 707 (28, 34), 709 (46),

710 (30, 52), 714 (5), 716 (18), 717 (18), *733*
Man'ko, V. I., 277 (128), *292*
Mannick, L., 364 (10f), *492*
Manos, D. M., 257 (72, 73), 258 (72, 73), *290*
Manz, J., 59 (255), 61 (264), 63 (264), *111,* 253 (60), 277 (124), 286 (138), *290, 292*
Marchetti, P., 89 (354), 94 (354), *114*
Marcus, R. A., 11 (147), 52 (202), 70 (302), 91 (147), *108, 110, 112,* 168 (44), 169 (48), 196 (70), 197 (70), *199,* 204 (20), 208 (25), 233 (67), 234 (80), *235, 236, 237, 238,* 241 (13), 243 (39), 255 (65), 257 (39), *289, 290,* 510 (28), *553*
Marling, J. B., 254 (61), *290*
Marsden, J., 202 (16), 206 (16), 209 (16), 210 (16), 219 (16), *235*
Mass, E., 641 (35), *659*
Matiuk, V. M., 244 (69), 257 (69), *290*
May, R. M., 202 (4), *235*
Maylotte, D. H., 626 (6)
Mead, A., 303 (42), *322*
Meagher, J. F., 511 (27a), *552*
Mei, C. C., 626 (8c, 8d), 633 (8d), 635 (8d), *638*
Meier, K., 637 (18), *638,* 702 (33), 703 (33), 707 (41), *711*
Merchant, V., 509 (1b), *551,* 661 (6), *677,* 714 (8), 716 (8), 729 (8), *733*
Merchant, V. E., 683 (14, 15), 694 (15), *710,* 729 (37), *733*
Messiah, A., 278 (129), *292*
Messing, I., 706 (37), *711*
Metiu, H., 85 (342), *113*
Michelson, A. A., 82 (325), *113*
Mies, F., 391 (39), *493*
Mikami, N., 6 (73), *107*
Miller, F. A., 663 (20), *677*
Miller, J. H., 37 (177), *109*
Miller, R. E., 330 (10), *361,* 366 (18), 453 (18), 462 (18), 491 (18), *493*
Miller, R. J. D., 293 (16), 294 (16), 298 (16), *321*
Miller, W. H., 162 (39), 165 (39), 169 (47), 171 (47), 176 (39), 182 (39), 197 (73), *199, 200,* 202 (2), 204 (19), *235, 236, 237,* 241 (15), 244 (43), 246 (43), 251 (43), 253 (43), 264 (43), 288 (43), *289, 290*
Mills, I. M., 296 (35), *322*
Milne, T. A., 104 (390), 365 (12a), *114, 493*
Mitropolski, Y. A., 149 (26), *199*
Mittleman, I. H., 577 (76), *581*
Mo, K. C., 220 (48), 221 (48), *236*
Moehlam, J. G., 293 (19), *321,* 511 (27c), *553*
Moehlmann, J. G., 244 (54), 248 (56), *290*
Molin, Yu, N., 639 (1d), *658*
Mollow, B. R., 556 (14, 16, 19, 24), 578 (14, 16), *579,* 616 (86), 617 (88), *624*
Montagutelli, J., 592 (58), 594 (58), 595 (58), *623*
Monts, D. L., 6 (71, 72), 44 (71), *106, 107,* 331 (13), 341 (29), *361,* 510 (24a), 543 (24a), 551 (24a), *552*
Mooradian, A., 509 (4a), 511 (4a), 533 (4a), 535 (4a), 550 (4a), *552,* 592 (20, 57), 598 (70), *622, 623*
Moore, C. B., 2 (18), 3 (42), 5 (42), 11 (42), 12 (42), 18 (42), 21 (42), 65 (273), 70 (42), 72 (42), 85 (335, 340), *105, 112, 113,* 244 (44), 246 (44), 251 (44), 253 (44), 264 (44), 288 (44), *290,* 364 (7d), *492,* 509 (56), 511 (56), 512 (56), 513 (56), 535 (56), *552,* 556 (5) *579,* 591 (5), 626 (8a, 8b, 8c, 8d), 633 (8d), 635 (8d), *638,* 639 (1b, 3), 640 (20b), 647 (1b), 655 (1b), *658,* 679 (3), *710,* 714 (11), 728 (35), *733*
Morales, D., 491 (72), *494*
Moret-Bailly, J., 588 (37), 592 (58), 594 (58, 65), 595 (58), *623*
Mori, H., 54 (219), *110,* 512(40a), 513 (40a), 514 (40a), 515 (40a), 525 (40a), 532 (40a), *553*
Morrell, J. A., 293 (5), 294 (5, 26a), 298 (5, 26a), 302 (26a), 305 (26a), 306 (26a), 307 (26a), 309 (26a), 310 (26a), 313 (26a), *321, 322*
Morse, M. D., 58 (244), *111*
Morsink, J., 510 (23c), 536 (23c), 551 (23c), *552*
Moser, J., 202 (16), 206 (16), 209 (16), 210 (16), 219 (16), *235*
Moskowitz, A., 303 (42), *322*

Mott, N. F., 19 (151), 91 (151), *108*
Moulton, P. F., 584 (12), 588 (12), 589 (12), 592 (20, 57), *622, 623*
Movshev, V. G., 244 (69), 257 (69), *290*
Moy, J., 63 (265), *111*
Muckerman, J. T., 24 (165, 268), *109, 111*
Muenter, J. S., 337 (28), *361*
Mukamel, S., 2 (13, 16), 3 (13, 16), 5 (13, 16), 14 (16), 15 (111), 17 (13, 16, 136), 31 (13), 32 (13, 16), 34 (13, 16), 39 (13, 16), 40 (13, 16), 41 (16), 42 (16), 43 (16), 52 (206), 58 (206, 246), 70 (289, 290, 312, 313), 72 (206, 289, 312, 313), 76 (312, 313), 77 (312, 313), 89 (1b), 90 (16), *105, 107, 108, 110, 111, 112,* 241 (9), 244 (9), *289,* 313 (52), 315 (55), 320 (58, 61), *322,* 388 (42), 396 (42), 403 (42), *493,* 509 (6, 15, 19a, 19b), 510 (20, 22), 512 (19a, 19b, 43, 44), 513 (19a, 19b, 20), 515 (19a, 19b), 516 (19a, 19b, 43, 44), 517 (19a, 19b), 520 (19a, 19b), 521 (43, 44), 525 (19a, 19b), 535 (19a, 19b), 536 (44), 538 (19b), 539 (19b), 540 (20, 54), 543 (19a, 19b), 551 (64), *552,* 556 (34, 35), 567 (34), 568 (34), *580,* 601 (80), *624,* 640 (22, 29), *658*
Mulliken, R. S., 342 (32), *362,* 444 (51), *494*
Mulloney, T., 171 (52), *199*
Munslow, W., 639 (12), 647 (12), *658*
Murrell, J. N., 301 (40), *322*
Muthukumar, M., 193 (65), *200*

Naaman, R., 5 (61), 51 (198), 52 (198), 81 (323), *106, 110, 113*
Nagy, P. J., 241 (18), *289*
Naj, C. T., *292*
Nayfeh, A. H., 147 (19), 148 (19, 23), *198, 199*
Nereson, N. G., 584 (11, 12), 588 (11, 12), 589 (11, 12, 47), 590 (47), 591 (11), *622, 623*
Nesbet, R. K., 261 (80), 262 (80), 277 (124), 284 (80), *291, 292*
Netzel, T. L., 91 (370), 98 (370, 371), 99 (370, 371), 100 (370, 371), 101 (370, 371), *114*
Newton, R. G., 202 (13), *235*
Nielsen, H. H., 588 (36), 593 (59), *623*
Nieman, J., 47 (191), 50 (191), 70 (191), *109,* 509 (11a), 510 (11a), *552,* 667 (25), 683 (18), *677,* 692 (18, 24), 693 (18, 24), *711*
Nikitin, E. E., 2 (4), *104,* 364 (7a), *492*
Nill, K. W., 588 (44), *623*
Nilsson, A., 639 (16), *658*
Nitzan, A., 14 (101, 107), 34 (101), 40 (181, 187, 188), 41 (188), 42 (190), 47 (193, 194), 49 (194), 50 (194), *107, 109,* 320 (62), *322,* 535 (47, 49), 548 (49), *553,* 556 (36), *580,* 674 (28), 676 (28), *677,* 692 (26), 693 (26), *711*
Noid, D. W., 70 (302), *112,* 168 (44), 169 (48), 196 (70), 197 (70), *199,* 204 (20), 208 (25, 80), 216 (25), *236, 238*
Nordholm, K. S. J., 54 (218), *110,* 162 (38), 187 (60), *199,* 204 (21), *236*
Nordstrom, R. J., 585 (28), 621 (28), *622*
Norris, P. E., 705 (36), *711*
Norrish, R. G. W., 626 (7), *638*
Novick, S. E., 335 (22, 23, 24, 25), 336 (26), *361, 493*
Nowak, A., 639 (16), *658*
Nowak, A. V., 592 (53, 54, 56), *623*
Nyi, C. A., 88 (347), *113*

Ochs, W., *292*
O'Dwyer, M. F., 89 (353), 94 (353), *114*
Oka, T., 574 (65), *580*
Olson, R. W., 82 (281), *112*
Omont, A., 572 (52), *580*
Oppenheim, I., 285 (137), *292,* 511 (30), 512 (43), 516 (43), 521 (43), *553*
Oref, I., 52 (204), 57 (204), *110,* 233 (64), *237,* 240 (1), 241 (6), 244 (1), 257 (6), *288,* 640 (30), *659*
Orenstein, A., 637 (20), *638,* 707 (39), *711*
Orlowski, T. E., 15 (125, 129), 66 (125, 129), 69 (125, 129), 82 (125, 129), 83 (129), 84 (129), 510 (23b), 536 (23b), 551 (23b), *552*

Ottinger, Ch., 282 (132), *292*
Ouhayonn, M., 584 (13), 587 (13), 589 (13), 598 (13), 600 (13), *622*
Oxtoby, D. W., 164 (42), 196 (42), *199,* 226 (55), 229 (55), *237,* 241 (12), *289*

Pacey, P. D., 626 (6), *638*
Pack, R. T., 426 (48b), 472 (66b), 473 (66b), 478 (48b), 479 (48b, 66b), *494*
Panfilov, V. N., 639 (1d), *658*
Pantell, R. H., 2 (23), *105*
Papulovski, V. P., 661 (2), *676*
Parker, D. H., 51 (199), *110*
Parker, G. A., 426 (48b), 478 (48b), 479 (48b), *494*
Parks, E. K., 231 (59), *237*
Parmenter, C. S., 65 (276), *112,* 233 (65), *237*
Parr, C. A., 189 (61), *200,* 241 (18), *289*
Parr, R., 59 (256), 60 (256), *111, 292*
Parson, J. M., 257 (72, 73, 74), 258 (72, 73, 74), 262 (74), *290, 291,* 511 (27b), *553*
Parson, W. U., 98 (372), 99 (372), 100 (372), 101 (372), *114*
Pasternack, L., 703 (35a), 709 (35a), *711*
Patel, C. K. N., 75 (316), *113,* 584 (1, 2), 585 (2), 591 (1), *622*
Patterson, I., 295 (34), *322*
Paul, M. A., 296 (35), *322*
Pauling, L., 364 (1), *492*
Pecelli, G., 205 (24), 219 (24), *236*
Pechukas, P., 169 (43), *199*
Peerlkamp, P. K., 663 (17), *677*
Penrose, O., 209 (27), *236*
Percival, I., 215 (176), *235*
Percival, I. C., 169 (45), 180 (45), *199,* 208 (26), *236*
Perel'man, N. F., 18 (140), *108*
Perez, M., 588 (40), *623*
Perlmutter, A., *622*
Perry, D. S., 61 (262), *111,* 244 (53), 248 (53), *290*
Perry, J. W., 293 (8), 294 (8), 298 (8), 299 (8), 305 (8), 307 (8), *321*
Person, W. B., 444 (51), *494,* 591 (50), *623*
Pestov, E. G., 578 (79), *581*
Peters, K., 9 (86), 98 (86, 373), 99 (86, 373), 100 (373), 101 (373), *107, 114*
Petersen, F. R., 598 (70), *623*
Petrov, Yu. N., 661 (3), *677*
Petuchowski, S. J., 573 (57), *580*
Pimentel, G. C., 59 (259), *111,* 242 (24), 244 (24), 247 (24), 253 (24), *289,* 626 (4), *638*
Pitman, E. J. G., 272 (110), *292*
Pitts, J. N., Jr., 12 (94), *107*
Plainchamp, P., 573 (60), *580*
Pliva, J., 296 (36), *322*
Plivac, J., 460 (57), *494*
Plum, C. N., 709 (43), *711*
Poincare, H., 133 (14), *198*
Polanyi, J. C., 58 (250), 59 (257, 258), 61 (262), 64 (268), *111,* 244 (53), 248 (53), 286 (139), *290, 292,* 626 (6), *638*
Polanyi, M., 625 (1), *637*
Pollak, E., 56 (230), 61 (263), 62 (263), *110, 111,* 241 (19), 242 (19), 244 (42), 283 (135), 286 (136), *289, 290, 292*
Pomphrey, N., 162 (40), 169 (45), 180 (40, 45), 196 (40), *199*
Porter, G., 626 (7), *638*
Porter, R. N., 241 (18), *289*
Potapov, V. K., 244 (69), 257 (69), *290*
Pottier, L. C. L., 364 (8a, 8b), *492*
Pradere, F., 573 (63), *580*
Prasad, P., 8 (81), 9 (81), *107*
Preses, J. M., 626 (10), 635 (10), *638*
Press, A., 186 (58), *200*
Preston, R. K., 70 (301), *112*
Priestley, E. B., 3 (52a), 73 (52a), *106,* 655 (45), *659*
Prigogine, I., 295 (33), *322*
Prince, R. C., 9 (87), 98 (87), 99 (87), 100 (87), *107*
Prior, Y., 709 (49), *711*
Pritchard, D. E., *361*
Pritchard, H. O., 641 (34), *659*
Procaccia, I., 260 (77, 78), 261 (77), 283 (77), *291*
Proch, D., 27 (172), 65 (274), 73 (172), 81 (321), *109, 112, 113,* 714 (2), 728 (38), *733*
Prokhorov, A. M., 661 (3), *677*

Pryor, A. W., 57 (235), *111*
Pullman, B., 2 (13), 3 (13), 5 (13), 17 (13), 31 (13), 32 (13), 34 (13), 39 (13), 40 (13), 59 (256), 60 (256), *105, 111,* 233 (67), *237*
Pullman, R., 543 (57), 544 (57), *553*
Pummer, H., 65 (274), *112*
Puretzkii, A., 640 (28), 655 (28), *658*
Puretzky, A. A., 602 (81), *624,* 681 (9), 709 (45), *710, 711*
Puthoff, H. E., 2 (23), *105*

Quack, M., 52 (203), 70 (314), 72 (314), 76 (314), *110, 113,* 241 (10), 243 (10), 244 (10), *289,* 512 (42), *553*
Quelly, J. J., 509 (11b), 510 (11b), *552,* 709 (48), *711*
Quick, C. R., 242 (28), 244 (28), *289,* 626 (9), 635 (9), *638,* 714 (10), *733*

Rabalais, J. W., 664 (22), *677*
Rabin, Y., 556 (12, 25, 26), 560 (25), 565 (26), 567 (26), 569 (25), 570 (25), 573 (12), 577 (26), 578 (26), *579*
Rabinovitch, B., 640 (30), *658*
Rabinovitch, B. S., 52 (204), 57 (204, 236), *110, 111,* 233 (64), *237,* 240 (1), 241 (2, 5, 6), 244 (1, 2, 5), 257 (6), *288,* 511 (27a), *552,* 641 (34), *659*
Rabinowitz, P., 3 (52b), 73 (52b), *106,* 639 (18), 644 (38), 647 (38), 656 (18, 38), 657 (18), *658, 659*
Rabitz, H., 54 (222), *110,* 467 (63), 476 (63), *494*
Ragone, A., 334 (11), *361*
Ramakrishna, B. L., 6 (65a), 36 (65a), 37 (65a), 44 (65a), 75 (65a), *106*
Ramaswarmy, R., 54 (222), *110,* 467 (62, 63), 469 (62), 476 (63), *494*
Ramsperger, H. C., 243 (38), *289*
Rank, D. H., 364 (9b, 9c, 9d), *492*
Rao, B. S., 364 (9b), *492*
Rao, K. N., 588 (41), 589 (41), *623*
Rapp, D., *493*
Rautian, S. G., 578 (79), *581*
Raymer, M. G., 578 (82), *581*
Raz, B., 88 (350), *113*
Rebane, K. K., 2 (28), 87 (346), 88 (346), *105, 113*
Rebane, L. A., 83 (328), *113*
Rebick, C., 241 (20), *289*
Reddy, K. V., 53 (212), 54 (212), *110,* 293 (6), 294 (6), 298 (6), 315 (6), 316 (6), *321*
Redfield, A. G., 557 (42), *580*
Reidel, D., 233 (67), *237,* 543 (57), 544 (57), *553*
Reilly, J. P., 244 (67), 257 (67), *290*
Reinhardt, W., 162 (41), 169 (41), 171 (41), *199,* 224 (52), 225 (52), *236*
Reinhardt, W. P., 233 (74), *237*
Reiser, C., 639 (7), *658*
Reisler, H., 637 (19), *638,* 684 (19), 686 (19, 20), 691 (20), 693 (19, 20), 694 (19), 695 (28, 29b), 696 (30), 700 (28), 701 (28), 702 (34), 703 (34), 705 (34), 707 (28, 34), 709 (30), 710 (30), *711,* 716 (18), 717 (18), *733*
Renhorn, I., *361*
Rentzepis, P. M., 3 (37), 4 (37), *105,* 9 (83, 86), 14 (101), 34 (101), 88 (83, 351), 90 (361, 362), 91 (370, 371, 373), 100 (370, 371, 373), *105, 107, 113, 114*
Requena, A., 471 (65), 476 (68), 478 (68), *494*
Reynaud, S., 512 (45b), 516 (45b), *553,* 556 (22), 577 (22), 578 (22), *579,* 617 (90), *624*
Rhodes, C. K., 584 (8), 591 (8), 614 (8), *622*
Rhys, A., 90 (366), *114*
Rice, O. K., 243 (37), *289*
Rice, S. A., 2 (8, 14), 3 (14), 5 (14), 6 (73), 7 (78), 17 (14), 22 (78), 29 (8), 31 (8), 34 (8, 14), 36 (8), 39 (8, 14), 40 (14, 180), 55 (225), 74 (318), 89(8), *104, 105, 107, 109, 110, 113,* 118 (6), 162 (38), 164 (42), 176 (38), 184 (57), 187 (60), 191 (62), 192 (64), 193 (65, 66), 194 (66), 196 (42), 197 (74), *198, 199, 200,* 202 (1), 204 (21), 205 (21), 215 (17d), 226 (55), 229 (55), 233 (71), *235, 236, 237,* 241 (12), 293 (1), 295 (33), 311 (51), *321, 322,* 356 (46), *362,* 369 (32), *493,* 510 (25), 511 (27b), 534 (48), *553*

Richardson, M. C., 509 (1a, 1b), *551,* 661 (4, 6), *677,* 681 (4), 683 (4, 14), 706 (14), *710,* 714 (8), 716 (8), 729 (8), *733*
Richardson, M. R., 3 (39), 19 (39), 21 (39), 70 (39), *105*
Rink, J., 509 (3), *552*
Riuuirok, Th. W., 548 (63), *553*
Robertson, G. N., 366 (23a, 23c), 369 (23a, 23c), 373 (23c), 374 (23c), 482 (23c), *493*
Robiette, A. G., 595 (67), *623*
Robins, E. J., 104 (391), *114,* 365 (12b, 12c), *493*
Robinson, C. P., 509 (3), *552*
Robinson, D. W., 65 (275), *112*
Robinson, G. W., 2 (12), 5 (12), 17 (12), 39 (12), 90 (357, 358), *105, 114,* 311 (48b), *322, 494*
Robinson, P. J., 56 (232), 58 (232), *110,* 452 (53), 466 (53), *494,* 510 (266), 511 (26b), *552,* 649 (41), *659*
Robinson, P. T., 720 (26), *733*
Rockley, M. G., 98 (372), 99 (372), 100 (372), 101 (372), *114*
Rockwood, S., 244 (67), 257 (67), *290*
Rockwood, S. D., 3 (33), *105,* 509 (3), *552*
Rod, D. L., 205 (24), 219 (24), *236*
Ron, A., 311 (50), *322*
Ronn, A. M., 47 (191), 50 (191), 70 (191), *109,* 509 (11a), 510 (11a), *552,* 667 (25), *677,* 683 (18), 692 (18, 24), 693 (18, 24), *711*
Root, J. W., 244 (51), 248 (51), *290*
Rose, M. E., 472 (67), *494*
Rosen, N., 22 (163), *109,* 366 (21), 385 (21), 388 (21), 409 (21), *493*
Rosenberger, A. T., 573 (57), *580*
Rosenfeld, R. N., 57 (233), *110*
Ross, J., 512 (43), 516 (43), 521 (43), *553,* 626 (2), *637*
Rossetti, R., 85 (338), 86 (338), *113*
Rossi, M., 81 (323), *113*
Rothe, K. W., 53 (210), *110,* 584 (13), 587 (13), 589 (13), 598 (13), 600 (13), *622*
Rothschild, W. G., 465 (59), *494*
Rousseau, D. L., 14 (108), *107*
Rowlinson, J. S., 274 (120), *292*
Russell, G. A., 97 (382), *114*
Russell, T., 350 (43), 354 (43), *361*
Russell, T. D., 467 (60), *494*
Ryabov, E. A., 12 (95), 70 (95), *107,* 509 (2a), *551,* 661 (5, 7), *677,* 681 (5, 6, 8), 682 (10), 714 (9), 716 (9), 717 (9), *733*
Rynbrandt, J. D., 241 (5), 244 (5), *288*

Saari, P., 87 (346), 88 (346), *113*
Sadowski, C. M., 706 (37), *711*
Saenger, K. L., 7 (79), 22 (79), *107*
Safron, S. A., 58 (238), *111,* 243 (40), *290*
Sage, M. L., 295 (28, 29), 296 (38), 303 (44), *322,* 304 (44), 305 (45), 307 (47), 313 (38), 314 (38, 47), 317 (38), 320 (29)
Salamo, G. J., 584 (9), 591 (9), 614 (9), *622*
Salmon, R., 202 (7), *235*
Salomon, C. Ch., 584 (13), 587 (13), 589 (13), 598 (13), *622, 623*
Salow, H., 311 (48a), *322*
Sanchez, A., 573 (58), *580*
Sander, R. K., 5 (60), *106*
Sanders, J. H., 570 (49), *580*
Sandri, G., 148 (24), *199*
Sargent, M., 2 (23), *105*
Sartakov, B. G., 586 (30), 618 (30), *622*
Satchler, G. R., 593 (62), 615 (62), *623*
Savolamer, J., 15 (130), 66 (130), 67 (130), *108*
Sazanov, V. N., 70 (299), *112*
Schaafsma, A., 663 (17), *677*
Schatz, G. C., 171 (52), *199,* 261 (79), *291,* 472 (66c), 473 (66c), 479 (66c), *494*
Scheidecker, J. P., 118 (4), 124 (4), 141 (4), *198*
Schek, I., 40 (186, 188), 41 (186, 188), 70 (186, 188, 306), 72 (186, 188, 306), 73 (306), 75 (186, 188), 76 (186, 188), 77 (186, 188), 78 (186, 188), 80 (188, 320), 81 (186, 188), *109, 112, 113,* 303 (44), 320 (60), 304 (44), *322*
Schieve, W. C., 210 (29), *236*
Schlag, E. W., 2 (9), 5 (9), 6 (70), 17 (9), 18 (139), 39 (9), 44 (70), *104, 106, 108,* 641 (34), *658*

Schmailzl, U., 65 (274), *112*
Schmatjko, K. J., 242 (36), *289*
Schmidt, S. P., 726 (31), *733*
Schmiedl, R., 707 (41), *711*
Schneider, S., 2 (9), 5 (9), 17 (9), 39 (9), *104*
Screiber, J. L., 58 (250), *111*
Schriefer, R., 9 (85), 91 (85), *107*
Schroder, H., 27 (172), 73 (172), *109,* 728 (38), *733*
Schultz, A., 6 (64b), 36 (64b), 37 (64b), 44 (64b), 75 (64b), *106*
Schultz, M. J., 70 (310), 72 (310), 74 (310), 76 (310), 81 (310), *113,* 640 (31), 648 (39), *659*
Schultz, P., 5 (62a), 12 (62a), *106*
Schultz, P. A., 5 (62b), 12 (62b), 52 (205), 53 (207), 58 (231), *106, 110,* 233 (63), *237,* 241 (8), 244 (8, 62, 63, 64), 254 (62, 63, 64), *288, 290,* 509 (7a, 7b), 512 (7a, 7b), 550 (7a, 7b), 551 (7a, 7b), *552,* 633 (15), 635 (15), *638,* 639 (4), 640 (24), 641 (4), *658,* 714 (1a, 1b, 1c), 716 (1a, 1b, 1c), 728 (1a, 1b, 1c), *732*
Schultz-Dubois, E. O., 3 (34), *105*
Schumacher, E., 3 (49), *106*
Schuster, G. B., 726 (31), *733*
Schutz, R., 366 (19), 453 (19), *493*
Schwinger, J., 512 (37), *553*
Scoles, G., 330 (10), *361,* 366 (18), 453 (18), 462 (18), 491 (18), *493*
Scott, G. W., 89 (355), *114*
Scott, J. F., 585 (26), *622*
Scott, L. F., *622*
Scotti, A., 214 (35a), 219 (47), *236*
Scully, M. O., 2 (23), 15 (121, 122), 66 (121, 122), *105, 108,* 584 (7), 585 (23, 26), 591 (7), *622*
Seaver, M., 65 (276), *112*
Secrest, D., 427 (49c), *493, 494*
Section, V. C., 55 (229), 56 (229), 58 (229), *110*
Seilmeier, A., 87 (345), *113*
Seitz, F., 90 (365), 91 (365), 97 (365), *114*
Seligson, D., 573 (58), *580*
Selmeier, S., 293 (20), *321*
Selwyn, L., 639 (9), *658*
Selzle, H. L., 18 (139), *108*
Setser, D., 639 (12), 647 (12), *658*
Setser, D. W., 58 (253), 59 (253), *111,* 241 (3), 242 (30, 31), 244 (3, 30, 31, 50, 51), 248 (50, 51), 249 (50), 253 (50, 51), *288,* 331 (13), *361,* 701 (32), *711*
Shaffer, W. H., 588 (36, 39), *623*
Shamah, I., 65 (277), *112*
Shank, C. V., 90 (363), *114*
Shannon, C. E., 274 (118), *292*
Shapiro, M., 58 (243), *111,* 233 (69), *237,* 396 (43b), 427 (43b), *494*
Shapiro, S. L., 2 (19), 3 (19), 4 (19), *105*
Sharfin, W., 365 (15e), 367 (15e), 370 (15e), 375 (15e), 422 (15e), *493*
Sharfin, W. F., 328 (8), 348 (38), 352 (46), 353 (49), 358 (49), *361, 362*
Sharp, T. E., *493*
Shaw, R. J., 89 (353), 94 (353), *114*
Shea, R. F., 15 (122), 66 (122), *108*
Shelepin, L. A., *622*
Shen, Y., 5 (62a, 62b), 12 (62a, 62b), *106*
Shen, Y. R., 52 (205), 53 (207), 58 (205, 231), *110,* 233 (63), *237,* 241 (8), 244 (8, 62, 63), 254 (62, 63), *288, 290,* 509 (7a, 7b), 512 (7a, 7b), 550 (7a, 7b), 551 (7a, 7b), *552,* 572 (53), *580,* 633 (15), 635 (15), *638,* 639 (4), 640 (24), 641 (4), *658,* 714 (1a, 1b, 1c), 716 (1a, 1b, 1c), 728 (1a, 1b, 1c), *732*
Shibanov, A. N., 707 (38), *711*
Shilnikov, L. P., 202 (8), *235*
Shirley, J. H., 558 (44), *580*
Shobatake, K., 511 (27b), *553*
Shoemaker, R. L., 15 (116, 117, 118), 66 (116, 117, 118), *108,* 510 (23a), 536 (23a), 551 (23a), *552,* 585 (26), *622*
Shore, B. W., 544 (58b), *553,* 556 (33), *580*
Shortridge, R. G., 244 (47, 57), 245 (47, 57), 246 (47), *290*
Shpolskii, E. V., 8 (80), 9 (81), 88 (80), *107*
Shuler, K. E., 511 (30), *553*
Siebrand, W., 2 (11), 5 (11), 17 (11), 39 (11), *105,* 293 (9), 294 (9), 298 (9), *321*
Silberstein, J., 257 (71), *290*
Silverman, B. D., 584 (6), 585 (6), 607 (6), 620 (6), *622*

Sinai, G., 212 (33), 213 (33), *236*
Sinha, M. P., 6 (64b), 36 (64b), 37 (64b), 44 (64b), 75 (64b), *106*
Sirkin, E. R., 244 (49), 247 (49), 248 (49), *290*
Sitaram, P., 364 (9c), *492*
Slater, N. B., 52 (200), *110,* 243 (38), *289*
Slater, R., 722 (27), *733*
Sloan, J. J., 59 (257), *111, 292*
Slusher, R. E., 75 (316), *113,* 584 (1, 2), 585 (2), 591 (1), *622*
Smalley, R. E., 6 (65a, 65b, 66, 67, 71, 72), 23 (66, 67), 24 (66, 67, 164a, 164b, 164c), 36 (65a, 65b), 37 (65a, 65b), 44 (65a, 65b, 66, 67, 71), 75 (65a, 65b, 66, 67), *106, 107, 109,* 326 (5), 327 (6), 328 (7), 331 (7, 11, 13), 334 (11), 340 (30, 31), 341 (29), 345 (35), 349 (40), *361, 362,* 365 (14, 15a, 15b), 366 (17), 367 (15a, 15b), 370 (15a, 15b), 371 (15a), 375 (15a, 15b), 422 (15a, 15b), 424 (15a), 427 (15a), 453 (17), 463 (17), *493,* 510 (240), 543 (24a), 551 (24a), *552*
Smith, A. L., 331 (13), *361*
Smith, D. D., 293 (18), *321*
Smith, E. W., 572 (52), *580*
Smith, I. W., 242 (31), 244 (31), *289*
Smith, I. W. M., 58 (253, 254), 59 (253), *111,* 364 (7c), *492*
Smith, J. H., 65 (275), *112*
Smith, S. D., 2 (20), 3 (33, 51), 5 (62b), 73 (51), 11 (93), 12 (99), 15 (134), 30 (93), 72 (134), 75 (134), *105, 106, 108,* 366 (20), 453 (20), *493,* 592 (20), 602 (82), *622, 624*
Snowden, R. G., 641 (34), *659*
Soep, B., 5 (60), *106*
Solarz, R. W., *112*
Sorbie, K. S., 301 (40), *322*
Spoliti, M., 662 (13), *677*
Sprandel, L. L., 295 (33), 322
Stafast, H., 70 (291), *112*
Stammreich, H., 663 (19), *677*
Stannard, P. R., 294 (25), 296 (25), *321,* 543 (56), *553*
Starace, A. F., 14 (106), *107*
Steadman, S. G., 264 (82, 83), 383 (83), *291*
Stechel, E. B., 233 (70, 75), *237, 238*
Steed, J. M., 336 (27), *361*
Stegun, I. A., 384 (36), 386 (36), *493,* 593 (60), *623*
Steiner, W., 311 (48a), *322*
Steinfeld, J., 639 (7), *658*
Steinfeld, J. I., 4 (59) (237), *106, 111,* 241 (7), 244 (7), 259 (7), *288,* 509 (11b), 510 (11b), *522,* 558 (41), 592 (53, 54, 55), *623,* 709 (48), *711*
Steinmetzer, H. C., 719 (21), 722 (21), 727 (21), *733*
Stel'makh, O. M., 661 (3), *677*
Stenholm, S., 570 (49), *580*
Stepanov, B. I., 366 (22), 369 (22), 463 (22), *493*
Stephenson, J. C., 63 (266), 81 (322), *111, 113,* 242 (29), 244 (29), *289,* 509 (8a, 8b), *552,* 633 (16), *638,* 707 (42), *711,* 714 (6), *733*
Stevenson, J. C., 639 (10), *658*
Stitch, M. L., 509 (5a), 511 (533 (5a), 535 (5a), *552,* 605 (84), *624*
Stoddard, S. D., 146 (18), *198*
Stoichoff, B. P., 3 (32), *105*
Stokseth, P., 509 (4a), 511 (4a), 533 (4a), 535 (4a), 550 (4a), *552*
Stone, J., 53 (210), 70 (285, 286, 295, 296, 309), 72 (309), 76 (309), *110, 112,* 509 (14), 511 (14), 516 (14), 537 (14), *552,* 602 (83), *624,* 633 (16), *638,* 639 (10), 640 (20, 25, 32, 33), 641 (32), 657 (32), *658, 659*
Stone, J. M., 295 (30), *322*
Stoneham, A. M., 97 (383), *114*
Stratt, R. M., 162 (39), 165 (39), 176 (39), 182 (39), *199,* 233 (68), *237*
Strauss, H. T., 293 (17), 294 (17), 298 (17), 307 (17), *321*
Street, R. A., 19 (151), 91 (151), *108*
Strelcyn, J. M., 213 (34), 214 (34), 233 (78), *236, 238,* 241 (17), 242 (17), *289*
Stroud, C. R., Jr., 556 (21), 572 (50), 569 (21), *579, 580*
Struve, W. S., 88 (351), *113*
Stryland, J. C., 364 (10f), *492*
Sturge, M. D., 19 (155), *108*
Sturrock, P. A., *199*

Sudbo, A. S., 233 (63), *237*, 244 (62, 63, 64), 254 (62, 63, 64), *290*, 639 (4), 640 (24), 641 (4), *658*
Sulkes, M., 7 (78), 22 (78), *107*, 352 (46), *362*, 369 (32), *493*
Sur, A., *361*
Sutton, D. G., 592 (53, 54), *623*
Svec, H. J., 664 (23), *677*
Swimm, R. T., 169 (49), 173 (49), *199*
Swinney, H. L., 202 (14), *235*
Swofford, R. L., 293 (2, 3, 4, 5), 294 (2, 3, 4, 5, 26a), 295 (26a), 298 (2, 3, 4, 5, 26a), 302 (26a), 305 (26a), 306 (26a), 307 (26a), 309 (26a), 310 (26a), 313 (26a), *321*, *322*
Szoke, A., 577 (77), 578 (82), *581*, 584 (8), 591 (8), 614 (8), *622*

Tabor, M., 169 (46), 176 (46), 196 (70), 197 (70), *199*, *200*, 204 (18, 22), 233 (62), 234 (62), *235*, *236*, *237*
Tamir, M., 58 (249), 70 (311), 72 (311), 76 (311), *111*, *113*
Tang, C. L., 584 (3, 6), 585 (6), 589 (3), 591 (3), 607 (6), 620 (6), *622*
Tang, K. Y., 65 (276), *112*
Tannenwald, P. E., 584 (6), 585 (6), 607 (6), 620 (6), *622*
Tardy, D. C., 57 (236), *111*
Tavares, Y., 663 (19), *677*
Taylor, E. H., 626 (2), *637*
Taylor, J. R., 268 (101), *291*, 414 (46), 416 (46), *494*
Tehver, I., 83 (332), *113*, *539*, *553*
Teller, E., *109*
Tellinghuisen, J., 331 (11), 331 (13, 14), 334 (11), 351 (44), 352 (45), 353 (47), *361*
Tellinghuisen, P. C., *361*
Temkin, R. J., 556 (10, 11), 573 (10), *579*
Thiele, E., 55 (228), 70 (284, 285, 286), *110*, *112*, 215 (39), *236*, 608 (85), *624*, 640 (20, 32), 641 (32), 657 (32), *658*, *659*
Thirtle, J. H., 662 (13), *677*
Tisone, G. C., 331 (13, 14), *361*
Tisone, G. G., 3 (31), *105*
Tlide, D. M., 9 (87), 98 (87), 99 (87), 100 (87), *107*
Toda, M., 146 (17), *198*, 221 (50), 223 (50), 234 (50), *237*
Toennies, J. P., 626 (5), *638*
Tokuyama, M., 512 (40b), 513 (40b), 514 (40b), 515 (40b), 525 (40b), 532 (40b), *553*
Tolman, R. C., 267 (94), 278 (94), 288 (94), *291*
Tomlinson, W. J., 75 (316), *113*
Torrey, H. C., 556 (13), *579*
Toschek, P., 556 (9), 569 (9), 573 (9), *579*
Toyozawa, Y., 83 (333), 90 (368), 91 (368), 93 (368), *113*, *114*, 547 (59), 548 (59), *553*
Tramer, A., 18 (146), *108*
Tredgold, R. H., 218 (43), *236*
Trevalion, P. A., 365 (12b), *493*
Treve, Y. M., 124 (8), 161 (8), *198*, 233 (61), *237*
Tribus, M., 264 (84, 86), 274 (116), 265 (86), *291*, *292*
Tric, C., 40 (182, 183), *109*
Troe, J., 52 (203), *110*, 241 (10), 243 (10), 244 (10), *289*
Trotman-Dickenson, A. F., 641 (34), *659*
Tsao, J., 709 (99, 50), *711*
Tsien, T. P., 426 (48b), 478 (48b), 479 (48b), *494*
Tully, J. C., 58 (238), *111*, 243 (40), *290*
Tumanov, O. A., 661 (5), *677*, 681 (5), *710*
Turner, J. S., 146 (18), *198*, 210 (29), *236*
Turro, N. J., 719 (21, 22), 722 (21, 22, 27), 727 (21), 728 (22), *733*
Tusa, J., 7 (78), 22 (78), *107*, 352 (46), *362*, 369 (32), *493*

Uhlenbeck, G. E., 511 (31), 535 (31), *553*
Ulenikov, O. N., 588 (35), *623*
Ulstrup, J., 19 (152), 91 (152), 94 (152, 379, 381), 98 (152), *114*
Umstead, M. E., 244 (47, 57, 58), 245 (47, 57, 58), 246 (47), *290*
Utling, B. D., 364 (4), *492*

Valentini, J. J., 7 (79), 22 (79), *107*

Vandergrift, A. E., 104 (390), *114*
Van der pol, B., 149 (25), *199*
Van der Werf, R., 38 (178), *109*, 728 (36), *733*
Van Hove, L., 76 (324), *113*, 194 (68), *200*
Van Kampen, N., 511 (29), *553*
Van Kranendonk, J., 334 (15), *361*, 536 (51), *553*
Van Lerberghe, A., 584 (13), 587 (13), 589 (13), 598 (13), 598 (71), 599 (71), 600 (13), *622*, *624*
Van Velsen, J. F. C., 219 (37), *236*
Varsanyi, F., 311 (49), *322*
Velazco, J. E., 331 (13), *361*
Venanzi, L. M., 644 (37), *659*
Vigue, J., 18 (138), *108*
Villaeys, A. A., 616 (87), *624*
Vodar, B., 364 (9a), *492*
Voelker, S., 83 (330), 87 (330), *113*
Volkov, V. V., 263 (81), *291*
Von Neumann, J., 267 (96), *291*
Von Zeipel, H., 149 (28), *199*

Wagner, A. F., 231 (59), *237*
Wald, G., 9 (84), *107*
Waldman, J., 556 (10), 573 (10), *579*
Waldman, M., 335 (22), *361*, *493*
Walker, G. H., 126 (11), *198*, 226 (46), 227 (46), *236*
Walker, R. B., 70 (301), *112*
Wallace, R., 294 (23), 298 (23), *321*
Wallace, S. C., 3 (32), *105*
Wallace, S. F., 3 (30), *105*
Walpole, J. N., 3 (36), *105*, 592 (57), *623*
Walther, H., 53 (210), *110*, 584 (13), 587 (13), 589 (13), 598 (13), 600 (13), *622*
Wang, C. W., 75 (316), *113*
Wang, M. W., 511 (31), 535 (31)
Wang, R. T., 9 (89), 98 (89), 99 (89), 100 (89), *107*
Watanabe, A., 326 (4), *360*, 364 (10a, 10b), *492*
Watson, J. K. G., 295 (30), *322*, 588 (33), *623*
Watson, K. M., 2 (24), 18 (141), 40 (24), *105*, *108*, 577 (73), *580*
Watson, T., 709 (46), *711*
Webster, C. R., 662 (16), 663 (16), *677*
Wei, J., 351 (44), *361*
Weinstein, N. D., 58 (238), *111*, 243 (40), *290*
Weisman, R. B., 18 (137), *108*
Weiss, C. O., 573 (59, 60), *580*
Weisshar, J. C., 728 (35), *733*
Weitz, F., 74 (317), *113*
Welge, K. H., 707 (40, 41), *711*, 714 (3), *733*
Welsh, H. L., 326 (4), *360*, 364 (10a, 10b, 10d, 10e, 10f, 10g, 10i), *492*
Wendell, K. L., 282 (133), *292*
Wenzel, H., 19 (157), *109*
West, G. A., 53 (208), *110*, 242 (27), 244 (27), 247 (27), *289*, 709 (51), *711*
Weston, R. E., Jr., 53 (208), *110*, 242 (27), 244 (27), 247 (27), *289*, 626 (10), 635 (10), *638*
Wharton, L., 6 (65a, 65b, 66, 67, 68, 74), 23 (66, 67), 24 (66, 67, 164a, 164b, 164c), 36 (65a, 65b), 37 (65a, 65b), 44 (65a, 65b, 66, 67, 68), 75 (65a, 65b, 66, 67), *106*, *107*, *109* 326 (5), 327 (6), 328 (7), 329 (9), 331 (11), 341 (30, 31), 345 (35), 346 (36), 348 (39), 349 (40), 352 (46), 353 (49), 358 (49), *361*, 365 (14, 15a, 15b, 15c, 15d, 15e), 366 (17), 367 (15a, 15b, 15d, 15e), 370 (15d, 15e), 375 (15a, 15b, 15d, 15e), 422 (15a, 15b, 15d, 15e), 427 (15a), 453 (17), 463 (17), *493*, 510 (24b), 551 (24b), *552*
White, R. M., 664 (23), *677*
White, W. B., 663 (20), *677*
Whitley, R. M,, 556 (21), 569 (21), 570 (21), *579*
Whittaker, E. T., 277 (127), *292*
Wichman, E., 264 (87), 265 (87), *291*
Wiersma, D. A., 15 (123, 124, 128), 66 (123, 124, 128, 280), 69 (123, 124), 82 (123, 124), 83 (329, 331), 84 (124, 331), 87 (280, 331), *108*, *112*, *113*, 510 (23c), 536 (23c), 551 (23c), *552*,

Wiesenfeld, J. M., 85 (340), *113*
Wieting, R. D., 82 (281), *112*
Wiggins, T. A., 325 (3), *360,* 364 (19b, 19c, 19d), *492*
Wightman, K. J., 215 (17f), *235*
Wigner, E. P., 593 (61), *623*
Wild, U., 664 (24), *677*
Williams, P. F., 14 (108), *107*
Willis, C., 639 (9), *658*
Wilson, A. D., 58 (248), *111,* 556 (38), 577 (38), *580*
Wilson, D. J., 215 (39), *236,* 241 (18), *289*
Wilson, E. B., 455 (55), 460 (55), *494*
Wilson, J. R., 65 (278), *112*
Wilson, K. R., 12 (94), 58 (248), *107,* 396 (43a), 427 (43a), *494*
Wilson, T., 719 (20), 722 (20), 724 (20), *733*
Windsor, M. W., 9 (88), 98 (88, 372), 99 (88, 372), 100 (88, 372), 101 (372), *107, 114*
Winterfeld, C., 568 (46), *580,* 714 (14), *733*
Wittig, C., 242 (28), 244 (28), *289,* 626 (9), 635 (9), 637 (9), *638,* 639 (8), *658,* 684 (19), 686 (19, 20), 691 (20), 692 (19), 693 (19, 20), 694 (19), 695 (28, 29a, 29b), 696 (30), 700 (28), 701 (28), 702 (34), 705 (34), 707 (28, 34), 709 (46), 710 (52), *711,* 714 (10), *733*
Wodarczyk, F. J., 626 (8a), *638*
Woerner, R. L., 90 (363), *114*
Wolfrum, J., 60 (260), *111,* 242 (36), *289,* 626 (3), *638*
Womlock, J. M., 3 (44), *106*
Wood, J. H., 664 (21), *677*
Woodin, R., 639 (17), 642 (36), 656 (17), *658, 659*
Woodin, R. L., 3 (53), 73 (53), *106,* 509 (5b), 511 (5b), 512 (5b), 533 (5b), 535 (5b), *552*
Woodruff, S. B., 470 (64), *494*
Woskoboinikov, P., 556 (10), 573 (10), *579*
Woste, L., 3 (49), *106*
Wright, K. R., 241 (13), *289*
Wright, M. D., 65 (278), *112*
Wulfman, C. E., 281 (130, 131), *292*
Wurzberg, E., 626 (11, 12, 13a, 13b), 628 (11, 12, 13a, 13b), 629 (11, 12, 13a, 13b), 632 (13a, 13b), 637 (11, 12, 13a, 13b), *638*
Wyatt, R. E., 176 (54), *199,* 231 (58a), 233 (73), 234 (58a), *237,* 261 (79), 277 (124), *291, 292*
Wyne, J. J., 3 (46), 18 (46), *106*

Yablonovitch, E., 3 (40), 4 (54), 5 (40), 11 (40), 18 (40), 21 (40), 30 (40), 52 (206), 58 (206), 70 (40, 310), 72 (40, 206, 310), 73 (40), 74 (40, 310), 76 (40, 310), 80 (40), 81 (40, 310), *105, 106, 110, 113,* 241 (9), 244 (9), *289,* 293 (22c), *321,* 509 (5c, 6a, 12b), 511 (6a), 533 (5c), 535 (5c), 536 (12b), 550 (5c), *552*
Yahav, G., 12 (96), 70 (96), *107,* 661 (11, 12), *677,* 683 (16, 17), 693 (16, 17), 694 (16, 17), *710,* 716 (19), 719 (19, 23, 24), 723 (19), 725 (19), *733*
Yakovlenko, S. I., 577 (69, 70), *580*
Yang, S. C., 58 (242), 59 (242), *111*
Yardley, J. T., 709 (44), *711*
Yee, T. K., 556 (39), *580*
Yen, Y. R., 366 (19), 453 (19), *493*
Yogev, A., 3 (47, 48), 70 (47, 48), *106,* 639 (14), *658*
Yoon, B., 516 (46), 551 (46), *553*
Young, J. F., 65 (278), *112,* 577 (74), *581*
Yu, M. H., 639 (8), *658,* 684 (19), 686 (19), 692 (19), 693 (19), 694 (19), 695 (29a, 29b), *711,* 714 (5), 716 (18), 717 (18), *733*
Yuan, J. M., 65 (279), *112,* 577 (76), *581*

Zaleskaya, G. A., 686 (21), 693 (21), *711*
Zamir, E., 53 (209), *110,* 242 (25), 244 (25), 247 (25), 254 (25), 255 (25), 282 (134), *289, 290, 292*
Zandee, L., 51 (198), 52 (198), *109,* 242 (23), 244 (68, 70), 256 (70), 257 (70), *289, 290*
Zapka, W., 573 (62), *580*
Zare, R. N., 3 (50), 5 (60, 61), 6 (64b), 12 (97), 15 (112), 36 (64b), 37 (64b), 44 (64b), 47 (191), 50 (191), 51 (198), 52 (198), 64 (270, 272), 70 (191), 81 (323),

106, 107, 109, 110, 111, 112, 113, 286 (139), *292,* 509 (11a), 510 (11a), *552,* 667 (25), *677,* 683 (18), 692 (18), 693 (18), *711*
Zaslavskii, G. M., 132 (13), *198,* 209 (28), 226 (28), 229 (28), *236*
Zdasink, G., 3 (30), *105*
Zembrod, A., 584 (4), *622*
Zenack, A. Z., 82 (326), *113*
Zewail, A., 118 (6), *198,* 203 (17d), 204 (20), 215 (17d), *235, 236,* 241 (12), *289,* 510 (23b), 536 (23b), 551 (23b), *552*
Zewail, A. H., 2 (21), 15 (125, 126, 127, 129), 51 (199), 55 (224, 225), 66 (125, 126, 127, 129), 69 (125, 126, 127, 129), 82 (125, 126, 127, 129), 83 (129), 84 (129, 334), *105, 108, 110, 113,* 293 (6, 8, 18), 294 (6, 8), 298 (6, 8), 299 (8), 305 (8), 307 (8), 315 (6), *321*
Ziegler, J., 575 (67), *580*
Zimmermann, I. H., 65 (279), *112*
Zittel, P., 364 (7d), *492*
Zittel, P. F., 65 (273), *112*
Ziurys, L. M., 331 (13), *361*
Zvijac, D. J., 255 (66), *290*
Zwanzig, R., 54 (218), 76 (324b, 324d, 324e), *110, 113,* 511 (36), 512 (36), 513 (36), 514 (36), 525 (36), 532 (36), *553*

SUBJECT INDEX

Absorption lineshapes, bond modes and, 315-321
Acetone (ACE), 641
Argon, complexes of iodine and, 358-360
Atom-molecule reactions, 625-638
  experimental, 628-629
  introduction to, 625-628
  limitations of, 633-637
  results of, 629-633

Benzene, 6
Biological systems, photoselective chemistry and, 9
Biophysics, photoselective chemistry and, 98-102
Bloch-Redfield equations, 557-559
Bond modes, 293-322
  absorption lineshapes, 315-321
  beyond normal coordinates, 295-298
  cooperative excitations, 309-313
  energetics, 298-302
  intensities, 302-306
  intramolecular dynamics (involving X—H bonds), 313-315
  introduction to, 293-295

Chromatium (photosynthetic bacterium), 98, 99
Classical dynamics, stochastic transition and, 205-214
  completely integrable systems, 206-208
  divergence of trajectories, 206
  K entropy, 212-214
  mixing systems, 208-210
  Poincaré surface of section, 205-206
  typical systems, 210-212
Classical mechanical descriptions (coupled nonlinear oscillator systems), 119-161
  numerical studies of dynamics, 133-147
  perturbation theory, 147-159
  properties of trajectories, 119-132
Coherence effects, energy acquisition and, 4
Coherent optical effects: experimental observables, 15-16
  for pedestrians, 65-69
Coherent propagation effects, SF_6 spectroscopy and, 603-617
  calculation of polarization density, 604-606
  electromagnetic field equations, 603-604
  sidebands at resonant molecular transition frequencies, 611-613
  spatial (M) degeneracy, 613-617
  spectral content of radiated field, 609-611
  thin-sample approximation, 606-609
Coherent pulse propagation effects, SF_6 spectroscopy and, 583-624
Colinear analytical models of triatomic vdW molecules, 372-413
  diabatic distorted wave treatment without linearization, 390-395
  diatomic and simplest golden-rule expression, 372-388
  general formalism for VP decay, 395-407
  matrix elements, 407-413
  Rosen's relative coordinates treatment, 388-390
Collisional activation, deactivation and, 257-261
Collisional effects, energy acquisition and, 6-7
Collisional processes, photoselective chemistry, 58-65
Complex molecules, infrared laser chemistry, 639-659
  model, 648-652
  molecules, 641-648
    laser dissociation experiments in molecular beam, 644-647
    number of photons absorbed, 647-648
  real-time reaction monitors, 652-655
Condensed phases, relaxation and dephasing in, 82-90
Cooperation excitations, bond modes, 309-313
Coupled nonlinear oscillator systems, 119-195
  classical mechanical description of, 119-161
    numerical studies of dynamics, 133-147

perturbation theory, 147-159
properties of trajectories, 119-132
quantum-mechanical description of, 161-195
differences between classical and, 184-186
ergodicity and reaction rate, 186-195
Nordholm-Rice analysis, 162-168
relationship between classical trajectories and, 176-183
semiclassical quantization, 168-175
Coupled oscillators, numerical experiments on, 214-219
Critical-point analysis, 225-226

Decay of excited states, 13-14
exponential, 13
nonexponential, 14
quantum beats in, 14
Dimers (vdW), intramolecular dynamics, 427-439
facts and correlations, 427-428
VP dynamics, 428-431, 439-453
local mode basis, 440-442
miniexciton basis, 442-443
normal mode basis, 443-453
V→T process, 431-435
V→V + T process, 435-439
Dimethyltetrahydrofuran (DMTHF), 641
Dimethysulfoxide (DMSO), 641
Dissociation, multiphoton, 254-257

Electronic predissociation, 27-28
Electronic relaxation, 40-46
inverse, 46-51
Energetics, bond modes, 298-302
intramolecular effects on, 306-309
Energy acquisition, photoselective excitation and, 2-9
biological systems, 9
coherence effects, 4
high energy, 3
high power, 3
matrices and mixed crystals, 7-9
molecules, 4-6
at low pressures, 4-5
in supersonic beams, 5-6
in thermal molecular beams, 5
selective collisional effects, 6-7
solutions, 9
spectral range, 3
tunability, 3
ultrashort duration, 4
Enrgy disposal, 12-13
Energy-resolved observables, 14-15
Energy storage, problem of, 9-11
excitation amplitudes, 10-11
initial conditions, 11
molecular level structure, 10
Entropy, information theoretic approach:
decline of, upon improved experimental resolution, 287-288
maximum, intermezzo on, 250-254
Excitation amplitudes, characterization of (energy storage), 10-11
Experimental observables, photoselective chemistry, 13-16
coherent optical effects, 15-16
energy-resolved, 14-15
time-resolved, 13-14

Formaldehyde, 6

Gases, vibrational relaxation in, 367

Heavy ion transfer reactions, 263-264
Helium, complexes of iodine and, 353-355
Heterodimers: VP dynamics of, 428-431
V→T process in, 431-435
V→V+T process in, 435-439
High energy, energy acquisition and, 3
High-order infrared multiphoton excitation (MPE), 70-82
dephasing (isolated, collision-free molecule), 81-82
probabilities for MPE, 77
random coupling model (RCM), 78-79
reduction schemes, 77-78
traditional energy level scheme, 71
vibrational quasicontinuum, 74-76
weak coupling approach, 76
High-order multiphoton excitation (MPE), 47
High power, energy acquisition and, 3
Homodimers, intramolecular dynamics of VP of, 439-453
Homogeneous relaxation, 557

Information theoretic approach, 239-292
background of, 240-244
decline of entropy on improved experimental resolution, 287-288

initial state, 269-275
intramolecular dynamics, 275-285
maximum energy formalism, 264-269
outlook for, 285-287
surprisal analysis, 244-264
Infrared laser chemistry of complex molecules, 639-659
model, 648-652
molecules, 641-648
laser dissociation experiments in molecular beam, 644-647
number of photons absorbed, 647-648
real-time reaction monitors, 652-655
Inhomogeneous broadening, origins of, 5
Initial state, 269-275
activation process, 271-272
maximum entrophy principle, 266-268
most conservative inference, 272-275
stationary state, 270-271
Intensities, bond mode, 302-306
Interstate electronic relaxation (ER), 29
Intramolecular dynamics, 117-506
bond modes, 293-322
information theoretic approach, 239-292
approximations, 283-284
constraints, 279-281
incomplete resolution of final states, 276-277
once is enough, 275-276
sum rules, 281-283
time-dependent constants of motion, 277-278
time evolution, 278-279
onset of statistical behavior, 201-238
transfer of vibrational energy, 117-200
van der Waals molecules, 323-506
Intramolecular and intermolecular processes, 16-17
classification of, 16-17
Intramolecular energy flow, 51-56
Intramolecular energy randomization (IVR), 640
Intramolecular vibrational energy transfer, 117-200
coupled nonlinear oscillator systems, 119-195
classical mechanical, 119-161
quantum-mechanical, 161-195
introduction to, 117-119
Intrastate vibrational energy redistribution (IVR), 29
Inverse electronic predissociation (IEP), 28
Inverse electronic relaxation (IER), 46-51
Inverted-Raman double-resonance, 574-575
Iodine complexes, van der Waals molecules, 353-360
and argon, 358-360
and helium, 353-355
and neon, 355-358

Ladder double resonance, 570-574
Linearized equation, 225
Low pressures, molecules at, 4-5
Luminescence, electronic, resulting from infrared multiple photon excitation, 679-711
chemiluminescent reactions, 694-707
$C_2(a^3\pi_u)$ +NO, 695-701
$C_2(X^1\Sigma_g^+, a^3\pi_u)$ +O_2, 701-705
other reactions, 705-707
collisionless and collisional production, 681-694
kinetics, 687-692
mechanisms, 692-694
survey of experimental results, 681-687
introduction to, 679-680
Luminescence of parent molecule, induced by multiphoton infrared excitation, 661-677
experimental, 666
introduction to, 661-666
results, 666-673
4000 A region, 672-673
5300 A region, 671-672
6300 A region, 668-671

Maximum entropy formalism, 264-269
algorithm, 265-266
initial state, 266-268
variational approximations, 268-269
Mixed crystals, energy acquisition and, 7-9
Molecular level structure, characterization of (energy storage), 10
Molecular multiphoton processes (MMP), 509-533
coherences and intramolecular dephasing, 540-546
complete incoherent driving, 544-546

early stages of I-coherent driving, 540-541
perturbative line-broadening limit, 541-544
quasicontinuum, 544
evaluation of higher order correlation functions, 527-535
Hamiltonian model and REM for the populations, 516-522
COP reduced equations of motion, 518-520
POP reduction scheme, 520-522
intramolecular dipole correlation functions, 546-551
introduction to, 509-513
REM expansion to second-order in the field, 522-527
REM for populations and coherences, 535-540
systematic reduction scheme, 513-516
Molecules, energy acquisition and, 4-6
low pressures, 4-5
supersonic beams, 5-6
thermal beams, 5
Mo's method, 220-221
variational equations approach and, 233-235
Multiphonon processes, photoselective chemistry, 90-98
Multiphoton-induced chemistry, 509-733
atom-molecule reactions, initiation of, 625-638
complex molecules (infrared laser chemistry), 639-659
electronic luminescence (resulting from IR MPE), 679-711
luminescence of parent molecule, 661-677
N-level multiple resonance, 555-581
reduced equations of motion, 509-553
SF_6 spectroscopy and coherent pulse propagation effects, 583-624
unimolecular multiple photon dissociation, 713-733
Multiphoton ionization (MPI), 51
Multiple resonance, *N*-level, 555-581
introduction to, 555-557
master equations, 567-569
open subsets, 564-567
steady-state solutions, 559-564
three-level double resonance, 569-579
inverted-Raman, 574-575
ladder, 570-574

Naphthalene, 6
Neon, complexes of iodine and, 355-358
Nonradiative electronic processes (NREP), 90-98
Nonreactive molecular processes, 29-40
accessibility criterion, 30-40
energy-levels model, 32
Hamiltonian formalism, 33
intermediate level structure, 36-40
interstate and intrastate level scrambling, 31
similarity criteria, 30
small molecular limit, 34-35
spectrocopic implications of interstate coupling, 37
statistical limit, 35-36
time-resolved emission spectrum, 38
transition from small limit to statistical limit, 39
Nordholm-Rice analysis, 162-168
Nuclear heavy ion transfer reactions, 263-264
Numerical methods of triatomic vdW molecules, 413-427
application to perpendicular VP, 420-427
comparison between distorted wave calculations and, 417-420
scattering formalism, 413-417

Onset of statistical behavior, *see* Statistical behavior, onset of (intramolecular energy transfer)
Ovalene, 6
fluorescence excitation spectrum of, 7

Pentacene, 6
absorption spectrum of, 8
fluorescence excitation spectrum, 6
Perturbation theory, 147-159
Photoselective chemistry, 1-114
biophysics and, 98-102
coherent optical effects for pedestrians, 65-69
collisional processes, 58-65
electronic relaxation, 40-46
inverse, 46-51
energy acquisition, 2-9
energy disposal, 12-13
energy storage, 9-11

experimental observables, 13-16
coherent optical effects, 15-16
energy-resolved, 14-15
time-resolved, 13-14
future potential of, 102-104
high-order multiphoton excitation (MPE), 70-82
intramolecular energy flow, 51-56
intramolecular and intermolecular relaxation, 16-19
meaning of, 2
multiphonon processes, 90-98
nonreactive molecular processes, 29-40
accessibility criterion, 30-40
similarity criteria, 30
predissociation, 19-28
relaxation and dephasing in condensed phases, 82-90
unimolecular reactions, 56-58
van der Waals molecules, 343-360
binding energies, 349-352
competing processes, 348-349
lifelines, 345-348
product-state distributions, 352-360
Phtalocyanine, 6
Poincaré surface of section, 205-206
Polyatomic molecules, vibrational predissociation, 453-464
background information, 453-464
electronically excited complexes, 463-464
model calculations (complexes), 454-463
Polyatomic species (van der Waals molecule), 334-343
infrared, microwave, and radiofrequency spectra, 334-337
visible and ultraviolet spectra, 337-343
Predissociation, photoselective chemistry, 19-28
electronic, 27-28
inverse electronic, 28
rotational, in a diatomic molecule, 20-21
vibrational, 24-26
Pyridine (PYR), 641

Quantum-mechanical description (coupled nonlinear oscillator systems), 161-195
differences between classical and, 184-186
ergodicity and reaction rate, 186-195
Nordholm-Rice analysis, 162-168
relationship between classical trajectories and, 176-183
semiclassical quantization, 168-175

Reduced equations of motion, 509-533
coherences and intramolecular dephasing, 540-546
complete incoherent driving, 544-546
early states of I-coherent driving, 540-541
perturbative line-broadening limit, 541-544
quasicontinuum, 544
evaluation of higher order correlation functions, 527-535
Hamiltonian model and REM for the populations, 516-522
COP reduced equations of motion, 518-520
POP reduction scheme, 520-522
intramolecular dipole correlation functions, 546-551
introduction to, 509-513
REM expansion to second-order in the field, 522-527
REM for populations and coherences, 535-540
systematic reduction scheme, 513-516
Rotational predissociation in a diatomic molecule, 20-21
Rotational states, distribution of, 261-263

Semiclassical quantization of coupled oscillator systems, 168-175
SF_6 spectroscopy, 583-624
coherent propagation effects, 603-617
calculation of polarization density, 604-606
electromagnetic field equations, 603-604
sidebands at resonant molecular transition frequencies, 611-613
spatial (M) degeneracy, 613-617
spectral content of radiated field, 609-611
thin-sample approximation, 606-609
introduction to, 584-586
ν_3 band, 586-603
fitting of spectroscopic parameters, 596-598
model, 601-603
transition moments, 598-601
vibration-rotation basis, 592-594
vibration-rotation Hamiltonian, 594-595
Simply-connected multiple resonances, 558

Spectral range, energy acquisition and, 3
Spectrascopy and structure, van der Waals molecules, 330-343
diatomic species, 331-334
polyatomic species, 334-343
Statistical behavior, onset of (intramolecular energy transfer), 201-238
classical dynamics and stochastic transition, 205-214
completely integrable systems, 206-208
divergence of trajectories, 206
K entropy, 212-214
mixing systems, 208-210
Poincaré surface of section, 205-206
typical systems, 210-212
exponentiating trajectories and, 231-233
introduction to, 201-203
models for regular to erratic transition, 220-231
critical-point analysis, 225-226
Mo's method, 220-221
overlapping resonances, 226-230
variational equations approach, 221-225
numerical experiments on coupled oscillators, 214-219
problems of interest, 203-205
Stochastic transition, classical dynamics and, 205-214
completely integrable systems, 206-208
divergence of trajectories, 206
K entropy, 212-214
mixing systems, 208-210
Poincaré surface of section, 205-206
typical systems, 210-212
Supersonic beams, molecules in, 5-6
Surprisal analysis, 244-264
$(C_3H_4O)^{\neq}$ and other elimination reactions, 245-248
collisional activation and deactivation, 257-261
distribution of rotational states, 261-263
F+RH, 248-250
heavy ion transfer reactions, 263-264
intermezzo on maximum entropy, 250-254
multiphoton dissociation, 254-257

Tetracene, 6
Tetrahydrofuran (THF) compound, 641
Thermal inhomogeneous broadening effects (TIB), 5
Three-level double resonance, 569-579
inverted-Raman, 574-575
ladder, 570-574
Time-resolved observables, 13-14
Trajectories: divergence of, 206
properties of, 119-132
statistical behavior and, 231-233
Triatomic van der Waals molecules, 334
Tributyl phosphate (TBP), 641
Trimethyl phosphate (TMP), 641
Tunability, energy acquisition and, 3

Ultrashort duration, energy acquisition and, 4
Unimolecular multiple photon dissociation, 713-733
collision-free LIMD chemiluminescent reactions, 716-719
introduction to, 713-716
experimental methods, 714-715
mechanism of LIMD, 715-716
SiF_4 experiments, 729-732
of tetramethyldioxetane, 719-729
introductory notes, 719-722
nature of emitting species, 726-729
time behavior, 722-727
Unimolecular reactions, photoselective chemistry, 56-58

van der Waals molecules, 6, 323-506
chemical synthesis and experimental probes, 325-330
dimers (intramolecular dynamics), 427-453
facts and correlations, 427-428
VP dynamics of heterodimers, 428-431, 439-453
V→T process in heterodimers, 431-435
V→V + T process, 435-439
energetics of iodine-rare gas, 467
intramolecular dynamics, 363-506
introduction to, 323-325, 364-370
dynamics of VP and vdW molecules, 367-370
vibrational relaxation in gases, 367
VP of ArHCl, 367
VP of hydrogen-bonded systems, 336-337
models (VP of triatomic vdW molecules), 370-427
colinear analytical, 372-413
facts to consider, 370-371
numerical methods, 413-427

photoselective chemistry, 343-360
binding energies, 349-352
competing processes, 348-349
lifelines, 345-348
product-state distributions, 352-360
quasiclassical trajectory studies, 470
rotation effects on VP rates, 470-482
scaling theoretical analysis of VP in H I, 467-470
spectroscopy and structure, 330-343
diatomic species, 331-334
polyatomic species, 334-343
vibrational predissociation, 23
VP of polyatomic molecules, 453-464
background information, 453-454
complexes, model calculations, 454-463
electronically excited complexes, 463-464
VP of vdW complexes (containing triatomic molecules), 482-492
Variational equations approach, 221-225
Mo's method and, 233-235
Vibrational predissociation (VP), 24-26
of ArHCl, 367
dynamics and vdW molecules, 367-370
energetics of the iodine-rare gas van der Waals molecules, 467
of hydrogen-bonded systems, 366-367
of linear vdW heteronuclear dimers, 428-431
models of triatomic vdW molecules, 370-427
quasiclassical trajectory studies, 470
recent progess in, 466-492
rotation, effects of on, 470-482
scaling theoretical analysis in the H I, 467-470
of vdW complexes containing triatomic molecules, 482-492

Water (H_2O), 641